STATISTICAL PROCESS ANALYSIS

The Irwin/McGraw-Hill Series
Decision Sciences

To Amir and Vanessa

PREFACE

The purpose in writing *Statistical Process Analysis* (*SPA*) is to give readers a strategy for effective use of statistics in the area of process management/analysis. There are a number of excellent books that treat this topic from the perspective of traditional coverage of statistical process control (SPC) methods. As such, *SPA* does not intend to provide another *purely* traditional approach to the area of SPC and its primary tool, the control chart. This does not imply traditional SPC methods and concepts will not be treated, they indeed are, and they even serve as unifying themes throughout the book. However, this book markedly differs in that SPC methods are treated within a broader data analysis perspective.

There are a number of efforts being made to change statistics curricula in business and engineering schools. In part, these efforts are motivated by signals from industry that universities have failed to train students with tools that are useful for analyzing real applications. Reports from the annual American Statistical Association conferences on "Making Statistics More Effective in Schools of Business" consistently reflect a need for instructors, curricula, teaching materials, and books to emphasize and integrate the themes of statistical thinking, quality management, and practical data analysis in conjunction with a user friendly software package. Furthermore, there is a need to speak to a general audience, not just those in manufacturing. *SPA* attempts to address all of these needs.

UNIQUE FEATURES

Using a variety of real world data sets, the book emphasizes the visual examination of data plots and the development of an intuitive understanding of statistical modeling. Since the data sets are real, the reader will experience a wide assortment of applications: some that allow for the appropriate use of traditional SPC methods while others challenge the appropriateness of the use of traditional SPC methods.

When SPC methods are not most appropriate, the book helps the reader develop the data skills to better understand the process at hand. This unique contribution of the book should be underscored. It is safe to say that nearly all SPC textbooks rely primarily on data sets that demonstrate the appropriateness of traditional SPC techniques, that is, the data sets are "textbook" examples. *SPA* shows the readers that the real world is much more interesting, with processes behaving in a variety of fashions that can not be properly controlled or analyzed by traditional methods. Most SPC books either give no coverage to real-world challenges, or, on the other hand, treat the issue as a special topic limited to one or two sections of some chapter.

To truly understand most processes encountered in practice, one needs to be armed with a broad range of data analysis skills. Still, the book resists the temptation to use overly sophisticated statistical models. Instead, *SPA* keeps the data analysis down to easily understood tools such as regression methods. All the statistical tools presented in the book have been successfully understood and implemented by students (undergraduate, MBA, and EMBA) from introductory statistics to elective quality courses taught by the author at the University of Wisconsin-Milwaukee as well as similar courses taught by colleagues at the University of Chicago.

Below is a summary of the unique features of *Statistical Process Analysis (SPA)*:

✓ *SPA* is self-contained, flexible, and informative for different audiences: those with limited statistical backgrounds, those with statistical and/or SPC training interested in developing skills beyond the standard treatment, students (undergraduate or graduate, business or engineering), and industry practitioners.

✓ *SPA* provides a detailed background discussion of the principal features of the modern management paradigm and the integral role of statistical thinking and problem solving within this paradigm.

✓ *SPA* provides thorough coverage of the traditional SPC methods, using real world examples and data, from elementary to advanced topics.

✓ *SPA* recognizes that process data analysis is applicable to *any* process not just manufacturing processes. Because of the origins of SPC methods, most books focus primarily on manufacturing processes; *SPA* looks at manufacturing and many other processes.

✓ *SPA* helps readers develop skills for general process data analysis. *SPA* keeps the data analysis tools understandable. This helps promote the use of the book material beyond the classroom to industry itself.

✓ *SPA* integrates a popular (both in education and industry) statistical software package, Minitab. This is not a trivial point. No currently available book focussed on the area of SPC *fully* integrates an all-purpose statistical package for general data analysis flexibility. Some books make "spotty" connections to a general statistical package, such as Minitab. However, most books either do not incorporate software (hand-generated plots) or incorporate SPC software narrowly designed for the given book.

✓ *SPA* uses real data sets throughout the book, both within the main text presentation and the exercises. These data sets are made available on a data disk in the Instructors Manual or at the McGraw-Hill Web site. (www.mhhe.com/bstat)

✓ *SPA* contains nearly 400 end-of-chapter questions. Each chapter has a balance of review questions (designed to check students' understanding of major concepts), solving problems (designed to help students develop their technical comprehension), and computer-based problems (designed to help students practice and hone their data analysis skills).

LEVEL AND AUDIENCE

To use the text, little mathematical expertise beyond basic college algebra is necessary; no knowledge of calculus is required. For those without formal statistics training, Chapter 2 serves as a self-contained introduction to the basic statistical concepts useful for subsequent developments.

SPA is intended to appeal to a wide variety of audiences. In a university setting, the book has natural opportunities in courses for management, engineering, or statistics students. For instance, *SPA* can serve as a textbook for junior-, senior-, or graduate-level courses in quality control or management often found in engineering schools (such as, industrial engineering) and business schools.

Beyond quality-related courses, there is an opportunity for the use of this book in a university setting. The book can be used as the principal text in application-oriented statistics courses focussed on diverse topics such as basic statistical methods, time-series analysis, statistical quality control, regression, forecasting, and, of course, data analysis. Or, as an alternative, the book can be used as a supplementary resource that enables students to achieve practical data analysis skills while the instructor focuses on more theoretical concepts. Indeed, most of the data sets and the corresponding data analysis found in this book serve as the core of an Executive MBA-required business statistics course I teach at the University of Wisconsin-Milwaukee. It should be noted that the only prerequisite for the course is a two-week refresher course on basic mathematics and algebra. Given the course's emphasis on practical data analysis, I am happy to report that the course has been a tremendous success with students often reporting being "hooked" on data analysis.

Beyond the university setting, *SPA* can serve as a resource for individual practitioners and for company training programs focussed on statistical quality control and continual process improvement. As noted in Chapter 1, many companies (such as, General Electric) are providing in-depth statistical training (often known as "black belt" training) for all their employees. My review of black belt training materials suggests that even though they provide excellent coverage of important statistical techniques, they still fall short on providing trainees with the important basic data analysis skills integrated throughout this book. As such, *SPA* can serve as valuable complement to many of the training materials currently in use.

USE OF MINITAB

This book is unique in its integration with the statistical computer package Minitab; in particular, this book is tied to Release 12. This integration emphasizes the computer as a practical tool for insightful data analysis. Access to a statistical package is essential and presumed for all but the most sophisticated readers. One can not learn data analysis by simply reading text; one should attempt to develop a degree of skill in actual computation.

Although there are many excellent statistical packages currently available, I chose to use Minitab because of its wide acceptance by educational institutions and its extensive accessibility around the world in business and government. It is currently used by more that 2,000 colleges and universities and in the business world, by companies of all sizes, from start-ups to major corporations, including Allied Signal, Ford Motor Company, 3M, General Electric, General Motors, and Lockheed Martin. In fact, the majority of Fortune 500 companies use Minitab as part of their routine operations.

Minitab, is a registered trademark of Minitab, Inc.

McGraw-Hill Higher Education

*A Division of The **McGraw-Hill** Companies*

STATISTICAL PROCESS ANALYSIS

Copyright © 2000 by The McGraw-Hill Companies, Inc. All rights reserved. Printed in the United States of America. Except as permitted under the United States Copyright Act of 1976, no part of this publication may be reproduced or distributed in any form or by any means, or stored in a data base or retrieval system, without the prior written permission of the publisher.

This book is printed on acid-free paper.

1 2 3 4 5 6 7 8 9 0 DOC/DOC 9 0 9 8 7 6 5 4 3 2 1 0 9

ISBN 0-256-11939-2

Vice president/Editor-in-chief: *Michael W. Junior*
Publisher: *Jeffrey J. Shelstad*
Executive editor: *Richard T. Hercher, Jr.*
Marketing manager: *Zina Craft*
Project manager: *Karen J. Nelson*
Senior production supervisor: *Heather D. Burbridge*
Designer: *Kiera Cunningham*
Supplement coordinator: *Susan Lombardi*
Compositor: *GAC Indianapolis*
Typeface: *10/12 Times Roman*
Printer: *R. R. Donnelley & Sons Company*

Library of Congress Cataloging-in-Publication Data

Alwan, Layth C.
    Statistical process analysis / Layth C. Alwan.
        p.   cm.   (Irwin/McGraw-Hill series in operations and
    decision sciences)
        ISBN 0-256-11939-2
        1. Process control -- Statistical methods.   2. System analysis
    -- Statistical methods.   I. Title.   II. Series.
    TS156.8.A45   2000
    658.5'62/015195 dc--21   99-27679

http://www.mhhe.com

STATISTICAL PROCESS ANALYSIS

Layth C. Alwan
University of Wisconsin—Milwaukee

Boston • Burr Ridge, IL • Dubuque, IA • Madison, WI • New York • San Francisco • St. Louis
Bangkok • Bogotá • Caracas • Lisbon • London • Madrid
Mexico City • Milan • New Delhi • Seoul • Singapore • Sydney • Taipei • Toronto

My choice is also based on the fact that Minitab can be learned quickly and easily, providing students with a powerful data analysis system that can also be used in other courses and in their professional careers. Minitab's intuitive user interface along with its context-sensitive online Help and simple tutorials make it easy for first-time users to perform statistical analysis without spending an inordinate amount of time learning the software.

Notwithstanding the ease of learning Minitab on one's own, I have incorporated two aids in this book. First, for any Minitab output that is first encountered, I have superimposed next to the output the pull-down menu choices required to produce the output. For instance, the first example of a scatter plot is in Figure 1.10. Referring to this figure, you will find "Graph >> Plot" superimposed next to the scatter plot. Since many outputs can easily be obtained by directly typing in a Session command, I have also shown for certain first-encountered output both the pull-down menu choices and the corresponding Session command, leaving to the user to decide which approach to use. For example, refer to Figure 2.14 to find that summary statistics are shown to be obtained by either the Session command "MTB> describe 'reading'" or by the pull-down menu choices of "Stat >> Basic Statistics >> Descriptive Statistics". As a second aid, in an end-of-book appendix, I have also provided a overview of Minitab along with a basic summary of some important utility functions (e.g., inputting data and saving data).

As a note to instructors, there are two versions of Minitab: the full (or professional) version and the student version. The professional version offers the complete range of statistical and quality-related capabilities. Many universities have a site license for Minitab which brings to the students the full power of Minitab. The student version of Minitab is a trimmed version of the full Minitab. I mention this version because Irwin/McGraw-Hill (the publisher of this book) can "bundle" a personal copy of the student version of Minitab with each copy of this book. Bundling is economically attractive because students get the textbook and the software at a price much lower than buying each separately.

With respect to the student version capabilities, it has all the basic statistical capabilities necessary to carry out the data analysis techniques of this book such as, descriptive statistics, graphing, time-series functions, and regression modeling. The student version retains a large subset of the full version's SPC capabilities. The only notable exception is the CUSUM control chart which is only available on the full version. This is not an insurmountable obstacle, one could perform the CUSUM computations by hand or spreadsheet (see Chapter 8) and then use Minitab's graphing options to create a CUSUM chart.

Even though Minitab is used extensively throughout the book, it should be emphasized that the reader is not bound to the software and is not disadvantaged using another software package. Because the computer output from most statistical packages is quite similar, the reader can safely use this book with programs other than Minitab. To allow for this flexibility, all data sets used in this book are provided both in Minitab and ASCII formats. An ASCII formatted data set is a text file, which can be universally read into any statistical or spreadsheet software. ASCII data files found in the supplied data disk are identified by the ".dat" filename extension while the Minitab data files are identified by the ".mtw" filename extension. To avoid confusion, a given data set will have the same prefix file name. For example, a data set found in a file named "response.dat" would be an ASCII file and the Minitab formatted file would be named "response.mtw."

For most software (e.g., Excel), you can simply open the ASCII file and the software will automatically recognize the number of columns (one or more) associated with the data set and then read them into the software appropriately. If, however, there is a need to know how many columns are associated with the data set prior to the input process, there is a simple process within Windows that will enable you to determine the content of the data file. In particular, you should first double click "My Computer" and then go to the directory in which the ASCII file resides and then double click the file itself. At this point, Windows will automatically open a utility known as WordPad (rudimentary word processing package) and show you the contents of the file. Now that you know the contents of the file, you can proceed to your software to instruct it to input the data appropriately.

A WORD ON STATISTICS

H. G. Wells wrote, "Statistical thinking will one day be as necessary for efficient citizenship as the ability to read and write." For industry that day has arrived. The foundation of "statistical thinking" is increasingly being recognized as *critical* to effective management (at all levels).

However, statistics is not generally perceived as a popular subject. There are many reasons for this perception which I shall not dwell on. I do offer one insight drawn from my many years of teaching. When statistics is approached from a data analysis perspective, students develop an eagerness for the subject and are surprised to find themselves viewing data analysis as useful, interesting, and even exciting. Once you become aware of what data analysis can do, you will see that opportunities for practical application are omnipresent, in business, engineering, and elsewhere. I, personally, have found data analysis skills to be professionally and personally rewarding. In fact, I regard data analysis as a hobby of sorts which I plan to continue well beyond my professional career. I hope that you will come to agree with me as you read this book.

ACKNOWLEDGMENTS

This book was written over a period of several years, and I owe thanks to many people. First, I must recognize the generous assistance from Irwin/McGraw-Hill. Foremost, I am grateful to my editor, Richard T. Hercher, Jr. He was not only enthusiastic about the ideas and approach of this book but he also provided professional guidance and long-term support that brought this project to fruition. I am thankful to Colleen Tuscher for her efforts in resolving some startup problems encountered in the earlier stages of the project. My thanks go to Nicolle Schieffer for her patience, organization, and cheerful attitude that resulted in pleasant culmination of the project. Karen Nelson deserves special recognition for her professional management of the production of the book.

The final stages of the book occurred while I was on sabbatical lecturing at Al Akhawayn University in Morocco as a Fulbright Scholar. Because the university is located in a remote town, receipt and delivery of book materials were not trivial tasks. I would like to thank Youssef Senhaji and Lhoussaine Oumaidi of the University Business Office for their tremendous efforts to get materials to and from an express courier service office located in a distant city.

I would like to thank Minitab Inc. for continually providing me, through its author assistance program, with the most recent release of its Minitab Statistical Software (Minitab for Windows). I am also grateful to the many individuals in Minitab's technical support staff who generously gave their time to resolve certain technical issues encountered during the development of this book.

I am greatly indebted to present and former MBA and Executive MBA students at the University of Wisconsin-Milwaukee for their constructive comments on the many data applications found in the book and for their contribution of real data applications from their places of work.

My sincere appreciation is extended to my colleague Timothy C. Haas for being available on numerous occasions to discuss various statistical issues with me. His comments lead to a number of improvements in the presentation of material found in this book. I am also grateful to William A. Berezowitz of General Electric Medical Systems for his many helpful insights and his data contributions.

I am thankful to all the reviewers involved in the review process. I am particularly appreciative of the following reviewers: Charles W. Champ, Georgia Southern University; Sudhakar D. Deshmukh, Kellogg School of Management-Northwestern University; Ben Huneycutt, Clemson University; Binshan Lin, Louisiana State University-Shreveport; Don Richter, Stern School of Business-New York University; Steve E. Rigdon, Southern Illinois University; and Athanasios Vasilopoulos, St. John's University. These reviewers made a multitude of excellent suggestions. I took every suggestion very seriously and I hope they can recognize their contributions within the text.

Given the theme of this book, I would be remiss if I did not acknowledge the profound influence that Professor Harry V. Roberts of the Graduate School of Business at the University of Chicago had on my development as a statistician. From him, I learned the potential and excitement of statistics. It was because of him that practical data analysis permeates my thinking and brings clarity on everything from personal to professional activities. I will forever be grateful for his mentoring.

Finally, I owe very special thanks to my family. I am deeply appreciative of my parents. It was my mother who instilled in me the passion for incessant inquiry for knowledge; her memory is always in my heart. With respect to my father, I owe much of my success and well being to his guidance and loving support from my childhood through my adult years. I am also thankful for his professional contributions to the development of this book. As a professor of statistics himself, he provided valuable comments and criticisms based on reading preliminary versions of the manuscript. Last but not least, I would like to thank my wife, Athena Maria, for her patience, understanding, and love in helping make this book a reality.

Layth C. Alwan

Brief Contents

Contents

C h a p t e r

1

Data Analysis and Process Management

CHAPTER OVERVIEW

The predominant focus of this book is on statistical methods relevant to the study and monitoring of processes. The word *process* tends to be associated with a manufacturing application, such as a machining process. As will be evident to the reader, we shall take a much broader view of process applications in this book. With such a view, we will discover the need to learn a wider variety of statistical concepts and methods than typically covered in statistical process control books. To appreciate the focus of this book, this chapter provides a backdrop for the necessity of process management activities and statistical methods in today's competitive environment.

1.1   QUALITY MANAGEMENT: A BRIEF OVERVIEW

In the 1980s, U.S. companies began to realize that they were playing a game very different from the one they were winning 15, 10, and even 5 years earlier. Companies were faced with the combined pressures of

- Fierce global competition, suddenly far stronger than most managers had realized.
- Rapidly changing customer demands for products and services.
- Increasing customer demand for product quality and customer service.

Now, there is a general recognition that to succeed in global competition, companies must make a commitment to customer responsiveness and continuous improvement in quickly developing innovative products and services that at once combine exceptional quality, fast and on-time delivery, and low prices and costs. It is clear that this competitive challenge is not a passing fad but rather a long-term, high-stakes battle in an ever-changing marketplace. Change is, and will continue to be, the only constant.

Companies have also learned that to meet the challenges of the competitive market, an organizational shift to a different paradigm of management thought and action is necessary. Simply stated, the new paradigm calls for managers at all levels to orient their thinking and actions toward developing an appropriate organizational system dedicated to providing superior customer value.

In the 1980s, this emerging paradigm was most commonly referred to as *total quality management (TQM)*. The novelty of TQM left many managers struggling and even disagreeing with its basic definition. In an attempt to bring clarity to the meaning of a total quality approach, participants in the Total Quality Forum sponsored by Procter & Gamble, a consortium of business leaders (chief executive officers of major corporations) and academic leaders (deans of business and engineering schools), arrived at the following consensus definition of a total quality management based system:[1]

[1] Procter & Gamble, Report to the Total Quality Leadership Steering Committee and Working Councils, Cincinnati, OH: Procter & Gamble, 1992.

> Total Quality (TQ) is a people-focused management system that aims at continual increase of customer satisfaction at continually lower real cost. This is a total system approach (not a separate area or program), and an integral part of high-level strategy. It works horizontally across functions and departments, involves employees, top to bottom, and extends backward and forward to include the supply chain and the customer chain. TQ stresses learning and adaptation to continual change as keys to organizational success.
>
> The foundation of TQ is philosophical: the scientific method. It includes systems, methods, and tools. The systems permit change; the philosophy stays the same. TQ is anchored in values that stress the dignity of the individual and the power of community action.

As can be seen in the above definition, TQ is another label of such a management system. In fact, there is clearly a recent trend in many companies away from using the TQM acronym to signify their quality management efforts. Other common alternative names include *total quality (TQ), total customer satisfaction (TCS), total quality assurance (TQA), continuous quality improvement (CQI),* and *business process improvement (BPI).* Some companies have even developed their own personalized versions; for example, Motorola, General Electric, and Allied Signal refer to their customer-oriented management systems as Six-Sigma. The most important point to recognize is that the substance that underlies the acronym is what truly matters. Indeed, all acronyms should be able to pass out of use without affecting the usefulness of the management system described here.

The ideas of a quality management system seem so appealing that many managers insist that they have been following the approach all along. However, upon close examination, quality management leads to management practices that are very different from traditional management practices. Points of divergence between quality and traditional management thinking are often itemized in quality management books and articles. Some of the more conspicuous differences, drawn and adapted from a much longer list found in Roberts and Sergesketter (1993), are as follows:

- It is no longer assumed that higher quality is attainable only at higher cost. When higher quality is attained by reduction of waste, lower cost is a consequence of higher quality.

- Unlike older programs of *quality control,* quality assurance and quality improvement are not tasks of professional specialists alone; they concern everyone in an organization. Management must initiate quality management, but everyone must be involved, not only in special team improvement projects but also in everyday work.

- Good managers must understand the specific *processes* of an organization; the idea that management skills are readily transferable to any kind of organization is misleading.

- Management must influence or control upstream causes in processes in order to influence quality, costs, or profits, which are only the downstream results of processes.

- Tools and techniques of quality improvement—including the use of statistics, teams, and small groups—are useful only if the organizational culture is sound, including especially the orientation toward pleasing customers.

- In managing a process, it is essential to understand the nature of statistical variation and to try to isolate the root causes of variation.

The quality management paradigm is not a newly founded body of knowledge. It is really an accumulation of ideas from many sources that have, in some shape or form, been successfully applied over time by many leading companies. This body of knowledge is evolving and continuously improving. For the interested reader, comprehensive discussions and reviews of the managerial aspects of the quality management paradigm can be found in Bounds et al. (1994), Evans and Lindsay (1999), Goetsch and Davis (1997), and Kolarik (1995).

1.2   QUALITY PIONEERS

The work of numerous individuals[2] has helped shape *contemporary* quality thinking. There is, however, a select group of individuals who have had a particularly profound impact. Since these individuals should be familiar to all interested in the quality area, we briefly review their key contributions.

WALTER A. SHEWHART

If one person could be regarded as the father of the modern quality philosophy, it would be, without question, Walter A. Shewhart (1891–1967). Even though Shewhart is most recognized for his development of the statistical control chart, his contributions to the quality field were much broader. Shewhart's view of quality issues is encapsulated in the following quotation from his landmark book in quality control, *Economic Control of Quality of Manufactured Product,* published in 1931.

> Looked at broadly there are at any given time certain human wants to be fulfilled through the fabrication of raw materials into finished products of different kinds. These wants are statistical in nature in that the quality of a finished product in terms of the physical characteristics wanted by one individual is not the same for all individuals.
>
> The first step of the engineer in trying to satisfy these wants is therefore that of translating as nearly as possible these wants into the physical characteristics of the

---

[2]As of 1999, The American Society for Quality (ASQ) has honored 18 individuals by conferring on them the status of Honorary Member. This status is "reserved for those who are so well-known and clearly preeminent in the [quality] profession that there should be almost no doubt of their being worthy." A detailed summary of the background and contributions of each Honorary Member can be found from an ASQ Web site (http://www.asq.org/about/history/honorary.html).

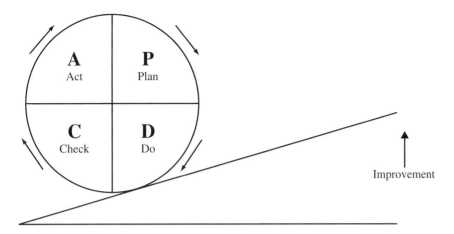

Figure 1.1   PDCA cycle and continuous improvement.

thing manufactured to satisfy these wants. In taking this step intuition and judgement play an important role as well as the broad knowledge of the human element involved in the wants of individuals.

   The second step of the engineer is to set up ways and means of obtaining a product which will differ from the arbitrarily set standards for these quality characteristics by no more than may be left to chance. (p. 54)

As can be seen, Shewhart was clear on the need for customer focus and developing means for satisfying customers which are among the basic tenets of any modern-day quality management system.

   The central contribution of Shewhart was his encouragement to tackle quality-related issues in a "scientific" manner. The foundation of Shewhart's notion of a scientific approach is based on two basic elements:

1. The understanding that variability is omnipresent, and thus statistical methodologies should be used to understand, monitor, and improve processes.

2. The belief that improvement of any process should be viewed as a systematic sequence of steps. Shewhart proposed a cyclical procedure to improvement which originally was referred to as the *Shewhart cycle*. Nowadays, the Shewhart cycle is typically referred to as the *plan-do-check-act (PDCA) cycle*[3] (see Figure 1.1).

The basic activities involved in PDCA are:

1. **Plan.**
    a.   Select a potential opportunity for improvement.
    b.   Describe the current situation surrounding the improvement opportunity.
    c.   Identify potential root causes underlying the situation.

---

[3]Other authors refer to the improvement cycle as plan-do-study-act (PDSA). The term *study* is used to emphasize the importance of learning from the improvement attempts.

    d. Select a solution and action plan, including targets for improvement.

2. **Do.** Implement the solution or change, typically as a small-scale pilot run.

3. **Check.** Evaluate the result of the change.

4. **Act.** Act on the evaluation from the check phase. If the experiment proved successful, implement changes as part of standard operating procedures, that is, making the changes a routine activity.

Upon completion of one PDCA cycle, there is then an opportunity for a new turn at the cycle. The continual "turning" of the PDCA cycle supports the philosophy of continuous improvement.

    Given Shewhart's pioneering role in the area of statistical process control with his development of the control chart, we will have many opportunities later in this book to delve more deeply into his perspective of the role of statistical methodology in the control of processes.

W. Edwards Deming

W. Edwards Deming (1900–1993) is widely recognized as the tireless and uncompromising advocate and teacher of quality methods for management who had profound influence on the start of and subsequent progress of what is sometimes called the postindustrial *quality revolution.*

    Much of Deming's perspective on quality can be traced back to his formative years. Deming was trained as a physicist, earning a Ph.D. from Yale University in 1928. While working as a physicist at the U.S. Department of Agriculture (USDA), Deming became responsible for courses in statistics at the USDA, and he invited Shewhart to lecture there. His affiliation with Shewhart had a profound effect on Deming shifting his interests to the area of statistics with application to quality. During the 1940s, Deming, along with other preeminent statisticians, applied statistical quality control principles at the U.S. Bureau of the Census, and he taught many short courses in statistical methods that had a substantial impact on the U.S. effort in World War II.

    In the 1950s, Deming was invited to war-ravaged Japan as an expert in statistics and quality control methods, to serve as a teacher and consultant to Japanese industry. Deming preached the importance of top-management leadership, customer-supplier partnerships, statistical analysis, goal setting, communication, and continuous improvement in product development and manufacturing processes. Japanese business leaders listened and embraced his ideas. More than 20,000 Japanese engineers were trained in statistical methods, and Japanese industry began to develop the world's most sophisticated assembly lines and processes. In 1991, *U.S. News & World Report* concluded that Japan's Deming-inspired miracle was one of the nine most important turning points in world history.[4]

---

[4]In the *U.S. News & World Report*'s rankings, Deming's hand in Japan's economic revolution (ranked ninth) is part of a list that includes The Apostle Paul, whose preaching led to mass acceptance of Christianity (ranked first) and the bubonic plague, which killed one-third of Europeans and resulted in the end of serfdom,

1. Create constancy of purpose toward improvement of product and service, with the aim to become competitive and to stay in business, and to provide jobs.

2. Adopt the new philosophy. We are in a new economic age. Western management must awaken to the challenge, must learn their responsibilities, and take on leadership for change.

3. Cease dependence on inspection to achieve quality. Eliminate the need for inspection on a mass basis by building quality into the product in the first place.

4. End the practice of awarding business on the basis of price tag. Instead, minimize total cost. Move toward a single supplier for any one item, on a long-term relationship of loyalty and trust.

5. Improve constantly and forever the system of production and service, to improve quality and productivity, and thus constantly decrease costs.

6. Institute training on the job.

7. Institute leadership. The aim of leadership should be to help people and machines and gadgets to do a better job. Leadership of management is in need of overhaul as well as supervision of production workers.

8. Drive out fear, so that everyone may work effectively for the company.

9. Break down barriers between departments. People in research, design, sales, and production must work as a team, to foresee problems of production and in use that may be encountered with the product or service.

10. Eliminate slogans, exhortations, and targets for the work force asking for zero defects and new levels of productivity. Such exhortations only create adversarial relationships, as the bulk of the causes of low quality and low productivity belong to the system and thus lie beyond the power of the work force.

11. *a.* Eliminate work standards (quotas) on the factory floor. Substitute leadership.

    b. Eliminate management by objective. Eliminate management by numbers, numerical goals. Substitute leadership.

12. *a.* Remove barriers that rob the hourly worker of his right to pride of workmanship. The responsibility of supervisors must be changed from sheer numbers to quality.

    b. Remove barriers that rob people in management and in engineering of their right to pride of workmanship. This means, *inter alia,* abolishment of the annual merit rating and of management by objective.

13. Institute a vigorous program of education and self-improvement.

14. Put everybody in the company to work to accomplish the transformation. The transformation is everybody's job.

Figure 1.2   Deming's 14 points.

Much of Deming's management philosophy is summarized in his famous 14 points (see Figure 1.2) in which he advocates a blueprint for overall management and organizational philosophy and strategy. Deming modified the specific wording of the points over the years, which accounts for the slightly different versions found in

making way for the Renaissance (ranked second). Some might feel that it is "farfetched" to include Deming's influence in a short list with such historic events. Any ranking is, of course, subjective and a potential source of debate. Personally, I take no position on the specific rankings of these events. My only perspective is that if a credible and influential source, such as the *U.S. News & World Report,* includes Deming's influence as part of a list of 10, 20, or, for that matter, 100 important events in history, then there is little argument about his profound historic impact.

various publications. Deming emphasized in later years that he regretted the numbering of the points since many people wrongly interpret the numbers as a priority listing or as a sequence of tasks to be done in progression. Instead, Deming advocated that the points should be viewed in whole as an interactive and synergistic system of management. Detailed elaboration of each point can be found in Deming (1986), Gitlow et al. (1995), and Scherkenbach (1988).

JOSEPH M. JURAN

Upon completing a bachelor of science in electrical engineering, Joseph M. Juran (1904–   ) joined Western Electric in 1924 to work in the inspection department of the historic Hawthorne Works plant in Chicago. In 1926, a Shewhart-led team of Bell Laboratory statisticians made a visit to the Hawthorne plant with the goal of applying some of the statistical tools and methods that they had been developing to operations at the plant. Recognized for his analytical ability, Juran was chosen as one of the 20 trainees and then as one of two engineers for a newly formed Inspection Statistical Department—one of the first such departments established in industry in the United States. This fortuitous experience set Juran firmly on the path toward his life's work in quality.

While at Western Electric, Juran authored a pamphlet called *Statistical Methods Applied to Manufacturing Problems.* This writing became an input to the pioneering book *AT&T Statistical Quality Control,* which is still in publication today and continues to serve as an important reference to quality professionals. In 1945, after a leave of absence from Western Electric for government service, Juran resigned from Western Electric to embark on a second career as an independent consultant, devoting the rest of his life to philosophize, consult, lecture, and write on quality management.

Like Deming, Juran was invited to Japan to teach quality principles in the 1950s. Both men emphasized that quality thinking and customer orientation were necessary ingredients for competitiveness. Juran echoed Deming's recommendations for top-management commitment to quality, the need for continual improvement, the use of data and quality control techniques, and quality training at all levels of an organization. Juran's contributions to Japan were not overlooked; Emperor Hirohito awarded him Japan's highest award that can be given to a non-Japanese, the Order of the Sacred Treasure.

Even though most aspects of Juran's and Deming's messages are quite similar, there are some points of divergence. Unlike Deming, Juran does not propose a radical organizational change like that laid out in Deming's 14 points. Juran, in fact, does not agree with all of Deming's points. For instance, Juran does not embrace the notion that management must drive out fear (point 8). From Juran's perspective, "Fear can bring out the best in people."[5] Recognizing that certain organizational changes are no doubt necessary, Juran's approach to improve quality is to work within the organizational system familiar to the managers. As a case in point,

---

| [5]J. Main, "Under the Spell of the Quality Gurus," *Fortune,* August 18, 1986, pp. 30–34.

he argued that successful implementation of a quality improvement program requires speaking the "language" of the employees. Juran stated that upper managers speak the language of money; front-line workers speak the language of things; and middle managers speak in both languages, translating between money and things. Hence, to capture the attention of financially oriented executives, Juran advocated that quality issues be cast in terms of financial figures.

In the pursuit of high quality, Juran prescribes a focus on three managerial processes, called *The Juran Trilogy*[6] (Table 1.1): (1) *quality planning*—the process of developing methods to be in tune with customers' needs and expectations; (2) *quality control*—the process of comparing products produced with goals and specifications; (3) *quality improvement*—the ongoing process of improvement necessary for continued success.

Table 1.1   The three universal processes of managing for quality

Quality Planning	Quality Control	Quality Improvement
Establish quality goals	Evaluate actual performance	Prove the need
Identify who are the customers	Compare actual performance to quality goals	Establish the infrastructure
Determine the needs of the customers	Act on the difference	Identify the improvement projects
Develop product features which respond to customers' needs		Establish project teams
Develop processes able to produce the product features		Provide the teams with resources, training, and motivation to:
		• Diagnose the causes • Stimulate remedies
Establish process controls; transfer the plans to operating forces		Establish controls to hold the gains

Reprinted with the permission of The Free Press, a division of Simon & Schuster, from *Juran on Quality by Design* by J. M. Juran. Copyright © 1992 by Juran Institute Inc.

As can be seen from Table 1.1, Juran approaches quality improvement by means of a project-by-project program. The approach calls for the identification of improvement needs, selection of appropriate projects, and creation of an organizational structure that guides the diagnosis and analysis of projects. Successful improvement efforts raise quality performance to unprecedented levels which Juran calls "managerial breakthroughs."

Finally, Juran is well known for espousing the *Pareto principle,* which states that whenever a number of individual factors contribute to some overall effect,

---

[6]The Juran Trilogy is a registered trademark of Juran Institute, Inc.

relatively few of those factors account for the bulk of the effect. For example, we might find that while a manufacturer produces 30 different products, 5 of those account for 80 percent of its customer sales, or 80 percent of defects are due to 20 percent of causes or sources. Since a quality improvement team cannot address all sources of a problem, Juran suggests that organizations concentrate on the "vital few" sources of problems and not be distracted by those of less importance, the "trivial many." Focusing on the vital few enables a team to achieve the highest return on the investment of resources and effort.

ARMAND V. FEIGENBAUM

Although not as well known as Deming and Juran, Armand V. Feigenbaum (1922–) has had profound influence on the evolution of quality management. In 1941, Feigenbaum began his career at General Electric, where he served as a manager of manufacturing and quality control. While working on his Ph.D. at the Massachusetts Institute of Technology, he wrote the now famous book *Total Quality Control*, first published in 1951 under the title *Quality Control: Principles, Practice, and Administration.*

With this seminal work, Feigenbaum was a pioneer in championing an organization-wide approach to quality. He viewed quality not as the sole responsibility of a small group of technical specialists, but rather as a strategic business tool requiring the involvement of all functional areas in the company. According to Feigenbaum (1983), "Total quality control is an effective system for integrating the quality-development, quality-maintenance, and quality-improvement efforts of various groups in an organization so as to enable marketing, engineering, production, and service at the most economical levels which allow for full customer satisfaction" (p. 6). As shown in Figure 1.3, he views total quality control as a horizontal, cross-functional concept stretching across the functional divisions of an organization. The Japanese embraced Feigenbaum's model and expanded it further to include the participation of the entire workforce in the management of quality, regardless of hierarchical level. In this form, quality management involved a complete mobilization of improvement and control efforts—including all functions and all levels—in one corporate effort.

Since the works of Deming, Juran, and Feigenbaum represent seminal perspectives on quality management, it is often asked if one approach reigns as a more appropriate approach to quality. Internationally regarded management theorist Tom Peters (1988, p. 74) has asked, "Which system? There's a lot of confusion here. Should you follow W. Edwards Deming, father of the Japanese quality revolution . . .? . . . Or Armand Feigenbaum's Total Quality Control? Or Joseph Juran? Or invent a system of your own?" His conclusion: "Frankly, it makes little difference which you choose, among the top half-dozen or so, as long as it is thorough and followed rigorously."

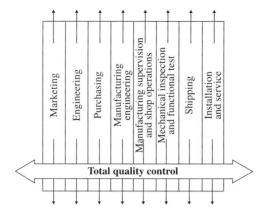

Figure 1.3   Feigenbaum's horizontal scope of total quality
control. Reproduced, with permission from
McGraw-Hill, from A. V. Feigenbaum (1983),
Total Quality Control, 3rd ed., p. 83.

KAORU ISHIKAWA

In the 1950s, Japan's economy was devastated not only by World War II but also
by a worldwide reputation for shoddy and poor-quality products. A deliberate
and concentrated effort was undertaken to rid Japan of this reputation. As discussed, part of this effort included the invitation of Deming and Juran to teach
methods for improving quality and competitive position. During this time, Kaoru
Ishikawa (1915–1989) initiated a campaign aimed at promoting the application
of simple tools useful for quality control and improvement. Ishikawa felt that all
individuals in an organization should become involved in quality problem solving. He advocated the use of six quality tools: check sheets, Pareto diagrams, histograms, cause-and-effect diagrams, scatter diagrams, and control charts. These
six tools plus the technique of flowcharting are often referred to as the *seven
basic tools.*[7] These tools will be covered in greater detail in Section 1.6.

Ishikawa is also credited with the creation of the *quality circle.* A quality circle is a team of employees who volunteer to meet regularly for the purpose of
identifying, recommending, and making workplace improvements. The use of
quality circles has become a genuine movement throughout the entire industrial-
ized world. The Japanese Union of Scientists and Engineers (JUSE) reports that
quality circle activities have spread to more than 50 countries worldwide. In
Japan alone, Main (1994) reports an estimated 743,000 circles in 1988, and JUSE
alone had 350,000 circles registered in 1992. One of the reasons for the success

---

[7]Interestingly, many authors mistakenly include the flowcharting technique as one of the tools originally advocated by Ishikawa. However, careful study of Ishikawa's writings will reveal no mention of flowcharting.

of quality circles is the sense of teamwork and cooperation that results as employees work together toward a common goal.

GENICHI TAGUCHI

Born in Japan, Genichi Taguchi (1924–) developed a keen interest in the area of statistics during World War II; it is reported that his interest in statistics was sparked by his receipt of a draft notice resulting from a random selection process. Immediately following the war, Taguchi honed his statistical skills while working for the Institute of Statistical Mathematics. In 1950, he joined the Electrical Communication Laboratory (ECL), where he was given the opportunity to apply methods of design of experiments and data analysis to improve the productivity and efficiency of various research and development projects. His experience at ECL set Taguchi on a path to develop a unique approach to quality design and improvement.

Taguchi's approach to quality has been characterized as an "engineering approach" to quality. Taguchi observed that variation in product characteristics, relative to their specifications, and inconsistency of product performance in the field are primary sources of customer dissatisfaction or *quality loss*. As such, he stresses the need to produce according to specifications with minimal product performance variation in the customer's environment. In Taguchi's terminology, *noise* means variation. Taguchi identifies three distinct types of noise factors:

1. *External noise*—variables in the environment (e.g., temperature, humidity, and barometric pressure) that affect product performance.

2. *Deterioration noise*—changes in product performance as a result of wear or storage.

3. *Manufacturing noise*—differences between individual units that are manufactured to the same specification.

With the objective of minimizing overall noise, Taguchi proposed the use of two approaches to quality control: on-line quality control and off-line quality control. On-line methods include various techniques for maintaining target values and controlling variation around the target during the production phase. These techniques include methods such as statistical control charts for process monitoring and adaptive control methods for on-line process adjustment based on observed deviations from target values. The discussion of statistical control charts will be the core of this book. Adaptive control methods will be discussed in detail in Chapter 10.

It is the development of the off-line quality control methods that makes Taguchi's contributions distinctive. Off-line methods are technical aids for quality and cost control prior to the product's going into production. Off-line activities include market research, product development, and process development. Once production is initiated, the on-line quality control activities come into action. Taguchi (1978) suggested that an off-line quality system should be based on three components: (1) system design, (2) parameter design, and (3) tolerance design.

System design is the process of applying scientific and engineering knowledge to determine both the product configuration and the design of the manufacturing process for the product in question. System design requires not only knowledge of the customers' needs but also an understanding of the manufacturing environment. Techniques employed in system design include methods for establishing customer requirements and translating those requirements to engineering terms; for example, brainstorming techniques and *quality function deployment (QFD)* are often utilized in system design.

Once the system is chosen, the parameters, or values of the system variables (product and process), need to be chosen. *Parameter design* is the process of identifying key product or process variables that affect product variation and then setting product and process parameters so that the least amount of variation will result in the product's function. To accomplish this, the parameters need to be chosen in such a way as to minimize the product's sensitivity to varying operating conditions. Taguchi calls this *robust design*. As an example, consider a cake mix product. Potentially, the cake mix could be used by consumers in low-, medium-, and high-altitude areas. Ideally, consumers want the cake to rise to a certain height with a certain degree of fluffiness. A cake mix with a given set of ingredients and a baking procedure is considered robust if the outputs of cake height and fluffiness are not greatly affected by the variations in altitude.

Finally, *tolerance design* calls for the determination of tolerances around parameter settings identified by parameter design. Often tolerances are set arbitrarily rather than by scientific reasoning. Tolerances that are too tight increase manufacturing costs unnecessarily, and tolerances that are too loose increase performance variation. Rather than tightening tolerances across the board, Taguchi calls for tolerance determination to be chosen based on economic considerations, that is, based on the tradeoff between manufacturing costs and customers' loss due to performance variation. Tolerance design also requires the knowledge of those factors which contribute most to the end-product variation. If a factor does not have much effect on the end-product performance, it can be specified with a wider tolerance.

An essential ingredient in Taguchi's off-line quality control scheme is the utilization of statistical *design of experiments (DOE)* methods, particularly during parameter and tolerance design phases. In general, DOE is a statistical approach for studying the effects of purposeful changes of input variables on some output variable. Taguchi's application of DOE techniques departs somewhat from the classical DOE approach in that he integrates concepts known as loss functions and signal-to-noise ratios (see Taguchi, 1986, for details). Although many companies (Japanese and U.S.) have reported success with Taguchi's approach, a number of investigators (e.g., see Montgomery, 1997, section 14-7) have revealed technical deficiencies to his DOE recommendations. Notwithstanding these criticisms, many of these same investigators agree that any such technical flaws should not undermine Taguchi's pioneering contribution to the quality arena and the importance of his broad philosophy of quality engineering.

GEORGE E. P. BOX

Regarded as one of the preeminent statisticians of the 20th century, George E. P. Box (1919–) actually began his studies in the area of chemistry. During World War II, he joined British Army Engineering where he studied the effects of poison chemicals on small animals. Recognizing that the analysis of his studies required the aid of statistics, he was forced to learn statistical techniques on his own since no statistician was available for consultation.

Following the war, Box pursued degrees (bachelor's and doctoral) in mathematical statistics. During this period of study, Box accepted a position at Imperial Chemical Industries (ICI), where he served as a practicing industrial statistician. His years at ICI generated in him a deep appreciation for the development of statistical techniques for product and process improvement and manufacturing applications in general. Box later accepted a professorial position at the statistics department at the University of Wisconsin.

His contributions to the field of statistics have been pioneering and cover a wide span of topics applicable in the quality sciences: design of experiments, time-series analysis, data analysis, and process control. We will utilize a number of his statistical contributions with this book. Given the importance of his statistical contributions and their relevance in the quality area, Box is, without question, a key leader in the evolution of the quality area. When Box was elected to the prestigious status of honorary member of the American Society for Quality (ASQ), Frank Caplan[8] summarized Box's qualifications by stating, "It should be apparent that George Box is, in the field of the quality sciences, the consummate 'Renaissance man' who has made significant and enduring contributions to the profession of quality control and the allied arts and sciences . . . [his] contributions encompass considerable scope and have already had lasting effect and, as a result, he fully deserves the honor of the highest level of membership that ASQ can bestow."

1.3   FOCUS ON PROCESSES

We have already used the word *process* a number of times. The notion of a process is fundamental with modern approaches to quality, and even management in general. Simply defined, a *process* is any collection of activities that are intended to achieve some result, typically to create added value for a customer. Specific business examples might be the manufacturing and assembly of a product; writing of computer code; purchasing, billing, and treating of patients in a hospital; and employee training. Some processes may be contained wholly within a particular business function, such as writing computer code within the MIS department. However, most processes (such as purchasing or product design) are cross-functional, spanning various functional areas on an organizational chart.

[8]Founding editor of *Quality Engineering*, an American Society for Quality–sponsored journal.

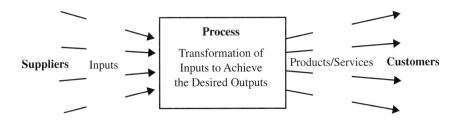

Figure 1.4   Basic scheme of a process.

The modern approach to management sees each employee in an organization as part of one or more processes. From this perspective, the responsibility of every employee is to receive input, add value to that input, and supply it to the next person in the process. Berwick, Godfrey, and Roessner (1990) call this responsibility the *triple role*—the employee as customer, processor, and supplier. Figure 1.4 is a linear representation of the basic elements of a process. Process inputs can include such things as material, equipment, information, people, policies, procedures, methods, and environment. The notion of supplier-customer relationships begins with external suppliers, progresses through the organization in a series of internal supplier-customer relationships, and ends with external customers.

Breaking down organizational activities into processes, or the steps required to perform activities, enables employees to focus on and identify exactly how work gets done. By understanding their own needs and carefully defining them, employees help suppliers improve the quality of the suppliers' work. In the same way, by understanding the needs of their customers and carefully defining these, employees improve the quality of their own work.

There are numerous ways to classify processes. Juran (1992) refers to large processes, typically spanning different departments and subsuming various departmental operations (steps, tasks, etc.), as *macroprocesses* as opposed to *microprocesses,* which represent the subprocesses embedded within macroprocesses.

Other classification schemes are based on the goals of the processes. For example, processes directed at producing goods are referred to as industrial, production, or manufacturing processes. Industrial processes are easiest to visualize since these processes produce things. As defined by Harrington (1991), an *industrial process* is "any process that comes in physical contact with the hardware or software that will be delivered to an external customer, up to the point the product is packaged (e.g., manufacturing computers, food preparation for mass customer consumption, oil refinement, changing iron ore into steel)." Another increasingly important type of process is commonly referred to as a *business* or *administrative process*. Business processes include all service processes and processes that support industrial processes (e.g., billing, payroll, and engineering changes). Business processes are notorious for frustrating customers (internal and external) more frequently than others. Harrington (1991) reports that external customers are 5 times more likely to abandon a company for its poor business processes than for its poor products.

Thus far, we have focused attention on processes common *within* an organization. However, the notion of a process is ubiquitous. Any repetitive phenomenon can be viewed as the result of a process, and as such, the idea of a process can range from our everyday personal lives, to external organizational processes, to even more general societal processes. At the personal level, process examples include cooking, playing golf, organizing one's daily activities, controlling weight, and studying. Or, we may be interested in studying and improving corporate sales, air pollution levels,[9] airline accident rates, or crime rates. One of the great contributions of the quality revolution is the realization that *any* process, not just the classical applications to manufacturing or service processes, can potentially be improved. In this book, we will try to adopt this wider view of processes in our presentation of data applications.

1.4   RELIANCE ON DATA

As a rule, process improvement cannot be achieved by armchair reasoning or inspiration. To understand how a process is performing and whether improvement has occurred requires data. As stated by a leader in the Japanese quality movement, "Data is a guide for our actions. From data we learn pertinent facts, and take appropriate actions based on these facts" (Kume, 1985). At General Electric, Hoerl (1998) reports that the emphasis on data for sound decision making and process improvement efforts is becoming ingrained in the organization.

> Similarly, vague statements, such as "We think we have that under control" or "Recent performance seems to be improving" are no longer accepted at any level. The typical response to these kinds of statements now is, "Show me the data!" (p. 39)

In a similar message, quality professionals have long quoted the classic credo "In God we trust, all others bring data."

The general advice to "bring data" is sound, but we must also be selective about the data we bring. Data are not good in and of themselves. If data are to guide process improvement, they must arise from meaningful measurements of relevant aspects of the process. Data that are irrelevant cannot be magically transformed by data analysis into a valuable product. Unfortunately, many companies accumulate mountains of useless data and fail to collect other potentially useful data. World-class companies have come to realize the importance of collecting data on key measures of operational performance. Examples of key measures include:

---

[9]The April 1991 issue of *Quality Progress*, the ASQ–sponsored magazine, was devoted to articles in which the environment is treated as a customer.

- Customer preferences and satisfaction.
- Complaints.
- Costs.
- Defect or scrap rates.
- Yields.
- Cycle times.
- Customer wait times.
- Measured quality characteristics (e.g., dimensions, weights, and concentrations).
- Breakdowns or failures.
- Work-in-progress inventories.
- Number of changes to work orders or methods after instruction.
- Days lost due to absenteeism or sickness.
- Accidents.

With the exception of customer satisfaction and complaints, the focus of the above-listed measures is on internal processes. The argument for this is that these internal processes drive organizational success. Notice also that most of the listed measures are not expressed in financial terms. Without denying the potential importance of internal processes on overall organizational success, there is a growing recognition that greater attention needs to be paid to measures of external processes and of the success of the overall application of process management methods for the organization as a whole. External measures reflect, in varying degrees, the aspects of quality, productivity, and profitability that are essential to the success of an organization. For any given organization, below is a listing of different external company measures that may be useful to monitor:

- Sales volume.
- Return on investment.
- Stock market rate of return.
- Net value added per employee: sales minus costs of material, suppliers, and work done by outside contractors, divided by total number of employees.
- Number of creditors and debtors.
- Net accounts receivable days.
- Loss of customers.
- New customers captured.

In selecting projects for the improvement of internal processes, companies are emphasizing the need to improve the internal nonfinancial metric and also produce bottom-line results. For instance, as reported by Snee (1998), General Electric views three elements as crucial to the success of process improvement projects:

1. The selection of projects that have well-defined measures.

2. The conversion of process improvement (e.g., defect and cycle time reduction) to significant dollars.

3. The definition of projects that can be completed swiftly—within 3 to 6 months.

The need for well-defined measures for process analysis is in accord with Deming's requirement for operational definitions. An *operational definition* is a set of criteria which enable one to translate a concept into communicable meaning. A requirement that a shirt be "clean" has no meaning unless a specific method is given to determine just what clean means. What is a clean shirt? Is it based on how it looks? Is the shirt "cleaner" if it smells a certain way? Without an operational definition, there is no way to communicate effectively. Balestracci (1998) provides a humorous scenario to explain the importance of an operational definition:

> One of my favorite Dilbert cartoons shows Dilbert sitting with his girlfriend, and he says, "Liz, I'm so lucky to be dating you. You're at least an eight," to which she replies, "You're a ten." The next frame shows them in complete silence, after which he asks, "Are we using the same scale?" to which she replies, "Ten is the number of seconds it would take to replace you." (p. 4)

An operational definition is not so much "true" as useful, for a particular purpose. Its purpose is to state or imply the same actions to everyone who uses it. Suppose, for example, a military unit is planning an attack. The commanding officer gives the order to the unit to "synchronize watches." Everyone accordingly sets his or her own watch to the time shown on the commanding officer's watch. This serves as a basis for an operational definition of the duration time of the mission. Whether or not it agrees with the true time, determined by the U.S. Naval Observatory Master Clock, does not matter.

Deming argues that an operational definition must be based on three elements: (1) a specific test of the product or service; (2) a criterion (or criteria) for judgment; and (3) a decision (yes or no) as to whether the product or service meets the criterion (or criteria). To demonstrate the development of an operational definition, Deming (1986) has the reader consider the meaning of a label on a blanket which reads "50 percent wool." Deming argues that the label has no operational meaning. For example, a manufacturer could produce a blanket that is all cotton on one half of the blanket (say, from one's head to waist) and all wool on the other half of the blanket (say, from one's waist to feet)! Such a blanket is indeed 50 percent wool. However, the typical consumer will interpret the label as meaning the wool is dispersed throughout the blanket. To operationalize this interpretation, Deming offers the following operational definition:

Test: Cut 10 random holes in the blanket, 1 or 1.5 centimeters in diameter. Number the holes 1 to 10, and hand 10 associated cutouts to a chemist for a test. The chemist will record x_i, the proportion of wool by weight of cutout i. Compute $\bar{x}$, the average of the 10 proportions.

Criteria: $\bar{x} \geq 0.50$ and $x_{\max} - x_{\min} \leq 0.02$

Decision: If the sample fails on either criterion, the blanket fails to meet the specification of "50 percent wool."

1.5 ROLE OF STATISTICS

Obtaining relevant data is the first step toward process understanding and improvement. However, it is not enough to collect data. Insightful statistical analysis of the data is essential. The application of statistical methods to quality and productivity improvement activities is well documented. In Japan, statistical concepts and methods not only are taught extensively to employees and used in companies but even have penetrated elementary and secondary education. As Box (1984) reports: ". . . many millions of people in Japan have some training in the use of statistical methods. These techniques permeate the whole of their industries and are the basic tools for a never-ending incentive towards improvement." Thurow (1993), a well-noted economist and former dean of MIT's Sloan School of Management, points out that the use of statistical methods clearly had a significant impact on manufacturing: "Without statistical process control, today's high-density semi-conductor chips cannot be built. They can be invented, but they cannot be built." Ishikawa (1985) summarizes the impact of statistics by stating: "The use of statistical methods, including the most sophisticated methods, has become deeply rooted in Japan. . . . Japan's advance in productivity cannot be dissociated from the use of statistical methods. It was through these that the quality level has risen, reliability has risen, and cost has fallen."

Insightful data analysis does not always require highly sophisticated tools. Quite often elementary tools, such as the seven basic tools (which will be presented in section 1.6), have proved to be very effective in practice. There are also many applications for which more sophisticated statistical tools are required to properly analyze the data. As established in the Preface, this book is about data analysis. Generally, data analysis follows a sequence that entails display of data, formulation of a tentative explanation or model for what is happening in the data, fitting of the model to data, and diagnostic checking of the adequacy of the fitted model. The results of data analysis can then be applied for purposes of process evaluation, prediction, and improvement.

Throughout this book we will emphasize the recurring theme that statistical tools are designed to be most effective under certain conditions. In particular, our focus will be on the standard techniques of *statistical process control (SPC)* and their effectiveness in various process scenarios. In Chapter 4, we will learn that standard SPC methods, known as control charts, were developed more than 70 years ago for the control of *manufacturing* processes. With manufacturing processes, the objective is to center the process on dimensional targets and to reduce variability around these targets. Accordingly, with certain additional assumptions (see Chapter 4), SPC techniques were developed to help monitor and control processes.

But, as discussed in the previous section, organizations nowadays are interested in understanding, tracking, and improving a wide variety of processes, not just target-based manufacturing processes. For instance, a company may wish to monitor its monthly sales volume process to see if certain management interventions had any effect on the process (positive or adverse). The analysis of a typical monthly sales data series will require the use of statistical methods to accommodate trends, seasonalities, and other time movements. Such an application requires the insight provided by statistical methods that go beyond the standard SPC techniques. Furthermore, we will learn that these broader methods may be required and very useful when studying manufacturing processes, which is the intended focus of standard SPC techniques.

In a thought-provoking article, Gunter (1998) strongly emphasizes the need to enhance the statistical toolbox for current and future challenges in process management:

> . . . the core principles underlying it—that variability is the issue and that it is important to distinguish between variability that is inherent to the system and the variability that is not—are as valid today as they were 70 years ago. But times have changed. . . . Control charts have had a long and successful run, but it is time to move beyond these now archaic and simplistic tools. . . . The reality of modern production and service processes has simply transcended the relevance and utility of this honored but ancient tool. . . . If the tools used by the quality community do not add value nor help solve real problems, then those who use them will be swept away. . . . But of this I am certain: If 70 years from now ASQ's quality engineer certification exam is still at its current level and concerned with its current subject matter, there will be no takers. The knowledge will be both obsolete and worthless. We simply have got to move on and up to the modern data analysis methodology required for today's quality reality. (pp. 113–119)

It should be underscored that *our view is not that standard SPC techniques— control charts— have no role in quality application.* Indeed, as we demonstrate throughout the book, standard SPC techniques continue to play an integral role in process management. However, we wholeheartedly agree with Gunter's sentiments that effective process management today can no longer rely on this one set of statistical tools. There is a need to broaden the statistical foundation with the tools of data analysis.[10] With such a base, we will learn when standard SPC techniques can be rewarding and when broader approaches are needed.

Our discussion of the importance of statistical tools needs to be placed in a proper framework. In the absence of a favorable organizational culture, statistical techniques—elementary or sophisticated—have limited potential at best. Leading writers (e.g., Deming, 1986; Juran, 1964; Ishikawa, 1985), have both emphasized that statistical applications cannot be made in an organizational vacuum and have helped to define an organizational culture conducive to effective

---

[10]As a technical note, Gunter points out that there is a need to handle *autocorrelation* (concept to be learned later in this book) and use techniques such as time-series methods. We will learn that without a fundamental understanding of standard SPC techniques (concepts and how to implement), we cannot effectively handle autocorrelation or use time-series methods.

statistical application. To summarize our earlier discussions, the keys to a favorable organizational culture include a shared dedication throughout the organization to process improvement and problem solving based on data rather than subjective opinion or emotion; the recognition that all employees can contribute to process improvement; and an organizationwide dedication to pleasing customers, internal and external. These conditions do not arise spontaneously, but must be shaped and promoted by top managers.

Unfortunately, even with good intentions to shape a customer-oriented culture, many U.S. companies, Hoerl (1995) has observed, still have "barely scratched the surface relative to tapping the vast potential bottom-line impact of statistical methods." In his view, there is an imbalance in the use and nature of statistical concepts and methodologies within the typical organizational hierarchy. He writes:

> A critical problem has been that statistically literate managers have emphasized new and better tools at the operational level almost exclusively, ignoring the need to understand statistical concepts at the strategic level and to integrate these concepts with the managerial systems at the tactical level. Well-meaning managers have also frequently insisted on statistical training of operational employees, while feeling no need to receive any training themselves. (p. 63)

Hoerl suggests that the promotion of statistical applications throughout an organization will be made more successful if the nature of statistical concepts and methods assimilated by employees is consistent with their routine responsibilities and activities. For example, he argues that for upper management an infusion of "statistical thinking" is more critical than training in any specific statistical tool. We turn to Roberts (1991) to bring out the salient points of statistical thinking:

> We shall see that statistical thinking requires familiarity with ideas such as the regression phenomenon, sample selection, location and variability, correlation and auto correlation, distributions, conditional probability, and the hazards of measurement error in data. These ideas are useful even in the absence of formal statistical evidence. . . . Statistical thinking is central to effective management: statistics is one of the glues that bind together the functional areas of business. Among the writers on quality and productivity improvement, Deming has most clearly spelled out the connection of statistical thinking and management. In particular, he emphasizes that understanding variability is central to management, and that statistical training is needed for proper understanding of variability. . . . Interpretation of the latest data point—last month's sales, the most recent closing stock price, or the yield of a production batch— brings out clearly the need for statistical thinking by managers. A manager not understanding statistical thinking can only guess and pontificate about what the point means. Only a manager who understands variability can understand how statistical analysis can help to put the new information into proper perspective. (pp. 8–9)

Even though many companies, as reported by Hoerl (1995), have yet to fully tap into the potential of broad-based application of statistical concepts and methodologies, there are leading companies making significant efforts to ingrain the use of statistics throughout their organizations. Most notably, General Electric, one of the largest companies in the world, is in the midst of a massive undertaking to become "by year 2000, a Six Sigma quality company, which means a company that produces virtually defect-free products, services and transactions."[11] As defined by Hoerl (1998), *six sigma* (denoted 6σ) is "a disciplined, quantitative approach for improvement of defined metrics in manufacturing, service, or financial processes." To support the conversion to a six sigma company, GE has committed to an ambitious training program of green belts and black belts. To become a black belt, an employee is required to complete a 4- or 5-week highly quantitative training program. The statistical topics integrated in the program are

- Basic statistics using Minitab (week 1).
- Process capability (week 1).
- Measurement systems analysis[12] (week 1).
- Statistical thinking (week 2).
- Hypothesis testing (week 2).
- Correlation (week 2).
- Simple regression (week 2).
- Design of experiments (week 3).
- Analysis of variance (week 3).
- Multiple regression (week 3).
- Statistical process control/advanced process control (week 4).

Indeed, with the exception of design of experiments and analysis of variance, we will have the opportunity to deal with each of these topics in this book. In addition, we will introduce many other particularly useful data analysis techniques that are not included in a typical black belt training program. The reader might have noticed and inferred from the above listing that GE has designated Minitab as the statistical software of choice. As explained in the Preface, Minitab also serves as the primary software vehicle for this book.

In summary, the concepts and methods of statistics not only are important for effective process improvement efforts, but also play a critical role in the making of an effective manager in general. We believe that this alone is a powerful motivation for delving into learning useful statistical topics such as those discussed in this book. However, there may actually be an even more motivating factor—career potential! As quoted in *The Wall Street Journal,* "Mr. Welch [CEO of GE] has told young managers that they haven't much future at GE unless they are selected to become Black Belts." As we have learned, this means that managers of the future at GE are expected to possess broad-based statistical skills in their

---

[11]"GE Quality 2000: A Dream with a Plan," speech presented by John F. Welch, Jr., Chairman and CEO of GE, at GE Company 1996 annual meeting, Charlottesville, VA, 1996.

[12]Measurement systems analysis is synonymous with gage R&R analysis, which will be presented in Chapter 9 of this book.

portfolio. We have focused on GE because of its position as one of the largest and most influential companies in the world. But note that GE is not the only large company to invest in six sigma type of initiatives. For example, Motorola and Allied Signal are two other major companies that actually pioneered and developed many of the 6σ concepts, before GE's adoption. Furthermore, the literally thousands of supplier companies to these large corporations are experiencing the need to make the transition to a 6σ approach if they wish to remain in the supplier chain. Thus, as we enter the next millennium, we can expect an ever-increasing interest in the utilization of statistical methods for the improvement of processes.

1.6   BASIC PROCESS IMPROVEMENT TOOLBOX

A *scientific approach* to process improvement is the key contribution of the quality management paradigm. As we have learned, this means that instead of management by impulse, or by exhortation, or by preconception, process improvement is based on the management of facts—facts about the process at work and about the root causes of less-than-desired process performance. To facilitate the collection, organization, and analysis of facts, several problem-solving methods or tools have been developed.

In this section, we provide an overview of a popular collection known as the seven basic tools: flowchart, check sheet, Pareto diagram, histogram, cause-and-effect diagram, scatter diagram, and control chart. For additional details, excellent resources for the presentation of elementary problem-solving tools include *The Memory Jogger II* (1994) and Ishikawa (1982). In general, problem-solving tools have a variety of uses in the improvement of processes. Certain tools are used primarily to obtain an understanding of the process and to identify possible causes of problems—for example, flowcharts, Pareto diagrams, and cause-and-effect diagrams. Other tools are used mainly to gather information, for example, check sheets. Still others are used to display information and evaluate improvement attempts—histograms, scatter diagrams, and control charts. Finally, certain tools are used mainly to monitor and control a process after a verifiable improvement has occurred, for example, control charts.

As you may recall, these activities of process understanding, root cause analysis, evaluation of improvement efforts, and implementing successful changes to the process are encompassed within the systematic plan-do-check-act (PDCA) cycle (introduced in Section 1.2). While certain tools are most useful at particular stages of the PDCA cycle, there is no exact correlation. Many tools, for example, control charts or histograms, can be useful in different ways at different stages in the cycle.

Many of the basic tools seem too simplistic to lead to any significant benefit. However, the quality literature is rich with examples of the rewards attained from the application of the most elementary problem-solving tools. Based on a long experience of promoting the use of basic quality tools in Japanese industry,

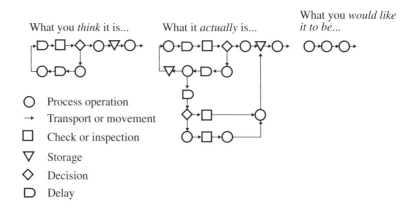

Figure 1.5   Flowchart perspectives.

Ishikawa (1990) contends that the vast majority of quality-related problems can be solved with the "skillful" application of the basic tools.

FLOWCHART

Using a variety of different symbols, a flowchart is a graphical representation of the sequence of events (e.g., tasks, decisions, and flow) in a process. It can be used to trace an entire process or a particular segment of a process. Flowcharts are universally applicable; for example, they can be used to study a manufacturing process, the process of admission to an emergency room, or even the steps in making a sale. Process improvement teams most commonly start their efforts with a flowchart since it provides a documentation of the existing process.

Quite often, individuals who work directly with a process are surprised to find out that their understanding of the process is markedly different from what is revealed by the flowchart. Day-to-day involvement with a process tends to create a false sense of familiarity, resulting in a biased perception that the process is much less complex than it is in actuality. As such, a flowchart can offer a fast start on process improvement since it can expose unexpected complexities (e.g., unnecessary loops and redundancies) or non–value added activities (e.g., an unnecessary inspection or waiting points that increase overall cycle time). Thus, the flowchart paves the way for process simplification by eliminating redundancy and unnecessary steps. Figure 1.5 summarizes our discussion.

CHECK SHEET

The collection of data is fundamental to process improvement. The need for data can occur at various stages during a problem-solving effort: (1) Data can be used to help select and define an opportunity for improvement. (2) Data can be used to reveal underlying causes of the problem under study. (3) Data can be used to check the effectiveness of a solution. (4) Data can be used to monitor and control a process to ensure the desired process performance. Data collection should not be performed in a haphazard fashion. The design of a relevant and effective data collection scheme requires one to address some of the following basic questions:

- What question do we wish to answer?
- What type of data will we need to answer the question?
- What data analysis tools do we envision using?
- Where can we find the data?
- Who can provide us with the data?
- How can we collect the data with minimum effort and with minimal chance of error?

A common format for collecting data is known as a *check sheet*. A check sheet is a device to collect data about a product, service, or process in a simple and organized manner. Strictly interpreted, a sheet is designed for entry of data by entering checkmarks or tallies. However, as many practitioners and authors do, we take the broader view that a check sheet can be any data-recording device designed for easy collection of relevant data.

There are three common types of check sheets: attribute check sheets, variable check sheets, and location check sheets. An attribute check sheet is designed to collect data on a countable phenomenon, such as the number of defects or number of lost invoices. Given that the data are in countable form, attribute check sheets are typically implemented by entering checkmarks or informative symbols in the appropriate places. A variable check sheet is any data collection form devoted to acquiring data measured on a continuous scale, such as size, weight, diameter, or waiting time. Finally, a location check sheet is a schematic drawing of a product which is marked on to reveal concentrations of defects by location. Figure 1.6 shows examples of these three types of check sheets.

It should be pointed out that check sheets are most often used for the collection of data on internal operational-level processes. The ability to easily input and retrieve reliable data on non-operational-level processes (e.g., sales data, cost data) is just as critical. For such processes, a combination of telecommunications, computer networks, and database management systems can be used to facilitate data entry and collection activities. Communication and cooperation among personnel involved in the maintenance of these systems are also important, to alleviate any unnecessary difficulties.

PARETO DIAGRAM

In the 19th century, the Italian economist Vilfredo Pareto (1848–1923) observed that about 20 percent of the Italian population controlled about 80 percent of the country's wealth. This observation led to what is known as the *Pareto principle,*

Defect Location Check Sheet   **Variable Check Sheet**

Project: Monitor Enamel Cracks
Location: Assembly
Dates: 3/1–3/6

Item _1056-91 Shaft_
Date _8/14/99_   Operator _Maria Dafnis_
Operation _Machining_
Characteristic _Diameter_
Units _Inches_

Sample Data

1.06	1.05	1.07	1.05
1.07	1.04	1.08	1.03
1.07	1.06	1.03	1.04
1.06	1.04	1.05	___
1.09	1.06	1.02	___

Average:	1.07	1.05	1.05	___
Range:	0.03	0.02	0.06	___

Refrigerator Model #A873Z

Attribute Check Sheet [*]

Product: _____

Manufacturing stage: final insp. _____

Type of defect: scar, incomplete, misshapen

Total no. inspected: 2530

Remarks: all items inspected

Date: _____
Factory: _____
Section: _____
Inspector's
  name: _____
Lot no.: _____
Order no.: _____

Type	Check	Subtotal
Surface scars	//// //// //// //// //// //// //	32
Cracks	//// //// //// //// ///	23
Incomplete	//// //// //// //// //// //// //// //// //// ///	48
Misshapen	////	4
Others	//// ///	8
	Grand total:	115
Total rejects	//// //// //// //// //// //// //// //// //// //// //// ////	
	//// //// //// //// //// /	86

Figure 1.6   Check sheet examples.

| *Adapted by permission from K. Ishikawa (1982), *Guide to Quality Control*, Tokyo: Asian Productivity.

Stat >> Quality Tools >> Pareto Chart

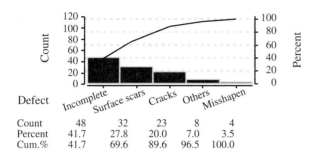

Defect	Incomplete	Surface scars	Cracks	Others	Misshapen
Count	48	32	23	8	4
Percent	41.7	27.8	20.0	7.0	3.5
Cum.%	41.7	69.6	89.6	96.5	100.0

Figure 1.7   Example of a Pareto chart.

or the *80-20 rule*. As noted in an earlier section, Juran found that the Pareto principle held true not just in economics but in a variety of industrial situations as well. Recall that Juran referred to the few items that account for the majority of an effect as the *vital few,* as distinguished from the *trivial many.* By performing Pareto analysis, process improvement efforts can be focused on the issues with greatest impact (the vital few) versus the less significant issues (the trivial many).

Pareto analysis is basically a two-step process: (1) Collect data on the contributing factors, and (2) display the data in a meaningful way. The most common way of displaying data to reveal the Pareto principle for a particular application is by means of a Pareto diagram. A Pareto diagram is simply a bar chart in which the frequencies (or relative percentages) of various factors on some effect are plotted in rank order (from highest to lowest).

Pareto diagrams are often used to analyze data collected in check sheets. As an example, Figure 1.7 shows a Pareto diagram for the attribute check sheet data on assembly defects found in Figure 1.6. This Pareto diagram was constructed by Minitab. Notice that the heights of the bars correspond to the frequency counts as read off the y axis scale on the left. The superimposed curve represents the cumulative percentage of the categories (from left to right) and is keyed to the y axis scale on the right. As can be seen, the highest category of defects is "incomplete," accounting for nearly 42 percent of the total. Furthermore, we see the Pareto principle in action by virtue of the fact that a small number of categories—incomplete, surface scars, and cracks—account for nearly 90 percent of defects. With the vital few highlighted, the next step is to focus on finding effective solutions to these most pressing problems.

CAUSE-AND-EFFECT DIAGRAM

Process improvement calls for not only the identification of problems but also the solution of these problems. Most of us have experienced instances in which problems we thought were corrected continued to occur. A recurrence of a problem is inevitably due to the fact that the attempted solution was a "Band-Aid" solution, and did not address the underlying root-level causes. Consider the following scenario given in Imai (1986, p. 50) to illustrate the use of continual questioning to expose the true cause for a machine stoppage:

Question 1:   Why did the machine stop?

Answer 1:   Because the fuse blew due to an overload.

Question 2:   Why was there an overload?

Answer 2:   Because the bearing lubrication was inadequate.

Question 3:   Why was the lubrication inadequate?

Answer 3:   Because the lubrication pump was not functioning right.

Question 4:   Why wasn't the lubricating pump working right?

Answer 4:   Because the pump axle was worn out.

Question 5:   Why was it worn out?

Answer 5:   Because sludge got in.

Hence, the questioning process revealed that the root cause of the machine stoppage is sludge in the lubricating pump. One reasonable solution would be to attach a strainer to the lubricating pump. Had there not been an attempt to get to the root level, the solution would have likely been an intermediate countermeasure, such as replacing the fuse. An intermediate solution only leaves the process prone for a likely recurrence of the problem.

The questioning process represents one means of identifying potential underlying causes. When a process improvement effort is team-based, a group problem-solving procedure, known as *brainstorming,* is most commonly used to generate various ideas and theories about the causes of a problem. With brainstorming, an environment clear of criticisms is created so as to encourage team members to use their collective thinking power to generate ideas and unrestrained thought. Brainstorming is useful not only for generating theories of possible causes of a problem, but also for finding solutions to a problem, and finding ways to implement these solutions. For a detailed discussion of the implementation of brainstorming and other team-based cooperative efforts, refer to Scholtes (1988).

As often is the case, potential causes of a problem are interrelated with some being subsumed by others. To aid in the organization and visualization of interrelated causes, Kaoru Ishikawa developed a graphical tool known as a *cause-and-effect diagram,* also known as an *Ishikawa diagram* or a *Fishbone diagram.* The *effect* represents the problem we are interested in studying, and the *causes* are the potential underlying factors influencing the effect. Depending on the application, Ishikawa (1982) suggested one of three types of cause-and-effect diagrams that can be constructed for the discovery of root causes: (1) dispersion analysis diagram, (2) cause enumeration diagram, and (3) process analysis diagram.

A *dispersion analysis diagram* is constructed around a set of generic elements or factors underlying most any system or process. For manufacturing systems, Ishikawa (1982) suggested the consideration of five generic elements known as the "five M's": material, machine, measurement, method, and manpower. For

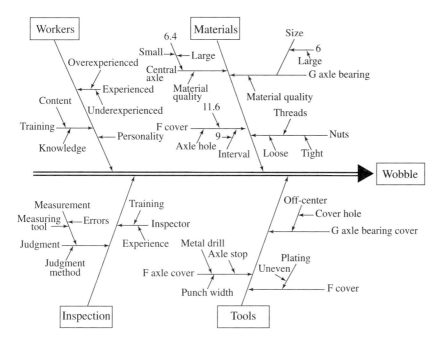

Figure 1.8   Cause-and-effect diagram for wobbling (dispersion analysis). Adapted by permission from K. Ishikawa (1982), *Guide to Quality Control,* Tokyo: Asian Productivity.

nonmanufacturing systems, Berwick, Godfrey, and Roessner (1990) propose a set of generic elements called the "five P's": patrons (those who use the system), people (those who work in the system), provisions (supplies), places (work environment) and procedures (methods and rules of work). To construct a dispersion analysis diagram using the five M's or five P's, a process improvement team assembles and organizes its brainstorming around the following questions:

- What materials, machines, measurement systems, methods, and manpower are used in the process? Or, what patrons, people, provisions, places, and procedures are used in the process?

- How might these materials, machines, measurement systems, methods, and manpower be the cause of the problem we are studying? Or, how might patrons, people, provisions, places, and procedures be the cause of the problem we are studying?

As an example, Ishikawa (1982) considers a situation where a wobble during a machine rotation is responsible for product defects. To eliminate the wobble, process analysts must understand its causes. Figure 1.8 illustrates how the effect (wobble) relates to a number of potential causes. In this case, Ishikawa organizes the potential underlying causes around four of the five generic elements: manpower (labeled *workers*), materials, measurement (labeled *inspection*), and machine (labeled *tools*). As can be seen, these four elements are associated with the main branches on the diagram. Potential causes are mapped onto the branches as the "twigs."

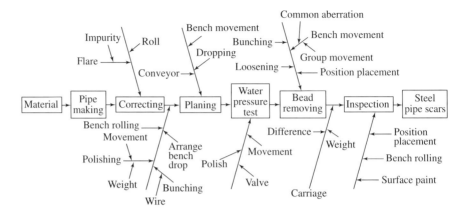

Figure 1.9   Cause-and-effect diagram for steel pipe scars (process analysis). Adapted by permission from K. Ishikawa (1982), *Guide to Quality Control,* Tokyo: Asian Productivity.

A *cause enumeration diagram* differs only from the dispersion analysis diagram in that the five M's or five P's are not required to serve as branches. In other words, cause enumeration is an unconstrained analysis of possible causes. Typically, a cause enumeration diagram is constructed from a wide-open brainstorming activity. Once a list of possible causes is generated, a team works on organizing the causes into natural main groupings that will then represent the main branches of the diagram. Next, the team decides on the interrelationship of the enumerated causes, within the identified groups, with the goal of mapping them onto the twigs of the diagram.

A *process analysis diagram* is, in a sense, a hybrid of a flowchart and a cause-and-effect diagram. The purpose of this diagram is to enable one to visualize where potential causes of an effect occur in the sequence of events making up a process. An example of a process analysis diagram is given in Figure 1.9.

Once a cause-and-effect diagram has been constructed, the process improvement team needs to collect data to confirm or rule out the theories on possible causes reflected by the diagram. A common mistake is to assume that the brainstormed causes shown on a cause-and-effect diagram accurately reflect reality. Rather, a cause-and-effect diagram should be viewed as a *creative* tool that helps team members plan for data collection so that they can pinpoint actual or most frequently occurring causes. In the end, most applications will be dominated by the Pareto principle in that only a few of the many potential causes will have the greatest influence on the effect.

SCATTER PLOT

A *scatter plot* (or *scatter diagram*) is a graph of data on one variable against another variable with the intent of showing whether the two variables are related in any systematic fashion. In situations where there is a potential cause-and-effect relationship between two variables (suggested possibly from a cause-and-effect diagram), a scatter plot serves as a first step toward gaining fundamental insight into the statistical nature of the causation. With this said, we must underscore a caution with scatter plots. By itself, a scatter plot that reflects a relationship between two variables does not imply there is a causal relationship. A scatter plot merely highlights whether a relationship exists and the strength of that relationship.

A poignant illustration of the utility of the scatter plot (and, more generally, the importance of statistical thinking) is given by the *Challenger* disaster. On January 28, 1986, the 25th mission (51-L mission) in the space shuttle program—flown by the *Challenger*—ended tragically with the loss of all seven crew members and the destruction of the vehicle when it exploded only 1 minute 13 seconds after liftoff. So troubling was the disaster to NASA and the public at large that President Reagan appointed an independent commission to investigate the circumstances surrounding the accident and to establish probable cause or causes of the accident. Chaired by William P. Rogers, Former Secretary of State under President Nixon, the Presidential Commission is typically referred to as the *Rogers Commission.*

As stated in the final report of the Rogers Commission, the loss of the space shuttle *Challenger* "was caused by a failure in the joint between the two lower segments of the right Solid Rocket Motor. The specific failure was the destruction of the seals that are intended to prevent hot gases from leaking through the joint during the propellant burn of the rocket motor." Furthermore, the commission concluded that the primary factor affecting the functionality of these seals, known also as *field joint O-rings,* is ambient temperature. In particular, it was determined that field joint O-rings used for the *Challenger* and all previous launches are substantially less resilient when ambient temperatures are low. The forecasted ambient temperature for the *Challenger* was about 23 degrees colder than that of any previous launch. Sadly, data were *readily* available prior to the launch which, if properly analyzed by either NASA or the supplier of the O-rings (Thiokol), would have made it quite likely (as concluded by the Commission) to cancel the launch, averting the national tragedy. As stated in the Rogers Commission report: "A careful analysis of the flight history of O-ring performance would have revealed the correlation of O-ring damage and low temperature. Neither NASA nor Thiokol carried out such an analysis; consequently, they were unprepared to properly evaluate the risks of launching the 51-L mission in conditions more extreme than they had encountered before."

Perhaps even more disturbing is that the most elementary data analysis would have revealed the association between O-ring and temperature. In particular, if available data for rocket engines recovered from 23 of the 24 previous launches had been organized statistically into a simple scatter plot, simple visual analysis would have suggested that the expected damage to field joint O-rings increased as launch temperature decreased. Figure 1.10 shows a Minitab-produced scatter plot of incidents of field joint O-ring damage versus temperature at launch (°F). To aid the visual analysis, we superimposed an "eyeball" fit of what might be the underlying relationship as suggested by the data. The scatter plot clearly indicates a

Graph >> Plot

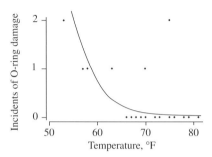

Figure 1.10   Scatter plot of incidents of O-ring
damage versus launch temperature.

negative correlation between the two variables; that is, higher incidents of O-ring damage are expected at lower launch temperatures, and lower incidents of O-ring damage are expected at higher launch temperatures. As it turns out, the launch temperature for the *Challenger* was 36°F. More refined analysis has since been conducted on the data (see Hoadley, Dalal, and Fowlkes, 1989) which, in the end, confirm the salient conclusion—don't launch—as suggested from a visual analysis of a simple scatter plot.

HISTOGRAM

In nature and in every process, variability exists. Managers without an understanding of variation and its omnipresence are in an unfavorable position to effectively manage the processes around them. This ultimately places the organization at risk in its ability to serve its customers (internally and externally). Collecting data on processes and organizing the data appropriately are first steps in bringing recognition to the presence and type of variation. One simple and useful tool for graphically representing variation in a given set of data is a *histogram*.

A histogram divides the range of data into intervals and shows the number, or percentage, of observations that fall into each interval. As an example, consider the case where a quality improvement team at an electric utility company servicing a large Midwestern metropolitan area was interested in studying the response time to emergency service calls (in particular, the time interval between the customer call-in and the arrival of a service crew at the scene). Table 1.2 shows 40 actual response time measurements (in minutes) collected by the team. (As we do throughout the book, we also provide individual data sets on the enclosed diskette. These data are found in the file "response.dat".) Although the table contains a great deal of information about the variation in response times, it is difficult to extract the information from a list of numbers alone.

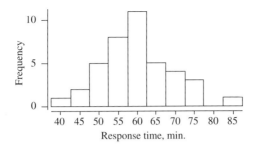

Figure 1.11   Histogram of response times.

Table 1.2   Emergency service response times (minutes) for an electric utility company

61	48	62	62	44	52	53	84	53	71
39	62	68	50	58	54	66	53	53	77
60	59	71	51	76	50	57	59	55	59
59	74	67	62	64	68	55	46	63	64

A histogram of the service response times (generated by Minitab) is shown in Figure 1.11. The height of each bar indicates the frequency with which the response time fell within the corresponding class interval. The numbers on the horizontal axis represent the midpoints of the class intervals. For example, the histogram shows that five service calls had response times that fell in the interval from 47.5 to 52.5 minutes (min.) (midpoint 50 min).

Of greater interest than the frequency of response times in any particular interval is the overall distributional picture the histogram reveals. For example, it can now be clearly seen that

- The variation in response time is wide, ranging from about 40 min to almost $1\frac{1}{2}$ hours (h).
- The highest concentration of response times is in the neighborhood of 50 to 70 min.
- The distribution is essentially symmetric around an average or mean time of about 60 min.

Thus, a histogram tells a much richer story about the data than gleaned from the data listing alone. Throughout the book, we will find that the histogram is an extremely valuable tool for general data analysis. In both Chapters 2 and 3, we provide detailed discussions on the role of histograms in data analysis and the cautions to be heeded when interpreting histograms.

CONTROL CHART

Given that this book is largely devoted to control chart techniques, traditional and enhanced versions, we necessarily limit our discussion at this point. Briefly, control charts were originally developed in the 1920s by Walter Shewhart to help managers and employees reduce the costs of inspection and rejection and to attain more uniform quality. By incorporating a measure of inherent variation in a process, a control chart serves as a basic data analysis tool to monitor a process for a change (unfavorable or favorable) as the result of some sort of process intervention (intended or unintended).

As will be explained in great detail in Chapter 4, control charts were originally designed to monitor manufacturing processes for which process outcomes are desired to be near a specified target with minimal variation around the target. The need to study and control such a class of processes is as vital today as it was during the era when control charts were first developed. Accordingly, we will devote a considerable amount of time to fleshing out the conceptual and technical details of the basic control chart technique as it was originally devised. Furthermore, we will provide numerous examples for which the basic control chart provides useful and appropriate insights.

However, let us not forget that from our earlier discussions (Sections 1.3 and 1.4) we concluded that processes are omnipresent (internally and externally) and are not restricted to a particular domain, such as target-based manufacturing processes. We will learn that there are times when the basic control chart is still applicable, and for a variety of reasons, there are times when it is not. Our goal is to arm the reader with the conceptual and technical understanding of basic data analysis techniques that will allow the reader to intelligently devise appropriate strategies for the study of the process in question. With this said, we are now prepared to embark on a data analysis adventure.

EXERCISES

1.1. How has the nature of competition changed over the last few decades?

1.2. Discuss the importance of quality in the national interest.

1.3. For the employees listed below, identify their suppliers and customers.
   a. Physician.
   b. Office secretary.
   c. College professor.
   d. Assembly worker.

1.4. For the products and services listed below, give an example of an external customer and an internal customer.
   a. Automobile.
   b. College education.
   c. In-patient surgery.
   d. Dairy products.

1.5. For the processes listed below, identify key inputs and outputs.
  a. Serving lunch at a restaurant.
  b. Delivering an overnight package.
  c. Designing a portable computer.
  d. Publishing a textbook.

1.6. To know what good quality is, is to recognize bad quality. Unfortunately, bad quality permeates our professional and personal lives. The objective of this exercise is to creatively release your frustrations by writing down your experiences with bad quality. Collect a handful of "That's not quality" examples with vivid details of the episodes. For each encountered scenario, put on a quality consultant "hat" to suggest *briefly* appropriate measures, and possibly methods, that might help the organization recognize and address the problem for future improvements.

1.7. Locate an issue of *Quality Progress* magazine (published by the American Society for Quality). Look through the issue and find a quality-related article of interest to you. Read the article and write a brief (say, $\frac{1}{2}$ page) critical review.

1.8. Discuss the fundamental concepts at the base of a total quality system.

1.9. Explain all the steps of the PDCA improvement cycle.

1.10. Consider the following discussion from Juran and Gryna (1993), *Quality Planning and Analysis* (New York: McGraw-Hill), regarding the notion of company quality culture:

> Employees in an organization have opinions, beliefs, traditions, and practices concerning quality. We will call this the company quality culture. Gaining an understanding of this culture should be part of a company assessment of quality.
>
> Formal approaches to assessing the quality culture are still evolving, but two can be identified, i.e., focused discussions with groups of employees and the use of questionnaires. With either of these approaches assessment of the quality culture can be done separately or as part of a larger survey of attitudes on many matters.
>
> Understanding a company's quality culture is important for assessment. Gaining that understanding has risks.

  a. Suppose you are a quality consultant. Develop a questionnaire to assess the quality culture of an organization. Characterize this organization as consisting of three levels—upper management, middle management, and front-line employees. State at least three questions that will be asked to each of these three levels to assess the current state of the quality culture. Each question should be clear and concise. Clearly state how the employees are to respond to each question, that is, by open-ended answers, rating scales, yes/no, etc. State how the answers will be analyzed and reported to management.

b. State the benefits and risks associated with implementing such a questionnaire. What resources are necessary to ensure successful implementation?

1.11. Deming, Juran, and Feigenbaum all believe in striving for world-class quality. Each, however, has his own set of guidelines for achieving such a quality level. Pick and discuss one aspect where all three agree.

1.12. Select one of the quality pioneers discussed in this chapter and briefly describe his contributions to quality management.

1.13. Match each name with the appropriate contribution to quality.

Shewhart	Robust design
Deming	Quality circle
Juran	Time-series analysis
Feigenbaum	Pareto principle
Ishikawa	14 Obligations for management
Taguchi	Statistical process control
Box	Cross-functional management

1.14. Below are excerpts from Deming's book *Out of the Crisis* as they relate to one of his 14 points. Match the quote with the appropriate point:

a. "Wrong way. A division in the civil service of a state has the task of preparing titles to automobiles. The foreman of the group described the mistakes that are made: misspelling of name of owner, mistakes in address, mistakes in serial number, in model, and many others. She estimated that only one in seven of the mistakes made ever comes back for correction, yet correction of these mistakes cost the state millions of dollars per year. She learned that she could purchase for $10,000 software that would indicate an inconsistency while a title is being typed. . . . A better way, in my opinion, would be to improve the forms for clarity and ease. . . ."

b. "Piece work is more devastating than work standards. Incentive pay is piece work. The hourly worker on piece work soon learns that she gets paid for making defective items and scrap—the more defectives she turns out, the higher her pay for the day. Where is her pride of workmanship?"

c. "People in design worked with people in sales and with engineers to design a new style. Salesmen, showing prototypes to wholesalers, piled up orders. The outlook was bright until bad news came—the factory could not make the item economically. Small changes in style and specifications turned out to be necessary for economic production. These changes caused delay in the factory. . . . The result was loss of time and loss of sales in a changing market. Teamwork with manufacturing people at the start would have avoided these losses."

d. "Do it right the first time. A lofty ring. But how could a man make it right the first time when the incoming material is off-gauge, off-color, or otherwise defective, or if his machine is not in good order, or the measuring instruments not trustworthy?"

e. "No one can put in his best performance unless he feels secure."

1.15. Consider an automobile brake system. Identify product variables associated with a brake design and environmental conditions that could result in variation in product performance in terms of stopping distance at a given speed. Explain why an antilock brake system (ABS) might be a production innovation leading to a more robust braking system.

1.16. Search the library or Internet to learn about Deming's "red bead experiment." What is the lesson learned from this experiment?

1.17. Visit the American Society for Quality (ASQ) Web site at http://www.asq.org and find information about a quality pioneer not discussed in this chapter. Briefly describe the contributions of the pioneer to quality management.

1.18. Provide three operational measures of the quality of a college course. Be specific about your measures.

1.19. Provide three operational measures of the quality of a delivered pizza. Be specific about your measures.

1.20. Provide three operational measures of overall company performance. Be specific about your measures.

1.21. Provide an operational definition for the following:
   a. Reliable computer chip.
   b. Cordless telephone with clear reception.
   c. Excellent teacher.
   d. High-quality color scanning.
   e. Poor country.
   f. Messy desk.

1.22. Create a flowchart that captures the sequence of possible events for registering for a class at your college.

1.23. Create a flowchart that captures the sequence of possible events for the process of an individual going from curbside at an airport to an airplane seat.

1.24. Below are the total numbers of points scored by Chicago Bulls players during the 1997–1998 playoff and championship series:

Player	Points	Player	Points
Brown	9	Kukoc	275
Buechler	11	Longley	142
Burrell	80	Pippen	353
Harper	142	Rodman	102
Jordan	680	Simpkins	16
Kerr	103	Wennington	44

      a. Create a Pareto chart for points scored by player.

      b. What percentage of the 12-player playoff roster is responsible for approximately 50 percent of points scored? What percentage is responsible for approximately 80 percent of points scored?

1.25. Perform a study of the applicability of the Pareto principle to any of the following: *(a)* population of countries; *(b)* annual sales of U.S. companies; *(c)* number of medals received by country in the 1996 Summer Olympics; *(d)* causes of human mortality in the United States

1.26. What are the four main branches of a dispersion analysis cause-and-effect diagram when applied to a manufacturing application? What are the four main branches of the diagram when applied to a service application?

1.27. It was mentioned that Ishikawa is credited with the development of the cause-and-effect diagram. Is the cause-and-effect diagram and its heavy dependence on theorizing (i.e., speculating) consistent with Ishikawa's philosophy of the need to always use "facts and data to make presentations—utilization of statistical methods"? Explain.

1.28. Consider the process of downloading files from the Web. Brainstorm as many possible causes for variation in download time.

1.29. Suppose your commute from home to school involves 3 miles of city street driving and 5 miles of highway driving. Brainstorm as many possible causes for variation in the time it takes to get from home to school.

1.30. Referring to Exercise 1.28, construct a cause-and-effect diagram to display the relationships of the enumerated causes.

1.31. Referring to Exercise 1.29, construct a cause-and-effect diagram to display the relationships of the enumerated causes.

1.32. Explain the difference, in structure and application, between a histogram and a Pareto diagram.

1.33. Explain the difference, in structure and application, between a histogram and a scatter plot.

1.34. Explain the difference, in structure and application, between a checklist and a scatter plot.

1.35. Consider data collected on the amount of scrap and the line speed of a production line. With amount of scrap on the vertical axis and line speed on the horizontal axis, create a scatter plot of the data and describe what relationship (or lack of) is reflected in the plot.

Scrap:	288	110	73	328	206	318	135	125	353	245
Speed:	249	126	104	278	177	281	156	159	285	216

1.36. Consider data collected on the annual number of days absent and the age of the employee. With days absent on the vertical axis and age on the horizontal axis, create a scatter plot of the data and describe what relationship (or lack of) is reflected in the plot.

Age:	37	41	62	65	55	27	54	57	20	42
Days:	4	3	3	1	3	5	4	2	7	3

REFERENCES

Balestracci, D. (1998): "Data 'Sanity': Statistical Thinking Applied to Everyday Data," Special Publication (Summer), Parsippany, NJ: American Society for Quality, Statistics Division.

Berwick, D. M., A. B. Godfrey, and J. Roessner (1990): *Curing Health Care: New Strategies for Quality Improvement,* San Francisco: Jossey-Bass.

Bounds, G., L. Yorks, M. Adams, and G. Ranney (1994): *Beyond Total Quality Management,* Burr Ridge, IL: McGraw-Hill/Irwin.

Box, G. E. P. (1984): Comment on "Present Position and Potential Developments: Some Personal Views. Industrial Statistics and Operational Research" by A. Baines, *Journal of Royal Statistical Society,* Series A, 148, pp. 325–326.

Deming, W. E. (1986): *Out of the Crisis,* Cambridge, MA: MIT Center for Advanced Engineering Study.

Evans, J., and W. M. Lindsay (1999): *The Management and Control of Quality,* 4th ed., Cincinnati, OH: South-Western College Publishing Co.

Feigenbaum, A. V. (1983): *Total Quality Control,* New York: McGraw-Hill.

Gitlow, H., A. Oppenheim, and R. Oppenheim (1995): *Quality Management: Tools and Methods for Improvement,* 2nd ed., Burr Ridge, IL: McGraw-Hill/Irwin.

Goetsch, D. L., and S. B. Davis (1997): *Introduction to Total Quality,* Upper Saddle River, NJ: Prentice-Hall.

Gunter, B. (1998): "Farewell Fusillade," *Quality Progress,* April 31, pp. 111–119.

Harrington, H. J. (1991): *Business Process Improvement,* New York: McGraw-Hill.

Hoadley, B., S. R. Dalal, and E. B. Fowlkes (1989): "Risk Analysis of the Space Shuttle: Pre-Challenger Prediction of Failure," *Journal of the American Statistical Association,* 84, pp. 945–957.

Hoerl, R. W. (1995): "Enhancing the Bottom-Line Impact of Statistical Methods," *Quality Management Journal,* 2, pp. 58–74.

_____ (1998): "Six Sigma and the Future of the Quality Profession," *Quality Progress*, June 31, pp. 35–42.

Imai, M. (1986): *Kaizen: The Key to Japan's Competitive Success,* New York: Random House.

Ishikawa, K. (1982): *Guide to Quality Control,* 2nd ed., Tokyo, Japan: Asian Productivity.

_____ (1985): *What Is Total Quality Control?* Englewood Cliffs, NJ: Prentice-Hall.

_____ (1990): *Introduction to Quality Control,* Tokyo, Japan: 3A Corporation.

Juran, J. M. (1964): *Managerial Breakthrough,* New York: McGraw-Hill.

_____ (1992): *Juran on Quality by Design,* New York: Free Press.

Kolarik, W. J. (1995): *Creating Quality,* Burr Ridge, IL: McGraw-Hill/Irwin.

Kume, H. (1985): *Statistical Methods for Quality Improvement,* Tokyo, Japan: 3A Corporation.

Main, J. (1994): *Quality Wars: The Triumphs and Defeats of American Business,* New York: Free Press.

The Memory Jogger II (1994): Methuen, MA: GOAL/QPC.

Montgomery, D. C. (1997): *Design and Analysis of Experiments,* 4th ed., New York: Wiley.

Peters, T. (1988): *Thriving on Chaos: A Handbook for a Management Revolution,* New York: Knopf.

Roberts, H. V. (1991): *Data Analysis for Managers,* Belmont, CA: Duxbury Press.

_____ and B. F. Sergesketter (1993): *Quality Is Personal: A Foundation for Total Quality Management,* New York: Free Press.

Scherkenbach, W. W. (1988): *The Deming Route to Quality and Productivity,* Rockville, MD: Mercury Press.

Scholtes, P. R. (1988): *The Team Handbook,* Madison, WI: Joiner Associates.

Shewhart, W. (1931): *Economic Control of Quality of Manufactured Product,* New York: D. Van Nostrand Co., reprinted by the American Society for Quality in 1980, Milwaukee, WI.

Snee, R. D. (1998): "Getting Better Business Results," *Quality Progress,* June 31, pp. 102–106.

Taguchi, G. (1978): "Off-line and On-line Quality Control Systems," *Proceedings of the International Conference on Quality Control,* Tokyo, Japan.

_____ (1986): *Introduction to Quality Engineering: Designing Quality into Products and Processes,* White Plains, NY: Kraus International, UNIPUB (Asian Productivity Organizations).

Thurow, L. C. (1993) *Head to Head,* New York: Warner Books.

c h a p t e r

2

A Review of Some Basic Statistical Concepts

CHAPTER OVERVIEW

The purpose of this chapter is to impart an understanding of some essential statistical principles useful for developments and discussions of subsequent chapters. We present basic visual and numerical means of summarizing process data. Additionally, we introduce some probability distributions which play critical roles in the development of process data analysis methods. These topics lead to a discussion of statistical estimation and inference which entails making statements about a population based on sample statistics. For many of the topics, our coverage is necessarily brief. If the reader has some statistical background, for the most part, this chapter should serve as a handy review. However, even for the more statistically versed, we strongly encourage more than a quick passing of the discussions pertaining to summarization of process data found in Sections 2.3 and 2.4. These sections discuss the ways of characterizing process behavior over time and the distinction between time and distribution behavior, topics not typically discussed in standard statistics texts.

2.1   POPULATIONS, PROCESSES, AND SAMPLES

Statistics is often described as a scientific discipline concerned with the collection, summarization, analysis, and interpretation of numerical data. When we study data, we are typically doing so because the data characterize some phenomenon of interest to us. In statistics, the ultimate target of our interest is called a *population*. Simply defined, a population is the set of collection of all possible observations of some specific characteristic. Depending on the nature of the phenomenon under study, populations can be classified in one of two ways: finite or conceptual.

FINITE POPULATIONS

Finite populations arise when we wish to examine some common observable characteristic of a set of individuals or objects taken at a *particular* time, sometimes referred to as existing populations. Finite or existing populations are "tangible," in the sense that all elements of these populations are identifiable and with enough effort information can be obtained on each and every element. In studying particular characteristics of a finite population, one has two choices: (1) conduct an exhaustive count or census of all the elements that comprise the given population or (2) study a subset of the population and then from the knowledge of the subset draw some conclusions about the population. Because of the fact that finite populations can potentially be completely enumerated, any statistical study of a finite population is often called an *enumerative study* (terminology originally due to W. Edwards Deming). Enumerative studies are also known as *cross-sectional studies* because the population is defined at a particular slice of time.

Making a complete enumeration may present many limitations of both time and money. Furthermore, in a number of cases, to obtain the needed information, the element has to be completely or partially destroyed, as, for example, in testing the strength of the glass envelope of a lightbulb. In this case, complete enumeration requires the destruction of the entire population. Since complete enumeration is often impractical, studying a subset of the population, called a *sample,* is done as an alternative. The aim, of course, is to know as much as possible about the total population. Since information is incomplete, any statement (or conclusion) about the population cannot be made with absolute certainty. Therefore, such generalizations from a particular sample to the whole population must be accompanied by some degree of uncertainty. Drawing conclusions from a sample about the population is called *statistical inference.* Hence, the task is to develop statistical methods or procedures for making uncertain but nevertheless valid and useful inferences about the population.

As an illustration, suppose we define a population as a crate of 300 parts made from an injector molding machine on a particular day. All the elements of the population exist, literally sitting in a crate. One measurement of interest, to the nearest 0.01 centimeter (cm), is the outer diameter of the part which is specified to within 5.00 ± 0.05 cm. If diameters on all 300 parts are obtained, then this population is fully characterized and there is no need for statistical methods. We can then answer any question that is posed about the population. For instance, what is the spread of diameters from narrowest to widest in the class? What percentage of the diameters in the crate is less than 4.95 cm? Or, what is the average diameter of the parts in the crate? Answers to these types of questions offer a variety of numerical summaries of the population at hand. In general, numerical summaries of a population are called *parameters.* Short of a complete enumeration, a sample from the crate can be gathered to make reasonable but less certain "guesses" to the types of questions posed above. Any numerical summary of a sample is referred to as a *sample statistic* (or just *statistic*).

In summary, the goal of enumerative studies is to simply characterize the existing population as well as possible, given the amount of information obtained. Enumerative studies can serve as a basis for practical actions, but these actions are limited exclusively to the population itself. For example, the manufacturer may want to obtain a sample of parts to estimate the percentage of defects in the crate to decide whether the crate of parts should be passed along to the next stage of production.

CONCEPTUAL POPULATIONS

With an enumerative study, we were only concerned with characterizing the population itself defined at a particular time that has occurred. There was no regard to the possibility that the particular population might be part of an ongoing process. Enumerative studies are limited to the confines of the existing population (e.g., the crate of parts) and bear no insights for other populations (past or

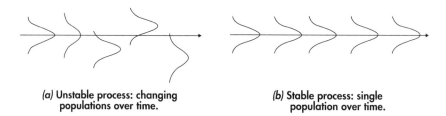

(a) Unstable process: changing
populations over time.

(b) Stable process: single
population over time.

Figure 2.1   Unstable versus stable process behavior.

future). In a sense, enumerative studies consider only a "snapshot" of some measurable phenomenon at one point in time. By contrast, our primary interest is to examine the behavior of some measurable phenomenon as time elapses, that is, to study the behavior of a process over time. Given the time dimension, studies of processes are frequently called *time-series studies.* Alternatively, Deming refers to such undertakings as *analytic studies.*

Unlike enumerative studies, the notion of a finite population no longer exists with analytic studies. When studying processes, we are considering ongoing processes which generate both past and future outcomes. No longer can we establish a listing or frame of the population elements since we cannot identify elements that do not exist currently. We need to think of populations associated with processes as *conceptual.* A conceptual population is the totality of observations that *might* occur from performing a particular operation in a particular way.

Another distinction that should be pointed out is that there is only one single population in question in an enumerative study; however, this may or may not be the case with an analytic study. As illustrated in Figure 2.1*a,* the underlying process is undergoing continual irregular change in terms of the populations at given time points. Thus, no single conceptual population can reasonably serve as a guide to represent past or future process behavior. This is in contrast to the process shown in Figure 2.1*b,* which is predictable, and its future outcomes can be safely inferred from past behavior from a single conceptual population. As will be discussed in subsequent chapters, *statistical process control (SPC)* methods—control charts—play an important role in both helping to determine the stability of the process and maintaining its stability.

There are basically two goals of any time-series (or analytic) study: (1) to gain an understanding of the past behavior through the analysis of sample data taken sequentially over time and (2) to predict how the process is likely to perform in the future based on knowledge derived from past data. By establishing a statistical baseline of the past and likely future behavior of a process, two additional opportunities arise. First, there is an opportunity for the discovery of surprises—unfavorable or favorable to the process. For example, an observed data point (from the past or in the future) may be inconsistent with past data or with predicted behavior. Such points are challenges for management to reveal underlying causes—special causes—for process improvement purposes by either eliminating or perpetuating the sources, depending on their nature. The second opportunity goes beyond the accidental discovery of inconsistent behavior to the purposeful intervention in the process to gain insights on the causal system so as to improve the process. Management can experiment with new methods or procedures and assess whether these new approaches have resulted in a process significantly different from predicted behavior. Control charts and data analysis techniques presented in this textbook are useful for helping highlight special causes and process changes due to either unknown or purposeful interventions. Thus, our primary focus is on statistical methods supporting analytic studies.

It is not only statistical factors that distinguish analytic from enumerative studies. Deming (1975) writes that an analytic study "is one in which action will be taken on a cause-and-effect system to improve performance of a product or a process in the future." Statistical methods are necessary to assess whether an action taken on a cause-and-effect system has had any significant impact. However, effective analytic studies cannot be performed in a statistical vacuum. The choice of reasonable actions for improvement and the implications from such improvement actions necessarily go beyond statistical procedures. In particular, knowledge of the process is a critical component in a continual effort to improve the process. Without knowledge of the process, efforts to understand and explain sources of variation — common or special—become futile. In summary, statistical theory is critical in analytic studies, but "One must never forget the importance of subject matter" (Deming, 1986).

2.2   TYPES OF PROCESS DATA

There are a variety of ways to characterize process-related data. A broad categorization can be made based on whether the data measurements are expressed numerically. If the information is nonnumerical, then the data are *qualitative*. If the information can be expressed numerically by either counting or measuring, then the data are *quantitative*. Below we briefly discuss these two basic types of data characterization.

Information that can fall into one of several categories is a common type of qualitative data. Examples of qualitative data include information identifying the supplier of a shipment, the determination of defective or satisfactory, the occupation of a survey respondent, and the color of a finished product. When studying categorical information, we usually want to know how many or what proportion of observations fall in each category. As introduced in Chapter 1, for process improvement studies, a popular way of presenting qualitative information is via the Pareto chart (see Figure 1.7). The Pareto chart of Figure 1.7 indicates the occurrence and relative frequency of five categories of product damage. Recognize that the categories themselves, such as "cracks," are qualitative information while the counts of the categories are quantitative data. Later, we will find that categorical information can be quite important in the development of statistical models to better explain process behavior. For example, data analysis of a particular process might reveal that output generated on Mondays is consistently high. As such, an appropriate statistical model should incorporate this information. To do so, a meaningful numerical value needs to be assigned to the category "Monday" versus "Tuesday" and so on. In Chapter 3, we will learn that categorical information can be represented by a set of variables, called indicator variables, that take on the values 0 or 1. There are many other forms of qualitative information in applications. For instance, data gathered from focal groups, such as comments and suggestions, are not easily quantifiable but are invaluable "voice of the customer" information that nearly every organization utilizes.

In this textbook, we are primarily concerned with statistical investigation of quantitative data that represent any sort of numerical recording of information. Quantitative data can be classified as *continuous* or *discrete*. Continuous data are associated with measurements on some interval scale of such characteristics as length, weight, volume, and time. Examples of continuous data include the waiting time in a service center, fill volume of fluid (in liters) in a bottled beverage, firing temperature of an oven used in the manufacture of ROM chips, value of the Dow Jones Industrial Index, and blood cholesterol level of an individual. In contrast to continuous data, discrete data cannot be associated with a continuum of values but rather assume specific values along the number line, with gaps between them. The most common form of discrete data is a count of some observable occurrence. Examples of count data include the number of blemishes on an automobile surface, the monthly number of dissatisfied customers, and the number of defective units in a batch. Count data can be expressed in terms of whole numbers or percentages. For example, it is equivalent to report 10 defects in a batch of 150 parts or that 6.67 percent ($= 10/150$) of the batch is defective.

Control charts apply primarily to quantitative data, both continuous and discrete versions. However, the terminology used to indicate the study of continuous versus discrete data is different from standard statistical practice. Namely, control charts for continuous quantitative data are referred to as variable control charts while control charts for discrete data are referred to as attribute control charts.

2.3 CHARACTERIZING PROCESS DATA: VISUAL DISPLAYS

Once process data have been collected, the challenge is to effectively allow the data to speak for themselves to facilitate meaningful interpretation. Here we address the basic means of describing process data which entails the presentation of pertinent visual displays along with useful numerical summaries. These descriptive methods will help the analyst develop a feel for the nature of the underlying process behavior.

TIME-SERIES DISPLAY

Depending on the nature of the data, the *first* step of any statistical investigation should be an appropriate graphical presentation. A visual examination of the data enables the analyst to quickly gain a feel or sense of what is happening. Graphs are a critical means for revealing any sorts of patterns in the data. No set of numerical summaries can compensate the pattern recognition potential of one's eyes. The old saying "A picture is worth a thousand words" is truly apropos in data analysis.

Table 2.1   Stopwatch reading

10.13	9.98	9.92	9.98	9.92	9.93	9.78	10.07	9.84
10.01	9.97	9.97	9.92	10.09	10.09	9.96	10.08	10.01
9.84	10.08	9.91	10.15	9.90	9.93	9.88	9.94	9.98
10.00	9.92	10.02	10.09	9.99	10.05	10.01	10.03	9.91
10.01	10.07	10.06	10.05	10.21	9.95	9.86	10.03	10.02
10.10	9.88	10.13	9.83	9.97				

Since process outcomes are naturally sequential over time, the presentation of a time-series (or time-sequence) plot would be the relevant starting point. A time-series plot is simply a plot of the data over time, with the horizontal axis scaled to time or sequenced observation numbers. In quality applications, a time-series plot of process data is known as a *run chart*. Since time-series plot, time-sequence plot, and run chart are all equivalent in meaning, we will take the liberty to use them interchangeably in subsequent discussions. To illustrate a time-series plot, consider a sequence of 50 observations directly gathered by the author. The experiment is to sequentially start and stop an electronic stopwatch as close as possible to 10 seconds (s).[1] Measurements were taken to the nearest 0.01s. The data are shown in Table 2.1 and can also be found in file "timing.dat."

As noted in the Preface, the statistical software package Minitab for Windows will be used as a vehicle for our data analysis. For purposes of graphing, there are two modes for graphical presentation in Minitab: high-resolution and character.[2] The high-resolution mode offers the user a more polished presentation with great flexibility for editing graphs with preferred choices of symbols, type-sets, and fonts. The character mode results in graphs that consist of basic keyboard characters, much like creating graphs by using a typewriter. Although it may seem that the high-resolution mode is always the clear choice, on some occasions, as we will illustrate in subsequent chapters, the low-resolution mode provides an invaluable perspective to the data analyst.

---

[1]This data-generating experiment was introduced to me by Professor Harry V. Roberts, University of Chicago. Professor Roberts refers to the associated data as a "lightning data set." Professor Roberts developed the concept of lightning data sets to allow students to quickly generate data that can easily be related to in order to gain basic insights in (1) data collection, (2) variability, and (3) data analysis.

[2]Mainframe versions and some of the earlier personal computer (PC) versions of Minitab only run in character mode.

Graph >> Character Graphs >> Time Series Plot

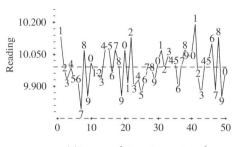

(a) **Low-resolution time-series plot.**

Graph >> Time Series Plot

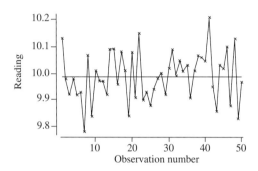

(b) **High-resolution time-series plot.**

Figure 2.2   Time-series plots of the stopwatch data series.

As a first experience, Figure 2.2 shows both a high- and low-resolution time-series plot of the stopwatch data. Starting with the digit 1, Minitab in the low-resolution mode (Figure 2.2*a*) repetitively uses the 10 digits to label the observations; this is the default which we will later have occasion to change. In contrast to the high-resolution plot (Figure 2.2*b*), the low-resolution plot has the unfortunate quality of *not* having the plotted points connected by straight lines. To gain maximum insight from a low-resolution plot, the "hassle" of manually connecting points by means of pen and straightedge—as I have—must not be avoided.[3] The benefit of this exercise is similar to that of the old children's game of connecting the dots which allows underlying pictures (for us, patterns) to come to the surface.

We should also point out that both plots have been enhanced with the addition of a horizontal line at the average of the data. In control chart jargon, this line is referred to as the *centerline*. The centerline serves two purposes. First, it gives a

---

[3]As an alternative, most Windows-based word processing software have a line-drawing capability. A line is created simply by dragging the mouse from one point to the next.

more tangible feel for the level of the process data. Second, it serves as a reference line to better aid the eyes in determining whether patterns exist in the data.

With the sequence plot along with superimposed centerline in place, we are ready to formulate some impressions about the time series. A good starting point is to ask the following question: Based on the 50 observations shown in the time-series plot, what would we guess the future observations (51, 52, and so on) to be? Such predictions would, of course, depend greatly on the nature of the observed data. If, for instance, the measurements were steadily trending upward, it would be sensible to project future observations, at least in the short run, to be at higher levels. Or, as another example, suppose observations tended to go up and down, say, from around 9.5 to around 10.5 and back down. If the 50th observation hovered near 10.5, we would naturally guess, based on this observed pattern, the next observation to be near 9.5.

Now consider either time-series plot of Figure 2.2. We might first recognize that the measurements exhibit clear variation around the centerline. My goal was to hit 10 precisely each and every time; however, even with my best efforts, the results varied. It is a fair bet that variability of human reflexes is the primary source of the variability in results. Momentary lack of concentration due to outside disturbances may be another contributor. This experiment has enabled us to witness the important phenomenon of errors in measurement. This phenomenon is worthy of a slight digression in our discussion.

A measurement error is any deviation from the true value of the dimension being measured. In our case, the true value is 10s, and the deviations from this value are the magnitudes of the measurement errors. In practice, measurements, even when repeatedly taken on the same item, will inevitably exhibit variation. In essence, measurement errors distort the true view of the processes under study. One goal is to have the errors introduced by a measuring system reflect no systematic biases; that is, the errors should not be consistently high or low relative to the true value but rather should average out to zero. In addition, the measurement errors should be as small as possible compared to the dimensions being controlled. In the area of metrology, the study of measurements, these two requirements are known as *accuracy* and *precision*, respectively.

In particular, the accuracy of the measurement system refers to the extent to which the average of repeated measurements on a same item differs from the true value. The difference between the average and the true value is known as the *systematic error* (or *bias*) of the measurement system. In our example, the average for the 50 readings is 9.989s, implying an estimated bias of -0.011s from the true value of 10 seconds. Practically and statistically, it is fair to say that this bias is small and not significantly different from zero. In other words, the measurements reflect an accurate measurement system. Stout (1985) recommends that the bias in a measurement system ideally be no more than 10 percent of the minimum dimension that needs to be assessed. For an inaccurate measurement system, the systematic error can be viewed as the correction needed—in the opposite direction—to calibrate the system.

For our example, even though the measurement system appears to be accurate, the system possesses a certain level of imprecision. *Precision* is a measure

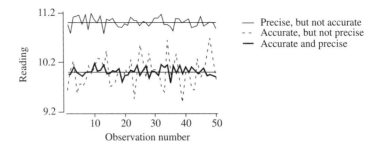

Figure 2.3   Relative comparisons of accuracy and precision.

of the ability of a measurement system to replicate its own measurements on a dimension of a single item. Precision is inversely related to the variation or dispersion of the replicated measurements. Hence, the more dispersed the measurements, the more imprecise the measurement system. Given the omnipresence of variability in any system (measurement or not), it is a truism to say that every measurement system is subject to a certain level of imprecision. For our stopwatch example, we "see" from Figure 2.2 the level of imprecision in terms of variability around the centerline. In Section 2.5, we will introduce some numerical summaries that quantify such process variability. Figure 2.3 summarizes the notions of accuracy and precision in a relative sense for three different sequences of measurements for the stopwatch experiment.[4] In Chapter 9, we will resume our discussion of measurement systems in a more detailed manner.

Let us now return to our impression of the stopwatch data as reflected by either of the time-series plots given in Figure 2.2. As we have discussed, the plot has enabled us to recognize the presence of variability in the observations. We can also notice that the data are not trending up or down but rather "hover" around the centerline. This is to say the average level of the process is constant or remains the same. Notice also that there are times when consecutive observations bounce from one side of the centerline to the other while at other times they do not, with no persisting tendency to do one or the other. In general, the sequence of observations is "patternless." One way of thinking of patternless behavior is to realize that the information that one or more points are above or below the centerline is of no value in predicting whether or not future points will be above or below the centerline. As a final characterization of the stopwatch data, we note that the dispersion of observations around the centerline appears to be about the same from the beginning of data collection to the end, going down to about 9.75 and up to about 10.25.

Collectively, the characteristics of constant process level, no systematic pattern of observations around the centerline, and constant process dispersion define a *random process*. In traditional SPC terminology, a random process is regarded

---

[4]For fun, conduct your own stopwatch experiment of measuring 10s to assess your personal measurement system. See if you can "beat" my results in terms of accuracy and precision!

as a process in a *state of statistical control*. Earlier we asked what would be reasonable guesses for observations 51, 52, and so on, based on the given observed behavior. For a random process, the best single guess for the future is intuitively a simple extension of the centerline, which in this case is 9.989. There is no reason to believe future observations will more likely fall above or below the centerline. Thus, the best we can do is to hedge our bets and guess the centerline.

Of course, we cannot expect future outcomes to be exactly as predicted. So it would be helpful to provide some range of likely future outcomes. Assuming basic process conditions remain the same, the observed variability around the centerline provides that basis. This enables us to say statements such as that our best guess for future observations 51, 52, and so on is 9.989 but we would not be surprised if the observations fell roughly between 9.75 and 10.25 (see Figure 2.2). In essence, we have just informally constructed a control chart. Simply speaking, a control chart is nothing more than a formal means of highlighting significant changes in the process relative to predicted process behavior; we will address the specific details of control chart construction in subsequent chapters. The indication that we would not be surprised if future observations fell roughly between 9.75 and 10.25 can be even further refined. For instance, we can add that it is more likely an observation will fall in the interval from 10.0 to 10.1 than in the interval from 10.1 to 10.2; similarly, it is more likely an observation will fall in the interval from 9.9 to 10.0 than in the interval from 9.8 to 9.9. These refinements give us a feel for how the observations are concentrated or *distributed*. In the next subsection, we will present a visual means to get a more complete impression of the distribution of the data.

Now that we have experienced a time-sequence plot of a random process, let us consider some time-sequence plots of some clearly nonrandom processes, shown in Figure 2.4. In attempting to assess whether a particular process is nonrandom, we need to be on the lookout for one or more of the following violations of a random process:

Violation 1:   The general or overall process level does not remain constant over time.

Violation 2:   There exists a persisting systematic pattern in the data. As an informal test, if the centerline cannot serve as the best prediction for any individual point in the future, there is a systematic pattern.

Violation 3:   The variation around the process level is not constant.

It is important to underscore that the use of the term *nonrandom process* is not meant to imply that the process is not subject to any random variability. All processes are subject to random variability in some shape or form. The term *nonrandom process* is simply a common means (particularly in the area of time-series analysis) to imply a process behavior other than "pure" random variation which is constant in mean and variance over time, that is, any process not exhibiting one of the violations listed above.

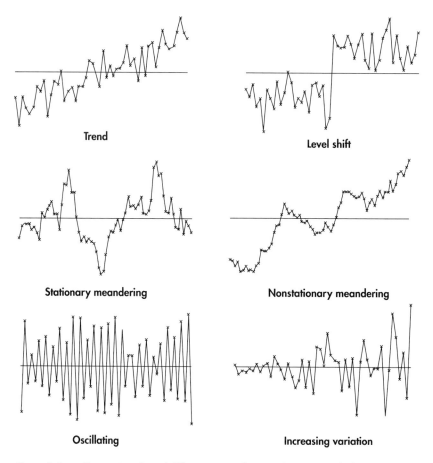

Figure 2.4   Time-series plots of different types of nonrandom process behavior.

In the case of the top left time-sequence plot of Figure 2.4, both violations 1 and 2 apply. The overall process level is not constant, but rather drifts away as time passes. Unlike the case of a random process, a horizontal centerline would not be a useful guide for predicting the future behavior of this process; instead, future observations are expected to be farther away. Notice that we did not indicate the occurrence of violation 3 for the illustrated trend process. The reason is that the variation *around* the systematic trend is similarly dispersed.

Adjacent to the trend series is a graph that shows a process for which the level has shifted, giving the appearance of a step. Since the variation is about the same around different process levels, violation 3 is not the culprit. However, violation 1 clearly applies to this situation since the overall process level has "jumped" up. Consequently, the centerline is consistently below the process after the shift, and thus it cannot represent a reasonable forecast for future process outcomes. We should point out that a sustained shift, as illustrated, is not the only type of process level change. The process level might, for instance, abruptly shift away for only a short time. Such transient behavior would likely result in the appearance of one or more extreme points (known as *outliers*) situated well beyond the general crowd of remaining observations. Regardless of the duration of the process level shift, the process would be deemed nonrandom.

The plots in the second row of Figure 2.4 are characterized as being *meandering*. The term *meandering* is meant to suggest the image of a winding stream or river. As a possible alternative image, I have compared this behavior to a crack in a sidewalk. The appearance of these images results because successive observations tend to be close together in value. As we will discuss in Chapter 3, closeness of successive observations is technically referred to as *positive autocorrelation*.

Even though both plots reflect meandering time series, the nature of their meandering is quite different. In the first case (left graph), the process wanders around the centerline; lingering above the centerline for a while and then below the centerline for a while. The net effect of this movement is a constant overall process level; hence, violation 1 is not the issue. Nevertheless, the process is extremely nonrandom based on violation 2, that is, the existence of a systematic pattern. The pattern is the persistent tendency for successive observations to be close together. Remember, the litmus test for detecting pattern is to ask yourself, Given the values of past observations, is the centerline the best guess for future individual observations? Here, the answer is definitely no. If, for instance, the most recently observed point is above the centerline, then the observed meandering behavior suggests that the next point would most likely also be above the centerline and *not* on the centerline; similarly, if the most recently observed point is below the centerline, then the next point would most likely also be below the centerline.

In contrast to the first meandering series, the second meandering series (right graph) appears to be "running away" rather than moving around the centerline. Recognize that the nature of the movement away from the centerline is quite different from that of the previously discussed trend process. In contrast to the meandering process, the trend process drifts away in a nearly straight-line fashion. One of the best ways to relate to the drifting meandering behavior is to think of the evolution of most any financial index, such as the Dow Jones Industrial Index which represents the general price level of the stock market. If you look at the Dow Jones index over time (daily, weekly, monthly, or yearly), you will see a time series for which adjacent observations are relatively close to one another (with the exception of market crashes!) but, in the long run, a series that continually meanders farther and farther away. Thus, for such processes, the general level is not constant, implying the occurrence of violation 1. In addition, similarly to the first meandering series, the centerline does not serve as a suitable guide in the prediction of future outcomes. Instead, predictions are better made by using the fact of a persistent pattern (violation 2) of close together successive observations.

In summary, both processes shown in row 2 of Figure 2.4 exhibit nonrandom meandering but predictable behavior. Additionally, for each of the processes, the variation around the meandering pattern remains about the same from beginning to the end of the time-sequence plot. The primary difference between the two processes lies in the long-run implications. Namely, the general level of the first process "averages out" to some constant level while the general level of the second process is not constant and is actually undefined, since the process is continually moving away from past short-term levels. If both the general level and the variation of a time series remain constant throughout, the time series is said to be *stationary*.[5] Hence, the first process represents a stationary process, and the second process represents a nonstationary process. With respect to previous examples, the random stopwatch process is stationary while both the trend and level shift processes are nonstationary. Generally speaking, a random process is a stationary process without a systematic pattern. However, a stationary process is not necessarily a random process.

The distinction between stationarity and nonstationarity of a process has great practical importance. For instance, if a stationary process is centered to some desired target value, we expect the process not to permanently drift away from the target value. Thus, stationarity implies a degree of *practical* stability relative to some target or specification values. We emphasized the word *practical* to contrast with the notion of statistical stability. A process can be classified as statistically stable if its future outcomes are predicable within limits. In that sense, predictable nonstationary processes such as the trend example can be viewed as statistically stable; we may not like the practical implications of the trend, but it nonetheless is a predictable process. On the flip side, the process level shift example and the process depicted in Figure 2.1*a* represent statistically unstable processes since their past behaviors do not provide reliable guidance for the prediction of future outcomes.

Continuing with the time-sequence plots of Figure 2.4, we see that the next graph reflects a process that tends to bounce from one side of the centerline to the other. Given the oscillating pattern to the data, we would designate the process as nonrandom[6] based on violation 2. The process is stationary since it appears that the average level remains constant at the centerline along with the fact that the scattering of points around the centerline seems fairly constant. In practice, oscillatory behavior is a fairly uncommon *natural* pattern. This behavior, however, can result when management or an operator overreacts to random variation and attempts to adjust (e.g., changes a machine setting) the process back to some target level. This back-and-forth "knob" turning has the perverse effect of increasing process variation and inducing an oscillatory pattern; in Chapters 4 and 10, we will study in greater detail the effects of counterproductive adjustments.

The last time-sequence plot reflects a process for which dispersion around the centerline is steadily increasing. This is not an uncommon occurrence in practice. Many measurement systems are known to lose precision over time. Or, a machine

---

[5]By using probability theory, a more technical definition for stationarity can be given. The interested reader should consult Box, Jenkins, and Reinsel (1994).

[6]As a statistics instructor, I have found that many students have a preconceived notion that oscillatory behavior is a case of extreme randomness. This incorrect perception quickly disappears following a little data analysis experience.

may become more "wobbly" through the wear and tear of constant use. Statistically speaking, this process is nonrandom based on a nonconstant variation violation, that is, violation 3. Note that neither violation 1 nor violation 2 is at hand since the process level is keeping steady and there is no systematic pattern in the short-term process levels; the systematic pattern is related to the dispersion, not the level of the process. Thus, the best guess for the future outcomes is the centerline but with increasing uncertainty as one projects farther into the future.

With Figure 2.4, we have illustrated only a handful of nonrandom processes. There are, of course, many more (infinite to be precise!) possible nonrandom scenarios. Some other examples can simply be created by combining two or more scenarios found in Figure 2.4. For instance, it is quite possible to have a trend process with steadily increasing variation around the trend line as time evolves. An important type of nonrandom behavior not illustrated in Figure 2.4 is seasonality. Seasonal behavior is a fairly regular pattern of movement within some time frame. For example, the consistently higher use of electricity from year to year in the months of July and August represents a seasonal pattern. In general, a *season* is defined as a period of time which within itself has a constant number of subperiods. Examples include 12 months of the year, 5 or 7 days of the week, and 3 production shifts per day. Since many quality applications (e.g., monthly customer satisfaction ratings or daily defect rates) are sampled within a defined season, the ability to identify and analyze potential seasonal behavior will prove very useful. In subsequent chapters, we will have several opportunities to encounter seasonality in real process data.

In summary, we have seen that a time-sequence plot is a fundamental tool for establishing a basic impression of underlying process behavior. With no further refined statistical analysis, this plot alone can serve as a basis for making reasonable predictions about the future process outcomes.

HISTOGRAM

In the last section, we saw how a time-sequence plot enables us to see the nature of process data over time. Among other things, a time-sequence plot reveals the general level variability of the data relative to the centerline. For the stopwatch example, we saw a random sequence of data all of which fell between 9.75 and 10.25s (see Figure 2.2). Even though a time-sequence plot has much to reveal, there are still several important process characteristics not easily extracted from this type of plot. For example, are the observations evenly spread in the range of 9.75 to 10.25s? That is, are the chances the same that a process outcome will fall in any interval from 9.75 to 10.25s? If the observations are not evenly spread, where do the data tend to group or be concentrated? In general, we are seeking to characterize the *distribution* of the data.

To gain such a perspective for the stopwatch data, consider the displays found in Figure 2.5. Starting at the top left, we present a time-sequence plot, in character mode, of the stopwatch readings along with a superimposed centerline. Suppose

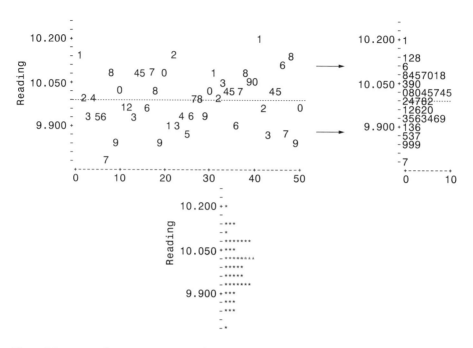

Figure 2.5   Transforming a time-series plot into a histogram.

now we push all the observations to the left so that they pile on the vertical axis which gives us the graph to the right of the time-sequence plot. Next, we delete the centerline and the horizontal axis and replace each numeral symbol with an asterisk. As a result of these activities, we have the bottom display given in Figure 2.5. By collapsing the values in the piles, we have basically created a display of the data with the time dimension eliminated. This graph, known as a *histogram*, makes clearer the dispersion of the measurements as well as the relative frequency of the various values. We see, for example, that more values occur in the middle with the number of values tapering off in the direction of the extremes.

Fortunately, to create a histogram on a routine basis we do not have to go through the word processing gymnastics shown in Figure 2.5! In either character or high-resolution mode, Minitab will automatically display a histogram. In Figure 2.6, we show both versions of the histogram for the stopwatch readings data.

Consider first the character version of the histogram seen in Figure 2.6*a*. Notice the resemblance to the histogram we "created" in Figure 2.5. Both graphs have no time dimension, and both show the frequency of observations for different measurement values. But the graphs are not identical for a couple of reasons. First, the values on the axis of one graph are flipped relative to the values on the axis of the other graph. In particular, the frequencies of the low values (less than 10s) are plotted on the bottom of the vertical axis for the histogram in Figure 2.5 while they are plotted on the top of the vertical axis for the histogram in Figure 2.6*a*, and vice versa for high values (greater than 10s). This difference has

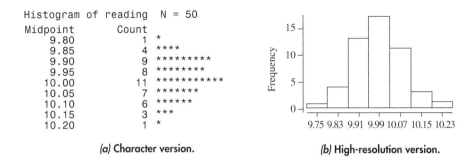

(a) Character version.   (b) High-resolution version.

Figure 2.6   Histograms of the stopwatch data.

no consequence for the interpretation of the histogram shape and data concentration, either way will work. The main reason for the difference in the two displays is related to the manner in which the data are grouped and stacked against the vertical axis. In the general construction of a histogram, the value axis is divided into adjacent and equal-width class intervals or cells. The number of observations that fall within each interval is then counted. For each interval, the next step is to plot a bar or a stack of symbols which is proportional in length to the number of observations in the interval.

Looking at Figure 2.6a, we see that Minitab—by default—divided the axis into nine class intervals of width 0.05. The boundaries of the class intervals can be inferred from the designated midpoints. In particular, starting from the top, the first class interval ranges from 9.775 to 9.825, the second class interval ranges from 9.825 to 9.875, and so on. Minitab does provide options to allow the user to select the width of the class intervals along with a desired starting point. (As a homework exercise at end of this chapter, the reader is asked to use these options to create a histogram identical to the one shown in Figure 2.5.) Both the number of observations and a stack of asterisks proportional to the number are displayed to the right of the midpoint designation. For example, one observation fell in the first class interval (9.775 to 9.825) while four observations fell in the second class interval (9.825 to 9.875).

With respect to the high-resolution histogram (Figure 2.6b), Minitab displays the graph at a 90° angle relative to the character version. Instead of stacks of asterisks, bars are associated with the frequencies. With this chart, the use of class intervals and frequency counts in the construction of a histogram becomes self-apparent. For purposes of visualizing the histogram's shape and data concentration, either version of the histogram will do the job. However, the character display has one nice advantage of providing the user with the exact frequency counts. Extracting exact counts from the high-resolution version is a bit tedious and prone to error. We will find that the handy availability of these counts can be

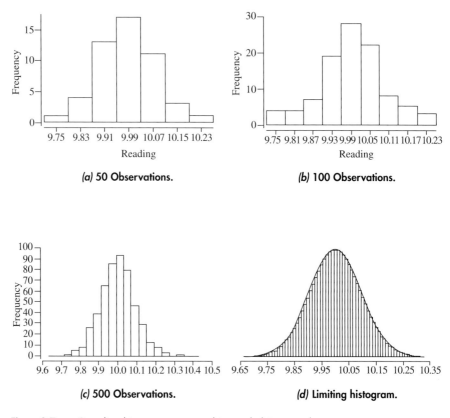

Figure 2.7   Sampling histograms approaching underlying population.

used in a simple check of how closely the data conform to a given theoretical distribution. For this reason and based on this author's personal preference, the character histogram will be used most often in this textbook.

Comparing Figure 2.6*a* (or 2.6*b*) with Figure 2.5, we notice how the grouping of data values helps smooth out some of the jaggedness, giving rise to a clearer indication of the underlying distribution of the data. Note that the default command was not used in creating the histogram in Figure 2.6*b*. If the default mode were used for the stopwatch data, a high-resolution histogram based on five class intervals would be created; interestingly, the default character histogram has nine class intervals. Based on my judgment, with only five class intervals, a bit too many observations are lumped together, producing too much smoothing and an unnecessary loss of detail in the distribution. As a rough but reasonable rule of thumb, the number of intervals should be approximately equal to the square root of the number of data points. In our case, we have 50 points, so about 7 intervals would be used.

In any case, we see that the tendency for observations to concentrate in the middle and to taper off is more evident from the histograms of Figure 2.6 than from our earlier created histogram of Figure 2.5. We also notice that the histograms

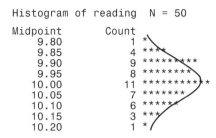

```
Histogram of reading   N = 50

Midpoint        Count
    9.80            1   *
    9.85            4   ****
    9.90            9   *********
    9.95            8   ********
   10.00           11   ***********
   10.05            7   *******
   10.10            6   ******
   10.15            3   ***
   10.20            1   *
```

Figure 2.8   Freehand sketch superimposed
on sample histogram.

are fairly *symmetric* in shape. This means that the upper half of the histogram is nearly a mirror image of the lower half of the histogram.

One of the main reasons for the construction of a histogram is that it helps us envision the shape of the underlying population from which the data were drawn. Through the stacks of asterisks or bars, the sample histogram forms a rough image of what the population might look like. Remember for analytic or time-series studies, we are dealing with conceptual populations, that is, the totality of observations that might occur if we were to perform some operation over and over. From another perspective, the conceptual population can be thought of as a very large sample—theoretically, infinite—of data from a particular process. By associating a single population to the process, it is critical to recognize that we are implicitly assuming that the underlying process remains unchanged over time. If the process level and/or dispersion changes in any fashion over time, the notion of a single population to explain the distribution of the data at any time is no longer applicable; this is such an important point that we will revisit it shortly.

Suppose, hypothetically, that we are able to generate more and more measurements for the stopwatch experiment as an ongoing process. As seen in Figure 2.7, if more and more observations are obtained, and the histograms are constructed by grouping the measurements into class intervals of smaller and smaller width, we will see the irregularities—bumps and valleys—become less and less noticeable. Eventually, the histograms will approach a smooth curve which represents the underlying population distribution of the measurements. In practical applications, we typically do not have the luxury of having so many data, enabling us to pretty much see the underlying smooth distribution curve. Instead, we need to infer the smooth image of the population histogram from the sample histogram.

One useful way of helping infer the population shape is to fit a freehand sketch of a smooth curve onto the sample histogram. To illustrate, in Figure 2.8, I sketched a smooth curve on the character histogram for the 50 stopwatch readings to suggest what the population histogram might look like. The key to good freehand sketching is to not overreact to the sample irregularities. As Roberts and Ling (1982) state, "The freehand fit to the histogram should not just skim over the peaks! It should level off the peaks and raise the valleys." Notice with only

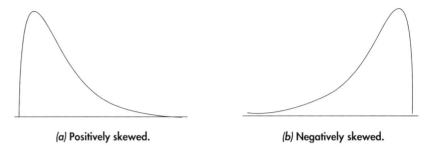

(a) Positively skewed.   (b) Negatively skewed.

Figure 2.9   Asymmetric distributions.

50 observations the histogram along with the freehand sketch indeed suggests a population curve compatible with the limiting histogram seen in Figure 2.7*d*. Both the shapes of the freehand curve and the limiting distribution closely resemble a mathematically generated curve known as the *normal* curve, which we will describe in greater detail in Section 2.8.

Even though the normal curve with its characteristic symmetric "bell" shape provides a good approximation for many data processes, not all histograms resemble this well-known shape. In theory, there are an infinite variety of distributional shapes other than the normal curve. Fortunately, however, most processes in practical applications are well approximated by a small handful of distribution shapes. The most common departure from the normal curve in practical applications is asymmetry, or skewness.[7] Figure 2.9 shows two types of skewness. When the tail of the distribution is stretched more to the right toward higher values, we say that the distribution is right- or positively skewed (Figure 2.9*a*). Similarly, a left- or negatively skewed distribution is one that stretches more to the left toward smaller values (Figure 2.9*b*). Of these two types of skewness, positive skewness is the more predominate distribution shape in practice. In summary, symmetric distributions closely resembling a normal distribution or positively skewed distributions account for a substantial proportion of process data; the practical data sets presented in this textbook will attest to this reality. Since most SPC methods assume that data approximately conform to the normal distribution, we will rely heavily on histograms, along with other diagnostic checks (see Section 3.2), to determine whether this basic assumption is being met.

In addition to the use of histograms for the validity of assumptions for the implementation of statistical methods, histograms are a wonderful means of assessing process capabilities relative to quality requirements, such as engineering specifications. For example, Figure 2.10*a* suggests a process that is centered on target and performing quite well relative to both upper and lower specification limits.[8] Figures 2.10*b* and *c* reflect less capable processes due to two different

---

[7]A distribution can be symmetric but not normal. For example, some statistics textbooks introduce symmetric distributions characterized as being platykurtic or leptokurtic. Platykurtic and leptokurtic distributions depart from the normal distribution by having either too little or too great concentration in the tails of the distribution.

[8]As will be illustrated and discussed in Section 2.4, we should warn that firm conclusions should not be drawn from histograms without initial inspection of a time-sequence plot of the observations.

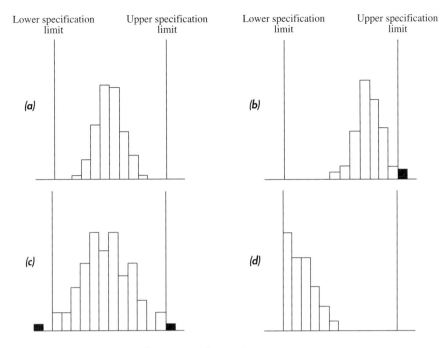

Figure 2.10   Histograms relative to specification limits.

troubles. In the first case (Figure 2.10*b*), the process is off target in the direction of the upper specification limit, resulting in a percentage of unacceptable product (shaded area). In the second case (Figure 2.10*c*), the process is centered on target, but the variation is so large that again unacceptable product results. The histogram of Figure 2.10*d* relates to an interesting experience of a certain customer who decided to assess incoming product from a specific supplier. As can be seen, all the product received is within specification limits. However, the histogram reveals quite clearly that the supplier must be sifting out unacceptable product prior to shipment. From the customer's perspective, such sifting implies substantial costs in terms of inspection and scrap incurred by the supplier, a price the customer will no doubt have to pay.

2.4   TIME VERSUS DISTRIBUTION BEHAVIOR

In the last two sections, we have seen how a time-series plot can shed light on how the process level and variation evolve over time relative to a centerline whereas a histogram provides insight about the data concentration around the centerline with no regard to a time dimension.

Using the stopwatch data, we constructed a histogram by "collapsing" sequenced observations into a single gathering of data (Figure 2.5). An implicit

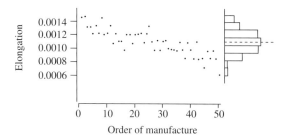

Figure 2.11   Time-series plot and histogram of elongation measurements. Figure adapted with permission from Deming (1986).

assumption in this collapse is that observations from different time points have been generated from a single identical conceptual population. It would make little sense to use a single histogram as a basis for judgment if the underlying distribution changed over time. To bring out the dangers of inappropriate judgment based on a histogram, consider the following passage from Deming's *Out of the Crisis* (1986):

> As an example, we turn attention to a distribution that appears to have all the good qualities that one would ask for, but which was misleading, not just useless. [Figure 2.11] shows the distribution of measurements made on 50 springs used in a camera of a certain type. Each measurement is the elongation of the spring under a pull of 20 g. The distribution is fairly symmetrical, and both tails fall well within specifications. One might therefore be tempted to conclude that the process is satisfactory.
>
> However, the elongations plotted one by one in the order of manufacture show a downward trend. Something is wrong with the process of manufacture, or with the measuring instrument.
>
> Any attempt to use the distribution in [Figure 2.11] would be futile . . . It would tell nothing about the process. . . . (pp. 312–313)

To further elaborate on Deming's discussion, the histogram based on the 50 spring observations is not a guide for the future, and thus it has no predictive value. If we were to rely on the histogram in Figure 2.11, we would be lulled into concluding that the underlying process level is around 0.0011 and the range of likely process outcomes is from 0.0006 to 0.0014. However, the best guess for the short-term future is not near 0.0011 but rather is somewhere below 0.0006. Moreover, in dramatic contrast to the implications of the histogram, it is quite unlikely that any future outcomes will even fall above the centerline.

Clearly, the difficulty stems from the fact that the spring elongation process is exhibiting a nonrandom downward trend which is impossible to infer from a histogram alone. As generally depicted in Figure 2.12*a*, the underlying conceptual population for a trend process is continually changing in terms of level. The implication is that no single population can be used to characterize the likely process outcomes over the course of time. To bring out this point from another perspective, if we collapse the time dimension, the underlying distributions do not overlay as a single distribution (see Figure 2.12*a*). But notice, however, that the distributions for the random variation *around* the trend line appear identical in shape; in other words, the distributions over and above the trend line resemble each other. To see that the distributions around the trend are indeed identical, we can "eliminate" the trend by simply laying the process flat on its side, which gives the graph of Figure 2.12*b*. Now, if the distributions are collapsed down, we are left with a single distribution describing the likely process outcomes for any point in time.

In general, the fact that distributions overlay exactly as one distribution is the result of two properties. First, the distributions do not move around, but rather they are all lined up in the same position. Hence, the process cannot exhibit any systematic patterns, such as trend, meandering, shift, and seasonality. Second, the distributions must all look exactly the same in terms of shape. Recognize that together these two properties define a random process. *Hence, interpretation of a histogram is appropriate only when the data exhibit random behavior.* As demonstrated by Deming's spring elongation data, analysis of histograms in the presence of nonrandom behavior is a futile and risky exercise. If data exhibit systematic patterns, the first order of business is to analyze the data from a time-series perspective. As suggested by Figure 2.12, histograms can come in later to summarize the nature of the data variability over and above the identified systematic patterns.

The above discussion invites us to introduce important time-series terminology for classifying different time-series behaviors. Namely, as an alternative to saying that a process is random, we can say that the process is *independent and identically distributed,* which is commonly abbreviated as *iid*. The term *independent* suggests that the individual process observations taken at different time points are unrelated.[9] Hence, knowledge that certain observations are high or low is not suggestive that other observations (past or future) are high or low. The other aspect of *identically distributed* means that the underlying distribution of process outcomes is exactly the same for each time period. Remember also that the traditional notion of a state of statistical control is associated with a random process. Therefore, the designations *random, iid*, and *in a state of statistical control* are all interchangeable.

The hypothetical process of Figure 2.12*b* (not Figure 2.12*a*) reflects an *iid* process; another example can be seen from Figure 2.1*b*. Since we deemed the stopwatch data as random, these data are inferred to have arisen from an underlying *iid* process. In all these cited examples of *iid* behavior, the underlying shape of the conceptual distribution was either associated with or inferred as a bell-shaped normal curve. Random or *iid* processes do not have to be associated with a normal curve. The requirement for *iid* behavior is that the distribution, whatever it may be, remain the same over the course of time. Generally, the notions

---

[9]For the more technically inclined, the property of independence implies that the joint probability of an observed sample is equal to the product of the probabilities of each individual observation in the sample.

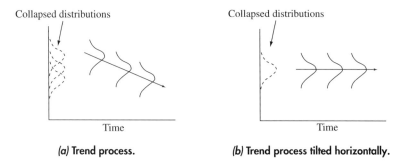

(a) **Trend process.**

(b) **Trend process tilted horizontally.**

Figure 2.12 Trend process in terms of underlying populations.

of randomness and distribution are distinct concepts; that is, randomness (or lack of) does not imply a particular distribution, and vice versa. As illustrated in Figure 2.13, a process can be random and normal, random and nonnormal, nonrandom and normal, or nonrandom and nonnormal. In the important case of a process that is *iid* and normal, the common convention is to refer to such a process as being *iidn*.

In summary, the first step in the analysis of process data must always be the construction of a time-series plot. If and only if no systematic patterns are evident, then a histogram can be constructed to gain further insights about the data. If, however, the data reflect systematic patterns, these patterns should be analyzed and summarized before histograms come into play. In Chapter 3, we will introduce statistical tools to help in the data summarization of patterned data.

2.5 SUMMARIZING PROCESS DATA: NUMERICAL MEASURES

In the last two sections, we found that graphical techniques substantially added to our understanding of process data. We now examine several important numerical measures that are used to describe and summarize observations. Summary measures can correspond to either populations or samples. As briefly noted earlier, numerical summaries of underlying populations or processes are referred to as *parameters* while numerical summaries of data sets are referred to as *sample statistics* (or, simply, *statistics*). In this section, we concentrate on a few common sample statistics. However, recognize that since a population can be thought of as a "large" data set (relatively speaking), the insights gained about many sample statistics directly apply to the associated population parameters.

The most common numerical descriptions of data sets are generally of two types:

1. Measures that indicate the approximate center of the data distribution, called *measures of central tendency or location*.

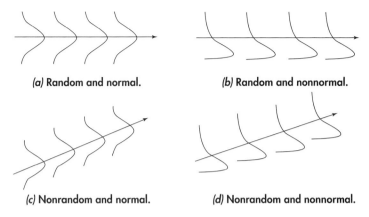

(a) Random and normal.   (b) Random and nonnormal.

(c) Nonrandom and normal.   (d) Nonrandom and nonnormal.

Figure 2.13   Distinction between time and distribution behavior.

2. Measures that indicate the scatter or spread of the data around the central lo-
    cation of the data distribution, called *measures of variability or dispersion.*

Note that there are numerical summaries of other aspects of a distribution. For
example, there is a numerical measure of skewness that is used to summarize the
amount of asymmetry in the data distribution. In our data analysis applications
throughout the textbook, we will rely on graphical analysis rather than a numerical measure to determine the presence of skewness.

MEASURES OF CENTRAL TENDENCY

When we calculate data summaries, the first consideration is typically to find a
value representing the general location or centrality of the observations. Since, in
most cases, observations tend to concentrate or bulk up in a middle area between
the two extremes, a summary value representing the center is (in a sense) a value
that is typical or representative of the whole data set. Here we will look at three
measures of central tendency: sample mean, sample median, and sample mode.

Sample Mean   By far the most common measure of central tendency is the sample *mean*, which is also referred to as the sample *average*. It is a measure that
uses all the data, it is easily computed, it is easily understood by all, and it has
many desirable statistical properties. Because of these reasons, the sample mean
is a dominant measure of central tendency in the area of SPC and in general
applications.

The sample mean is simply computed by adding all the observations and dividing the total by the number of observations. For convenience, we can express
this computation and others to follow symbolically. It is conventional to use the
letter x to generically represent an individual measurement and the letter n to denote the number of observations in the sample. To distinguish the individual

Stat>>Basic Statistics>>Display Descriptive Statistics

```
MTB > describe 'reading'

Descriptive Statistics

Variable          N        Mean   Merdian     TrMean      StDev    SE Mean
reading          50      9.9890    9.9850     9.9889     0.0918     0.0130

Variable    Minimum    Maximum        Q1         Q3
reading      9.7800    10.2100    9.9200    10.0625
```

Figure 2.14   Summary statistics for the stopwatch data.

measurements, numeral subscripts are used. Thus, the individual measurements are given by $x_1, x_2, x_3, \ldots, x_n$. The sample mean or average is defined as follows:

$$\bar{x} = \frac{x_1 + x_2 + \ldots + x_n}{n} = \frac{\sum\limits_{i=1}^{n} x_i}{n} \tag{2.1}$$

where the Greek symbol Σ is used to more concisely represent the summation operation. To signify the averaging operation, it is customary to place a bar over the letter associated with the measurements, which in the case of Equation (2.1) gives us the notation $\bar{x}$ (pronounced "x bar").

To illustrate the computation of the sample mean, recall the stopwatch example with the 50 observations given in Table 2.1. For these data, we have

$$\bar{x} = \frac{10.13 + 9.98 + 9.92 + \ldots + 9.83 + 9.97}{50} = \frac{499.45}{50} = 9.989$$

Using Minitab, we find from Figure 2.14 the value of 9.989 for the sample mean along with values of other summary statistics reported; we will soon refer to this output for some of the other displayed values.

The sample mean not only represents a centrally located value of data but also can be uniquely interpreted as a balance point for the data. As depicted in Figure 2.15, if a wedge (known as a fulcrum) is placed at the mean with the histogram sitting above on a flat plane, then the histogram will perfectly balance. Equivalently, the mean can be thought of as the center of gravity of the distribution. In some statistical contexts, the intuitive appeal of a balancing point measure can have a downside. Namely, extreme points or outliers can substantially throw off the balance point relative to the centrality of the data. Consider the following data:

$$0 \quad 0 \quad 0 \quad 0 \quad 0 \quad 1 \quad 1 \quad 1 \quad 1 \quad 50$$

The sample mean is 5.4. This value clearly does correspond to the bulk of the data distribution and, as such, cannot be interpreted as (or close to) a "typical value."

Is the sensitivity of the sample mean to outliers a detrimental factor in the implementation of SPC methods that are based on this statistic? Actually, the

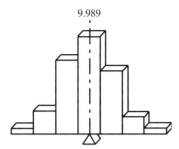

Figure 2.15   The mean as a balancing point.

answer is not so obvious. We need to recognize that there are two fundamental phases in the implementation of any control chart procedure. The first phase involves a retrospective analysis of observed process data. The goals of this phase are to gain an understanding of the process and to establish a statistical benchmark (or model) for the likely future process outcomes based on the data at hand. In the case of the stopwatch data, we found the 50 observations to be random. Furthermore, none of the observations appeared unusual or extreme; in other words, nothing "stuck out like a sore thumb." Given the nature of the data, we intuitively argued that the best guess (model) for the future is an extension of the centerline which, as you will recall, was simply the average of the data, that is, the sample mean. Imagine now that some substantial outliers did exist among the 50 stopwatch observations. Given the general sensitivity of the sample mean to outliers, there is a potential that these outliers might unduly influence the value of the centerline, rendering it less reliable as a guide for future outcomes. As will be noted in subsequent discussions on control charts, if special explanations can be given to demonstrate that extreme points are indeed unrepresentative of the process, then the points should be deleted from the data set. Thus, it is during this first phase of process modeling that we must be wary of outliers and their detrimental effects on the sample mean and other related statistics.

The second phase in control chart implementation involves observing the process in real time and comparing the new process data with the benchmarks established in the first phase of data analysis. A primary goal during this phase is to monitor the process for any fundamental changes that require management and/or operator intervention. As we will learn later, many control charts entail collecting a sample of two or more observations at specific points in time. In turn, statistics, such as the average of the data, are computed and plotted over time. Since changes often result in unusual or extreme data values, it is particularly important that these points not be camouflaged by the statistical computations. Hence the sensitivity of the sample mean statistic to extreme points becomes an advantage during the phase of on-line monitoring and detection.

Sample Median   In cases where we want to protect against the influence of outlying points, the sample *median* can be used as an alternative statistic. The median

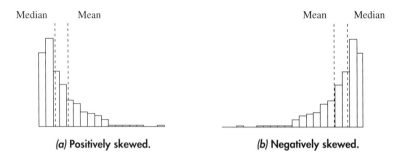

Figure 2.16   Median versus mean for asymmetric data distributions.

is identified by finding the most central value of the data when arranged in ascending order. If the number of observations in the data set is odd, the median is simply the middle item in the ordered data set. If the number of observations in the data set is not odd (even), the convention is to calculate the median as the average of the two middle items. As a result, the median is a value that divides the data set into two parts, with at least half the data values being less than or equal to this value and at least half the data values greater than or equal to this value. Symbolically, the median is denoted as $\tilde{x}$ (pronounced "x tilde").

Consider the following data where the sample mean was unduly influenced:

$$0 \quad 0 \quad 0 \quad 0 \quad 0 \quad 1 \quad 1 \quad 1 \quad 1 \quad 50$$

Since there are an even number ($n = 10$) of observations, the median is the average of the fifth and sixth observations, which is $(0 + 1)/2 = 0.5$. This value is clearly more representative of the bulk of the data than the sample mean value of 5.4. Furthermore, if the outlying value of 50 were made even more extreme, say, 500, the sample median would remain unchanged. Hence, the sample median is insensitive or robust to outliers. This robustness allows the median to remain more centrally located. However, with no further information, the median value by itself does not flag the possibility of extreme points and hence potentially conceals critical information. As discussed earlier, this insensitivity can be viewed as a downside for the purposes of control charting.

With respect to the stopwatch data, the median computed by Minitab (Figure 2.14) is seen to be 9.985. The nearly identical value of the median and the sample mean ($\bar{x} = 9.989$) reflects a basic property of these two statistics. Namely, if the data distribution is symmetric, the values of the sample median and mean will coincide. In contrast, these two statistics will differ when the data distribution is asymmetric, as depicted in Figure 2.16. In both cases, notice that the mean has been "pulled" in the direction of the elongated tail away from the bulk of the distribution. In many applications, such as reliability and lifetime testing studies, positively skewed distributions are commonplace. In such situations, the median might be used to provide a fairer representation of the centrality of the data.

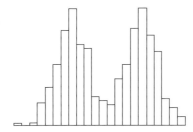

Figure 2.17   Bimodal data distribution.

Sample Mode   The *mode* is the value that occurs most frequently in a set of data. When data are grouped to form a histogram, the mode is associated with the class interval with the most observations. For grouped data, the mode can be arbitrary since it can change by a simple reclassification of the class intervals.

In practice, reporting of the value(s) for the mode is infrequently done. In fact, most statistical software—including Minitab—do not routinely provide this statistic. The notion of the mode is more useful in describing the general shape of the data distribution. A distribution that has an identifiable single peak or "hump" is described as being *unimodal.* In contrast, the data distribution shown in Figure 2.17 is *bimodal* due to its two distinct peaks. The appearance of multiple modes is often indicative of two or more fundamentally different sources of variation being mixed together, for example, data from different operators, times, or machines. Since there is more than one underlying population, the sample mean or median statistics have little applicability and are actually quite misleading. Once multiple modes have been identified, it is most appropriate to segregate the different sources. Thereupon, traditional statistics can be applied to the data from the segregated sources.

MEASURES OF VARIABILITY

The measures of central tendency help us understand where the data distribution is generally located, but they tell us nothing about the variation among the observations. To clarify the nature of the data distribution, we must obtain some measure of variability. Alternative terms for *variability* include *dispersion, spread, scatter*, *precision*, and *uncertainty*. Even though there are several measures of variability, we will address only two basic types which are dominant in the quality control area: sample range and sample standard deviation.

Sample Range   The simplest and most easily understood measure of variability is the *range*, commonly denoted by *R*, which is the difference between the largest and smallest values in the data set. From Figure 2.14, the largest and smallest values in the stopwatch data set are 10.21s and 9.78s, respectively. Therefore, the range is $10.21 - 9.78 = 0.43$s.

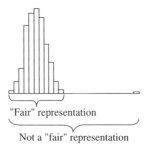

"Fair" representation

Not a "fair" representation

Figure 2.18   Undue influence of an outlier on
the sample range statistic.

The range has one clear limitation. Namely, it considers only the two observations and fails to account for the variability of the remaining observations in the data set. Thus, the majority of the information contained in the data set is ignored by the range. As the number of observations increases, the percentage of available information used becomes progressively smaller. Generally, for sample sizes larger than 25, the range is one of the least efficient measures of variability; this implies that in repeated samples from the same source, the ranges exhibit greater variation (that is, less stability) from sample to sample than will the other measures. However, for smaller sample sizes ($n \leq 10$), the range actually fares quite well as a measure of variability relative to other common measures.

The fact that the range is defined by the extreme values can be viewed as either a disadvantage or an advantage depending on the application. In the context of an enumerative study where the purpose is to infer the variability of the underlying population, there is the danger that an outlying point might mislead the analysis. As illustrated in Figure 2.18, all it takes is one "bad apple" to completely throw off the range statistic in fairly representing the spread of the preponderance of the data. For small samples, outlying observations are not easily detected, hence increasing the likelihood of misinterpretation of the true underlying variability. Enumerative studies are especially susceptible since inference is typically based on a single sample with no comparison to other samples, which would allow one to assess if a particular sample were unusual or not. However, in analytical studies, sequential data samples are gathered and compared with one another to help reveal any unusual values reflecting process disturbances. Since extreme points suggest the need for process investigation, the sensitivity of the range statistic to such points can be put to good use.

In summary, the range is easy to calculate, reasonably efficient in small samples, and sensitive to unusual observations. Because of these factors, the sample range statistic is used extensively in the area of SPC.

Sample Variance and Standard Deviation   Rather than base a measure of variability on only two points (largest and smallest), there are alternative and more comprehensive measures that involve all the observations of the data set. The most common and well known are the sample *variance* and the square root of the sample variance—the sample *standard deviation.*

Sample variance (and standard deviation) is based on the notion that a measure of variability should be small when observations are clustered closely around the mean and should be large when observations are scattered widely around the mean. Thus, we are considering distances of individual observations from the mean of the data set. For a data set consisting of n observations x_1, x_2, $x_3, \ldots, x_n$, the differences of the observations from the mean are given by $x_1 - \bar{x}$, $x_2 - \bar{x}, \ldots, x_n - \bar{x}$. These differences are more commonly referred to as *deviations* from the mean.

At first glance, it might seem that the sum of deviations $\Sigma(x_i - \bar{x})$ would be at the heart of defining a measure of variability. However, the sum of deviations will always equal zero. The reason is that the sum of the positive deviations from the mean offsets the sum of the negative deviations; we know intuitively they must offset because the mean is the balancing point of the data distribution. One way to combat the canceling effect is to eliminate the minus signs by squaring the individual deviations prior to the summing operation. (We should note that taking absolute values would also eliminate the cancellation effect, but it is a mathematical operation that makes certain theoretical developments difficult.) Hence, we consider the following:

$$\sum_{i=1}^{n}(x_i - \bar{x})^2 \tag{2.2}$$

This quantity represents the sum of squared deviations from the mean; for simplicity, it is also referred to as the *sum of squares*. Notice that the sum of squares by itself provides a measure of variability. If all the observations are the same, the sum of squares is 0, consistent with the fact of no variability in the data. If the observations are tightly clustered around the mean, the sum of squares will tend to be small; but if the observations are widely dispersed about the mean, the sum of squares will be relatively larger. The only difficulty with a sum of squares measure in its present form is that it increases with sample size. Thus, for two data sets with essentially the same amount of variability, the data set with more observations will tend to have a higher sum of squares. For equal-size data sets, the sum of squares measure can be used to fairly differentiate the levels of variability.

Given the confounding effect of sample size on the sum of squares, it will prove to be more convenient to compute the following statistic:

$$s^2 = \frac{\sum_{i=1}^{n}(x_i - \bar{x})^2}{n - 1} \tag{2.3}$$

This measure is known as the *sample variance*. The sample variance can be interpreted as being the "average" squared deviation from the mean. We place the word *average* in quotes since the notion of average is commonly equated with dividing some sum by the number of terms added up in the numerator. Here, this would suggest that an average squared deviation would be more precisely associated with the sum of squares divided by n, not $n - 1$. If our sole purpose is to descriptively compare levels of variability in different data sets, then a divisor of $n, n - 1$, or any other constant makes no difference. The justification of $n - 1$ instead of n stems from the potential use of the sample variance statistic in inferring or estimating the underlying population variance parameter. Based on a criterion that will be discussed in Section 2.10, this slight modification improves s^2 as an estimate of the population variance.

From Equation (2.3) notice that the variance is not expressed in the same units as the original data. For example, if the original data are measurements taken in feet, then the deviations from the mean are measured in feet and the variance is measured in feet squared. Since it may be difficult to relate to squared units, the positive square root of the sample variance is often computed to return to original units. The square root of the sample variance is known as the *sample standard deviation* and is given by

$$s = \sqrt{\frac{\sum_{i=1}^{n}(x_i - \bar{x})^2}{n - 1}} \tag{2.4}$$

As a computational example, consider again the stopwatch data for which we know the sample mean is $\bar{x} = 9.989$. Substituting this sample mean value along with the individual measurements (see Table 2.1) into Equation (2.3), we calculate the sample variance as

$$s^2 = \frac{(10.13-9.989)^2 + (9.98-9.989)^2 + \ldots + (9.83-9.989)^2 + (9.97-9.989)^2}{50 - 1} = 0.00843$$

The sample standard deviation is simply $s = \sqrt{0.00843} = 0.0918$. Referring to the Minitab output shown in Figure 2.14, we find the value of 0.0918 is provided under the heading "StDev."

2.6   RANDOM VARIABLES AND PROBABILITY DISTRIBUTIONS

We have discussed and illustrated how the histogram is a tool for showing the frequency distribution of a sample of data. We have also discussed the notion of a population as the totality of possible observations that either exist (enumerative study) or might occur (analytic study). Ideally, "large" samples can be collected to construct the frequency distribution which essentially represents the population. A frequency distribution based on such an extensive data set can, in turn, be used to make judgments on whether current or future observations would be considered typical, unusually (significantly) large, or unusually (significantly) small. However, it is generally not practical to obtain extensive data sets. Alternatively, we can attempt to find a mathematical function that can reasonably approximate the large-sample frequency distribution, that is, the population. Such a theoretical model could then serve as a basis for judgment of observed data instead of reliance on a large-sample frequency distribution.

Depending on the application, the process of finding a reasonable model or representation of some statistical phenomenon incorporates, to varying degrees, a combination of information on the sampling and experimental conditions that underlie the data generation, theory (such as physical laws), experience, and pertinent data. (*Note*: The term *experiment* is meant to define any procedural method that leads to certain observed results when completed.) Suppose we are interested in a model associated with the number of possible heads obtained from randomly flipping a two-sided coin a certain number of times. Given these experimental conditions, the laws of probability theory can be employed to deduce that a certain theoretical probability distribution (known as the binomial distribution, see Section 2.7) would most likely explain the random phenomenon. Fortunately, the development of statistical theory over the last 150 years has led to a set of general paradigms that help the data analyst, in many cases, narrow the search for a model to adequately describe the underlying distribution for the population in question.

However, before we proceed in confidence with any general model, collection of data on the random phenomenon is critical. There are two fundamental purposes served from the incorporation of data in model development. First, data should be analyzed to verify the adequacy of the specified model. It may be that the assumed conditions underlying the chosen theoretical model have not been or cannot be met; hence, the model cannot provide reliable guidance for judgment or decision making. If the prespecified model is found to be inadequate, alternative models must be sought. One of the basic themes of this textbook is to not blindly rely on idealized models (or their derivatives) without running some checks of consistency from the observed data. Once a reasonable theoretical model can be accepted and verified, the second use of data is to estimate the "particulars" of the model. For instance, in the coin-tossing experiment, if the above-mentioned binomial distribution is deemed as a satisfactory theoretical basis, this theoretical distribution needs to be made more specific by providing an estimate of the probability of obtaining a head for any given toss.

To develop a mathematical framework for presenting theoretical frequency distributions, as a basic starting point we need to establish a symbolic representation for the possible outcome values from an experiment. First, we define a *random variable* to be a rule or function that associates a single numerical value with each outcome of an experiment. Customarily, capital letters (such as X) are used to denote random variables. Since a random variable entails outcomes that are uncertain, it is also referred to as a chance variable or a stochastic variable.

A batch of n items is inspected and items are classified as either acceptable or defective. Let X denote the number of defective items in the given batch. Therefore, X is a random variable whose possible values are: $\{0, 1, 2, \ldots, n\}$.

Example 2.1

| **Example 2.2** | Let X be the number of customer complaints in a given month. Hence, X is a random variable taking on the following possible values: $\{0, 1, 2, 3, \ldots\}$. |

| **Example 2.3** | Consider an experiment of recording the life of a lightbulb. A bulb is left lighted until it burns out, and its life is recorded. Let X represent the set of values associated with this experiment. In this case, X can theoretically assume any value greater than or equal to 0. |

As previously discussed in Section 2.2, quantitative data, and thus random variables, may be classified as discrete or continuous. The random variables associated with Examples 2.1 and 2.2 are discrete while the random variable associated with Example 2.3 is continuous. Given the notion of uncertainty with random variables, it should be clear that a random variable induces the assignment of probabilities for the possible outcomes of an experiment. The function that relates the set of probabilities to the set of random variable values is called a *probability distribution*.

For discrete random variables, one denotes the probability of the random variable taking a particular value x by $P(X = x)$, $p(x)$, or $pr(x)$. Every discrete probability distribution must satisfy the following rules:

1. It is nonnegative, $P(X = x) \geq 0$, for any value x.
2. The sum of probabilities over all possible values of random variable X must be equal to 1, that is,

$$\sum P(X = x) = 1$$

where the summation is over all possible values of X.

| **Example 2.4** | Let X denote the total number of tails in tossing a fair coin 2 times. The possible values of X are 0, 1, and 2. The probabilities of these outcomes are |

$$P(X = 0) = \tfrac{1}{4} \quad \text{(H, H)}$$

$$P(X = 1) = \tfrac{1}{2} \quad \text{(H, T) and (T, H)}$$

$$P(X = 2) = \tfrac{1}{4} \quad \text{(T, T)}$$

Notice that $\sum_{i=0}^{2} P(X = i) = 1$.

Since continuous random variables are defined over continuous intervals that contain an uncountable infinite number of points, one cannot assign a positive probability to every possible point and still have the probabilities sum to 1. Imagine asking the probability that the next person you encounter weighs 150 pounds (lb)? In common parlance, this is usually taken to mean approximately 150 lb. However, technically speaking, it is impossible for a person to weigh exactly 150 lb, since this implies 150.00000 . . . out to an infinite number of decimal places. So the probability that a continuous random variable equals any given point is 0. Of course, there is no measurement device that can provide infinitely precise measurements. Practically speaking, every measured variable is discrete since it is rounded to a certain finite number of digits. But even so, we will still want to treat most measurement variables as strictly continuous since it is inconvenient to list all values of X along with their associated probabilities. Furthermore, in most cases, the probability that a given specific measurement will occur is small; thus the continuous framework well approximates the situation. Actually, the continuous framework can be quite flexible to satisfactorily approximate the distributions of many discrete variables. In fact, some of the classic control charts for discrete variables (known as p and c charts, see Chapter 7) are constructed and interpreted based on an approximating continuous distribution function.

Since we cannot assign probability values to each possible outcome on a continuum, statements are made regarding the probability that a continuous random variable will assume a value within some defined interval. For example, we may want to know the probability that the next person we encounter weighs between 147 and 153 lb rather than some specific value. A *probability density function* (pdf) is used to establish the link between probabilities and continuous intervals. The function, denoted by $f(x)$, is represented by a smooth curve which lies entirely above or on the x axis. A pdf is designed in such a manner that probabilities are associated with areas under its graph. Specifically, the probability that the random variable is in the interval (a, b) is the area under the curve between the points $X = a$ and $X = b$. Thus, the total area under the curve must equal 1. To summarize,[10]

1. $f(x) \geq 0 \qquad$ for all x

2. $\int_{-\infty}^{\infty} f(x)dx = 1$

3. $\int_{a}^{b} f(x)dx = P(a < X < b) = P(a \leq X \leq b)$

The reason that it does not matter if a "greater than" sign or a "greater than or equal to" sign is used in property 3 is due to the fact that probability at specific points is 0 for continuous random variables.

The time in minutes that an individual talks on a long-distance phone call can be considered a random variable with the following probability density function:

| **Example 2.5**

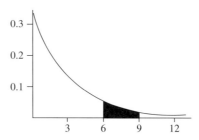

Figure 2.19   Probability density function for telephone example.

$$f(x) = \begin{cases} \frac{1}{3}e^{-x/3} & x \geq 0 \\ 0 & \text{otherwise} \end{cases}$$

where the symbol e is the well-known constant approximately equal to 2.71828. Figure 2.19 shows the graph for this exponential pdf function. The reader can verify that the function is indeed a density function by showing that the total area under the curve equals 1. Suppose we are interested in knowing the probability that an individual will talk between 6 and 9 min. This probability corresponds to the shaded area in Figure 2.19. To determine the area under the exponential curve, we can employ the integration technique shown below:

$$
\begin{aligned}
P(6 \leq X \leq 9) &= \int_6^9 \frac{1}{3}e^{-x/3}dx \\
&= e^{-x/3}\Big|_6^9 \\
&= e^{-2} - e^{-3} \\
&= 0.1353 - 0.0498 \\
&= 0.0855
\end{aligned}
$$

Hence, the probability is 0.0855, or 8.55 percent, that an individual will talk between 6 and 9 min on a long-distance phone call.

Since we will not assume knowledge of the integration technique in this textbook, it would be useful to seek an alternative means of determining the required probability. For many important distributions, tables are available that enable one to approximate fairly accurately the desired probabilities. In this textbook, we include tables for four distributions that are particularly important for SPC and data analysis applications, namely, the normal distribution, the t distribution, chi-square distribution, and the F distribution. These distributions will be discussed in detail later. As an alternative to the use of tables, most statistical software provide the capability of computing probabilities for a variety of distributions (both discrete and continuous). For instance, Minitab can compute the probability of a random variable's being less than or equal to some specified value—known as the *cumulative probability*—when the random variable follows one of a number of common distributions (including the exponential).

Calc>>Probability Distributions>>Exponential>>Cumulative Probability

```
MTB > cdf 6;
SUBC> exponential 3.

Cumulative Distribution Function

Exponential with mean = 3.00000

      x    P( X <= x)
 6.0000       0.8647

MTB > cdf 9;
SUBC> exponential 3.

Cumulative Distribution Function

Exponential with mean = 3.00000

      x    P( X <= x)
 9.0000       0.9502
```

Figure 2.20   Cumulative probabilities from Minitab.

To use Minitab to determine cumulative probabilities, Minitab requires the input of the parameter(s) that fully describe the distribution of interest. For the exponential distribution, there is only one parameter involved, namely, the expected value or mean of the distribution. As we will show in the next subsection, the mean of the exponential distribution in this example is 3. From Figure 2.20, we find the following probabilities:

$$P(X \le 6) = 0.8647 \qquad P(X \le 9) = 0.9502$$

The desired probability can then be determined as follows:

$$P(6 \le X \le 9) = P(X \le 9) - P(X < 6)$$
$$= P(X \le 9) - P(X \le 6)$$
$$= 0.9502 - 0.8647$$
$$= 0.0855$$

PROPERTIES OF RANDOM VARIABLES

In Section 2.5, we discussed how summary statistics such as the sample mean, sample variance, and sample standard deviation can be used to concisely summarize key features of sample data distribution. In a similar fashion, the intent behind these summary statistics can be extended to develop analogous measures that concisely summarize key features of a probability distribution. Remember that a probability distribution for a random variable is nothing more than a theoretical representation of the underlying population which can essentially be viewed as a sample data distribution based on "lots" of data. Any measurable characteristic of a population or its theoretical representation is known as a

Graph >> Chart

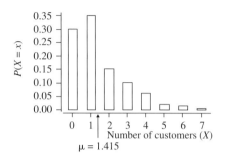

Figure 2.21   Probability distribution for customer example.

parameter. We introduce now two of the most important properties of a random variable: its expected value and variance.

Expected Value   The expected value of a random variable represents a measure of the center of probability distribution. Similar to the sample mean, the expected value of a random variable can be interpreted as the center-of-gravity point of the distribution, which in this case is the probability distribution as opposed to the histogram or data distribution. Thus, the expected value is also known as the *population mean parameter.* For a random variable X, the expected value of X is denoted by $E(X)$ or the Greek symbol μ. (To avoid confusion, at times we will put the symbol for the random variable in the subscript, say, μ_X or μ_Y.) Depending on whether X is discrete or continuous, the *expected value* is defined as

$$\mu = E(X) = \sum_{\text{all } i} x_i P(X = x_i) \qquad \text{discrete case} \qquad \text{(2.5)}$$

$$\mu = E(X) = \int_{-\infty}^{\infty} x f(x)\, dx \qquad \text{continuous case} \qquad \text{(2.6)}$$

Thus, expected value is the weighted average of all the possible outcomes of the random variable, with the probabilities being the weights reflecting the likelihood of each outcome.

Example 2.6   Let X be a random variable representing the number of customers arriving at a checkout counter during a 5-min period. Suppose the probability distribution of this discrete random variable is given by

x	0	1	2	3	4	5	6	7
$P(X = x)$	0.30	0.35	0.15	0.10	0.06	0.02	0.015	0.005

By using Equation (2.5), the expected value $E(X)$ is found to be

$$E(X) = 0(0.30) + 1(0.35) + 2(0.15) + 3(0.10) + 4(0.06) + 5(0.02) + 6(0.015) + 7(0.005)$$

$$= 1.415$$

A graph of the probability distribution along with its mean is shown in Figure 2.21. Literally speaking, the words *expected value* can be a bit misleading since 1.415 customers are not to be expected as a value of X—only the integers 0 to 7 are possible. The point is that $E(X)$ should be interpreted in terms of an average. It is the average (mean) number of customers that we would expect to see if we were to observe a very large number of 5-min periods, that is, if the "experiment" were repeated over and over.

Example 2.7

To determine the expected value for a continuous random variable, consider the probability density function given in Example 2.5. From (2.6), the expected value is given by[11]

$$E(X) = \int_0^\infty x(\tfrac{1}{3}e^{-x/3})\,dx = (-xe^{-x/3} - 3e^{-x/3})\big|_0^\infty = (0 - 0) - (0 - 3) = 3$$

Thus, the mean time for a long-distance phone call is 3 min.

Variance   Recall that the sample variance s^2 measures the "average" squared deviation of the sample data values from the sample mean. This sample statistic provided a numerical measure of the spread of the data distribution. To measure the spread of a probability distribution, a natural extension would be to determine the average squared deviation of the random variable X from its population mean μ. Given that the expected value is a general measure of the average, what we just described is known as the *variance* of a random variable and is given by

$$\sigma^2 = \text{Var}(X) = E[(X - \mu)^2]$$

For calculating variances, an equivalent and more convenient version of the above expression is

$$\sigma^2 = \text{Var}(X) = E(X^2) - \mu^2 \tag{2.7}$$

Since variance is measured in squared units, we take the square root of the variance, represented by σ, to return to original units and obtain the standard deviation of the random variable.

Example 2.8

To determine the variance of the random variable given in Example 2.6, we first find $E(X^2)$ as follows:

$$E(X^2) = \sum_{\text{all } i} x_i^2 p(x_i)$$

$$= 0^2(0.30) + 1^2(0.35) + 2^2(0.15) + 3^2(0.10) + 4^2(0.06) + 5^2(0.02) + 6^2(0.015) + 7^2(0.005)$$

$$= 4.095$$

---

[11]Again, readers unfamiliar with the integration technique can completely ignore the mathematical details with no risk. For those familiar with and interested in the details, the integral for this problem is solved by using the standard integration trick known as integration by parts.

We found earlier that $\mu = 1.415$, hence the variance is

$$\text{Var}(X) = 4.095 - (1.415)^2 = 2.093$$

Therefore, the standard deviation of X is $\sigma = \sqrt{2.093} = 1.447$.

We close this section by providing some useful properties of expectation and variance:

1. $E(c) = c$, where c is a constant.
2. $E(cX) = cE(X)$, where c is a constant.
3. $E(X \pm Y) = E(X) \pm E(Y)$, where X and Y are random variables.
4. $\text{Var}(c) = 0$, where c is a constant.
5. $\text{Var}(cX) = c^2 \text{Var}(X)$, where c is a constant.
6. $\text{Var}(X \pm Y) = \text{Var}(X) + \text{Var}(Y)$, where X and Y are independent random variables.

2.7   SOME USEFUL DISCRETE DISTRIBUTIONS

There are a number of discrete probability distributions that have received particular names; have wide applications in business, engineering, and science; and are of great theoretical importance. Below we will treat three discrete distributions that are particularly important to the area of SPC.

BERNOULLI DISTRIBUTION

In a common situation an experiment is characterized by the repetition of trials in which there are just two possible outcomes. Manufactured items may be individually tested to determine whether they are acceptable or defective; a firing at a target results in either a hit or a miss; a coin toss results in a head or a tail; and a customer may be classified as satisfied or dissatisfied.

A random variable that can take on only one of two numerical values is said to have a *Bernoulli distribution*. It is convention to generically designate the two possible outcomes as "success" and "failure." The meanings of these two terms should not be taken literally. As it turns out, there is a popular control chart (*p* chart, see Chapter 7) for the analysis of percentage of defectives where the outcome of a "defective item" is defined as a success! The term *success* should be viewed as only a means to indicate when one of the two possible outcomes has occurred. As long as there is consistency, the assignment of the labels is quite arbitrary.

To quantify the two possible outcomes, it will be convenient to code one outcome with the value 1 and the other outcome with the value 0. These two possible outcomes are complements, and if p denotes the probability of outcome value 1, then the probability of outcome value 0 is $1 - p$. By using Equations (2.5) and (2.7), the expected value, variance, and standard deviation of a Bernoulli random variable can easily be found to be, respectively,

$$\mu = E(X) = p$$
$$\sigma^2 = \text{Var}(X) = p(1 - p) \qquad \text{(2.8)}$$
$$\sigma = \sqrt{p(1 - p)}$$

That the expected value is p for a Bernoulli random variable can be intuitively reasoned. Recall that one interpretation of the expected value is that it is the average value of the observations if the associated experiment is repeated a very large number of times. Since the sum of 0s and 1s in a sequence of results is just the number of 1s, average of such a sequence is simply the proportion of 1s. If a very large number of trials is observed, we would naturally expect the proportion of 1s to be nearly equal to the probability of getting a 1 (success) on any given trial, namely, p.

BINOMIAL DISTRIBUTION

In practical applications, we are not typically interested with a single Bernoulli trial. Rather, most situations deal with the observations of a sequence of Bernoulli trials. Suppose that a Bernoulli experiment is repeated such that the trials are independent and that the probability of either outcome remains unchanged for each trial. Experiments of this kind are called *binomial experiments,* and the sequence of Bernoulli random variables is known as a *Bernoulli process.*

   If n Bernoulli trials are conducted, one may be interested in computing the probability of obtaining a specific number of successes, say x, from these n trials. Denote this random variable by X. The desired probability $P(X = x)$ is a joint probability of n independent trials. By using the fact that the joint probability of independent random variables equals the product of the individual probabilities, the probability that x successes and $n - x$ failures are obtained in n trials is

$$\underbrace{p \cdot p \cdot p \cdots p}_{x \text{ successes}} \cdot \underbrace{(1 - p) \cdot (1 - p) \cdots (1 - p)}_{(n - x) \text{ failures}} = p^x (1 - p)^{n-x}$$

But this is just one particular sequence or order of successes and failures. There are a number of sequences having the same probability. The number of sequences of x successes and $n - x$ failures in n trials is given by

$$\binom{n}{x} = \frac{n!}{x!(n - x)!}$$

where $n!$ is read "n factorial" and is defined by

$$n! = n(n - 1)(n - 2) \cdots (2)(1)$$
$$1! = 1 \qquad \text{and} \qquad 0! = 1$$

Thus, the desired probability (probability of x successes in n independent trials) is the sum of the probabilities of all such possible sequences. That is,

$$P(X = x) = \binom{n}{x} p^x (1 - p)^{n-x} \qquad x = 0, 1, \ldots, n \qquad (2.9)$$

where $n > 0$ and $0 < p < 1$. A random variable having this probability distribution is said to have a *binomial distribution*.

Example 2.9

A machine is known to consistently produce 10 percent defective items. A sample of 10 items is taken.

1. What is the probability that none of the items is defective?

$$p = 0.1 \qquad 1 - p = 0.9 \qquad n = 10 \qquad \text{and} \qquad x = 0$$

Thus, $P(X = 0) = \binom{10}{0}(0.1)^0(0.9)^{10} = 1 \cdot (0.9)^{10} = 0.3487$.

2. What is the probability that precisely one item is defective?

$$P(X = 1) = \binom{10}{1}(0.1)^1(0.9)^9 = 10(0.1)(0.9)^9 = 0.3874$$

3. Using Equation (2.9), we can find the complete distribution (probabilities rounded to the fourth decimal place):

x	0	1	2	3	4	5	6	7	8	9	10
$P(X = x)$	0.3487	0.3874	0.1937	0.0574	0.0112	0.0015	0.0001	0.0000	0.0000	0.0000	0.0000

4. What is the probability that at most two items are defective?

$$P(X \le 2) = P(X = 0) + P(X = 1) + P(X = 2) = 0.3487 + 0.3874 + 0.1937 = 0.9298$$

5. What is the probability that at least three items are defective?

$$P(X \ge 3) = P(X = 3) + P(X = 4) + \cdots + P(X = 10)$$

Notice,

$$P(X \ge 3) = 1 - P(X < 3) = 1 - P(X \le 2)$$

Thus, using the previous computation, we have $P(X \ge 3) = 1 - 0.9298 = 0.0702$.

We should point out that manual computation of binomial probabilities can be avoided through the use of Minitab. Minitab provides the user with options to compute the probability of a specific outcome or a cumulative probability for many of the important distributions discussed in this textbook. To illustrate, below are the commands (session and menu) and output for determining the answer for part 2 of Example 2.9:

Graph >> Chart

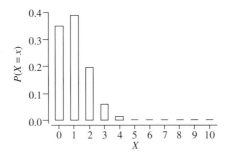

Figure 2.22   Binomial distribution with $p = 0.10$ and $n = 10$.

```
MTB > pdf 1;                    Calc>>Probability Distributions>>Binomial>>Probability
SUBC>   binomial 10 .1.
Probability Density Function
Binomial with n = 10 and p = 0.100000
    x       P( X = x)
  1.00        0.3874
```

Using Minitab, the cumulative probability for part 4 of Example 2.9 is found as follows:

```
MTB > cdf 2;            Calc>>Probability Distributions>>Binomial>>Cumulative probability
SUBC> binomial 10 .1.
Cumulative Distribution Function
Binomial with n = 10 and p = 0.100000
    x     P( X <= x)
  2.00        0.9298
```

To visualize the nature of the binomial distribution, Figure 2.22 portrays graphically the probability distribution for the binomial random variable described in Example 2.9. Notice that the distribution for this particular example is skewed. In general, the binomial distribution will be positively skewed for p less than 0.5 and negatively skewed for p greater than 0.5; thus, symmetry corresponds to $p = 0.5$. The mean, variance, and standard deviation, respectively, of a binomial random variable X are given by

$$\mu = E(X) = np$$
$$\sigma^2 = \text{Var}(X) = np(1 - p) \qquad \text{(2.10)}$$
$$\sigma = \sqrt{np(1 - p)}$$

In SPC applications, we are frequently interested in the proportion or percentage of "successes" in a given sequence of trials. For example, in a process of producing items that may be classified as defective or nondefective, the interest may be to report the proportion of the defective items produced by the process. If we obtain X successes in n trials, then the sample proportion of successes, denoted by $\hat{p}$, is defined as $\hat{p} = X/n$. By using a couple of the properties of expectation and variance noted at the end of Section 2.6 along with Equation (2.10), it can easily be shown that the mean, variance, and standard deviation of the sample proportion random variable $\hat{p}$ are, respectively,

$$\mu_{\hat{p}} = E(\hat{p}) = p$$

$$\sigma_{\hat{p}}^2 = \text{Var}(\hat{p}) = \frac{p(1-p)}{n} \qquad \qquad \textbf{(2.11)}$$

$$\sigma_{\hat{p}} = \sqrt{\frac{p(1-p)}{n}}$$

Knowledge of the properties of a binomial random variable and a proportion random variable given in Equations (2.10) and (2.11), respectively, will prove to be vital in the construction of binomial-based control charts detailed in Chapter 7.

POISSON DISTRIBUTION

There are a number of random situations in which the probability of a success on a single trial is small and the number of trials is large. Computing probabilities by direct use of the binomial distribution for a large number of trials can be a long and tedious task. On the other hand, a new distribution with less demanding computations can be used as an approximation to the binomial distribution. The following probability distribution, known as the *Poisson* distribution, is often used for that purpose,

$$P(X = x) = \frac{e^{-\lambda}\lambda^x}{x!} \qquad x = 0, 1, 2, \ldots \qquad \textbf{(2.12)}$$

where λ is some fixed positive real number, called the parameter of the distribution. The symbol λ is conventional in nearly all statistical applications; however, we should point out that this symbol is sometimes replaced by the letter c in the area of SPC.

The Poisson distribution can be viewed as a limiting form of the binomial distribution when n approaches infinity ($n \rightarrow \infty$) and p approaches zero ($p \rightarrow 0$) in such a way that their product (np) is some fixed number λ. In other words, as the number of trials n gets larger and as p gets smaller in such a way that $np = \lambda$, the binomial distribution approaches the Poisson distribution. In practical applications, n and p do not have to be very extreme for the Poisson distribution to suitably approximate the binomial distribution. As a rule of thumb, the Poisson distribution is an adequate approximation when $n > 20$ and $p < 0.05$, with the approximation improving as n gets larger and p smaller.

Consider a sample of 100 items drawn from a large number of items produced by a machine that is known to produce 2 percent defectives. What is the probability that precisely 5 items are defective? Using the binomial distribution, we have

$$p = 0.02 \qquad 1 - p = 0.98 \qquad n = 100, \qquad \text{and} \qquad x = 5$$

$$P(X = 5) = \binom{100}{5}(0.02)^5(0.98)^{95}$$

Computing this quantity by *hand* can be slightly tedious; if done, we find the exact probability to be 0.0353 (or 3.53 percent). Consider, however, the Poisson distribution where we have

$$\lambda = np = (100)(0.02) = 2 \qquad x = 5$$

Hence, by applying Equation (2.12), we find

$$P(X = 5) = \frac{e^{-2}(2^5)}{5!} = 0.0361 \text{ or } (3.61\%)$$

As can be seen below, Minitab can be employed to arrive at the above probability value,

```
MTB > pdf 5;              Calc>>Probability Distributions>>Poisson>>Probability
SUBC> poisson 2.
Probability Density Function
Poisson with mu = 2.00000
       x      P( X = x)
    5.00       0.0361
```

Example 2.10

In addition to being an approximating distribution to the binomial, the Poisson can be viewed as a distribution in its own right arising from certain random processes called Poisson processes. Events that are generated by a Poisson process require a given interval (of time, length, or space) that can be divided into small subintervals, each of which can be considered as a trial. The subintervals should be sufficiently small that no more than one success can occur in a subinterval. As in the case of the binomial, the trials (subintervals) are assumed to be independent. For example, in the study of waiting lines, one may assume that arrivals coming for service per unit of time (say, per minute) constitute a Poisson process; it is safe to assume that there is some subinterval, say, 1s, in which it is unlikely that more than one arrival can occur. For Poisson processes, the parameter λ represents the average number of occurrences of the event over the given interval of time, length, or space.

Flaws in the plating of manufactured sheet metal occur at random, on the average of one per 20 square feet (ft^2). What is the probability that a sheet measuring 5 ft by 10 ft will have at

Example 2.11

least two flaws? To determine this probability, first we need to identify λ. Since one flaw per 20 ft^2 occurs on average, we would expect 2.5 flaws for a 5-ft by 10-ft sheet (or 50 ft^2); thus, $\lambda = 2.5$. The probability of observing at least two flaws is given by

$$P(X \geq 2) = P(X = 2) + P(X = 3) + \ldots$$

Computationally, the difficulty is that the above probability is expressed as an infinite sum. Alternatively, we can recognize that the event of getting "at least two flaws" is the complement of the event of getting "zero flaws or one flaw." Hence,

$$P(X \geq 2) = 1 - [P(X = 0) + P(X = 1)] = 1 - \left[\frac{e^{-2.5}(2.5^0)}{0!} + \frac{e^{-2.5}(2.5^1)}{1!} \right] = 0.713$$

For future reference, it will be helpful to know that the mean, variance, and standard deviation of a Poisson random variable X are given, respectively, by

$$\mu = E(X) = \lambda$$

$$\sigma^2 = \text{Var}(X) = \lambda \qquad \textbf{(2.13)}$$

$$\sigma = \sqrt{\lambda}$$

We complete our discussions of the Poisson by pointing out a useful property of the distribution. In particular, if $X_1, X_2, \ldots, X_k$ represent k independent Poisson random variables with parameters $\lambda_1, \lambda_2, \ldots, \lambda_k$ respectively, the distribution of

$$Y = X_1 + X_2 + \ldots + X_k$$

is also Poisson with mean and variance equaling

$$\lambda = \lambda_1 + \lambda_2 + \ldots + \lambda_k \qquad \textbf{(2.14)}$$

2.8   NORMAL DISTRIBUTION

In the previous section, a set of useful discrete distributions was presented. We now turn our attention to the continuous case. There are a number of continuous distributions that are important in the general analysis and modeling of process data along with the construction and interpretation of control charts. However, in this section, we focus on only one continuous distribution: the *normal* distribution. Some other useful continuous distributions are more naturally presented in the context of statistical inference and the distribution of sample statistics, to be discussed in the following sections.

Without question, the normal distribution is one of the best-known continuous probability distributions. It is also known as the *Gaussian* distribution, named in honor of Carl Friedrich Gauss (1777–1855), a mathematician and astronomer who used the distribution extensively to explain data from many physical phenomena. The normal distribution is quite important in SPC for three distinct reasons:

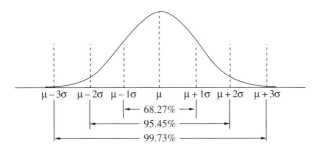

Figure 2.23   Areas under the normal curve.

1. Measurements generated from many random processes follow closely this distribution; for example, recall the histogram for the stopwatch measurements.

2. The normal distribution can often provide a close approximation for a number of discrete distributions such as the binomial and Poisson.

3. Distributions of statistics such as the sample mean and sample proportion are approximately normal, regardless of the distribution of the individual measurements used in the calculation of the statistic.

Because of these reasons, the starting point in the construction of most control charts is an assumption of normality for the controlled measurements (or statistics).

Recall that probabilities for continuous random variables are associated with the areas under a mathematical curve known as the probability density curve. For the normal distribution, the mathematical curve is represented by the density function

$$f(x) = \frac{1}{\sqrt{2\pi\sigma^2}} e^{-(x-\mu)^2/(2\sigma^2)} \qquad -\infty < x < \infty \qquad (2.15)$$

where π is the familiar constant $3.14159\ldots$, e is the constant 2.71828, and x is any real value. Consistent with previously established notation, the constants μ and σ^2 represent the mean and variance, respectively, of the distribution. In "shorthand" notation, a normal random variable X with mean μ and variance σ^2 is written as $X \sim N(\mu, \sigma^2)$.

As reflected by Figure 2.23, a typical graph of the normal density function is a symmetric bell-shaped curve. Also from Figure 2.23, we indicate the areas under the curve relative to the benchmarks of ± 1, ± 2, and ± 3 standard deviations around the mean. Namely, 68.27 percent of the distribution is between $\mu \pm 1\sigma$, 95.45 percent of the distribution is between $\mu \pm 2\sigma$, and 99.73 percent of the distribution is between $\mu \pm 3\sigma$. Awareness of the area associated with 3 standard deviations around the mean is particularly important because nearly all control chart procedures involve a decision rule that calls for investigation when data follow beyond 3 standard deviations from the mean. The rationale is that outlying points are unlikely (about 3 in 1000) if a process is in control and generating observations that are normally distributed (or approximately so); thus, if an outlying point does occur, it more likely suggests a fundamental departure or change in the underlying process.

At times it is useful to compute the probability under the normal curve for intervals other than the ones noted above. Since probability is associated with an area under the density function, the mathematical technique of integration is required to find probabilities for specific intervals. To make matters more difficult, the integral for a normal curve cannot be written in terms of elementary functions. Instead, it must be approximated by lengthy numerical methods. Fortunately, these computations have been carried out to create tables that allow us to approximate the area under the normal curve for selected intervals. Since different combinations of μ and σ give rise to different normal curves, it is obviously impossible to tabulate an infinite number of normal distributions. And yet any statistics text one looks at will have a table for just one particular normal distribution. The reason for this is that any normal distribution can be transformed easily into a particular normal distribution known as the standard normal distribution or the Z-distribution. Assuming that a random variable is distributed normally with mean μ and standard deviation σ, the transformation to the standard normal distribution is accomplished as follows:

$$Z = \frac{X - \mu}{\sigma} \tag{2.16}$$

It can be shown that Z has a mean equal to 0 and a variance equal to 1. Thus, the standardization of any normal random variable $X \sim N(\mu, \sigma^2)$ results in another normal random variable $Z \sim N(0, 1)$. By standardizing a particular value $X = x$, the resulting $Z = z$ value has a very special interpretation. Namely, it represents how many standard deviations above or below the mean the original x value is located.

Appendix Table I is an upper-tail probability table for the standard normal distribution. Namely, each entry in the table represents the probability that the standard normal random variable is greater than or equal to a given value z, that is, $P(Z \geq z)$. Since the table provides only probabilities to the right of positive values of z, we can utilize the following facts to determine probabilities for general intervals:

1. $P(z_1 \leq Z \leq z_2) = P(Z \leq z_2) - P(Z \leq z_1)$
2. $P(Z \geq z) = 1 - P(Z \leq z)$
3. $P(Z \leq -z) = P(Z \geq z)$

Note also that the appearance (or lack) of an equals sign in the probability statement has no bearing on the computation since the probability that a continuous random variable equals a specific value is 0.

The standard normal probability table can also be used in "reverse" to determine a specific z value above which lies some specified percentage of the distribution. In terms of notation, we can define z_a as the point along the standard normal distribution such that $P(Z \geq z_a) = a$; this point can also be referred to as the $100(1 - a)$ percentile point. The following examples illustrate the use of the standard normal table.

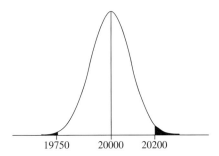

Figure 2.24   Probability of a defective resistor.

Suppose that it has been determined that the resistance of a particular resistor is normally distributed with $\mu = 20{,}000$ ohms (Ω) and $\sigma = 90\ \Omega$. If an acceptable resistance range is from 19,750 to 20,200 Ω, what is the expected number of rejected resistors in a production run of 10,000 resistors? The answer to this question is predicated on the assumption that the process associated with the manufacture of the resistors will remain stable and in control. As we learned from Deming's spring elongation example (Section 2.4), if the process is out of control and/or exhibits systematic behaviors such as trend, then a single data distribution or population cannot serve as a meaningful guide. Assuming a stable, in-control process, we are interested in determining the probability that X, a normal random variable with mean 20,000 and standard deviation 90, is less than 19,750 or exceeds 20,200 (see Figure 2.24). To find this probability, we do the following:

Example 2.12

$$P(\text{rejected resistor}) = P(X < 19{,}750) + P(X > 20{,}200)$$

$$= P\!\left(\frac{X - \mu}{\sigma} < \frac{19{,}750 - 20{,}000}{90}\right) + P\!\left(\frac{X - \mu}{\sigma} > \frac{20{,}200 - 20{,}000}{90}\right)$$

$$= P(Z < -2.78) + P(Z > 2.22)$$

$$= 0.00271794 + 0.0132094 \qquad \text{From Appendix Table I}$$

$$= 0.01592734$$

Thus, 1.59 percent of 10,000 resistors (or 159 resistors) is expected to be beyond acceptable specifications. With Minitab (or most any statistical software), the need for standardization and tables can be altogether avoided. As seen below, we can find the necessary probabilities from the unstandardized normal distribution:

```
MTB > cdf 19750;        Calc>>Probability Distributions>>Normal>> Cumulative probability
SUBC> normal 20000, 90.

Cumulative Distribution Function

Normal with mean = 20000.0 and standard deviation = 90.0000
         x     P( X <= x)
   1.98E+04       0.0027
```

Note: In the session output for the 'cdf' command, Minitab expresses x-values of magnitude 1000 or larger in scientific notation.

```
MTB > cdf 20200;
SUBC> normal 20000, 90.

Cumulative Distribution Function

Normal with mean = 20000.0 and standard deviation = 90.0000
          x     P( X <= x)
    2.02E+04       0.9869
```

Given the reported probabilities, we would make the following computations:

$$P(\text{rejected resistor}) = P(X < 19{,}750) + P(X > 20{,}200)$$

$$= P(X < 19{,}750) + 1 - P(X \leq 20{,}200)$$

$$= 0.0027 + 1 - 0.9869$$

$$= 0.0158$$

The slight disparity between the table-based result of 0.0159 and the above result of 0.0158 is due to the fact that Minitab rounds to the fourth decimal place when the results are reported to the user on the terminal screen. If the option is used to output and store the results into the worksheet, Minitab will report the results more accurately to the sixth decimal place.

Example 2.13	A courier service guarantees absolutely that priority overnight packages will be delivered the next day by 10:30 a.m. Suppose management has found that delivery times are normally distributed with the mean delivery time at 10:00 a.m. and a standard deviation of 9 minutes. By what time can management expect 90% of packages to be delivered? Essentially, what we are seeking is a 90th percentile point on the delivery distribution. To simplify the problem, we can equate the mean delivery time of 10:00 a.m. with $\mu = 0$. In terms of our established notation, the 90th percentile point is associated with the value $x_{0.10}$ that satisfies $P(X \geq x_{0.10}) = 0.10$ where $X \sim N(0, 9^2)$ or $N(0, 81)$. To do so, we begin by finding $z_{0.10}$ on the standard normal distribution. From Appendix Table I, we find the z value closest to giving 0.10 is $z_{0.10} = 1.28$. Using the standardizing transformation (2.16), we have:

$$z_{0.10} = \frac{x_{0.10} - 0}{9} \qquad \text{implying} \qquad x_{0.10} = 9z_{0.10} = 9(1.28) = 11.52$$

Hence, we can expect 90% of packages to be delivered by 11.52 minutes after 10:00 a.m. or 10:11:31 a.m. (hr:min:sec). The problem can also be solved directly using Minitab:

```
MTB > invcdf .90;   Calc>>Probability Distributions>>Normal>> Inverse cumulative probability
SUBC> normal 0, 9.

Inverse Cumulative Distribution Function

Normal with mean = 0 and standard deviation = 9.00000

P( X <= x)          x
   0.9000      11.5340
```

Here the disparity (11.52 versus 11.534) is due to Minitab's being able to more accurately determine the 90th percentile than can be determined from the normal table.

2.9   SAMPLE STATISTICS AS RANDOM VARIABLES

In our development, we have suggested that the individual observations in a data set can be viewed as individual samplings generated from a larger picture known as a population. Since probability distributions can be used to model populations, we can equivalently think of individual data observations as the outcomes of a repeated experiment for some defined random variable that is characterized by a probability distribution. For example, the 50 stopwatch observations appear to be outcomes of a random variable X (defined as a single start or stop time measured in seconds) which is normally distributed with some mean μ and variance σ^2. One question we might ask is what are the "true" values for parameters μ and σ^2? In the area of quality control, there is often a need to estimate underlying population parameters. For example, if a measurable characteristic of a manufactured part is a prespecified—engineering—target requirement, a statistical investigation can be made to collect process data, estimate the underlying population mean, and compare the estimate with the desired target level. In the stopwatch experiment, the desire was to have the underlying mean μ be at a target value of 10. Estimation of pertinent population parameters along with analysis of the 50 stopwatch measurements can suggest whether it is reasonable to believe that the mean is (or is close to) 10s.

Quite obviously, since any sample is by definition incomplete information, statements about population parameters are always subject to a level of uncertainty. Thus, our goal is to extract pertinent information from the sample to provide a reasonable estimate of the population parameter of interest. Herein lies the basic process of *inferential* statistics, namely, the generalization of descriptions of sample data to make statements about the values of population parameters.

Typically, information derived from a data set for the purposes of estimation takes the form of a sample statistic calculated from the sample observations. For example, the population mean μ is usually estimated from the computed value of $\bar{x}$, the population variance σ^2 from s^2, and the probability of success from the sample proportion $\hat{p}$. In all reality, an estimate based on a given sample will not equal the true value of the population parameter. Some difference between the estimate and the corresponding parameter, known as *sampling error*, is likely to occur. And if we were to obtain another sample, the amount of sampling error would most likely be different from that of the previous sample. In general, if we were to take several samples of a given size and calculate a particular sample statistic for each, we would obtain a set of different values for the statistic. This is intuitively obvious since the individual observations on which a sample statistic is based are outcomes of some random variable. Hence, sample statistics themselves can be viewed as random variables. As is the case for any random variable, a sample statistic can be associated with some probability distribution. In statistics, it is common practice to refer to the probability distribution for a statistic as its *sampling distribution*.

By knowing the sampling distribution for a statistic, we can get a sense of how exact (or off) an estimate of a population parameter might be. This opens the way for comparing the relative performances of competing statistics, that is, enabling us to determine what statistic is preferable in the estimation of a particular population parameter. In Section 2.10, we present some criteria that will help in the relative assessment of sample statistics as estimators of a given population parameter.

In addition to the development of useful estimators, the knowledge of the sampling distribution of a statistic is important because it gives us a general understanding of how we may expect values of a sample statistic to vary from sample to sample. In SPC applications, such an understanding is particularly important since a primary goal is to identify unrepresentative values potentially reflecting process change and the need for action. Not knowing how a statistic generally varies makes it difficult to assess when a particular sample statistic value is indeed unusual. We now present the sampling distributions useful for the development of SPC methods and the general analysis of process data.

SAMPLING DISTRIBUTION OF THE SAMPLE MEAN $\overline{X}$: σ KNOWN

Assume X is any random variable with an expected value of μ and a variance of σ^2. Now suppose that the sample mean $\overline{X}$ represents the average of a sample of n independent observations taken on X. [*Note:* When we are studying the sample mean as a random variable, the capital letter X is used as opposed to the small letter x found in Equation (2.1), which represents the value of the sample mean for a particular sample when there is no longer uncertainty.] As discussed in the previous section, $\overline{X}$ (or any sample statistic) is a random variable and, as such, has a probability distribution. As you might imagine, the mathematical form of the probability distribution of $\overline{X}$ depends to a certain extent on two factors: (1) the probability distribution of X and (2) the sample size n used to compute the sample mean. However, there are some characteristics about the probability (or sampling) distribution of $\overline{X}$ that can be derived regardless of the shape of the underlying distribution of X. Namely, using the properties of expected values of variances found in Section 2.6, we find

$$E(\overline{X}) = E\left(\frac{1}{n}\sum_{i=1}^{n} X_i\right) = \frac{1}{n}\sum_{i=1}^{n} E(X_i) = \frac{1}{n}(n\mu) = \mu \tag{2.17}$$

and

$$\sigma_{\overline{X}}^2 = \text{Var}(\overline{X}) = \text{Var}\left(\frac{1}{n}\sum_{i=1}^{n} X_i\right) = \frac{1}{n^2}\sum_{i=1}^{n} \text{Var}(X_i) = \frac{1}{n^2}(n\sigma^2) = \frac{\sigma^2}{n} \tag{2.18}$$

Accordingly, the standard deviation of the sample mean statistic is denoted by $\sigma_{\overline{X}}$ and given by

$$\sigma_{\overline{X}} = \frac{\sigma}{\sqrt{n}} \tag{2.19}$$

In other words, if repeated random samples of size n are drawn from any population (discrete or continuous) with mean μ and variance σ^2, the sampling distribution of the corresponding sample means $\bar{X}$ has mean μ, and variance σ^2/n. It should not be too surprising and quite clear intuitively that if the expected value of X is μ, then the distribution of $\bar{X}$ values is also centered at μ. The variance result indicates that the variance of $\bar{X}$ decreases with increasing sample size; in part, this is due to a canceling effect of small and large values in the averaging operation and with increased sample size the impact of more extreme X values is further lessened (or diluted). Now that we know two key parameters of the sampling distribution of $\bar{X}$, what can we say about the shape of the distribution? To answer this question, we consider two general scenarios depending on whether or not the individual observations arise from a normal distribution.

Sampling from Normal Populations: Exact Sampling Distribution   If we are sampling from a normal population (that is, X is a normal random variable), then it can be shown that $\bar{X}$ will also follow a normal distribution with mean μ and variance σ^2/n. In terms of the notation for normal distributions, we can summarize the result by writing

$$\bar{X} \sim N(\mu, \sigma^2/n) \qquad \text{(2.20)}$$

Recall from Equation (2.16) that if we are to express a random variable X in standard units such that its mean is 0 and its variance is 1, we transform the variable from X to $(X - \mu)/\sigma$. With this in mind, the fact summarized in Equation (2.20) can be expressed as follows:

$$Z = \frac{\bar{X} - \mu}{\sigma / \sqrt{n}} \sim N(0, 1) \qquad \text{(2.21)}$$

Sampling from Nonnormal Populations: Central Limit Theorem   We have established that if the distribution underlying the individual measurements is normal, then the resulting sampling distribution of $\bar{X}$ is also normal. But quite often the samples are drawn from populations that are nonnormal, for example, skewed distributions. Theoretically, given the infinite number of possible departures from normality, it may seem that there is little hope of arriving at some general statements about the sampling distribution of $\bar{X}$. However, there is a remarkable result, known as the *central limit theorem*, that allows us to focus on statements made about the sampling distribution of $\bar{X}$ even when X follows any nonnormal distribution.

Loosely speaking, the central limit theorem states that sums or averages of a "larger" number of independent random variables from any population follow approximately the normal distribution. This approximation improves as the number of variables involved increases. More specifically for our purposes, it says that for a large enough n, the sampling distribution of $\bar{X}$ is approximately normally distributed *regardless* of the probability distribution of X. Since we know that the mean and variance of $\bar{X}$ are μ and σ^2/n, respectively, the central limit theorem allows us to state that for "large enough" n,

$$\bar{X} \overset{\sim}{} N(\mu, \sigma^2/n) \tag{2.22}$$

where the symbol $\overset{\sim}{}$ stands for "approximately distributed." In terms of standardized units, we can equivalently state that for "large enough" n,

$$\frac{\bar{X} - \mu}{\sigma / \sqrt{n}} \overset{\sim}{} N(0, 1) \tag{2.23}$$

To be practical, how large is "large enough"? Of course, this depends on the amount of departure from nonnormality in the X distribution. If the distribution for X is severely nonnormal, we may require a large n, say, $n > 50$ (or even 100), for adequate approximation to normality. However, if the underlying distribution does not differ too greatly from normality, the distribution of $\bar{X}$ will often appear normal for n as small as 4 or 5. Fortunately, most practical applications lean toward this latter scenario.

Approximating the Binomial and Poisson Distributions   Recall that a binomial random variable X is defined as the number of successes in n independent Bernoulli trials. Equivalently, X can be viewed as the sum of n Bernoulli random variables which take on the value 1 or 0 depending on whether the outcome is a success or failure, respectively.

Since the central limit theorem applies to either sums or averages of independent random variables, the binomial distribution can be viewed as a special application of the theorem. Thus, for large enough n, a binomial random variable has a distribution that approaches the normal distribution. From Equation (2.10), the mean and variance of a binomial random variable are np and $np(1 - p)$, respectively. For a binomial random variable X, this implies that for large enough n,

$$X \overset{\sim}{} N[np, np(1 - p)] \tag{2.24}$$

In most quality control applications, p is associated with the probability of a defective item. Since such a probability is usually quite small, the associated binomial distribution for the number of defects in a sample of n items will be markedly skewed. In Chapter 7, we provide some general rules of thumb to determine, for a given probability p, what sample size n is required to give a satisfactory approximation of the binomial distribution by the normal distribution.

In Section 2.7, we also presented the random variable $\hat{p} = X/n$ which can be interpreted as the average number or the sample proportion of successes in a sequence of n Bernoulli trials. Since $\hat{p}$ is a sample average, the central limit theorem again applies. Using the facts given in Equation (2.11), we can state for large enough n,

$$\hat{p} \overset{\sim}{} N\left[p, \frac{p(1 - p)}{n}\right] \tag{2.25}$$

As noted above, guidelines for the size of n required will be discussed in Chapter 7.

Our focus in this special application of the central limit theorem has been on the binomial distribution. But recall from Section 2.7 that the Poisson distribution is a theoretical extension of the binomial distribution. Hence, it comes as no surprise that the approximation of the binomial by the normal extends to the Poisson. It turns out that as the mean of the Poisson distribution (λ) increases, the Poisson distribution approaches the normal distribution. Since the mean and variance of a Poisson random variable X are both equal to λ, we can state that for large enough λ

$$X \sim N(\lambda, \lambda) \tag{2.26}$$

As a rule of thumb, if λ exceeds 5, then the normal distribution serves as an adequate approximation.

SAMPLING DISTRIBUTION OF THE SAMPLE MEAN $\overline{X}$: σ UNKNOWN

In the previous section, we explored the distribution of sample means $\overline{X}$ when samples are selected from normal or nonnormal populations. In both cases, we were able to summarize the sampling distribution results through the statistic

$$\frac{\overline{X} - \mu}{\sigma / \sqrt{n}}$$

which was found to follow either exactly or approximately the standard normal distribution depending on whether X was a normal random variable. As can be seen in the denominator of the statistic above, we assumed that the value of standard deviation σ is known.

In most applications, the value of σ is unknown. In such cases, we need to obtain an estimate for σ from the sample data. One reasonable possibility is to replace the *population* standard deviation with the *sample* standard deviation, which was defined in Equation (2.4). We now have the statistic

$$T = \frac{\overline{X} - \mu}{S / \sqrt{n}} \tag{2.27}$$

Since we are concerned with the sample standard deviation as a random variable, we use the capital letter S not s as found in Equation (2.4). The distribution of this random variable T cannot be easily derived because the denominator in Equation (2.27) is not constant but rather is random from sample to sample. It can be shown that if X is a normal random variable, then T follows a distribution known as the *Student's t distribution*, or more simply the *t distribution*. The initial discovery of this distribution is due to William Sealy Gosset (1876–1937) who studied the statistical problem of estimating a population mean based on smaller samples. At the time, Gosset was employed by Guiness Brewery of Ireland which had a policy prohibiting employees from publishing research results under real names. As a result, Gosset was forced to publish under a pseudonym which he chose to be Student.

The *t* distribution is a continuous distribution and is similar to the standard normal distribution: It is bell-shaped and symmetric with mean zero and extends from minus infinity to plus infinity. It depends on one parameter, called the number of

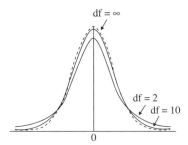

Figure 2.25   Student's *t* distribution for different degrees of freedom (df).

degrees of freedom (df). The number of degrees of freedom in this case is $n - 1$ (the sample size minus 1). However, the dispersion of the *t* distribution is greater than that of the standard normal distribution; that is, the variance is greater than 1.

The variance of a *t* distribution is a function of the number of degrees of freedom; as the number of degrees of freedom becomes large, the variance approaches 1. Hence, as depicted in Figure 2.25, the *t* distribution approaches the standard normal distribution as *n* gets larger. As a rough rule, the *t* distribution and the standard normal distribution are nearly identical for $n \geq 30$; thus, statistical analysis can be based on either distribution for $n \geq 30$. Since the shape of the *t* distribution depends on the number of degrees of freedom, it is not practical to tabulate the entire distribution. Selected percentile points of the *t* distribution, for various degrees of freedom, are given in Appendix Table II. Minitab can be invoked to find an even wider range of percentile points.

In prefacing the *t* distribution, we indicated that the population from which the samples are selected is assumed to be normal. If the population is not severely nonnormal and the sample size is larger (say, $n \geq 30$), then the central limit theorem effect comes into play, allowing us to approximate the distribution of the *T* statistic given in Equation (2.27) by the standard normal distribution. However, if the population is nonnormal and the sample size is too small, neither the *t* distribution nor the standard normal distribution adequately describes the distribution behavior of the random variable *T*.

In the development of standard control chart methods, the assumption is made that the sample size and total number of samples are both large enough to base judgment on the normal distribution. However, our discussion of the *t* distribution is not without a purpose. We will find that the *t* distribution is quite useful in interpreting the results of statistical modeling procedures (e.g., in regression modeling) used in the general analysis of process data. The use of the *t* distribution in statistical modeling is due to the fact that this distribution applies to random variables which are more broadly defined than is suggested by Equation (2.27). In particular, consider the general statistic

$$T = \frac{\hat{\theta} - \theta}{S_{\hat{\theta}}}$$

(2.28)

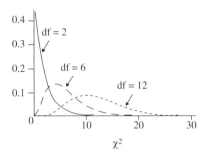

Figure 2.26   Chi-square distribution for different degrees of freedom (df).

where $\hat{\theta}$ is a sample statistic used to estimate some population parameter θ, and $S_{\hat{\theta}}$ is the sample estimator of the true standard deviation of $\hat{\theta}$, that is, $\sigma_{\hat{\theta}}$. Under certain conditions, statistics of the form found in Equation (2.28) follow a t distribution with the number of degrees of freedom depending on sample size and the context in which the statistic is being used.

SAMPLING DISTRIBUTION OF THE SAMPLE VARIANCE S^2

In the last two sections, we considered various underlying conditions to explain the distribution behavior of the sample mean statistic $\bar{X}$. Let us now consider the probability distribution of the sample variance statistic S^2 defined, in terms of random variables X and $\bar{X}$, as

$$S^2 = \frac{\sum_{i=1}^{n}(X_i - \bar{X})^2}{n - 1}$$

If the population from which the individual observations are generated is normal, the sample-to-sample randomness of S^2 is explained through the following random variable:

$$\chi^2 = \frac{(n - 1)S^2}{\sigma^2} \tag{2.29}$$

The random variable χ^2 follows a distribution known as the *chi-square distribution*. Even though the derivation of this statistic is based on the normality of the X variable, the results will hold approximately as long as the departure from normality is not too severe. The chi-square distribution is not symmetric with the tendency to tail off toward the upper end (that is, it is positively skewed). As with the t distribution, the chi-square distribution is a one-parameter distribution. The parameter of the chi-square distribution is the number of degrees of freedom (df), as was the case with the t distribution. Not accidentally, the number of degrees of freedom in this case is $n - 1$, as it was with the T statistic defined by Equation (2.27). Figure 2.26 illustrates the shapes which a chi-square distribution may assume. Notice that as the degrees of freedom increase, the chi-square curves become more symmetric, approaching the shape of a normal curve.

As with the t distribution, detailed tables showing all areas for all degrees of freedom are not feasible. Rather, we provide a table of selected percentiles of the chi-square distribution for various degrees of freedom in Appendix Table III. Additionally, Minitab can be used to find areas and percentile points for a wide variety of chi-square distributions.

2.10   ESTIMATION OF PARAMETERS

As was pointed out, estimation is one major area of statistical inference. Generally, there are two types of estimation procedures: (1) point estimation and (2) interval estimation. A *point estimate* is a single value obtained from the observed values of a sample to approximate the value of the population parameter in question. In this kind of estimation, there is no way of measuring our degree of confidence in the estimate. In order to measure this confidence in the estimate, a range of values, rather than a single value, is calculated. This is called an *interval estimate* (or confidence interval). An interval estimate, therefore, is an interval specified by two numbers calculated from the observed values of a sample.

In our subsequent discussions, it is useful to clarify the distinction between the terms *estimator* and *estimate*. The accepted convention is to use the word *estimator* to refer to a function of the sample values and the word *estimate* to refer to the value of the function where the observed sample values have been inserted. Hence, an estimator is nothing more than a sample statistic.

CRITERIA FOR CHOOSING AN ESTIMATOR

First recall that the values of an estimator (sample statistic), when computed from all possible samples, form a population known as a sampling distribution. Thus, an estimator is a random variable that has its own distribution, expected value, and variance.

A number of criteria may be used to judge or evaluate the relative merits of different estimators. Among these criteria are the properties that an estimator should be *unbiased, consistent*, and *efficient*. These terms are explained next.

A estimator is said to be *unbiased* if the expected value of the estimator is equal to the value of the parameter being estimated. To put it differently, an estimator is unbiased if the average of all possible values of such an estimator (or the average value of the estimator calculated from all possible samples drawn from the population under consideration) is equal to the population parameter. For example, suppose we consider the sample mean statistic $\overline{X}$ to estimate the population mean μ of the random variable X. From Equation (2.17), we found that $E(\overline{X}) = \mu$. Thus, $\overline{X}$ is qualified to be called an unbiased estimator of μ. On the other hand, if the sample variance is computed as the average of the squared deviations

$$\nu^2 = \frac{\sum_{i=1}^{n}(X_i - \bar{X})^2}{n}$$

then ν^2 is a biased estimator of σ^2 because it can be shown that

$$E(\nu^2) = \sigma^2 \frac{n-1}{n} \neq \sigma^2$$

Therefore, a correction in computing the sample variance is made in order to obtain an unbiased estimator of σ^2. The correction is made by multiplying ν^2 by the ratio $n/(n-1)$, which is the same as computing the sample variance as

$$S^2 = \frac{n\nu^2}{n-1} = \frac{\sum_{i=1}^{n}(X_i - \bar{X})^2}{n-1}$$

This is the version of the estimator shown earlier in this chapter. We should point out that even though $E(S^2) = \sigma^2$, $E(S) \neq \sigma$. Thus, S is a biased estimator of σ. Fortunately, the bias does diminish as the sample size increases, allowing one to use S with minimal consequence. But, to be precise, we present in Chapter 5 a correction factor for S that gives us an estimator S which is actually unbiased.

An estimator is said to be *consistent* if, as the sample size n gets large, the probability approaches 1 that the estimator will be arbitrarily close to the value of the parameter. The sample mean $\bar{X}$, for instance, is a consistent estimator of μ. This can be recognized by looking at the variance of $\bar{X}$, which is σ^2/n. As n gets arbitrarily large, the variance of $\bar{X}$ approaches 0; hence, the probability that $\bar{X}$ will be close to its expected value μ approaches 1.

An estimator of a parameter having the smallest variation (minimum variance) among a specified group of estimators is an *efficient* estimator. The variance of an estimator measures its precision—the smaller, the better. If two unbiased estimators of a parameter were available, one would prefer the estimator with the smaller variation. As an example, the sample mean $\bar{X}$ and sample median $\tilde{X}$, are unbiased estimators of the mean μ of the normal population. That is, $E(\bar{X}) = \mu$ and $E(\tilde{X}) = \mu$. However, the variances are

$$\text{Var}(\bar{X}) = \frac{\sigma^2}{n}$$

$$\text{Var}(\tilde{X}) = \frac{\pi}{2}\frac{\sigma^2}{n} \qquad \text{(approximately for large samples)}$$

Thus, $\text{Var}(\tilde{X}) > \text{Var}(\bar{X})$, and therefore the sample mean is preferred to the sample median as an estimator of μ because the sample mean has minimum variance. In fact, it may be shown that $\bar{X}$ has minimum variance among all unbiased estimators of μ if the underlying distribution is normal; such an estimator is referred to as the minimum-variance unbiased estimator, or the most efficient unbiased estimator.

POINT ESTIMATION

Now that we covered some guidelines for choosing estimators, let us summarize which estimators are good candidates for estimating some key population parameters. In terms of the population mean μ, we have found that the sample mean $\overline{X}$ is an unbiased estimator of μ; this is true regardless of the nature of X (discrete or continuous). Additionally, we have learned that the sample variance S^2 is an unbiased estimator of the population variance σ^2.

In the special application of a Bernoulli random variable, recall from Equation (2.8) that the population mean is the probability of success p. An unbiased point estimate of p is the sample average of a sequence of Bernoulli observations, that is, $\overline{x} = x/n$ (where x is the number of successes observed and n is the sample size or number of trials). But recall that this special average was earlier called the sample proportion of successes and labeled $\hat{p}$. For the binomial distribution, the mean is $\mu = np$ [see Equation (2.10)]. Since $\hat{p}$ is an unbiased estimator of p, it is straightforward that $n\hat{p}$ is an unbiased estimator for the mean of a binomial distribution. With respect to the Poisson distribution, both the mean and variance equal the parameter λ; thus a sample average of a set of Poisson observations serves as an unbiased point estimate of both μ and σ^2.

INTERVAL ESTIMATION

Because any sample is an incomplete look at the population, we cannot expect a point estimate to equal the corresponding population parameter. The point estimate may or may not be close to the parameter in question. Therefore, a point estimate by itself conveys no information about the potential sampling error involved, that is, about the accuracy of the estimation procedure. For this reason, it is often desirable to obtain an interval of possible values in which the parameter is likely to be. Such interval is called an *interval estimate* or *confidence interval* for a parameter. Construction of confidence intervals requires knowledge of the sampling distribution of the estimator used in obtaining the point estimate. Below, we show the construction for some of the more commonly used intervals.

Interval Estimate for the Population Mean: σ Known   Suppose a random sample of size n is taken from a population with a known variance σ^2 (or known σ). Consider the problem of estimating the unknown mean of the population μ. We learned that the sample mean $\overline{X}$ is an unbiased point estimator of μ and that its sampling distribution is normally distributed (or approximately so from the central limit theorem) with variance σ^2/n. When the standard normal distribution was introduced in Section 2.8, we learned that the area to the right of $z_{\alpha/2}$ would be $\alpha/2$ and, based on symmetry arguments, the area to the left of $-z_{\alpha/2}$ would also be $\alpha/2$. Thus, the area between these two percentile points is $1 - \alpha$. With all these facts in mind, we can write

$$1 - \alpha = P\left(-z_{\alpha/2} \leq \frac{\bar{X} - \mu}{\sigma / \sqrt{n}} \leq z_{\alpha/2}\right)$$

$$= P\left(-z_{\alpha/2}\frac{\sigma}{\sqrt{n}} \leq \bar{X} - \mu \leq z_{\alpha/2}\frac{\sigma}{\sqrt{n}}\right)$$

$$= P\left(\bar{X} - z_{\alpha/2}\frac{\sigma}{\sqrt{n}} \leq \mu \leq \bar{X} + z_{\alpha/2}\frac{\sigma}{\sqrt{n}}\right)$$

The last probability statement suggests that a $100(1 - \alpha)$ percent interval estimate or confidence interval for a given set of observations is computed by

$$\bar{x} - z_{\alpha/2}\frac{\sigma}{\sqrt{n}} \leq \mu \leq \bar{x} + z_{\alpha/2}\frac{\sigma}{\sqrt{n}} \qquad (2.30)$$

It is important that the probability statement

$$P\left(\bar{X} - z_{\alpha/2}\frac{\sigma}{\sqrt{n}} \leq \mu \leq \bar{X} + z_{\alpha/2}\frac{\sigma}{\sqrt{n}}\right) = 1 - \alpha$$

be interpreted correctly. In this statement, the parameter μ should not be viewed as a random variable; rather it is an unknown constant. Thus, we should not say that the probability that μ falls in the range such and such is $1 - \alpha$. However, $\bar{X}$ is a random variable that varies from sample to sample. Hence, if we take different samples of size n from a single population,[12] we will potentially get different values for $\bar{X}$. As such, the endpoints of the interval vary from sample to sample. So the probability statement is correctly interpreted by saying, "If we were to gather a sample and construct an interval estimate, the probability that this random interval would include within it the actual population mean μ is $1 - \alpha$." Once we have taken the sample, we say that we are $100(1 - \alpha)$ percent confident that our interval contains the population parameter, because if we performed the same experiment a large number of times, we know that in $100(1 - \alpha)$ percent of the cases the interval would contain the population parameter.

Implicit in our above discussion, the value α indicates the proportion of times that we will be incorrect in assuming that an interval contains the population parameter. As we will discuss in the next chapter, this very value also represents the type I error rate used in testing hypotheses.

Suppose that the amount of liquid fill dispensed by a bottling machine is normally distributed with $\sigma = 0.9$ ounce (oz) for whatever machine setting is established. The interest is to estimate the underlying mean level of fill volume for a particular production run with its particular

Example 2.14

---

[12]We use the adjective *single* in front of the word *population* to imply that the population being sampled is the same from sample to sample. As we will discuss in Chapter 4, one of the goals of the SPC methods is to detect changes in the underlying population sample over time.

machine settings. So a sample of $n = 16$ filled bottles is randomly selected from the output of the machine on a given production run. If the average fill for the 16 bottles is 15.1 oz, what is a 95 percent interval estimate for μ? To construct the interval, we first compute the standard deviation of the sample mean, which is in this case $\sigma/\sqrt{n} = 0.90/\sqrt{16} = 0.225$. In terms of the notation given in Equation (2.30), a 95 percent confidence interval translates into $\alpha = 0.05$. Thus, we need $z_{\alpha/2} = z_{0.025}$ which from Appendix Table I is 1.96. The desired confidence interval is

$$15.1 - 1.96(0.225) \le \mu \le 15.1 + 1.96(0.225)$$

$$14.659 \le \mu \le 15.541$$

Thus, we are 95 percent confident that this interval (14.659, 15.541) covers the true mean. We cannot be sure that this particular interval truly contains the population mean, but if the sampling process were repeated over and over, about 95 percent of the intervals computed by the above procedure would contain the population mean.

Interval Estimate for the Population Mean: σ Unknown   The assumption of having a known population variance is often not very realistic. Instead, the population variance can be estimated from the sample values. Recall that if the population from which samples are taken is normally distributed and the sample standard deviation s is used in place of σ, then the sampling distribution of $\overline{X}$ is the t distribution with $n - 1$ degrees of freedom. Following the framework of Equation (2.30) but with s replacing σ and t percentile points for z percentile points, we see that the $100(1 - \alpha)$ percent confidence interval for the true mean is

$$\overline{x} - t_{\alpha/2, n-1}\frac{s}{\sqrt{n}} \le \mu \le \overline{x} + t_{\alpha/2, n-1}\frac{s}{\sqrt{n}} \tag{2.31}$$

where $t_{\alpha/2, n-1}$ denotes the percentile point on the t distribution with $n - 1$ degrees of freedom such that $P(t \ge t_{\alpha/2, n-1}) = \alpha/2$.

Example 2.15	**A** producer of lightbulbs claims that the average life of its bulbs is 2000 h. A government agency took a sample of 10 bulbs, and sample results were $\overline{x} = 1680$ h and $s = 215$ h. Assuming that bulb life is normally distributed, find the 99 percent confidence interval for true average bulb life. This level of confidence implies $\alpha = 0.01$, and thus $\alpha/2 = 0.005$. Since $n = 10$, we need the value of $t_{0.005, 9}$ which can be obtained from the t distribution table and found to be 3.25. Hence, the 99 percent confidence interval is

$$1680 - 3.25\left(\frac{215}{\sqrt{10}}\right) \le \mu \le 1680 + 3.25\left(\frac{215}{\sqrt{10}}\right)$$

$$1459.04 \le \mu \le 1900.96$$

The confidence interval allows us to make judgments of the producer's claim of $\mu = 2000$ h. By virtue of the fact that the above confidence interval does not encompass the value of 2000, the claim does not appear plausible or consistent with the observed data. So we tend to discredit the producer's claim. In general, a check of whether observed data are consistent with particular parameter values is a major area of statistical inference known as *hypothesis testing*. The use of confidence intervals is an intuitive means of determining if conjectured parameter values are reasonable, given the data at hand.

Interval Estimate for Population Proportion   It was mentioned earlier that, for sufficiently large n, the binomial distribution may be approximated by the normal distribution. Often in quality control applications, one is interested in estimating the parameter p, the population proportion of successes (typically, defects) in a sequence of n items. We found that an unbiased estimator of p is $\hat{p}$, the sample proportion of successes. Furthermore, we know from Equations (2.11) and (2.25) that the standard deviation of $\hat{p}$ is $\sqrt{p(1 - p)/n}$. Assuming that n is large enough for normality to hold, if we follow the same general procedure discussed previously for the construction of confidence intervals, then a $100(1 - \alpha)$ percent confidence interval for p is given by

$$\hat{p} - z_{\alpha/2} \sqrt{\frac{p(1 - p)}{n}} \leq p \leq \hat{p} + z_{\alpha/2} \sqrt{\frac{p(1 - p)}{n}}$$

The difficulty with the above interval is that the computation of the standard deviation requires knowledge of p — the one thing we do not know! A reasonable alternative is to substitute the unbiased estimator $\hat{p}$ for p, which gives

$$\hat{p} - z_{\alpha/2} \sqrt{\frac{\hat{p}(1 - \hat{p})}{n}} \leq p \leq \hat{p} + z_{\alpha/2} \sqrt{\frac{\hat{p}(1 - \hat{p})}{n}} \tag{2.32}$$

Example 2.16

In an attempt to measure certain aspects of customer satisfaction, the service department of a large metropolitan automobile dealer randomly selected 250 customers who recently had some form of service or repair done on their automobiles. It was found that 220 out of the 250 customers (88 percent) were pleased with the service performed. What is a 90 percent confidence interval for the proportion of pleased customers? From the standard normal table, we find by linear interpolation that $z_{\alpha/2} = z_{0.10/2} = z_{0.05} = 1.645$. Thus,

$$0.88 - 1.645 \sqrt{\frac{0.88(0.12)}{250}} \leq p \leq 0.88 + 1.645 \sqrt{\frac{0.88(0.12)}{250}}$$

or

$$0.846 \leq p \leq 0.914$$

With 90 percent confidence, the interval between 0.846 and 0.914 will cover the true proportion of pleased customers.

Interval Estimate for Population Variance   In many manufacturing applications, we are concerned with not only the mean level but also the variation of process produced by different methods or procedures. Hence, we now consider the construction of confidence intervals for population variance.

Suppose that the random variable X is normal (or approximately so) with unknown mean μ and unknown variance σ^2. We have learned that an unbiased estimator of σ^2 is S^2 and that the sampling distribution of $(n-1)S^2/\sigma^2$ is a chi-square distribution with $n-1$ degrees of freedom. To find a $100(1-\alpha)$ percent confidence interval for σ^2, we note the following probability statement:

$$P\left[\chi^2_{1-\alpha/2,\,n-1} \leq \frac{(n-1)S^2}{\sigma^2} \leq \chi^2_{\alpha/2,\,n-1}\right] = 1-\alpha$$

where $\chi^2_{\alpha/2,n-1}$, and $\chi^2_{1-\alpha/2,n-1}$ are the $100(1-\alpha/2)$ and $100(\alpha/2)$ percentile points of a chi-square distribution with $n-1$ degrees of freedom, respectively. After terms are rearranged, this statement is equivalent to

$$P\left[\frac{(n-1)S^2}{\chi^2_{\alpha/2,\,n-1}} \leq \sigma^2 \leq \frac{(n-1)S^2}{\chi^2_{1-\alpha/2,\,n-1}}\right] = 1-\alpha$$

So, for a given sample of data, the $100(1-\alpha)$ percent confidence interval for σ^2 is constructed as follows

$$\frac{(n-1)s^2}{\chi^2_{\alpha/2,\,n-1}} \leq \sigma^2 \leq \frac{(n-1)s^2}{\chi^2_{1-\alpha/2,\,n-1}} \tag{2.33}$$

Example 2.17

To illustrate the computation of a confidence interval for σ^2, consider the data from Example 2.15. In that example, we were told that $n = 10$ and $s = 215$ h which implies $s^2 = 46{,}225$ hours^2 (h^2). For a 95 percent confidence interval, we look up in Appendix Table III the values of $\chi^2_{0.025,9}$ and of $\chi^2_{0.975,9}$, which are 19.02 and 2.70, respectively. By using Equation (2.33), the 95 percent confidence interval for σ^2 is

$$\frac{9(46{,}225)}{19.02} \leq \sigma^2 \leq \frac{9(46{,}225)}{2.70}$$

$$21{,}873.03 \leq \sigma^2 \leq 154{,}083.33$$

We may be tempted to question the sensibility of such "large" numbers. However, we need to keep in mind that we are dealing with squared units for numbers (deviations from the sample mean) that are in the hundreds or possibly thousands. If we rescaled the data to be in units of thousands (for example, 2000 h is equated to the value 2), then the 95 percent confidence interval for σ^2 would be $0.021873 \leq \sigma^2 \leq 0.154083$.

EXERCISES

2.1. List the two types of statistical studies. Specify a rule that distinguishes these studies.

2.2. Using the first example below as a guide, fill in the missing parts of the table found below:

Phenomenon	Experimental Units	Population	Population Type
a. Patient satisfaction of hospital stay in this month	Patients staying in hospital this month	Set of satisfaction ratings of all patients who stayed in the hospital this month	Finite
b. Diameter of pipes produced on an assembly line			
c. Gross monthly sales of a company			
d. Current percentage of ISO 9001 certified companies in Wisconsin			

2.3. Explain why inferential statistics is not required if we have a census of cross-sectional data.

2.4. Explain why inferential statistics is always required if we are studying a process.

2.5. Identify the following information as qualitative or quantitative:
   a. Gender of a customer.
   b. Age of a machine.
   c. Pipe length (inches).
   d. Customer's response to a statement (strongly disagree, disagree, neither agree nor disagree, agree, strongly agree).
   e. Income level of a customer.
   f. Month of the year.

2.6. Classify the following quantitative data as discrete or continuous:
   a. Stopping distances of cars moving 60 miles per hour (mi/h).
   b. Daily number of employees absent.
   c. Processing times of mortgage loans.
   d. Number of air bubbles found in a glass windshield.
   e. Pass/fail data on manufactured items.
   f. Weekly closing prices for the Dow Jones Industrial Index.

2.7. You must make a decision regarding the procurement of hydraulic pumps from two suppliers. The target for the diameter of the pump body interior is 2.25 inches (in). Purchasing has requested a production run of 50 sequential observations from each supplier (see below; read across, then down). The data are also supplied in a file named "piston.dat." In addition, we show below the summary statistics and histograms for each data set.

```
piston1
    1.94   2.09   2.09   2.03   2.26   2.13   2.17   1.93   2.12   2.24
    2.07   2.18   2.03   2.23   2.22   2.16   2.15   2.28   2.06   2.17
    2.13   2.28   2.33   2.14   2.32   2.22   2.35   2.19   2.26   2.43
    2.24   2.37   2.16   2.40   2.41   2.34   2.44   2.12   2.40   2.35
    2.53   2.27   2.27   2.46   2.29   2.47   2.61   2.49   2.45   2.46

piston2
    2.35   2.16   2.29   2.09   2.10   2.30   2.21   2.49   2.49   2.04
    2.70   2.36   2.24   1.83   2.32   2.47   2.54   2.29   2.05   1.92
    2.13   2.43   2.46   2.48   2.35   2.14   2.22   2.21   2.39   2.38
    2.21   2.54   2.41   2.44   2.10   2.30   1.95   2.26   2.37   2.33
    2.67   2.04   2.12   2.34   2.03   2.32   2.42   2.49   2.38   2.28
```

Variable	N	Mean	Median	TrMean	StDev	SE Mean
piston1	50	2.2546	2.2500	2.2545	0.1551	0.0219
piston2	50	2.2886	2.3100	2.2914	0.1877	0.0265

Variable	Minimum	Maximum	Q1	Q3
piston1	1.9300	2.6100	2.1375	2.3775
piston2	1.8300	2.7000	2.1375	2.4225

Given the data, what recommendations would you make regarding the supplier selection decision?

2.8. Consider the stopwatch reading data set introduced in Section 2.3 and found in file "timing.dat." Using the Minitab character graphs option, create a character histogram of the data which resembles the histogram shown in Figure 2.5. (*Note:* The histogram that you will create will be upside down relative to the histogram shown in Figure 2.5. The goal is to create a histogram with the same class intervals' associated frequencies.)

2.9. Indicate whether each of the underlying processes depicted below is iid. If you conclude that the underlying process to be not iid, explain why.

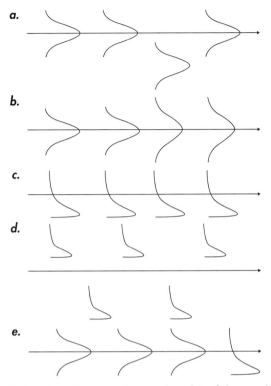

2.10. Below is a character time-series plot of the number of monthly power interruptions over a 3-year period for an electric utility company located in
the south central region of the United States. (1 represents January, 2 represents February, . . . , 9 represents September, 0 represents October, A
represents November, and B represents December.)

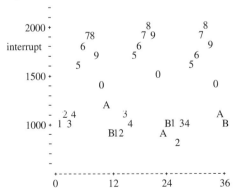

Do the data appear to arise from an underlying random process? Explain.

2.11. Refer to the character histogram below:

```
Midpoint    Count
  110.00        2  **
  116.00        8  ********
  122.00        7  *******
  128.00        8  ********
  134.00        2  **
  140.00        1  *
```

   a. How many observations are in the data set?
   b. What is the class width?
   c. What are the endpoints of the first class?

2.12. The thickness in millimeters (mm) of eight randomly selected card-boards is shown below:

   2.47   2.44   2.52   2.51   2.52   2.42   2.57   2.62

   a. Compute the sample mean.
   b. Compute the sample median.
   c. Compute the sample range.
   d. Compute the sample standard deviation.

2.13. The yield strengths, in pounds per square inch (lb/in^2, or psi), for a random sample of nine high carbon steel masts used in the assembly of large antenna systems are measured:

   86,427 86,894 87,095 87,864 85,834 86,852 86,446 87,254 85,483

   a. Compute the sample mean.
   b. Compute the sample standard deviation.

2.14. Consider the data in Exercise 2.13.
   a. Compute the sample median.
   b. How much can the value of the smallest observation increase without changing the sample median?

2.15. Let $x_1, x_2, \ldots, x_n$ represent the values of a sample of n observations. Suppose the sample observations are arranged by value in ascending order, that is, smallest to largest. Denote the ordered values by $x_{(1)}, x_{(2)}, \ldots, x_{(n)}$, where $x_{(1)} \le x_{(2)} \le \cdots \le x_{(n)}$.
   a. Using the order statistics notation, define the sample range
   b. Using the order statistics notation, define the sample median
   c. If $x_{(1)} = x_{(n)}$, what does this imply about the values of the observations?

2.16. Refer to the Minitab output shown below:

Variable	N	Mean	Median	TrMean	StDev	SE Mean
x	50	20.772	20.723	20.711	4.079	0.577

Variable	Minimum	Maximum	Q1	Q3
x	13.488	29.162	17.085	24.304

    a. How many observations are in the data set?
    b. What is the sample mean?
    c. What is the sample range?
    d. What is the sample variance?

2.17. The data shown below are the times (in minutes) that successive customers had to wait for service at a division of a motor vehicles facility (read across, then down).

2.97	4.54	3.98	2.54	2.09	5.82	3.22	4.25	3.48
4.81	2.81	4.79	2.98	6.64	9.16	2.03	2.03	0.62
2.80	6.32	3.11	2.43	2.47	3.46	2.32	2.24	1.14
4.08	8.05	6.20	2.37	3.44	3.20	6.58	3.47	4.68
4.13	5.59	0.76	2.06	10.02	2.06	0.53	4.37	

    a. Hand-compute the sample mean and standard deviation.
    b. Hand-construct a time-series plot of the observations. Comment on the time-behavior of the process as reflected by the time-series plot.
    c. Hand-construct a bar-type histogram, using the rough rule of thumb of choosing the number of class intervals approximately equal to the square root of the sample size. Start out with the first midpoint positioned at 1.5. Comment on the shape of the histogram.

2.18. Consider the waiting time data in Exercise 2.17. These data are given in a file named "motor.dat." Use Minitab to answer parts *a, b,* and *c* of Exercise 2.17.

2.19. In assemblies where contact to metal surfaces is critical, surface smoothness becomes a critical quality dimension. A profilometer is one of the traditional instruments used to measure surface roughness. The profilometer "stylus" of a finite radius is run against the metal surface to produce a voltage output which is proportional to the height of the surface. Surface roughness measurements, in microinches (μin), for a random sample of polished metal bearings are shown below:

214.86	200.30	213.96	192.94	196.14	205.50	208.94	201.80
199.18	191.97	205.48	191.21	208.03	203.01	211.02	206.97
199.52	199.32	198.32	196.93	207.53	202.12	210.48	207.43
195.08	217.95	203.79	198.46	220.55	204.15	212.56	201.11
204.14	205.47	208.95					

    a. Hand-compute the sample mean and standard deviation.
    b. Hand-construct a time-series plot of the observations. Comment on the time-behavior of the process as reflected by the time-series plot.
    c. Hand-construct a bar-type histogram, using the rough rule of thumb of choosing the number of class intervals approximately equal to the square root of the sample size. Start out with the first midpoint positioned at 190. Comment on the shape of the histogram.

2.20.   Consider the surface roughness data in Exercise 2.19. These data are given in a file named "roughness.dat." Use Minitab to answer parts *a*, *b*, and *c* of Exercise 2.19.

2.21.   In section 2.5, we intuitively argued that $\sum_{i=1}^{n}(x_i - \bar{x}) = 0$. Algebraically demonstrate this fact to be true.

2.22.   The sample *percentile* is a measure of the relative standing of a value within a data set. For example, the median represents the 50th percentile; that is, 50 percent of the observations are less than or equal to the median. In general, the *p*th percentile is the value x_p such that *p* percent of the observations are less than or equal to x_p. Below is the general procedure for calculating sample percentiles:

   (1) Arrange the *n* observations in ascending order. The smallest observation is given rank 1, the second smallest is given rank 2, and so on.
   (2) Calculate the index $i = (p \times n)/100$.
   (3) If *i* is an integer, the *p*th percentile is the average of the values having ranks *i* and $i + 1$. If *i* is not an integer, the *p*th percentile is equal to the value having a rank equal to the next integer value greater than *i*.

   a. For the waiting time data of Exercise 2.17, use the outlined procedure to determine the 10th and 90th percentiles.
   b. The lower quartile (also called the first quartile) is the 25th percentile of the data and is denoted by Q_1. The upper quartile (also called the third quartile) is the 75th percentile of the data and is denoted by Q_3. Find Q_1 and Q_3 for the waiting time data.

2.23.   Refer to Exercise 2.22 for the general procedure for finding percentiles and the definitions of lower and upper quartiles (Q_1 and Q_3). Determine Q_1 and Q_3 for the surface roughness data of Exercise 2.19.

2.24.   A pair of balanced dice are tossed. If *X* equals the total number of spots showing on the dice, then, for $k = 2, 3, \ldots, 12$, find
   a. $P(X = k)$ *b.* $P(X \le k)$ *c.* $P(X \ge k)$

2.25.   Find the expected value and variance of the random variable defined in Exercise 2.24.

2.26.   Find the mean and variance for the following distribution:

x	0	1	2	3	4
$P(X = x)$	0.20	0.10	0.10	0.40	0.20

2.27.   If *X* is a random variable whose domain is $1, 2, 3, \ldots, n$ and whose probability function $P(X = x) = ax$, determine the value of *a*. Find the expected value and variance of *X*.

2.28. Consider the following function:

$$f(x) = \begin{cases} \frac{5}{2}x & 0 < x \le c \\ 0 & \text{otherwise} \end{cases}$$

a. What value of c will make $f(x)$ a probability density function?
b. Given the appropriate value for c, find the mean and variance of the distribution.

2.29. The *geometric distribution* describes the number of trials until a "success" is encountered. Suppose independent trials are performed and the probability of a success at each trial is p. If X is the number of the trial on which the first success occurs, then

$$P(X = x) = p(1 - p)^{x-1} \qquad x = 1, 2, 3, \ldots$$

a. If the probability that an inspected item is defective is 0.02, what is the probability of having to look at 15 items before encountering a defective one?
b. Again, if the probability that an inspected item is defective is 0.02, what is the probability 10 or more items need to be inspected before encountering a defective one?

2.30. By a series of tests of a certain type of electric relay, it has been determined that in approximately 1.5 percent of the trials, the relay will fail to operate under certain specified conditions. What is the probability that in 15 independent trials under these conditions the relay will fail to operate 3 or more times?

2.31. Refer to the Minitab output given below to answer the questions:

```
MTB > pdf;
SUBC> binomial 5,.07.

Probability Density Function

Binomial with n = 5 and p = 0.0700000
            x      P( X = x)
            0        0.6957
            1        0.2618
            2        0.0394
            3        0.0030
            4        0.0001
            5        0.0000
```

a. What is the probability of success?
b. What is the sample size?
c. What is the probability of two successes?
d. What is the probability of at most one success?

2.32. Suppose a production process produces 1 percent defective items. Assuming independence of items, what is the probability that the sample proportion defective ($\hat{p}$) on a random sample of 100 items is less than or equal to 2 percent?

2.33. Determine the mean and standard deviation of the binomial distribution used in Exercise 2.30.

2.34. Determine the mean and standard deviation of the binomial distribution used in Exercise 2.32.

2.35. Suppose a quality control procedure is to take 10 items from production during each shift and stop production if one or more defectives are found. Assume that production is stable and that the probability of a defective item is p. The probability that an inspection does not stop production is called that *operating characteristic function* of the inspection procedure. Derive an expression for this function and plot it.

2.36. Derive the result that $E(X) = np$ for the general binomial distribution $P(X = x)$ given by Equation (2.9) by writing out $\sum_{x=0}^{n} xP(X = x)$, then factoring out the common factor np, and finally recognizing that the remaining summation is the sum of all possible binomial outcomes when the sample size is $n - 1$.

2.37. Use the Poisson distribution to approximate the exact binomial probability computed in Exercise 2.30.

2.38. We will learn in Chapter 9 that a 6σ process is equated with a process producing 3.4 per million nonconforming items. Suppose 10,000 items are tested using a high-speed computer-automated inspection device. Using the Poisson distribution as an approximation for the binomial distribution, what is the probability of observing zero nonconforming items? What is the probability of observing one nonconforming item?

2.39. Suppose that the number of e-mails arriving to a particular user during business hours is random at a rate of 0.025 per minute. Assuming the Poisson distribution, what is the probability that the user will get the following?
   a. Five e-mails during any given business hour
   b. At least one e-mail during any given business hour

2.40. Splices in a certain manufacturing tape occur at random, on the average of 1 per 2500 feet (ft). Assuming the Poisson distribution, what is the probability that a 4000-ft roll of tape has
   a. No splices?
   b. At most two splices?
   c. Two or more splices?

2.41. Refer to the Minitab output given below to answer the questions:

```
MTB > cdf;
SUBC> poisson 1.4.

Cumulative Distribution Function

Poisson with mu = 1.40000

        x      P( X <= x)
        0          0.2466
        1          0.5918
        2          0.8335
        3          0.9463
        4          0.9857
        5          0.9968
        6          0.9994
        7          0.9999
        8          1.0000
```

    a. What is value of parameter λ?
    b. What is the probability of observing at least one occurrence?
    c. What is the probability of observing exactly three occurrences?

2.42. Derive the result that $E(X) = \lambda$ for the general Poisson distribution $P(X = x)$ given by Equation (2.12).

2.43. The resistances of carbon resistors of 1500 Ω nominal value are normally distributed with $\mu = 1500$ and $\sigma = 200$.
    a. What proportion of resistors is expected to have resistance greater than 1150 Ω?
    b. Within plus and minus what amount around the mean do 80 percent of resistances fall?

2.44. A review of a field service technician's records indicates that the time taken for a service call can be represented by a normal distribution with mean 50 min and standard deviation 10 min.
    a. What proportion of the service calls takes more than 30 min?
    b. What proportion of the service calls takes less than 1 h?

2.45. The percentage of moisture content of a popular breakfast cereal is normally distributed with mean 1.5 percent and standard deviation 0.7 percent. Suppose this quality characteristic is subject to a maximum specification of 3.0 percent. What is the probability that a sample of cereal selected at random will conform to specifications?

2.46. The service life of a certain product is normally distributed. Suppose 94 percent of the items have lives exceeding 2000 h and 3.51 percent have lives exceeding 16,240 h. Find the mean and standard deviation of the service life.

2.47. The length of a bolt made by a machine parts company is normally distributed with a standard deviation σ known to be 0.02 mm. The lengths of six randomly selected bolts are 15.02, 14.97, 14.95, 14.94, 14.98, and 15.01 mm.

    a. Construct a 95 percent confidence interval for the mean bolt length.

    b. Specifications for the process require a mean length μ of 15.00 mm for the population of bolts. Do the data indicate conformance to this specification?

2.48. A particular vacuum cleaner is specified to produce suction of 9 lb/in^2. Assume that suction strengths are normally distributed with a standard deviation σ known to be 1.2 lb/in^2. A random sample of 10 vacuum cleaners has an average suction strength of 8.3 lb/in^2.

    a. Construct a 95 percent confidence interval for the true mean suction strength.

    b. Do the data indicate conformance to the mean target specification?

2.49. The length of time between the billing and the receipt of payment was recorded for a random sample of 20 clients of a legal firm. The sample mean and sample standard deviation for the accounts were 23.4 days and 7.1 days, respectively. Length of time can be assumed to be normally distributed. Construct a 95 percent confidence interval for the true mean cycle time.

2.50. Consider the data in Exercise 2.17. Construct a 95 percent confidence interval for the mean waiting time.

2.51. Consider the data in Exercise 2.19. Construct a 95 percent confidence interval for the mean surface roughness.

2.52. In our discussion of the stopwatch data set, we stated (in Section 2.3) that "it is fair to say that this bias is small and not significantly different from zero." In other words, we are unable to reject the possibility that the true mean equals 10s. To confirm such a conclusion, refer to the summary statistics for the data set given in Figure 2.14 and construct a 95 percent confidence interval for the true mean.

2.53. Suppose a random sample of 100 critical parts for a newly developed airplane were subjected to a severe stress test. In the test, 8 parts failed. Find a 95 percent confidence interval for the population proportion of parts that would fail the same test.

2.54. A pharmaceutical company tests 700 pain relief medication bottles for their resistance to tampering. Among those bottles, 11 were found to be sealed incorrectly, rendering them prone to possible tampering. Construct a 99 percent confidence interval for the population proportion of bottles that are incorrectly sealed.

2.55. In the chapter, we developed confidence intervals for the following parameters: μ, p, and σ^2. Consider now the Poisson parameter λ, that is, the expected number of occurrences for a unit interval.

    a. Suppose the numbers of occurrences for n randomly selected unit intervals are counted; label them $x_1, x_2, \ldots, x_n$. Given these observed numbers, develop a $100(1 - \alpha)$ percent confidence interval for λ.

    b. What assumptions must be made for the proper interpretation of your proposed confidence interval?

2.56. In an automated process, a machine is used to fill a container with a liquid product. If the average amount is different from the specified target value, the machine can be adjusted to correct the mean. If, however, the variance of the filling process is too high, the machine needs to be shut down and repaired. Thus, there are regular checks of the variance of the filling process. A random sample of 35 containers gives an estimate $s^2 = 3.451$. Give a 90 percent confidence interval for the population variance σ^2.

2.57. Consider the data in Exercise 2.17. Construct a 95 percent confidence interval for the variance of waiting time.

2.58. Consider the data in Exercise 2.19. Construct a 95 percent confidence interval for the variance of surface roughness.

REFERENCES

Box, G. E. P., G. M. Jenkins, and G. C. Reinsel (1994): *Time Series Analysis: Forecasting and Control,* 3rd ed., Englewood Cliffs, NJ: Prentice Hall.

Deming, W. E. (1975): "On Probability as a Basis for Action," *The American Statistician*, 29, pp. 146–152.

————— (1986): *Out of the Crisis*, Cambridge, MA: MIT Center for Advanced Engineering Study.

Roberts, H. V., and R. F. Ling (1982): *Conversational Statistics with IDA*, New York: The Scientific Press/McGraw-Hill.

Stout, K. (1985): *Quality Control in Automation*, Englewood Cliffs, NJ: Prentice-Hall.

3

Modeling Process Data

CHAPTER OVERVIEW

In Chapter 2, we introduced the concept of randomness along with the corresponding notion of a process in a state of statistical control. We also introduced the idea of a distribution and emphasized its distinction from the concept of randomness. One point to keep in mind is that the development and interpretation of many standard statistical process control (SPC) methods rest on an assumption that a stable process is one that generates observations (or related statistics) which are random and normally distributed. The reality is that processes encountered in practice are often nonrandom and/or nonnormal but are still quite predictable.

As we discussed in Chapter 2, time-series plots and histograms are extraordinarily valuable in the discovery of time and distribution behavior of processes. In this chapter, we present additional tools that will enable us to recognize departures from randomness and/or normality. If departures are identified in a given process, then we need to consider statistical methods to describe or "model" the nature of the process behavior.

By means of real data, we will demonstrate how methods of data transformation and regression modeling provide a powerful, but easily understood, body of techniques for the general description of most process data. Even though our aim in this chapter is to provide a self-contained presentation of data analysis techniques, it is more useful to view the material of the chapter as a first exposure and to recognize that the applications offered in subsequent chapters will fully reinforce the necessary concepts.

For readers who wish additional resources, there are a wide variety of textbooks written on regression modeling. Among the well regarded are those by Draper and Smith (1981), Neter et al. (1996), and Ryan (1996). For a time-series emphasis, excellent textbooks include, but are not limited to, Abraham and Ledolter (1983); Box, Jenkins, and Reinsel (1994); and Cryer (1986). However, many of these texts assume a mathematical level beyond a realistic technical training level of many people involved in process monitoring. A few authors attempt to provide a practical, computer-oriented approach to data analysis and regression modeling. These include Cryer and Miller (1991), Menzefricke (1995), and Roberts (1991).

3.1   DIAGNOSTIC CHECKS FOR RANDOMNESS

Recall one of the lessons learned from Deming's spring elongation example presented in Section 2.4. Namely, the first stage in the analysis of process data should be a study of the observations sequenced over time. By studying process observations over time, we can hopefully assess the general conditions of the underlying process. One of the most basic questions we might ask is, Is a process random over time? Recall also that a random process is a predictable process from past data in the sense that its general level (or location) and variability

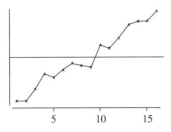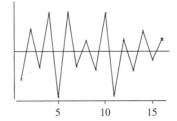

Figure 3.1   Two extreme nonrandom series ($n = 16$)

around that level remain nearly constant over time. Furthermore, a random process is characterized as a process absent of any sustaining patterns.

For a given set of process data, visual examination of the time-series plot of the observations clearly plays an important role in *diagnosing* the time behavior of a process. However, there can be situations in which the underlying patterns in a process are subtle; hence it is useful to consider other diagnostic checks to supplement the visual inspection of a time-series plot. In the subsections below, we present two additional checks useful in the assessment of time-series randomness. In addition, the first of these checks—the runs test—will provide a nice opportunity to detail the basic concepts of statistical hypothesis testing useful for our process data analysis explorations.

Runs Test

A simple numerical check of randomness of a series is a runs test. The runs test classifies observations as being above ($+$) or below ($-$) some reference line, usually the sample mean, which is the default value in Minitab. Based on the plus and minus symbols, a *run* is defined as a sequence of consecutive symbols of one kind preceded and followed by a symbol of another kind. For example, in the sequence $- - - + + - - - - + + + - - + +$ there is a run of three minuses, followed by a run of two pluses, followed by a run of four minuses and so on. Hence, there are a total of 6 runs in the 16 observations. For this hypothetical sequence, what do the 6 runs suggest in terms of concluding randomness or nonrandomness? To understand how a runs count can be useful for making this distinction, let us consider two extreme situations, depicted in Figure 3.1.

The first plot reflects a time series which is distinctly nonrandom, characterized by a trend or drift. Converting this series to a sequence of plus and minus symbols gives us the sequence $- - - - - - - - - - + + + + + + +$, implying a runs count of 2 in the 16 observations. On the other extreme, the second plot of Figure 3.1 reflects a nonrandom oscillating series. The corresponding sequence for a runs count is $- + - + - + - + - + - + - + - +$ which produces a total of 16 runs in 16 observations. It appears that too few runs or too many runs suggest nonrandom behavior. Since a random process has neither the persisting tendency to stay (or cluster) on one side of the mean nor the persisting tendency to oscillate around the mean, we might intuitively expect the number of runs for a random process to be somewhere in between the number of runs produced from these opposite scenarios. Thus, if a sequence of 16 observations comes from a random process, the expected number of runs is *about* $\frac{16}{2} = 8$. More precisely, the expected number of runs is found as follows:

$$\text{Expected number of runs} = \frac{2m(n - m)}{n} + 1 \qquad \textbf{(3.1)}$$

where n is the number of observations and m is number of pluses. For the hypothetical sequence $- - - + + - - - - - + + + - - + +$ for which we counted 6 runs, the expected number of runs for this sample is

$$\text{Expected Number of Runs} = \frac{2(7)(9)}{16} + 1 = 8.875$$

Our general approximate rule of one-half the number of observations (8) is not too far off. Of course, for any given sequence of observations we would not observe 8.875 runs. The point is that we would expect the average runs count to be 8.875 over many repeated samples of 16 observations having 7 pluses and 9 minuses.

The question is now whether 6 observed runs are close enough to 8.875 expected runs to deem the underlying process random. The following statistical fact is particularly useful for this determination. Namely, if the underlying process is random, the observed number of runs with a given number of pluses and minuses over many samples will approximately follow the normal distribution with mean given by Equation (3.1) and standard deviation given below:

$$\text{Standard deviation of runs} = \sqrt{\frac{2m(n - m)[2m(n - m) - n]}{n^2(n - 1)}} \qquad \textbf{(3.2)}$$

Applying (3.2) to our example, we compute

$$\text{Standard deviation of runs} = \sqrt{\frac{2(7)(9)[2(7)(9) - 16]}{16^2(16 - 1)}} = 1.9$$

To accommodate the variability in the actual runs count due to chance, we can provide an interval (or range) of possible runs count values that might occur when the process is truly random. An interval which is commonly reported is one that encompasses 95 percent chance given an underlying random process. Since runs counts are approximated by the normal distribution, we would expect about 95 percent of the runs counts to lie within 2 standard deviations of the mean[1]. For our example, this implies that if the process is random, about 95 percent of the runs counts (with 7 pluses and 9 minuses) will be within $8.875 \pm 2(1.9)$, that is,

---

[1] The choice of the whole number 2 is a matter of convenience since 95 percent of the area of the normal distribution is *precisely* associated with ± 1.96 standard deviations. Technically, a $2-$standard-deviation interval corresponds with slightly over 95 percent (actually, 95.45 percent).

between 5.075 and 12.675. Since 6 runs fall within this 95 percent interval, we are inclined to conclude that the associated process is random. In contrast, the 2 runs for the trended sequence of Figure 3.1 (which also has 7 pluses and 9 minuses) falls well below the limits of the 95 percent interval which suggests that this process is nonrandom.

The above discussions pertain to what is more formally known as *statistical hypothesis* testing. Any assertion or assumption concerning the nature of a process and/or the form of the probability distribution underlying a process is called a *statistical hypothesis*. A procedure or a rule to be used for deciding whether to accept a hypothesis is called a *statistical test* of a hypothesis. The hypothesis to be tested is defined as the *null hypothesis* and is commonly denoted by H_0. Once the null hypothesis has been established, the next step is to establish the *alternative hypothesis*, denoted by H_1. The null hypothesis and alternative hypothesis together make up all possibilities that might occur for the process in question. Thus, we are confronted with the problem of testing (or choosing) one possible hypothesis H_0 against the alternative hypothesis H_1.

We began this section by asking whether a specific process can be deemed random. To place the problem in a formal hypothesis-testing framework, we consider the following competing hypotheses:

Null hypothesis	H_0: Process is random
Alternative hypothesis	H_1: Process is nonrandom

To make a choice between these two possible states of nature, we need a decision rule that specifies the conditions under which one hypothesis is more plausible than the other. At the core of any statistical decision rule is some relevant sample information (called a *test statistic*) appropriate to help choose between the competing hypotheses. For our case, we have demonstrated that the number of runs observed from a process is one possible test statistic for judging whether a process is random. Earlier, we learned that if the underlying process is truly random (if H_0 is true), the number of runs is a random variable which is approximately normally distributed with mean and standard deviation given in Equations (3.1) and (3.2), respectively. If the process is not random (if H_1 is true), then the expected number of runs will be different from Equation (3.1). Consequently, if H_1 holds true, the observed number of runs will tend to be either large or small relative to what would be expected if the process were truly random.

With the observed number of runs as the test statistic, extreme values (large or small) suggest that we reject H_0 in favor of H_1. How extreme? To answer this question, we must recognize that there are two possible errors in hypothesis testing, as illustrated in Figure 3.2. Based on the observed data, if the null hypothesis is accepted when H_0 is true, then clearly a correct decision has been made. Similarly, if the null hypothesis is rejected, when H_1 is true, the decision is also correct. There is, however, a chance that sampling variation will lead us to incorrect decisions. We may reject the null hypothesis when it is true or accept the null hypothesis when it is false. These two types of incorrect decisions are called

State of nature

Decision	H_0	H_1
Accept H_0	Correct decision	Type II error
Reject H_0	Type I error	Correct decision

Figure 3.2   Consequences of hypothesis testing.

type I and *type II* errors, respectively.[2] The probabilities of committing these errors are denoted by the Greek letters α (alpha) and β (beta), that is,

$$\alpha = P[\text{type I error}] = P[\text{reject } H_0 \mid H_0 \text{ is true}]$$

$$\beta = P[\text{type II error}] = P[\text{accept } H_0 \mid H_1 \text{ is true}]$$

Ideally both the probabilities of a type I error and a type II error would be made as small as possible in a hypothesis test. However, we must contend with a basic tradeoff in hypothesis testing; for a given sample size, by lowering the type I error risk α, we must assume a higher type II error risk β, but lowering β will result in a higher α. In general, the three quantities α, β, and sample size n are interrelated in such a manner that both α and β can be made small only by increasing n.

Since, for a given sample size, α and β cannot be simultaneously lowered to arbitrary levels, the standard approach is to specify and control α to some small value, thereby ensuring that the probability is low that we will incorrectly reject a true null hypothesis. Why do we specify a value for α rather than β? The problem is that type II error is less subject to our control. Unlike the type I error, which statistical tests permit us to control by our selection of α, the probability of making a type II error is dependent on the underlying magnitude of the departure of a process parameter from the hypothesized value. Since this magnitude is typically unknown, β is accordingly unknown.

The probability of type I error is also called the *level of significance*. There is nothing "etched in stone" when it comes to choosing a level of significance; we generally just want α to be low. Albeit arbitrary, the most commonly used values for α are 0.01, 0.05, and 0.10. Probably, the most prevalent practice followed in general statistics is to design a decision rule such that the probability of rejecting H_0 when it is true is 0.05, or 5 percent. In fact, as we will soon see, the 5 percent level of significance is so common that most statistical computer packages (including Minitab) default many of their statistical hypothesis-testing routines at this value.

---

[2] In Chapter 4, we will point out a connection between hypothesis testing and control chart procedures. Thus, familiarity with hypothesis-testing concepts and terminology is useful for later developments.

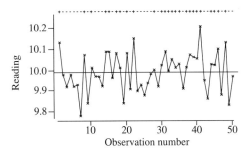

Figure 3.3   Counting number of runs in stopwatch data series.

For the above reasons and because of personal experience, we will be generally guided, although not too rigidly, in our data analysis explorations by a 5 percent level of significance as a rule of thumb. This is indeed what we did earlier when we made a judgment of randomness based on an interval constructed such that it encompassed 95 percent of possible run count values if the underlying process is random. An observed run count that falls outside this interval would be viewed as extreme or *statistically significant,* persuading us to believe that the process was not random; more precisely, we say that the observed run count is "significant at the 0.05 level." As explained, there is a small chance (5 percent) that we will incorrectly deem the process nonrandom even though the process is random, but our premise is that a more tenable explanation for a significant runs count is that the underlying process is in fact nonrandom.

As it stands, we can assess randomness by constructing an interval that encompasses 95 percent of possible runs count values, and then report whether the observed runs count is significant at the 0.05 level. But this approach does not allow us to answer the question, If an observed runs count is insignificant at the 0.05 level, is it also insignificant at the 0.10 level? Neither can we answer the question, If an observed runs count is significant at the 0.05 level, is it also significant at the 0.01 level? One way to answer these questions is to construct intervals that encompass 90 and 99 percent of possible run count values.

Recalculating the intervals for different desired levels of significance is a rather cumbersome and clumsy approach. Alternatively, we can compute a single value, called a *p value*, which allows us to determine the significance of an observed test statistic for any desired level of significance. A *p value* of a hypothesis test is defined as the probability of obtaining an observed sample value that *deviates* as far, or farther, from the expected value of the test statistic when the null hypothesis H_0 is true. If this probability is small, this suggests that the observed test statistic lies far from the expected value and we would be inclined to reject the null hypothesis since the deviation is probably too large to be explained by chance alone. On the flip side, if this probability is large, we do not reject the null hypothesis and do conclude that the deviation is due to chance, that is, sampling variation.

To further assimilate the concept of a *p value*, we illustrate its calculation for the runs test. Recall from Chapter 2, the stopwatch data set which consisted of

50 consecutive start and stop measurements of 10. Figure 3.3 shows the time-sequence plot of the data along with a string of plus and minus signs, indicating whether a point is above $(+)$ or below $(-)$ the sample mean of 9.989. If we count the number of distinct sequences of pluses and minuses, we find that there are 24 observed runs with 25 pluses and 25 minuses. If 25 pluses and 25 minuses are observed and the underlying process is truly random, then applying Equations (3.1) and (3.2) gives us the following expected number of runs and standard deviation of runs:

$$\text{Expected number of runs} = \frac{2(25)(25)}{50} + 1 = 26$$

$$\text{Standard deviation of runs} = \sqrt{\frac{2(25)(25)[2(25)(25) - 50]}{50^2(50 - 1)}} = 3.499$$

So, if the underlying process is random, the runs count for a sequence of 50 observations with 25 pluses and 25 minuses is expected to be 26 with a standard deviation of 3.499. Thus, the observed number of runs (24) deviated by 2 runs from the expected number of runs (26). To calculate the p *value*, we need to find the probability that the runs count statistic deviates by 2 or more from the expected value of 26, assuming the process is indeed random. This is equivalent to determining the probability that the runs statistic is less than or equal to 24 $(= 26 - 2)$ or greater than or equal to 28 $(= 26 + 2)$.

As stated earlier, runs counts are approximately normal when the process is random. Since the normal distribution is symmetric around the mean, P(number of runs ≤ 24) is equal to P(number of runs ≥ 28). Hence, the required probability is double either one of these probabilities. Below, we invoke Minitab to compute the probability that a normal random variable with a mean of 26 and standard deviation of 3.499 is less than or equal to 24:

```
MTB > CDF 24;
SUBC> Normal 26 3.499.

Cumulative Distribution Function

Normal with mean = 26.0000 and standard deviation = 3.49900
        X     P( X <= x)
   24.0000        0.2838
```

Hence, the p value is computed to be $2(0.2838) = 0.5676$. This means that for a random process, more than 56 percent of the time we would see as big as or bigger deviation from the expected number of runs than the one observed here. One can think of a p value as a measure of how unusual or surprising the data are if the null hypothesis is true. In this case, the large p value suggests that a runs count that deviates by 2 from the expected number of runs is not unusual for a random process. Thus, there is little reason to believe the underlying process is not random.

Before we introduced the concept of the p value, we explained that if the observed number of runs falls beyond the interval of $\pm$ 2 standard deviations around the expected number of runs, the sample value is deemed significant at the 0.05 level. Such an interval is constructed on the basis that the probability is 0.05 that a sample value from a random process deviates 2 or more standard deviations from the expected number of runs. Since the probability beyond the limits is at most 0.05, the p value associated with a sample value beyond 2 standard deviations must be less than 0.05. On the other hand, the p value associated with a sample value's falling within the limits is necessarily greater than 0.05. In general, for a specified value of α, the p value can be used to choose between H_0 and H_1 in the following manner:

If p value $\geq \alpha$, accept H_0.

If p value $< \alpha$, reject H_0.

With the p value, we know if H_0 is rejected for any specified α level. Suppose, for instance, the p value is 0.07. We know that the null hypothesis would be rejected at a 10 percent level of significance but not a 5 percent level of significance. The p value can be viewed as the minimum level of significance for α that would result in the rejection of the null hypothesis. Reporting the p value allows analysts to make their own decisions about the results of the experiment in terms of the stated hypotheses. For our purposes, we will be generally guided by the following rules of thumb to describe the strength of the sample evidence:

- If the p value is greater than 0.10, the sample results are *not significant,* that is, they are consistent with the null hypothesis. For the runs test, the process would be viewed as random.

- If the p value is between 0.05 and 0.10, the sample results might be regarded as *marginally significant.* For the runs test, this level of p value raises our suspicions and is suggestive of nonrandomness, but the evidence cannot be viewed as overwhelming.

- If the p value is between 0.01 and 0.05, the sample results are regarded as *significant;* the null hypothesis is discredited. For the runs test, the process would be viewed as nonrandom.

- If the p value is less than 0.01, the sample results are regarded as *highly significant*; the evidence against the null hypothesis is quite convincing. For the runs test, the evidence very strongly suggests the process is nonrandom.

Beyond the runs test, these rules of thumb can be universally applied to *any* hypothesis-testing procedure for which a p value is reported.

When we illustrated the runs test on the stopwatch data set, we exerted a lot of manual effort to count the observed number of runs (24), to count the number of pluses and minuses (25 and 25), to calculate the expected number of runs (26), and ultimately to compute the p value for the test (0.5676). Fortunately, as you might suspect, Minitab performs these computations in one shot. Figure 3.4 shows the Minitab output for the runs test. Minitab has also counted 24 runs

Stat>>Nonparametrics>>Runs Test

```
MTB > runs 'reading'

Runs Test

  reading

  K =     9.9890

  The observed number of runs =   24
  The expected number of runs =   26.0000
  25 Observations above K    25 below
            The test is significant at   0.5676
            Cannot reject at alpha = 0.05
```

Figure 3.4   Runs test for stopwatch data series.

around the mean, denoted by K, and, as we did, found 25 observations above the mean and 25 observations below the mean. Confirming our calculation, Minitab shows the expected number of runs to be 26. On the next-to-last line, Minitab says "the test is significant at 0.5676." This is Minitab's way of reporting the p value which we found to be the same.

On the last line, Minitab says, "Cannot reject at alpha = 0.05." The translation is that the runs test cannot reject the null hypothesis that the process is random at the 0.05 level of significance, the default α value in Minitab. This is because the p value is greater than $\alpha = 0.05$. If the p value is less than 0.05, Minitab reports the p value and does not state the last line, implicitly suggesting that the runs count is significant.

AUTOCORRELATION FUNCTION

To facilitate discussions in this and subsequent sections, it is convenient to establish a notation system for sequenced observation. To introduce time order, we place an index (1, 2, 3, . . .) in the subscript of the measurement variable. Typically, the index is mapped to equally spaced time periods; for example, days, weeks, months, and years. When time periods are not equally spaced, the index is interpreted to represent the order of the observation in sequenced time series. Thus, x_1 denotes the value of a random variable X for the first period (or the first observed value), x_2 denotes the value of a random variable X for the second period (or the second observed value), and so on. In general, x_t represents the value of the time series at time period t (or the tth observed value).

We have, on a number of occasions, defined a random process as a patternless process with no tendencies for observations to appear below or above the mean because prior observations have been below or above the mean. In Chapter 2, we also referred to a random process as an independent process to suggest that the observations taken at different time points are independent or unrelated with one another. However, if a process exhibits a persisting pattern over time, observations will reflect some sort of association or dependency relative to prior observations. To check whether associations exist, we need a way of relating observations made at different time periods.

Consider again the 50 sequential stopwatch readings of Chapter 2. By using the established system of notation, this time series can be denoted as $reading_t$, where $t = 1, 2, \ldots, 50$. To compare observations against prior observations, we need now to create new variables that take on earlier values of the original variable ($reading_t$). This is accomplished by a process known as *lagging*. By lagging a variable, we are creating another variable by arranging the data so that original observations are lined up with prior observations from a certain number of periods in the past.

Table 3.1   Stopwatch data series and its lags

t	$reading_t$	$reading_{t-1}$	$reading_{t-2}$	$reading_{t-3}$	$reading_{t-4}$
1	10.13	*	*	*	*
2	9.98	10.13	*	*	*
3	9.92	9.98	10.13	*	*
4	9.98	9.92	9.98	10.13	*
5	9.92	9.98	9.92	9.98	10.13
6	9.93	9.92	9.98	9.92	9.98
7	9.78	9.93	9.92	9.98	9.92
•					
•					
•					
46	10.10	10.02	10.03	9.86	9.95
47	9.88	10.10	10.02	10.03	9.86
48	10.13	9.88	10.10	10.02	10.03
49	9.83	10.13	9.88	10.10	10.02
50	9.97	9.83	10.13	9.88	10.10

To illustrate the concept of lagging, we present in Table 3.1 the stopwatch data along with lagged variables going one, two, three, and four "periods" back.[3] Adjacent to the column of original data ($reading_t$), we have a column of data for a variable labeled $reading_{t-1}$. This variable is known as the first lag of $reading_t$, or more generally, a lag 1 variable. Thus, for a given time period, a lag 1 variable takes on the value of the immediately preceding observation. Notice that the entry for the first row of $reading_{t-1}$ is the symbol *; this denotes a "missing value" since there is no observation $reading_t$ prior to the first observation. The next column in the display is the second lag of the time series ($reading_{t-2}$). This variable

[3] In this application, think of a certain number of periods in the past as a certain number of observations in the past.

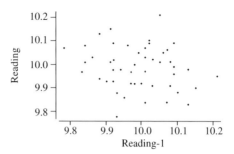

Figure 3.5   Plot of reading$_t$ versus reading$_{t-1}$.

contains values of reading$_t$ from two prior periods. Following this logic, we may define the kth lag of any time series x_t as a variable x_{t-k} which contains values of x_t from k periods back.

Once lagged variables are created, we can more easily investigate the presence of (or lack of) associations between observations over time. One visual means to demonstrate associations is to produce a two-dimensional plot in which each observation (on the vertical axis) is plotted against an observation a certain number of periods prior (on the horizontal axis); that is, to plot a time-series variable versus its lag. For example, in Figure 3.5, we plot each stopwatch value (reading$_t$) observed at time t, against the previous value (reading$_{t-1}$), observed at *time $t-1$*, for t = 2, 3, . . . , 50. Hence, there are 49 points on the graph representing the 49 available pairs (reading$_t$, reading$_{t-1}$). From this graph, also known as a *scatter plot*, there seems to be an ever so slight hint of a negative relationship between reading$_t$ and reading$_{t-1}$, which simply might be the result of sampling variation. What we can say is that there is clearly no substantial evidence of an association between the two variables, that is, between successive observations over time.

Contrast the stopwatch reading scatter plot of Figure 3.5 with the scatter plots found in Figure 3.6. The scatter plot of Figure 3.6a provides clear evidence that successive observations are negatively, or inversely, related since lower values of x_{t-1} tend to imply higher values of x_t, and higher values of x_{t-1} also tend to imply lower values of x_t. In other words, if a given observation is low, the subsequent observation will tend to be high, and vice versa. From a time-series perspective, this is indicative of a process that tends to go up and down (oscillate). For an example of an oscillating series, refer to the second plot of Figure 3.1.

On the other hand, the scatter plot of Figure 3.6b reflects a positive relationship between successive observations. This implies that if a given observation is high, the subsequent observation will tend to be high, and if a given observation is low, the subsequent observation will also tend to be low. Hence, observations close together in time tend to be close together in value. Processes that exhibit trend and/or meander show this tendency. In practical applications, positive

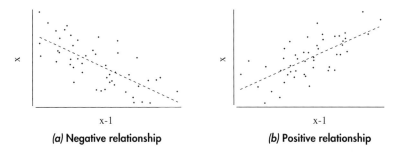

(a) Negative relationship   (b) Positive relationship

Figure 3.6   Different scatter plots for x_t versus x_{t-1}.

association between successive observations is, by far, the most common form of nonrandomness.

In both cases, the associations represented in Figure 3.6 are said to be *linear*. A linear association means that if we draw a smooth line on the scatter plot to best represent the tendencies between the two variables, the line will be a straight line—as we have done with the two dashed lines for the two scatter plots of Figure 3.6. In many situations the relationship between two variables is nonlinear; that is, the tendency between the two variables is best represented by a curved line. For example, the relationship between world population and time is one that is curving up at a faster and faster rate. Even though nonlinear relationships are abundant in application, they rarely occur when one is comparing a time-series variable versus its lags. Typically, either there is no association between a time-series variable and its lags, or the association is closely linear.

Beyond visual inspection of a scatter plot, it is useful to have some measure calculated from the data that enables us to more formally test for the existence of a linear association. The summary statistic known as the *sample correlation coefficient*, denoted by r, provides such a measure. The sample correlation coefficient is a single value that describes the degree of linear association between two variables. In our case, we are concerned with the correlation between a time-series variable and its first lag, that is, the correlation between successive observations. Since we are not looking at a correlation between two arbitrary variables but rather at a correlation computed from the observations of a single time series, a sample correlation coefficient is commonly referred to as a sample *auto*correlation ("self" correlation) coefficient. More specifically, the sample correlation coefficient computed between a time-series variable and its first lag is called the lag 1 sample autocorrelation coefficient, denoted r_1 where the 1 in the subscript signifies the order of the lag. For purposes of brevity, it is common practice to say lag 1 sample autocorrelation instead of lag 1 sample autocorrelation coefficient.

As seen from Figure 3.7, r_1 (or any sample correlation coefficient) can theoretically take on values ranging from a minimum value of -1.0 to a maximum value of $+1.0$. The extreme values of -1.0 and $+1.0$ for r_1 reflect a perfect linear association between successive observations, one extreme being a perfect

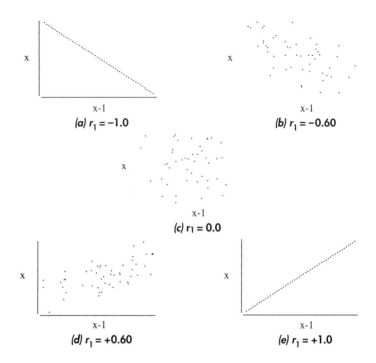

Figure 3.7   Scatter plots for different lag 1 autocorrelation values.

negative linear association and the other being a perfect positive linear association. Between the two extremes, a computed sample autocorrelation of 0 for r_1
indicates that there is no linear relationship between successive observations at
all. In practice, a computed autocorrelation rarely, if ever, equals 0 even if the
successive observations are truly independent since sampling variation will generally result in a small amount of correlation between observations.

Up to this point, our attention has been focused on assessing the extent of
linear association between successive observations. We may be interested also in
studying the relationship between observations two periods apart. In terms of the
scatter plot method, this entails plotting a time-series variable (x_t) versus its second lag (x_{t-2}). Computing the sample correlation between these two variables
gives us the lag 2 sample autocorrelation r_2. Similarly, scatter plots can be made
for data three periods apart, four periods apart, and so on. The corresponding
sample autocorrelation coefficients are r_2, r_3, and so on. In general, we define the
lag k sample autocorrelation r_k as the sample correlation based on observations
taken k periods apart.

Minitab provides a routine that computes r_k for several lags ($k = 1, 2, 3, \ldots$)
and plots the respective sample autocorrelations against k in the form of a bar
graph. This resulting graph is known as the sample *autocorrelation function
(ACF)*. The ACF for the stopwatch data series (reading$_t$) is presented in

Stat>>Time Series>>Autocorrelation

```
MTB > acf 'reading'

Autocorrelation function

ACF of reading

                -1.0 -0.8 -0.6 -0.4 -0.2  0.0  0.2  0.4  0.6  0.8  1.0
                +----+----+----+----+----+----+----+----+----+----+
    1  -0.184                         XXXXXX
    2   0.185                              XXXXXX
    3  -0.070                           XXX
    4   0.065                              XXX
    5  -0.101                          XXXX
    6   0.015                              X
    7   0.165                              XXXXX
    8  -0.168                         XXXXX
    9  -0.104                         XXXX
   10  -0.021                            XX
   11  -0.090                          XXX
   12  -0.097                          XXX
```

Figure 3.8   Autocorrelation function for stopwatch data series.

Figure 3.8. As can be seen, the lag 1 sample autocorrelation is -0.184, the lag 2 sample autocorrelation is 0.185, the lag 3 sample autocorrelation is -0.070, etc.

Since a sample autocorrelation provides a measure of association between observations of a time series, it seems natural to use ACF as a diagnostic tool to check for randomness. The ACF is indeed one of the most important tools for checking for randomness. The following key facts serve as the basis for the development of the ACF as a diagnostic check. If a process is truly random, the true (or theoretical) autocorrelation at any lag k is equal to 0. Of course, even if the process is truly random, the sample autocorrelations will most certainly not equal 0 due to chance variation. It can be shown that sample autocorrelations computed from a random process will behave approximately as normal random variables with mean equal to 0 and standard deviation equal $1 / \sqrt{n}$, where n is the sample size.

Recall that 95 percent of the normal distribution is within 2 standard deviations of the mean. Thus, given randomness, the probability is about 95 percent that the individual sample autocorrelations r_k will fall between $\pm 2(1 / \sqrt{n})$ around 0. A test that rejects the null hypothesis, that the true autocorrelation at lag k equals 0 when r_k falls beyond $-2 / \sqrt{n}$ or $+2 / \sqrt{n}$, is based on a significance level of approximately $\alpha = 0.05$.

For the stopwatch data series, the sample size n is equal to 50, and thus the approximate standard deviation of the sample autocorrelations is $1 / \sqrt{50} = 0.141$. Therefore, at the 5 percent significance level, the randomness hypothesis cannot be rejected if the sample autocorrelations are all within the interval $\pm 2(0.141) = \pm 0.282$ around 0. In Figure 3.8, vertical dashed lines positioned at

± 0.282 have been superimposed on the ACF. Note that Minitab did not place these superimposed lines.[4] The lines can be produced by manual drawing with pen and straightedge or by line-drawing options found in most word processing software. Notice that the lag 1 sample autocorrelation of -0.184 falls well within the cutoff limits. Thus, the slight negative association between reading$_t$ and reading$_{t-1}$ originally seen in the scatter plot (Figure 3.5) cannot be viewed as a significant departure from zero correlation. In fact, all 12 sample autocorrelations reported by Minitab are within 2 standard deviations of 0, providing no suspicion against randomness, which confirms both our initial visual inspection of the data and the conclusion of the runs test.

A word of caution is required when following the guidance of $\pm 2(1 / \sqrt{n})$ limits to *simultaneously* test several r_k values. The problem is that although the probability of a false rejection is 0.05 when one is testing any r_k, the chances are collectively higher that at least one of the several sample autocorrelations tested will fall outside the limits. Thus, if the process is truly random, the probability is greater than 5 percent that the randomness hypothesis is falsely rejected. Background knowledge of the process under investigation along with common sense goes a long way to help minimize the possibility of overreacting. If one sample autocorrelation falls beyond significance limits at some "oddball" lag but all the remaining sample autocorrelations are insignificant, there is probably good reason to resist the temptation to reject randomness.

For example, let us assume hypothetically that only the lag 8 sample autocorrelation for the stopwatch data is significant, and accordingly we decide to reject randomness. By doing so, we are concluding that the data are nonrandom with a persisting pattern in the form of some relationship between observations taken 8 "periods" apart. How reasonable is it to believe that a measurement associated with the experiment of starting and stopping a stopwatch is somehow related to a measurement taken 8 readings prior? Not very. It would be more prudent to reserve judgment and conclude, at least for now, that the so-called pattern is a passing aberration. On the other hand, if sample autocorrelations at lags 1, 2, or possibly 3 are significant, it is not farfetched to interpret the correlations as being due to some carryover related to the experimenter's memory of the most recent measurements. As an example, if the most recent measurement is below the target of 10, there may be a tendency for the experimenter to self-correct on the next trial by lengthening the time between the start and stop actions, or shortening the reaction time if the most recent measurement is above 10. This behavior would result in successive observations that were negatively autocorrelated, most likely giving rise to a significant lag 1 sample autocorrelation deserving of legitimate attention.

Finally, by combining the ACF diagnostic checks with a visual inspection of the data along with a runs test diagnostic, we can safeguard ourselves against overdependence on one method for drawing an appropriate judgment.

---

[4] Minitab has a routine that creates a high-resolution ACF graph. With this routine, Minitab superimposes lines associated with a 5 percent level of significance based on a slightly more refined estimate of the standard deviation of the sample autocorrelations. In addition, Minitab reports a more sophisticated test for randomness known as the *modified Box-Pierce Q test* (for details, see Farnum and Stanton, 1989).

3.2   DIAGNOSTIC CHECKS FOR NORMALITY

In the development of many statistical process control and data analysis methods, some assumptions are usually made regarding the distribution of the variables being analyzed. For many statistical methods, it is assumed that the data in question are well approximated by the normal distribution. In many applications, the normal distribution is indeed a good approximation. However, there is no law of nature that establishes the normal distribution as an appropriate distribution for real data under all circumstances. Hence, some diagnostic checks for normality are essential. In Chapter 2, we introduced the histogram as an important tool for assessing the form of the underlying distribution. In the next subsection, we will suggest that the histogram be constructed after rescaling the data in a particular way.

STANDARDIZED DATA

When one is characterizing the normal distribution, three facts are often recited. Namely, the areas under the normal curve between ± 1, ± 2, and ± 3 standard deviations around the mean are 68.27, 95.45, and 99.73 percent, respectively. Given these facts, a simple check for normality would be to count the number of data observations that fall between ± 1, ± 2, and ± 3 *sample* standard deviations around the *sample* mean relative to the total number of observations. The process of counting the number of observations within these intervals can be greatly simplified by converting the measurement units of the data to a more convenient scale. In particular, the data can be rescaled so that each observation represents the number of sample standard deviations that it deviates from the sample mean.

Conceptually, this rescaling process, known as *data standardization*, is analogous to the conversion of a normal random variable to a standard normal random variable studied in Chapter 2, except that data standardization is done with the sample mean $\bar{x}$ and sample standard deviation s_x rather than the population mean μ and population standard deviation σ. If x_i represents the ith observation in a sample of n observations, the corresponding standardized value y_i is computed as follows:

$$y_i = \frac{x_i - \bar{x}}{s_x} \qquad i = 1, 2, \ldots, n \qquad \textbf{(3.3)}$$

As will be noted in Chapter 5, there are alternative computations of standardized values based on slightly different estimates of the population standard deviation. We will stick with Equation (3.3) mainly because the default of Minitab's built-in command "Center" standardizes data by using this computation.

Suppose that we standardize the stopwatch data using (3.3). Below are the summary statistics for both the original data and the standardized values (labeled "stdread" in Minitab):

```
MTB > histo 'stdread' −3.25  .5

Histogram

Histogram of stdread   N = 50

Midpoint  Count
 −3.250     0
 −2.750     0                                                          −3
 −2.250     1    *
 −1.750     3    ***                                                   −2
 −1.250     3    ***
 −0.750    10    **********                                            −1
 −0.250     8    ********
 −0.250    10    **********
  0.750     7    *******
  1.250     4    ****                                                  +1
  1.750     3    ***
  2.250     1    *                                                     +2
  2.750     0
  3.250     0                                                          +3
```

Figure 3.9   Histogram of standardized observations.

Variable	N	Mean	Median	TrMean	StDev	SEMean
reading	50	9.9890	9.9850	9.9889	0.0918	0.0130
stdread	50	0.000	-0.044	-0.001	1.000	0.141

Variable	Min	Max	Q1	Q3
reading	9.7800	10.2100	9.9200	10.0625
stdread	2.276	2.407	-0.752	0.801

The sample mean of the standardized data is precisely equal to 0, and the sample standard deviation is precisely equal to 1. This will always be true for any standardized data computed from Equation (3.3).

Just how extreme is the maximum value of 10.21? Looking more closely at Minitab's reported summary statistics, we begin to see the convenience of this rescaling. In terms of original units, the answer to the question is not immediate, but on the standardized scale we learn in a flash that this observation is 2.407 standard deviations above the mean.

Let us now see an advantage of having standardized values to assess the compatibility of the data with the normal distribution. In Figure 3.9, we present a low-resolution histogram of the standardized data, created from Minitab with a customized option. Specifically, we requested that Minitab start the midpoint of the lowest interval at −3.25 and make all the class intervals 0.5 wide. By doing so, we will be able to easily count the number of observations within very specific intervals. Technically, the lowest class interval midpoint can be set to any value as long as the fractional part is either at 0.25 or 0.75; for example, −4.25

or -3.75 works equally well. One caution: Make sure the lowest interval is placed far enough out to ensure that the most extreme negative observations are shown.

Notice that we superimposed on the histogram horizontal dashed lines[5] positioned at ± 1, ± 2, and ± 3. Given the design of the class intervals, we know that *precisely* 35 ($= 10 + 8 + 10 + 7$) standardized observations fall within ± 1 around 0, which implies that 35 of the unstandardized observations fall within ± 1 sample standard deviation around their sample mean. Therefore, 70 percent ($= 35/50$) of the data distribution is contained within ± 1 sample standard deviation around the sample mean, which is about what is expected if the underlying process distribution is normal. Below are the comparisons between sample results and what is expected given normality for all three intervals:

Interval Around Mean	Sample	Expected
± 1 standard deviation	$\frac{35}{50}$, or 70%	68.27%
± 2 standard deviations	$\frac{48}{50}$, or 96%	95.45%
± 3 standard deviations	$\frac{50}{50}$, or 100%	99.73%

For each of the three intervals, the observed counts are quite compatible with normality. The counting comparisons provide us with a simple "quick and dirty" diagnostic for normality. To allow the reader to perform the counting diagnostic easily, all histograms presented for the remainder of this book will be constructed in character-graphics mode[6] for standardized observations with the appropriate midpoints and class interval width. We do this also because the standardized scale has the appeal of more easily relating the magnitude of deviations of individual observations from the sample mean, which is especially convenient in the identification of unusual values.

NORMAL PROBABILITY PLOT

When studying a histogram for conformity to normality, we are attempting to assess how well the data distribution matches in shape with a *mentally* visualized theoretical normal curve. Rather than trying to imagine an idealized curve, there is a graph, called a *normal probability plot,* that explicitly compares the data with the theoretical normal distribution.

---

[5] To show the $+3$ line, we had to modify Minitab's output by adding the intervals with midpoints 2.750 and 3.250 which have no observations. Minitab automatically presents positive intervals no farther than the interval encompassing the largest positive observation.

[6] My choice of presenting character-based histograms in this book is based on personal preference. The use of high-resolution histograms is a reasonable alternative that the reader may prefer.

Table 3.2   Values for constructing the normal probability plot

reading$_t$	i	reading$_{(i)}$	$y_{(i)}$	$\dfrac{i - 0.375}{n + 0.25}$	$P(Z \leq y_{(i)})$
10.13	1	9.78	−2.27637	0.012438	0.011412
9.98	2	9.83	−1.73178	0.032338	0.041656
9.92	3	9.84	−1.62287	0.052239	0.052309
9.98	4	9.84	−1.62287	0.072139	0.052309
9.92	5	9.86	−1.40504	0.092040	0.080005
9.93	6	9.88	−1.18720	0.111940	0.117575
9.78	7	9.88	−1.18720	0.131841	0.117575
10.07	8	9.90	−0.96937	0.151741	0.166181
9.84	9	9.91	−0.86045	0.171642	0.194771
10.01	10	9.91	−0.86045	0.191542	0.194771
9.97	11	9.92	−0.75153	0.211443	0.226167
9.97	12	9.92	−0.75153	0.231343	0.226167
9.92	13	9.92	−0.75153	0.251244	0.226167
10.09	14	9.92	−0.75153	0.271144	0.226167
10.09	15	9.93	−0.64261	0.291045	0.260239
9.96	16	9.93	−0.64261	0.310945	0.260239
10.08	17	9.94	−0.53370	0.330846	0.296775
10.01	18	9.95	−0.42478	0.350746	0.335499
9.84	19	9.96	−0.31586	0.370647	0.376054
10.08	20	9.97	−0.20694	0.390547	0.418028
9.91	21	9.97	−0.20694	0.410448	0.418028
10.15	22	9.97	−0.20694	0.430348	0.418028
9.90	23	9.98	−0.09803	0.450249	0.460954
9.93	24	9.98	−0.09803	0.470149	0.460954
9.88	25	9.98	−0.09803	0.490050	0.460954

Let $x_1, x_2, \ldots, x_n$ represent the observations from a random sample of size n. For convenience of presentation, using Equation (3.3), suppose that these observations are converted to standardized values $y_1, y_2, \ldots, y_n$. Let now $y_{(1)}, y_{(2)}, \ldots y_{(n)}$ represent the standardized observations that have been ordered from smallest to largest. For example, $y_{(2)}$ is the second-smallest standardized observation in the data set, and $y_{(n)}$ is the largest standardized observation in the data set. It can be shown (see Blom, 1958) that $y_{(i)}$ serves a reasonable estimate of the point on the underlying unknown distribution such that the probability of being to the left of this point is equal to

$$\frac{i - 0.375}{n + 0.25} \tag{3.4}$$

Table 3.2 Values for constructing the normal probability plot (concluded)

reading$_t$	i	reading$_{(i)}$	$y_{(i)}$	$\dfrac{i - 0.375}{n + 0.25}$	$P(Z \leq y_{(i)})$
9.94	26	9.99	0.01089	0.509950	0.504344
9.98	27	10.00	0.11981	0.529851	0.547683
10.00	28	10.01	0.22873	0.549751	0.590460
9.92	29	10.01	0.22873	0.569652	0.590460
10.02	30	10.01	0.22873	0.589552	0.590460
10.09	31	10.01	0.22873	0.609453	0.590460
9.99	32	10.02	0.33765	0.629353	0.632186
10.05	33	10.02	0.33765	0.649254	0.632186
10.01	34	10.03	0.44656	0.669154	0.672403
10.03	35	10.03	0.44656	0.689055	0.672403
9.91	36	10.05	0.66440	0.708955	0.746782
10.01	37	10.05	0.66440	0.728856	0.746782
10.07	38	10.06	0.77332	0.748756	0.780333
10.06	39	10.07	0.88223	0.768657	0.811173
10.05	40	10.07	0.88223	0.788557	0.811173
10.21	41	10.08	0.99115	0.808458	0.839193
9.95	42	10.08	0.99115	0.828358	0.839193
9.86	43	10.09	1.10007	0.848259	0.864348
10.03	44	10.09	1.10007	0.868159	0.864348
10.02	45	10.09	1.10007	0.888060	0.864348
10.10	46	10.10	1.20899	0.907960	0.886666
9.88	47	10.13	1.53573	0.927861	0.937698
10.13	48	10.13	1.53573	0.947761	0.937698
9.83	49	10.15	1.75356	0.967662	0.960247
9.97	50	10.21	2.40707	0.987562	0.991959

In other words, $y_{(i)}$ is an estimate of the point for the underlying unknown distribution such that the *cumulative* probability to the left of this point equals Equation (3.4).

The idea now is to compute the theoretical cumulative probability to the left of $y_{(i)}$, given the assumption of normality. If the data came from a normal distribution, we would expect, for a given value of $y_{(i)}$, the cumulative probability given by (3.4) and the theoretical cumulative probability assuming normality to be nearly equal. Let us demonstrate the comparison with the stopwatch data. In Table 3.2, we show in the first column the observations in original units and in original order. The third column has the observations arranged in ascending order. The index of the rank order i is given in the second column. In the fourth

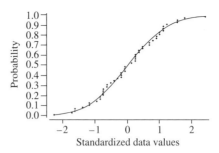

Figure 3.10   S-shaped normal probability plot.

column, we show the ordered observations converted to standardized values $y_{(i)}$. Based on the index of the rank order and the sample size ($n = 50$), the cumulative probabilities derived from (3.4) are given. For example, if $i = 1$, then the associated cumulative probability is computed to be

$$\frac{1 - 0.375}{50 + 0.25} = 0.012438$$

In the fifth column of Table 3.2, we give all the cumulative probabilities derived from (3.4). Finally, the sixth column of Table 3.2 gives the associated cumulative probabilities $y_{(i)}$ assuming the data arise from the normal distribution. Since the data are in standardized units, we calculate the cumulative probability from the standard normal distribution. Consider, for example, the first ordered value, which is $y_{(1)} = -2.27637$. Using either Minitab's CDF command or the normal probability tables, we can find that the cumulative probability associated with this value is P($Z \leq -2.27637$) = 0.011412.

Comparing columns five and six of Table 3.2, we see that the values in any given row appear to be close. To aid the comparison, we can simultaneously plot the cumulative probabilities in column five versus $y_{(i)}$ and the normal-based cumulative probabilities in column six versus $y_{(i)}$ (see Figure 3.10). Notice that the normal-based cumulative probabilities sketch out a perfectly smooth S-shaped curve. The cumulative probabilities derived from the rank order plot fairly consistently along the smooth benchmark. We should resist the temptation to react to slight bumps or gentle undulations about the curve. These deviations can simply be explained away as sampling variation. Thus, the plot (which we can call a normal probability plot) suggests that the underlying distribution generating the data can be safely assumed to be the normal distribution.

The normal probability plot we just created requires us to compare a scattering of points relative to an S-shaped curve. However, for most of us, it is much easier to judge how closely a plot agrees with a *straight* line. By adjusting the distances between hash marks on the vertical axis, the theoretical S-shaped curve can be mapped into a perfectly straight line. Thus, if the plotted points lie nearly along a straight line, we can conclude the data are compatible with the normal distribution. However, if the plotted points deviate systematically from a straight line, nonnormality is indicated. In Minitab, the normal probability plot is auto-

Graph>>Probability Plot

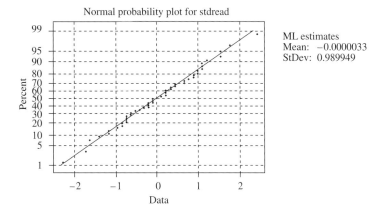

Figure 3.11   Straight-line-based normal probability plot.

matically created with the appropriate vertical scaling to allow for comparison with a straight line. In Figure 3.11, we give the normal probability plot for the stopwatch data as created by Minitab.[7] If you compare the normal probability plots of Figures 3.10 and 3.11, you should find that relative positions of plotted points versus the S-shaped curve are identical to the relative positions of plotted points versus the straight line. Thus, we again can conclude the data are compatible with the normal distribution because the plotted points of Figure 3.11 are fairly consistent with the straight-line benchmark.

3.3   DEALING WITH NONNORMAL DATA

Often data do not conform to the standard assumption of normality. Consequently, there is a danger of misleading conclusions resulting from the interpretation of statistical methodology based on this standard assumption. For example, a substantial departure from normality can throw off the proper interpretation of a statistical test. We might think we are rejecting a null hypothesis at the level of significance of 0.05 when in fact the true p value is larger than 0.05. Generally, making judgments based on the normal model in the presence of nonnormality increases the risks of deeming an unacceptable process as acceptable or deeming an acceptable process as unacceptable.

   If process data reflect substantial nonnormality, it may be useful to investigate the process to see if there are any nonstatistical explanations for the nonnormal distribution pattern. A histogram that reflects two distinct humps—modes—is not likely to be a natural distribution pattern. Rather, the unusual distribution pattern is more likely to have resulted from a mixture of two underlying distributions. In

[7] Minitab offers two different ways to create a normal probability plot: (Graph >> Probability Plot) or (Stat >> Basic Statistics >> Normality Test).

common manufacturing situations this occurs when data are obtained from the combined output of presumably identical machines which in reality produce output with different mean levels. In such a case, the machines need to be calibrated, or another possibility is to perform separate statistical analyses for the output from each machine.

Most often nonnormality is inherent in the process and not the result of some mixture of different sources. Bissell (1994, p. 257) indicates that there are a number of well-known manufacturing processes that give rise to intrinsically nonnormal distributions, which include measures of distortion (eccentricity, warping, surface finish), electrical phenomena (capacitance, insulation resistance), low levels of substances in materials (trace elements, impurities), and other physical properties (maximum strengths, time to failure). Nonnormality is not restricted to manufacturing processes. Customer waiting times for service, daily mileage logged by a salesperson, and times for a computer system to recover after a system crash are a few measures typically found to be nonnormal.

When nonnormality is inherent in a process, there are several statistics-based actions that might be entertained. One strategy is to base the analysis on a more suitable theoretical distribution. For instance, distributions such as the exponential, Weibull, and gamma have been developed to explain many phenomena such as the waiting time for the arrival of the next customer, the life span of a single component or of a complex system, and the life span of a mechanism in which eventual failure is the result of accumulated damage. Determining which of these distributions is the most appropriate for a specific application is beyond the scope of this book. For interested readers, a practical but comprehensive coverage of this subject can be found in Hahn and Shapiro (1967).

Another possibility is to arrange the data collection such that averages instead of individual observations are dealt with. The idea is to tap into the central limit theorem effect that assures us that sample averages are approximately normal regardless of the underlying distribution if the sample size is large enough. If averages are already being used, then it may be advisable to increase the sample size to improve the approximation to normality.

It may be possible to consider the use of other statistical methods, known as *nonparametric methods*, that do not rely on the assumption of normality or, for that matter, on any distribution. The runs test we encountered earlier is actually an example of a nonparametric method: The only assumption under the null hypothesis is that the observations are randomly generated from some continuous distribution whose form need not be specified. The statistical literature of nonparametric methods has grown rapidly, but there have been limited developments in the SPC area. For interested readers, Gibbons (1990) provides an excellent overview of nonparametric methods along with an extensive reference list. The limited application of nonparametric methods in SPC is primarily due to the fact that these methods are typically robust or insensitive to extreme points such as outliers. Hence, nonparametric methods will tend to repel rather than highlight unusual observations. For general statistical applications, robustness is appealing since the statistical summarization of the majority of the data is not unduly "thrown" off by a few unrepresentative observations. But, in SPC applications, statistical methods need to react to the presence of unusual observations since such observations can be important signals for needed improvement or corrective actions.

NORMALIZING TRANSFORMATIONS

A popular data analysis strategy for dealing with nonnormality is to transform the data to achieve approximate normality. The idea is then to analyze the transformed data by using the wide variety of standard statistical and SPC techniques developed under the assumption of normality. A very common approach to "normalizing" a data set is to take all the data values and raise each of them to some power λ, that is, x^λ. Within this class of transformations, known as *power transformations*, the square root ($\sqrt{x}$, or $x^{1/2}$), the cube root ($\sqrt[3]{x}$, or $x^{1/3}$), and the inverse ($1/x$, or x^{-1}) are among the most frequently used transformations to yield approximate normality. However, there is one very special and useful transformation that is defined within the class of power transformation that does not involve raising the data to a power per se! In particular, the logarithmic transformation corresponds to the power of $\lambda = 0$; for an explanation, see Tukey (1977). Logarithms can be taken to any base; however, the most common practice is to choose e as the base, where $e = 2.71828 \ldots$. Logarithm to base e is known as the *natural logarithm*, denoted by ln.

The process of determining the most appropriate power of λ for the given data is typically a process of trial and error. This process can be made a bit more systematic by recognizing how changing the value of λ up or down might impact the data distribution. Since raising a number to the power 1 gives the same number, doing nothing to the data corresponds to a power transformation with $\lambda = 1$; this would be most appropriate if the data conform well to the normal distribution and there is no need to transform. As we move away from $\lambda = 1$, there will be an impact from the power transformation. It turns out that values of $\lambda < 1$ correspond to power transformations useful for normalizing positively (or right) skewed data while values of $\lambda > 1$ correspond to power transformations useful for normalizing negatively (left) skewed data. The more extreme the degree of skewness, the more extreme the value of λ chosen relative to the pivotal value of 1.

A number of procedures have been developed to reduce the number of iterations of decreasing and increasing the value of λ until the resulting observations appear close to the normal distribution. One of the more popular procedures is the Box-Cox procedure. Based on the method of maximum likelihood estimation (see Box and Cox, 1964), the Box-Cox procedure automatically identifies a possible choice for λ. The full-blown version of Minitab, as opposed to the Student Edition of Minitab, includes the Box-Cox procedure as an option.

If software is not available to implement the Box-Cox method, Hines and Hines (1987) developed a graphical method for determining directly a reasonable value of λ. Their method requires the computation of the sample median and certain sample percentiles. Recall that to find the sample median, we need to order

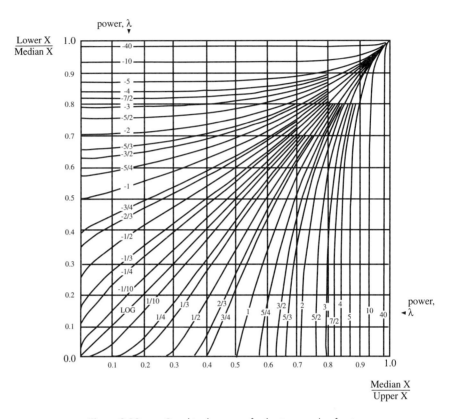

Figure 3.12   Graphical means of selecting a value for λ.

the observations from smallest to largest and find a middle value such that at least 50 percent of the observations are less than or equal to it and at least 50 percent are greater than or equal to it. A sample percentile is a value such that a certain percentage of the observations fall below it. For example, the sample median is the 50th percentile of the sample.

We need, in addition to the sample median, two percentiles symmetrically positioned above and below the sample median. For example, we might consider the 10th and 90th percentiles or the 25th and 75th percentiles. These pairs are called the *lower* and *upper quartiles*. If a sample distribution is roughly symmetric, then the difference between the sample median and the lower quartile will be roughly equal to the difference between the sample median and the upper quartile. So the method of Hines and Hines (1987) is to find a value of λ such that these two differences are nearly equal, which will tend to make the data distribution symmetric and more closely compatible with the normal distribution. Using Figure 3.12 (kindly supplied by W. G. S. Hines and R. J. O'Hara Hines), one plots the ratio of the lower quartile to the sample median on the vertical axis against the ratio of the sample median to the upper quartile. The suggested value of λ is read off the closest curve.

```
Descriptive Statistics

Variable          N       Mean     Median    TrMean     StDev    SE Mean
calorie          60       1832       1562      1728       852        110

Variable    Minimum    Maximum        Q1        Q3
calorie        1020       5640      1292      2014
```

Figure 3.13   Summary statistics for daily caloric intake data.

To illustrate the normalization of data, consider a personal process observed by a former executive masters of business administration student. This student was interested in studying her daily weight measurements along with her daily caloric intake. Using guidance from food labeling and a nutrition fact book, she was able to measure quite accurately the total daily calories down to multiples of 5 calories. In Table 3.3, we provide the student's calorie measurements for 60 consecutive days. The data are stored in the file "calories.dat." Since these data constitute a time series, the first step should be to analyze the time behavior of the data. We do not perform the random checks (time-series plot, runs test, and ACF) here but encourage the reader to do so. If done, the random diagnostics will show no evidence of nonrandom behavior.

Table 3.3   Daily caloric intake

1240	1740	1020	1620	1175	2015	1685	1765	2545	1410	1260
3170	1294	2660	1630	1480	1860	1265	1700	1292	1565	1725
1450	2845	3250	1730	1460	1100	1100	1730	1460	1145	2140
1140	1850	2010	1330	2190	1495	5640	1360	1105	1560	1100
1590	2550	4500	3070	1375	2685	1350	1230	1215	1340	1515
1695	1225	1365	2810	3075						

Given the data are random, we can appropriately proceed with the distributional checks of the data. Before we do so, notice from Figure 3.13 that the sample mean is 1832 and the sample median is 1562. The fact that the sample mean appears to be substantially pulled above the sample median is a hint that data distribution is positively skewed; this phenomenon was discussed in Section 2.5 and illustrated by Figure 2.16. The histogram of the standardized data along with a normal probability plot is shown in Figure 3.14. Without question, both plots reflect a data distribution that is highly skewed. It would be greatly misleading to base any sort of judgment on the overall process or individual observations using the guidance of the normal distribution. As an example, we will learn that standard SPC methods deem observations beyond ± 3 standard deviations around the mean as signals of unrepresentative events requiring special investigation. This symmetric rule, known as the "3-sigma" or 3σ rule, is supported by an assumption

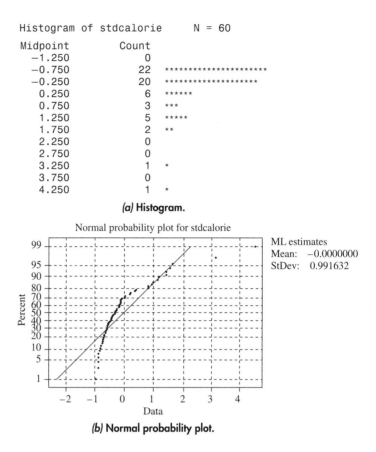

```
Histogram of stdcalorie     N = 60
Midpoint              Count
  -1.250                0
  -0.750               22    **********************
  -0.250               20    ********************
   0.250                6    ******
   0.750                3    ***
   1.250                5    *****
   1.750                2    **
   2.250                0
   2.750                0
   3.250                1    *
   3.750                0
   4.250                1    *
```

(a) **Histogram.**

Normal probability plot for stdcalorie

ML estimates
Mean: −0.0000000
StDev: 0.991632

(b) **Normal probability plot.**

Figure 3.14   Normality checks applied to standardized calories.

that the data approximately follow the symmetric normal distribution. Notice from Figure 3.14*a* that if we applied the 3σ rule to the calorie data, two observations (both on the high end) would be labeled as unusual points. The difficulty with accepting such a conclusion is that these two observations are not sitting outside a data distribution that can be generally regarded as consistent with the normal distribution. But rather, these two points seem to be *consistent* with the overall skewness exhibited by the whole data set.

Let us now consider a power transformation to normalize the data. Given that the data are positively skewed, the most suitable value of λ would be less than 1. From Figure 3.13, the median caloric intake is 1562, the 25th percentile (Q1) is 1292, and 75th percentile (Q3) is 2014. Consistent with a nonsymmetric distribution, the median is not halfway between the lower and upper quartiles. If we plot $1292/1562 = 0.83$ on the vertical axis and $1562/2014 = 0.78$ on the horizontal axis of Figure 3.12, we find the plotted point to be closest to the $\lambda = -3/2$ line. Hence, the suggested transformation is to raise all the data to the -1.5 power, that is, $x^{-1.5}$ or $1/x^{1.5}$.

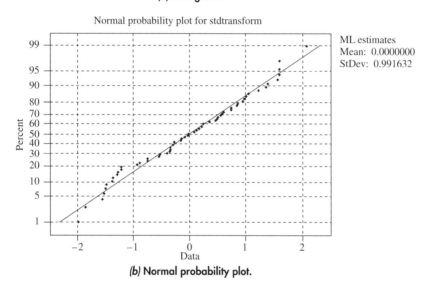

```
Histogram of stdtransform        N = 60
Midpoint            Count
  -2.250               0
  -1.750               4    ****
  -1.250               8    ********
  -0.750               6    ******
  -0.250              12    ************
   0.250               9    *********
   0.750              11    ***********
   1.250               5    *****
   1.750               4    ****
   2.250               1    *
```

(a) Histogram.

(b) Normal probability plot.

Figure 3.15   Normality checks applied to standardized transformed calories.

In Minitab, if column 1 (c1) contains the original data and if one wishes to put the transformed data in another column, say c2, the transformation is easily carried out by typing the following command: let $c2 = c1^{**}(-1.5)$. Where the double asterisk stands for "raised to the power of." To check whether the transformed data resemble the normal distribution, we do the usual first step of standardizing the data and then proceeding with normality checks.

In Figure 3.15, we present the histogram and normal probability plot of the transformed values on a standardized scale. As can be seen, the effect of the transformation is remarkable. Unlike the original observations, the transformed values are quite compatible with the normal distribution. Now the normal distribution can provide proper guidance in the analysis of the data. For instance, the 3σ rule can be safely applied, and it tells us that there are no unusual observations warranting special explanation. Given the two "seemingly" unusual observations on the original scale, one might feel that the transformation has somehow

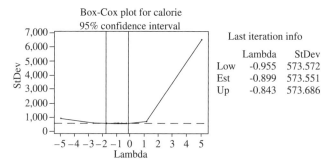

Stat>>Control Charts>>Box-Cox Transformation

Figure 3.16   Box-Cox plot for daily caloric intake data.

camouflaged these values. This is not the case. Normalizing transformations do not intrinsically change the relative positioning of the data values. What they do is to reexpress the data, while preserving the rank order, to a scale that allows the normal distribution to serve as a benchmark for interpretation and judgment. In our example, the transformation enables us to conclude that the two high values are not unusual relative to the underlying skewed distribution since the corresponding transformed values are not unusual relative to the normal distribution.

The alternative approach to determining a value for λ is to invoke the Box-Cox method. In Figure 3.16, we show the computer output resulting from implementing the procedure by Minitab. Notice that the point estimate for λ is -0.899. In addition, Minitab computes a 95 percent confidence interval for the true value of λ. Notice that this interval includes the value of -1.5 which was arrived upon using the method of Hines and Hines (1987). In an end-of-chapter exercise, the reader is asked to evaluate transforming the data based on $\lambda = -0.899$ versus $\lambda = -1.5$.

Implicitly, in the discussions of the power transformations, we have assumed that the data set to be transformed contains all positive values. The difficulty with nonpositive values is that some transformations cannot be performed on such values; for example, the square root of a negative number and the logarithm of 0 or a negative number are undefined. When nonpositive numbers are present, a positive number that is at least larger in magnitude than the minimum value in the data set can be added prior to taking a transformation. Finally, it should be pointed out that not every data set can be made to conform well to the normal distribution through data transformations. If the smallest value in the data set is also the most commonly occurring value, it is nearly impossible to transform such data to a normal distribution. We should emphasize that the problem occurs when the smallest individual value is the most frequent and not necessarily when the leftmost class interval of a histogram has the most observations. As long as the most frequently occurring individual observation is in the interior of the data distribution, the normalizing strategies discussed in this section should produce a data distribution that appears approximately normal.

3.4 CONCEPTUAL FRAMEWORK FOR PROCESSES

One of the primary goals in the study of any process is prediction. Prediction is fundamental for management planning; for example, prediction of future sales helps in the planning of inventory and procurement levels. Prediction also provides management with the opportunity to recognize surprises (good or bad) inconsistent with predictions. Such surprises are opportunities for discovery of underlying events that have impinged on the process. The information gained from the study of these events can be used to maintain or improve overall process performance. Finally, prediction can be used to assess whether a purposeful management intervention to improve the process has succeeded. Without a baseline of expected process performance, there is no way to determine if a new method or procedure really has had any impact on process performance. To develop methods for the prediction of a process, it is useful to establish a general model framework for processes. In the subsections below, we begin our development by considering the simplest of process behaviors, namely, a random process (i.e., an iid process).

RANDOM PROCESS MODEL

Recall that a random process is a process with constant underlying mean and variance levels. Consider, for a moment, an idealized situation in which there is no variation. In such a situation, all the process observations are equal to the mean, say μ. However, in reality, observations do not fall exactly on the true mean. They get randomly "bumped" up and down away from the true mean.

Conceptually, data generated by random process can be represented by the following equation:

$$\text{Observed data value} = \text{Mean level} + \text{Random deviation} \qquad \textbf{(3.5)}$$

The random deviation component describes the random phenomenon that might push what is actually observed above the mean level or below it. Therefore, random deviations can be positive or negative. When a random deviation is positive, the associated observation will reveal itself above the mean; and when a random deviation is negative, the associated observation will reveal itself below the mean.

Even though a particular random deviation may be positive or negative, it is reasonable to assume that the average of a given set of random deviations should be close to zero; theoretically, this average equals zero exactly for a very large (infinite) number of deviations. In the development of statistical modeling techniques, it is quite common to assume that random deviations follow the normal distribution. The rationale for this assumption is based on an informal translation of the central limit theorem (see Section 2.9). Namely, the central limit theorem suggests that observed measurements that arise from the cumulative effect of many small but unidentified independent sources should closely resemble the normal distribution. This cumulative notion is indeed the standard interpretation of random deviations. That is, random deviations reflect the effects of many independent sources where no one of these sources is dominant; the nearly un-avoidable phenomenon of measurement error is a good example of one possible source. Given this interpretation, the normal distribution is a good assumption to start with, but not something we should take for granted.

For a more compact presentation, the above discussion can be formulated by using a convenient notation. If we denote the random deviation at time t by ε_t, the general model for a random process X_t is given by

$$X_t = \mu + \varepsilon_t \quad t = 1, 2, \dots \qquad (3.6)$$

where μ is the underlying mean level. The random deviations ε_t are also com-monly referred to as *random errors* or *disturbances*. The term *error* is not meant to imply a mistake or blunder; instead, it is a standard statistical term to represent a deviation of the observation from the underlying level of the process.

For the purposes of the development and interpretation of standard data analy-sis techniques, the following assumptions about random errors are necessary:

1. The random errors are random or independent. Since the random errors oc-cur sequentially over time, this assumption is equivalent to saying that ran-dom errors are generated from a random process.

2. The random errors have an expected value (or mean) of 0

$$E(\varepsilon_t) = 0$$

where $E(\cdot)$ denotes the mean or expected value of a random variable (refer to Section 2.6).

3. The random errors have a common variance of σ_ε^2. This specification is of-ten called the *assumption of constant dispersion*. The idea is that the vertical scatter of points about the systematic pattern tends to be the same every-where.

4. The random errors are normally distributed random variables. Using the no-tation first introduced in Section 2.8, this specification along with assump-tions 2 and 3 above implies that $\varepsilon_t \sim N(0, \sigma_\varepsilon^2)$.

The above assumptions define an ideal scenario for standard statistical mod-eling techniques. One should never presume that these assumptions will be auto-matically met in all applications. Instead, we need to be on guard for substantial violations from the standard assumptions and prepared to rectify the situation ac-cordingly. As an example, a particular application might suggest the underlying distribution for the random errors to be nonnormal (i.e., a violation to assump-tion 4). For such a case, the remedy might be to simply reexpress the data by us-ing one of the normalizing transformations discussed in Section 3.3.

GENERAL PROCESS MODEL

In the previous subsection, we developed a general model for a random process. However, in practice, processes often exhibit nonrandom patterns. These pat-

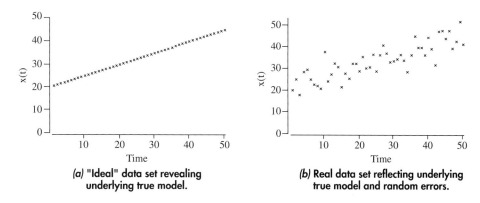

(a) "Ideal" data set revealing underlying true model.

(b) Real data set reflecting underlying true model and random errors.

Figure 3.17   Effects of random errors.

terns are typically persisting, and they are not merely explained away as passing phenomena due to some isolated special events. We began our development of the general model for a random process by considering an idealized situation of no variation around the underlying mean level. Similarly, modeling nonrandom processes would be rather straightforward if the observations emerged exactly in some crystal-clear pattern. For example, if a process followed a nonrandom pattern in the form of a straight-line upward trend, it would be wishful thinking to expect the data to appear as plotted in Figure 3.17a. For this series, modeling the data would be ridiculously easy. We can pick off the intercept ($= 20$) from the graph and compute the slope of the trend line from any two points, and we are done. If all processes generated data in this fashion, there would be no work for statisticians!

   More realistically, a trending process might reveal itself in the manner displayed in Figure 3.17b. The reason again is that all processes are subject to some level of random variation, which in this case randomly bumped some observations above the underlying trend and others below the trend. Random variation "muddies" our ability to see the ideal picture, shown in Figure 3.17a. In general, we will never see the idealized systematic pattern persisting in a process.

   To generalize our earlier development of a model for a random process to include a process with underlying systematic patterns, we can refer to the underlying component which the observations randomly deviate around as the "true model." In the case of a random process, the true model is simply the mean parameter μ. As we will explain in Section 3.6, the true model that is underlying a simple trend can be written as $\beta_0 + \beta_1 t$, where t is the time index ($t = 1, 2, \ldots$), β_0 is an intercept parameter, and β_1 is the slope parameter. Thus, referring to Equation (3.6), we can write the general process model as

$$X_t = \text{true model} + \varepsilon_t \qquad t = 1, 2, \ldots \qquad (3.7)$$

where ε_t is the random error variable based on the same assumptions detailed in the previous subsection.

3.5   MODEL FITTING BASED ON LEAST-SQUARES ESTIMATION

We have learned that observations from a process can be broken down or decomposed into two parts; one part is the contributions from the true model, and the other part is the contributions from random variation. In practice, we witness the combination of these two effects, never knowing the exact magnitude of the contribution of either part. Clearly, to make useful predictions of future process outcomes, the challenge rests greatly on our ability to understand and estimate the true model component.

When estimating the true model component of a process, we aim to summarize it by an equation or model. An equation that represents an estimate of the true model behavior is commonly referred to as the *estimated model* or *fitted model*. Since the true model cannot account for all the variation in the data observations, neither can we expect the fitted model to account for all the data variation.

The question is, How do we determine an appropriate fitted model for a given set of data? To answer this question, let us consider the stopwatch data series which we have demonstrated to be consistent with a random process. Figure 3.18 offers two possible fitted models for the data series. Clearly, the fitted line shown in Figure 3.18*a* is not a good choice, given its placement well above the data. In contrast, the fitted line shown in Figure 3.18*b* is a more reasonable choice, given that it is nestled "inside" the scatter of points. In general, there are an infinite number of fitted lines that can be superimposed on the data series. The challenge is to select the line that "best" describes the data.

An informal approach to model fitting is to simply draw a direct freehand sketch of a plausible fitted line. For simple nonrandom patterns such as trends, one can usually approximate such behaviors quite well by means of "eyeball" fitting. There is a lot to be said about informal model fitting. Since it is unlikely that all individuals in any given organization can be trained to routinely use more formal statistical methods, one possible remedy is to encourage some users to plot time-series plots (also known as run charts) and to train them in direct visual analysis of time-series plots. Even if the aim is to develop control charts (time-series plots with superimposed "warning" limits) later, such preliminary visual analysis of time-series plots can be very helpful. It can bring out the importance of interpreting time-series plots for basic understanding of specific processes under surveillance.

The most obvious drawback of informal model fitting is the guesswork. To avoid subjective judgment in establishing the fitted line, we need a formal criterion that defines a "best"-fitting line. If we look again at the different fitted lines in Figure 3.18, we can get some clues on how to define a useful criterion. We ask, "What makes the fitted line in panel *a* such a poor choice relative to the fitted line in panel *b*?" Clearly, it is because the first fitted line *deviates* greatly from all the observations relative to the second fitted line. In general, a *residual* defines the deviation between an observation and a fitted model:

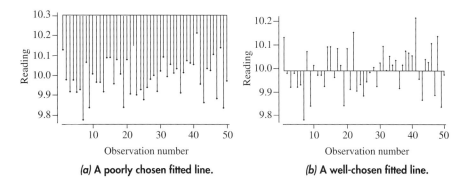

(a) **A poorly chosen fitted line.**   (b) **A well-chosen fitted line.**

Figure 3.18   Different fitted lines for the stopwatch data series.

$$\text{Residual} = \text{Observed data value} - \text{Fitted model} \qquad \textbf{(3.8)}$$

Previously, we denoted the observed value at time t by x_t. For more convenient transition into regression modeling (to be presented in the next section), we denote the observed value at time t by y_t. Accordingly, let $\hat{y}_t$ be the fitted model value at time t. Finally, we denote the residual value at time t by e_t. With this notation, Equation (3.8) can be rewritten as

$$e_t = y_t - \hat{y}_t \qquad \textbf{(3.9)}$$

Given the respective fitted models, the residuals are visually represented in Figure 3.18 by the vertical line segments. Intuitively, if the fitted line is well placed, the residual deviations will be minimized, not just for one point but collectively for all the points in the plot. An immediate and natural reaction might be to find a fitted line that minimizes the sum of residuals, that is, $\sum_{i=1}^{n} e_i$, where n is the number of data points. The difficulty with this criterion is that residuals can be positive or negative; a positive residual occurs when the fitted line is below the actual observation, and a negative residual occurs when the fitted line is above the actual observation. With this in mind, the sum of residuals in Figure 3.18a will be less than the sum of residuals in Figure 3.18b, implying the fitted line in the left graph is better than the fitted line in the right graph!

What we need is a criterion that makes the residuals small, regardless of whether they are positive or negative. One possibility is to find the fitted line that minimizes the sum of absolute values of the residuals, $\sum_{i=1}^{n} |e_i|$. Even though this criterion is quite sensible and could be adopted, it is not used routinely in practice. The primary reason is that the computation procedure for finding a fitted line satisfying this criterion is relatively complex. Another problem is that minimizing the sum of absolute residuals, in some cases, might not lead to a unique fitted line; that is, more than one fitted line might meet this criterion. For these reasons, very few statistical software packages offer model estimation procedures based on this criterion.

An alternative criterion that avoids the problem of plus and minus residuals is to define the best-fit model as the model for which the sum of squared residuals is minimized, that is, $\sum_{i=1}^{n} e_i^2$ is minimized. This procedure is called the *method of least squares.* Minimizing the sum of squared residuals is by far the most widely used guiding principle for statistical model fitting. The sum of squared residuals is typically denoted by the acronym SSE, which stands for *sum of squared errors.* The use of the word *errors* in SSE can be misleading because it could suggest that we are looking at the sum of squared random errors, which is not the case. In general, we have

$$\text{SSE} = \sum_{i=1}^{n}(y_i - \hat{y}_i)^2 = \sum_{i=1}^{n} e_i^2 \tag{3.10}$$

In the case of fitting a model to random data, we simply want to fit a horizontal line at some value b, that is, $\hat{y}_i = b$. The method of least squares requires us to find the value of b that minimizes

$$\text{SSE} = \sum_{i=1}^{n}(y_i - b)^2 \tag{3.11}$$

The minimum of SSE can be found analytically by using the techniques of calculus. In particular, we need to take the derivative of SSE with respect to b, set the derivative equal to zero, and then solve for b. When done,[8] we would find that that $b = \bar{y}$; that is, the fitted model ($\hat{y}_i$) is simply the sample mean ($\bar{y}$). Thus, for the stopwatch data series, the fitted model is a horizontal line positioned at $\bar{y} = 9.989$ (refer to Figure 2.14 to find the sample mean of the data). Prior to this formal result, we have used the sample mean, based on intuitive arguments, as an appropriate guess for future observations from a random process. We now have the formal justification to do so.

The method of least squares is a universal approach that can be used to fit processes with systematic patterns. We turn to such applications in subsequent sections.

3.6   MODELING A SIMPLE LINEAR TREND PROCESS

In this section, we consider the modeling of a simple departure from random behavior that is called a *trend*. In our earlier discussions of nonrandom behavior, we displayed a time series in Figure 3.17*b* which reflected an apparent trend. Visual analysis of this plot suggested a tendency for the process level to systematically shift up as time passes. Furthermore, the rate of increase appears to be relatively constant. These two properties characterize a process which is following a *simple linear trend.* As we will see in a number of applications, one may encounter trend behavior that is curving rather linear.

---

[8]For readers with a background in calculus, we leave the details of the derivation as an end-of-chapter exercise.

In practice, trends are commonplace and of great practical importance. Typically, company sales and, more generally, many economic variables have underlying trends. In a variety of machining operations, it is also common to witness dimensions of machined parts reflecting a steady trend due to tool wear, aging machinery, or the cumulative effect of the lack of maintenance. Continual quality improvement efforts often impact processes in the form of a trend, for example, steadily decreasing defect levels. Roberts (1998) illustrates that trend models nicely track personal fitness processes where the measurements are consecutive lap times or split times associated with working out at some constant effort level. For these processes, the trend effect reflects slowing of performance due to accumulated fatigue. We now turn to an application driven by simple trend behavior.

In Section 2.4, we discussed an application taken from Deming (1986) in which the data set displayed nonrandom behavior. The example was originally introduced to discuss the misleading consequences of constructing a histogram in the presence of such behavior. Let us take the application one step further by seeking a statistical model that will account for the systematic behavior evidenced in the data.

Table 3.4   Elongation measurements

0.0014541	0.0014682	0.0013200	0.0013129	0.0012071	0.0013341
0.0012212	0.0014541	0.0012000	0.0012141	0.0013341	0.0010729
0.0012071	0.0011012	0.0011012	0.0009741	0.0012000	0.0012000
0.0010729	0.0012000	0.0012071	0.0012141	0.0010871	0.0012212
0.0013341	0.0010941	0.0009741	0.0011153	0.0009741	0.0011012
0.0011082	0.0009953	0.0009882	0.0009812	0.0009741	0.0010941
0.0009812	0.0008612	0.0010871	0.0009812	0.0008541	0.0011012
0.0008400	0.0008541	0.0007200	0.0009741	0.0008541	0.0009741
0.0008471	0.0006212				

Recall, the observations are 50 consecutive measurements (from successive production runs) of the elongation of a camera spring made under a pull of 20 grams (g). These values[9] are provided in Table 3.4 and can be found in the file "deming.dat." In Figure 3.19, we show the time-series plot of the data again but produced from Minitab (see Figure 2.11 for the original plot reproduced from Deming, 1986). The downward trend is clearly visible. Underlying the random variation, the process level appears to decrease systematically in a straight-line fashion. Given our visual impression, it is reasonable to tentatively identify the underlying model as a simple linear trend.

---

[9]Deming (1986) provides only a time-sequence plot of the observations and does not explicitly give the measurement values. The numerical readings were estimated through the use of a very finely scaled measuring instrument applied to the time-sequence plot.

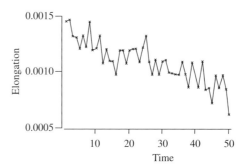

Figure 3.19   Time-series plot of elongation measurements.

At this stage, it is helpful to know that the mathematical equation for a straight line has the general form $Y = \beta_0 + \beta_1 X$, where Y is the variable associated with the vertical axis and X is the variable associated with the horizontal axis. The constant β_0 represents the value of Y when $X = 0$ (also known as the *intercept*), and the constant β_1 represents the slope of the line. The *slope* of a line refers to the rate at which the line rises or falls, that is, the change in Y for a given change in X. Such an equation represents a perfect relationship between Y and X. In practice, even if Y and X were related in this manner, (3.7) suggests that each observation y_i for variable Y, given a value x_i for variable X, arises from the following scheme:

$$Y_i = \beta_0 + \beta_1 x_i + \varepsilon_i \tag{3.12}$$

where ε_i is a random error term. The above relationship is known as a *simple regression model,* where the term *simple* signifies that the value of the dependent variable Y is predicted from the value of one and only one independent variable X. When two or more independent variables are used in the equation to predict Y, the relationship is called a *multiple regression model.* For now, we will focus our attention on the simple regression model, but we will quickly move into applications requiring a multiple regression framework.

Let us return to our study of the elongation observations plotted in Figure 3.19. As with any time-series plot, the elongation measurement values are plotted vertically while the horizontal axis represents time order by the numbers 1, 2, 3, ... Rather than produce, as we did, the plot found in Figure 3.19 by using Minitab's automatic time-series routines, consider an alternative route. Namely, imagine that one column in Minitab contains the elongation observations in order while another column contains the sequence of numbers 1, 2, . . . , 50. These two columns represent two variables: the elongation measurements (elongation$_t$) and a time index t. Using Minitab's Y-versus-X plot command, we can plot directly (elongation$_t$) on the vertical axis against t on the horizontal axis, as shown in Figure 3.20.

Recall that a plot of a variable measured along the vertical axis versus a variable measured along the horizontal axis is called a scatter plot. A scatter plot is a basic but important graphical tool that enables us to visualize the nature of the relationship between two variables. Notice that the scatter plot in Figure 3.20 and

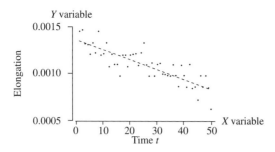

Figure 3.20   Scatter plot of elongation measurements versus time index.

the time-series plot in Figure 3.19 are essentially identical except the points are not connected in the scatter plot. Even though both plots provide the same information, let us focus for now on the scatter plot with the points not connected. This allows us to momentarily ignore the plot as being a time-sequence plot and rather to view it as a generic Y-versus-X plot. In fact, we labeled the vertical axis of the plot as the Y variable and the horizontal axis as the X variable. From this perspective, we see that there is negative association between the two variables. Furthermore, inspection of the scatter plot suggests that the underlying relationship between the two variables is linear. Thus, it is reasonable to believe that these variables are conceptually related in a manner expressed by (3.12).

Since the underlying parameters (β_0 and β_1) that define the underlying linear model are unknown, we need to estimate these model coefficients by using the available sample information. As discussed in the previous section, an equation that represents an estimate of the underlying model is known as a fitted model. When one is estimating a linear relationship, the general equation for the fitted model can be written as

$$\hat{y}_i = b_0 + b_1 x_i \tag{3.13}$$

The value of b_0 is the sample estimate of parameter β_0, and the value of b_1 is the sample estimate of parameter β_1. The value $\hat{y}_i$ is the fitted value or predicted value of variable Y when $X = x_i$.

To find a reasonable fitted line, we employ the method of least squares discussed in the previous section; that is, we find the fitted line that minimizes the sum of squared residuals. Substituting the fitted equation given by (3.13) into the general definition of SSE given by (3.10), we obtain

$$\begin{aligned}
\text{SSE} &= \sum_{i=1}^{n}(y_i - \hat{y}_i)^2 \\
&= \sum_{i=1}^{n}[y_i - (b_0 + b_1 x_i)]^2 \\
&= \sum_{i=1}^{n}(y_i - b_0 - b_1 x_i)^2
\end{aligned}$$

By using again the techniques from calculus, it can be shown that least-squares estimates for the slope and intercept terms are:

$$b_1 = \frac{\sum_{i=1}^{n}(x_i - \bar{x})(y_i - \bar{y})}{\sum_{i=1}^{n}(x_i - \bar{x})^2} \tag{3.14}$$

$$b_0 = \bar{y} - b_1\bar{x} \tag{3.15}$$

It is important to note that estimators based on the least-squares method have all the desirable properties discussed in Section 2.10. In particular, least-squares estimators are the most efficient unbiased estimators of underlying regression model parameters. In the case of a simple regression model, this implies that if the underlying model is indeed of the form $Y = \beta_0 + \beta_1 X + \varepsilon$, then the least-squares estimators (b_0 and b_1) associated with the fitted line $\hat{y}_i = b_0 + b_1 x_i$ will be unbiased, that is, $E(b_0) = \beta_0$ and $E(b_1) = \beta_1$. Furthermore, these estimators have the lowest variance of all possible unbiased regression estimators. The bottom line is that the method of least squares gives us what we would hope for, namely, estimates which are "close" to the true underlying model parameters.

The benefits of the properties of least-squares estimators are reaped only if the nature of the fitted line is consistent with the nature of the underlying model; for example, the underlying relationship between Y and X is a straight line, and the fitted model is a straight line. However, if the underlying relationship is curved and one fits the data with a straight line, the fact that the least-squares method was used to compute the fitted line will not save a bad situation. Generally, least-squares estimates can be computed even if the fitted model is inappropriate for the data studied. For this reason, it is essential that all fitted models be taken through diagnostic checks for model appropriateness. We will soon illustrate the diagnostic checking procedure for the modeling of the elongation data set.

Given the obvious linear relationship between the vertical axis and horizontal axis variables as seen in Figure 3.20, we can compute a least-squares line by asking Minitab to regress the Y-variable on the X-variable. But before we do so, let us be more specific with our variable names. In particular, the Y-variable (dependent variable) represents the elongation measurements which we earlier denoted by the name elongation$_t$. Looking at the scatter plot, we see that the X variable is simply a variable that takes on values 1, 2, 3, . . . , 50 which represents time or time order. Thus, the independent variable is the time index t.

Notice that we stated that this index represents "time or time order." In some applications, we might be studying a process sampled from period to period in which the intervals between periods are equal (e.g., shifts, days, weeks, months, years). For such cases, the regression coefficient in front of the time index is interpreted as the average growth (or decline) of the process from period to period. In some situations, observations are not equally spaced, such as observations taken from production run to production run or from operation to operation. The time index coefficient can safely be interpreted as the average change in the

Stat>>Regression>>Regression

```
MTB > regr 'elongation' 1 't'

The regression equation is
elongation = 0.00137 −0.000011 t

Predictor        Coef            StDev            T        P
Constant   0.00136522       0.00002914        46.84     0.000
t          −0.00001066       000000099       −10.72     0.000

S = 0.0001015    R − Sq = 70.5%    R − Sq(adj) = 69.9%

Analysis of Variance

Source             DF             SS              MS        F          P
Regression          1      1.18428E-06     1.18428E-06   114.96     0.000
Residual Error     48      4.94479E-07     1.03017E-08
TOTAL              49      1.67876E-06
```

Figure 3.21   Regression output for estimated trend line.

process from one measurement period to the next. A good example is the monitoring of a process subject to tool wear. Since it is the actual machining operation that contributes to the deterioration, not the time between operations, the time index can be appropriately used to estimate the rate of deterioration from operation to operation.

But we have to be careful when a process is sampled at irregular time intervals and there has been activity at some subinterval between the sampled intervals. For instance, if a data series of company sales includes observations which are spaced apart by days, weeks, and years with no regularity, the interpretation of the time index applied to these data has little meaning. In general, any data analysis interpretation is made less reliable under such erratic sampling schemes. We must always recognize that effective process understanding is a careful blend of data analysis and the design of a process data collection scheme.

The Minitab regression output for the two variables (elongation vs. t) is displayed in Figure 3.21. Naturally, the first question is, What is the fitted model? Scanning the details of the output, we find the key result printed under the heading "The regression equation is." Furthermore, the regression coefficients are summarized, with more significant digits, under the column labeled "Coef." In particular, we find that

$$\text{Fitted elongation}_t = \hat{y}_t = 0.00136522 - 0.00001066t$$

Assuming that this model is deemed appropriate, this relationship says that on average, elongation measurements are decreasing at a rate of $b_1 = 0.00001066$ unit from one production run to the next. There may be a temptation to conclude that 0.00001066 is for practical purposes 0, thus negating the existence of the trend. But be careful; as we shall soon demonstrate, this would be a mistaken conclusion. We must remember that the original measurements are *small*, in particular, to three and four significant digits.

The b_0 coefficient of 0.00136522 is the estimated constant or y intercept. Technically, we can interpret this value as the height of the line for $t = 0$. In this case, the y intercept has no interest in and of itself; just think of it as a necessary value to position the line appropriately within the data scatter.

Table 3.5   Data display of actual observations, fitted values, and residuals

```
MTB > print 'elongation' 'fitted' 'residual'

Data Display
    Row      elongation          fitted          residual
      1       0.0014541       0.0013546         0.0000995
      2       0.0014682       0.0013439         0.0001243
      3       0.0013200       0.0013332        -0.0000132
      4       0.0013129       0.0013226        -0.0000097
      5       0.0012071       0.0013119        -0.0001048
      6       0.0013341       0.0013012         0.0000329
      7       0.0012212       0.0012906        -0.0000694
      8       0.0014541       0.0012799         0.0001742
      9       0.0012000       0.0012692        -0.0000692
     10       0.0012141       0.0012586        -0.0000445
      .           .               .                 .
      .           .               .                 .
      .           .               .                 .
```

With the estimated regression model, fitted values and residuals can be computed. Table 3.5 provides a printout of the fitted values and residuals corresponding to the original observations. To illustrate how these values relate, consider the first observation (row 1), for which $t = 1$. Substituting the value 1 for the independent variable t in the estimated regression equation, we compute the fitted value for the first period:

$$\hat{y}_1 = 0.00136522 - 0.00001066(1) = 0.00135456$$

This is indeed the fitted value found in row one of the data display. This fitted value of 0.00135456 is less than the actual first-period observation of 0.0014541. The difference between these values represents the residual value for the first period, in particular,

$$e_1 = y_1 - \hat{y}_1 = 0.0014541 - 0.00135456 = 0.0000995$$

All the remaining entries of fitted values and residuals are related in exactly the same manner. Reviewing the residuals in the other displayed rows, we see that some are positive and others are negative. This is an indication that the fitted line

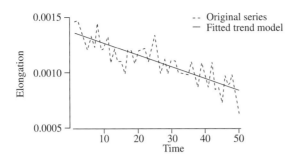

Figure 3.22   Fitted line superimposed on original data series.

is placed in the "midst" of the data scatter. It turns out that any fitted line which minimizes the sum of residuals squared will always have negative and positive residuals; furthermore, these residuals will always average out to zero.

Given that each observed value has a corresponding fitted value and given that the observed values are in time order, the fitted values are in time order. As such, we can plot the fitted values as a time-series plot. In Figure 3.22, we present a time-series plot of the original observations along with the fitted values superimposed. As can be seen, the fitted line (estimated trend line) seems to track well the underlying evolution of the process.

Is this process to be viewed as a process in or out of control? This all depends on one's perspective. In Chapter 2, we briefly noted that the traditional perspective is to associate a process in a state of statistical control with a purely random process. Based on this definition, the observed trend process would be deemed an out-of-control process. However, the term *out-of-control* can sometimes be a bit of a misnomer since it tends to suggest a process which is completely unpredictable. This is not the case here. A trending process is indeed quite predictable, at least for short periods into the future. Namely, extending the estimated trend line shown in Figure 3.22 serves as our "best guess" for future outcomes. We will soon illustrate the computations of prediction.

It is critical to recognize that the systematic variation present in this application affects all the outcomes of the process, not just some isolated observations. Recognizing this reality is an important first step to better understand the process which will, hopefully, lead to more effective strategies for process improvement. The fact that this process is predictable does not imply that it is acceptable or that we should be content. As Deming (1986, p. 313) states, ". . . the elongations plotted one by one in order of manufacture show a downward trend. Something is wrong with the process of manufacture, or with the measuring instrument." An investigation should be directed to search out the sources of the systematic variation with the objective of reducing or removing their effects. If this were done, the overall process variation would be reduced, thus making possible higher levels of quality through process stability.

DETAILED BREAKDOWN OF REGRESSION OUTPUT

Beyond the estimated fitted line provided in Minitab's computer output, there are a number of other reported statistics which deserve our attention. Below, we discuss in some detail the regression output, and we provide explanations of particularly important aspects. Even though our discussions are directed to the current application, the ideas are universally applicable to subsequent regressions (simple and multiple) found in the text.

1. To start, let us look at the top part of the Minitab output originally shown in Figure 3.21:

```
The regression equation is
elongation = 0.00137 -0.000011 t

Predictor          Coef        StDev          T        P
Constant       0.00136522    0.00002914     46.84    0.000
t             -0.00001066    0.00000099    -10.72    0.000
```

As we have already learned, the entries under the "Coef" heading are the estimated regression intercept and slope coefficients (b_0 and b_1). These estimates are combined in equation form to give the fitted line which is conveniently reported under "The regression equation is."

Since the values of b_0 and b_1 will vary from sample to sample, these estimators are random variables. As is the case for any random variable, each of these sample statistics can be associated with some probability or sampling distribution.[10] Earlier, we noted that the means (or expected values) of the sampling distributions for b_0 and b_1 are β_0 and β_1, respectively. The standard deviations for these distributions are denoted by σ_{b_0} and σ_{b_1} respectively.

In practice, these standard deviations are unknown and need to be estimated. We denote the estimated standard deviations by s_{b_0} and s_{b_1}. Formulas exist to compute s_{b_0} and s_{b_1} (see Neter et al., 1996), however, we need only be concerned with Minitab's final computations. Under the heading "Stdev" (which is in the box above), Minitab reports the estimated values for s_{b_0} and s_{b_1} which we find to be 0.00002914 and 0.00000099, respectively. These values are also referred to as *standard errors of the coefficients.*

Think of the standard error of a coefficient as a measure of the uncertainty in the estimated regression coefficient. In an idealized scenario in which the data between two variables fall on a perfectly straight line (see Figure 3.17a), there is no uncertainty regarding the intercept and slope of the regression line, implying standard errors of zero. As the scatter of observations increases, there is a corresponding higher level of uncertainty in how close the estimated coefficients are to the true underlying model parameters.

With the standard errors of the coefficients available, we can go beyond a best guess for an underlying regression parameter to the reporting of interval

---

[10]For a review of the notion of sample statistics as random variables with probability distributions, refer to Section 2.9.

estimates for the underlying model parameters (β_0 and β_1). Given that the fitted line is appropriate for a given set of data and the underlying assumption of normality for the error terms is satisfied, it can be shown that the sampling distribution for the regression estimators is approximately normal. (We will soon see that all the standard assumptions of the regression model are indeed satisfied.) However, because the true standard deviation of the random error term is unknown and needs to be estimated, the sampling distribution is technically a t distribution (see Section 2.9). When using a t distribution, we need to know the appropriate degrees of freedom. When we make inferences about the model parameters of a simple regression, the degree of freedom used is $n - 2$, as opposed to $n - 1$ when we make inferences about the population mean of a single variable. In general, when we use the t distribution in making inferences about a regression model, the number of degrees of freedom used is equal to the sample size n minus K, where K equals the number of predictor variables plus 1.

In our example, suppose we wish to construct a 95 percent confidence interval for the slope parameter β_1. Since the sample size is 50, we need $t_{0.025, 48}$ which can be found to equal 2.0106. Accordingly, the 95 percent confidence interval for β_1 is

$$-0.00001066 - 2.0106(0.00000099) \le \beta_1 \le -0.00001066 + 2.0106(0.00000099)$$

$$-0.00001265 \le \beta_1 \le -0.00000867$$

The above interval suggests with a high degree of confidence that the true slope is negative. The ranges of values do not include the value of zero. Hence, we can be very confident that the true regression slope is not zero. The implication of this conclusion is extremely important, as we will see below. In the same fashion, interval estimates can be made for the intercept coefficient β_0, estimated by $b_0 = 0.00136522$ with $s_{b_0} = 0.00002914$.

2.   One of the most fundamental questions we can ask in the construction of a regression model is, How useful is a particular independent variable in the prediction of the dependent variable? To answer this question, first recall that for a simple regression the true model is set within the framework $Y = \beta_0 + \beta_1 X + \varepsilon$. Notice that if $\beta_1 = 0$, the values of the independent variable X are of no use in predicting Y. Thus, the question of the usefulness of an independent variable in predicting the dependent variable boils down to a question of whether we would conclude $\beta_1 - 0$. This question can be placed in a hypothesis-testing framework with the following competing hypotheses:

$$H_0: \beta_1 = 0 \qquad \text{versus} \qquad H_1: \beta_1 \ne 0$$

If we accept the null hypothesis, we are concluding that there is no importance to the independent variable whereas rejection of the null hypothesis suggests that the independent variable is providing some intrinsic value in the prediction of the dependent variable.

The confidence interval just earlier constructed for β_1 can be used to test the competing hypotheses. Because the interval did not include the value $\beta_1 = 0$, we reject the null hypothesis $\beta_1 = 0$ based on a 5 percent level of significance. Minitab, however, does not construct confidence intervals for its regression output. Instead, Minitab reports a t ratio and p value associated with the above competing hypotheses. Clearly, choosing between these two hypotheses depends greatly on the value of b_1, which for this current application is -0.00001066. But this value alone is relatively meaningless. We also need to consider the estimated standard deviation ($s_{b_1} = 0.00000099$) because it will enable us to assess just how far away the slope coefficient value is from zero. With this in mind, consider the following computation:

$$\frac{-0.00001066 - 0}{0.00000099} = -10.77$$

Hence, the sample slope coefficient is 10.77 standard deviations below zero. This number is referred to as the t *ratio* and can be found below in the regression output. (*Note:* The slight difference between -10.77 and -10.72 is due to fact that Minitab computes the t ratio based on values for b_1 and s_{b_1} to more significant digits than is printed out in the session output.)

Predictor	Coef	Stdev	t-ratio	p
Constant	0.00136522	0.00002914	46.84	0.000
t	-0.00001066	0.00000099	-10.72	0.000

The name t *ratio* is not accidental. Assuming the null hypothesis is true ($\beta_1 = 0$), Student's distribution or the t distribution accounts for all possible t ratio values. From Section 2.9, we learned that the t distribution is close in interpretation to the standard normal distribution which is symmetric with mean equal to 0 and standard deviation equal to one. In our example, the t ratio falls more than 10 standard deviations away from the mean of the distribution. Such a sample result is too extreme to be written off as the result of sampling variation alone. It is more reasonable to conclude that the t ratio is statistically significant; that is, the null hypothesis is rejected, and the slope coefficient is considered different than zero. The implication is that the independent variable—time index t—is significantly contributing to the prediction of the elongation measurements. In other words, there is a significant downward trend in the elongation measurements.

Minitab also reports the p value associated with the computed t ratio statistic. For the slope coefficient, the p value is found to be 0.000, that is, 0 to three decimal places, meaning the p value is less than 0.0005. The concept of a p value as a measure of "significance" was first introduced in the context of a runs test. The interpretation of a p value is no different here and is universal to all hypothesis-testing applications. In particular, the p value of 0.000 translates to the following statement: *If* the true slope coefficient is zero, the chances are less than 5 in 10,000 that we will observe a sample slope coefficient falling at least as far from 0 as 0.00001066. The unlikeness of such a sample result if the null hypothesis were true persuades us to reject the null hypothesis ($\beta_1 = 0$) in favor of the alternative hypothesis ($\beta_1 \neq 0$).

In general, the standard convention is to deem a sample regression coefficient as significantly different than zero when the p value is less than 0.05. This is a rule of thumb that we will follow in helping decide whether a specific independent variable should be retained in the model. However, there is no law requiring absolute rigidity to this threshold. Readers can make their own decisions about what p value constitutes a significant result.

3. Let us now turn to a summary statistic (denoted R-sq or R^2) which is reported directly below the detailed output pertaining to the regression coefficients:

```
S = 0.0001015    R-Sq = 70.5%    R-Sq(adj) = 69.9%
```

To see how this statistic is computed, we need to look at the part of the regression output related to the analysis of variance:

```
Analysis of Variance
Source          DF          SS            MS          F       P
Regression       1    1.18428E-06   1.18428E-06   114.96   0.000
Residual Error  48    4.94479E-07   1.03017E-08
Total           49    1.67876E-06
```

Under the column labeled "SS," we find three sum-of-squares quantities referred to as the *sum-of-squares total (SST), sum-of-squares for regression (SSR), and sum of squared errors (SSE)*. Basically, these three quantities represent measures of variation and are computed as follows:

$$SST = \sum_{i=1}^{n}(y_i - \bar{y})^2 \tag{3.16}$$

$$SSR = \sum_{i=1}^{n}(\hat{y}_i - \bar{y})^2 \tag{3.17}$$

$$SSE = \sum_{i=1}^{n}(y_i - \hat{y}_i)^2 \tag{3.18}$$

Before we extract meaning out of these quantities, let us remember the goal of the regression model. Namely, our goal is to predict y_i which is equivalent to saying that our goal is to explain the variation in the y data. Looking at the definition of SST, we see that it is a measure of variation of the y_i values around their mean $\bar{y}$. Notice that SST is actually the numerator of the fraction used to compute the sample variance of y and thus is interpreted as the total variation in y.

Enter now the regression model. Clearly, there will be a part of the total variation in the y_i values that will be explained by the fitted model and a part of the total variation in the y_i values that will not be explained by the regression model. Recall that a residual is the part of the observation not explained by the fitted model. Furthermore, recall that SSE is the sum of squared residuals. Thus, SSE is seen to be the amount of variation *not* explained by the fitted model. Accordingly, SSR must be the amount of variation explained by the fitted model. Hence, the three of sum-of-squares terms are related as

$$SST = SSR + SSE \tag{3.19}$$

Since one goal is to find a model that helps explain the movements or variation in the dependent variable, it is natural to ask how much of this variation is indeed accounted for by the model relative to the overall variation. Given the breakdown of variations above, an intuitive statistical measure is to simply divide the variation in the fitted values by the variation in the observations, that is, to determine what percentage of the total "pie" is explained by the fitted model. Accordingly, we define

$$R^2 = \frac{SSR}{SST} \tag{3.20}$$

The R^2 (R-sq) statistic—also called the *coefficient of determination*—is the most commonly reported single-valued measure of the "success" of the fitted model. The statistic ranges from 0 (model accounts for none of the data variability) to 1 (model accounts for all the data variability).

From Figure 3.21, we find SSR = 1.18428E-06 and SST = 1.67876E-06. Substituting these values into (3.20), we find

$$R^2 = \frac{1.18428E\text{-}06}{1.67876E\text{-}06} = 0.705 \qquad \text{or} \quad 70.5\%$$

As you can notice, this is precisely the number that Minitab reports. It indicates that the fitted trend model accounts for 70.5 percent of the variability in the data set of 50 elongation measurements.

The practical implication of R^2 can be quite important. In many quality control applications, any additional process variation is viewed as an unwanted evil since it undermines the process's ability to produce a consistent product or service. So R^2 provides a measure of the percentage of total variation attributed to some observed systematic pattern which is summarized in a fitted model. If the systematic sources of variation can be removed, R^2 can be interpreted as a measure of opportunity for process improvement efforts in terms of variation reduction.

We close our discussion about R^2 with a few cautionary comments. Many regression users inappropriately rely solely on R^2 as an overall measure of the quality of the fitted model. For example, if R^2 is a "big" number, such as 80 or 90 percent, there is a great temptation to assess the regression model as a good or an excellent fit. The problem is that the process of checking the validity of a proposed fitted model involves many steps, as we will outline soon, which cannot be summarized by any single-valued statistic. For example, the value of R^2 tells us nothing about whether the assumption of underlying normality is satisfied.

It is also important to know that R^2 never decreases as additional independent variables are added. This will happen even if the added variables have no significant relationship with the dependent variable. One can easily inflate R^2 toward 100 percent with a large set of "nonsense" independent variables. Even though an existing data set can be made to be perfectly fit, a model built on this basis would be useless for the purposes of future prediction. Hence, on the surface, a large R^2 can be deceptively appealing. To combat misleading representations by R^2, there is a variant called the *adjusted R^2* that adjusts for the number of variables in the regression model. Its value is reported immediately to the right of the unadjusted R^2 value. Basically, this measure works on the following idea: Adjusted R^2 will decrease if the added independent variable adds little to the ability of the model to explain the variation in the dependent variable; otherwise adjusted R^2 will increase.[11]

Given this built-in safeguard, adjusted R^2 is particularly useful for comparing competing fitted models which are based on different numbers of independent variables; this is an activity that we will not pursue in detail in this text. Since we will not report a particular fitted model as adequate if it incorporates any nonexplanatory independent variables, we will be content with reporting and interpreting the unadjusted R^2. In any case, for well-chosen fitted models, unadjusted R^2 and adjusted R^2 can be expected to be relatively close in value.

4.   The last aspect of the regression output which we will note at this point is highlighted below:

$$\boxed{\text{S = 0.0001015}} \qquad \text{R-Sq = 70.5\%} \qquad \text{R-Sq(adj) = 69.9\%}$$

The quantity S is the standard deviation of the residuals. Since residuals represent the unexplained variation around the fitted line, the standard deviation of the residuals serves as an estimate of the true standard deviation of the random error component σ_ε.

To know how the standard deviation of the residuals is precisely estimated, we first write the basic definition of variance as applied to the random error variable:

$$\sigma_\varepsilon^2 = E(\varepsilon_i - \mu_\varepsilon)^2 \tag{3.21}$$

Now recall that the expected value (or mean) of the random errors is assumed to be zero. Thus, (3.21) becomes

$$\sigma_\varepsilon^2 = E(\varepsilon_i^2) \tag{3.22}$$

This implies that the variance of the random errors is the expected value (or mean) of the random error variable squared. One reasonable strategy, then, is to estimate this population average of squared errors with the sample average of the squared residuals. Since the sum of squared residuals is SSE, we are considering the following estimate:

$$\hat{\sigma}_\varepsilon^2 = \frac{\text{SSE}}{n} \tag{3.23}$$

However, it turns out that $\hat{\sigma}_\varepsilon^2$ is a biased estimate of σ_ε^2. To correct for the bias, we simply need to change the denominator a bit:

$$s_\varepsilon^2 = \frac{\text{SSE}}{n - K} \tag{3.24}$$

where K again is the number of predictor variables plus 1. The above quantity is also known as the *mean squared error*, or *MSE* for short. The square root of MSE is then an estimate for the standard deviation of the random errors and is what we earlier called the standard deviation of the residuals.

---

[11]For the interested reader, adjusted R^2 is computed as

$$R_{adj}^2 = 1 - \frac{n-1}{n-K}\frac{\text{SSE}}{\text{SST}}$$

From Figure 3.21, MSE is found in the analysis-of-variance table under the column labeled "MS" cross-referenced with the row labeled "Residual Error." We find MSE is equal to 1.03017E-08. The square root of 1.03017E-08 is 0.0001015 which is the value reported next to "S=" in the regression output.

The standard deviation of the residuals (0.0001015) is less than the standard deviation of the original y observations (0.0001851). This is due to the fact that the regression equation has explained some of the original variation in the elongation data. Generally, the smaller the standard deviation of the residuals is relative to the standard deviation of the original data, the greater the explanatory power that is attributable to the fitted model.

The standard deviation of residuals has an important interpretation for quality improvement efforts. If the nonrandom pattern summarized by the fitted model is deemed an unwanted extra variation affecting the overall process, then the removal of this variation will result in a new process affected only by the unexplained variation which is represented by the residuals. A process removed of its systematic component have a level of overall variation summarized by the standard deviation of the residuals.

DIAGNOSTIC CHECKING OF FITTED MODEL

We have just examined many of the principal outputs of the regression computation routinely printed out by Minitab. However, the fact that the regression calculations can be performed does not necessarily imply that the fitted model will be adequate for the intended application. In this section, we direct our attention to basic diagnostic checks of model adequacy.

From our earlier discussions, observed data can be decomposed in two ways:

1. Observed data value = True model + Random error
2. Observed data value = Fitted model + Residual

Notice that if the fitted model is close to the true model, then the residuals will be close to the random errors. Herein lies the basic guiding principle in assessing the adequacy of a fitted model in describing the nature of a process. Namely, if the residuals behave as random errors, the fitted model is viewed as close to the true model. We have specified random errors to be a sequence of independent and normally distributed observations. Using the randomness and normality diagnostics discussed earlier in this chapter, we have the tools to check how closely the residuals conform to randomness and normality which, in turn, enables us to make judgments about the appropriateness of the fitted model.

Diagnostic checking of the residuals (or any data) should always start with the random checks, *not* with the normality checks. As we have warned on a number of occasions, normality checks can be misleading in the presence of nonrandomness. So, we proceed to normality checks on the residuals only if the residuals are deemed random. If the residuals are found to be nonrandom, we should "bail" out of the diagnostics and attempt to seek an improved fitted model by incorporating the information about the pattern found in the residuals.

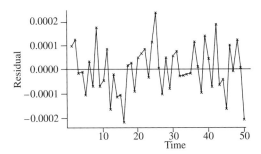

Figure 3.23   Time-series plot of the residuals.

To check for randomness, we begin by studying a time-series plot of the residuals which is given in Figure 3.23. The centerline for the time-series plot is the mean of the residuals which is always equal to zero, a fact associated with fitted models based on the method of least squares. Unlike the original elongation series, the visual impression of the residuals is consistent with randomness. More specifically, there are no patterns in the level of the residual process, and the dispersion of the residuals appears constant through time. During diagnostic checking, we should also be on the lookout for the presence of extreme observations or outliers inconsistent with the rest of the data. Such points should be investigated in the hope of finding special explanations which, in turn, may be useful information for process improvement purposes. Formal criteria for the detection of unusual points will be developed in our presentation of control charts in the chapters to follow. Without the need for formality, it is clear from Figure 3.23 that no points "stick out" unusually.

As supplements to the time-series plot, the runs test and ACF are provided in Figure 3.24. The runs test shows 28 observed runs versus 25.96 expected runs for a random process. The p value of 0.5594 is substantially larger than the conventional cutoff of 0.05; thus we do not reject the null hypothesis of randomness. Finally, the ACF with superimposed 95 percent limits ($\pm 2 / \sqrt{50}$) shows no significant autocorrelations, which further confirms the randomness of the residuals. With these three checks, we can conclude that the residuals are compatible with a random process. The fitted model has successfully passed the first phase of diagnostic checking. We must now explore the distributional behavior of the residuals.

In Section 3.3, we suggested that when we check a set of data for compatibility with the normal distribution, it is more desirable to rescale the observations to a standardized scale. As a matter of convenience, Minitab provides the user with the opportunity to request the computation and storage of standardized residuals when invoking the regression command. In Figure 3.25a, we present a histogram of the standardized residuals ("stres"). The histogram reflects a symmetric data distribution that is consistent with the assumption of normality. Given the convenient design of the class intervals, we can also make the following count comparisons:

```
residual

K =  −0.0000

The observed no. of runs = 28
The expected no. of runs = 25.9600
24 Observations above K     26 below
            The test is significant at 0.5594
            Cannot reject at alpha = 0.05
```

(a) Runs test.

```
ACF of residual
          −1.0 −0.8 −0.6 −0.4 −0.2  0.0  0.2  0.4  0.6  0.8  1.0
          +----+----+----+----+----+----+----+----+----+----+
  1  −0.062                         XXX
  2  −0.006                          X
  3   0.068                         XXX
  4  −0.047                         XX
  5   0.000                          X
  6   06040                         XX
  7   0.084                         XXX
  8  −0.272                 XXXXXXXX
  9  −0.031                         XX
 10  −0.136                       XXXX
 11   0.012                          X
 12   0.015                          X
```

(b) ACF.

Figure 3.24   Randomness checks of the residuals.

Interval Around Mean	Sample	Expected
±1 Standard deviation	$\frac{32}{50}$, or 64%	68.27%
±2 Standard deviations	$\frac{47}{50}$, or 94%	95.45%
±3 Standard deviations	$\frac{50}{50}$, or 100%	99.73%

Overall, the observed counts appear compatible with normality. As a final check of normality, the normal probability plot for the standardized residuals is presented in Figure 3.25*b*. The plot appears nearly linear, indicating that the distribution of the standardized residuals is close to a normal curve.

In summary, the residuals are consistent with both randomness and normality, implying a "clean bill of health" for the fitted model. Hence, the estimated simple trend model is an appropriate summary for the underlying nonrandom behavior of the elongation process. Additionally, there is no evidence of isolated unusual observations relative to the trend behavior.

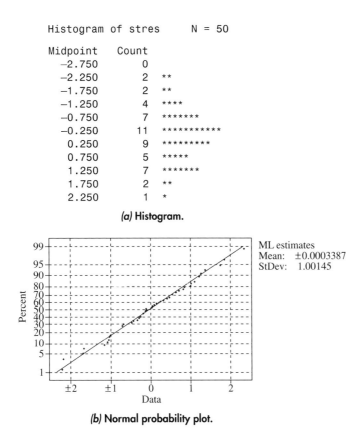

```
Histogram of stres      N = 50

Midpoint    Count
  -2.750      0
  -2.250      2    **
  -1.750      2    **
  -1.250      4    ****
  -0.750      7    *******
  -0.250     11    ***********
   0.250      9    *********
   0.750      5    *****
   1.250      7    *******
   1.750      2    **
   2.250      1    *
```

(a) **Histogram.**

ML estimates
Mean: ±0.0003387
StDev: 1.00145

(b) **Normal probability plot.**

Figure 3.25   Normality checks of the standardized residuals.

PREDICTION OF FUTURE OUTCOMES

We have demonstrated how regression modeling can be used to summarize systematic patterns revealed in a sequence of data observations. In itself, the regression summarization is valuable to a better understanding of the general nature of the process. The regression model can also be used for predicting future observations of the dependent variable.

Recall that if a process exhibits random and normal behavior, the "best guess" prediction of future outcomes is simply an extension of the centerline positioned at the sample average. For a simple trend process, the centerline is tilted to represent the estimated trend line (refer to Figure 3.22). Best-guess predictions are then obtained by extending the tilted centerline into the future.

Predictions are simple to compute when the fitted model is a simple linear trend. Since the independent variable is the time index ($t = 1, 2, 3, \ldots$), simply insert the appropriate value of the index for the time period to be forecast. To illustrate, if we wish to predict the next observation for period 51, we substitute $t = 51$

into the regression equation $0.00136522 - 0.00001066t$. When done, the prediction for the next observation is $0.00136522 - 0.00001066(51) = 0.00082156$.

Often, we are interested in going beyond a best guess of a future outcome (also called a *point* prediction) to the reporting of a range of likely future outcomes. Such a range is commonly referred to as a *prediction interval*. As we discussed, the standard deviation of the residuals (0.0001015) provides a measure of the level of uncertainty around the fitted line. Since we have found the residuals to be approximately normal, we construct the following intervals regarding the future outcome at $t = 51$:

1. An approximate 68 percent prediction interval for a future process outcome is $0.00082156 \pm 1(0.0001015)$, or 0.00072006 to 0.00092306.

2. An approximate 95 percent prediction interval for a future process outcome is $0.00082156 \pm 2(0.0001015)$, or 0.00061856 to 0.00102456.

3. An approximate 99.7 percent prediction interval for a future process outcome is $0.00082156 \pm 3(0.0001015)$, or 0.00051706 to 0.00112606.

The interpretation of prediction intervals, like that of confidence intervals, is based on a repeated-sampling concept. Namely, if we take many samples of size n from the process and construct prediction intervals from each of these samples, we expect a certain percentage (say, 68, 95, or 99.7 percent) of these intervals to include the future outcome.

Using the "Predict" subcommand of the regression command, Minitab can be used to compute a point prediction and a prediction interval for any desired probability level. For example, below is Minitab's computation of a 95 percent prediction interval for $t = 51$:

```
Predicted Values
     Fit   StDev Fit        95.0% CI             95.0% PI
 0.000821    0.000029   (0.000763,0.000880)  (0.000609,0.001034)
```

Under the heading "Fit," Minitab provides the same point prediction value (0.000821) as we calculated. However, Minitab's 95 percent prediction interval (0.000609, 0.001034) is slightly wider than the interval we calculated (0.000619 to 0.001025). This is because of two factors. First, the formal calculation of a prediction interval is based on multiples from the t distribution rather than our approximate choices based on the normal distribution. Given that our sample size is "large" (≥ 30), the difference between a normal-based multiple and the corresponding t-based multiple is slight. Second, Minitab uses a more refined measure for the uncertainty, known as the *standard error of prediction*. The standard error of prediction makes allowance for two sources of uncertainty:

1. The first source is due to the variation of the process outcomes around the true line. This source of uncertainty is estimated by the standard deviation of the residuals.

2. The second source is the uncertainty about the fitted line itself, that is, the uncertainty about the location of the true line relative to the fitted line.

For a moderately large sample size (say, $n \geq 30$), the second source of uncertainty is small relative to the uncertainty associated with the variation of individual observations around the regression line. As a consequence, the standard deviation of residuals is a reasonable approximation of the precise measure of prediction uncertainty.

Generally speaking, since most applications in process analysis offer sufficiently large sample sizes, we can be content with summary measures based on the residuals as suitable guides for future process behavior. We can also proceed on this basis because it parallels the traditional implementation of control chart methods to be discussed in the following chapters. With the assumption that the process at hand is random, standard control charts use the sample mean as a point prediction for the future. In addition, prediction limits, called control limits, are computed based on *only* the variation of the observations around the sample mean. Hence, typical control charts are based on *only* the first source of uncertainty described above. Control chart procedures can be thought of as a special case of regression modeling in which the fitted model is the sample mean and the residuals are the observations around the sample mean.

The elongation data set has allowed us to delve deeply into many concepts of regression analysis. This leaves us in a good position to introduce more applications posing different data analysis challenges. We now turn to a data series which exhibits trended behavior but is not reconciled by a simple linear trend model as found in the elongation series.

3.7   MODELING A NONLINEAR TREND PROCESS

Quality management and continuous process improvement are more than buzzwords at a large metropolitan medical center located in the southeastern region of the United States. The medical center offers a broad spectrum of health-related services, including general medical and surgical services, pharmaceutical services, home care, and occupational health services. The foundation of the medical center's quality management initiative was laid over a decade ago (circa 1985) when the incoming hospital president supported a culture of process data gathering, high respect for individuals (customers and employees), systems thinking, and cross-functional teams.

Over the years, there have been numerous improvements of administrative processes (e.g., patient admissions and billings) and clinical processes (e.g., patient care and laboratory testing). In this section, we look at data collected on a process identified as being in dire need of substantial improvement. In particular, there was great concern about the inpatient medication management system. An initial pilot study of the system revealed an astonishing statistic: Nearly 35 percent of ordered medications were not given (missing), were wrong, or were correct but of a wrong dosage strength. A dedicated process improvement team discovered that the overall process of proper medication delivery from physician to patient hinged on several key processes, including, but not limited to, the communication processes among various medical staff (physicians, nurses, and pharmacists), the process of filling and delivering medication orders from medical wards to hospital pharmacy, and the process of delivering medication doses from hospital pharmacy back to intended medical wards and ultimately to the patients.

To monitor the number of medication errors, the improvement team adapted a simple, but effective, means of data collection originally devised by a team tackling the same issue of medication management problems at an Air Force medical center.[12] The method is to attach to all medication carts three canisters labeled *missing medication, wrong medication,* and *incorrect dosage.* If a nursing staff person discovered a medication error during patient rounds, he or she would drop a bead in the appropriate canister. Using this collection method, data are gathered and tracked by using run charts (time-series plots) on a daily, weekly, and monthly basis.

Table 3.6   Weekly medication errors

110	79	83	49	45	76	66	28	82	40
33	52	42	49	19	45	84	58	32	32
52	23	36	15	22	15	7	13	41	17
55	20	12	10	25	59	18	24	24	3

In Table 3.6, we present total weekly medication error counts over the course of 40 weeks roughly beginning at the onset of the process improvement efforts. The data can also be found in file "mederror.dat." The time-series plot of these data (Figure 3.26) gives a visual impression of a downward trend which, if genuine, is good news given the context of this application. Careful inspection of the plot suggests that the downward trend is not a straight shot (that is, not linear); rather, the trend effect appears to gently curve. During the tracking of these weekly counts, there was concern by hospital personnel that "setbacks" requiring special explanations may have occurred (see the highlighted observations). Our data analysis will attempt to address these various issues raised about the data series. First, we turn to the nonrandom features reflected in the data series.

There are a number of ways to model a curving trend, also called a *curvilinear trend.* One way to capture the curving trend is to create a trend variable equal to the square of the time index *t.* Another possibility is to create an independent variable which is the inverse of the time index; this is particularly useful when the trend appears to be flattening out. Of course, there are an infinite number of

---

[12]The quality improvement experiences of this Air Force medical center along with those of other hospitals are reported in a publication of the Joint Commission on Accreditation (1992).

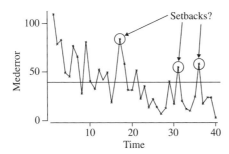

Figure 3.26   Time-series plot of weekly medication errors.

possible functions of the time index that can be entertained. Fortunately, in most practical applications, trending behavior can be adequately described by using only a handful of possible trend variables.

In this case, let us entertain three trend variables: time index t, time index squared t^2, and inverse of the time index $1/t$. Remember that we will check the chosen fitted model for appropriateness of fit. So, if these variables are not sufficient for modeling the data series, we will be alerted during our study of the residuals.

When we are confronted with a set of potential independent variables, the decision is not necessarily to choose one variable to the exclusion of others. It is quite possible that some combination of the available independent variables is required to appropriately capture the nonrandom features of the data. One strategy in exploring possible regression models is to try every possible combination of the three independent variables; the reader can verify that this will entail the computation of seven possible regression models. This is not a horrendous task with three independent variables, but the number of possible regression models increases rapidly as the number of variables increases. In particular, if k independent variables are considered, there are $2^k - 1$ possible regressions. Fortunately, we can use Minitab to make the process of model searching a more expeditious affair.

There are a number of time-saving methods which have been developed to help the analyst find good subsets of predictors without requiring the fitting of all possible subset regressions. Minitab offers three selection methods: the forward selection method, backward elimination method, and stepwise regression method. We will now provide only a brief explanation of these methods. Readers interested in more detailed discussion should refer to Afifi and Clark (1996).

The forward selection method is a sequential procedure that adds variables one at a time to the model. At each step, this method examines the set of independent variables not already in the model and selects the variable that makes the greatest contribution to the overall fit for the dependent variable. Once a variable enters the model, it is never dropped. The process continues until no variable is outside the model or a criterion can be chosen to decide when to stop adding variables.

The backward elimination method works in opposite fashion to the forward selection method. This method begins with all independent variables in the fitted model and proceeds by eliminating or weeding out the least useful variables one at a time. Variables that are dropped can never reenter the model. In Minitab, the process stops when all the variables included in the model have *t* ratios with absolute values of 2 or better. Recall that an absolute-value *t* ratio of 2 is closely associated with a 5 percent level of significance.

The stepwise regression procedure is probably the most commonly used searching technique. The procedure combines the ideas of both forward and backward selection methods. Stepwise regression proceeds in sequential manner, as does the forward selection method, by examining the set of independent variables not already in the model and selects the variable that makes the greatest contribution to the overall fit for the dependent variable. Once a new variable is added to the model, the stepwise regression procedure does not immediately proceed to the next step. Instead, the idea of backward elimination is applied. Namely, stepwise regression evaluates all the variables currently in the model to determine if any variable should be dropped. The process of adding and dropping continues until no remaining variables are deemed worthy of entering the model and no previously entered variables are chosen for removal. Afifi and Clark (1996) discuss the general criteria for entering and removing variables.

In the default mode of stepwise regression, Minitab adds and drops variables in such a manner that all the variables at the final step have absolute *t* ratios exceeding 2, which equates to all the variables being approximately significant at the 5 percent level (that is, *p* values < 0.05). For the most part, we will use this default mode of stepwise regression to facilitate the process of screening out contributing from noncontributing predictor variables.

Automatic procedures can be quite valuable for quick exploration of candidate variables, saving the analyst considerable time spent having to look at numerous individual regression runs. Given their ease of implementation, there can be a dangerous temptation to throw at these procedures a slew of independent variables. Such a mindless strategy is bound to pick some "nonsensical" variable that, by sheer chance, is correlated with the particular sample but is not really characteristic of the underlying process. This danger is known as model *overfitting* or *overspecification*.

In time-series studies, model overfitting is more prone to occur when one attempts to predict a given time series from information found in other time series; in Chapter 10, we will illustrate the modeling of a process based on data from other processes. In an exercise of reckless abandon, I have taken the daily closing prices of a particular company's stock as a dependent variable and run it in stepwise regression procedure, using many independent variables, to find, in the end, that the independent variables of Dow Jones performance (makes sense) and my daily weight were chosen as significant predictors of the company's stock performance! The moral of this story is that regression and automated procedures are no substitutes for good sense. Background knowledge of the situation being studied should be used in establishing a relevant set of candidate predictor variables. Thereafter, automated procedures can aid the analyst

Stat>>Regression>>Stepwise

```
MTB > step 'mederror' 't' '^t2' '1/t'

Stepwise Regression
  F-to-Enter:  4.00  F-to-Remove:  4.00
  Response is mederror on  3 predictors, with n =    40

      Step        1      2
  Constant    70.04   53.07

  t           -1.47   -0.95
  T-Value     -5.71   -3.20

  1/t                    58
  T-Value              2.91

  S            18.8    17.2
  R-Sq        46.14   56.18
```

Figure 3.27   Stepwise regression of mederror, on three trend variables.

in sifting through the variables to determine what combination might be included in the fitted model.

Let us now illustrate the use of stepwise regression in the modeling of the medication errors series. The dependent variable is the medication errors over time, which we denote by $mederror_t$. Recall that based on our visual study of the data series, we tentatively identified three trend variables: t, t^2, and $1/t$. Figure 3.27 shows the Minitab stepwise regression output of the dependent variable against the three independent variables. At step 1, Minitab explores the three variables and chooses the time index t as the best first choice. Looking vertically from top to bottom under the "1" heading, we see that Minitab provides a summary of the regression statistics if $mederror_t$ is regressed on t. In particular, the estimated regression model is "Fitted $mederror_t = 70.04 - 1.47t$" with the t ratio for b_1 equaling -5.71 and $R^2 = 46.14$ percent.

At step 2, Minitab decided to retain the time index t and add the inverse of time index $1/t$ to the model. At the end of step 2, Minitab terminated the stepwise procedure. The implication is that the remaining independent variable t^2 would not significantly contribute to the overall fit of $mederror_t$. Minitab makes this judgment based on the fact that if t^2 were included in the regression equation, its absolute t ratio would be less than 2. (The reader can verify that the t ratio for t^2 would be less than 2 by explicitly running a regression of the dependent variable on all three independent variables.) Thus, based on a 5 percent level of significance, we cannot reject the hypothesis that the regression coefficient in front of t^2 equals 0, implying no significant predictive contribution from t^2.

Hence, a promising model appears to include the two independent variables t and $1/t$. From the stepwise output, we can find that the corresponding estimated model is given by

$$\text{Fitted } (mederror_t) = 53.07 - 0.95t + 58 \, (1/t)$$

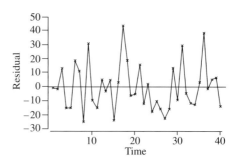

Figure 3.28   Time-series plot of the residuals for the first fitted model.

```
Histogram of stres        N=40
Midpoint   Count
  -1.750     0
  -1.250     4   ****
  -0.750    12   ************
  -0.250     7   *******
   0.250     7   *******
   0.750     4   ***
   1.250     2   **
   1.750     2   **
   2.250     1   *
   2.750     1   *
```
(a) Histogram.

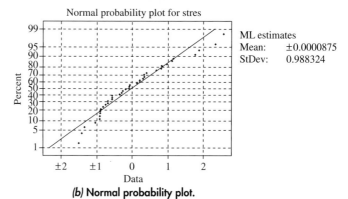

(b) Normal probability plot.

Figure 3.29   Normality checks of the standardized residuals for the first fitted model.

The individual t ratios for predictors t and $1/t$ are given by -3.20 and 2.91, respectively, and R^2 equals 56.18 percent. To investigate the merits of this fitted model, we need to check the behavior of the associated residuals.

The time-series plot of the residuals is shown in Figure 3.28. Unlike the original observations, the residuals appear to be consistent with random behavior. We

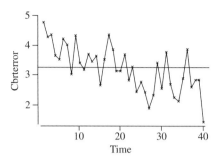

Figure 3.30 Time-series plot of transformed weekly medication errors.

leave it to the reader to verify that the runs count and ACF of the residuals are also compatible with randomness. We now turn to the question of normality. We display the histogram and normal probability plot of the standardized residuals in Figure 3.29. The histogram reveals a substantial evidence against normality, in the form of a positively skewed distribution. The systematic "bowing" of the normal probability plot is further evidence of the departure from normality. Thus, the standard regression assumption of underlying normality is violated, which implies that the fitted model does not pass the diagnostics of model appropriateness. In a sense, we must go back to the drawing board to pursue an alternative path for modeling the data series.

Given the discussions of Section 3.3, one reasonable remedy is to seek a normalizing power transformation of the medication errors data series. One might think that the transformation should be applied to the residuals since they are the data exhibiting the nonnormality. This would be an incomplete solution because even though the transformed residuals might be more normally distributed, the fitted model would still be based on the original observations and their underlying nonnormality. Since the interpretation of some aspects of the fitted model (such as t tests) depends on the approximate underlying normality, we must solve the problem up front by transforming the original observations. Thus, the approach is not to transform the residuals directly, but rather to transform the original observations so that a fitted model estimated from these transformed observations will have normally distributed residuals.

The graphical aid (Figure 3.12) for determining a possible power transformation is more suitable for directing the application of the transformation to nonnormal but random process data. When data are nonrandom, as is the case with the medication errors data series, we need to resort to a process of trial and error. Basically, we try a transformation on the original observations and check whether the resulting residuals are approximately normal; if they are not, we try another transformation.

When the residuals exhibit positive skewness, the square root ($x^{1/2}$), the cube root ($x^{1/3}$), the logarithm ($\ln x$) and the inverse (x^{-1}) are all good starting candidates. As it turns out, the cube root works best for this application; the reader is

```
The regression equation is
cbrterror = 4.17 - 0.0448 t

Predictor          Coef          StDev            T            P
Constant         4.1720         0.1794        23.25        0.000
t              -0.044795        0.007627       -5.87        0.000

S = 0.5568    R - Sq = 47.6%    R - Sq(adj) = 46.2%

Analysis of Variance

Source              DF            SS            MS           F           P
Regression           1        10.695        10.695       34.50      0.0000
Residual Error      38        11.782         0.310
Total               39        22.477
```

Figure 3.31   Regression output of estimated trend line for transformed data series.

encouraged to try other transformations to see how they work out. Figure 3.30 presents a sequence plot of the cube root medication errors data, that is, (mederror$_t$)$^{1/3}$; in Minitab, we call this variable cbrterror. The series still trends down, but the trend now seems approximately linear. Consistent with our visual impression of linearity, when a stepwise regression of the transformed data on the three trend variables t, t^2, and $1/t$ is performed, only the simple time index t is reported as significant.

The regression output for the simple linear trend fit is shown in Figure 3.31. We find that the estimated equation is given by

$$\text{Fitted (mederror}_t)^{1/3} = 4.172 - 0.044795t$$

The contribution of the trend term is highly significant as reflected by a t ratio of -5.87 which corresponds to a p value smaller than 0.0005. Statistically speaking, the highly significant negative coefficient for the trend variable suggests that this independent variable contributes significantly to the overall fit. The practical implication is that the improvement in the process is genuine.

You will also notice that the R^2 statistic is 47.6 percent for this fitted model. For the first regression run on the untransformed data, R^2 was found to be 56.18 percent. Be wary of making comparisons since R^2 does not reflect the appropriateness of the fitted model. We should not forget that an important underlying regression assumption—normality—was materially violated in the earlier analysis. Let us now investigate how the newly estimated model stacks up diagnostically.

We leave it to the reader to verify that the residuals of the simple trend model, estimated from the transformed data, are compatible with randomness; remember random diagnostics include a time-series plot, runs test, and ACF. The normality checks applied to the standardized residuals are given in Figure 3.32. In contrast to the residuals from the previous fit, these residuals exhibit approximate normality. Furthermore, it seems clear that no points are outstanding

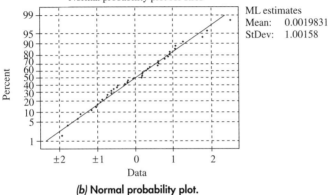

```
Histogram of stres    N = 40
Midpoint   Count
 -2.250      0
 -1.750      3    ***
 -1.250      3    ***
 -0.750      8    ********
 -0.250      6    ******
  0.250      7    *******
  0.750      7    *******
  1.250      3    ***
  1.750      2    **
  2.250      1    *
```

(a) **Histogram.**

(b) **Normal probability plot.**

Figure 3.32   Normality checks of the standardized residuals for the second fitted model.

relative to the data distribution. This suggests that the earlier noted concerns of setbacks are not supported. In summary, the fitted model, estimated from the cube root–transformed data, passes all the standard diagnostic checks. Hence, this fitted model provides an adequate summary of the underlying evolution of the process.

To visualize the fitted model, we show in Figure 3.33*a* a time-series plot of the transformed data along with the sequence of fitted values. Based on the transformed scale, there has been a steady linear decrease—improvement—in medication errors over time. By untransforming the fitted values (raising them to the third power), we can gain a perspective of the implications of the model in terms of the original units (Figure 3.33*b*). Notice that the untransformed fitted values nicely adapt to the gentle curvature evident in the data series. In the original units scale, we see from the fitted line that continuous improvement efforts are steadily paying off, but at a decreasing rate as higher levels of quality are attained. Furthermore, we have learned from the data analysis that none of the weeks had unusually high or low medication errors relative to the steady improvement.

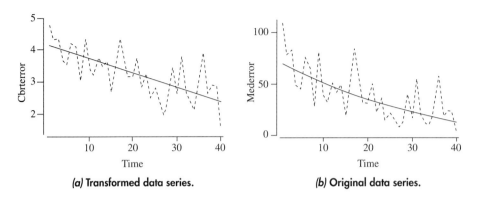

Figure 3.33   Fitted values in transformed and original units.

3.8   MODELING A MEANDERING PROCESS

In the last two applications, we learned how to estimate trend effects by using regression analysis. In this section, we turn to a data series which exhibits a non-random pattern that is not accountable by a trend-based regression model.

Table 3.7   Daily weight measurements

182.0	181.2	180.6	181.5	181.2	181.3	182.0	181.0	180.7
180.9	181.0	181.5	183.0	181.9	182.8	184.2	184.0	183.4
183.4	182.0	180.4	181.1	181.1	182.5	181.8	181.0	179.4
180.0	178.8	178.3	178.6	178.4	179.5	179.2	181.4	181.6
182.5	182.6	183.0	182.4	181.6	180.4	181.9	182.2	182.2
182.1	180.8	181.6	181.5	182.4	184.1	184.1	183.9	182.9
182.6	182.7							

We shall use an application pertaining to the self-monitoring of a personal process. In particular, the data are daily weight measurements of a former MBA student (at the University of Wisconsin-Milwaukee) over the course of 8 weeks, starting on a Monday, that is, for 56 successive days. Using an electronic scale (it provides weights to the nearest 0.1 lb.), the measurements were obtained under comparable conditions, defined as every morning at 8:00 a.m., before breakfast and/or exercise activities. Furthermore, the student reports that he maintains a very regimented lifestyle; for example, his daily eating consumption and sleeping patterns remain fairly consistent. The observations are both given in Table 3.7 and in the file "weight.dat."

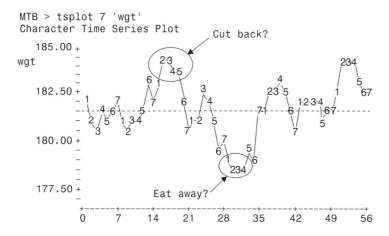

Figure 3.34   Time-series plot of daily weight measurements.

The ideas and developments of the quality and productivity have taught us that *any* process can be improved, not just a manufacturing or a service process in an organization. They are also applicable to the improvement of personal processes (perhaps, the most important of all processes), such as athletic performance, health conditions, and efficiency in daily tasks. Because personal processes are easily related to and are genuinely interesting, they provide wonderful opportunities to experience actual data collection, to perform data analysis, and to experiment with process improvement interventions. For example, one former student monitored her morning and evening blood pressure (systolic and diastolic) measurements with the intent of quitting smoking halfway into the semester. Through simple data analysis, she was able to conclude that the planned experiment resulted in statistically significant improvement in the process, that is, lower blood pressure levels.

In fact, there are reportedly physicians proposing the use of run charts and statistical process control (SPC) methods by patients in the management of certain chronic heath problems; examples include the management of diabetes, hypertension, blood coagulation problems, and weight control. As we will find out with the weight series, many personal health processes are not in statistical control in the traditional sense of random behavior. Often, the processes exhibit nonrandom, but *predictable* behavior. Since standard SPC methods are grounded in the assumption of random variation, there is a great risk of misleading conclusions being drawn from the application of these methods to many personal health processes. The broader perspective offered by data analysis provides a more appropriate perspective for process monitoring.

We begin, our analysis, as usual, with a time-series plot, shown in Figure 3.34. Instead of plotting the graph in high resolution, it is a good idea to use Minitab's character graphics mode, which enables us to label the days of the week, where 1 signifies Monday, 2 signifies Tuesday, and so on.[13] Such labeling makes it easier to see seasonal patterns, such as a tendency for Mondays (or any other day) to be consistently high or low. In Chapter 6, we will show an example of a manufacturing series which when plotted in *default* high-resolution mode, camouflages a consistent seasonality; hence, a better understanding of the process is sacrificed. As mentioned in Chapter 2, one downside with the character plot is that the points are not connected. So, as done in Figure 3.34, it is necessary to manually connect successive points to facilitate visual analysis.

From the time-series plot, we see a form of nonrandomness that is unlike the earlier trend example. Here, the data series appears to "snake" or "meander" around the centerline positioned at the sample mean of 181.61 lb. As we discussed in Chapter 2, the appearance of meandering is the result of successive observations tending to be close together in value. In Chapter 2, we also introduced the concept of a stationary process. A stationary process is one that has no tendency to run away from the overall mean level *in the long run*. This seems to be the case for the weight series. The process is meandering around the centerline. Each time the points wander away from the overall mean level (above or below), they appear to revert back soon in the direction of the mean.

To confirm the nonrandom nature of the weight series, the runs test and ACF of the data are given in Figure 3.35. As can be seen, there are only 13 runs above and below the mean, which is significantly less than the expectation of about 29. The reported p value for the runs test is 0 to four decimals. ACF along with 95 percent limits reveals substantial *positive* autocorrelations at the first three lags, providing further evidence of the presence of systematic nonrandom effects. The positive autocorrelations reflect the tendency for observations close in time to be close in value. For this reason, meandering series are also referred to as *positively autocorrelated series.*

Imagine that your daily weight followed a random process. What an uncomfortable ("shaky") existence that would be! Thankfully, the laws of nature are kind to us, allowing our weight levels to flow closely together over time.

This application also teaches a wonderful lesson for the need of "statistical thinking." One of Deming's great contributions is his insights on the connection of statistical thinking to effective management. He emphasizes that in the absence of statistical thinking, managers lack the appreciation of random variation. Subsequently, these managers tend to overreact to individual process outcomes that are due to chance causes not attributable to special circumstances. Such overmanaging of processes leads to organizational waste and confusion and can even result in more poorly performing processes.

---

[13]Data labeling can be done in the high-resolution mode. In particular, the user would need to create a column of labels and then use the "Data Labels" option within the "Annotation" option of the time-series plot dialog. The advantage of the high-resolution mode over the character graphics mode is that the points are connected, and thus there is no need for manual connection. The disadvantage of the high-resolution mode is that plotted points are restricted to a plot of fixed horizontal width. If the data series is longer, all the plotted points, along with their data labels, are "squeezed" together, resulting in a time-series plot that is quite difficult to visually analyze. Regardless of data-series length, plotted points of character-based plots are kept a fixed distance, making for easy visual analysis.

```
wgt

K  =  181.6107

The observed number of runs = 13
The expected number of runs = 28.9643
27 Observations above K     29 below
         The test is significant at 0.0000
```

(a) Runs test.

```
ACF of wgt
       −1.0  −0.8 −0.6 −0.4 −0.2  0.0  0.2  0.4  0.6  0.8  1.0
       +----+----+----+----+--.-+----+----+----+----+----+
 1   0.810                         XXXXXXXXXXXXXXXXXXXXX
 2   0.628                         XXXXXXXXXXXXXXXXX
 3   0.418                         XXXXXXXXXXX
 4   0.229                         XXXXXXX
 5   0.107                         XXXX
 6   0.032                         XX
 7  −0.046                        XX
 8  −0.114                       XXXX
 9  −0.158                       XXXX
10  −0.221                      XXXXXXX
11  −0.199                       XXXXXX
12  −0.199                       XXXXXX
13  −0.220                       XXXXX
14  −0.277                      XXXXXXXX
```

(b) ACF.

Figure 3.35   Randomness checks of the weight series.

Deming's arguments for the importance of statistical thinking in understanding randomness are equally applicable to nonrandom processes. In the daily weight application, not understanding the statistical implications of a weight process can also lead to unnecessary tampering. For example, the student would be overreacting during the period of days 15 to 18 (circled on Figure 3.34) in concluding there is an increase in weight requiring immediate action, such as a carrots and water diet! On the other side, the lower weight levels for days 28 to 31 are no excuse for a binge on cakes and sweets! Both situations are overreactions since, given the nature of the stationary meandering pattern, the weight levels will naturally revert to the mean without special interventions. Assuming that the overall average weight level is at a desirable level, the best action is no action.

The notion that successive observations are close together in value is a clue as to how we might model the nonrandom pattern in the weight series. It suggests that the values of previous periods can be used for the prediction of future periods. If we denote the dependent variable by weight$_t$, this implies lagged variables (weight$_{t-1}$, weight$_{t-2}$, etc.) would serve as independent variables.

A relevant question is, How far back do we have to go? The ACF in Figure 3.35 would lead us to conclude that the prediction of any given period requires knowledge of the previous three periods. However, this may be more lags than we truly need. If adjacent observations are close in value, then it stands to reason that

```
MTB > pacf 'wgt'                        Stat>> Time Series>>Partial Autocorrelation

Partial Autocorrelation Function

PACF of wgt
                 −1.0 −0.8 −0.6 −0.4 −0.2  0.0  0.2  0.4  0.6  0.8  1.0
                  +----+----+----+----+----+----+----+----+----+----+
    1    0.810                               XXXXXXXXXXXXXXXXXXXXX
    2   −0.083                           XXX
    3   −0.193                        XXXXXX
    4   −0.087                           XXX
    5    0.057                             XX
    6    0.027                             XX
    7   −0.131                          XXXX
    8   −0.093                           XXX
    9    0.011                             X
   10   −0.112                          XXXX
   11    0.136                              XXXX
   12   −0.102                          XXXX
   13   −0.158                         XXXXX
   14   −0.190                        XXXXXX
```

Figure 3.36   Partial autocorrelation function of weight series.

observations two apart will tend to be close together. Thus, the lag 2 autocorrelation is in part due to the lag 1 autocorrelation. Similarly, the lag 3 autocorrelation is affected by the lag 1 and lag 2 autocorrelations. Hence, the ACF, while useful for testing for nonrandomness, tends to provide an inflated viewpoint as to the number of useful lags needed for modeling the nonrandom effects effectively.

Because of the confounding effects of autocorrelations at certain lags on autocorrelations at other lags, an alternative version of correlation known as *a partial autocorrelation coefficient* can be computed for each lag, to correct for these effects. The partial autocorrelation coefficient of lag k can be thought of as the correlation between observations k units apart, after the correlations at intermediate lags have been adjusted "for." This adjustment is made to see if there is any intrinsic correlation between observations k units apart over and above the correlation induced by intermediate observations. A plot of partial autocorrelation coefficients against the lag values is known as *a partial autocorrelation function (PACF)*.

The PACF for the weight series is shown in Figure 3.36. As was the case with the ACF, the sampling distribution of a sample partial autocorrelation coefficient is approximately normal with mean $= 0$ and standard deviation $= 1 / \sqrt{n}$ when the underlying process is random. Thus, as with the ACF, we can superimpose 95 percent cutoff limits at $\pm 2 / \sqrt{n}$ to highlight statistically significant partial autocorrelations. We see that the first partial autocorrelation can be viewed as a significant departure from zero. Notice that the value of the lag 1 partial autocorrelation (0.810) is exactly the same as the value of the lag 1 simple autocorrelation found in Figure 3.35*b*. This is always true since at lag 1 there are no intermediate or shorter lags to adjust for.

After taking the lag 1 correlation into account, notice that the lag 2 correlation, which was significant on the ACF, falls by the "wayside" on the PACF.

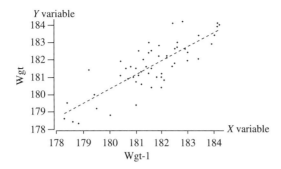

Figure 3.37   Scatter plot of weight versus weight lag 1.

Actually, for all lags above 1, none of the partial autocorrelations significantly depart from zero.

The results of the PACF suggest that the lag 1 variable weight$_{t-1}$ alone should serve as a useful predictor for the variable weight$_t$. To visualize how successive values of the weight series are related, consider the scatter plot of weight$_t$ versus weight$_{t-1}$ given in Figure 3.37. Ignoring for a moment the actual variable names, the scatter plot suggests that the vertical-axis variable (Y variable) is linearly related to the horizontal-axis variable (X variable). This plot immediately orients us to see how regression can be used for the modeling of the weight data. Namely, the scatter plot clearly directs us to fit a linear function between the Y variable (weight$_t$) and the X variable (weight$_{t-1}$). By fitting such a model, we are assuming that the data are generated by the following model:

$$\text{weight}_t = \beta_0 + \beta_1 \, \text{weight}_{t-1} + \varepsilon_t$$

where β_0 and β_1 are the true underlying intercept and slope coefficients and, as usual, ε_t represents the random error component. Since this model suggests regressing a time-series variable on previous values of the same time series, the terms *autoregression* and *autoregressive—self regression*—are used in describing such models. Since only a lag 1 term is used in the above equation, the model is said to be an *autoregressive model* of order 1, or AR(1) for short. A time series which is dependent on both lag 1 and lag 2 values is represented by an AR(2) model, and so on.

Autoregressive models fall in a special class of time-series models called *autoregressive integrated moving-average* (ARIMA) models. ARIMA models are also known as *Box-Jenkins models,* named after the authors of a now classic text *(Time Series Analysis: Forecasting and Control)* on time-series modeling, first published in 1970. In their original work, the authors developed a structured approach to model building that emphasizes simple or parsimonious models. If a data series is modeled by an autoregressive model requiring many lagged terms, the Box-Jenkins approach attempts to find an alternative model based on fewer independent variables. Even though there are some appealing aspects to Box-Jenkins modeling, regression-based modeling provides a less technical but equally effective means of summarizing most process applications. For more details on Box-Jenkins modeling, refer to the revised edition of the original treatise by Box, Jenkins, and Reinsel (1994).

Given what we have learned, we seem ready to estimate a regression equation for predicting weight$_t$ based on the one independent variable weight$_{t-1}$. However, before doing so, we need to address an issue hinted at earlier but not followed up in detail. Namely, the time-series plot shown in Figure 3.34 is constructed with seasonal labels for the day of the week, which begs the question of whether there are any tendencies for certain days to be higher or lower relative to other days of the week. Even with the aid of labeling and careful visual study of the sequence plot, it is to hard to discern if seasonality exists or not. The problem is that the observations are tightly bound to the meandering pattern, making it difficult to visually extract possible seasonal patterns.

To draw a more definitive conclusion on the possible existence of day-of-the-week effects, we can proceed in a couple of directions. One possibility is to ignore the seasonality issue and fit the autogressive lag 1 model. Presumably, this fitted model will account for most, of, if not all, the meandering pattern. After we fit the model, we can look at a time-series plot of the residuals, which are free of the meandering behavior, to identify any lingering seasonal patterns. If these residuals do not reflect seasonality, then it is safe to say that no seasonality exists in the original data. If these residuals do reflect seasonality, then we must go back and refine the model to capture the seasonal pattern along with the meandering pattern.

Another approach is more of an "up-front" strategy. We can create independent variables which are dedicated to capturing seasonal patterns. If we find that a seasonal variable significantly contributes to the overall fit, then seasonality has been detected and incorporated in the fitted model. As a learning experience, let us implement this strategy for our application.

There are a variety of possible seasonal patterns. For this example, we will explore the simplest, but quite common, case in which seasonality is reflected by certain days tending to move up or down a regular and fixed amount relative to the current underlying level of the process, as determined by the other possible nonrandom features in the process.

To begin, we create a coded variable in Minitab called "Day" which takes on the numbers 1, 2, 3, 4, 5, 6, and 7 for the seven days of the week. These numbers are placed in the "Day" column of Minitab so that we know on what day any given weight measurement has occurred. Typically, we will not be interested in the "Day" variable per se as a possible independent variable. Instead, its role is to allow us to create other variables that will be used to capture seasonal regularities. In particular, we apply Minitab's "Indicator" command to the "Day" variable, which creates seven new variables (called indicator variables) that take on the value of 0 or 1, according to the day of the week.

From the Minitab printout given in Table 3.8, we can see the results of the described actions. In general, an indicator variable indicates whether an observation falls into a certain category; 1 means that it does, and 0 means that it does not. We see, for example, that the indicator variable "mon" is 1 for weight measurements taken on Mondays and 0 otherwise. Since the categories represent seasons (days of the week), the associated indicator variables are more specifically referred to as *seasonal* indicator variables.

Table 3.8   Data display of actual observations along with seasonal indicator variables

Row	wgt	day	mon	tues	wed	thur	fri	sat	sun
1	182.0	1	1	0	0	0	0	0	0
2	181.2	2	0	1	0	0	0	0	0
3	180.6	3	0	0	1	0	0	0	0
4	181.5	4	0	0	0	1	0	0	0
5	181.2	5	0	0	0	0	1	0	0
6	181.3	6	0	0	0	0	0	1	0
7	182.0	7	0	0	0	0	0	0	1
8	181.0	1	1	0	0	0	0	0	0
9	180.7	2	0	1	0	0	0	0	0
10	180.9	3	0	0	1	0	0	0	0
.									
.									
.									

To see the implications of seasonal indicator variables in a fitted model, suppose that the fitted model is given by $\hat{y}_t = 100 + 30(\text{mon}) - 50(\text{wed})$. This implies that the process level is estimated to be

$100 + 30(1) - 50(0) = 130$ on Mondays

$100 + 30(0) - 50(1) = 50$ on Wednesdays

$100 + 30(0) - 50(0) = 100$ on Tuesdays, Thursdays, Fridays, Saturdays, and Sundays

Thus, a seasonal indicator variable found in a fitted model adjusts the prediction of the process level at the associated season either up or down, depending on the value and the sign of the regression coefficient.

With the seasonal variables in hand, we can now press ahead on the model fitting of the weight data series. The candidate independent variables include the first lag of the weight series (weight$_{t-1}$) and the seven seasonal indicator variables; in Minitab, we denote the lag 1 variable "wgt−1" since subscripts are not possible. Rather than try all combinations of these eight independent variables, we can screen out contributing variables by using a stepwise selection procedure. Rather than allow Minitab to stop when no remaining variables are viewed as significant,

```
MTB > step 'wgt-1' 'mon' 'tues' 'wed' 'thur' 'fri' 'sat' 'sun';
SUBC> fenter =0;
SUBC> fremove=0.
```

Stepwise Regression

F-to-Enter   0.00   F-to-Remove: 0.00

Response is wgt   on 8 predictors, with N = 55
N(case with missing observations) =   1 N(all cases) = 56.

Step	1	2	3	4	5	6	7
Constant	32.98	32.30	32.85	30.87	31.50	30.90	30.90
wgt-1	0.818	0.822	0.819	0.829	0.825	0.828	0.828
T-Value	10.23	10.28	10.19	10.22	10.08	9.99	9.87
tues		0.35	0.40	0.46	0.51	0.56	0.56
T-Value		1.07	1.19	1.36	1.45	1.49	1.28
wed			0.27	0.33	0.38	0.43	0.43
T-Value			0.81	0.97	1.09	1.15	0.99
mon				0.35	0.40	0.45	0.45
T-Value				0.96	1.06	1.13	0.99
thur					0.21	0.26	0.26
T-Value					0.59	0.69	0.59
sun						0.16	0.16
T-Value						0.41	0.36
fri							0.00
T-Value							0.00
S	0.854	0.852	0.855	0.856	0.862	0.869	0.878
R-Sq	66.37	67.10	67.52	68.11	68.33	68.45	68.45

Figure 3.38   Stepwise regression of weight$_t$ on weight$_{t-1}$ and seasonal variables.

we will use stepwise regression based on the forward selection method, which brings in variables one at a time until all variables have been entered. This will allow us to see the big picture in terms of the relative contributions of all the independent variables, insignificant as this may be.

The stepwise regression results are shown in Figure 3.38. At step 1, the lag 1 variable weight$_{t-1}$ was chosen as the most contributing independent variable of the eight possible choices. The associated t ratio for this lagged variable is 10.23, which is clearly significant at any reasonable level of significance. At step 2, the indicator variable for Tuesday ("tues") observations entered as the next-best choice; however, it enters with a t ratio of only 1.07, which is substantially less than the conventional cutoff value of 2. Similarly, the t ratios of the seasonal variables at all the remaining steps have low values; as the saying goes, "There is

```
The regression equation is
wgt = 33.0 + 0.818 wgt-1

55 cases used 1 cases contain missing values

Predictor          Coef        StDev           T           P
Constant          32.98        14.53        2.27       0.027
wgt-1           0.81846      0.08002       10.23       0.000

S = 0.8537      R-Sq = 66.4%       R-Sq(adj) = 65.7%

Analysis of Variance

Source              DF           SS          MS           F           P
Regression           1       76.249      76.249      104.61       0.000
Residual Error      53       38.630       0.729
Total               54      114.879
```

Figure 3.39   Estimated lag 1 regression equation.

nothing to write home about." Thus, it appears that there is no seasonal pattern to the weight observations. The only independent variable that needs to be entertained is the lag 1 variable.

You might have noticed from Figure 3.38 that stepwise regression stopped one step short (at step 7) of including all eight independent variables. The reason stems from the nature of the seven daily indicator variables. Note that knowing the values of *any* set of six indicator variables defines exactly the value of the remaining indicator variable. For example, if mon = tues = wed = thurs = fri = sat = 0, then it must be the case that sun = 1. At step 7, since six indicator variables (mon, tues, wed, thur, sat, and sun) are included, to go one more step to include the remaining variable (fri) would add no useful information; in other words, fri is a redundant variable after step 7. Computationally, it is impossible to include a redundant variable in a regression equation. It would force Minitab to attempt the impossible task of dividing by zero, hence, the reason for no step 8.

Based on the results of our stepwise regression, we estimate the lag 1 autoregressive model (see Figure 3.39). From the details of the output, we find that the fitted model is given by

$$\text{Fitted weight}_t = 32.98 + 0.81846 \text{ weight}_{t-1}$$

Thus, to obtain a fitted value for any given time period, multiply the previous observation by 0.81846 and add 32.98 to the result. As recognized from the stepwise regression output, the contribution of the lag 1 independent variable is highly significant; here we see that the p value is less than 0.0005. The model explains away around 66 percent (R^2) of the variation observed in the original data series.

Generally, the lag 1 regression coefficient might be interpreted as a measure of "carryover" from one period to the next. Here, the interpretation is that on top

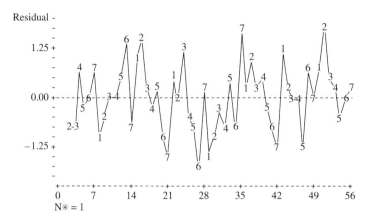

Figure 3.40   Time-series plot of residuals.

of the baseline constant of 32.98 lb, about 82 percent (0.81846) of the individual's weight carries over into the next day. For processes dependent on only their lag 1 observations, it can be shown that stationary processes correspond to a lag 1 regression coefficient that is less than 1 in absolute value. Imagine if the lag 1 coefficient were equal to 2. That would suggest that the individual's weight was essentially doubling each day! Such a process is nonstationary, since there is no tendency for it to revert to some overall mean level.

Before firmly concluding the appropriateness of the estimated lag 1 model, we need to perform diagnostic checks on the residuals. The time-series plot of the residuals labeled by the day of the week is provided in Figure 3.40. Notice below the plot Minitab prints out $N*=1$. This means that there is one missing value for the residual series. In particular, there is no residual for the first period. This is because of the lag 1 variable in the fitted model. To obtain a fitted value for the first period, we need to know the individual's weight at $t = 0$, that is, before data were collected. Since no value is known for the zero period, we cannot compute a fitted value for the first period, which in turn means that we cannot compute the residual for the first period. So any plot of residuals or fitted values will necessarily start at period 2.

The sequence plot shows the residuals to be compatible with randomness free of the meandering behavior underlying the original observations. There is also no tendency for any of the numerals 1 through 7 to be consistently high or low, implying the lack of a seasonality or day-of-the-week effects. None of the observations seem out of place relative to the general level of variation of the residuals. Hence, we would not deem any weight readings over the 56 days as particularly high or low.

The runs test and the ACF (Figure 3.41) also confirm the randomness conclusion. (*Note:* To be accurate, the placement of $\pm 2 / \sqrt{n}$ limits on the ACF would be based on $n = 55$, the number of residual values.) With respect to normality, the histogram of standardized residuals (Figure 3.42) is well described

```
residual

K = −0.0000

The observed no. of runs  =   30
The expected no. of runs  =   28.4909
28 Observations above K      27 below
          The test is significant at 0.6812
          Cannot reject at alpha = 0.05
```

(a) Runs test.

```
ACF of residual
        −1.0 −0.8 −0.6 −0.4 −0.2  0.0  0.2  0.4  0.6  0.8  1.0
          +----+----+----+----+----+----+----+----+----+----+
  1    0.074                            XXX
  2    0.134                            XXXX
  3   −0.032                           XX
  4   −0.131                          XXXX
  5   −0.096                           XXX
  6    0.049                            XX
  7    0.003                            X
  8   −0.070                           XXX
  9    0.045                            XX
 10   −0.253                      XXXXXXX
 11    0.027                            XX
 12    0.021                            XX
 13    0.086                            XXX
```

(b) ACF.

Figure 3.41   Randomness checks of the residuals.

```
        −2.750    0
        −2.250    1   *
        −1.750    2   **
        −1.250    6   ******
        −0.750    8   ********
        −0.250   10   **********
         0.250   12   ************
         0.750    7   *******
         1.250    3   ***
         1.750    4   ****
         2.250    2   **
```

Figure 3.42   Histogram of the standardized residuals.

by the normal distribution. We leave the construction of the normal probability plot to the reader which, when done, will also reflect compatibility with normality.

Given the passing grades on all the diagnostic checks, we can conclude that the systematic nonrandom pattern is well captured by the fitted model. To view the estimated nonrandom pattern over time, we show in Figure 3.43 a sequence plot of connected fitted values, denoted by the symbol Z, along with the original observations, denoted by the numerals. When the fitted value and actual observation are very close in value, Minitab is unable to print both a Z and a numeral on the plot, so instead the plus symbol is printed. As can be seen, the fitted values

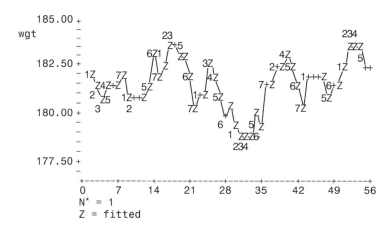

Figure 3.43   Plot of fitted values along with original observations.

provide a more focused picture of the meandering pattern affecting all the process.

In addition to summarizing the systematic effects that are operating throughout the data set, the fitted model can be put to use for the prediction of future values. For example, to predict the weight one period in the future (weight$_{57}$) from period 56, we would substitute the weight value of period 56 (weight$_{56}$) into the fitted model. Referring to Table 3.7, we find that weight$_{56}$ = 182.7 lb. Hence, the point forecast of period 57 is weight$_{57}$ = 32.98 − 0.81846(182.7) = 182.51 lb.

Recall that the uncertainty of the point forecast is approximately measured by the standard deviation of the residuals, which from Figure 3.39 is 0.8537. With this standard deviation, we construct a 95 percent prediction interval for a future weight reading to be 182.51 ± 2(0.8537) and a 99.7 percent prediction interval for a future weight reading 182.51 ± 3(0.8537). If the actual weight falls outside such intervals, it can be viewed as evidence of an unusual increase or drop in the individual's weight, likely due to some special circumstances. As we will soon learn, the conceptual idea of using a fitted model for predicting and monitoring ongoing observations from the process stems directly from the control chart scheme.

3.9   MODELING A COMBINATION OF PROCESS EFFECTS

In our previous applications, we witnessed a couple of processes reflecting *only* trend (linear and nonlinear) effects and a process reflecting *only* lagged effects. We also discussed the possibility of processes exhibiting seasonal effects. In many applications, processes are not isolated to just one of these nonrandom effects. Many realistic situations are dictated by two or more nonrandom patterns simultaneously. For example, it is quite common to see a process exhibit meander-ing behavior *around* a trend. In this section, we study a process driven by several systematic effects.

Without question, one of the most critical processes that any organization would be concerned about is the number of current customers over time. De-pending on the nature of the organization (profit or not-for-profit), changes in the organization's customer base have a direct impact on key performance measures, such as market share, revenue, and profit.

Consider the membership totals for the Statistics Division of the American Society for Quality (ASQ). ASQ is a leading organization with a mission of com-municating and promoting all aspects of quality-related principles and concepts. Established in 1979, the Statistics Division is the second largest of ASQ's 21 di-visions, comprising nearly 10 percent of ASQ's total membership. As stated in the ASQ Statistics Division Newsletter (1995), "One measure of our success should be increased Division membership. The ASQ Statistics Division is a re-source organization which serves to promote Statistical Thinking and the correct use of statistical methods for quality and productivity . . ." (p. 19).

Table 3.9   Monthly membership totals (ASQ, Statistics Division)

14,668	14,720	14,644	14,919	15,099	15,264	15,461	15,579	13,482
13,893	14,112	14,479	14,666	15,090	14,831	14,874	14,822	14,936
15,153	15,985	13,934	14,309	14,496	14,736	14,800	14,870	14,392
14,387	14,414	14,424	14,514	14,561	12,363	12,588	12,728	12,987
13,075	13,280	13,180	13,189	13,120	13,105	13,144	13,237	11,322
11,558	11,795							

To track the membership success of the division, the end-of-the-month mem-bership totals are collected and reviewed. Analysis and discussion of member-ship totals data series are published periodically in the division's newsletter. In Table 3.9, the monthly membership totals (member$_t$) are from February 1991 to December 1994, that is, 47 observations.[14] The data are also available from file "asq.dat."

The time-series plot of the data series is given in Figure 3.44. The 12 in the Minitab command causes Minitab to label the months of the year—1, 2, 3, 4, 5, 6, 7, 8, 9, 0, A, B—where, for example, 1 means January, 8 means August, and A means November. The data series is clearly nonrandom. For one, the member-ships appear to be trending down over time. Furthermore, the trend seems to be curved downward, suggesting an increasing rate of membership decline. Beyond

[14]The author is grateful to Rick Lewis (Division Chair) and Bob Mitchell (Membership Committee Chair) for making available these data. The author wishes to also acknowledge the staff at the membership department of the national ASQ headquarters for providing insights to certain movements found in the data series.

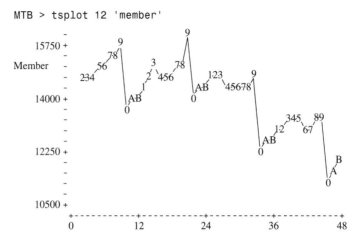

Figure 3.44   Time-series plot of monthly membership series.

the trend, there is a strong suggestion of seasonality. The most obvious evidence is the consistent drop in membership levels from September (9) to October (0).

The lower levels in Octobers are due to an annual membership adjustment made by ASQ every September-October to purge unpaid renewals. Every year, the deadline for renewing membership falls in the beginning of April. During this period, members are asked to either forward a renewal payment or indicate their intent not to renew. Members who do not respond to either of these choices are ultimately dropped from the membership database during the September-October purge. Those who wish to continue as ASQ members are also given the opportunity to continue, drop, or add affiliation(s) with one or more of the 21 divisions (which includes the Statistics Division).

There may also be lagged effects in the data series. One way to notice the possibility of lagged effects is to look at three "mini-series" starting in October and ending in September of the following year. Within each of these mini-series, the successive observations are close to one another, giving rise to a meandering appearance for the string of observations.

Given the possibility of lagged effects, we use the PACF (Figure 3.45) to see how many lags might be needed. As can be seen, the coefficients at lags 1 and 13 fall well beyond the benchmarks of $\pm 2 / \sqrt{47} = 0.292$, the lag 1 coefficient in the positive direction and the lag 13 coefficient in the negative direction.

If the PACF is interpreted literally, we create a lag 1 variable ($member_{t-1}$) and a lag 13 variable ($member_{t-13}$). The lag 1 variable is a reasonable candidate independent variable since most of the successive observations are closely bound. To believe that a lag 13 variable should be incorporated into the fitted model is to believe that all observations 13 months apart are inversely related. Intuitively, such a relationship over time makes little sense in the context of the application.

Should we ignore the significant correlation and dismiss it as some sort of aberration? The answer lies in recognizing that lag 13 is curiously close to a lag

```
PACF of member
              -1.0  -0.8 -0.6 -0.4 -0.2  0.0   0.2   0.4   0.6   0.8   1.0
               +----+----+----+----+----+----+----+----+----+----+
 1    0.770                              XXXXXXXXXXXXXXXXXXXX
 2    0.027                              XX
 3   -0.036                            XX
 4    0.193                              XXXXXX
 5    0.035                              XX
 6    0.106                              XXXX
 7   -0.122                          XXXX
 8    0.036                              XX
 9    0.073                              XXX
10    0.021                              XX
11    0.069                              XXX
12    0.201                              XXXXXX
13   -0.574                 XXXXXXXXXXXXXXX
14   -0.030                              XX
15   -0.025                              XX
16   -0.052                              XX
```

Figure 3.45   PACF of membership series.

number that has greater meaning for the application. Namely, the number 13 is adjacent to the number 12, which represents the number of seasonal periods (12 months in a year). Recall that our earlier visual inspection of the data suggested evidence of seasonality. What is occurring here is a compounding of different nonrandom effects (seasonality, lag 1 effects, and trend) to produce significant lag 13 correlation.

With this in mind, we will initially try only the lag 1 variable. If one lag is not sufficient, we should presumably be alerted later during diagnostic checking. To capture the seasonal effects, we create 12 seasonal indicator variables. Given the appearance of a curving downward trend, we also create two plausible trend variables: the time index t and the time index squared t^2. To sort through these 15 candidate independent variables, we invoke the stepwise procedure.

From the output shown in Figure 3.46, we see that five variables (t^2, member$_{t-1}$, apr, sept, and oct) have been screened out as contributing significantly to the overall fit. Patching together the coefficients reported in step 5, we find the estimated fitted line to be given by

Fitted member$_t$ = 2529 − 0.251t^2 + 0.843 member$_{t-1}$ − 332apr + 240sept − 2042oct

The absolute values of all independent variable t ratios exceed 2, the approximate 5 percent threshold. This model also explains away a considerable amount of the variation in the membership data as reflected by a very high R^2 of 98.53 percent.

Given our preliminary visual examination, the inclusion of the trend variable, the lag variable, and the October indicator variable is no surprise. The coefficient of −332 in front of the April indicator variable implies an estimated decrease of 332 members from March to April. Looking at Figure 3.44, we see that this statistically significant seasonality is too subtle to visually discern. In light of some earlier noted background information, the April seasonality can be

```
Response is member on 15 predictors, with N =  46
   N(cases with missing obs.) =  1 N(all cases) =  47
```

Step	1	2	3	4	5
Constant	15095	15176	2218	1837	2529
t^2	−1.339	−1.308	−0.213	−0.188	−0.251
T-Ratio	−9.79	−11.40	−2.81	−2.97	−4.08
oct		−1211	−2054	−2115	−2042
T-Ratio		−4.45	−18.80	−23.15	−23.39
member-1			0.863	0.891	0.843
T-Ratio			17.13	21.16	20.18
apr				−371	−332
T-Ratio				−4.52	−4.36
sept					240
T-Ratio					2.97
S	621	520	186	154	141
R-Sq	68.55	78.45	97.30	98.20	98.53

Figure 3.46   Stepwise regression of member, on lagged, trend, and seasonal variables.

explained. Namely, April is the deadline for membership renewal. This would mean that slightly more than 300 current ASQ members are explicitly choosing either not to continue as ASQ members outright or to continue as ASQ members but not to remain affiliated with the Statistics Division.

The final variable selected by the stepwise procedure is the September indicator variable with an associated positive coefficient of 240. This implies that there is an annual expected increase of 240 members from August to September. Rather than attempt to seek an explanation for this result, we will find it rewarding to perform diagnostic checks of the model. In Figure 3.47, we present the most basic checks for randomness and normality, a time-series plot of the residuals and a histogram of the standardized residuals. Looking at either of these two plots, we are immediately struck by a substantial outlier residing in the residual data. More than 4 standard deviations above the mean of the residuals, the outlier is associated with September 1992. (This outlier falls well beyond the standard control chart scheme which alerts users to observations 3 standard deviations away.) Bear in mind that the fitted model expects September observations to be higher, and still this outlier appears.

When we look back at the sequence plot of the original data (Figure 3.44), it can be seen that the membership total for this September jumped up substantially relative to the movements of the other September totals. Explanation should be sought by the organization to discover if there are any special circumstances which led to this membership surge.

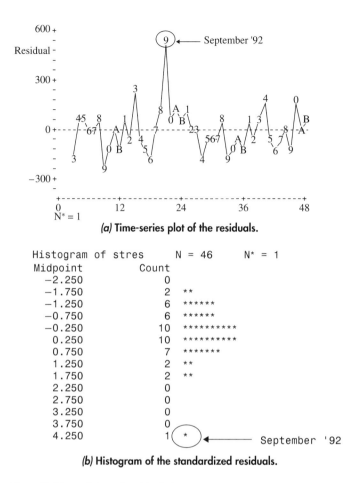

(a) Time-series plot of the residuals.

```
Histogram of stres     N = 46      N* = 1
Midpoint          Count
  -2.250            0
  -1.750            2   **
  -1.250            6   ******
  -0.750            6   ******
  -0.250           10   **********
   0.250           10   **********
   0.750            7   *******
   1.250            2   **
   1.750            2   **
   2.250            0
   2.750            0
   3.250            0
   3.750            0
   4.250            1   *
```
September '92

(b) Histogram of the standardized residuals.

Figure 3.47   Selected residual plots.

One might suspect that the unusual observation is somehow a database "blip" associated with the unpaid-renewal purging process performed during the September-October time frame. Is it possible that there was no "real" surge in membership, but rather the larger September total resulted from a late start of the 1992 purging process? This argument would be convincing if the difference between the membership total prior to September and the membership total after purging, which is reflected in the October total, were relatively consistent from year to year. Studying the original series (Figure 3.44), reveals that the differences between the August level (8) and the October level (0) for 1991, 1993, and 1994 are quite close in value. However, this August/October difference for 1992 is nearly one-half that of the other years. Thus, it seems unlikely that the source of the substantial outlier is related to a mistimed purging process; rather it appears to be an unusually high *addition of new members.* The challenge to the ASQ and the Statistics Division is to pinpoint the factors underlying the favorable effect on membership; if these factors are found, they may offer insights for future improvements.

   We need to react to outliers not only for their organizational implications, but also for their statistical implications. A primary statistical concern with outliers is that they might unduly influence the estimated coefficients of the regression equation and supporting summary statistics (such as t ratios). In Chapter 2, we discussed a similar concern with extreme points on the sample mean estimate. Because of the potential misleading impact of outliers on the estimated model, some analysts recommend tossing out all extreme points, regardless of whether explanations can be given, and reestimating the model without these points. The difficulty with this policy is that improbable observations that are part of the natural underlying variability will be discarded. We must recognize that regression analysis assumes an underlying normal distribution and the likelihood (albeit small) of extreme points purely due to chance causes. By basing models only on "cleansed" data, there is a serious consequence of underestimating the level of the underlying random variation. Prediction limits of future observations will be too tight; as a result, there is a greater risk of future points falling outside limits and being incorrectly deemed unusual.

   A much more conservative approach is to discard an outlier only if it can be traced to a specific cause not associated with the general workings of the process. With this philosophy, a worst-case scenario would be to retain a substantial outlier if no explanation could be found for it, leaving us "stuck" with the observation and its distorting effects. A reasonable stance is to come to some compromise between the two extreme approaches. For example, discard an outlier either if there is a special explanation for it or if the fitted model estimated from the data with the outlier is dramatically different from the fitted model estimated from the data without the outlier. As Anscombe (1973) states, "We are usually happier about asserting a regression equation if the relation is still appropriate after a few observations (any ones) have been deleted—that is, we are happier if the regression relation seems to permeate all the observations and does not derive largely from one or two."

   In the above discussions, we speak of "tossing out" or "discarding" outliers. For some data sets, this can be done literally by simply deleting the observation from the data set; in Minitab, this is accomplished by deleting the row in the spreadsheet associated with the outlier. However, for time-series data, especially nonrandom data, deleting observations in this manner is problematic. Suppose, for instance, that we delete the September 1992 data point (observation number 20). This action will result in the August 1992 and October 1992 observations being adjacent to each other when they really should be two periods apart. The application and interpretation of a lag 1 independent variable would be inappropriate at this time period. Furthermore, the data series will be out of synchronization one month after the 19th observation.

   Rather than directly delete an outlier, we should seek an alternative way of taking out the effects of the outlier which preserves the correct relative positioning of the data observations. There are two ways of doing this. One way is to replace the outlier observation with a missing value. In Minitab, a missing value is signified by the * symbol, the asterisk. For the membership series, we would then have the 19th value being the August 1992 observation, the 20th value being the * symbol, and the 21st value being the September 1992 observation. Since the relative contributions of the independent variables might change, it is important to treat the modified data series as a completely new data set.

A second way of handling an outlier is to create a special independent variable and model the effects of the outlier with this variable. To do so, we set up an indicator variable which takes on the value 1 in the 20th position and 0 for all other places, as shown below:

```
sept92
0    0    0    0    0    0    0    0    0    0    0    0    0    0    0
0    0    0    0    1    0    0    0    0    0    0    0    0    0    0
0    0    0    0    0    0    0    0    0    0    0    0    0    0    0
0    0
```

This variable is added to the original group of candidate independent variables, and a new fitted model can be found. It turns out that by including sept92 in a fitted model, observation 20 is fit perfectly, leaving a residual of zero. In other words, the effects of the September 1992 observation are completely absorbed by the special indicator variable. This is conceptually equivalent to deleting the observation without the problem of disturbing the relative positions of the data observations.

To carry through the analysis with the special indicator variable, we performed a stepwise regression on all the original independent variables plus the new one. The stepwise procedure (not shown) selected the following five variables: t^2, $member_{t-1}$, apr, oct, and sept92. The detailed regression output based on these independent variables is shown in Figure 3.48. All the t ratios are highly significant with all p values less than 0.0005. The model explains away more than 99 percent of the observed variation in the data series.

For comparison, the fitted models with and without the effects of the outlier, respectively, are as shown:

$$\text{Fitted } member_t = 2529 - 0.251t^2 + 0.843member_{t-1} - 332apr + 240sept - 2042oct$$

$$\text{Fitted } member_t = 2389 - 0.222t^2 + 0.852member_{t-1} - 337apr - 2058oct + 772sept92$$

Notice that the removal of the outlier resulted in certain adjustments of the estimated regression coefficients. The major change after removal of the effects of the outlier is the dropping out of the September indicator variable. Unlike the first model, the new model suggests that there is no consistent year-to-year seasonal rise in September membership. This means that the membership total for September 1992 was so large that it unduly influenced the September indicator variable to become significant, giving the impression of a significant year-to-year effect.

By removing the outlier effect with an independent variable instead of replacing the observation with a missing value, we have the advantage of explicitly summarizing the impact of the outlier. In particular, we learn from the

```
The regression equation is
member = 2389 - 0.222 t^2 + 0.852 member-1 - 337 apr - 2058 oct + 772 sept 92

46 cases used 1 cases contain missing values

Predictor        Coef       StDev          T         P
Constant       2389.2       424.7       5.63     0.000
t^2           -0.22239     0.04199      -5.30     0.000
member-1       0.85213     0.02835      30.05     0.000
apr           -336.66       54.42       -6.19     0.000
oct          -2057.79       60.93      -33.77     0.000
sept92         772.5        105.4        7.33     0.000

S = 101.8     R-Sq = 99.2%      R-Sq(adj) = 99.1%

Analysis of Variance

Source            DF         SS          MS          F         P
Regression         5    5351532    10703065    1033.21     0.000
Residual Error    40     414360       10359
Total             45   53929683

Source       DF     Seq SS
t^2           1   36968213
member-1      1    3266335
apr           1      28118
```

Figure 3.48   Final estimated regression equation for membership series.

regression coefficient for sept92 that there was an increase of more than 770 members in September 1992 over and above the general movement of the process as summarized by the trend, lag one, April, and October variables.

The t ratio of 7.33 indicates that the estimated increase is more than 7 standard deviations away from what is expected by the estimated systematic pattern. This is a much more dramatic assessment of the outlier effect which was earlier found to be slightly more than 4 standard deviations away from what is expected by the fitted model. This is because the removal of the outlier gave us a more accurate estimate of the systematic nonrandom pattern, and hence we have a more appropriate baseline to determine the extent of the departure from what is expected. The p value for the sept92 variable reports just how likely it is to observe a membership total deviating more than 770 from that expected purely due to chance causes. The p value of 0.000 (< 0.0005) indicates deviation is unlikely to be the result of random variation. For the curious, the exact probability is around 7 in 1 billion! Without question, some exceptional event did in fact happen in September 1992.

To complete the analysis, we should look at the standard checks of the residuals. A time-series plot of the residuals and a histogram of the standardized residuals are offered in Figure 3.49. Based on these two basic checks, the residuals appear to be random and approximately normal. The reader can confirm

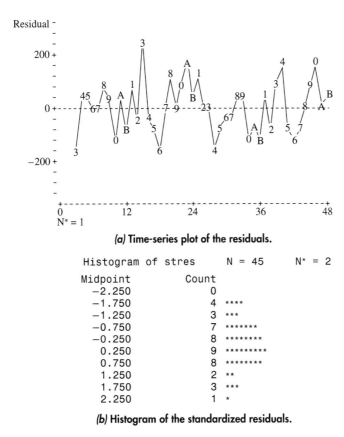

(a) **Time-series plot of the residuals.**

```
Histogram of stres      N = 45      N* = 2
Midpoint          Count
  -2.250            0
  -1.750            4   ****
  -1.250            3   ***
  -0.750            7   *******
  -0.250            8   ********
   0.250            9   *********
   0.750            8   ********
   1.250            2   **
   1.750            3   ***
   2.250            1   *
```

(b) **Histogram of the standardized residuals.**

Figure 3.49   Selected residual plots for final estimated model.

that the other checks (runs test, ACF, and normal probability plot) substantiate this conclusion.

Finally, we visually summarize the way in which the fitted model tracks the data. Figure 3.50 gives a time-sequence plot of the fitted values (solid line) and the actual membership totals (dashed line). Clearly, the fitted model has captured the essence of the process tendencies over time.

The average losses of 337 and 2058 members for the April and October months, respectively, coupled with the curving trend downward summarized by the t^2 component do not bode well for the Statistics Division. In the Statistics Division's Newsletter (1996), it is stated that ". . . we are continuing to lose members, the 'Assessing Member Needs' tactical plan seeks to develop a customer satisfaction measurement for the Statistics Division." The fitted model can serve as a litmus test for the success of any attempts to improve the membership process. If membership totals basically remain within prediction intervals (say, 95 or 99.7 percent) computed from the fitted model, then improvement efforts cannot be judged as having any significant impact.

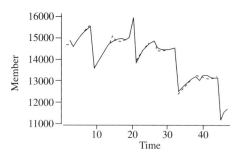

Figure 3.50   Time-series plot of fitted values along with original membership series.

3.10   SUMMARY OF PROCESS MODELING STEPS

We have covered the basic data tools for assessing the time and distribution behavior of process data. When data behave differently from the standard assumptions of randomness and normality, we learned how the techniques of data transformation and regression modeling can help address such departures. The examples of the chapter illustrated the wide variety of possible process behaviors in real applications. We found that processes can exhibit various combinations of random or nonrandom behavior along with normal or nonnormal behavior. As a brief recap, we witnessed the following combinations:

1. A process (stopwatch readings) that exhibits *random* and *normal* behavior. Such a process is amenable to standard statistical process control techniques, to be discussed in the following chapters of this book.

2. A process (daily caloric intake) that exhibits *random* and *nonnormal* behavior. We learned that a data transformation can produce random and normally distributed data. Statistical techniques appropriate for random and normal data can then be applied to the transformed data.

3. A process (elongation measurements) that exhibits *simple linear trend* behavior with *normally distributed* deviations around the pattern. We learned that a regression model using a simple time index can appropriately summarize the process behavior.

4. A process (medication errors) that exhibits *nonlinear trend behavior* with *nonnormally distributed* deviations around the pattern. We learned to transform the process data and then fit a regression model to the transformed data.

5. A process (daily weight) that exhibits *meandering behavior* with underlying *normality*. We learned that lagged variables can be incorporated into a regression model to capture meandering behavior.

6. A process (monthly membership total) that exhibits a nonrandom pattern resulting from a combination of effects: *trend, meandering,* and *seasonality*

with underlying *normality*. This example also provided us with the opportunity to witness and deal with an outlier in the context of a nonrandom process.

In conclusion, if conducted in a systematic manner, most process data series can be satisfactorily summarized by the data analysis methods developed in this chapter. For basic guidance, follow the general steps for analyzing the process behavior of any process series x_t:

1.   Determine whether x_t data are random. The checks include a time-series plot, runs test, and ACF. If random, proceed to step 2. If not, proceed to step 4.

2.   Determine whether x_t data are compatible with normality. This step is only appropriate when the data are deemed random. The checks include a histogram of the standardized data; counts of number of observations between ± 1, ± 2, and ± 3 standard deviations; and a normal probability plot.

   If the random data are consistent with the normal distribution, the sample mean serves as the fitted model; no further steps listed here are required. If nonnormality is detected which cannot be attributed to "nonstatistical" explanations (e.g., mixed processes), proceed to step 3.

3.   Seek a normalizing transformation for the random but nonnormal data. Explore the family of power transformations, possibly guided by Figure 3.12 or the Box-Cox method. Once an appropriate transformation is found, the transformed data can be treated in the same manner as any other set of random and normal data; *no further steps listed here are required.*

4.   Since the data are found to be nonrandom, we explore regression modeling to summarize process behavior. Begin with x_t as the dependent variable. Next, gather plausible independent variables. If the time-series plot reflects trend behavior, consider creating time trend variables (say, t and t^2). Meandering behavior suggests the creation of lagged variables (x_{t-1}, x_{t-2}, . . .). PACF is a good guide for determining how many lagged variables to try. If the data series is sampled on seasonal periods, consider creating seasonal indicator variables.

   When many candidate independent variables are involved, stepwise methods can be used to screen out the useful variables. At this stage, estimate the regression equation based on the promising independent variables. Proceed to step 5.

5.   Begin the diagnostics for model appropriateness by checking whether the residual data are random over time. The checks include a time-series plot, runs test, and ACF. If the residuals are found to be nonrandom, return to step 4 to consider an alternative model. If the residuals are random, proceed to step 6.

6.   Determine whether residuals are compatible with normality. This step is only appropriate when the residuals are deemed random. The checks include a histogram of the standardized residuals; counts of number of observations between ± 1, ± 2, and ± 3 standard deviations; and a normal probability plot. If residuals exhibit discernible nonnormality, apply normalizing transformations to the original x_t and then return to step 4.

Look out for substantial outlying residuals. If special explanations are discovered, create special indicator variables, then return to step 4. If no special explanations can be found but the outliers appear so substantial that they might have a disproportionate effect on the regression equation, consider again the creation of special indicator variables; if subsequent analysis reveals little impact of the unexplained outliers on the fitted model, then *do not remove* the outliers from the data.

If these final sets of diagnostics are satisfactory, the estimated regression equation can serve as an appropriate summary of the observed process behavior as well as a means for future prediction. Thus, *no further steps listed here are required.*

EXERCISES

3.1. For parts *a* to *c* below, use the given sequence of pluses and minuses (plus implies original observation above the mean and minus implies original observation below the mean) to determine the values of n and m and the observed number of runs.
  a. $+ - + + + - - - - + - -$
  b. $- - - + + - - - - -$
  c. $+ - - + + - + + - - - + + + - - - - - + - - + +$

3.2. In parts *a* to *e* below, we give a sequence of pluses and minuses (plus implies original observation above the mean, and minus implies original observation below the mean) that is associated with a random process, simple trend process, oscillating process, or stationary meandering process. Indicate the most likely match between the sequence and one of four possible time behaviors.
  a. $+ + + - - + - + - -$
  b. $+ - + - + + - + - + - - + - + - +$
  c. $- - - - - - - + - + + + + + + + + + + + + + + +$
  d. $- - - - + + + + + - - - - - - - - + + + + + + + +$
      $- - - - + + + + - - - - - -$
  e. $+ + + + + + - + + - - - - - - - - - - - - - - -$
      $- - -$

3.3 The thicknesses of glue (in millimeters) applied to cardboard are listed below for samples randomly selected once each half hour.[15]

0.159	0.137	0.163	0.141	0.150	0.162	0.157	0.145	0.151
0.144	0.135	0.151	0.146	0.147	0.156	0.157	0.146	0.151
0.162	0.149							

  a. Determine the values of n and m and the observed number of runs.
  b. Assuming the underlying process is random, what is the expected number of runs?

[15]All sequenced data provided in the exercises are read across and then down.

    c. If the underlying process is random, about 95 percent of the runs counts, given n and m from part *a*, will fall within what interval?

    d. Based on results from part c, test the null hypothesis of randomness at a 5 percent level of significance.

3.4. Consider the glue thickness data in Exercise 3.3. Show the details of the computations to determine the p value for the runs test applied to these data.

3.5. Weekly sales (in thousands of dollars) for a Web-based retail store are shown below (data are also found in "websales.dat"):

```
61.6361   62.9236   66.7807   64.7094   64.6682   66.2040   68.9796
63.4278   66.8801   70.7408   65.0917   67.5841   67.6864   66.5981
70.5323   73.5983   69.2186   73.6375   71.7811   72.2896   76.8067
78.5283   77.7869   75.1795   78.6190   81.8842   79.8196   80.6729
76.9433   80.3274
```

    a. Determine the values of n and m and the observed number of runs.

    b. Assuming the underlying process is random, what is the expected number of runs?

    c. If the underlying process is random, about 95 percent of the runs counts, given n and m from part *a*, will fall within what interval?

    d. Based on results from part c, test the null hypothesis of randomness at a 5 percent level of significance.

3.6. Consider the sales data in Exercise 3.5. Show the details of the computations to determine the p value for the runs test applied to these data.

3.7. Determine the value of the missing part of the runs test output applied to a data series of length 100:

```
MTB > runs 'x'

Runs Test
     x

K = −0.0218

The observed number of runs = 49
The expected number of runs = 50.2800
44 Observations above K    56 below
       The test is significant at   ????
       Cannot reject at alpha = 0.05
```

3.8. For three distinct data series, below are three runs tests (labeled R1, R2, and R3) and three ACFs (labeled ACF1, ACF2, and ACF3). Indicate which runs test corresponds with which ACF. Justify your match.

```
R1

K = −0.0460

The observed number of runs = 70
The expected number of runs = 50.9200
52 Observations above K    48 below
       The test is significant at   0.0001
```

```
R2

K =  0.0435

The observed number of runs = 53
The expected number of runs = 50.9200
52 Observations above K    48 below
        The test is significant at    0.6754
        Cannot reject at alpha = 0.05

R3

K =  −0.1773

The observed number of runs = 29
The expected number of runs = 51.0000
50 Observations above K    50 below
        The test is significant at    .0000

ACF1

        -1.0  0.8   0.6   0.4   0.2   0.0   0.2   0.4   0.6   0.8   1.0
        +----+----+----+----+----+----+----+----+----+----+
  1    0.589                             XXXXXXXXXXXXXXXX
  2    0.322                             XXXXXXXXX
  3    0.195                             XXXXXX
  4    0.169                             XXXXX
  5    0.027                             XX
  6    0.029                             XX
  7    0.013                             X
  8    0.010                             X
  9   -0.048                             XX
 10   -0.038                             XX

ACF2

        -1.0 -0.8  -0.6  -0.4  -0.2   0.0   0.2   0.4   0.6   0.8   1.0
        +----+----+----+----+----+----+----+----+----+----+
  1   -0.010                             X
  2    0.057                             XX
  3   -0.151                          XXXXX
  4    0.132                             XXXX
  5   -0.081                           XXX
  6    0.057                             XX
  7   -0.137                          XXXX
  8   -0.065                           XXX
  9   -0.016                             X
 10   -0.103                          XXXX
```

ACF3

```
              -1.0 -0.8 -0.6 -0.4 -0.2  0.0  0.2  0.4  0.6  0.8  1.0
               +----+----+----+----+----+----+----+----+----+----+
    1   -0.531                 XXXXXXXXXXXXXX
    2    0.266                               XXXXXXXX
    3   -0.243                    XXXXXXX
    4    0.294                               XXXXXXXX
    5   -0.299                   XXXXXXXX
    6    0.206                               XXXXXX
    7   -0.129                      XXXX
    8    0.133                          XXXX
    9   -0.153                     XXXXX
   10    0.133                          XXXX
```

3.9. Consider the glue thickness data in Exercise 3.3. Obtain the ACF for the data series. Based on the ACF, would you deem the process random?

3.10. Consider the sales data in Exercise 3.5. Obtain the ACF for the data series. Based on the ACF, would you deem the process random?

3.11. Consider the following histogram on standardized data:

```
Histogram of stdx    N = 60
Midpoint   Count
  -2.750      0
  -2.250      1    *
  -1.750      2    **
  -1.250      6    ******
  -0.750     11    ***********
  -0.250      8    ********
   0.250     16    ****************
   0.750      8    ********
   1.250      4    ****
   1.750      2    **
   2.250      1    *
   2.750      1    *
```

a. What percentage of the observations falls between ± 1, ± 2, and ± 3 standard deviations around the mean?

b. What percentage of the observations falls beyond 2 standard deviations above the mean?

c. What percentage of the observations falls beyond 1.5 standard deviations below the mean?

3.12. Suppose a data set of 5 observations is standardized to obtain the following numbers:

    1.21298   0.62939   −0.55256   −1.35039   0.06058

If the sample mean and standard deviation of the unstandardized observations are, respectively, 8.798 and 1.354, what are the values of the unstandardized observations?

3.13. Consider the daily caloric intake data series introduced in Section 3.3. Transform the data, using the Box-Cox recommended value of $\lambda = -0.899$. Compare the results of this transformation with $\lambda = -1.5$ as determined by Hines and Hines (1987) and illustrated in the section.

3.14. Consider the daily caloric intake data series introduced in Section 3.3 (data given in Table 3.3 and found in the file named "calories.dat"). Based on a power transformation of $\lambda = -1.5$, provide a 95 percent prediction interval for a future caloric intake value.

3.15. ComputerTeck Co. is a manufacturer of several computer series, including the CYBER 120 series. Whenever changes to the series are made, documentation must be revised. Such revisions are initiated by change orders. A study was undertaken to understand how change order processing times vary over time. Data were collected for 65 sequentially timed orders during the period from July 1995 to October 1997. The time (in working days) it took for each change order to be approved was recorded. The data are shown below and also available in the file "change.dat."

29	21	17	9	49	20	34	66	16	28
6	44	15	15	63	30	21	21	11	16
32	14	10	15	10	28	59	32	24	125
27	23	22	28	74	39	8	16	23	31
20	23	40	12	72	33	51	4	16	42
23	29	37	24	62	48	10	9	15	8
46	6	7	5	48					

a. Perform randomness checks on the data series. What do you conclude?

b. Perform normality checks. If the data are found to be nonnormal, seek an appropriate normalizing transformation.

c. Provide a 95 percent prediction interval for future change orders in terms of the number of working days.

3.16. What are the steps that should be used in evaluating the appropriateness of a regression model? Write down each step in the order in which it should be evaluated, and provide a brief explanation of its importance.

3.17. Consider the stopwatch data series introduced in Section 2.3. Given that the data series is fitted with the sample mean, what are the residual values for the data set? What is the value of SSE?

3.18. If you are familiar with basic techniques of calculus, show that if you wish to fit a horizontal fitted line to the data series, then $\bar{y}$ is the least-squares fitted model.

3.19. If you are familiar with basic techniques of calculus, show that Equations (3.14) and (3.15) are indeed least-squares estimates for a simple linear regression model.

3.20. Show algebraically that $\sum_{i=1}^{n} e_i = 0$ when a data set is fitted with the least-squares model $\hat{y}_i = b_0 + b_1 x_i$.

3.21. Below is a portion of a regression output based on a sample size of $n = 25$:

```
Predictor      Coef     StDev      T       P
Constant    41.0264      (1)    48.65   0.000
x               (2)   0.9922     2.01     (3)

S = (4)        R-Sq = (5)

Analysis of Variance

Source           DF       SS      MS
Regression      (6)    56.66   56.66
Residual Error  (7)      (8)     (9)
Total          (10)   380.63
```

Determine the values for the 10 missing values.

3.22. What is the general relationship between SST and the sample variance of the dependent variable s_y^2?

3.23. *Substandard product loss,* referred to as *SSPL,* is a major concern of any organization. At a particular organization with annual revenue of approximately $100 million dollars, SSPL accounts for nearly 7 percent of sales, or $7 million annually. Given this tremendous dollar loss, efforts are underway to continually reduce SSPL. Below are data on the number of incidents of SSPL by week during the year 1998 (data are also found in "sspl.dat"):

28	33	25	24	28	29	31	27	26	27
29	27	22	24	25	26	25	26	24	23
27	23	23	25	22	20	25	19	19	20
22	17	20	19	17					

  a. Time-series plot the 35 consecutive observations.
  b. Perform the runs test on the data series. Based on a 5 percent level of significance, does the runs test deem the process to be random?
  c. Obtain the ACF for the data series. Based on the ACF, would you deem the process as random or not?

3.24. Consider the SSPL data in Exercise 3.23. Fit the data series with a simple trend model. Perform and show all model diagnostic checks. Given the fitted model, what is the rate of improvement in SSPL? Provide a 90 percent prediction interval for SSPL in week 36.

3.25. Consider the medication error series introduced in Section 3.7. Based on the final fitted model presented in the section, provide a 95 percent prediction interval for the number of medication errors for week 41 in original units.

3.26. The number of customers arriving at Franklin Valley Bank each day is recorded for 5 consecutive weeks (data are also found in "bank.dat"):

	Monday	Tuesday	Wednesday	Thursday	Friday	Saturday
Week 1	112	99	100	97	130	96
Week 2	120	109	104	104	126	106
Week 3	114	106	99	107	118	102
Week 4	114	97	104	101	125	104
Week 5	112	100	103	97	124	109

a. Make a time-series plot of the 30 consecutive observations in default high-resolution mode (without data labels).

b. Make a time-series plot of the 30 consecutive observations, i.e., labeling the observations by day of the week. Comment on this plot in comparison to the plot from part *a*.

c. Perform the runs test on the data series. Based on a 5 percent level of significance, does the runs test deem the process to be random?

d. Obtain the ACF for the data series. Based on the ACF, would you deem the process to be random?

3.27. Consider the bank customer data in Exercise 3.26. Use seasonal indicator variables to find an appropriate fitted model for the data series. Perform all model diagnostic checks. Given the fitted model, how many customers are expected to arrive at the bank on each of the days of the week?

3.28. Consider the fitted model for the daily weight series shown in Figure 3.39. We noted in section 3.8 that if the lag one regression coefficient is less than 1 in absolute value, then the process is stationary. Construct a 95 percent confidence interval for the true lag one regression coefficient (β_1). Based on the resulting confidence interval, is it reasonable to conclude that the underlying weight process is stationary?

3.29. Consider the daily weight series introduced in Section 3.8. Provide a 95 percent prediction interval for the individual's weight on day 57.

3.30. An article in *Journal of Quality Technology* (Cryer and Ryan, 1990, Vol. 22, No. 3, p. 189) presents data from a chemical process in which the measurement variable is a color property. The data are shown below and also given in "color.dat:"

0.67	0.63	0.76	0.66	0.69	0.71	0.72	0.71	0.72	0.72
0.83	0.87	0.76	0.79	0.74	0.81	0.76	0.77	0.68	0.68
0.74	0.68	0.69	0.75	0.80	0.81	0.86	0.86	0.79	0.78
0.77	0.77	0.80	0.76	0.67					

a. Make a time-series plot of the 35 consecutive observations.

b. Perform the runs test on the data series. Based on a 5 percent level of significance, does the runs test deem the process to be random?

c. Obtain the ACF for the data series. Based on the ACF, would you deem the process to be random?

3.31. Consider the color data in Exercise 3.30. Obtain the PACF for the data series. Based on the information of the PACF, find an appropriate fitted model for the data series. Perform and show all model diagnostic checks. Given the fitted model, what is your prediction of the 36th color observation?

3.32. In a file named "boxja.dat" is the well-known data series referred to as "Series A" from Box, Jenkins, and Reinsel (1994). The data series consists of 197 concentration readings from a chemical process, readings taken every two hours.

a. Perform all randomness checks on the data series.

b. Obtain the PACF for the data series. Which lags are associated with significant correlations based on superimposed 95 percent limits on the PACF?

c. Given the information from part *b*, find an appropriate fitted model for the data series. Perform all model diagnostic checks.

d. Given the fitted model from part *c*, what is your prediction of the 198th concentration reading?

3.33. Consider the ASQ membership series introduced in Section 3.9. Provide a 95 percent prediction interval for the membership total one month into the future, that is, the 48th observation.

3.34. With concerns of global warming and the environment, the global temperature process has become a process receiving great attention from many scientists, politicians, and citizens at large. Starting from 1900 and ending in 1997, below are annual readings (in degrees Celsius) of the global temperature index (data are also given in "globaltemp.dat").

-0.14	-0.20	-0.33	-0.46	-0.48	-0.40	-0.28	-0.50	-0.45	-0.43
-0.41	-0.40	-0.36	-0.35	-0.21	-0.12	-0.31	-0.44	-0.33	-0.25
-0.28	-0.19	-0.30	-0.29	-0.29	-0.24	-0.08	-0.16	-0.17	-0.33
-0.11	-0.06	-0.13	-0.21	-0.15	-0.18	-0.12	0.02	0.06	-0.03
-0.08	-0.01	0.03	0.03	0.11	0.03	-0.05	-0.12	-0.09	-0.08
-0.15	-0.05	0.02	0.05	-0.16	-0.17	-0.25	0.02	0.09	0.03
-0.01	0.00	-0.02	0.02	-0.20	-0.18	-0.05	-0.08	-0.12	0.03
-0.05	-0.22	-0.08	0.06	-0.19	-0.15	-0.22	0.03	-0.05	0.05
0.07	0.09	0.06	0.25	0.03	-0.01	0.08	0.27	0.24	0.16
0.33	0.28	0.10	0.10	0.23	0.33	0.20	0.42		

Note: The data represent average global surface air temperatures and are relative to the temperature level in the year 1961. These data (and other climatic data) can be retrieved from the Website of the National Climatic Data Center (NCDC): http://www.ncdc.noaa.gov. NCDC is part of the Department of Commerce, National Oceanic and Atmospheric Administration (NOAA).

a. Perform all randomness checks on the data series.

b. Fit a simple trend model to the data series. In an NOAA press release, it states that "with the new data [1997] factored in, global temperature warming trends now exceed 1.0 degree F (0.55 degrees C) per 100 years." Is this statement consistent with the fitted simple trend model?

c. Perform all model diagnostic checks. Is a simple trend model an appropriate fit for the data series?

3.35. Consider the temperature data in Exercise 3.34. Go beyond the simple trend model to find an appropriate fitted model for the data series. *Hint:* Consider using a trend variable and lag(s) of temperature. Perform all model diagnostic checks. Given the fitted model, what is your prediction of the global temperature level for the year 1998?

REFERENCES

Abraham, B., and J. Ledolter (1983): *Statistical Methods for Forecasting,* New York: Wiley.

Afifi, A. A., and V. Clark (1996): *Computer-Aided Multivariate Analysis,* 3rd ed., London: Chapman & Hall.

Anscombe, F. J. (1973): "Graphs in Statistical Analysis," *American Statistician,* 27, pp. 17–21.

Bissell, D. (1994): *Statistical Methods for SPC and TQM,* London: Chapman & Hall.

Blom, G. (1958): *Statistical Estimates and Transformed Beta Variates,* New York: Wiley.

Box, G. E. P., and D. R. Cox (1964): "An Analysis of Transformation," *Journal of the Royal Statistical Society,* Series B, 26, 211–243.

———, G. M. Jenkins, and G. C. Reinsel (1994): *Time Series Analysis: Forecasting and Control,* 3rd ed., Englewood Cliffs, NJ: Prentice Hall.

Cryer, J. D. (1986): *Time-Series Analysis,* Boston: PWS-KENT.

——— and R. B. Miller (1991): *Statistics for Business: Data Analysis and Modeling,* 2d ed., Boston: Duxbury Press.

Deming, W. E. (1986): *Out of the Crisis,* Center for Advanced Engineering Study, Cambridge, MA: MIT.

Draper, N., and H. Smith (1981): *Applied Regression Analysis,* 2d ed., New York: Wiley.

Farnum, N. R., and L. W. Stanton (1989): *Quantitative Forecasting Methods,* Boston: PWS-KENT.

Gibbons J. D. (1990): "Nonparametric Statistics," in *Handbook of Statistical Methods for Engineers and Scientists,* ed. H. M. Wadsworth, New York: McGraw-Hill.

Hahn, G. J., and S. S. Shapiro (1967): *Statistical Models in Engineering,* New York: Wiley.

Hines, W. G. S., and R. J. O'Hara Hines (1987): "Quick Graphical Power-Law Transformation Selection," *American Statistician,* 41, pp. 21–24.

Joint Commission on Accreditation (1992): *Striving toward Improvement,* Oakbrook Terrace, IL: Joint Commission on Accreditation.

Menzefricke, U. (1995): *Statistics for Managers,* Belmont, CA: Duxbury Press.

Neter, J., M. H., Kutner, C. J., Nachtsheim, and W. Wasserman (1996): *Applied Linear Statistical Models,* 4th ed, Burr Ridge, IL: McGraw-Hill/Irwin.

Ryan, T. P. (1996): *Modern Regression Methods,* New York: Wiley.

Roberts, H. V. (1991): *Data Analysis for Managers with Minitab,* 2d ed., South San Francisco, CA: Scientific Press.

———— (1998): *Total Quality Management and Statistics,* Cambridge, MA: Blackwell Publishers.

Tukey, J. W. (1977): *Exploratory Data Analysis,* Reading, MA: Addison-Wesley.

Introduction to the Control Chart Concept

CHAPTER OVERVIEW

The recognition that process and product variations are omnipresent is not new. Concerned with the consistency of products manufactured for the Bell Telephone System, Walter A. Shewhart formulated a statistically based approach for studying and controlling manufacturing variation back in the 1920s. This approach has been popularized as the area of statistical process control (SPC). In an unpublished memorandum dated May 16, 1924, Shewhart made the first known sketch of a *control chart,* the distinctive tool of SPC. Thereafter, Shewhart continued to refine the concept and technique of the control chart, eventually leading to his publication in 1931 of the now classic book titled *Economic Control of Quality of Manufactured Product.*

In this chapter, we introduce the conceptual framework for the control chart technique devised by Shewhart and for SPC methods in general. Like any other statistically based technique, the control chart is designed to be most effective given certain underlying assumptions. We will discuss these assumptions along with the implications on standard SPC methods when these assumptions are materially violated in practical application.

4.1 APPROACHES TO PROCESS CONTROL

To appreciate the role of SPC methods, it is important to understand first what is meant by *process control.* Recall that a process can be simply defined as a combination of steps or operations through which inputs are converted or transformed to desired outcomes. Processes are typically repetitive, and the outcomes can be measured and recorded.

The term *control* is defined as a feedback loop based on the following four steps: (1) establishment of a standard, (2) measurement of actual performance, (3) comparison of actual performance with some standard, and (4) corrective action, if needed, addressing the discrepancy between actual and standard. Thus, process control, in general, can be defined by these steps when applied to a measurable process outcome.

Different interpretations and applications of this general feedback loop have led to two major subdivisions of process control, namely, *statistical process control* and *adaptive process control.* Box and Jenkins (1963) capture the thrust of each approach when they outline three important objectives of process control:

1. The detection of changes in process performance.
2. The identification of assignable causes of variation.
3. The adjustment of relevant process input variables so as to maintain a performance criterion in some desirable neighborhood.

Based on the pioneering work of Walter A. Shewhart, an emphasis on the development of statistical techniques to detect process changes and identify assignable causes of variation defines the general route of statistical process control. The word *control* in the acronym *SPC* may be confusing since it may convey the false impression that the process is continually adjusted or regulated. *Statistical process surveillance* or *statistical process monitoring* more accurately reflects the general intent of SPC.

The third objective—process adjustment to maintain desirable performance—is the main concern of adaptive process control (sometimes referred to as dynamic, on-line, or engineering control), as opposed to SPC. Here the word *control* accords more closely with everyday use, in the sense of *adjustment* of a process to offset systematic deviations of the process from a desired target level. For parts manufacturing processes, adjustment may mean changing the machine setting. For continuous processes, such as chemical production, adjustment may mean changing the level of an input which has a known effect on process output. Adaptive control ideas are even applicable with personal processes, for example, adjusting caloric intake to attain desired weight levels or adjusting medication dosages to attain desired levels of particular health indicators (blood pressure, hormone activity, etc.).

4.2   COMMON- AND SPECIAL-CAUSE VARIATION: BASIC BREAKDOWN

Among the great contributions of Shewhart is the lesson that all processes exhibit variability. After studying many manufacturing processes, Shewhart concluded that causes of process variation can be attributed to *chance or random causes* or to *assignable causes.* Deming (1986) refers to these two sources of variation as *common causes* and *special causes*, respectively: these terms were first introduced in Section 2.1.

Variation due to common causes, called *common-cause variation,* reflects the natural variation inherent in every process. This natural variability results from myriad minor, but inevitable, variations in process-related factors such as raw materials, machines, tools, training, measurement devices, and environment. Relatively speaking, although any one of these factors—common causes—may contribute a minute amount of variation, their combined effect yields a substantial amount of variation. Since it is assumed that common causes represent many minor *random* perturbations to which all process outcomes are exposed, their cumulative effect is by definition assumed to be random in nature.

Deming also refers to common causes as *system faults*, since they represent problems with the overall system itself. The notion of common-cause variation applies to *all* processes in *all* areas of business, not just internal processes. For instance, common-cause variation in corporate sales, customer satisfaction levels, or changes in stock prices is no doubt induced by an uncountable number of forces, such as competitors, customer perceptions, and macroeconomic variables. On the personal level, amount of sleep, stress, weather, and caloric consumption are among myriad factors inducing common-cause variation in health and fitness processes.

A special cause can be thought of as any factor impinging on a process resulting in variation inconsistent with common-cause variation. Unlike common-cause variation, whose individual sources are often difficult to isolate, special-cause variation can generally be traced to some identifiable source or event. Situations such as a bad batch of raw material, operator error, power outage, and a faulty setup are examples of special causes. As other examples, corporate merger, order of a regulatory agency, political disturbance, lawsuit, catching the flu, intake of wrong medication, and injury reflect special-cause events.

For manufacturing processes in which the goal is to reduce variability around product targets, extra variation due to special causes is viewed with negative connotations. However, special causes do not always impact negatively on a process. For example, Roberts (1991) reports a situation where in a study of a gasoline blending process, one particular blend gave substantially higher octane ratings than expected, given the overall pattern of the data. This outlier was found not to be a mistake, but rather a serendipitous discovery of a good thing which led to the issuance of a patent and increased profits for the company.

MANAGERIAL IMPLICATIONS

In Chapter 1, we discussed the fundamental importance of statistical thinking for effective management. One important component of statistical thinking that managers must possess is a clear understanding of the distinction between common and special causes of variation. To appreciate this notion, consider the following quote from Deming (1967):

> Confusion between common causes and special causes is one of the most serious mistakes of administration in industry, and in public administration as well. Unaided by statistical techniques, man's natural reaction to trouble of any kind, such as an accident, high rejection rate, stoppage of production (of shoes, for examples, because of breakage to thread), is to blame a specific operator or machine. Anything bad that happens, it might seem, is somebody's fault, and it wouldn't have happened if he had done his job right.
>
> Actually, however, the cause of trouble may be common to all machines, e.g., poor thread, the fault of management whose policy may be to buy thread locally or from a subsidiary. Demoralization, frustration, and economic loss are inevitable results of attributing trouble to some specific operator, foreman, machine, or other local condition, when the trouble is actually a common cause affecting all operators and machines, and correctable only at a higher level of management. The specific local operator is powerless to act on a common cause. He cannot change specifications of raw materials. He cannot alter the policy of purchase of materials. He cannot change the lighting system. He might as well try to change the speed of rotation of the earth.

Managers who do not understand the distinction between common- and special-cause variation are prone to view most process outcomes as the result of some special cause. As a result, individual workers, supervisors, or machines are often blamed for unsatisfactory performance which actually stemmed from common causes. Management must recognize that common-cause variation is system variation, and thus management, not individual players within the system, has responsibility.

There are two types of mistakes that may result from the confusion of common and special causes:

1. A variation is ascribed to a special cause when in fact the cause belongs to the system (common causes).

2. A variation is ascribed to the system (common causes) when in fact the cause is special.

Both mistakes are costly to an organization. According to Deming, the first mistake is predominant in management systems. The time and resources wasted in attributing common-cause variation as special-cause variation are enormous in light of the fact that system variation accounts for the majority of process variation. Many highly regarded professionals estimate that about 85 percent of process variation is due to common causes, while special causes account for the remaining 15 percent. Deming believes that system variation actually accounts for much more, well above 90 percent of total variation.

The second mistake might be viewed as the case of missed opportunities. By failing to recognize the occurrence of a special cause, management has lost an opportunity to learn about certain underlying sources which substantially impacted (favorably or unfavorably) the normal workings of the process. If the special cause is unfavorable, its effects can be corrected or removed and policies set forth to minimize a recurrence. If the special cause is favorable, its effects can be retained and policies set forth to promote or perpetuate the underlying sources of the special cause.

In order to control or improve any process, it is important to know who is in a position to act, that is, who has the authority and responsibility to take corrective action. Action on common-cause variation, by its very nature, demands management's attention. Only management can implement a fundamental change in the process required to alter the operation of common causes, for example, change technology, operating procedures, or the working environment.

The fact that management is in a position to alter common-cause variation does not mean individual employees have no role in improving the nature of common-cause variation. Such a perspective "can be devastating, since an organization needs everyone contributing to improvement . . . Management may be the only ones who can really do something about the opportunity for improvement, but they cannot act if they do not know about it" (Scherkenbach, 1986, p. 101). Often, local employees directly connected with the process are in a better position to identify potential sources of common-cause variation. Such employees can pass along suggestions and insights to management for corrective action. Thus both individual employees and management have critical roles in the altering of common-cause variation.

With respect to special-cause variation, its discovery is typically connected with the individual(s) responsible for the operation that yields data for analysis.

In some cases, especially shop floor level processes, action on special causes can be dealt with effectively by workers or by the immediate level of supervision, for example, adjusting a machine setting or fixing a jammed machine. However, in most cases, special causes have implications for management action. For example, an adverse change in process behavior due to raw material change may be detected by an operator, but the local operator usually has no control over the procurement of raw material. It is management's job to take action with suppliers to improve the quality of incoming materials. Similarly, unusual process behaviors due to special events in sales or customer satisfaction processes are clearly vital information for management-level actions.

4.3   STATE OF STATISTICAL CONTROL: HISTORICAL PERSPECTIVE

As we have learned, Shewhart envisioned the following dichotomy for process variation:

$$\text{Total variation} = \frac{\text{Common-cause}}{\text{random variation}} + \frac{\text{Special-cause}}{\text{variation}} \tag{4.1}$$

Bear in mind that Shewhart conceptualized this breakdown with his attention directed to processes related to the manufacture of parts and products. For such manufacturing processes where the objective is to produce items as close as possible to engineering targets or specifications, variation is viewed as undesirable. Hence, management's objective should be to reduce total process variation.

Reduction of total variation can be attained through the reduction of the common-cause component and/or the special-cause component. Ideally, both sources of variation can be acted upon. However, given limited resources of an organization, Shewhart recognized the need to devise a systematic but economical approach for the reduction of process variation. He recommended that the first line of attack on a process be to detect and remove special causes. Since any one special cause may result in substantial variation inconsistent with the random pattern of common causes, special causes are generally easy to detect. One of the great contributions by Shewhart is the development of a simple statistical device known as a *control chart*, which is used to help sort out the effects of special causes from common causes.

From Equation (4.1), we see that the removal of the special-cause component leaves a process governed only by common causes. Said equivalently, a process void of special causes exhibits random variation only. From Shewhart's perspective the attainment of a random process is a desired goal. The reason for this goal is again due to the nature of the process applications which Shewhart was interested in. Specifically, his focus was on devising statistical methods for the monitoring of the manufacture of parts and products. For such manufacturing processes, the objective is to produce products consistent with specified engineering targets. However, because of inherent variation in all processes, process

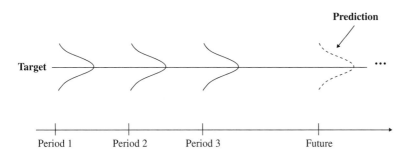

Figure 4.1   Random process centered at target.

outcomes will vary in spite of our best efforts to produce the same thing each and every time. Given this reality, what characteristics of the process might we hope for? Deming (1944) gives us an idea in the following passage:

> There is no such thing as constancy in real life. There is, however, such a thing as a constant-cause system. The results produced by a constant-cause system vary, and it may vary over a wide band or a narrow band. They vary, but they exhibit an important feature called stability. Why apply the terms constant and stability to a cause system that produces results that vary? Because the same percentage of these varying results continues to fall between any given pair of limits hour after hour, day after day, so long as the constant-cause system continues to operate. It is the distribution of results that is constant or stable.

A random process indeed offers the properties of stability and constancy in distribution. As illustrated in Figure 4.1, if centered to a target level, a random process is highly desirable since future outcomes are predicted to remain centered on the target. In contrast, a process which is unstable or erratic relative to target, as shown in Figure 4.2a, gives management no sense of future process performance. Thus, there is no assurance that product requirements will be satisfactorily met. Instability in process behavior is not the only undesirable scenario. Consider, for instance, the trend process depicted in Figure 4.2b. Even though this process is quite predictable, the implication of this predictable behavior is a steady departure away from the target level. Management should no doubt seek and remove the sources of the trend effect.

When a process exhibits only random variation, Shewhart classified the process as being in a *state of statistical control* (or simply in control), terminology first introduced in Section 2.3. From his classic treatise (1931), we find that prediction is a key element in Shewhart's definition of statistical control:

> A phenomenon will be said to be controlled when, through the use of past experience, we can predict, at least within limits, how the phenomenon may be expected to vary in the future. Here it is understood that prediction within limits means that we can state, at least approximately, the probability that the observed phenomenon will fall within the given limits. (p. 6)

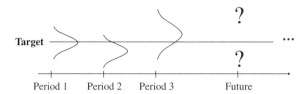

Target

| Period 1 | Period 2 | Period 3 | Future |

(a) Erratic (unstable) process relative to target

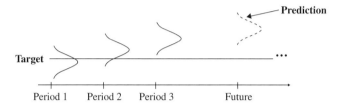

Target

| Period 1 | Period 2 | Period 3 | Future |

(b) Predictable process relative to target

Figure 4.2   Undesirable manufacturing processes.

In light of our applications and discussions in Chapter 3, we recognize that many process behaviors are accommodated by Shewhart's definition of statistical control. However, as implemented by Shewhart and his colleagues and successors, this general definition has been specialized to imply only random behavior, that is, iid behavior. Shewhart (1939) wrote: "We start with the assumption that when the operation of production is random . . . it is in a state of statistical control" (p. 26).

It is not surprising that Shewhart did not view other predictable behaviors as being in control. For example, the trend process of Figure 4.2*b* is predictable, but clearly not tolerable. In his framework, processes whose behaviors (erratic or predictable) do not consistently meet product requirements result in unnecessary waste and lower productivity, ultimately diminishing the organization's ability to satisfy customers' wants. Only a random process offers management a process that is predictable with the expectation to remain on target. In this light, Shewhart's implementation of a random process as an in-control process makes eminent sense.

It is important to note that the process depicted in Figure 4.1 is not taken to be in statistical control because the mean level of the process is centered at the engineering target. Statistical control is a purely statistical concept, not an engineering or specification-related notion. As long as the process exhibits random behavior around some mean level, it is deemed an in-control process. In Section 4.8, we will elaborate on the distinction between statistical control and process capability.

In Figure 4.1, we depicted the underlying distribution with a normal curve. Recall that randomness and distributional behavior are separate concepts. Randomness, for instance, does not imply normality. Randomness suggests only that the measures of process performance are independent and identically distributed

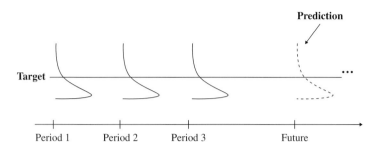

Figure 4.3   In-control process with underlying nonnormal distribution.

(iid), normal or otherwise. Identically distributed is what Deming referred to in the earlier quote as "distribution of results that is constant." Thus, as does the process in Figure 4.1, the iid process with underlying nonnormal distribution shown in Figure 4.3 represents a process in a state of statistical control.

4.4   STATISTICAL NATURE OF SPECIAL-CAUSE VARIATION

Given the earlier breakdown of variation between common- and special-cause variation along with the assumption that common-cause variation is random variation, any behavior not attributable to random variation is assumed to be the result of special causes. Shewhart (1939) summarized the idea by saying that "Our clue to the existence of assignable [special] causes is anything that indicates nonrandomness" (p. 26).

Based on Shewhart's work, a process influenced by special causes is deemed an *out-of-control* process. As implied by the pejorative adjective *out-of-control,* Shewhart viewed a process in an out-of-control state as unacceptable. He contended that special causes present in a process need to be eliminated so that the process is ultimately brought to a state of statistical control, that is, a stable system of random variation. Deming (1975) echos this same philosophy by stating "But a [stable system] is not a natural state. . . . It is a state of achievement, arrived at by elimination, one by one, by determined effort, of special causes of variation" (p. 5).

To detect special causes, it is important to have a more concrete notion of the statistical nature of special-cause variation. From our discussion, we see that Shewhart's framework suggests that any statistical behavior not consistent with independent and identically distributed (iid) behavior is a potential signal of a special cause. Theoretically, departures from iid behavior encompass an infinite number of possibilities.

Included within these possibilities are predictable processes with underlying systematic patterns, such as trends, seasonality, and lagged effects. Systematic patterns are persistent changes affecting all, or nearly all, the process outcomes.

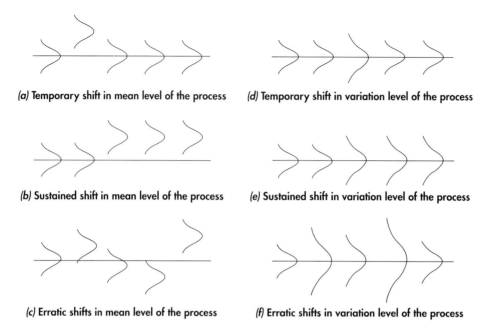

(a) Temporary shift in mean level of the process

(d) Temporary shift in variation level of the process

(b) Sustained shift in mean level of the process

(e) Sustained shift in variation level of the process

(c) Erratic shifts in mean level of the process

(f) Erratic shifts in variation level of the process

Figure 4.4   Out-of-control process behaviors.

Special-cause variation can also manifest itself as completely unexpected process behavior. Figure 4.4 illustrates a variety of unstable behaviors with respect to mean or variation levels. In Figure 4.4*a,* the mean level is abruptly, but temporarily, shocked away from the general level of the process. This might arise from some fleeting circumstance, for example, holiday, sickness, temporary staff, or exceptional weather conditions. Figure 4.4*b* shows a situation in which the mean level has experienced a sustained shift with no hint of returning to the original mean level. Sustained shifts often reflect an abrupt change in conditions which will remain in place until some corrective actions are pursued, for example, equipment malfunction, implementation of different method, or a change in raw material vendor. Figure 4.4*c* reflects a chaotic state for the process mean; namely, the mean level is shifting around in an erratic fashion, that is, in unpredictable ways.

The remaining processes portrayed in Figure 4.4 illustrate out-of-control scenarios in which the variation level changed over time. In Figure 4.4*d,* the process has experienced a transitory change in the variation level while in Figure 4.4*e* the change in the variation level is lasting. Figure 4.4*f* illustrates a process whose variation level is constantly changing in an erratic fashion. These figures illustrate out-of-control situations in either the mean level or the variation. As depicted earlier in Figure 4.2*a,* the possibility exists for a process to be out of control in terms of both the mean and the variation level.

In general, since a process in a state of statistical control is associated with an underlying distribution's remaining identical over time, any change in distributional shape, location, or spread is considered an out-of-control condition. In formal statistical language, an out-of-control condition is associated with a change in any distributional parameter (e.g., mean, variance, or skewness). By sampling a process and estimating the parameters of its statistical distribution, changes in the distribution can often be revealed when these parameter estimates are plotted against time. In subsequent sections, we will expand on this idea.

Implicit in our presentation of out-of-control processes, special causes are viewed as undesirable disruptions undermining the ability of the process to operate in an economical or efficient fashion. Farnum (1994) insightfully points out that exercising effective control over undesirable special causes rests on three factors: a mechanism for detecting the existence of special causes, the ability to trace back and find the underlying source contributing to the disruption, and the ability to rectify problems once they are found. These factors are equally important in cases where special causes have favorable effects, for example, a shift up in monthly sales or customer satisfaction ratings. In such cases, there is no need to rectify problems, but rather to learn from the underlying sources to promote their effects.

Statistical methods, such as control charts, provide a mechanism for sorting out special-cause effects. The potential benefits of these methods, in general, cannot be realized simply by executing the appropriate arithmetic calculations. Much depends on the skill of the person(s) following up the statistical signals that suggest the occurrence of special causes. Resolving the effects of special causes also depends on the nature of the data system. For manufacturing processes, Schonberger (1986) elaborates:

> The person who records data is inclined to analyze, and the analyzer is inclined to think of solutions. Success depends on recording the right kind of data at the right time. One approach is for operators to record a piece of data each time there is a work slowdown or stoppage. The vital piece of data to be captured is the cause of the slowdown or stoppage. (pp. 18–19)

In situations where detailed documentation does not uncover special causes, the use of basic process improvement tools such as flowcharts, cause-and-effect diagrams, and brainstorming can play a vital role in the search for explanations.

Finally, it cannot be overstated that organizational culture is a key ingredient for the successful control of special causes. Management is responsible for the company culture within which statistical methods are used. Management must provide an environment that actively encourages the use of statistical methods. And lastly, management must support actions necessitated by the discovery of special causes.

4.5   HAZARDS OF OVERCONTROL

When a process is in a state of statistical control, it is futile to seek special explanations for individual random deviations. Such overreacting can lead to actions which can have the perverse effect of decreasing process quality and increasing costs. As Deming (1986) states: "If anyone adjusts a stable process to try to compensate for a result that is undesirable, or for a result that is extra good, the output that follows will be worse than if he had left the process alone" (p. 327). Making ad hoc interventions when a process is under the influence of purely random variation is often called the problem of *overcontrol* or *tampering*. Consider the following 20 observations taken[1] from an in-control random and normally distributed process with an underlying mean μ of 10 and standard deviation σ of 0.1:

```
9.969  10.032  9.958 9.943   10.170  9.986  9.834  10.092  10.011   9.776
9.765  10.171  9.999 10.016  10.044 10.172 10.063  10.067   9.890 10.093
```

As it turns out, the sample mean $\bar{x}$ is 10.003, and the sample standard deviation s is 0.119.

Suppose an operator is instructed to keep the process outcomes as close as possible to a target value of 10.000. With good intentions, the operator attempts to do so by adjusting the process (say, by changing a machine setting) after each observed process outcome. In the beginning, assume the machine setting is 0. However, after outcome 1, the operator adjusts the setting up by 0.031 (= 10.000 − 9.969). As a result, the next outcome then becomes 10.063 (= 10.032 + 0.031). At this point, the operator figures another setting adjustment is required, this time down by 0.063 (= 10.063 − 10.000). So, after two periods, the machine setting is set at +0.031 − 0.063 = −0.032, which means the third process outcome becomes 9.926 (= 9.958 − 0.032). Continuing with this mindset, the operator "creates" the following process values:

```
9.969  10.063  9.926 9.985   10.227  9.816  9.848  10.258   9.919   9.765
9.989  10.406  9.828 10.017  10.028 10.128  9.891  10.004   9.823 10.203
```

Below are the summary statistics from Minitab for the original values versus the values from the adjusted or tampered process:

Variable	N	Mean	Median	Tr Mean	StDev	SE Mean
original	20	10.003	10.014	10.006	0.119	0.027
tamper	20	10.005	9.987	9.996	0.169	0.038

Variable	Min	Max	Q1	Q3
original	9.765	10.172	9.947	10.086
tamper	9.765	10.406	9.859	10.112

Notice that the average levels of both processes are on target; however, the sample standard deviation of the tampered process has increased from 0.119 to 0.169, an increase of 42 percent. Thus, the attempts to improve the process led to an undesirable increase in the level of process variation. In general, it can be shown that such an adjustment policy applied to a random process will theoretically

[1]These data were randomly generated by using Minitab's "Random" command. The "Random" command enables the user to specify one of a variety of distributions and the desired parameter values of the chosen distribution. In this case, Minitab randomly generated 20 observations from a normal distribution with $\mu = 10$ and $\sigma = 0.1$.

increase the standard deviation of the process by a factor of $\sqrt{2}$, that is, a 41 percent increase.

THE FUNNEL EXPERIMENT—TIME-SERIES PERSPECTIVE

By means of an example, we illustrated that a certain adjustment policy intended to improve a process actually worsened performance in terms of process variation. In general, *any* adjustment policy based on individual outcomes of an in-control random process will result in increased process variation.

To demonstrate the effects of overcontrol, Deming popularized a physical experiment based on the use of a funnel.[2] The experiment requires the following: a funnel, a marble that will fall through the funnel, a flat surface (e.g., tabletop), and a funnel holder. With these materials, participants repetitively drop a marble through the funnel, suspended a fixed distance over the flat surface. The goal is to have the marble come to rest as close as possible to a designated target marked on the flat surface. Participants are given the opportunity to move the funnel along the surface (i.e., make adjustments) prior to each marble drop. Deming has found that most people are tempted by one of the following rules of adjustment:

Rule 1   Leave the funnel set over the target without adjustment.

Rule 2   Assume for the kth drop ($k = 1, 2, 3, \ldots$) the marble will come to rest a distance z_k measured from the target. Move the funnel a distance $-z_k$ *from its last position*. Note, a flat surface is a two-dimensional space, hence moving the funnel entails movement in the x direction and y direction.

Rule 3   Set the funnel over point $-z_k$ *from the target*.

Rule 4   Set the funnel at each drop over the spot where the marble comes to rest.

Rule 1 corresponds to the case of making no adjustment while rules 2, 3, and 4 are different ways of intervening with the process.

In Figure 4.5, we see the tabletop scatter of points resulting from several repetitions of each of these rules. We can see that rule 1 results in the least scatter of points while rules 2, 3, and 4 add unnecessary variation to the distribution of points. The results of Figure 4.5 clearly demonstrate the counterproductive effects of adjustment to a random process on overall process performance as measured by variation around the target. However, the nature of the summarization of results ignores the fact that the process measurements arise in a time-ordered sequence. It would be particularly informative to see the impact of the funnel experiment rules from a time-series perspective.

---

[2]The use of the funnel to illustrate the effects of overcontrol is originally due to Lloyd S. Nelson, and thus often is referred to as the Nelson funnel experiment. There are other physical experiments that can be conducted to bring out the notion of overcontrol. For example, one might use a Galton quincunx board ("statistical pinball machine") which is a device for creating distributions by allowing beads to drop through rows of pins. A nice discussion of overcontrol and the quincunx board can be found in Gitlow et al. (1995).

Figure 4.5   Results of funnel experiment recorded on tabletop (from Deming 1986, reprinted with permission).

For a time-series perspective, rather than thinking of moving a funnel, think of adjusting a machine setting for a process over time. Initially, we assume that the process is in control, centered at the target with the machine setting fixed at zero. Rule 1 says to do nothing, that is, to leave the machine setting alone. Rule 2 is associated with the operator's turning the machine setting away from its current setting by an amount equal to the deviation between the last outcome and the target. The hypothetical example of the previous subsection, in which we showed that the operator increased the process variation by 42 percent by continually adjusting the machine setting by deviation between the last outcome and the target, was an example of a rule 2 intervention.

Rule 3 is associated with the operator's establishing the machine setting above or below the zero setting, opposite to the deviation between the last outcome and the target value. Rule 3 is a much more dramatic adjustment policy than rule 2 since the operator is tending to "swing" the machine settings from one side to the other side of the target setting. Rule 4 applies where the operator "tries to achieve uniformity by attempting to make every piece like the last one" (Deming 1986, p. 329). Thus, if the last outcome is 10 units above the target, the machine setting is set to 10.

To illustrate the induced behaviors from the four rules, we generated 100 sequential observations from an iid process centered at zero. For comparative purposes, we leave the first 50 observations alone while making adjustments to the last 50 observations, using each of the respective policies.

Again, rule 1 (Figure 4.6a) is seen to be the best choice. By understanding the distinction between random and special-cause variation, management recognizes that tampering with an in-control system is unnecessary and wasteful. Individual measurements are best treated with benign neglect.

Careful examination of Figure 4.6b reveals that the process has deteriorated after observation 50 in terms of greater variation around the target. Deming indicates that this rule will produce a "stable" output around the target, but the variance will be double under rule 2 what it is under rule 1. In time-series terminology, the process induced by rule 2 is *stationary* (see Section 2.3); that is,

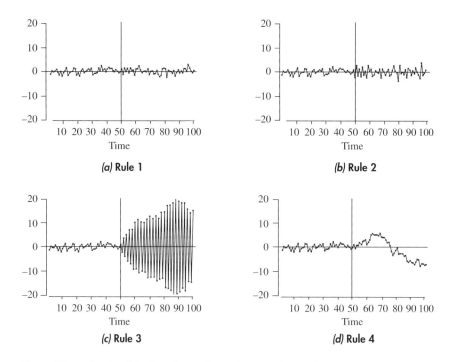

Figure 4.6   Results of the funnel experiment viewed in time order.

the overall process mean and variance will remain fairly constant over time. Even though the process under rule 2 is stationary or stable, it is not a random process. A rule 2 process follows a systematic pattern[3] characterized by a subtle tendency for successive observations to oscillate from one side of the target to the other. Relative to the induced behaviors by rules 3 and 4, the detrimental effects of rule 2 are not as apparent. In practice, there is a danger that the process deterioration resulting from rule 2 tampering might completely escape the notice of individuals responsible for the process in question.

The result of rule 3 tampering is quite dramatic. There is an immediate explosive increase in process variation. Furthermore, a rule 3 induced process will exhibit extreme period-to-period oscillation, with the magnitudes of the deviations from the target value ever increasing with time.

Under rule 4, we see in Figure 4.6d a process that is wandering away from the target value; in this case, the process eventually moves away from the target in the negative direction. In general, a rule 4 induced process could wander away in any direction with no tendency to revert to target.

In our discussion of overcontrol of processes, we have focused on an operator adjustment scenario in which an operator makes ill-advised attempts to improve the process based on the observation of individual process outcomes.

[3]The time-series pattern is known as a *moving-average order 1*, MA(1), *process*. This fact is discussed in detail in Alwan (1991).

However, the problem of overcontrol is not exclusive to machine setting applications. Deming (1986) and Gitlow et al. (1995) illustrate that these rules of intervention are common at all levels of an organization and even pervade our daily lives. Suppose, for example, your golf scores through time are in statistical control but you cannot resist the temptation to make adjustments to stance, grip, or stroke after any shot perceived to be bad. Gitlow et al. (1995) nicely point out that setting the current period's goal based on last period's overage or underage is an example of rule 3. For example, a sales quota policy stating that if you are short of this month's goal by a certain amount, you must increase next month's goal by that amount is rooted in rule 3 thinking.

Deming (1986) notes that on-the-job training is "a frightening example of rule 4 . . . where people on a job train a new worker. This worker is then ready in a few days to help to train a new worker. The methods taught deteriorate without limit. Who would know?" (p. 330). Rule 4 is associated with situations in which there is an attempt to have the new process outcome be close to the previous outcome. Engineering changes to a product based on the latest version of the design with no regard to the original design are an example of rule 4. Eventually, the current design will have no resemblance to the original design.

All these examples of tampering stem from a lack of understanding of random variation. More specifically, tampering implies acting on random variation as if it were special-cause variation. As long as the process is exhibiting in-control random behavior around a desired target level, any adjustments or reactions based on individual process outcomes waste time, quality, and resources and obscure real opportunities to make the process better. Short of fundamental changes to the system, no action is the best action.

It is important to underscore that the no-action policy is the appropriate policy for a very specific process behavior, namely, a random process void of special-cause disturbances. For nonrandom predictable processes or random processes with special causes, actions (such as process adjustment) based on individual process outcomes may indeed be quite helpful. Failure to respond to legitimate opportunities for process improvement represents the other extreme of overcontrol, namely, undercontrol. In Chapter 10, we will demonstrate how adjustments to an autocorrelated process can improve the process in terms of levels of variability. In general, thoughtful data analysis can help minimize the hazards of overcontrol and undercontrol. We now turn to the control chart which under certain conditions can provide proper guidance on decisions of whether to intervene.

4.6   BASIC STATISTICAL FRAMEWORK FOR CONTROL CHARTS

The idea of a control chart is simple but ingenious. In his brilliance, Shewhart recognized that it would be desirable and possible to set limits upon the natural variation of any process, so that variation within these limits is likely due to

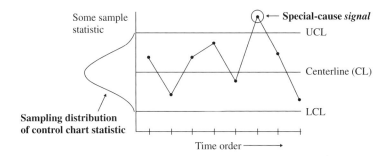

Figure 4.7   General structure of a control chart.

chance causes, but variation outside these limits would likely indicate a change in the process due to some special cause(s).

All control charts work on the basis that successive samples (also called *subgroups*) of a given size n (≥ 1) are taken from a process at more or less regular intervals. From each of these samples, a number of sample statistics may be computed depending on the nature of the individual process measurements, for example, the sample mean $\bar{x}$, the sample standard deviation s, or the sample range R. As illustrated in Figure 4.2a, different distribution parameters can potentially change over time; thus there is a need to look at different sample statistics, each dedicated to the monitoring of different aspects of the distribution.

If the individual process measurements are generated from a random process with identical distributions over time, then any quantity based on the measurements will also be random with identical sampling distributions. (*Note:* The distribution of the individual measurements and the distribution of the sample statistic are not necessarily the same.) As an example, suppose successive samples of size n are drawn and sample mean $\bar{x}$ is computed for each sample. If the individual measurements are randomly generated from a constant population over time with mean μ and standard deviation σ, then the fluctuations of sample means will be random, described by a constant sampling distribution over time with mean μ and standard deviation $\sigma/\sqrt{n}$ (see Section 2.9). Furthermore, the central limit theorem tells us that this sampling distribution will be approximately (if not, exactly) normal, depending on how closely the underlying distribution for the individual measurements conforms to the normal distribution and the size of n.

The basic idea of a control chart is illustrated in Figure 4.7. A control chart displays process measures over time, whether individual readings or summary statistics of samples. The time period or sample number is conventionally laid out on the horizontal axis, with the data scale on the vertical axis. Superimposed on the time-series plot are typically three horizontal lines. One line is the centerline (CL). Typically, the centerline represents the underlying mean of the sampling distribution for the sample statistic being charted. In practice, this underlying mean is never known. So instead, the centerline is taken to be the average value of the control chart observations.

The two other superimposed lines are called the *upper control limit* (*UCL*) and *lower control limit* (*LCL*). The idea is to place these limits so that if the process is in control, essentially all control chart observations will fall within limits. Given such a placement, it stands to reason that if any observation falls outside the control limits, there is reasonable evidence that a special cause is in the works.

From Figure 4.7, we see that the control limits are placed equidistant above and below the centerline. This common practice is based on an assumption that the sampling distribution for the control chart statistic is symmetric, as depicted in Figure 4.7. Most control charts are designed on the basis that not only is the sampling distribution symmetric but also it is well approximated by the normal distribution.

Normality is justified on the grounds that it often arises from the cumulative effect of many small, independently acting, sources of variation. Since the dominant component of process variation—common-cause variation—is indeed viewed as such a cumulative effect, normality would appear to be a reasonable assumption. Furthermore, due to the central limit theorem effect, many sample statistics (e.g., the sample mean) are well approximated by the normal distribution, regardless of the underlying distribution for the individual measurements.

For now, we will go along with the standard assumption of normality in the development of the control chart. However, we need to be prepared for the possibility of nonnormality when challenged by real applications. Fortunately, the data analysis tools presented in earlier chapters will enable us to reconcile the standard charting methods with data exhibiting nonnormal behavior.

HYPOTHESIS-TESTING CONNECTION

An outlying observation does not guarantee the presence of a special cause; it only serves as a credible statistical *signal* of the possibility of its existence. It is theoretically possible that pure random variation—void of special causes—produced the outlying observation. In such situations, the process would be incorrectly judged as out of control and a search for special causes would be done in vain. Similarly, there is a risk that observations which are truly the result of special causes will result in no action because they fell within control limits.

The notion of making these two possible false judgments might sound familiar in light of our discussion of hypothesis testing in Chapter 3. In fact, a control chart can be viewed as a statistical device for testing the hypotheses:

Null hypothesis H_0: Process is in a state of statistical control

Alternative hypothesis H_1: Process is *not* in a state of statistical control

Observations falling outside the control limits reject the null hypothesis of statistical control, while observations within control limits fail to reject the null hypothesis of statistical control. The comparison to a hypothesis-testing framework *must be made with care* because the standard hypothesis-testing situation

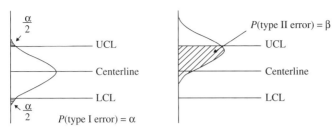

Figure 4.8   Type I and II error probabilities.

envisages a single test, not a *dynamic* test applied on an ongoing basis to each observation that is plotted, as is the case with a control chart procedure.

As with any hypothesis-testing problem, we can speak of the probability of type I error (concluding that the process is out of control when it is really in control) and the probability of type II error (concluding that the process is in control when it is really out of control). Using the notation from Chapter 3, the probabilities of type I and II errors are denoted by α and β, respectively. Figure 4.8 illustrates both of these probabilities.

In the first case, the process is assumed to be in control. There is, however, a risk of incorrectly inferring that the process is out of control if an observation falls beyond either control limit. Since the control limits of Figure 4.8 are symmetric around the centerline, there is an equal chance of a control chart observation incorrectly signaling an out-of-control condition. The total risk of type I error (α) is given by the sum of the two tail areas of the distribution outside of the control limits as shown in Figure 4.8a. Since a type I error is a false alarm of an out-of-control condition, the risk of type I error is also referred to as the *false-alarm rate*.

Suppose now the process of Figure 4.8a goes out of control, resulting in a mean shift of the sampling distribution. In such circumstances, the hope is that a control chart observation will fall outside the limits. However, there is a risk of an observation's falling within the control limits (exhibited by the shaded area in Figure 4.8b), leading one to incorrectly conclude that the process is in control. For this particular illustration, the risk of type II error (β) is substantial; in fact, it is greater than the probability of getting an observation which would lead to the correct conclusion that the process is out of control. For more substantial shifts, the opportunity to catch the out-of-control condition is greater, and accordingly the β risk is smaller.

The placement of the control limits is clearly a major influencing factor in determining the size of the risks of type I and II errors. As the control limits are placed farther apart, there is a smaller probability that an observation due to random causes will fall outside the limits, and thus a smaller probability of type I error. Placed widely enough, the α risk can be made negligible. Simply stated, one cannot make a mistake of concluding that the process is out of control when it is really in control if no observation is allowed to signal an out-of-control condition! But, by doing so, the probability of type II error is increased; that is, true out-of-control conditions will more likely escape necessary attention. The trade-off goes both ways. By narrowing the gap between control limits, we gain a greater sensitivity to out-of-control conditions, but also must assume a greater risk of unnecessary special-cause searching.

Given the tradeoff, it would seem that we are left with two possible strategies. Namely, place control limits which give some prespecified level of risk of making a type I error (α) or some prespecified level of risk of making a type II error (β). Even though either approach is theoretically possible, one practical difficulty with selecting β is that it depends on the magnitude of the process shift. Hence, to prespecify β, one must have in mind a *particular* process shift size which one would like to detect with some prespecified probability. Only in very specialized industries, typically sophisticated, is there a stated desire to detect a specific process shift size among an infinite array of possible departures.

Given the practical difficulty of prespecifying the level of β, common practice is to design control chart limits such that the risk of false alarms (α) is maintained at some desired level; that is, specify α in advance. In the next section, we consider levels of α most commonly selected in practice.

THREE-SIGMA CONTROL LIMITS

We are now in a position to summarize the control chart in terms of a general model. Let Y represent the control chart statistic of interest. As was noted, the standard control chart implicitly assumes that this control chart statistic Y is approximated by the normal distribution. Suppose that μ_Y represents the mean, or expected value, of Y, and σ_Y represents its standard deviation.

In Chapter 2, we introduced the notation $z_{\alpha/2}$ as the percentile of the standard normal distribution such that $P(Z \geq z_{\alpha/2}) = \alpha/2$. In general, $z_{\alpha/2}$ can be interpreted as the number of standard deviations around the mean needed to encompass $100(1 - \alpha)$ percent of the normal distribution. Hence, the probability that the observed value of Y will fall between $\mu_Y \pm z_{\alpha/2}\sigma_Y$ is $1 - \alpha$ as a result of *purely chance causes*. What we actually have here is the general form of the control chart proposed by Shewhart, namely,

$$\text{UCL} = \mu_Y + z_{\alpha/2}\sigma_Y$$

$$\text{CL} = \mu_Y \qquad\qquad (4.2)$$

$$\text{LCL} = \mu_Y - z_{\alpha/2}\sigma_Y$$

In his honor, control charts derived from the principle of control limits being a certain number of standard deviations around the centerline are often referred to as *Shewhart control charts*.

Since the probability is $1 - \alpha$ that a process outcome will randomly fall between the Shewhart control limits, the probability is α that a process outcome will randomly fall *outside* the limits. As discussed in the previous section, this probability represents the risk of type I error, that is, the risk of incorrectly concluding that the process is out of control. To determine specifically the risk at hand, a value for $z_{\alpha/2}$ must be chosen.

Historically, two general approaches have evolved in the selection of $z_{\alpha/2}$. One approach, especially common in the United Kingdom and Europe, is to select limits with a very specific α value in mind, typically a very small value on the order of 2 or 3 in 1000. Suppose, for instance, α is specified to be 0.002. This would imply the need to determine $z_{0.002/2} = z_{0.001}$ which is the value of z on the standard normal distribution such that the area to the right of it is 0.001, or equivalently, the area to its left is 0.999. Below, we use Minitab to find the desired z value:

```
MTB > invcdf 0.999;
SUBC>  Normal 0.0 1.0.

Inverse Cumulative Distribution Function

Normal with mean = 0 and standard deviation = 1.00000

P( X <= x)              x
   0.9990           3.0902   ←  Z0.001
```

Accordingly, the Shewhart control chart with an α risk of 0.002 is given by

$$\text{UCL} = \mu_Y + 3.09\sigma_Y$$

$$\text{CL} = \mu_Y \qquad\qquad \textbf{(4.3)}$$

$$\text{LCL} = \mu_Y - 3.09\sigma_Y$$

Limits of this type are known as *probability limits*. In this case, the limits would be called 0.002 probability limits.

As an alternative to specifying *first* an α level to imply a specific $z_{\alpha/2}$ multiple, many practitioners follow a somewhat reverse route; namely, first select the $z_{\alpha/2}$ multiple, typically an easily remembered number, which in turn implies the α level. Based on much experimentation, Shewhart recommended a convenient multiple value of 3 for $z_{\alpha/2}$. As a result, control limits for a sample statistic y are as follows:

$$\text{UCL} = \mu_Y + 3\sigma_Y$$

$$\text{CL} = \mu_Y \qquad\qquad \textbf{(4.4)}$$

$$\text{LCL} = \mu_Y - 3\sigma_Y$$

These limits are called *3-sigma* or 3σ limits and serve as a basis for most control charts found in practice. For an underlying normal distribution, the probability is $\alpha = 0.0027$ (roughly 3 in 1000) that an observation will randomly fall outside control limits. Or, equivalently, the probability is 0.9973 that an observation will randomly fall within the control limits.

There is a bit of controversy among some quality professionals as to whether Shewhart based his choice of 3 given an assumption that the process follows the normal distribution. In his book *Economic Control of Quality of Manufactured Product*, discussion of the normal distribution is a dominant theme. However, it appears that he justified the choice of 3 by a theorem known as *Tchebycheff's theorem.* This theorem states that for any underlying distribution (normal or not), the proportion of observations that lie within k standard deviations from the mean is at least $1 - 1/k^2$. For $k = 3$, this implies that the proportion of observations that lie within 3 standard deviations from the mean is at least $\frac{8}{9}$ (or, approximately 0.89). A proportion of 0.89 is substantially less than 0.9973 when the assumption can be made that the process follows the normal distribution. But let us consider Shewhart's perspective, as found in his other classic book, *Statistical Method from the Viewpoint of Quality Control*, published in 1939. He writes: "The control limits as most often used in my own work have been set so that after a state of statistical control has been reached, one will look for assignable causes when they are not present not more than approximately three times in 1000 subsamples . . ." (p. 36) He later comments: "For this purpose let us choose $1 - p' = .9973$ because this is about the magnitude customarily used in engineering practice. Of course, if we knew $\overline{X}'$ and σ', the desired range would be $\overline{X}' \pm 3\sigma'$" (p. 62). Thus, with these quotations, we see Shewhart's intimacy with the normal distribution and the design of the control limits.

Regardless of the original justification for 3σ limits, it can be stated that extensive industrywide experience has borne out their usefulness, given appropriate conditions. Such experience has demonstrated that investigation of 3σ out-of-control observations very often leads to the discovery of special causes. Furthermore, if a process is in control, unnecessary searches for special causes will be done very seldom, only about 3 for every 1000 control chart observations given an assumption of normality. For these reasons, nearly all control charts constructed for applications studied in this book will be based on a 3σ framework.

Our discussion up to this point has assumed that the parameters of the sampling distribution are known, in particular, μ_Y and σ_Y. Of course, in practice, this is rarely (if ever) the case; hence these parameters must be estimated. Let us denote the estimates for the mean and standard deviation by $\hat{\mu}_Y$ and $\hat{\sigma}_Y$, respectively. Substituting these estimates in the general 3σ limits of Equation (4.4), we have

$$\text{UCL} = \hat{\mu}_Y + 3\hat{\sigma}_Y$$

$$\text{CL} = \hat{\mu}_Y \qquad\qquad (4.5)$$

$$\text{LCL} = \hat{\mu}_Y - 3\hat{\sigma}_Y$$

For nearly all Shewhart-type control charts, $\hat{\mu}_Y$ (or the centerline) is simply computed as the overall average of all the observed control chart values. As will be seen throughout the text, the estimated standard deviation $\hat{\sigma}_Y$ can be computed in a variety of ways depending on the nature of the data and the control chart statistic.

It should be pointed out that 3σ limits based on estimated parameters cannot be *precisely* interpreted as having an associated 0.0027 false-alarm rate. Neither should 0.002 (or any value) probability limits be precisely interpreted as having an associated 0.002 false-alarm rate. These exact numbers are based on the strict assumptions of the parameters being known and an underlying sampling distribution which is exactly normal.

We learned in Chapter 2 that even if sample statistics, such as the sample mean, follow the normal distribution when the standard deviation parameter is known, this is not the case when the standard deviation parameter needs to be estimated. In particular, the Student t distribution more precisely summarizes sampling variability. However, in control chart applications, the sampling variation due to parameter estimation is ignored and the normal distribution is used without the refinements offered by the t distribution. Fortunately, for many applications, the rough-and-ready approximation offered by the normal distribution serves well for practical guidance. Nevertheless, since no sample statistic will exactly follow the normal distribution, or for that matter the Student t distribution, we should not take the probability 0.0027 too literally, but rather should appreciate that the 3σ limits offer a very small risk of unnecessary special-cause searching.

To make more tangible our conceptual discussions, we now illustrate the construction of a control chart. Consider again the stopwatch series of 50 consecutive measurements studied in Chapters 2 and 3. Using basic data analysis techniques, we learned that the data are well described by the normal distribution. Furthermore, we found the overall sample average $\bar{x}$ to be 9.989 and the sample standard deviation s to be 0.0918. Here, the plotted statistics are the *individual* stopwatch measurements. In other words, the sample or subgroup size n is 1. Substituting the estimates of $\bar{x}$ and s into the 3σ limits specified by (4.5), we obtain

$$\text{UCL} = 9.989 + 3(0.0918) = 10.264$$

$$\text{CL} = 9.989$$

$$\text{LCL} = 9.989 - 3(0.0918) = 9.714$$

As we will discuss in Chapter 5, there is an alternative procedure for estimating the true standard deviation, based not on the sample standard deviation s but on a measure of dispersion calculated from the average of the absolute changes from one observation to another (known as moving ranges). In Minitab, this latter approach—based on moving ranges—is the default calculation of control limits when one is dealing with control charts for a series of individual measurements.

In Figure 4.9, we present 3σ control charts for the stopwatch data based on both computational methods. For all practical purposes, the computational results for the control limits are the same. Examination of either control chart helps convey some important features about the process:

1. The observations are randomly scattered around the centerline set at the mean level. There is no hint of systematic behavior, such as trend, relative to the centerline. Such a process is said to have the property of *constant level.*

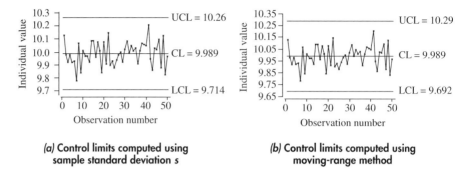

(a) **Control limits computed using sample standard deviation s**

(b) **Control limits computed using moving-range method**

Figure 4.9   Three-sigma control charts for stopwatch data series.

2. Even though the magnitudes of the deviations relative to the centerline are random, the average level of variability appears to be about the same throughout the plot. Such a process is said to have the property of *constant variation.*

3. No observations breach control limits. Hence, there is no suggestion of any disruption to the routine operation of the process due to special causes.

Thus, the process is in a state of statistical control operating under a stable set of chance causes; that is, the process is exhibiting iid behavior. Since the data also conform to the normal distribution, we can state even more precisely that the process is exhibiting iidn behavior.

By analyzing a given set of data through the use of control charts or other data analysis tools, we are actually looking back *retrospectively* on the performance of the process. When used retrospectively for these purposes, the control chart is said to be used *as a judgment.*

A control chart used as a judgment is only the beginning. If a process is found to be in statistical control (or brought to statistical control through the removal of special causes), control limits can be projected into the future so that the process can be monitored in real time to decide whether a process change has occurred, so that the process can be maintained in a state of statistical control. When used *prospectively* for these purposes, the control chart is said to be used *as an ongoing operation.*

For the stopwatch series, since the process is in a state of statistical control, we can extend horizontally the control limits and centerline into the future for prediction purposes. Of course, *prediction* does not mean an exact guess of the future; this is not possible for any process. Instead, we expect future observations to deviate randomly around the same level (centerline) as in the past, and with the same variability as in the past. Furthermore, we expect nearly all future process outcomes to vary within the projected control limits. Any future observation falling outside of the control limits should trigger a search for finding a special explanation.

4.7   AVERAGE RUN LENGTH CONCEPT

As discussed in the previous section, the false-alarm rate α serves as a measure of the performance of a control chart. Another measure of performance is the *average run length* (*ARL*). ARL is the average (or expected) number of points that must be plotted before an out-of-control signal is given. For Shewhart-type charts, the ARL for an in-control process (typically labeled ARL_0) is given by

$$ARL_0 = \frac{1}{\alpha} \qquad (4.6)$$

where α is the false-alarm rate. For example, the ARL for a 3σ Shewhart control chart applied to an iidn process is $1/0.0027 = 370.4$. In other words, for a stable in-control normally distributed process, we expect, on average, 370.4 plotted control points before an out-of-control signal is given.

If, however, the process has shifted, then the probability of an out-of-control signal increases, as we would hope. Recall from the previous section that the probability of not getting an out-of-control signal if the process has shifted is β. Thus, the probability of getting a signal (hence, detecting the shift) is $1 - \beta$; this probability is referred to as the *power* of the statistical procedure. It can be shown that the ARL for a Shewhart control chart when the process is out of control (typically labeled ARL_1) is

$$ARL_1 = \frac{1}{1 - \beta} \qquad (4.7)$$

As an example, suppose a Shewhart control chart is applied to individual observations X from an iidn process with known mean μ_0 and standard deviation σ. Assume now that after the implementation of the control chart, the process mean has shifted up 2 standard deviations; that is, the new mean level is $\mu_1 = \mu_0 + 2\sigma$. The probability of getting an out-of-control signal is

$$1 - \beta = P(X < \text{LCL}) + P(X > \text{UCL})$$

$$= P(X < \mu_0 - 3\sigma) + P(X > \mu_0 + 3\sigma)$$

$$= P\left(\frac{X - \mu_1}{\sigma} < \frac{\mu_0 - 3\sigma - \mu_1}{\sigma}\right) + P\left(\frac{X - \mu_1}{\sigma} > \frac{\mu_0 + 3\sigma - \mu_1}{\sigma}\right)$$

$$= P\left[Z < \frac{\mu_0 - 3\sigma - (\mu_0 + 2\sigma)}{\sigma}\right] + P\left[Z > \frac{\mu_0 + 3\sigma - (\mu_0 + 2\sigma)}{\sigma}\right]$$

$$= P(Z < -5) + P(Z > 1)$$

We can use Minitab (or Appendix Table I) to find that $P(Z < -5) = 0.000000$ and $P(Z > 1) = 0.158655$. Thus, we have

$$1 - \beta = 0.000000 + 0.158655 = 0.158655$$

Applying (4.7), we find the average run length to be

$$ARL_1 = \frac{1}{0.158655} = 6.3$$

Thus, for a Shewhart control chart applied to a series of individual measurements, a process mean shift of 2 standard deviations to an iidn process would be detected, on average, within 6 or 7 samples after the shift had occurred.

The ARL is a useful measure for the design of control charts. In particular, if one is interested in detecting mean shifts of a certain magnitude (typically, expressed in number of standard deviations), the ARLs for different control charts can be computed and compared. Given two control charts that have the same in-control properties, the control chart which detects the out-of-control situation more quickly is more desirable. In terms of ARLs, for a given shift size, the control chart with a smaller out-of-control ARL is more desirable. In Chapter 6, we will provide more details on the ARL measure for Shewhart control charts. We will also revisit the ARL concept in Chapter 8 when we discuss control charts based on a different conceptual scheme from the Shewhart design.

4.8 CONTROL LIMITS VERSUS SPECIFICATION LIMITS: DISTINCT CONCEPTS

Control limits are often confused with *specification* or *tolerance limits*. Specification limits represent the boundaries of acceptable product or service performance. What the customer wants, expects, or needs from the product or service is established and incorporated during the product or service design phase. Specification limits may be two-sided with a lower and upper specification limit (denoted, respectively, by LSL and USL). For example, a shaft diameter requirement of 25.00 ± 0.05 mm is a two-sided specification limit. Specification limits can be one-sided with only a lower or upper limit. A filling requirement of $(355 - 5)$ milliliters (mL) is a one-sided limit for which the lower specification limit is 350 mL and the target is 355 mL. A pizza delivery service which guarantees delivery within 1 h is establishing an upper specification limit of 1 h; any delivery in less than 1 h is acceptable, while any delivery exceeding 1 h is unacceptable performance.

Specification limits can be thought of as the voice of the customer; that is, they describe the *desired* process performance. Control limits, on the other hand, attempt to indicate the *actual* process performance.

How a process is viewed relative to one type of limits bears no insight on how the process would be viewed relative to the other type. To illustrate, consider the various process scenarios given in Figure 4.10. In Figure 4.10*a* and *b,* the associated process is statistically in control, in the sense that the process outcomes appear random around the centerline and no observations breach control limits (LCL and UCL). But statistical control does not mean the process is satisfactory in relation to the specification limits (LSL and USL). The process of

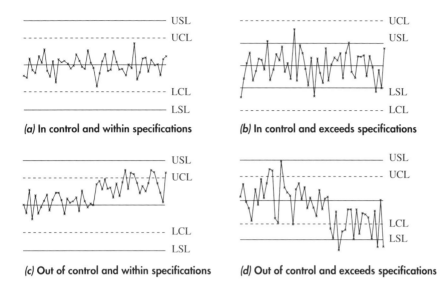

Figure 4.10   Control limits versus specification limits.

Figure 4.10*a* is "safely" meeting specifications; however, the process of Figure 4.10*b* is producing defective outcomes. How a process performs relative to specification limits is commonly referred to as the *capability* of the process. Figure 4.10*a* reflects a capable process while Figure 4.10*b* reflects an incapable process. The lesson here is that capability (or lack of) does not imply statistical behavior in terms of being in control or out of control.

Recall from Section 4.5, tampering with an in-control process will not help matters. Thus, the situation reflected in Figure 4.10*b* calls for fundamental change to the system (e.g., improved procedures, training, and technology). As depicted in Figure 4.10*c*, a process which is within specifications is not necessarily a process that is in control. For such a case, the fact that the process is within specifications should not be taken as any form of reassurance. The evidence of lack of control is a danger sign. The forces behind the out-of-control behavior could possibly push future process outcomes to unacceptable levels beyond specification limits. The situation shown in Figure 4.10*d* is the worst of the cases. The process is neither in control nor capable. Immediate steps should be taken to get the process in control and stable, by seeking and eliminating the effects of underlying special causes.

It is particularly important to emphasize that the plotted observations in Figure 4.10 represent measurements for individual items. In a typical application, specification limits are established for determining whether individual items conform to requirements. (There are some exceptions, e.g., a requirement that the average content weight of a package of items must be between certain specified values.) However, control limits can be applied to individual items and to statistics based on a sample of individual items, for example, a control chart which monitors the average of measurements for the individual items. A common mistake is to compare the sample average with specification limits. If the sample averages fall within the specification limits, this does not imply that the individual measurements are meeting specifications. It is quite possible that no individual measurements are within specifications, but because of canceling effects the average value is within specification limits.

In general, given the many different hazards associated with confusing specification limits and control limits, the best advice is to avoid superimposing both limits on any given sequence plot of process measurements. In Chapter 9, we will return to the issue of process performance in relation to specification limits by introducing various indices that quantify and summarize process capability. Until then, our focus will remain on the statistical performance and behavior of the process in question.

4.9   SUPPLEMENTARY RULES FOR DETECTING SPECIAL-CAUSE VARIATION

A single extreme point outside of 3σ control limits represents one possible statistical signal for the presence of special causes. Special causes can also give rise to unusual variation *within* control limits. To help detect such situations, other statistical rules have been developed. These rules, known as *runs* or *pattern rules*, single out patterns in successive observations that occur only with very small probability when the process is truly in a state of statistical control. Unlike the runs test (scc Section 3.1) which examines the overall randomness of a data series, runs rules search out short sequences of observations embedded within the overall data series which are inconsistent with randomness. Embedded nonrandom patterns provide suspicion that special causes have affected the natural workings of the process. An excellent discussion of the role of runs rules can be found in Nelson (1984). Additionally, Westgard and Barry (1997) provide a detailed discussion of the use and interpretation of multiple runs rules applied in a medical laboratory setting.

In the quality control literature, the recommended set of runs rules varies slightly from author to author. For example, Grant and Leavenworth (1996) suggest that an out-of-control condition is present if 7 successive points on the control chart are all on one side of the centerline, while Nelson (1984) requires 8 successive points, and then other authors are known to require 9 or 10 successive points. Often in conjunction with runs rules, the control chart is divided into zones or bands (labeled *A, B,* and *C*), each of width 1σ, starting from the centerline as shown in Figure 4.11. Below we list and describe the default set of runs rules incorporated in the statistical software package Minitab[4].

---

[4]In the full version of Minitab, the user has the option of changing any one of the default tests for special causes (Stat >> Control Charts >> Define Tests). For example, the user may wish to signal out of control if a point falls outside of 2.5 standard deviations from the mean rather than the conventional 3 standard deviations. Or, the user might require a signal if 8 successive points fall on one side of the centerline rather than Minitab's default of 9 successive points.

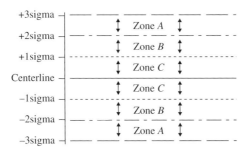

Figure 4.11   Control chart with identified zones.

Rule 1   Signal out of control if a single point falls beyond zone *A*.

Or, equivalently, signal out of control if any point falls outside 3σ control limits. This rule is by far the most common criterion for special-cause detection. As previously discussed, the probability that a point will fall outside these limits if the process is exhibiting iidn behavior is very small (theoretically, about 0.0027). Figure 4.12*a* shows the occurrence of a 3σ extreme point. (*Note:* When a sequence of points violates a run rule, Minitab places a numeral depending on the run violation next to the last point of the sequence in question.)

Rule 2   Signal out of control if 9 successive points fall on one side of the centerline.

For an iidn process, the numbers of points above and below the centerline are expected to be equal. Furthermore, if we are looking at a truly random process, the expected length of a run on one side of the centerline will be close to 2. Remember, an expected value is an average concept; we do not expect all runs to be of length 2. If you inspect a time-sequence plot of random observations, it is not unusual to find runs with lengths ranging from 1 to, possibly, 5 or 6. Refer to a sequence plot of the random stopwatch data to confirm these possibilities (see Figure 4.9).

However, a run of 9 points on one side of the centerline is fairly unlikely given random conditions. In fact, if the process is truly random with an equal chance of an observation's falling on one side or the other of the centerline, then the probability of 9 consecutive points falling on one side of the centerline is $(0.5)^9 = 0.00195$, or less than 2 in 1000. Similar to the probability of 0.0027 for 3σ control limits, the probability of 0.00195 can be interpreted as a probability of false alarm, that is, the probability of concluding the presence of a special cause when none exists. Individually, each runs rule is designed to provide a very small probability of this wrong judgment. Figure 4.12*b* illustrates the detection of a rule 2 violation. Apparently, the process mean has shifted downward but not enough to be detected by the 3σ limits. Regardless, this violation provides sufficient evidence for further investigation.

Rule 3   Signal out of control if 6 successive points are all increasing or all decreasing.

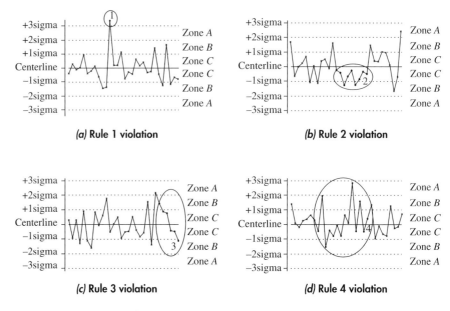

Figure 4.12   Examples of runs rules violations.

This rule attempts to detect a steadily rising or falling movement in the process which reflects, in most cases, the onset of process deterioration. An example of such a situation is shown in Figure 4.12c. Some authors regard this rule as a means of identifying a systematic linear trend. However, this rule does a very poor job of identifying linear trends, in general. For example, if this rule were applied to Deming's spring elongation data, which we found to be well fitted by a linear trend model (see Section 3.6), no rule 3 signals would be offered. For most trends, even though the overall tendency of the process is to move in a certain direction, we do not necessarily expect to witness a string of observations where each observation is strictly larger (or smaller) than all the previous observations. Hence, what this rule tends to pick up is very persistent or determined drifts in the process.

> **Rule 4**   Signal out of control if there are 14 successive points
> alternating up and down.

This rule attempts to highlight oscillatory or sawtooth behavior, as illustrated in Figure 4.12d. As illustrated in Section 4.5, oscillatory behavior can be indicative of *overcontrol* by process operators or by automated process adjustment mechanisms.

> **Rule 5**   Signal out of control if 2 of 3 successive points on one side of
> the centerline fall in zone A or beyond.

Here an out-of-control signal arises if 2 of 3 successive points fall beyond 2 standard deviations on one side of the centerline. Figure 4.12e shows an example of such a run violation. Rule 5 attempts to detect moderate-size shifts in the process

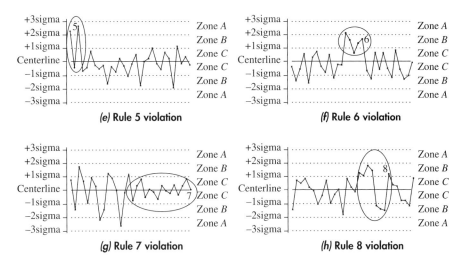

Figure 4.12   (cont'd)-Examples of runs rules violations.

which may not be picked up by conventional 3σ limits. In Europe, it is common practice to designate and superimpose on the control chart a pair of "warning limits" (LWL and UWL) at ± 2 standard deviations around the centerline. Thus, rule 5 violation is a situation where 2 of 3 successive points fall beyond one of the warning limits.

Rule 6   Signal out of control if 4 of 5 successive points on one side of the centerline fall in zone *B* or beyond.

Rule 6 constitutes another test for the detection of process mean shifts not substantial enough to give rise to extreme points. In particular, if 4 of 5 successive points fall beyond 1 standard deviation on one side of the centerline, there is a signal of an out-of-control condition, as illustrated in Figure 4.12*f*. This test is intended to detect small to medium-size shifts in the process mean.

Rule 7   Signal out of control if 15 successive points on either side of the centerline fall within zone *C*.

As seen from Figure 4.12*g*, this rule attempts to highlight process behavior in the form of unusually tight variation around the centerline. An out-of-control signal is flagged if 15 successive points fall within ± 1 standard deviation of the centerline. This type of behavior is sometimes referred to as *hugging of the centerline* since points on the control chart stick close to the centerline. Such a condition can arise from a decrease in process variability typically due to efforts to improve the process. If the control chart points are summary statistics derived from samples of more than one observation, hugging can be reflective of a sampling problem. For example, observations within the sample could be systematically sampled from two different sources in such a manner that the observations consistently offset each other and, as a result, there is less variability in the plotted sample statistics.

Rule 8   Signal out of control if 8 successive points on either side of the centerline fall beyond zone *C*.

To have a rule 8 violation, 8 successive points must avoid falling in zone *C,* that is, the region of ± 1 standard deviation around the centerline, as illustrated in Figure 4.12*h.* This rule could be signaling the independent influence of two different sources of variation on the control chart statistic. One source results in process outcomes to be high while the other source results in process outcomes to be low. Hence, control chart points show up at separate levels. We noted that the influence of different sources can result in hugging of the centerline. In that case, the different sources are mixed within the sample observations. In contrast, a rule 8 violation is more likely to result when one source completely influences all the observations of a given sample and the other source completely influences all the observations of another sample. In other words, the mixture of sources is between samples not within samples.

One should be careful when using several out-of-control criteria because, in hypothesis-testing language, we have multiple tests, implying that the associated false-alarm rate would be calculated by the compounding of the individual false-alarm rates for each test. As a result, the risk of false alarms may be increased greatly. Montgomery (1996) shows that the overall risk α_{overall} of using k criteria, each with false-alarm rate α_i, would be approximately

$$\alpha_{\text{overall}} = 1 - \prod_{i=1}^{k}(1 - \alpha_i) \tag{4.8}$$

The above equation is an approximation because it assumes that all the control criteria are independent of one another, when in fact they are not. However, the equation serves the purpose of illustrating the compounding effects of the simultaneous application of several control criteria. For instance, by simultaneously applying all the noted supplementary runs rules along with the standard 3σ rule, the overall false-alarm rate is found to be approximately 0.0195, which is more than 7 times the false-alarm rate of 0.0027 associated with the application of only the 3σ control limits.

Some investigators have noted that the concept of the false-alarm rate is not clearly defined when studying control criteria such as supplementary runs rules. The difficulty is that the probability of a false alarm depends on the number of samples since the start-up of the control chart. For example, on the very first sample, only the 3σ rule can cause a signal. At this stage, there is no possible false alarm for any of the other rules. For this reason, Champ and Woodall (1987) argue that the average run length measure is a more appropriate measure for comparing control criteria. Recall from Section 4.7, ARL is the average number of samples until a signal occurs. These investigators developed an innovative approach for determining ARLs (in control and out of control) for any combination of supplementary runs rules. They demonstrated that the addition of supplementary runs rules to a basic Shewhart control chart does improve the ability of the control scheme to detect smaller shifts, but the in-control ARL is substantially reduced. For example, applying all the supplementary runs rules to the Shewhart control chart results in an in-control ARL of about 90 as opposed to 370 for the Shewhart control chart alone. A shorter in-control ARL is not desirable because it implies more frequent signaling of out of control when the process is actually in control.

In the end, to decide whether to use additional criteria, the analyst must weigh the gained sensitivity to potential special causes against the cost of increased false signals and the inherent complexity of additional rules.

4.10   CLASSIFICATION OF STANDARD CONTROL CHARTS

Many types of control charts will be discussed in great detail in subsequent chapters of this book. Each of these chapters presents a certain type of control chart based on some rule of classification. Control charts can be classified in many ways, according to the type of the data, sample size(s), and type of control. Understanding these various classifications is essential to selecting an appropriate control chart for a particular application.

In Chapter 2, we noted that process data and random variables can be classified as continuous or discrete. Recall that continuous data are obtained from a continuous scale of measurement, such as weight, length, time, temperature, flow rate, and viscosity. Control charts devised for continuous data are known as *variable control charts*.

Discrete data typically result from counting. Examples of counting data are the number (or percentage) of defective electronic components in a production run, the number of complaining customers per month, the number of airplane arrivals in a given hour, the number of on-time deliveries, and the number of invoice errors. In quality applications, discrete data are known as *attribute data*. Hence, control charts devised for discrete data are called *attribute control charts*.

Within the broad categories of variable and attribute control charts, there can be further breakdowns. For variable data, one is often interested in monitoring the process for possible changes in the level and/or variability. To this end, there are variable control charts for monitoring the level of the process and variable control charts for monitoring the variability of the process. Each of these two types requires the computation of different sample statistics. For instance, the monitoring of the process level might involve sample statistics such as the sample mean ($\bar{x}$) or sample median ($\tilde{x}$) while the monitoring of process variability might be based on the sample range (R), sample standard deviation (s), or sample variance (s^2).

For classifying attribute control charts, there is a main dividing line based on the type of attribute data being dealt with. One type of attribute data is a count of nonconforming (or defective) units in a sample lot, or similarly, the percentage of nonconforming items in a sample. Control charts called *p charts* and *np charts* are designed for this type of attribute data. The other type of attribute data is a count of the number of nonconformities per unit, where a nonconformity is some defined occurrence. Control charts called *c charts, u charts,* and *D charts* are designed for this type of attribute data. To see the distinction between these charts, consider the following examples: If the number of defective—not working—computer chips in a batch of 100 chips is counted, then one considers *p charts* or *np charts.* In contrast, if the number of glass bubbles is counted on a windshield (not the number of defective windshields), then the latter type of control charts needs to be considered.

Another distinction made about control charts concerns the sample size *n* used at a particular sampling period. In general, all control charts involve some statistic computed from a sample of one or more process observations collected at a particular sampling period. In the SPC area, a sample of observations collected at a particular sampling period is often referred to as a *subgroup* or a *subsample*. How a subgroup of observations is formed is not an arbitrary procedure, as we will discuss in detail in Chapter 6. For variable control charts, subgroup sizes (*n*) typically range from 1 to 25. If subgroup sizes greater than 1 are used, the application necessarily lends itself to gathering up a group of observations during a relatively short time. For example, in many manufacturing scenarios, a "handful" of manufactured parts produced, close to each other in time, can be sampled and measured for critical characteristics and summarized as a single sample of information. However, there are processes for which the time between observations generated may be too long, making it less sensible to treat a set of observations as a single subgroup. For example, if one is interested in tracking monthly customer satisfaction ratings, it would not seem reasonable to group several months as a single subgroup. To take $n = 12$ would imply one subgroup observation per year! The alternative is to study the process as a data series of single or individual observations. Such a data series can be thought of as the special case of subgrouping where $n = 1$. Since there is really no grouping involved, some authors and practitioners prefer to view variable control charts for the $n = 1$ case as *control charts without subgrouping* and variable control charts for the $n > 1$ case as *control charts with subgrouping*.

For *p* chart applications, the subgroup sizes tend to be quite large (for example, *n* can be 100 or 500). Large sample sizes are required so that the sample proportion statistic can be well approximated by the normal distribution: In Chapter 7, we provide various recommended rules of thumb for the determination of subgroup sizes for *p* charts. Quite often with attribute charts, the amount of data sampled from period to period varies. For example, varying subgroup sizes occur when the chart is based on a complete audit of units from one production run to the next. When the amount sampled varies, there are a number of ways to handle the problem. Since, as we will see, the computation of control limits depends on the subgroup size for the given sampling period, one possible solution is to allow the control limits to vary from period to period. This approach along with other strategies will be discussed in detail in Chapter 7. At this point, it is important only to recognize that attribute charts can be classified by knowing whether subgroup sizes vary over time.

The notion of a subgroup is a bit different when one is dealing with many c chart–related applications. For both variable control charts and p charts, a subgroup comprises a certain number of items n, where n is some whole integer value (1 or 2 or 3 or . . .). In the case of c chart–related applications, the "items" inspected are viewed as "units" defined over some continuous interval, such as area, space, or time. For instance, a unit may be defined as 10 square feet (ft^2) of carpeting that is cut from each manufactured roll of carpeting. For this unit, there may be an interest in counting the number of disconnected woven loops. Monitoring the number of disconnected woven loops from manufactured roll to manufactured roll would fall under the context of a c chart–related application. Suppose for a particular sampling period, 12 ft^2 instead of 10 ft^2 was sampled. What is the subgroup size? The continuous nature of the sampling unit allows us to consider the subgroup size as 1.2 units. Such a fractional value is not possible for variable charts and p charts. One cannot sample 1.2 lightbulbs and ask how many work or are defective!

There is another important means of classifying control charts based on the type of control the chart is expected to exert. Farnum (1994) classifies two basic types of control: threshold and deviation control. Threshold control is concerned with detecting large shifts in the process while deviation control is concerned with detecting small shifts in the process. The basic 3σ Shewhart control chart introduced earlier is regarded as a threshold control chart. Control limits of Shewhart control charts can have difficulties signaling very small shifts in the process. Recall that we introduced runs rules in Section 4.9 to supplement the Shewhart control chart when we wish to detect departures which are not sufficiently large enough to give rise to points outside 3σ limits. Hence, Shewhart control charts supplemented with an appropriate combination of runs rules move the level of control from threshold to deviation type.

For some applications, a strategy of implementing a Shewhart control chart plus runs rules still does not provide adequate sensitivity for detecting small shifts. So, as an alternative, control charts have been developed with greater efficiency in the detection of very small shifts, that is, for deviation control. One alternative control chart is known as the *cumulative-sum (CUSUM) chart,* originally introduced by E. S. Page in the 1950s. The CUSUM approach differs from the standard Shewhart approach in that the statistic plotted on the CUSUM chart is based on the sum of current and past values of a statistic (such as individual measurements, $\bar{x}$, p, c, R, s, or s^2), not the individual values of these statistics. The idea is that while a small change in the process may not lead to a single point outside standard Shewhart control limits, the presence of this change may be detected if the effect accumulates over several samples. The *exponential weighted moving-average (EWMA) chart* is another control chart which combines past with current values to help detect small shifts more quickly. These non-Shewhart control schemes will be discussed in Chapter 8.

With all the above noted control charts, we are studying data on a single quality characteristic. As illustrated in Chapter 10, there are practical situations for which there is a requirement to study two or more quality characteristics simultaneously. Control charts based on the simultaneous monitoring of more than one quality characteristic are known as *multivariate* control charts. Multivariate control charts have been developed based on Shewhart and non-Shewhart control schemes for both variable and attribute data. In Chapter 10, we will present only Shewhart-based multivariate control charts for variable data. Figure 4.13 summarizes the classification of standard control charts to be covered in this book.

4.11 FIVE BASIC ASSUMPTIONS UNDERLYING USE OF STANDARD CONTROL CHARTS

Recall that as originally envisioned by Shewhart, the aim of control charting is to remove the effects of special causes of process output. The ultimate goal is to stabilize the process to a state of statistical control which is identified with a random process, that is, an iid process. Once a state of statistical control is attained, a control chart is used as a surveillance for any evidence of future departures. One then hopes to explain these departures in terms of special causes, and then move from out of control to in control by correcting or removing these special causes. We can extract from this summary some basic underlying assumptions in the implementation of all standard control charts[5] (variable and attribute):

Assumption 1: Standard control charts can reliably signal statistical behavior not attributable to random behavior.

Assumption 2: If properly identified, statistical behavior not attributable to random behavior is indeed undesirable, and thus the goal is its removal.

Assumption 3: If all statistical behaviors not attributable to random behavior are indeed undesirable, they can indeed be removed. Hence, it is possible to move the process to a state of statistical control defined by an iid process.

Other assumptions are entailed in the design of standard control charts. In particular, additional assumptions about the *distribution* of the monitored statistics are made:

Assumption 4: For variable control charts, it is assumed that individual quality measures, or most statistics computed from subgroups of these measures, are well approximated by the normal distribution.

Assumption 5: For attribute charts, a binomial distribution is assumed for p charts and a Poisson distribution for c charts. In addition, practitioners take it for granted that the normal

[5]The phrase *standard control charts* broadly encompasses basic Shewhart charts, non-Shewhart charts, and other standard control criteria, such as supplementary runs rules. Alternative phrases might be *traditional control charts* or *conventional control charts*.

Control Chart	Sample Statistic	Subgroup Size	Basis of Control	Chapter
Variable Control Charts				
x chart	Individual reading	$n = 1$	Shewhart	5
MR chart	Range	$n = 1$	Shewhart	5
$\bar{x}$ chart	Mean	$n > 1$	Shewhart	6
R chart	Range	$n > 1$	Shewhart	6
s chart	Standard deviation	$n > 1$	Shewhart	6
s^2 chart	Variance	$n > 1$	Shewhart	6
MA chart	Based on any of the Shewhart statistics	$n \geq 1$	Non-Shewhart	8
EWMA chart	Based on any of the Shewhart statistics	$n \geq 1$	Non-Shewhart	8
CUSUM chart	Based on any of the Shewhart statistics	$n \geq 1$	Non-Shewhart	8
Multivariate chart	Vector and matrices of summary statistics	$n \geq 1$	Shewhart	10
Attribute Control Charts				
p chart	Percent nonconforming	$n > 1$, fixed or varying	Shewhart	7
np chart	Number of nonconforming	$n > 1$, fixed	Shewhart	7
c chart	Number of nonconformities	$n = 1$, fixed	Shewhart	7
u chart	Average number of nonconformities	$n > 1$, fixed or varying	Shewhart	7
D chart	Weighted average number of nonconformities	$n > 1$, fixed	Shewhart	7
EWMA chart	Based on p- or c-chart type data	$n \geq 1$	Non-Shewhart	8
CUSUM chart	Based on p- or c-chart type data	$n \geq 1$	Non-Shewhart	8

NOTE: For c, u, and D charts, n is the number of unit intervals sampled (see Chapter 7 for details).

Figure 4.13   Classification of standard control charts covered in subsequent chapters.

distribution is an adequate approximation to the binomial and Poisson distributions.

As we will demonstrate throughout the book, there are many applications for which the listed assumptions are appropriate, and thus standard control charts provide an appropriate basis for process monitoring. However, there are also many practical applications for which the basic assumptions underlying the proper interpretation of standard control charts are challenged. In the next section, we outline common practical scenarios that can reduce the effectiveness of standard control charts.

4.12   LIMITATIONS OF STANDARD CONTROL CHARTS

In quantitative analysis, it is often said that "the model is only as good as its assumptions." Control chart methods are not immune to this statement. In the last section, we listed the basic assumptions underlying the standard implementation of control charts. In accordance with those listed assumptions, we discuss below the implications of practical challenges or violations to them.

CHALLENGES TO ASSUMPTION 1

Assumption 1 deals with the presumption that standard control charts can reliably sift nonrandom variability from random variability. Underlying all standard control chart procedures is a view of reality that envisages just two possibilities: a state of statistical control versus everything else. "A state of statistical control" is a very distinct concept—random behavior—while "everything else" is not. "Everything else" is any one of the infinite number of possible nonrandom departures from a state of statistical control. It might be a sporadic shift in the process level or variation due to a specific worker or machine or some special fleeting circumstance. Or, it might be a predictable pattern persisting always and affecting all process outcomes.

Earlier in Section 4.6, we discussed that checking for a state of statistical control—randomness—can usually be regarded as a test for a null hypothesis. An out-of-control state is any hypothesis alternative to this null hypothesis. Restating the hypothesis breakdown, we have

> Null hypothesis H_0: Random process
> Alternative hypothesis H_1: Everything else

To expect control chart limits or any set of control criteria to detect everything and anything under the "sky" defies common sense. Rather, we need to realize that any given control criterion is a statistical test designed to be sensitive to some *particular* departure from randomness. For instance,

1.   A Shewhart-type chart is reasonable if one believes that an important alternative hypothesis to the hypothesis of statistical control is the possibility that sudden and substantial shocks or shifts (in level and/or variability) may impinge on the common-cause system underlying a process that has previously been in control. So, the dichotomy between "randomness" and "everything else" is made in more concrete terms:

> Null hypothesis H_0: Random process
> Alternative hypothesis H_1: A shift in the random process

A shift may be either temporary or sustaining. As illustrated by Figure 4.14, Shewhart control charts are well designed to detect substantial shifts in processes which are otherwise random. In general, to see the types of out-of-control scenarios falling under the specific alternative hypothesis of a shift to a random process, refer to Figure 4.4. Against an alternative hypothesis of shifts, 3σ Shewhart control limits provide sufficient power to detect major shocks without sounding frequent false alarms in the absence of such shocks.

2.   CUSUM and EWMA charts have proved reasonable procedures for the detection of small to moderate *sustaining* shifts affecting an otherwise random process. So, in terms of hypotheses,

> Null hypothesis H_0: Random process
> Alternative hypothesis H_1: A small to moderate sustaining shift in
>   the random process

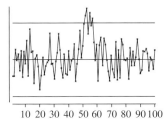

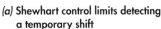

(a) Shewhart control limits detecting
a temporary shift

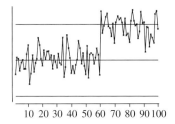

10 20 30 40 50 60 70 80 90 100

(b) Shewhart control limits detecting a
sustaining shift

Figure 4.14   Examples of substantial shifts to an otherwise random process.

As an example, consider the random process of Figure 4.15*a* which has shifted by a small amount around observation number 60. Notice that the shift is not large enough to allow Shewhart control limits to detect the out-of-control condition.

However, the CUSUM control chart (Figure 4.15*b*) does detect the shift, as evidenced by the numerous CUSUM observations falling above the upper limit; details of constructing CUSUM charts will be provided in Chapter 8. Even though the CUSUM control chart outperformed the Shewhart control chart in this example, it can be shown to be slower in the detection of sudden large shifts. As we have said, each type of control chart is designed to be most effective for particular departures from randomness.

3.   Runs rules are criteria for helping detect a variety of alternative hypotheses. For instance, some runs rules help detect small to moderate shifts not substantial enough for 3σ limits. Other runs rules look for unusual localized patterns not expected from a random process, for example, a string of oscillating process outcomes.

Thus, any given control criterion should not be viewed as a test of the null hypothesis of randomness versus the alternative hypothesis of "everything else." Instead, it should be viewed as a procedure to highlight a specific departure within the infinite set of possibilities lumped together as "everything else." For the most part, standard methods tend to focus on the detection of some sort of shift: (1) either localized or sustaining, and (2) small, moderate, or large. If, in reality, processes generally conform to randomness, with occasional shifts, then standard methods should work well. Unfortunately, such departures are not the only kind of violations that can occur.

The truth of the matter is that random processes with occasional unpredictable shifts represent only a slice of the possibilities that may occur in practice. It is very common for a process to be nonrandom but nonetheless a predictable process, one that is not affected by shifts or shocks. Underlying all predictable processes is a systematic pattern. As discussed in Chapters 2 and 3, there are a variety of systematic patterns frequently encountered in practice. For example, careful statistical analysis often yields evidence of trends (linear and

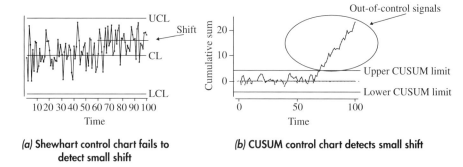

(a) Shewhart control chart fails to detect small shift

(b) CUSUM control chart detects small shift

Figure 4.15   Shewhart and CUSUM control chart applied to same data series.

nonlinear), seasonal variation, meandering or autocorrelated behavior, or some combination of these effects.

The recognition of systematic behavior—particularly, autocorrelation—in SPC applications has been extensively noted in the literature. Greatly due to the pioneering efforts of Professor George E. P. Box, autocorrelation in the chemical-processing industry has long been realized to be extensive. Montgomery and Friedman (1989) find that autocorrelation frequently occurs in SPC data from computer-integrated manufacturing environments. In the automated clinical laboratory setting, Alwan and Bissell (1988) found substantial autocorrelation in routine clinical chemistry SPC measurements. Similar findings in the continuous-process and high-technology industries have been cited, among others, by Baxley (1990); Berthouex, Hunter, and Pallesen (1978); Ermer, Chow, and Wu (1979); Harris and Ross (1991); Hunter (1990); and MacGregor and Harris (1990). In a classic paper on measurement processes, Eisenhart (1962) notes:

> Experience shows that in the case of measurement processes the ideal of strict statistical control that Shewhart prescribes is usually very difficult to attain, just as in the case of industrial production processes. Indeed, many measurement processes simply do not and, it would seem, cannot be made to conform to this ideal of producing successive measurements of a single quantity that can be considered to be "observed values" of independent identically distributed random variables. (p. 167)

Wetherill (1977) suggests that nonrandom autocorrelated behavior may be a pervasive phenomenon:

> For example, in the production of chocolate bars, it is not possible to control the mean weight in the Shewhart sense. The mean weight seems to wander, rather like a first-order autoregressive process, but with assignable causes of variation superimposed. . . . This sort of behavior has been observed by the author in a *number* of cases, and frequently either little is known about the causes of the oscillations of the process mean, or else very little can be done about them. (p. 65)

In an empirical study, Alwan and Roberts (1995) conservatively estimated that the rate of systematic patterns in SPC application is at least 80 percent in prac-

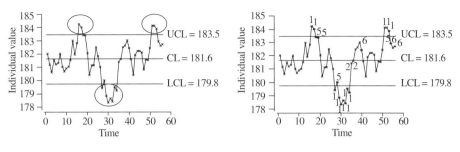

(a) Shewhart control limits only   (b) Shewhart control limits supplemented with runs rules

Figure 4.16   Misleading application of standard methods to daily weight series.

tice. They based their study on an analysis of a large sample of process application (variable and attribute) from a variety of industries (manufacturing and service). The implication is that non-iid behavior appears to be the rule rather than the exception in practice.

The logical question to ask is, What are the consequences of systematic patterns for the implementation and interpretation of standard SPC methods? Montgomery (1996) makes it quite clear that the consequences are detrimental:

> The most important of these assumptions [required for control charts] is that of independence of the observations, for conventional control charts do not work well if the quality characteristic exhibits correlation over time. Specifically, these control charts will frequently give misleading results if the data are correlated. (p. 375)

The conventional (or standard) control charts considered by Montgomery include the Shewhart control chart, CUSUM charts, and other procedures, such as EWMA charts. To help develop insights, we can illustrate the potential misleading problems by means of concrete examples.

Daily Weight Series   Consider the daily weight series studied in Chapter 3. We found that the daily weights followed a systematic meandering pattern which is predictable and stationary. Furthermore, we found no shifts, shocks, or any unusual departures over and above the natural workings of the weight process. Suppose we apply the standard control criteria to the weight series. In Figure 4.16a, a Shewhart control chart for the data is presented. (*Note:* This control chart was produced in Minitab's default mode using the method of moving ranges; we will discuss the details this method and others in Chapter 5.) From Figure 4.16a, we see 12 observations falling outside the control limits in three different clusters. Because the emphasis of standard methods is on detecting shifts in random processes, there is an immediate temptation to view the daily weight process as a sequence of three isolated shift episodes, each with its own special cause.

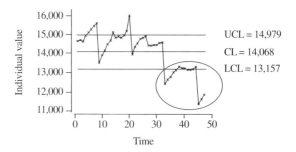

Figure 4.17   Misleading application of standard methods to the ASQ membership series.

Matters are made much worse when supplementary runs rules are applied to the data series, as shown in Figure 4.16*b*. Minitab flags 21 special-cause signals for a data series of 56 observations! Standard control criteria lead one to conclude that the process is wildly out of control, riddled with numerous special causes. Preoccupation with nonexistent special causes diverts attention from understanding the true workings of the process. The fact of the matter is that the process is affected only by two influences: a stable meandering pattern and in-control random variation around the pattern.

Standard methods tend to lock one into a mind-set that out-of-control signals must be associated with some specific localized special cause, diverting one away from seeing the big picture. This problem is not limited to inexperienced or untrained users. As demonstrated by the next couple of examples, the most experienced experts can fall into the trap of mechanically relying on the guidance of standard methods that should not have been used in the first place.

ASQ Membership Series   The ASQ Statistics Division customer membership series from Chapter 3 is an example of experienced users being potentially misled by standard control chart methods. Recall we found that this series is dominated by a beginning-to-end systematic pattern comprising seasonal, trend, and lagged effects. Additionally, we found that one particular month (observation 21) substantially stands out over and above these nonrandom effects. In Figure 4.17, we superimpose standard Shewhart control limits on the membership data series. It can be noticed that 18 out of the 47 observations fall outside control limits.

In a similar fashion to Figure 4.17, it was reported in the division newsletter (Winter, 1995) that standard Shewhart control limits were applied to the data series. In the newsletter, it was pointed out that a number of the observations at the end of the data series fall below the lower control limit, as highlighted in Figure 4.17. The following speculation was given to this set of "out-of-control" observations: "Suspected 'special causes' of this latest trend include a down-turned economy and the wider assortment of ASQ divisions (21) to choose from" (p. 19). Applying and interpreting the standard control chart moved the perspective from

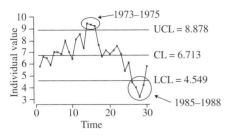

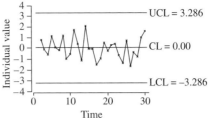

(a) Shewhart control limits applied to savings rate series

(b) Shewhart control limits applied to period-to-period changes in savings rate

Figure 4.18   Looking for special causes in savings rate series.

the process as a whole to bits and pieces of the process. Out-of-control signals are given special explanations while in-control signals are given less attention. As we have learned from the detailed data analysis, the *trend* in the process is not isolated to the last group of observations; rather it has a global nature related to the overall process as a whole, and it is pervasive from the beginning to the end of the data series. Attention should have been directed to understanding this pervasive process behavior. The horizontal control limits superimposed on the data series in Figure 4.17 serve only as unnecessary distractions.

With respect to the localized special-cause signal revealed by data analysis at observation 21 (September 1992), this observation is deemed an out-of-control signal by the standard control limits. However, there is no mention of this legitimate signal in the division newsletter. In all likelihood, the importance of this one observation was lost in the sea of many other out-of-control signals.

Savings Rate Series   Let us consider another example of misleading conclusions resulting from the standard control chart method. In a popular SPC textbook written by a deservedly well-known author[6] and consultant, the author demonstrates the construction of a control chart on the annual savings rate data (from 1960 to 1989); for the interested reader, the data can be found in "savings.dat". Figure 4.18*a* gives a Minitab version of the control chart constructed by the author; control limits are computed using the moving-range statistic (refer to Chapter 5). In discussing the results of the control chart, the author pointed out that the savings rate process was impacted by two distinct special causes during the years of 1973 to 1975 and 1985 to 1988, as highlighted on Figure 4.18*a*. The author then pointed out that individuals (e.g., government officials, economists) concerned with the saving process need to direct specific attention to these two periods to search for special-cause explanations.

---

[6]I do not mention the author since my objective is not to single out an individual's oversights. In fact, I applaud the author's attempt to show the use of SPC methods for less typical applications.

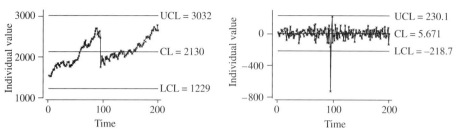

(a) **Shewhart control limits applied to weekly closing prices of the Dow Jones**

(b) **Shewhart control limits applied to week-to-week changes in Dow Jones**

Figure 4.19   Revealing special-cause shock in Dow Jones series.

As with the ASQ example, this is a good example of providing ad hoc explanations based on mechanical application and belief in the standard method. Economists and financial experts have long known that many economic and financial indices tend to follow certain systematic patterns. In particular, there is considerable evidence that a specific nonrandom process, known as a random walk, can adequately describe many economic and financial time series. A random walk is a time series for which the changes from period to period are random. In Figure 4.18b, we show a plot of the year-to-year changes of the savings rate process. As can be seen, the changes appear quite random and stable. Applying standard 3σ limits to these data suggests that there were no special causes for the savings rate process in the sense of shocks (up or down) that are incompatible with a state of statistical control.

Dow Jones Series   As we have seen, standard control limits in the presence of systematic variation can mislead users into believing that isolated points are to be associated with some special cause(s). The other danger is that standard control limits in the presence of systematic variation can camouflage legitimate special-cause signals. In Figure 4.19a, we superimpose standard 3σ limits[7] on 200 consecutive weeks of the weekly closing prices of the Dow Jones Industrial Index; for the interested reader, the data can be found in "dow.dat". Some users of control charts might naively conclude that since all points are within the limits, the process is in control or no special causes are at work. Another naïve approach would be to supplement control limits with other standard control criteria, such as runs rules; this lesson was learned with the weight process (see Figure 4.16b). For fun, if we had asked Minitab to invoke the full set of supplementary runs rules, we would have reported that 177 of the 200 weeks were associated with a special-cause signal!

The difficulty with this approach is that standard control criteria (whether horizontal control limits or runs criteria) are unable to provide appropriate

[7]For this example, we compute the standard control limits based on the sample standard deviation, s. The alternative is to use the moving-range statistic. Both methods will be presented in Chapter 5.

guidance in the presence of the evident nonrandom pattern in the Dow Jones series. As with the savings rate process, a very simple application of data analysis can provide the proper perspective. Again, the random walk model appropriately describes the time behavior of the Dow Jones series. As such, we expect the week-to-week changes in the Dow Jones Index to be random. In Figure 4.19*b*, we present a time-sequence plot of the weekly changes along with standard 3σ limits.

As can be seen, the changes are, for the most part, random around the centerline positioned at the average value of 5.59. The real issue which is brought to bear is the substantial out-of-control signal associated with week 95. Only this major shock deserves the title of out of control and justifies the search for a special cause. As it turns out, week 95 is the week of the famous October 19, 1987 ("Black Monday") stock market crash. Among the special-cause culprits, it was suspected that an initial downturn in the market was accelerated by automatic computer on-line systems which bought and sold stocks in a frenzied state.

The moral of this example is that standard control chart methods applied to the original data series (Figure 4.19*a*) did not isolate a truly legitimate special-cause signal embedded in a process exhibiting systematic variation. Once the systematic variation is accounted for, the standard control chart is able to discriminate the effects of the special-cause disturbance away from the remaining random variation.

With the above examples, we have gained a more specific sense of the dangers of interpreting standard methods in the presence of systematic nonrandom patterns. Namely, the simple interpretation offered by standard methods can often lead one to

1. Search for special causes when the data, properly interpreted, suggest that no special causes are present.

2. Fail to search for special causes when the data, properly interpreted, suggest that special causes are present.

3. Fail to identify systematic variation in the process, such as trend, seasonal, and meandering variation.

A natural solution to these problems is to go beyond the reliance on standard control charts to a broader approach which draws upon practical data analysis techniques, such as those detailed in Chapter 3. It should be underscored that standard SPC methods are not precluded from a broader data analysis approach. In fact, we will demonstrate that they are a *very* important part of the overall toolkit necessary for effective process data analysis. Let us now turn our attention to assumptions 2 and 3.

CHALLENGES TO ASSUMPTIONS 2 AND 3

Assumptions 2 and 3 presume that any nonrandom variation is undesirable and that its effects should (and can) be removed from the process. This traditional thinking is appropriate for some applications but not for others.

In manufacturing a product to a specified target, process improvement is equated with variation reduction. If systematic variation or sporadic variation can be properly identified, then actions should be sought for their possible removal with the ultimate goal of bringing the process to a state of statistical control in terms of being iid.

As we have just discussed, standard control chart methods often fail to properly distinguish the different types of variation. The goal of broader data analysis techniques (such as statistical modeling) is to make the proper sifting of variation so that appropriate actions can be taken. Hoerl and Palm (1992) articulate an important point that the goal of process study is not statistical modeling for the sake of statistical modeling; but rather the goal should be the understanding of the process so that actions can be taken to bring the process to a state of statistical control. They write:

> How could those knowledgeable in statistics be so misinformed about Shewhart charts? We believe that one root cause is a deep and fundamental difference in approach between statistical modeling of processes and the Shewhart method. The modeling approach might be characterized by the statement, "Fit the model to the process." . . . The Shewhart approach is just the reverse—namely, "Fit the process to the model." The model in this case is a series of independent random observations from a single distribution. Whatever causes the process to act in a way different from this is considered to be trouble (special cause) and should be permanently eliminated. (p. 268)

The ability to bring a process to a state of statistical control depends on understanding the process. So statistical modeling is not the end all. Instead, statistical modeling is a beginning step for process understanding which leads to improvement actions.

As Hoerl and Palm note, it is ideal to find the causes (sporadic or systematic) of nonrandom variation and permanently eliminate them. In some cases, the removal of causes of systematic variation is quite feasible. For instance, a machining process which exhibits a systematic trend might reflect some sort of tool wear or deterioration. Replacement of machine parts or a routine maintenance program might be needed actions to eliminate the trend effect. However, there are many manufacturing processes in which elimination of the underlying causes of systematic variation is just not possible. With respect to autocorrelated meandering behavior, Wetherill (1977, p. 65) writes, "This sort of [systematic] behavior has been observed by the author in a *number* of cases, and frequently either little is known about the causes of the oscillations of the process mean, or else very little can be done about them."

The fact that elimination of some causes of systematic behavior can be difficult or impossible does not preclude improvement actions in the spirit of variation reduction. In Section 4.1, we noted that there is an area of process control known as adaptive process control (APC). APC is aimed at improving the process by "predicting" how the process will wander off target if left alone, and specifying the adjustment called for to prevent systematic deviation from target.

For target-oriented processes, both SPC and APC have the same objective of reducing variability, leaving only iid or simple random variation. However, their routes to this goal are different; in particular, the intent of SPC is to detect, identify, and remove special causes of variation, while the intent of APC is to offset predictable deviations from target. SPC might be thought of as preventive medicine and APC as curative medicine. It is usually a question of economics as to whether it is less costly to seek out causes and remove them, or to tolerate the causes and compensate for them. With this said, there is no conflict between these two approaches and no reason why both approaches cannot be applied simultaneously for certain applications. In Chapter 10, we will illustrate some simple cases of APC.

Our discussion to this point, along with the comments by Hoerl and Palm, falls under the traditional application of process control, namely, manufacturing a product to target. One of the great developments of the quality management movement is the recognition that the ideas of process monitoring and improvement can be applied to any process, not just a manufacturing process.

For many processes, the traditional Shewhart mind-set of eliminating variability is not appropriate. In fact, forcing a traditional Shewhart mind-set to any and every process is potentially damaging. Would management wish to eliminate the causes and effects of an upward trend in the company sales process? Or, would management wish to eliminate the causes and effects of a downward trend in product defects over time? In both cases, the answer is, of course, no. Additionally, management will no doubt want to monitor these processes for any significant changes (good or bad, unexpected or purposeful) over and above the systematic effects. Since standard control charts tend not to do a good job under such process scenarios, a broader approach to process monitoring is required.

Systematic variation is often a natural and nonremovable component of process behavior. For example, we encountered the lagged effects of a personal daily weight process. Namely, a given day's weight tended to be relatively close to the weight of the previous day. We do not wish to nor can we remove such a natural phenomenon. Similarly, we found that the closing prices of the Dow Jones Industrial Average follow a particular meandering pattern. Again, this systematic behavior is nonremovable. There are many process applications in which the meandering tendency is present and must be dealt with.

Lagged effects and trends are not the only systematic effects intrinsic in many processes. Seasonality is also a common effect in many business processes. Retail, food, sporting equipment, entertainment, and utility industries have pervasive seasonal effects. One organization that recognized the need for organizationwide process monitoring and the inability of standard control charts to do the job is Florida Power and Light (FPL). FPL is the third-largest investor-owned utility in the United States, with the distinction of being the first non-Japanese company to be awarded the coveted Deming Prize for quality.

FPL reportedly monitors hundreds of business processes, such as monthly customer satisfaction levels and number of power outages or interruptions per month. As might be expected, FPL's processes are highly seasonal; for instance, summer months are systematically and significantly different on nearly all measures from other months. Such systematic effects cannot be removed and are inherent to FPL's business. Since standard control charts can mislead in the presence of systematic effects, such as seasonality, FPL uses time-series analysis to estimate the monthly seasonal levels. This information is used to deseasonalize the data, that is, to remove the seasonal component from the process data. Control charts (referred to as *seasonal-adjusted* control charts by FPL) are then applied to the deseasonalized data.

FPL's idea of deseasonalizing the data and then applying standard control charts is novel. But this approach works well if seasonality is the only source of systematic variation. A more flexible approach is to construct a statistical model for the process data, allowing one to incorporate seasonality and/or any other systematic effects that might be concurrently affecting the process.

CHALLENGES TO ASSUMPTIONS 4 AND 5

We have discussed the implications of nonrandom time behavior on the effectiveness of standard control criteria. We must also remember that certain distribution assumptions are made for proper interpretation of standard control criteria. For variable charts, it is assumed that the monitored statistic approximately follows the normal distribution. For attribute charts, a binomial distribution is assumed for proportions (p chart) and a Poisson distribution for counts (c chart). It is further assumed that the binomial or Poisson distribution is adequately approximated by the normal distribution.

In many practical applications, these distribution assumptions are challenged by the data, resulting in misleading conclusions. Positive skewness is probably the most common departure from normality encountered in practice. For certain control chart statistics, the response to nonnormality might simply be to increase the subgroup sample size. The idea is to allow the central limit theorem effect to take hold so that the sample statistic can be approximated by the normal distribution. However, increasing the sample sizes is not always a practical solution. Sampling extra units may be too costly to justify. In other cases, the application naturally lends itself to collecting a limited number of observations per sampling period; for example, Chapter 5 is devoted to applications for monitoring series of individual measurements.

To see the effect of nonnormality on the interpretation of standard control charts, consider the daily caloric intake data series studied in Section 3.3. We found this series of individual measurements to be random but highly nonnormal in the form of positive skewness. A blind application of 3σ Shewhart control limits to these data is shown in Figure 4.20a. The temptation is to view the two highlighted points above the upper control limit as special-cause signals. The difficulty with this viewpoint is the fundamental lack of adherence of the data distribution to the underlying assumption of normality. As demonstrated in Chapter 3, a simple power transformation produces data closely approximated by the normal distribution. Since the transformed data are random and approximately normal, application of 3σ Shewhart control limits to the transformed data (see Figure 4.20b) can safely be interpreted for possible special-cause signals. But as can be seen from the second

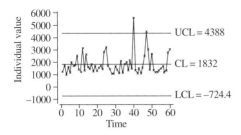

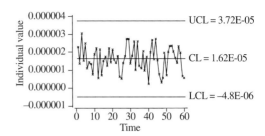

(a) Shewhart control limits applied to original
calorie series

(b) Shewhart control limits applied to transformed
calorie series

Figure 4.20   Nonnormality and control charts.

control chart, there is absolutely no evidence of exceptional or unusual caloric observations anywhere during the observation period.

There is another type of distribution departure unique to attribute charts. It is quite possible that attribute data are random and approximately normal but standard control limits still do not provide appropriate guidance for special-cause searching. The problem is that variation in the attribute data might not be due to the assumed underlying attribute distribution. Namely, proportions may not be purely binomial, or counts may not be purely Poisson. In such cases, the binomial and Poisson models do not serve as an appropriate basis for sifting out the unusual observations from inherent variation. In Chapter 7, we offer some basic statistical tests, with examples, for judging whether standard attribute limits should be applied to the attribute data series.

4.13   CONTROL CHARTS AND PDCA

We have learned that control charts are devices for detecting change over and above natural process variability. Because of its origin in manufacturing applications, the control chart is most often viewed as a statistical tool for detecting changes which are unexpected and carry bad consequences. By unexpected change we mean a significant change in process behavior which is the result of some unintended event's affecting the process. In this sense, unexpected change can be viewed as unintended change. There have been several instances in our discussion where we have alluded that not all changes are bad or unintended. This is a point worth reemphasizing with the following delineation:

1.   A change can be the result of an unintended event with bad consequences for the process.

2.   A change can be the result of an unintended event with good consequences for the process.

3.   A change can be the result of a purposeful intervention with bad consequences for the process.

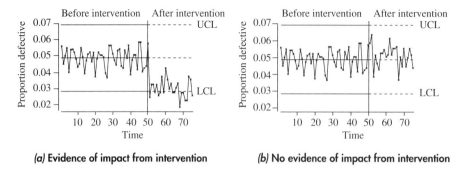

(a) **Evidence of impact from intervention**   (b) **No evidence of impact from intervention**

Figure 4.21   Effective use of standard control charts for assessing the impact of an intervention.

4. A change can be the result of a purposeful intervention with good consequences for the process.

As noted, the first case of unintended change with bad consequences is the traditional perception of change with the typical application of a control chart. In the second case, the unintended change can have favorable implications to the process, for example, a significant reduction in process defects. Such situations can be described by the word *serendipity,* which refers to the accidental discovery of good events. The discovery of useful surprises provides the organization with an opportunity to learn from the serendipitous event and to transfer the gained knowledge to future process improvements.

To go beyond passive observation, management can deliberately intervene in an attempt to improve the process. Examples of deliberate actions include a promotional campaign, a change in personnel training, and the introduction of new technology. Intervention ideas should not be blind guesses; instead, there should be appropriate planning and sound reasoning to suggest that potential interventions might have favorable prospects. The development of and planning for interventions are the primary activities of the plan stage of the plan-do-check-act (PDCA) cycle. Once an intervention is chosen and done (the do stage), there is a need to check or observe the effects of the intervention.

Under appropriate conditions, control charts can be quite helpful for judging whether the intervened process is any different from the process prior to the intervention. For instance, Gottman (1981) notes, "When the data are not correlated, a Shewhart chart can be used as a graphical technique to assess the effectiveness of an intervention" (p. 56). To illustrate, suppose the proportion of product defects in a given production run is monitored by using standard control charts, namely, p charts (see Chapter 7). Suppose the sample proportions are well approximated by the normal distribution and are random around an overall average close to 5 percent defective. With the goal of reducing the overall percentage defective, management felt that changing the primary supplier of raw material would have a significant beneficial impact. Figure 4.21 shows how the standard control chart can help assess the impact of the intervention for two possible

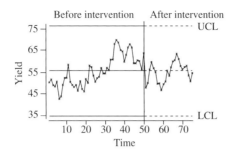

Figure 4.22   No evidence of impact from intervention using standard control chart.

scenarios. In the first case (Figure 4.21*a*), the intervention clearly had an impact, as reflected by the significant shift in process outcomes, many of which fall below the lower control limit. If one wishes, other standard control criteria, such as runs rules (see Section 4.9), can also be used to more quickly detect changes from past behavior. However, in the second case (Figure 4.21*b*), because the process outcomes *after* the intervention fall randomly within the control limits established *prior* to the intervention, this serves as clear evidence that the intervention had no impact on the process.

As we have learned, the standard control chart and other standard criteria are designed to detect changes in process behavior that is random and approximately normally distributed. However, the ability of standard methods to assess the impact of an intervention can be limited in the presence of other types of process behavior. For example, if the observations were autocorrelated, Gottman (1981) explains, ". . . then the problem is slightly more difficult than the use of Shewhart charts" (p. 56) He also notes, "Sophisticated examinations of the effect of even moderate autocorrelation in the data show that the estimates of effect size that assume no autocorrelation [such as Shewhart charts] may be badly biased . . ." (p. 56).

To illustrate the difficulties with using standard methods for assessing the impact of an intervention, consider Figure 4.22. The process measurements are yields, and as such, higher levels are desirable; for the interested reader, the data can be found in "yield.dat". Careful analysis of the preintervention data would reveal that the series follows a meandering pattern around an upward linear trend; in Chapter 5, we will illustrate the modeling of such a systematic pattern with another application. If we were to apply mechanically standard Shewhart control limits to the data series and project them out, we would find that the postintervention data fell well within the limits. Furthermore, it is difficult to determine if there was a shift in the data series after the intervention or if the postintervention data were simply a continuation of the meandering evidenced during the preintervention period.

One of two approaches might be used to assess *properly* the impact of an intervention. One way would be to create an independent variable which reflects

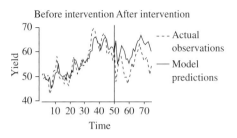

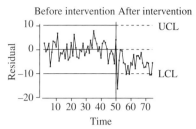

(a) Comparison of observed data series and preintervention model predictions

(b) Evidence of impact from intervention revealed

Figure 4.23   Appropriately assessing impact of an intervention in presence of systematic pattern.

the timing of the intervention. For example, we might create an indicator variable which simply takes on the value 0 for every period prior to the intervention and 1 for every period after the intervention. This intervention variable is then added to the regression model to estimate the intervention effect as part of the statistical model. If the regression coefficient for the intervention variable is viewed as significantly different from zero, then this is evidence supporting the conclusion that the intervention had a significant impact. In Chapter 7, we will provide a detailed example of this modeling approach.

Since the intervention variable spans the pre- and postintervention periods, the intervention variable model approach requires the analyst to fit a regression model to the whole data series, that is, pre- and postintervention data. There is another approach which is more akin to the standard control chart procedure. With the control chart approach, control limits are established from the preintervention data. Technically, these limits serve as the statistical model for the preintervention data. By projecting out these control limits, we are asking whether postintervention data differ significantly from what is expected by the preintervention model. If the data are random and normally distributed during the preintervention period, then the control limits serve as a proper statistical model. If, however, the preintervention data, as in this case, do not satisfy the conditions which are suitable for the application of standard control limits, then the data must be modeled appropriately.

As we mentioned, data analysis of the preintervention data would lead one to fit a model that captures a meandering effect and a trend component. In Figure 4.23a, we simultaneously plot the entire data series along with predictions of the process based solely on the preintervention model. Notice that the model is quite compatible with the preintervention data; however, there is a sustaining difference between the postintervention data and the model's predictions. Recall that the difference between any given observation and the model's prediction is defined as a residual. Hence, our comparison between the model and the actual data series can be equivalently reflected by plotting the residuals in time order.

Now, since the preintervention model is an appropriate fit for the preintervention data, the residuals behave as a random data series. In this case, the residuals are also well approximated by the normal distribution. This implies that a standard Shewhart control chart serves as an appropriate model for monitoring the preintervention residuals. If the postintervention residuals continue to behave in a manner consistent with preintervention limits, this implies the *absence* of a significant intervention effect. For our example, the postintervention residuals are clearly shifted down relative to the preintervention residuals (see Figure 4.23b). In fact, several residuals fall below the lower Shewhart control limit. The conclusion is that the purposeful attempt to improve the yield process actually resulted in a significant negative effect. But, to recap, a control chart applied directly to the original nonrandom yield series (Figure 4.22) was unable to properly assess the impact of the intervention. The standard application of the control chart actually misled us into believing there was no intervention effect when, in fact, there is a serious problem calling for management attention. Thus, we have again an application which calls for a broader data analysis perspective for proper understanding and monitoring of the process in question.

4.14   SUMMARY

In conclusion, under the proper conditions, standard control charts are ingenious tools for detecting special causes from an unintended event or a purposeful intervention. We, however, also recognize that practical challenges (e.g., systematic variation and/or nonnormality) to the appropriate interpretation of the standard SPC methods are frequent. Our presentation of examples that are not well suited for the application of standard SPC methods is not to suggest that they have no role or benefit in practice. Indeed, in *each* of the subsequent chapters, we will provide detailed coverage of the construction of many different types of standard control charts and will demonstrate their usefulness for gaining appropriate insight about the process in question. But given the frequency of practical challenges, we believe any coverage of process analysis based exclusively on standard SPC methods falls substantially short of what is needed to deal with reality.

EXERCISES

4.1. Describe the difference between chance and special causes.

4.2. How should a manager use variation to manage a group of people? Why should a manager be aware of common and special causes?

4.3. Refer to Figure 3.26. The possible reactions of believing that the highlighted points are "setbacks" are akin to what issues discussed in this chapter?

4.4. Refer to Figure 3.34. The possible reactions of "cut back" or "eat away" are akin to what issues discussed in this chapter?

4.5. From your personal experience, provide some examples of the two kinds of mistakes that can be made in interpreting variation.

4.6. Why is management's commitment so important to implementing SPC?

4.7. What is meant by the term *process stability*?

4.8. Why is a random process important for a target-based manufacturing process?

4.9. What is meant by this statement? "The process is in a state of statistical control." Does the statement imply anything about the type of distribution underlying the process?

4.10. Consider the following excerpt from a *New York Times* article (April 19, 1988) titled "Hot Hands Phenomenon: A Myth?" Explain in detail the issues of this chapter that directly relate to the excerpt.

> The gulf between science and sports may never loom wider than in the case of the hot hands. Those who play, coach or otherwise follow basketball believe almost universally that a player who has successfully made his last shot or last few shots—a player with hot hands—is more likely to make his next shot. An exhaustive statistical analysis led by a Stanford University psychologist, examining thousands of shots in actual games, found otherwise: the probability of a successful shot depends not at all on the shots that come before.
>
> To the psychologist, Amos Tversky, the discrepancy between reality and belief highlights the extraordinary differences between events that are random and events that people perceive as random. When events come in clusters and streaks, people look for explanations; they refuse to believe they are random, even though clusters and streaks do occur in random data.
>
> "Very often the search for explanation in human affairs is a rejection of randomness," Dr. Tversky said.

4.11. Suppose an operator is instructed to keep the process outcomes as close as possible to a target value of 50.00. Below are the values of the process outcomes had the operator done *nothing* (data are also given in "adjust1.dat").

```
51.28  49.42  48.75  48.00  50.45  48.70  51.31  49.66  50.21
49.58  51.70  49.70  50.09  50.15  47.89  49.76  49.39  50.98
49.87  49.68  50.81  49.05  49.38  50.78  50.49  49.67  50.66
50.32  51.40  50.62
```

Suppose that instead of leaving the process alone, the operator adjusts the process, by changing a machine setting, after each observed outcome, using rule 2 as described in Section 4.5. Assume the machine setting is initially set at zero.

a. Perform all random diagnostics on the observations from the unadjusted process. What is your conclusion?

    b. Determine now the values of the process that would have resulted given the noted adjustment policy.

    c. By what factor did the sample standard deviation change?

    d. Perform all random diagnostics on the observations from the adjusted process. What is your conclusion?

    e. Did the adjustment policy improve or worsen the process? Explain why.

4.12. Suppose an operator is instructed to keep the process outcomes as close as possible to a target value of 100.00. Below are the values of the process outcomes had the operator done *nothing* (data are also given in "adjust2.dat"):

94.36	92.28	95.66	94.40	96.55	97.74	93.72	95.75
102.38	105.29	102.03	98.37	101.23	102.52	102.67	101.89
102.20	107.38	105.14	105.40	99.63	101.23	101.39	100.34
102.80	100.51	103.68	101.71	101.23	100.35		

Suppose that instead of leaving the process alone, the operator adjusts the process, by changing a machine setting, after each observed outcome, using rule 2 as described in Section 4.5. Assume the machine setting is initially set at zero.

    a. Perform all random diagnostics on the observations from the unadjusted process. What is your conclusion?

    b. Determine now the values of the process that would have resulted given the noted adjustment policy.

    c. By what factor did the sample standard deviation change?

    d. Perform all random diagnostics on the observations from the adjusted process. What is your conclusion?

    e. Did the adjustment policy improve or worsen the process? Explain why.

4.13. Explain how and under what conditions a standard control chart can be a valuable aid for counteracting the problem of overadjustment.

4.14. In the classic book *Out of the Crisis*, Deming writes: "Some [business] leaders forget an important mathematical theorem that if 20 people are engaged on a job, 2 will fall at the bottom 10 percent, no matter what. It is difficult to overthrow the law of gravitation and laws of nature. The important problem is not the bottom 10 percent, but who is statistically out of line and in need of help."

    In another part of the book, Deming argues with vigor that management by objectives, management by numbers, and management by fear are one and the same. He writes: "It leaves people bitter, crushed, bruised, battered, desolate, despondent, dejected, feeling inferior, some even depressed, unfit for work for weeks after receipt of rating, unable to comprehend why they are inferior. . . . a merit rating is a meaningless predictor of performance."

> **a.** How do these separate quotations relate to each other?
>
> **b.** Deming argues that annual merit rating should be completely abolished. Assuming meaningful numeric ratings can be attached to employees, how can management use these numbers in a way that might appease Deming? Your suggested performance procedure should be based on some statistical rule, for example, rewarding only those falling above the sample average.

4.15. Describe the difference between variable and attribute data. What types of control charts are used for each type of data?

4.16. Explain the connection between a control chart and statistical hypothesis testing.

4.17. Define and explain type I and type II errors in the context of control charts. How are they related?

4.18. Discuss the implications of control limit placement in terms of type I and type II errors.

4.19. Determine the z multiple if you want to construct a Shewhart chart based on 0.005 probability limits.

4.20. Explain the notions of using control charts retrospectively versus prospectively.

4.21. Suppose 3σ control limits are applied to individual measurements from an iid process with an underlying exponential distribution with known mean parameter μ equal to 5. To answer the questions below, use the fact that for the exponential distribution, its mean equals its standard deviation (σ).

> **a.** What are the values of the lower and upper control limits?
>
> **b.** Given the control limits found in part *a*, use Minitab's CDF command to compute the false-alarm rate. How does the rate compare to the rate associated with an iidn process?

4.22. Consider the following Minitab-based simulation experiment:

> Step 1   Generate 1000 random observations from a specified probability distribution (Calc >> Random Data).
>
> Step 2   Apply Minitab's default standard 3σ control chart to the series of individual measurements (Stat >> Control Charts >> Individuals).
>
> Step 3   Count the number of observations falling outside the control limits.
>
> Step 4   Repeat the above three steps 10 times over. (*Note:* An alternative is to run the experiment once based on the random generation of 10,000 observations. However, if the Student Edition of Minitab is being used, there is a limitation of 3000 observations in the worksheet.)

Step 5 Estimate the average number of points per 1000 observations falling outside the control limits.

a. Perform the experiment on data arising from a normal distribution. For convenience, choose the Minitab default of the standard normal distribution.

b. Perform the experiment on data arising from a nonnormal distribution. In particular, select the default exponential distribution in Minitab (namely, an exponential with mean equal to 1).

c. Given the results from parts a and b, what did you learn from the experiment?

4.23. Discuss the relationship between ARL and type I and type II error rates. What information does the ARL immediately convey that the error rates do not?

4.24. Below are 60 consecutive observations in standardized units:

2.24	2.66	−0.24	3.09	2.40	1.94	1.22	3.26	2.12	2.11
2.91	0.09	3.74	2.72	0.45	2.24	2.47	1.69	1.78	2.73
1.17	2.79	2.55	2.62	0.09	3.15	2.24	2.81	1.32	2.93
1.77	1.53	2.69	2.70	1.70	2.34	3.62	1.25	1.48	2.70
0.79	3.35	2.34	2.04	3.22	1.45	2.86	3.24	2.73	2.01
2.07	1.88	3.30	1.25	3.24	2.34	1.12	2.91	2.03	3.76

a. The *run length* is the number of sample points until an out-of-control signal. Using only a 3σ criterion, find the observed run lengths in the provided data. (*Note:* Once an out-of-control signal occurs, a new run length is determined.)

b. What is the estimated ARL for the data series?

4.25. Suppose a 3σ Shewhart control chart is applied to a sample statistic following an iidn process with known mean μ and standard deviation σ. Assume now that after the implementation of the control chart, the mean of the statistic has shifted down by 1 standard deviation. What is the expected number of samples observed before the shift is detected?

4.26. Suppose that Shewhart control limits are applied to a sample statistic which follows an iidn process. Sometime after implementation of the control limits, assume that the process experiences a sustaining mean shift which the control chart has probability $1 - \beta$ of detecting on any given sample.

a. What is the probability the shift will be not be detected by the first sample but will be detected by the second sample after the occurrence of the shift?

b. What is the probability the shift will be not be detected until the kth sample after the occurrence of the shift?

c. What is the expected number of samples observed before the shift is detected?

Handwritten margin notes:

* a C chart has been centered @ $\bar{C} = 10$. Upper and lower control limits have been calculated to be 19.5 and .5. Assume that the process mean goes down to 5.0. What is the prob. that this shift beneficial shift will be detected b/c the new sample is below the lower control limit?

$\bar{x} = 30.0$
$\bar{R} = 2.0$

$\bar{x} \& R$ chart is in control; with n = 5; with sometime after sometime after implementation of the control limits, assume that the process experiences a sustaining mean shift, the new shift is 29.0. mean is 29.0. What is the prob. that the shift will not be detected?

d. What is the probability that at least one of k samples following the shift will signal out of control?

4.27. Describe the difference between control limits and specification limits.

4.28. If an in-control process goes out of control, does this necessarily imply the associated units produced will fall outside specification limits?

4.29. What is meant by a *run* within the context of control charts?

4.30. Suppose that the process in question is iidn. Assume also that there are enough sample points to give an out-of-control signal for any one of the runs rules discussed in Section 4.9. Given these assumptions, compute, for each runs rule, the probability of a false alarm.

4.31. Below are four different data series consisting of 20 consecutive observations in standardized units:

Observation Number	Series 1	Series 2	Series 3	Series 4
1	−1.48	−0.32	0.45	−0.48
2	0.95	0.64	−0.02	0.58
3	0.60	−0.51	−0.82	−0.40
4	−0.38	2.07	0.32	−0.28
5	0.23	2.37	0.12	0.74
6	−0.75	−1.04	0.62	−1.02
7	−1.79	0.08	0.72	−0.42
8	−0.90	−0.35	0.73	−2.17
9	−0.73	−0.53	0.30	0.21
10	−0.63	−0.21	1.16	−0.36
11	1.31	−1.00	0.34	1.65
12	1.04	−0.10	1.51	0.73
13	1.12	1.05	−1.95	1.10
14	−1.07	−1.95	−1.09	−0.80
15	1.56	−0.33	−1.59	1.69
16	−0.24	−0.42	1.45	−0.10
17	1.03	0.54	0.01	0.77
18	−0.22	−0.28	−1.39	−0.97
19	−0.51	0.09	0.23	−1.21
20	0.91	0.23	−1.09	0.76

For each of the data series, apply the eight control criteria of Section 4.9. In each case, are any of the criteria signaling the potential presence of special-cause variation?

4.32. Suppose you are given data from an in-control random process that might have been affected by special-cause variation in the form of a process mean shift. Once given the data, you soon learn that the time order of the data observations is not given and is no longer retrievable. Given the eight control criteria of Section 4.9, what is the most you can extract from the data set?

4.33. Suppose 3σ control limits are applied to a data series and "numerous" points signal out of control. You perform random diagnostics and find randomness cannot be rejected by any of the tests. Suggest a plausible explanation for the numerous out-of-control points.

4.34. Consider the savings rate series studied in Section 4.12. The data are given below and can also be found in a file named "savings.dat".

5.8	6.6	6.5	5.9	7.0	7.0	6.8	8.0	7.0	6.4
8.1	8.5	7.3	9.4	9.3	9.2	7.6	6.6	7.1	6.8
7.1	7.5	6.8	5.4	6.1	4.4	4.0	3.2	4.2	5.8

  a. Perform all randomness and normality checks on the changes in savings rate to confirm that the changes are in a state of statistical control. As a matter of convenience, the changes can be directly computed with Minitab's "differences" command (Stat $>>$ Time Series $>>$ Differences).
  b. Given that the changes are consistent with an iidn process, provide a 95 percent prediction interval for the change in savings rate from period 30 to 31.
  c. Provide a 95 percent prediction interval for the savings rate in period 31.

4.35. In Section 4.12, we introduced a time-series process known as a *random walk*.
  a. Based on the descriptive definition of a random-walk process given in that section, write down the general model for a random-walk process. Define all variables.
  b. Suppose you have observed T consecutive observations from a random-walk process. Sitting at time period T, what is an appropriate forecast for a future observation at time $T + 1$? Still sitting at time period T, what is an appropriate forecast for a future observation at time $T + 2$? Write down a general expression for the forecast k periods into the future.

4.36. Delineate the variety of possible process conditions that potentially undermine standard SPC methods' ability to provide proper guidance in the monitoring of the process in question.

4.37. Explain how and under what conditions a standard control chart can be a valuable aid in assessing whether an attempt to improve a process was successful.

REFERENCES

Alwan, L. C. (1991): "Time-Series Effects and Degradation of Control Chart Performance Resulting from Overadjustment," *Total Quality Management,* 2, pp. 99–112.

———— and M. G. Bissell (1988): "Time-Series Modeling for Quality Control in Clinical Chemistry," *Clinical Chemistry*, 34, pp. 1396–1406.

———— and H. V. Roberts (1995): "The Pervasive Problem of Misplaced Control Limits," *Applied Statistics*, 44, pp. 269–278.

Baxley, R. V. (1990): Discussion of "EWMA Control Schemes: Properties and Enhancements," *Technometrics*, 32, pp. 13–16.

Berthouex, P. M., W. G. Hunter, and L. Pallesen (1978): "Monitoring Sewage Treatment Plants: Some Quality Control Aspects," *Journal of Quality Technology,* 10, pp. 139–149.

Box, G. E. P., and G. M. Jenkins (1963): "Further Contributions to Adaptive Control: Simultaneous Estimation of Dynamics: Non-Zero Costs," *Bulletin International Statistical Institute, 34th Session*, 943, Ottawa, Canada.

Champ, C. W., and W. H. Woodall (1987): "Exact Results for Shewhart Control Charts with Supplementary Runs Rules," *Technometrics*, 29, pp. 393–399.

Deming, W. E. (1944): "Some Principles of the Shewhart Methods of Quality Control," *Mechanical Engineering*, 66, pp. 173–177.

———— (1967): "What Happened in Japan?" *Industrial Quality Control*, 24, pp. 89–93.

———— (1975): "On Probability as a Basis for Action," *The American Statistician*, 29, pp. 146–152.

———— (1986): *Out of the Crisis*, MIT Center for Advanced Engineering Study, Cambridge, MA.

Eisenhart, C. (1962): "Realistic Evaluation of the Precision and Accuracy of Instrument Calibration Systems," *Journal of Research of the National Bureau of Standards-C*, 67, pp. 161–187.

Ermer, D. S., M. C. Chow, and S. M. Wu (1979): "A Time-Series Control Chart for a Nuclear Reactor," in *Proceedings of the 1979 Annual Reliability and Maintainability Symposium*, Institute of Electrical and Electronic Engineers, pp. 92–98.

Farnum, N. R. (1994): *Statistical Quality Control and Improvement*, Duxbury, Belmont, CA.

Gitlow et al. (1995): *Quality Management: Tools and Methods for Improvement,* 2d ed., Irwin, Burr Ridge, IL.

Gottman, J. M. (1981): *Time-Series Analysis,* Cambridge University Press, Cambridge, England.

Grant, E. L., and R. S. Leavenworth (1996): *Statistical Quality Control,* 7th ed., McGraw-Hill, New York.

Harris, T. J., and W. H. Ross (1991): "Statistical Process Control Procedures for Correlated Observations," *Canadian Journal of Chemical Engineering*, 69, pp. 48–57.

Hoerl, R. W., and A. C. Palm (1992): "Discussion of Integrating SPC and APC," *Technometrics*, 34, pp. 268–272.

Hunter, J. S. (1990): "Discussion of EWMA Control Schemes: Properties and Enhancements," *Technometrics*, 32, pp. 21–22.

MacGregor, J. F., and T. J. Harris (1990): Discussion of "EWMA Control Schemes: Properties and Enhancements," *Technometrics*, 32, pp. 23–26.

Montgomery, D. C. (1996): *Introduction to Statistical Quality Control,* 3d ed., Wiley, New York.

——— and D. J. Friedman (1989): "Statistical Process Control in a Computer-Integrated Manufacturing Environment," in *Statistical Process Control in Automated Manufacturing*, eds. J. B. Keats and N. F. Hubele, Marcel Dekker, New York.

Nelson, L. S. (1984): "The Shewhart Control Chart—Tests for Special Causes," *Journal of Quality Technology*, 21, pp. 287–289.

Roberts, H. V. (1991): *Data Analysis for Managers with Minitab,* 2d ed., Scientific Press, South San Francisco, CA.

Scherkenbach, W. W. (1986): *The Deming Route to Quality and Productivity: Road Maps and Roadblocks,* CEEP Press, Washington, D. C.

Schonberger, R. J. (1986): *World Class Manufacturing*, Free Press, New York.

Shewhart, W. (1931): *Economic Control of Quality Manufactured Product*, D. Van Nostrand Co., New York; reprinted by the American Society for Quality Control in 1980, Milwaukee, WI.

——— (1939): *Statistical Method from the Viewpoint of Quality Control*, Graduate School of the Department of Agriculture, Washington, D. C.; reprinted by Dover Publications in 1986, Mineola, NY.

Westgard, J. O., and P. L. Barry (1997): *Cost-Effective Quality Control: Managing the Quality and Productivity of Analytical Processes*, American Association for Clinical Chemistry, Washington, D. C.

Wetherill, G. B. (1977): *Sampling Inspection and Quality Control,* 2d ed., Methuen, London, England.

Monitoring Individual Variable Measurements

CHAPTER OVERVIEW

In this chapter, we consider variable control charts for individual measurements. Charting procedures based on individual measurements have been developed because it is often not possible or practical to form samples of observations at a given time. This can occur for many reasons. The production rate may be slow, and monitoring of the process may be required before the time needed to form a subsample. A process may intrinsically generate one measurement representing a condition for a given time, as in continuous processes (e.g., chemical processing and pulp processing) or service-related processes. Companywide measures are typically monitored as a series of individual measurements, for example, daily, weekly, or monthly sales; inventory levels; or customer satisfaction levels. Control charts for individual observations are also commonly used in computer-integrated environments where every process item can be analyzed. In the medical laboratory industry, control charts for individual observations are the norm and are referred to as *Levey-Jennings charts* (see Westgard and Barry 1997). Finally, personal performance (work or fitness) and health processes (e.g., daily weight) fall under the category of processes generating individual measurements.

Since a series of individual measurements can theoretically be thought of as a series of subsamples of size 1, control charts for individual observations are ordinarily presented in many texts as a special case after presentation of the more general subsample-based control charts. However, to build the data analysis skills necessary for studying sequential observations, it is natural to proceed with the single-observation case first, which will lead us into the general case.

5.1 STANDARD CONTROL LIMITS FOR INDIVIDUAL MEASUREMENTS: X CHART

As discussed in Chapter 4, standard control chart theory is grounded in the assumptions that the underlying process behaves randomly, has constant variance, and is normally distributed. Recall that such a process is referred to as being independently and identically distributed normal (iidn). Based on these assumptions, control limits for the data are appropriately placed 3 standard deviations above and below the mean. Using the notation of Chapter 3, we let the random variable X_t represent the quality characteristic at time t. Given X_t has mean μ and standard deviation σ, control limits of $\mu \pm 3\sigma$ would be superimposed on the process data. However, in practice, these parameters are unknown and must be estimated.

For a set of n consecutive observations $(x_1, x_2, \ldots, x_n)$, the estimate of the process mean μ is simply given by the sample average $\bar{x}$. In practical application, the estimate of the process standard deviation σ may be obtained in two ways. In the first case, the sample standard deviation s (introduced in Chapter 2) may be used, where

$$s = \sqrt{\frac{\sum_{t=1}^{n}(x_t - \bar{x})^2}{n-1}} \tag{5.1}$$

Based on statistical theory, s can be shown to be a *biased* estimate of σ implying that, on average, s would be consistently off the mark in estimating the true standard deviation. To correct for this phenomenon, we divide s by an adjustment factor called c_4. Using $\bar{x}$ and s/c_4 as estimates for μ and σ, the control chart for single observations is given by

$$\bar{x} \pm 3\left(\frac{s}{c_4}\right) \tag{5.2}$$

For $2 \le n \le 24$, the values of c_4 are tabulated in Appendix Table V. For sample sizes in excess of 25, Montgomery (1996) indicates that the following approximation is suitable

$$c_4 \cong \frac{4n-4}{4n-3} \tag{5.3}$$

An alternative estimate of process variability is based on the variability observed between successive observations. For a series of n observations, the moving ranges are defined as

$$MR_t = |x_t - x_{t-1}| \qquad t = 2, 3, \ldots . \tag{5.4}$$

The moving range statistic MR is simply a special case of the general range statistic R which is the largest observation minus the smallest observation in a given sample. In this case, successive observations are paired off to form samples of size 2, allowing for the moving range to be computed based on (5.4). Notice that for a series of n observations, there will be $n-1$ moving ranges, implying that the mean moving range $\overline{MR}$ is given by

$$\overline{MR} = \frac{\sum_{t=2}^{n} MR_t}{n-1} \tag{5.5}$$

By using the mean moving range, it can be shown that an unbiased estimate for the process standard deviation is given by $\overline{MR}/d_2$, where the value of d_2 depends on the number of observations used in determining the individual range values. Hence, control limits based on $\overline{MR}$ are given by

$$\bar{x} \pm 3\frac{\overline{MR}}{d_2} \tag{5.6}$$

The control chart based on these limits is referred to as the *X chart* or *I chart*. For a sample size of 2 used in the computation of the moving ranges, we find from Appendix Table V that $d_2 = 1.128$, implying the control limits to be $\bar{x} \pm 2.66$ $\overline{MR}$. In practice, the moving-range–based limits appear to be more frequently used than the sample standard deviation–based limits. One indication of this fact

Table 5.1   Location measurements

−0.2302	0.9196	1.2809	0.9288	0.0430	0.4074	1.9086	0.1009
1.1807	−0.1985	0.2714	−0.1458	−0.1269	0.0350	1.5164	1.1236
−0.0814	−0.3408	2.1323	0.3683	0.1120	0.3759	−0.2727	0.2610
1.2097	0.3638	1.3688	2.0566	0.5649	0.7998	0.9407	0.3447
0.8107	0.1849	0.4069	0.2707	0.0701	−0.0498	−0.6768	0.5909
0.3270	0.3255	0.7791	1.5059	0.4731	1.1318	0.5256	0.8239
−0.1833	1.3530	−0.2356	−0.9623	0.8188	−0.0955	−0.3808	0.3889
0.7145	0.2056						

is that most computer software (including Minitab), by default, compute individual chart control limits based on the moving range.

5.2   EXAMPLE OF APPROPRIATE X CHART APPLICATION

The first example of single-observation charts is based on data taken from a large automobile assembly plant.[1] In one of the plant processes, each individual unpainted vehicle body passes through an automated optical measuring station containing 48 laser sensors. Ninety-five body dimensions are recorded using these sensors. One measurement (mm) is the fore/aft (forward/rearward) location of a hole for the attachment of front-wheel components. Since the location of this hole is related to vehicle handling stability, it was targeted as an important dimension to be monitored. Measurements were taken relative to design nominal, where negative (positive) values indicate a location forward (rearward) relative to this nominal. The data set consists of 58 individual measurements which are shown in Table 5.1 and found also in the file "vehicle1.dat".

In Figure 5.1, we report the basic summary statistics for the data. Before we construct a control chart, the data must be analyzed to determine whether the fundamental control chart assumptions of randomness and normality are satisfactorily met. As a first step, the data are plotted in time sequence along with the centerline in Figure 5.2. The time-series plot of the location data provides no evidence of nonrandom behavior. Furthermore, there does not appear to be any indication of systematic nonconstant variance. To confirm the seemingly iid behavior, we apply the randomness checks of the runs test and the ACF. The results of these diagnostics are provided in Figure 5.3.

[1]The author thanks Darrell Radson, University of Wisconsin-Milwaukee, for supplying the data for this section along with the data for Sections 5.3 and 5.6. The data found in this section are also studied in Radson and Alwan (1995).

```
Descriptive Statistics
Variable           N       Mean     Median     TrMean      StDev     SE Mean
loc               58     0.4886     0.3721     0.4666     0.6611      0.0868

Variable     Minimum    Maximum         Q1         Q3
loc          -0.9623     2.1323     0.0138     0.9219
```

Figure 5.1   Summary statistics for the location data.

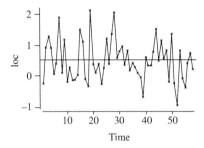

Figure 5.2   Time-series plot of the location series.

The runs test, with associated p value of 0.9699, provides no significant evidence against the null hypothesis that the data are random. The ACF, along with superimposed limits at $\pm 2/\sqrt{58}$ ($= \pm 0.263$), displays no autocorrelations that are significantly different from zero. In summary, all the standard diagnostic tests for randomness conclude that the data behave in a random fashion over time.

Before we apply the control limits given in either (5.2) or (5.6), it must be confirmed that the data satisfactorily conform to normality. In Figure 5.4, we provide the histogram and normality probability plot for the standardized values of the original data. As seen from both plots, there appears to be no severe departure from normality. There is a subtle hint of positive skewness; however, for all practical purposes, the working assumption of normality is reasonable. The randomness diagnostics along with the normality checks suggest that the underlying process is independent and identically distributed normal. With these assumptions satisfactorily met, we can now turn to the standard computations of the control limits.

From Figure 5.1, we find the estimated sample mean $\bar{x}$ and standard deviation s to be 0.4886 and 0.6611, respectively. Using (5.3) with $n = 58$, we find c_4 is 0.9956. As a result, we have from (5.2):

$$\text{UCL} = \bar{x} + 3\left(\frac{s}{c_4}\right) = 0.4886 + 3\left(\frac{0.6611}{0.9956}\right) = 2.481$$

$$\text{LCL} = \bar{x} - 3\left(\frac{s}{c_4}\right) = 0.4886 - 3\left(\frac{0.6611}{0.9956}\right) = -1.503$$

```
                    loc

                    K = 0.4886

                    The observed number of runs = 29
                    The expected number of runs = 29.1379
                    24 Observations above K      34 below
                         The test is significant at 0.9699
                         Cannot reject at alpha = 0.05
```

(a) **Runs test**

```
        ACF of loc

                -1.0 -0.8 -0.6 -0.4 -0.2  0.0  0.2  0.4  0.6  0.8  1.0
                 +----+----+----+----+----+----+----+----+----+----+
   1   0.021                              XX
   2  -0.031                              XX
   3   0.164                              XXXXX
   4   0.042                              XX
   5  -0.182                         XXXXXX
   6  -0.061                             XXX
   7  -0.193                         XXXXXX
   8  -0.158                          XXXXX
   9  -0.037                             XX
  10  -0.102                            XXXX
  11  -0.199                         XXXXXX
  12   0.190                              XXXXXX
  13   0.193                              XXXXXX
  14  -0.147                          XXXXX
```

(b) **ACF**

Figure 5.3   Randomness checks of the location series.

Alternatively, we can compute control limits based on (5.6). Table 5.2 provides the values for the moving ranges. Notice that there are 57 moving ranges in comparison to the original 58 observations. This results from the fact that the first original observation has no predecessor, thus no moving range can be computed. Hence, as seen in Table 5.2, there is an asterisk symbol representing a missing observation for the first moving-range value. To determine the moving-range-based limits, we find from Figure 5.5 the sample average of the 57 moving ranges $\overline{MR}$ to be 0.7426. The control limits are, therefore,

$$\text{UCL} = \bar{x} + 3\left(\frac{\overline{MR}}{d_2}\right) = 0.4886 + 3\left(\frac{0.7426}{1.128}\right) = 2.464$$

$$\text{LCL} = \bar{x} - 3\left(\frac{\overline{MR}}{d_2}\right) = 0.4886 - 3\left(\frac{0.7426}{1.128}\right) = -1.486$$

```
Histogram of stdloc     N = 58
Midpoint      Count
  -2.250          1    *
  -1.750          1    *
  -1.250          7    *******
  -0.750         10    **********
  -0.250         15    ***************
   0.250          8    ********
   0.750          6    ******
   1.250          5    *****
   1.750          2    **
   2.250          3    ***
```

(a) Histogram

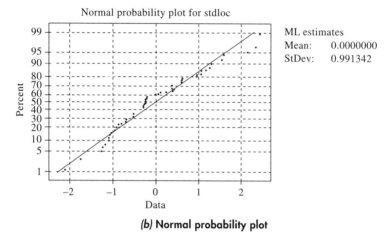

(b) Normal probability plot

Figure 5.4   Normality checks of the standardized location data.

Table 5.2   Moving ranges associated with the location data

*	1.1498	0.3613	0.3521	0.8858	0.3644	1.5012	1.8077
1.0798	1.3792	0.4699	0.4172	0.0189	0.1619	1.4814	0.3928
1.2050	0.2594	2.4731	1.7640	0.2563	0.2639	0.6486	0.5337
0.9487	0.8459	1.0050	0.6878	1.4917	0.2349	0.1409	0.5960
0.4660	0.6258	0.2220	0.1362	0.2006	0.1199	0.6270	1.2677
0.2639	0.0015	0.4536	0.7268	1.0328	0.6587	0.6062	0.2983
1.0072	1.5363	1.5886	0.7267	1.7811	0.9143	0.2853	0.7697
0.3256	0.5089						

Comparing the moving-range-based limits of $(-1.486, 2.464)$ with the sample standard deviation–based limits of $(-1.503, 2.481)$, we see little difference in the estimated limits. This is not surprising since the data generally conform

Variable	N	N*	Mean	Median	TrMean	StDev
mr(loc)	57	1	0.7426	0.6258	0.7084	0.5465

Variable	SE Mean	Minimum	Maximum	Q1	Q3
mr(loc)	0.0724	0.0015	2.4731	0.2918	1.0563

Figure 5.5   Summary statistics for the moving ranges of the location data.

well to the standard assumption of iidn which is the theoretical foundation for the interchangeability of the two approaches. However, note that Cryer and Ryan (1990) demonstrate that the moving-range-based estimator is less efficient than the standard deviation–based estimator in the estimation of σ when the underlying process is iid. In other words, on a repeated basis, the moving-range estimates would have greater fluctuation around the true process standard deviation than the sample standard deviation estimates. Despite this inefficiency, Rigdon, Cruthis, and Champ (1994) recommend using $\overline{MR}/d_2$ instead of s/c_4 to estimate the process standard deviation. Their recommendation is based on their demonstration of a significant bias in s/c_4 if the process was not in control when the preliminary data to construct the X chart were collected.

It is important to bear in mind that the preceding data analysis is by definition *retrospective* analysis. The primary purpose of the retrospective analysis is to gain an understanding of the process so that a benchmark can be established for future control (prospective analysis). Prospective analysis is the monitoring of newly generated process data to determine whether process change has occurred. The determination of change is relative to the process behavior understood through the retrospective analysis. Effective future control would naturally depend on having reliable estimates of the underlying process parameters. In the case of an underlying iidn process, this implies determining appropriate estimates for the process mean μ and standard deviation σ. The appropriateness of estimates rests greatly on whether the data used are representative of general process behavior.

If, for example, outliers or any other out-of-control signals are found to be special causes, then these points should be removed and process estimates along with control limits would then be recomputed using the remaining points. It is possible that the recomputed control limits would indicate "new" out-of-control signals warranting special-cause investigation. As a result, estimates and control limits may be recomputed. The revision of estimates and control limits continues until the process is viewed as being in control void of any special causes. Realize that there may be out-of-control signals that cannot be associated with special explanations. For instance, observations are expected to fall (with small probability) outside the control limits as a result of natural variation. In such a situation, the convention is to keep the observation(s) in the data set, using the computed estimates and control limits as appropriate for future process monitoring.

Of course, good judgment is called for when deciding about the retention of observations with no special-cause explanation. If an observation, with no

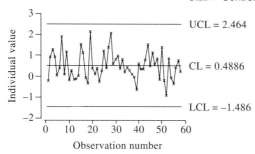

Figure 5.6 X-chart for the location series based on MR/d_2.

explanation, is very far beyond the 3σ limits, then its inclusion can dramatically affect the estimates and control limits, rendering them useless. The prudent action would be to set such an observation aside.

At this stage, we construct the control chart to examine the location data for possible special-cause signals. The above computations show the control limits for the individual measurements are virtually the same for limits obtained using (5.2) and (5.6). We show, for instance, the retrospective control chart with limits computed from (5.6) in Figure 5.6.[2] Observe that all points fall within the 3σ limits, implying that there is no suggestion for special-cause searching. In other words, the process is in a state of statistical control and is operating under a stable set of chance causes or common causes.

It can be pointed out that the realization of no outliers in the data could have been noticed earlier in the histogram of standardized values (see Figure 5.4). Even though this is true, the importance of plotting the data sequentially with limits cannot be overstated for several reasons. First, if an outlier does exist in the sequence, the histogram provides no information as to which observation is out of control and hence when the signal has occurred. Second, sequential plotting of data allows the control chart analyst to apply other special-cause detection criteria, such as runs rules (introduced in Section 4.9).

For ease of application of runs rules, the original data can be standardized around zero such that the value 1 is associated with 1 standard deviation above the mean, the value -1 is associated with 1 standard deviation below the mean, the value 2 is associated with 2 standard deviations above the mean, and so on. As noted in Section 3.2, if $\hat{\sigma}$ is an estimate of σ, the original observations x_t are standardized as follows:

$$y_t = \frac{x_t - \bar{x}}{\hat{\sigma}} \tag{5.7}$$

---

[2]By default, Minitab uses the labels *3SL* and $-3SL$ to signify *UCL* and *LCL*, respectively. Since these labels are not conventional and can mislead one into interpreting *SL* as meaning specification limit, we will use Minitab's graphics editor to replace the labels with the conventional labels of *UCL* and *LCL*.

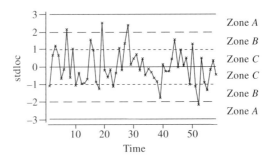

Figure 5.7   Zone chart for the standardized location values.

Table 5.3   Standardized location data based on $\overline{MR}/d_2$ as an estimate for process standard deviation.

−1.09190	0.65472	1.20355	0.66869	−0.67690	−0.12335	2.15707
−0.58894	1.05134	−1.04375	−0.32994	−0.96369	−0.93498	−0.68905
1.56129	0.96461	−0.86587	−1.25991	2.49689	−0.18274	−0.57208
−0.17120	−1.15646	−0.34574	1.09540	−0.18958	1.33708	2.38189
0.11590	0.47273	0.68677	−0.21859	0.48929	−0.46134	−0.12411
−0.33100	−0.63573	−0.81786	−1.77032	0.15540	−0.24518	−0.24776
0.44129	1.54534	−0.02355	0.97706	0.05621	0.50934	−1.02066
1.31308	−1.10011	−2.20401	0.50160	−0.88729	−1.32067	−0.15145
0.34316	−0.42990					

Based on developments in Section 5.1, either s/c_4 or $\overline{MR}/d_2$ might serve as estimates of σ. As a matter of consistency, it is recommended that the estimator used in the standardization be the same as that used in the construction of the 3σ control limits. In our case, each data point would have 0.4886 ($\bar{x}$) subtracted from it, and in turn, this quantity divided by 0.6583 ($\overline{MR}/d_2$). It is important to note that standardization does not intrinsically change the nature of the data; it simply is a rescaling of the data to different units. The standardized values, based on this calculation, can be found in Table 5.3. Inspection of the sequenced standardized values reveals no violations with respect to any one of the eight runs rules listed in Section 4.9. For example, the fact that there are no sequences of observations such that nine or more in a row are all positive or all negative implies no violation with respect to rule 2. We could have also plotted the standardized observations with limits placed ± 1, ± 2, and ± 3 (the boundaries of the zones) as shown in Figure 5.7. As concluded above, the zone chart for the data reveals no specific runs violations.

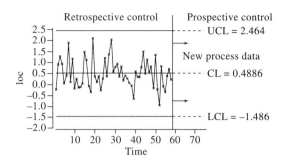

Figure 5.8   Extension of prospective control limits for the location series.

The data diagnostics, along with the application of control criteria, suggest that the location data behave as a simple random process with no indications of special causes. In such a case, the best prediction for future observations is simply the sample mean $\bar{x}$, that is, an extension of the control chart's centerline. Similarly, we can project the retrospective upper and lower control limits which provide a prediction of the variability around the centerline. By extending the control limits found in Figure 5.6, we find the prospective control chart in Figure 5.8.

New observations (59, 60, etc.) would be plotted in real time on the prospective chart. The new data would be monitored to see if the process was in control, as defined in the retrospective analysis. Out of control is indicated if any future individual observations fall outside control limits. Additionally, other control criteria, such as supplementary runs rules, may be used to identify potential special causes manifested as unusual patterns relative to random behavior.

5.3   MONITORING A SIMPLE LINEAR TREND PROCESS

In practice, processes often exhibit trend behavior reflected by a systematic level change (upward or downward) in the quality measurement as time evolves. A common situation arises when a monitored characteristic associated with machining operations drifts due to tool wear or deterioration. Trends might also reflect continuous quality improvements to processes such as yield (upward trend), defects (downward trend), and scrap (downward trend). In all situations, the recognition and understanding of trends have great practical implications.

In this section, we analyze a different set of data taken from the same automobile assembly plant considered in the previous section. In this case, in each production period, one completed automobile was randomly selected and its camber angle was measured (in minutes) off-line. A camber angle is a critical dimension for front-end stability. The data set consists of 59 consecutive measurements (*camber$_t$*) which are shown in Table 5.4 and found also in the file "vehicle2.dat".

Table 5.4   Camber measurements

−50.4000	−47.4000	−43.8000	−43.0000	−47.6000	−42.6000	−44.0000
−42.6000	−44.0000	−38.2000	−45.2000	−42.2000	−46.8000	−45.6000
−47.6000	−54.5000	−42.8000	−44.2000	−44.0000	−46.6000	−52.0000
−47.6000	−48.4000	−38.6700	−52.2000	−50.4000	−46.2000	−50.4000
−43.5000	−48.0000	−57.0000	−48.5000	−46.8000	−48.8000	−54.8000
−59.0000	−60.0000	−51.2500	−56.0000	−53.0000	−56.3300	−47.6000
−49.2500	−48.2000	−47.6000	−57.2000	−63.6700	−50.8000	−56.8000
−58.5000	−56.4000	−53.0000	−61.2000	−61.8000	−62.8000	−50.0000
−58.2500	−47.6000	−52.0000				

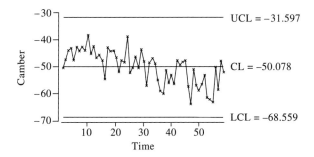

Figure 5.9 X chart for the camber series based on s/c_4.

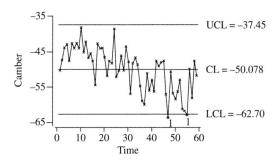

Figure 5.10 X chart for the camber series based on $\overline{MR}/d_2$.

In Figures 5.9 and 5.10, we find the time-series plot of the data with the application of control limits as determined by (5.2) and (5.6), respectively. The plots clearly show that the process is out of control in the sense of not being a random process. The runs test and ACF, shown in Figure 5.11, further confirm the nonrandom behavior. Specifically, there appears to be a systematic downward trend. Interpreting the control limits literally would be a misguided task. The sample standard deviation–based limits suggest no signals for special causes

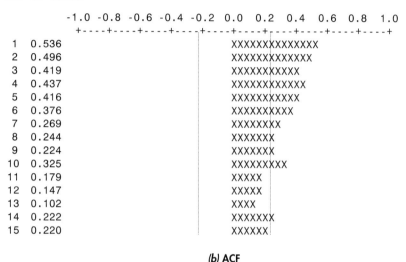

```
camber

K =    -50.0783

The observed number of runs   =     19
The expected number of runs   =     30.0847
33 Observations above K   26 below
       The test is significant at    0.0031
```

(a) Runs test

```
ACF of camber

            -1.0 -0.8 -0.6 -0.4 -0.2  0.0  0.2  0.4  0.6  0.8  1.0
             +----+----+----+----+----+----+----+----+----+----+
  1  0.536                            XXXXXXXXXXXXXX
  2  0.496                            XXXXXXXXXXXXX
  3  0.419                            XXXXXXXXXXX
  4  0.437                            XXXXXXXXXXXX
  5  0.416                            XXXXXXXXXXX
  6  0.376                            XXXXXXXXXX
  7  0.269                            XXXXXXXX
  8  0.244                            XXXXXXX
  9  0.224                            XXXXXXX
 10  0.325                            XXXXXXXXX
 11  0.179                            XXXXX
 12  0.147                            XXXXX
 13  0.102                            XXXX
 14  0.222                            XXXXXXX
 15  0.220                            XXXXXX
```

(b) ACF

Figure 5.11   Randomness checks of the camber series.

while the moving-range-based limits give signals at observations 47 and 55. The search for special causes is made even more confusing when one considers the use of supplementary runs rules. Application of these additional criteria suggests 10 "out-of-control" occurrences, each inviting special explanations, over and above the isolated indications revealed by the moving-range-based limits. All these methods misdirect the SPC efforts. The standard deviation–based limits suggest no particular search for special causes, while the moving-range-based limits and runs rules suggest numerous searches for special causes. Furthermore, no method provides real guidance in the identification of systematic behavior or in understanding what is really happening.

The differing results are due to the significant disparity in width between the standard deviation–based limits and the moving-range-based limits. It is worthwhile to point out that the typical statistical properties associated with the sample standard deviation defined in (5.1) are grounded in the assumption that the data are the result of random drawings from some population. In a time-series framework, this implies that the data are generated from a random (uncorrelated) process. In the case of trended data, this assumption is obviously not met. In general, the sample standard deviation does not provide a reliable estimate of process variability in the presence of nonrandom behavior.

To understand why the moving-range limits are seemingly too tight, first consider the case where X_t follows an underlying random process given by

$$X_t = \mu + \varepsilon_t \tag{5.8}$$

where μ is the process mean and ε_t is random deviation with $E(\varepsilon_t) = 0$ and $\mathrm{Var}(\varepsilon_t) = \sigma_\varepsilon^2$. If we define the random variable D_t as the difference between successive observations, that is, $D_t = X_t - X_{t-1}$, it follows that

$$D_t = X_t - X_{t-1}$$
$$= \mu + \varepsilon_t - (\mu + \varepsilon_{t-1}) \tag{5.9}$$
$$= \varepsilon_t - \varepsilon_{t-1}$$

Now, consider instead of a process with random errors around a horizontal mean μ, a process with the same random error process around a simple trend line. Recall from Chapter 3 that such a process is given by

$$X_t = \beta_0 + \beta_1 t + \varepsilon_t \tag{5.10}$$

where β_0 is the underlying intercept and β_1 is the underlying slope of the trend line. For this trend process, D_t is given by

$$D_t = X_t - X_{t-1}$$
$$= \beta_0 + \beta_1 t + \varepsilon_t - [\beta_0 + \beta_1(t - 1) + \varepsilon_{t-1}] \tag{5.11}$$
$$= \beta_1 + \varepsilon_t - \varepsilon_{t-1}$$

Alwan and Radson (1992) show that, for any time-series process, $E(\overline{MR}/d_2) = \sqrt{\mathrm{Var}(D_t)/2}$. (Recall that $\overline{MR}/d_2$ is an estimate of the standard deviation of the process X_t.) Comparing (5.9) and (5.11), we can see that the only difference between D_t for the random process and the trend process is the constant β_1. Since the variance of a constant plus a random variable is simply the variance of the random variable, $\mathrm{Var}(D_t)$ is the same for each process. This would mean that $E(\overline{MR}/d_2)$ is identical for each process, implying that the estimator $\overline{MR}/d_2$ "views" the variability of the two processes as being the same. Clearly, the observed variability of the two processes would not be the same; the variability of the random process would be exclusively due to the random component ε_t while the overall variability of the trend process would result from the random component ε_t plus the variability associated with the trend component. It should be pointed out that given the nonstationary nature of a trend process (recall the discussion in Chapter 2), the variance in the long run is actually infinite. So, when we speak of variability, we mean the short-term process variation. Therefore, $\overline{MR}/d_2$ underestimates the true short-term variability associated with a trend process, resulting in control limits placed relatively tightly around the centerline, as seen in Figure 5.10.

It has been suggested by some authors (e.g., see Wadsworth, Stephens, and Godfrey, 1986) that the tendency for the moving-range-based limits to be too narrow in the presence of nonrandom effects, such as trend, is an advantage over the application of sample standard deviation–based limits. The logic behind this suggestion is the fact that the moving-range-based limits will tend to signal more out-of-control points, alerting the prudent control chart analyst to question the appropriateness of the random assumption, rather than look for potentially nonexistent special causes at these particular points.

Even though the method of moving ranges may have some sensitivity to departures from randomness, we hesitate to recommend the application of these control limits for two primary reasons. First, given the great practical importance of the trend itself, it would seem more reasonable that emphasis should be placed on a detailed analysis of the data to help identify the systematic behavior. Second, in the presence of systematic nonrandom behavior such as trend, standard control limits are not intrinsically useful for detecting special causes. It must be remembered that the intent of control limits is to identify unusual behavior *relative* to the general workings of the process. For instance, when the process is random, one detects unusual variation relative to the natural variability around the process mean. Given that the horizontal projection of the sample mean (centerline) is the best prediction (or forecast) for the output of a random process, control limits placed around the sample mean will also be *horizontal* projections. However, for a nonrandom process a horizontal forecast projection would not appropriately describe the underlying evolution of the process level, which, in turn, implies the inappropriateness of horizontal control limits.

TREND CONTROL CHART

The inappropriateness of horizontal control limits has been recognized in the analysis of processes exhibiting trend behavior due to tool wear. Mandel (1969) developed a regression-based control chart which calls for centerline and control limits to be superimposed as parallel trend lines on the process data. As we have shown, the moving-range statistic is generally insensitive to the trend component. This allows us to perform the moving-range calculations on the original data to estimate the variability around the underlying trend. If the estimated trend line is given by $b_0 + b_1 t$, the trend control chart limits are given by

$$b_0 + b_1 t \pm 3 \frac{\overline{MR}}{d_2} \tag{5.12}$$

To illustrate this control chart, let us return to the trend process, camber$_t$, plotted in Figures 5.9 and 5.10. Clearly, the single value of the sample mean (-50.078) neither retrospectively nor prospectively provides a good "guess" for the underlying movement of the process level. A better descriptor for the evolution of the process level would appear to be a trend line which can be estimated by regressing camber$_t$ on the time index t. As seen from the regression output in Figure 5.12, the estimated regression line is given by $-42.8 - 0.241t$. The p value of 0.000 (that is,

```
The regression equation is

camber = - 42.8 - 0.241 t

Predictor          Coef        StDev            T          P
Constant        -42.844        1.204       -35.59      0.000
t              -0.24114      0.03490        -6.91      0.000

S = 4.565      R-Sq = 45.6%       R-Sq(adj) = 44.6%

Analysis of Variance

Source             DF          SS          MS          F          P
Regression          1      994.92      994.92      47.75      0.000
Residual Error     57     1187.73       20.84
Total              58     2182.65
```

Figure 5.12   Regression output for the fitted trend model for the camber series.

< 0.0005) for the trend variable is clearly significant. As we will soon see, the residuals for the fitted model appear to be random and normally distributed, implying that the simple trend model adequately describes the process data.

The average of the moving ranges for the original trended data can be found to be 4.748. By using (5.12), the trend-based control limits would be set at

$$\text{UCL} = -42.8 - 0.241t + 3\left(\frac{4.748}{1.128}\right) = -30.2 - 0.241t$$

$$\text{LCL} = -42.8 - 0.241t - 3\left(\frac{4.748}{1.128}\right) = -55.4 - 0241t$$

The trend control chart, based on these computed control limits, is shown in Figure 5.13. As can be seen, there are no points beyond the control limits, suggesting no special causes beyond the trend. This is in contrast to the naïve interpretation of the standard control limits (see Figure 5.10) which suggests two possible special causes at observations 47 and 55.

Prospectively, the control limits can be projected along the trend lines to monitor for any change from the predictable trend behavior. Unlike in random processes, the fact that the process is predictable with no evidence of special causes should provide no comfort. The implication of the process, as captured by the model, is that the process is systematically wandering away through time. It is essential that individuals, knowledgeable about the process, concentrate their efforts on seeking an explanation for the trend, in order to potentially stop or revert its direction. Prospective control limits can play a critical role in determining whether attempts to affect the trend have been successful. In this case, it is hoped that improvement efforts would result in newly generated data appearing above or drifting toward the prospective upper control limit.

FITTED-VALUES CHART AND SPECIAL-CAUSE CHART

The trend control chart nicely demonstrates that control limits are appropriately placed 3 standard deviations above and below the fitted model that estimates the

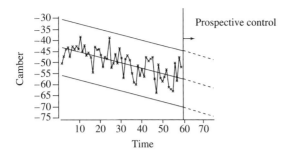

Figure 5.13   Trend control chart applied to the camber series.

underlying systematic behavior. Thus, control limits serve as guidance in the detection of any abnormal deviations from the fitted model. As defined in Chapter 3, deviations from the fitted model are commonly referred to as *residuals*. This implies that one could equivalently apply control limits to the residuals from a fitted model for the detection of abnormal variation relative to the general workings of the process.

In general, statistical modeling permits the decomposition of the process data into fitted values and residuals. A well-fitted model is measured by the fact that the residuals from the model will be uncorrelated over time and approximately normal with mean zero and constant variance. Except for possible local departures reflecting special causes, residuals from a well-fitted model will behave as a process in a state of statistical control in the standard sense. This implies that the standard control chart framework would apply for the detection of special causes.

Thus, the basic idea is to associate the systematic nonrandom effects with the fitted component of a statistical analysis of the process data and that special causes should be sought in the plot of the residuals, instead of mechanically applying standard SPC methods to the original data. So for purposes of process control and improvement, the standard control chart can be replaced by two charts: the *fitted-values chart,* a time-series plot of the fitted (retrospective) or predicted (prospective) values without control limits; and a *special-cause chart,* a standard control chart for the residuals (retrospective) or prediction errors (prospective). The fitted-values chart provides a visual representation of the systematic nonrandom effects underlying the process. The special-cause chart displays the underlying random variation with possible special causes which may be identified by basic standard SPC tools.

To illustrate, let us consider again the camber data series which exhibited trend behavior. Even though the trend model "appears" to be a suitable choice, it is important for effective process control to ensure that the estimated model is appropriate. This is accomplished by diagnostic checking of the residuals for conformity to a normally distributed random process with constant variance.

The first step in diagnostic checking would be to visually inspect the time-series plot of the residuals, which is given in Figure 5.14. A study of the residuals

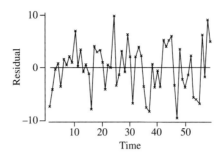

Figure 5.14   Time-series plot of the residuals from the fitted trend model for the camber series.

```
residual

K =     0.0000

The observed number of runs  =  30
The expected number of runs  =  30.4237
31 Observations above K 28 below
     The test is significant at 0.9112
     Cannot reject at alpha  =  0.05
```

(a) Runs test

```
ACF of residual

            -1.0 -0.8 -0.6 -0.4 -0.2 0.0  0.2  0.4  0.6  0.8  1.0
            +----+----+----+----+----+----+----+----+----+----+
 1   0.115                           XXXX
 2  -0.001                           X
 3  -0.096                         XXX
 4  -0.058                          XX
 5  -0.042                          XX
 6  -0.032                          XX
 7  -0.155                        XXXXX
 8  -0.163                        XXXXX
 9  -0.148                        XXXXX
10   0.141                           XXXXX
11  -0.073                         XXX
12  -0.093                         XXX
13  -0.083                         XXX
14   0.201                           XXXXXX
15   0.194                           XXXXXX
```

(b) ACF

Figure 5.15   Random checks of residuals from the fitted trend model for the camber series.

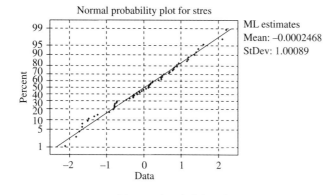

```
Histogram of stres      N = 59
Midpoint       Count
 -2.250          1    *
 -1.750          5    *****
 -1.250          3    ***
 -0.750          9    *********
 -0.250         10    **********
  0.250         13    *************
  0.750          9    *********
  1.250          6    ******
  1.750          1    *
  2.250          2    **
```

(a) Histogram

(b) Normal probability plot

Figure 5.16   Normality checks of the standardized residuals from the fitted trend model for the camber series.

plot reveals neither systematic nonrandom behavior nor nonconstant variance. The runs test of the residuals in Figure 5.15a gives a p value of 0.9112; thus we are unable to reject randomness of the residuals. The ACF of the residuals with significance limits (Figure 5.15b) shows no significant autocorrelations. The histogram of the standardized residuals, along with the corresponding normal probability plot, found in Figure 5.16 gives no apparent indication of nonnormality. Thus, the model diagnostics for the fitted trend model indicate that the residuals resemble a normally distributed random series, implying that the fitted model is satisfactory and adequately accounts for the nonrandom behavior.

Given that the residuals satisfy the basic requirements for the standard control chart theory, we can proceed to compute control limits for the data. We find that the average of the moving ranges for the residuals is 4.731. Notice that this value is quite close to the average of the moving ranges for the original trended data which was previously shown to be 4.748. This closeness is consistent with the results, shown earlier in this section, that the expected moving range is invariant to the trend component. Recognizing that the average of the residuals

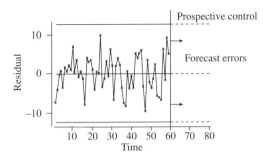

Figure 5.17   Special-cause chart for the camber series.

from a regression fit is always equal to zero, the control limits for the residuals are given by

$$\text{UCL} = 3\left(\frac{4.731}{1.128}\right) = 12.58$$

$$\text{LCL} = -3\left(\frac{4.731}{1.128}\right) = -12.58$$

Applying these limits, we construct the residual control chart (or special-cause chart) which is shown in Figure 5.17. As can be seen, there are no points breaching the control limits. Thus, there is no suggestion of special causes over and above the trend. This result should not be surprising given the constructed trend control chart shown in Figure 5.13. Except for a slight difference in the average of the moving ranges, the residual chart and the trend control chart are equivalent. One can think of the residual chart as basically the trend control chart laid horizontally.

Even though the two control charts will provide equivalent results in terms of highlighting individual outliers as possible indications of special causes, other control criteria are not easily reconciled with the trend control chart. As we saw earlier, runs rules will be misguided (10 special-cause violations are flagged) if applied to the trended data. Runs rules are better suited for the residuals which have been freed of any persisting nonrandomness. For instance, applying the set of runs rules to the residuals of this example gives no indications of special-cause variation.

It should be recognized that the special-cause control chart (Figure 5.17) and its counterpart trend control chart (Figure 5.13) are retrospective charts. As explained and illustrated earlier, prospective control limits for the trend control chart are simply obtained by extending the upper and lower trend control limit lines into the future. For the special-cause chart, the retrospective limits are horizontally extended, just as was done in Section 5.2 for the standard case (see Figure 5.8). Here, however, future points plotted are the forecast errors, which are the differences between the actual observed values and the predictions based on

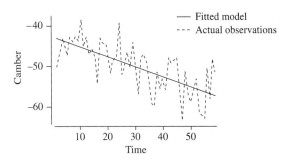

Figure 5.18   Fitted-values chart for the camber series.

the fitted time-series model. For example, consider the next period in the future, that is, period 60. By substituting the value of 60 for the independent variable t in the fitted model $(-42.8 - 0.241t)$, we would predict the value of the next period value to be -57.26. The forecast error then is given by $camber_t$ − prediction $= camber_{60} + 57.26$, where $camber_{60}$ is the actual observed value for the camber process at time period 60. If the computed forecast error falls outside the limits of $(-12.58, 12.58)$, then an out-of-control signal is flagged. Further forecast errors (for periods 61, 62, . . .) can be plotted for an on-line application of other control criteria such as runs rules.

It is important to underscore that this notion of retrospective (prospective) control vis-à-vis residuals (forecast errors) is conceptually equivalent to the standard case. Recognize that for a random process, such as the one studied in Section 5.2, the best fitted model is simply the sample mean $\bar{x}$, implying that the fitted-values chart is a straight line at the process average. The residuals (forecast errors) are the original (new) observations minus the sample mean. The sample average of these residuals, as for any other regression-based residuals, is equal to zero. Since all the fitted values are equal to a constant value, the residuals are equal to the original observations minus this constant, and thus the estimated variability of the residuals will be equal to the estimated variability of the original observations. Hence, the special-cause chart, which calls for the plotting of the residuals (forecast errors) with control limits around zero, is equivalent to plotting the original (new) observations with control limits around the sample mean.

The other chart that is a by-product of the time-series framework is the fitted-values chart. In Figure 5.18, we show the fitted values (solid line) along with the original observations (dashed line). This fitted-value plot gives a view of the process level and its evolution through time. In this situation, the plot clearly reflects that the process is being affected by persistent effects resulting in a downward trend, as opposed to transient, special causes. Given the undesirable implications of the downward trend, efforts should be made to find and remove the sources contributing to the trend.

In general, the two-chart implementation (special-cause chart and fitted-values chart) clearly isolates two distinct tasks. First, it provides a means (special-cause chart) for improved SPC searching for any possible special causes in the form of outliers or runs not consistent with the general process behavior. Second, it provides a means (fitted-values chart) for better understanding of the underlying process behavior, and thus provides useful clues for process improvement. This isolation of these two important tasks makes the two-chart implementation approach a more appealing and potentially more effective strategy than the single-chart method. In either case, it is necessary to capture the underlying systematic behavior by means of data analysis techniques such as time-series modeling.

Table 5.5   Time between computer failures

83.483	86.267	331.750	17.783	37.967	126.417	38.917
32.533	50.467	64.534	38.700	51.267	170.390	100.640
19.683	28.683	74.817	27.667	85.500	54.083	217.583
113.550	168.200	5.867	44.400	142.600	12.567	95.917
40.883	68.933	13.500	84.000	624.819	99.150	49.083
13.083	14.450	83.883	36.550	40.950	58.750	61.917
103.050	30.283	270.000	1.233	97.183	86.883	28.717
81.817	3.800	55.483	15.633	15.417	4.833	1.000
78.400	37.683	73.467	32.617	43.833	86.650	29.350
24.000	42.000	1.500	97.500	65.750	34.083	39.167
8.750	5.250	75.917	22.483	88.100	54.500	4.667
12.233	1.183	9.667				

5.4   MONITORING A NONRANDOM AND NONNORMAL PROCESS

In our next application, we study the process of time between computer failures. Specifically, the process measurements are the times (in hours) between failures of a computer system at a large Midwestern bank. The data set consists of 80 consecutive individual measurements which are shown in Table 5.5 and found also in the file "failure.dat".

By applying mechanically the moving-range control limits of (5.6) to the failure data (failure$_t$), the control chart shown in Figure 5.19 is obtained. Based on this control chart, there is an initial temptation to seek explanations for observations 3, 33, and 45. However, careful examination of the time-series plot suggests that the data may not be generated from an iidn process; thus the credibility of the control limits is suspect. In particular, the majority of the data appear

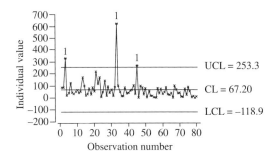

Figure 5.19 X chart for the computer failure series based on $\overline{MR}/d_2$.

"bunched" below and around the centerline with some observations floating upward. This suggests that the data are positively skewed. To confirm this suspicion, we construct the histogram and normal probability plot of the standardized data in Figure 5.20.

Both of these normality diagnostics clearly demonstrate that the data are severely nonnormal in the sense of being positively skewed. The fact that the times between failures are positively skewed is not too surprising. It is a known fact that if the number of occurrences of an event (e.g., a failure) in a period of time has a Poisson distribution, then the distribution for the time intervals between occurrences is an exponential distribution. The exponential distribution is characterized by its strong right skewness (e.g., see Figure 2.19) along with the property that the theoretical mean and standard deviation of this distribution are equal. For the computer failure data, the sample mean is 67.20 and the sample standard deviation is 85.58. The bit higher standard deviation relative to mean may be a hint of a departure from an underlying random Poisson process and, in turn, the exponential process.

Since the standard control methods (and regression methods, if needed) are based on the approximate normality of the data, we must seek alternative approaches in our analysis of the data. To determine the appropriate normalizing transformation, we might use the methods of Hines and Hines (1987) or of Box and Cox (1964), as described in Section 3.3. However, these methods work best when the data are generated from a random process. As we shall see, there is an underlying systematic pattern to the data which will require us to use regression modeling. In fitting regression models, attention then turns to getting the residuals to have close conformity to normality, not necessarily the original observations. By trial and error, we found that raising the data to the 0.2 power (i.e., fifth root) will, in the end, give excellent results.

In Figure 5.21, we display a time-series plot of the fifth root transformed data, that is, $(failure_i)^{0.2}$. Besides the more symmetric scatter in the data, the transformation revealed an important underlying phenomenon in the process not easily discerned in the original scale. Namely, the process is demonstrating a

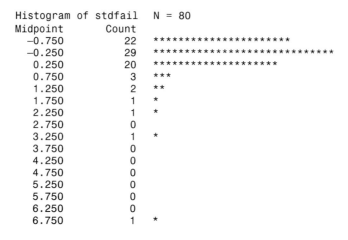

```
Histogram of stdfail  N = 80
Midpoint       Count
  -0.750         22    **********************
  -0.250         29    *****************************
   0.250         20    ********************
   0.750          3    ***
   1.250          2    **
   1.750          1    *
   2.250          1    *
   2.750          0
   3.250          1    *
   3.750          0
   4.250          0
   4.750          0
   5.250          0
   5.750          0
   6.250          0
   6.750          1    *
```

(a) **Histogram**

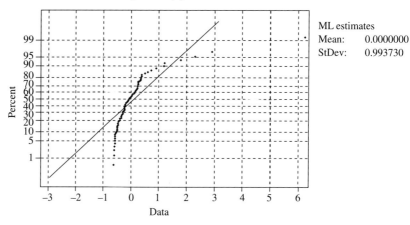

(b) **Normal probability plot**

Figure 5.20   Normality checks of the standardized computer failure data.

clear downward trend. To assess the significance of this apparent trend, we fit a trend model to the transformed data.

The regression output can be found in Figure 5.22. As evidenced by the p value (0.001) for the trend component being significant at the $\alpha = 0.05$ level, the trend is not illusionary. To check the adequacy of the fitted model, we should perform the standard residual diagnostics. We leave the randomness checks to the reader. If done, we would find the residuals to be generally compatible with randomness. To check for conformity to the normal distribution, we provide in Figure 5.23 the histogram and normal probability plot for the standardized residuals. In dramatic contrast to the original data (see Figure 5.20), these tests for normality indicate no substantial evidence against normality.

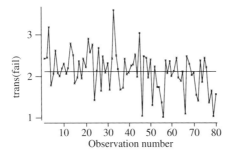

Figure 5.21   Time-series plot of the transformed computer failure series.

```
The regression equation is
trans(fail) = 2.43 - 0.00770 t

Predictor        Coef        StDev          T          P
Constant       2.4321       0.1032      23.57      0.000
t             -0.007699     0.002213     -3.48      0.001

S = 0.4572      R-Sq = 13.4%      R-Sq(adj) = 12.3%

Analysis of Variance

Source              DF          SS         MS          F          P
Regression           1      2.5289     2.5289      12.10      0.001
Residual Error      78     16.3022     0.2090
Total               79     18.8311
```

Figure 5.22   Regression output for the fitted trend model for the transformed computer failure series.

Thus, unlike the original series that was nonrandom *and* nonnormal, the residuals appear to be random and approximately normal, allowing for standard control chart analysis. Accordingly, we show in Figure 5.24 a control chart for individuals based on the moving-range estimate. We also supplement the control limits with the set of runs rules discussed earlier.

There appears to be only one distinct episode warranting search for special causes. Namely, there is a 3σ violation occurring at observation 33, as opposed to mechanical application of limits which suggested observations 3, 33, *and* 45 were indications of special causes. If a special-cause explanation can be provided for the favorable effect associated with observation 33, this knowledge would hopefully be incorporated for future system improvements.

REGRESSION MODEL REVISION FOR PROSPECTIVE CONTROL

In Section 5.2, we discussed the updating of process parameter estimates and control limits due to the presence of unrepresentative process data reflecting special causes. This previous discussion was specifically directed to the standard case where the underlying process is iidn, implying the need to update only the process

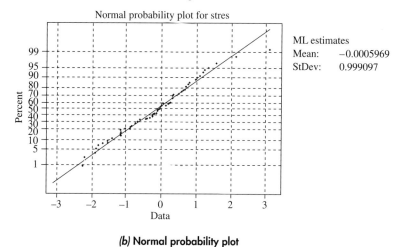

```
Histogram of stres     N = 80
  Midpoint      Count
   -2.250          2    **
   -1.750          4    ****
   -1.250          8    ********
   -0.750         11    ***********
   -0.250         13    *************
    0.250         17    *****************
    0.750         14    **************
    1.250          8    ********
    1.750          1    *
    2.250          1    *
    2.750          0
    3.250          1    *
```

(a) Histogram

Normal probability plot for stres

ML estimates
Mean: −0.0005969
StDev: 0.999097

(b) Normal probability plot

Figure 5.23   Normality checks of the standardized residuals from the fitted trend model for the transformed computer failure series.

level estimate $\hat{\mu}$ and process variability measure $\hat{\sigma}$. Similarly, if special causes are identified in the broader context of a fitted time-series model, their effects should be accounted for in order to more accurately estimate the underlying parameters that describe the general behavior of the process. For instance, if the isolated outlier in the computer system failure series proves to be a legitimate special cause, the related point needs to be "set" aside so that the estimates of the parameters associated with the trend line and variability around the trend line can be reestimated.

For the standard case, we simply delete the associated observations and proceed to revise the model estimates ($\hat{\mu}$ and $\hat{\sigma}$) based on the remaining data. In general, deletion of data values will result in observations being adjacent to each other which were not previously so. This problem was initially discussed with the ASQ example in Section 3.9. In particular, if x_i was deleted from a data set of n

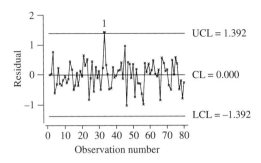

Figure 5.24   Special-cause chart for the transformed computer failure series.

sequential observations, the remaining data were given by $x_1, \ldots, x_{i-1}, x_{i+1}, \ldots,$ x_n. For time-series modeling purposes, the net effect of such deletion of points can be problematic given the fact that the ith data point is not x_i but really x_{i+1}. However, for random processes, the time dimension has no consequence for the estimation of parameters; observations can simply be viewed as random samplings from a single population.

In developing a strategy to remove or offset the effects of special causes in the modeling of nonrandom processes, it is useful to consider the framework of intervention analysis introduced by Box and Tiao (1975). An intervention is regarded as either a deliberate action imposed on the process or any other non-repetitive event (known or unknown) affecting the process. A special cause, for instance, reflecting a sudden shock in the process level, can be thought of as an intervention. In general, the effect of interventions might manifest itself in several different ways. For instance, an intervention could result in a change in the level of the time series, either immediately or after some delay; or it could induce more complicated time-series behaviors.

If the time of occurrence of interventions is known, intervention models can be utilized to account for their effects. Box and Tiao (1975) developed a procedure for modeling the effects of interventions on the general time-series behaviors. Their procedure is based on *transfer function analysis,* which is used to determine whether there is enough evidence that the series has actually changed and, if so, to determine its nature and magnitude. The basis of the intervention modeling is that interventions can be modeled by an appropriate set of indicator variables. For example, in the case of a special cause in the form of a temporal sudden shock, one might use a pulse indicator variable that takes the value 1 in the particular time period(s) and 0 in all others. On the other hand, if the form of the intervention is a sudden permanent change in level, one would use a step indicator variable that takes the value 0 prior to intervention period and the value 1 from that period on. From the time-series perspective, these indicator variables can be added to the time-series model to account for the effects of the sudden shocks, reflecting true special causes.

```
The regression equation is
trans(fail) = 2.40 - 0.00744 t + 1.47 special

Predictor          Coef         StDev            T          P
Constant        2.40333       0.09724        24.71      0.000
t              -0.007442      0.002079        -3.58      0.001
special          1.4659        0.4322         3.39      0.001

S = 0.4292     R-Sq = 24.7%       R-Sq(adj) = 22.7%

Analysis of Variance

Source             DF            SS           MS          F          P
Regression          2         4.6482       2.3241      12.62      0.000
Residual Error     77        14.1829       0.1842
Total              79        18.8311
```

Figure 5.25   Regression output for the fitted trend model with special-cause variable for the transformed computer failure series.

However, in practice the timing and even the occurrence of an intervention or special cause are generally unknown. If an intervention occurs at an unknown time—preventing the inclusion of indicator variables to account for the intervention—and a time-series model is fit to the data, then the result is a tendency for outlying residuals to occur. These outliers indicate a need for a search for special causes. Thus, looking for extreme residuals in the special-cause chart resulting from the estimated time-series model is a reasonable strategy for special-cause detection.

To illustrate the use of indicator variables to offset the effects of special causes, let us assume that observation 33 of the computer system failure series was found to have a special explanation. For example, suppose it was discovered that a power loss, which caused the prior system failure (i.e., the 32nd), resulted also in the configuration where several other communication devices were lost. In response to this situation, third-party computer consultants were brought in to reconfigure the equipment, using all the standard protocols required for the current system. Since it is generally known that incorrect configurations can lead to intermittent system failures, it can be reasonably inferred that the reestablishment of the correct configurations during this period also had a bearing on the unusually longer time to the next failure, as reflected by the 3σ outlier at observation 33. For purposes of process improvement, the lessons learned from this special-cause investigation should be considered. It might be advisable for company management to formally institutionalize a configuration verification program supported by documented procedures detailing the frequency of the configuration checks for a particular device, along with its configuration protocol.

With an explanation at hand for observation 33, we create an indicator variable which takes on the value 1 for that observation and the value 0 elsewhere. The next step is to estimate a new fitted model which incorporates the indicator variable along with the trend component. The results of this regression can be found in Figure 5.25. After we account for the special cause, the intercept and

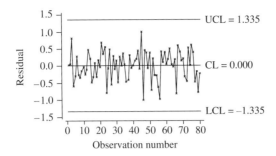

Figure 5.26   Revised special-cause chart for the transformed computer failure series.

slope of the trend line are slightly adjusted from (2.43, −0.00770) to (2.40, −0.00744). As would be true for any fitted model, the appropriateness of the fit needs to be checked. Performing all the standard diagnostic checks for randomness and normality on the residuals reveals that the fitted model is indeed satisfactory. The reader is encouraged to verify this conclusion. As a technical note, analysis of the residuals will reveal that the 33rd residual has value equal to 0. This is because the indicator variable perfectly accounts for the observation, as will be seen in the fitted-values chart. The implication is that the observation has been essentially "deleted" without affecting the proper sequencing of the other observations. (For certain technical reasons, the reader should be cautioned that Minitab assigns a missing observation value "*" to the standardized residual associated with the observation perfectly accounted for, in this case, the 33rd standardized residual. The user can simply substitute that missing value for the number 0, which is the value assigned to the raw residuals.)

Based on the new fitted model, the revised residuals along with control limits are given in Figure 5.26. The control chart provides no indications of special causes in the form of 3σ outliers or short runs. It would then be appropriate to project the residual limits of (−1.335, 1.335) for prospective control of the forecast errors. Given that the model is based on transformed data, the forecast errors would be the difference between the fifth root of the actual observation and the trend prediction line of $2.40 - 0.00744t$.

Now that the data analysis framework has enabled us to monitor appropriately for legitimate special-cause signals, attention needs to be focused on a more fundamental issue occurring with the process, namely, the systematic behavior as captured by the fitted model. The fitted-values chart based on the revised fit along with the transformed data can be found in Figure 5.27a. Notice that the model fully accounts for observation 33. Hence, for all practical purposes, the observation has been deleted as intended. Beyond this particular point, the interpretation of the model is that the times between failures are progressively getting shorter. Based on a fifth root scale, the intervals between failures decrease in a linear fashion. To gain a perspective in terms of the original

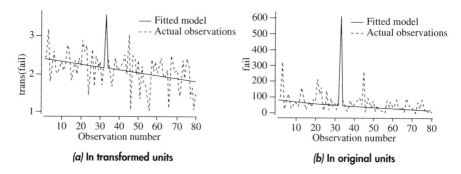

(a) **In transformed units**   *(b)* **In original units**

Figure 5.27   Fitted-values charts for the computer failure application.

units of measurement, one can raise the fitted values to the fifth power. A time-series plot of the fitted values in original units is given in Figure 5.27*b*. Again, we clearly see a systematic downward trend in the up times which drop from expected 60 h between failures in the beginning periods of observation to an expected 25 h between failures in the later periods. The computer system process is systematically deteriorating on a steady route toward unacceptable performance. Along with the monitoring of special causes as developed in this section, primary attention should be paid to understanding the trend process for purposes of systematic process improvement.

Table 5.6   Percentage of phosphorus series

0.045123	0.039692	0.042499	0.039905	0.039580	0.044515
0.042653	0.050871	0.051175	0.046192	0.045998	0.045287
0.043689	0.045428	0.037653	0.037126	0.038197	0.041416
0.039248	0.038642	0.042793	0.041774	0.041212	0.042322
0.043960	0.043397	0.049041	0.041046	0.040260	0.044297
0.036117	0.039916	0.044106	0.046558	0.041433	0.046357
0.039078	0.042592	0.043579	0.045991	0.045503	0.045529
0.043794	0.043867	0.040168	0.038482	0.041719	0.045132
0.039334	0.039720	0.038269	0.045140	0.044500	0.044594
0.046301	0.042862	0.043058	0.041517	0.042217	0.035953
0.038042	0.039301	0.043625	0.045408	0.047051	0.045239
0.041177	0.042900	0.048665	0.048620	0.050647	0.050802
0.046519	0.045999	0.046880	0.042557	0.039337	0.037700
0.038583	0.038924	0.041805	0.043915	0.038175	0.041224
0.041572	0.042356	0.037447	0.039595	0.046171	0.045421

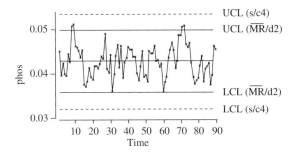

Figure 5.28   *X* chart for the phosphorus series with standard control limits.

5.5   MONITORING A PROCESS INFLUENCED BY LAGGED EFFECTS ONLY

In this section, we consider an application from the steelmaking industry. The data consist of 90 consecutive individual measurements of the percentage of phosphorus in periodic batches of iron. The data are found in Table 5.6 and in the file "steel1.dat". Since this chemical element is considered important for the quality of the iron produced, it is industry practice to monitor closely phosphorus measurements using standard control charts. It will interest the reader to know that this data set comes from a North American steelmaking company where it was indeed analyzed using the standard SPC methods with limited success.

In Figure 5.28, we present the time-series plot of the data, $phos_t$, with both standard control limits–based sample standard deviation (5.2) and moving range (5.6). In practical application, these limits are often calculated and mechanically applied to data that generally resemble this series. It is clear that the sample standard deviation–based limits are substantially wider than the moving-range-based limits.

Since no points breach the standard deviation–based control limits, control chart analysts might be tempted to conclude that the process is in control and that there is no need to search for special causes. In sharp contrast, there are four 3σ signals based on the moving-range-based limits. An equally hazardous temptation is to view the data like those shown above as a series of loosely connected episodes, each inviting special explanations. To highlight this point, consider the application of supplementary runs rules. Application of these additional criteria suggests 10 additional "out-of-control" episodes, implying a totality of 14 signals when coupled with outlier signals from the moving-range-based limits.

The process of achieving statistical control is sometimes pictured as finding special explanations for each episode, attempting a correction, observing the process predictively to see if the correction worked, discovering more special explanations for further episodes, etc., until "control" is finally reached. With this mind-set, in our example, the 14 identified out-of-control signals could potentially invite 14 different special explanations with associated actions. Since the 14 episodes include a total of 34 distinct observations, nearly 38 percent (= 34/90) of the series could be potentially labeled with ad hoc explanations!

By viewing the process as several loosely connected episodes begging unique explanation and action, the control chart practitioner is on a path of self-delusion to believe that work on each of these will change the underlying system to an inherent state of statistical control. Preoccupation with these numerous and potentially illusionary special causes distracts the user from the reality of *one* fundamental issue at hand, namely, the underlying systematic nonrandom behavior. The culprit is not some local, special, or sporadic occurrence but rather is a *persisting* systematic variation.

In particular, the time-series plot illustrates the presence of positive autocorrelation as reflected by the tendency of successive observations to be close together, giving the appearance of a "meandering" pattern. Unlike in a random series, knowledge that the most recent observations are above or below the centerline and by how much is useful for future prediction of such series. In particular, for a positively autocorrelated process, we would expect a new observation to be close to the current and recent observations. A characteristic of a positively autocorrelated process is the appearance of peaks and troughs which might tempt one to mistakenly identify them as unique up and down shifts in the process mean.

Based on the visual inspection of the data series, it is clear that the nonrandom behavior is *stationary*. This implies that neither the process mean nor the variance is drifting (up or down) away over time. The processes studied in Sections 5.3 and 5.4 reflected nonstationary behavior due to the systematic drift in their means as reflected by the fitted trend models. The notion of stationarity for the phosphorus series is important; it suggests that the process level will meander away but only temporarily, since there is a tendency for the observations to revert to the underlying mean. In a broader sense, the implication of stationarity is that the process behavior is stable. As will be discussed in Chapter 9, the distinction between stationarity and nonstationarity is useful in the determination of process capability and the development of appropriate capability measures.

To continue with the randomness diagnostics, the runs test and ACF with limits for the phosphorus series are provided in Figure 5.29. The runs test indicates significantly too few observed runs relative to what is expected under the assumption of randomness. Specifically, the low runs count of 27 for the positively autocorrelated series compared with the 45.98 expected count for a random process confirms the systematic tendency for observations to cluster. The associated p value of 0.0001 clearly rejects the null hypothesis of randomness for any reasonable level of significance. The ACF for the phosphorus data reveals significant autocorrelations at many of the lags.

From the random diagnostic checks, it was clear that the phosphorus series is nonrandom. The next step is to find and fit an appropriate time-series model that can properly account for this nonrandomness. In addition to being a useful tool for the detection of nonrandom behavior, the ACF serves as an important guide to identify potential models for the data. As discussed in Section 3.8, another aid in the search for candidate models is the *partial autocorrelation function (PACF)*,

```
phos

K = 0.0428
The observed number of runs  =  27
The expected number of runs  =  45.9778
44 Observations above K  46 below
      The test is significant at 0.0001
```

(a) Runs test

```
ACF of phos

          -1.0 -0.8 -0.6 -0.4 -0.2 0.0  0.2  0.4  0.6  0.8  1.0
           +----+----+----+----+----+----+----+----+----+----+
 1    0.518                        XXXXXXXXXXXXXX
 2    0.294                        XXXXXXXX
 3    0.128                        XXXX
 4   -0.013                         X
 5   -0.101                      XXXX
 6   -0.125                      XXXX
 7   -0.201                     XXXXXX
 8   -0.222                    XXXXXXX
 9   -0.203                     XXXXXX
10   -0.224                    XXXXXXX
11   -0.185                     XXXXXX
12   -0.172                      XXXXX
13   -0.090                       XXX
14   -0.035                        XX
15    0.010                         X
16    0.082                        XXX
17    0.078                        XXX
18    0.078                        XXX
19   -0.006                         X
20   -0.071                       XXX
21   -0.069                       XXX
22   -0.132                      XXXX
```

(b) ACF

Figure 5.29   Randomness checks of the phosphorus series.

which provides the "corrected" correlation between observations k periods apart after removing the effects of correlations at all intermediate lags.

For our study, the PACF of the phosphorus data set, with significance limits, is presented in Figure 5.30. Notice that the PACF reveals a significant correlation at lag 1 while all other lags are insignificant. The implication is that $phos_{t-1}$ alone might serve as a useful predictor variable of $phos_t$. To estimate the AR(1) model, we regress $phos_t$ on $phos_{t-1}$ with the results shown in Figure 5.31. The t ratio of 5.69 for the predictor variable $phos_{t-1}$ is highly significant with an associated p value < 0.0005. Notice that the estimated coefficient value of 0.521 for the lag 1

```
PACF of phos

               -1.0 -0.8 -0.6 -0.4 -0.2 0.0  0.2  0.4  0.6  0.8  1.0
               +----+----+----+----+----+----+----+----+----+----+
  1    0.518                              XXXXXXXXXXXXXX
  2    0.035                              XX
  3   -0.050                              XX
  4   -0.095                              XXX
  5   -0.074                              XXX
  6   -0.028                              XX
  7   -0.131                              XXXX
  8   -0.077                              XXX
  9   -0.045                              XX
 10   -0.113                              XXXX
 11   -0.043                              XX
 12   -0.092                              XXX
 13    0.016                               X
 14   -0.026                              XX
 15   -0.028                              XX
 16    0.042                               XX
 17   -0.056                              XX
 18   -0.016                               X
 19   -0.139                              XXXX
 20   -0.106                              XXXX
 21   -0.018                               X
 22   -0.155                              XXXXX
```

Figure 5.30   PACF of the phosphorus series.

variable is between -1 and $+1$, which provides further evidence that the underlying process is stationary, as discussed in Section 3.8.

As always, it is important to check the appropriateness of the fitted model. Study of the time-series plot (Figure 5.32) of the residuals reveals no apparent systematic meandering behavior as opposed to the original data. The runs test of the residuals in Figure 5.33a gives a p value of 0.3058; thus we are unable to reject the randomness of the residuals. The ACF of the residuals (Figure 5.33b), together with significance limits, shows no significant autocorrelations. Finally, the study of the histogram of the standardized residuals and normal probability plot in Figure 5.34 indicate that it is reasonable to assume normality for the residuals.

In summary, except for some possible local departures, the model diagnostics for the fitted AR(1) indicate that the residuals generally resemble an iidn process. Thus, the fitted model is satisfactory in capturing the systematic nonrandom behavior found in the original phosphorus series. In Figure 5.35, we present the special-cause chart for the residuals from the fitted AR(1) model. No points breach the control limits to signal a legitimate need for special-cause investigation as opposed to the four signals from the mechanically applied moving-range-based limits. In addition, supplementary runs rules suggest no indications

```
The regression equation is
phos = 0.0205 + 0.521 phos-1

89 cases used 1 cases contain missing values

Predictor            Coef        StDev             T       P
Constant         0.020486     0.003931          5.21   0.000
phos-1            0.52124      0.09158          5.69   0.000

S = 0.003039    R-Sq = 27.1%    R-Sq(adj) = 26.3%

Analysis of Variance

Source              DF          SS           MS        F       P
Regression           1    0.00029917   0.00029917   32.39   0.000
Residual Error      87    0.00080345   0.00000924
Total               88    0.00110261
```

Figure 5.31   Regression output for the fitted lag 1 model for the phosphorus series.

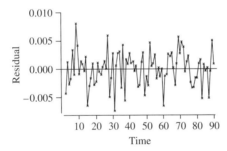

Figure 5.32   Time-series plot of the residuals from the fitted lag 1 model for the phosphorus series.

of special-cause variation. In comparison with the application of the runs rules to the original series, we see that for the original series 14 runs violations occurred versus 0 runs violations for the residual chart. Runs rules applied to the original series signal numerous illusionary special causes, suggesting that the process was nearly continually out of control, sometimes on the high side and sometimes on the low side. Applied to the residuals, runs rules suggest no need for special investigation.

The plot of fitted values along with the original values is given in Figure 5.36. The fitted plot gives us a view of the evolution of the estimated process level through time and shows us more clearly than does the plot of the original values the underlying systematic behavior of the process. Ideally, we would like to find, study, and remove the sources of the time-series effects reflected by the fitted chart. If the autocorrelation accounted for by the fitted model could be

```
residual

K =      -0.0000

The observed number of runs  =  50
The expected number of runs  =  45.2247
48 Observations above K   41 below
       The test is significant at 0.3056
       Cannot reject at alpha = 0.05
```

(a) Runs test

```
ACF of residual

        -1.0 -0.8 -0.6 -0.4 -0.2 0.0  0.2  0.4  0.6  0.8  1.0
        +----+----+----+----+----+----+----+----+----+----+
 1 -0.007                              X
 2  0.055                              XX
 3  0.030                              XX
 4 -0.046                              XX
 5 -0.087                             XXX
 6  0.000                              X
 7 -0.125                            XXXX
 8 -0.119                            XXXX
 9 -0.030                              XX
10 -0.116                            XXXX
11 -0.046                              XX
12 -0.105                            XXXX
13 -0.021                              XX
14  0.009                              X
15 -0.012                              X
16  0.090                              XXX
17  0.021                              XX
18  0.092                              XXX
19 -0.002                              X
20 -0.065                             XXX
21  0.030                              XX
22 -0.056                              XX
```

(b) ACF

Figure 5.33   Randomness checks of the residuals from the fitted lag 1 model for the phosphorus series.

removed, there is a potential for a 27.1 percent (i.e., estimated R^2 of the fitted model) improvement in the process, as measured by process variance reduction.

It may be the case that the underlying nonrandom behavior is not fully understood or that it is uneconomical to remove it. In such a situation, the fitted model provides information that might be utilized to determine an adjustment policy to compensate for the nonrandom behavior. As we have learned in Section 4.5, if a process is in a state of statistical control in the strict sense of iid, there is

```
Histogram of stres      N = 89  N* = 1
Midpoint        Count
 -2.250          3    ***
 -1.750          5    *****
 -1.250          6    ******
 -0.750         13    *************
 -0.250         14    **************
  0.250         24    ************************
  0.750         13    *************
  1.250          5    *****
  1.750          5    *****
  2.250          0
  2.750          1    *
```

(a) Histogram

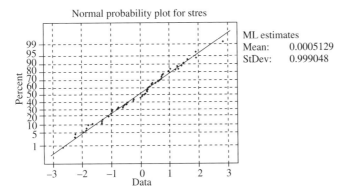

(b) Normal probability plot

Figure 5.34   Normality checks of the residuals from the fitted lag 1 model for the phosphorus series.

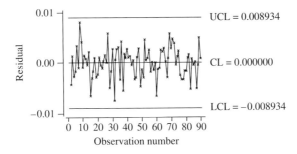

Figure 5.35   Special-cause chart for the phosphorus series.

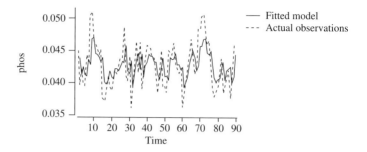

Figure 5.36   Fitted-values chart for the phosphorus series.

little to gain and much to lose by ad hoc intervention to improve the process based on its recent performance. However, in the case where data suggest auto-correlated behavior, process adjustment potentially can reduce process variation.

To illustrate the type of control decisions that could be based on the fitted model, suppose that the most desirable level of the phosphorus process is 0.045, and that increasing deviation from that level entails increasing economic loss from less-than-optimal product. Suppose further that at certain known cost it is possible at any time to recenter the process to 0.045. Then one can make an economic calculation to balance the expected loss of bad product over some specified time against the cost of recentering. This calculation will define action limits both below and above 0.045 at which the process should be recentered. Note that these action limits are conceptually different from control limits. They are not signals that it is time to look for special causes; rather, they are signals that a specific corrective action is needed. In Chapter 10, we will illustrate in greater detail the implementation of an adaptive control scheme.

Finally, it is interesting to point out that one could construct a control chart for the phosphorus series in much the same way as the trend control chart was developed for the camber series in Section 5.3. From (5.12), it can be seen that the basic idea is to superimpose control limits around the fitted values (or forecast function), which in this case implies control limits placed at

$$0.0205 + 0.521 \; \text{phos}_{t-1} \pm 3\frac{\overline{MR}_{\text{res}}}{d_2}$$

where $\overline{MR}_{\text{res}}$ is the average of the moving ranges for the residuals. For the case of a simple trend process, the estimate of the underlying random variation can be based on the moving ranges of the original data or the residuals. This is due to the invariance of the moving ranges to the trend component. However, for other non-random processes, such as an AR(1), the invariance does not apply, and thus moving-range calculations need to be performed on the residuals and not on the original data.

In Figure 5.37, we superimpose the time-varying control limits to the original phosphorus data. This single chart with variable 3σ limits and the special-cause chart with fixed 3σ limits (Figure 5.35) will provide equivalent results in

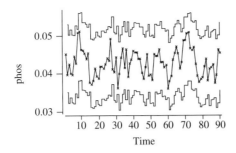

Figure 5.37   Phosphorus series with time-varying control limits.

terms of highlighting individual outliers as possible indications of special causes. However, from a pragmatic point of view, the special-cause chart gives a cleaner and sharper visual aid for the highlighting of individual outliers. Furthermore, other control criteria are not easily reconciled with the single-chart implementation. For instance, runs rules will be misguided by the underlying time-series effects reflected in the original data. As we saw earlier, runs rules should be applied to the residuals which are free of any persisting nonrandomness. Thus, for purposes of special-cause detection, the single chart is limited to detection of isolated outliers while the two-chart implementation opens the way for more general standard control criteria.

Table 5.7   Kleft measurements

−1.16	−0.49	−0.73	−0.63	−0.51	−1.17	−0.52	−0.31	−0.55	0.00
−0.61	−0.53	−0.41	−0.65	−0.66	−1.62	−0.84	−0.77	−1.05	−0.27
0.03	0.40	0.77	−1.29	−1.14	−1.59	−0.61	0.11	−1.49	−0.85
−0.03	−1.19	−0.87	−1.31	−1.04	−0.89	−0.96	−0.76	−1.07	−1.21
−1.66	−1.17	−1.38	−1.65	−1.75	−1.63	−1.85	−0.90	−0.84	−1.00
−1.14	−1.71	−0.75	−0.75	−0.44	−0.63	−0.57			

5.6   MONITORING A PROCESS INFLUENCED BY TREND AND LAGGED EFFECTS

The nonrandom processes studied in previous sections were found to be appropriately modeled with a single independent variable representing either a trend or a lagged effect. In Section 3.9, we illustrated with the ASQ data set that processes can be influenced by more than one type of systematic pattern. In this section, we

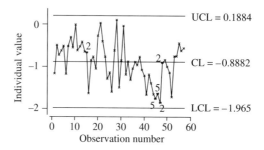

Figure 5.38 X chart for the kleft series based on MR/d_2 along with the application of supplementary runs rules.

provide another example of a process influenced by a combination of systematic effects. The data to be considered come from the automobile component process described in Section 5.2. In particular, the measurement kleft$_t$, is another critical dimension related to vehicle handling stability. The data set consists of 57 individual measurements which are shown in Table 5.7 and found also in the file "vehicle3.dat".

We begin by constructing the standard individual control chart based on the moving range. Such a control chart along with highlighted supplementary runs rules violations is shown in Figure 5.38. The control limits alone suggest no indications of special causes, tempting the user to deem the process in a state of statistical control. On the other hand, supplementary runs rules highlight a handful of out-of-control signals begging special explanations. Again, the control criteria are useless and distract from the real issue at hand, namely, the underlying nonrandom behavior.

The runs test and ACF with significance limits (see Figure 5.39) clearly confirm our suspicions of nonrandomness. The PACF (not shown) reveals one significant positive correlation at lag 1, implying an AR(1) model. However, inspection of the time-series plot hints of the possibility of a linear trend in conjunction with meandering behavior, reflecting the lagged effects.

Accordingly, we create two independent variables: one for the simple trend (t) effect and the other for lag 1 effect (kleft$_{t-1}$). At this stage, it is possible to regress the dependent variable (kleft$_t$) on all possible combinations of the two independent variables. Even though this task is quite manageable with so few independent variables, we use a stepwise regression procedure (forward) to facilitate the exploration of possible regressions. As can be seen from the stepwise regression output (Figure 5.40), both variables seem to contribute significantly in the prediction of kleft$_t$. Accordingly, we estimate a regression model based on the two components and provide the results in Figure 5.41.

The trend and lag effects appear to be genuine given that their respective p values are both smaller than the conventional cutoff of 0.05. The runs test of the residuals (not shown) finds an observed number of runs to be 31 versus an

```
kleft

K =   -0.8882

The observed number of runs  =   20
The expected number of runs  =   29.2807
31 Observations above K  26 below
      The test is significant at 0.0124
```

(a) Runs test

```
ACF of kleft
        -1.0 -0.8 -0.6 -0.4 -0.2 0.0  0.2  0.4  0.6  0.8  1.0
          +----+----+----+----+----+----+----+----+----+----+
 1   0.390                          XXXXXXXXXX
 2   0.236                          XXXXXXX
 3   0.265                          XXXXXXXX
 4   0.031                          XX
 5   0.162                          XXXXX
 6   0.151                          XXXXX
 7   0.207                          XXXXXX
 8   0.171                          XXXXX
 9   0.036                          XX
10   0.201                          XXXXXX
11   0.075                          XXX
12  -0.084                       XXX
13  -0.013                          X
14  -0.161                     XXXXX
```

(b) ACF

Figure 5.39   Randomness checks of the kleft series.

```
Stepwise Regression

F-to-Enter:    4.00         F-to-Remove:    4.00

Response is kleft on 2 predictors, with N = 56
N(cases with missing observations) = 1 N(all cases) = 57

        Step        1          2
    Constant    -0.5381    -0.3860

           t    -0.0117    -0.0084
     T-Value      -3.19      -2.17

      kleft-1                 0.28
     T-Value                  2.11

           S      0.444      0.431
        R-Sq      15.83      22.33
```

Figure 5.40   Stepwise regression output for the kleft series.

```
The regression equation is
kleft = -0.386 + 0.279 kleft-1 - 0.00842 t

56 cases used 1 cases contain missing values

Predictor          Coef        StDev           T         P
Constant        -0.3860       0.1399       -2.76     0.008
kleft-1          0.2785       0.1323        2.11     0.040
t            -0.008419     0.003888       -2.17     0.035

S = 0.4308      R-Sq = 22.3%      R-Sq(adj) = 19.4%

Analysis of Variance

Source             DF           SS          MS         F         P
Regression          2       2.8267      1.4133      7.62     0.001
Residual Error     53       9.8346      0.1856
Total              55      12.6613
```

Figure 5.41   Regression output for the fitted trend, lag model for the kleft series.

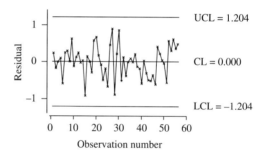

Figure 5.42   Special-cause chart for the kleft series.

expected number of 28.85 with a corresponding p value of 0.5614. Furthermore, the ACF (not shown) reveals no substantial autocorrelations. Additionally, the standard normality checks of the residuals are satisfactory. Overall, the diagnostic checks for the fitted model are quite favorable, indicating that the model is suitable for prediction.

The special-cause chart (Figure 5.42) reveals no outliers or short sequences of observations inconsistent with random behavior, which is in contrast to the results based on the mechanical application of the standard control criteria. The conclusion is that a search for isolated special causes would not be rewarding.

To visualize the effects of the systematic behavior, we display the fitted-values chart along with the original data in Figure 5.43. The chart more clearly shows that the process is meandering around a downward trend line. For purposes of process improvement, in terms of variation reduction, it is desirable to remove both of the nonrandom effects (trend and lag) captured by the model. However, given the implications of each nonrandom component, it would be sensible to

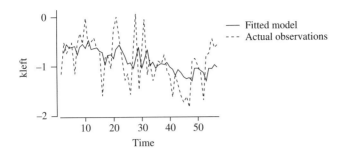

Figure 5.43   Fitted-values chart for the kleft series.

place higher priority on acting upon the trend behavior. Removing only the carry-over behavior associated with the lag 1 variable still leaves an unacceptable process that is systematically drifting away. On the other hand, if the causes of the trend can be eliminated, the resulting process could be a stable stationary auto-correlated process which, relatively speaking, is potentially tolerable.

5.7   MONITORING VARIABILITY FOR INDIVIDUAL MEASUREMENTS

In conjunction with a control chart for the original observations, it is common practice to use the moving ranges for the detection of process dispersion changes. In this section, we explore different strategies for monitoring moving-range data.

STANDARD MOVING-RANGE CHART

The standard approach is to plot the moving ranges with the following upper and lower control limits:

$$\text{UCL} = D_4\overline{MR}$$
$$\text{LCL} = D_3\overline{MR}$$

(5.13)

where D_4 and D_3 are general constants based on number of observations in the sample used to determine the range. The plot of the moving ranges with the above control limits is called a *moving-range chart*. The control limits are supposedly designed so that interpretation of the chart is equivalent to a standard 3σ control chart. For the case of the moving range, the number of observations in the sample is equal to 2 (i.e., the successive pairs of observations), and it can be shown that

$$D_4 = 1 + 3/2\sqrt{2\pi - 4}$$
$$D_3 = 1 - 3/2\sqrt{2\pi - 4}$$

(5.14)

Since D_3 is negative, the lower control limit is set to zero, and thus moving ranges are only compared to the upper control limit.

The concept of detecting changes in variation is, of course, very important. For this reason, Wheeler and Chambers (1992) stress the indispensability of the moving-range chart. Minitab,[3] along with all known SPC and statistical software, offers the moving range as seemingly an appropriate means for monitoring individual measurements for changes in process variability. However, the standard implementation of the moving-range chart in its standard form presents several problems. We see three fundamental problems.

First, the setting of the lower control limit to zero inhibits the chart from detecting reductions in variation. Technically, the occurrence of an out-of-control point on the lower end is possible, yet highly unlikely. An out-of-control point can be obtained only when two successive process points are equal, resulting in a moving range of zero and a point plotted on the lower control limit. Theoretically for continuous random variables, this event has an associated probability of zero. Practically, it is also unlikely when measurements are precisely taken to many significant digits, or when the variance of the process is inherently large. Given that an important goal of continual process improvement efforts is the reduction of variation, it is, therefore, important to have control charts that are capable of detecting such reductions.

Second, because of the overlapping of the moving ranges, successive moving-range values will be correlated. If the underlying process for the original measurements is iidn, it can be shown that, using the results in Cryer and Ryan (1990), the correlation between successive moving ranges is approximately 0.22. Even though the correlation is "slight," runs rules might still unnecessarily signal nonexistent special causes. Minitab gives the user the option to implement runs rules on the moving-range series. Given that implementing the moving-range chart alone is problematic, we believe that the runs rules option is ill advised and is simply compounding the problem.

Third, the underlying distribution of the moving ranges is highly skewed and non-negative. Thus, it makes little sense to superimpose the standard control limits which are based on the assumption of approximate normality of the data. The inconsistency between the true underlying distribution and the implied distribution associated with the control limits undermines the appropriate interpretation of the limits. Radson and Alwan (1995) demonstrate that the underlying departure from normality results in a type I error rate of 0.00915, which is greater than the implied rate of 0.0027 associated with 3σ limits.

To illustrate the standard moving-range chart, along with its deficiencies, consider the process studied in Section 5.2 which was deemed in control based on the analysis of the original measurements (loc_t). The individual moving ranges were given in Table 5.2, and their average ($\overline{MR}$) was found to be 0.7426.

---

[3]Minitab offers users two options: presentation of the moving-range chart alone (Stat >> Control Charts >> Moving Range) and presentation of the individuals chart (X chart) with the moving-range chart (Stat >> Control Charts >> I − MR).

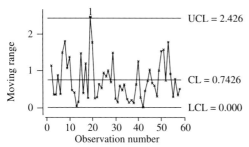

Figure 5.44   Standard moving-ranging chart for the locaton series.

By using this value along with (5.13) and the constants of (5.14), the control limits are as follows:

$$UCL = (1 + 3/2 \sqrt{2\pi - 4})(0.7426) = 2.426$$

$$LCL = 0$$

The moving-range chart for the location process is provided in Figure 5.44. Based on the upper control limit, observation 18 would be interpreted as an out-of-control signal suggesting a special-cause investigation of a variance increase. However, before we seek explanations, let us examine the distribution of the moving ranges to help us assess the appropriateness of such an action. Figure 5.45 gives the histogram and normal probability plot of the standardized moving-range values. The moving-range data are clearly nonnormal and positively skewed. In general, it can be shown that the underlying distribution of the moving-range statistic is the half-normal distribution (see Radson and Alwan, 1995).

In light of the deficiencies of the standard moving-range chart, some investigators argue against the implementation of the method; for example, see Nelson (1982, 1990); Roes, Does, and Schurink (1993); and Rigdon, Cruthis, and Champ (1994). Nelson (1982) argues that detection of variability changes should be handled by using the X chart alone. He states that "the chart of the individual observations actually contains all the information available." Based on analytical studies, Roes, Does, and Schurink (1993), and Rigdon, Cruthis, and Champ (1994) support the objections of Nelson. In particular, Rigdon, Cruthis, and Champ (1994) used thc ARL measure to demonstrate that the moving-range chart added little value in the detection of variance shifts over and above the implementation of the X chart alone. We agree with these investigators that the standard moving-range chart should not be implemented. However, we believe that there may be intermediate solutions short of totally dispensing with the moving ranges from the study. For example, Amin and Ethridge (1998) provide a procedure for choosing the multiples used in the design of the X chart and moving-range charts in such a manner that they believe the implementation of the two charts can be justified. These investigators, however, still allow the "redesigned"

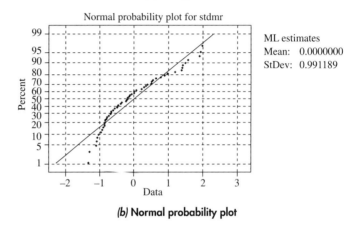

```
Histogram of stdmr    N = 57  N* = 1
Midpoint          Count
  -1.250             6  ******
  -0.750            17  *****************
  -0.250            12  ************
   0.250             7  *******
   0.750             5  *****
   1.250             5  *****
   1.750             4  ****
   2.250             0
   2.750             0
   3.250             1  *
```

(a) Histogram

(b) Normal probability plot

Figure 5.45   Normality checks of the standardized moving ranges computed from the location series.

moving-range chart to have no lower control limit; thus it is still incapable of detecting variation reductions.

In the next two subsections, we suggest alternative approaches to the standard moving-range chart that, in our opinion, salvage the use of moving ranges. Our premise is that data-analytic approaches can be used to directly attack the noted concerns of the standard moving-range chart.

HALF-NORMAL-BASED CHART

Given that the moving-range distribution is characterized by its asymmetry and skewness, the value of the standard 3σ symmetric limits is clearly questionable. One possibility is to use the correct probability limits from the true underlying distribution of the moving-range statistic. As mentioned above, this true distribution is the half-normal distribution. Using this distribution, one can establish $1 - \alpha$ control limits where α is the type I error rate. In the development of control limits, the standard convention is to associate the upper and lower control limits with $\alpha/2$ area from the upper and lower tails of the underlying distribution, respectively. Using a known relationship between the half-normal distribution and the normal distribution, Radson and Alwan (1995) showed that the upper and lower control limits, based on the half-normal distribution, are:

$$\text{UCL}_{\text{HN}} = \sqrt{2}\sigma_x z_{\alpha/4}$$
$$\text{LCL}_{\text{HN}} = \sqrt{2}\sigma_x z_{1/2-\alpha/4}$$

(5.15)

where $z_{\alpha/4}$ and $z_{1/2-\alpha/4}$ are the $100(1-\alpha/4)$ and $100(\frac{1}{2}+\alpha/4)$ percentiles of the standard normal distribution, respectively, and σ_x is the standard deviation of the original process measurements. If desired, the type I error rate can be controlled at the 3σ implied rate at $\alpha = 0.0027$, with $z_{\alpha/4} = z_{0.000675} = 3.20515$ and $z_{1/2-\alpha/4} = z_{0.499325} = 0.00169$.

In practical applications, one must first obtain an estimate of σ_x based on the original process data observations which, as discussed in Section 5.2, might be either $\overline{MR}/d_2$ or s/c_4. By using, for instance, the moving-range-based estimate for variation, the control limits of (5.15) are estimated by

$$\text{UCL}_{\text{HN}} = \frac{\sqrt{2z_{\alpha/4}}\,\overline{MR}}{d_2}$$
$$\text{LCL}_{\text{HN}} = \frac{\sqrt{2z_{1/2-\alpha/4}}\,\overline{MR}}{d_2}$$

(5.16)

To illustrate the half-normal-based control chart and make comparisons with the standard procedure, we choose $\alpha = 0.00915$ which, as noted earlier, is the true type I error rate for the standard moving-range chart. Following (5.16), the half-normal (denoted by subscript HN) control limits for the moving ranges computed from location data are

$$\text{UCL}_{\text{HN}} = \frac{\sqrt{2}(2.835)(0.7426)}{1.128} = 2.639$$

$$\text{LCL}_{\text{HN}} = \frac{\sqrt{2}(0.00573)(0.7426)}{1.128} = 0.0053$$

Superimposing these limits on the moving-range data results in the half-normal-based control shown in Figure 5.46.

In contrast to the standard procedure, the alternative limits do not isolate observation 18, yet they isolate a signal not originally revealed. Namely, observation 41 is below the lower control limit, suggesting a special cause in terms of a variability reduction. After finding observation 41 beyond the lower control limit, one should proceed to investigate the existence of a special cause associated with this signal. If a special-cause explanation is found, this observation should be deleted from the data and the retrospective limits recalculated.

This example illustrated nicely certain inadequacies of the standard procedure in special-cause detection. With the standard implementation, practitioners can be tempted to search for special causes that may not exist and hence may not

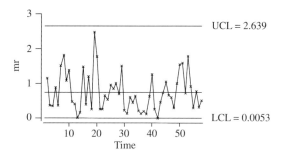

Figure 5.46   Half-normal-based control chart for the moving ranges computed from the location series.

recognize true special causes which warrant search. However, the half-normal based procedure appeared to provide a realistic judgment of upper-end observations and a possibility for greater detection of lower-end signals. Additionally, it should be recognized that the out-of-control signal at observation 41 was not revealed with the X chart (see Figure 5.6).

DATA ANALYSIS APPROACH TO MONITORING MOVING RANGES

As is true for all control chart procedures, proper implementation and interpretation of a given charting method rest on its underlying assumptions. For instance, we have demonstrated that the performance of the standard moving-range chart is hampered by the fact that its underlying assumption of normality for the moving-range statistic is unrealistic. To resolve this inappropriate assumption, an alternative method based on the half-normal distribution was developed in the previous section. However, it should be recognized that even this alternative method is grounded in certain assumptions necessary for proper implementation. The critical assumption for the development of the half-normal control chart is that the underlying distribution for the original process measurements is the normal distribution. If the distribution for the process measurements were found to be nonnormal, the performance of the alternative method would be questionable.

Consistent with the theme established throughout this book, a potentially more flexible approach is one based on direct data analysis. This framework allows not only for normalizing the data but also detecting systematic patterns in the data useful for general process understanding. To illustrate the data analysis approach for the monitoring of process variation, we consider two examples.

As a first example, consider the just analyzed moving-range data for the location series. As was seen in Figure 5.45, these data have an underlying positively skewed distribution. In light of this nonnormality, consideration should be given to find appropriate normalizing transformations. To determine the appropriate normalizing transformation, we might use the methods of Hines and Hines

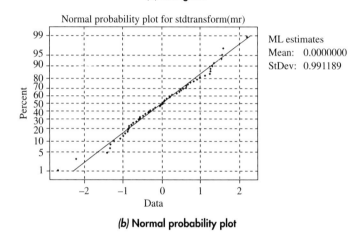

```
Histogram of stdtransform(mr) N = 57 N* = 1
Midpoint            Count
 -2.750                 1  *
 -2.250                 1  *
 -1.750                 0
 -1.250                 5  *****
 -0.750                12  ************
 -0.250                 9  *********
  0.250                10  **********
  0.750                 9  *********
  1.250                 6  ******
  1.750                 3  ***
  2.250                 1  *
```

(a) Histogram

Normal probability plot for stdtransform(mr)

ML estimates
Mean: 0.0000000
StDev: 0.991189

(b) Normal probability plot

Figure 5.47   Normality checks of the standardized transformed moving ranges computed from the location series.

or Box and Cox, as described in Section 3.3. These methods derive an estimate for λ based solely on the observed sample information. However, in the case of the moving-range statistic, we know its distribution (half normal), and thus we can utilize this fact to determine the "true" λ. For half-normal random variables, Alwan and Radson (1993) used simulation methods to find that a power transformation of approximately $\lambda = 0.4$ should be used.

In Figure 5.47, we show a histogram and normal probability plot for the standardized values of moving ranges raised to the 0.4 power. As we can see, the transformed moving ranges conform closely to the normal distribution. Given this close conformity, it would be more appropriate to apply 3σ limits to these data than to the untransformed moving-range data. By treating the transformed moving-range data as individual measurements, we can apply the standard control limits for individual measurements given by (5.2) or (5.6). By invoking the

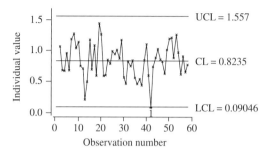

Figure 5.48   *X* chart applied to the transformed moving ranges computed from the location series.

default control chart for individual measurements on the transformed data from Minitab, we obtain the control chart given in Figure 5.48. Unlike in the control chart for the original moving ranges (see Figure 5.44), the scatter of the transformed data is clearly more in line with the symmetric limits. Similar to the half-normal-based chart, the transformed chart revealed observation 41 as a potential special cause and did not mistakenly highlight observation 18, as did the standard method.

Table 5.8   Fill measurements

353.86	352.40	353.02	352.03	352.63	353.53	353.58	352.86
352.90	353.53	352.96	353.78	353.37	353.82	353.57	352.55
353.15	352.67	353.43	353.86	353.74	353.61	353.72	353.66
353.58	354.02	353.96	353.50	353.78	353.79	354.14	354.02
353.69	354.10	353.82	353.31	353.96	353.76	353.71	353.52
353.49	353.46	353.81	353.48	353.87	353.68	353.20	353.33
354.18	353.96	353.87	353.91	353.70	354.41	353.97	353.89
353.99	353.75	353.83	353.85				

In our second example, consider the time-series plot of process measurements taken from a filling process given in Figure 5.49. The 60 observations are liquid fill levels, measured in milliliters at regular intervals. The data are shown in Table 5.8 and given also in the file "fill.dat". The time-series plot of the data clearly indicates that the process is trending upward. Additionally, it appears that variability around the systematic nonrandom pattern is decreasing; that is, the observations are progressively becoming tighter around the underlying pattern. The standard approach is to construct a moving-range chart, shown in Figure 5.50.

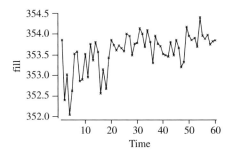

Figure 5.49   Time-series plot of the fill series.

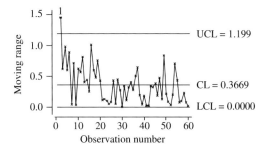

Figure 5.50   Standard moving-range chart for the fill series.

The first moving-range value is seen to be out of control, suggesting isolated special-cause explanation. However, careful visual inspection of the moving-range values reveals a systematic downward curved trend to the series. To model this trend, one might regress the moving-range data on the time index (t) and its square (t^2). Even though this regression model can be found to be generally satisfactory, a better fit can be obtained by considering the inverse of the time index ($1/t$) as an independent variable.

The regression output for this model is provided in Figure 5.51. The associated t ratio (5.88) of the independent variable $1/t$ is highly significant (p value < 0.0005). Except for some local departures, the regression diagnostics for this model are generally satisfactory. Specifically, the runs test for the residuals is compatible with randomness with an observed number of runs equal to 30 versus an expected 30.42. Furthermore, the ACF of residuals shows no significant correlation at any of the lags. The standard normal diagnostics of the residuals suggest that the normal distribution is a safe working assumption.

Given that the model suitably captures the underlying systematic component to the moving-range data, the residual control chart or special-cause chart is presented in Figure 5.52. The control chart reveals no 3σ outliers; however, the supplementary runs rules bring to our attention an unusual sequence of up and down

```
The regression equation is
mr   = 0.224 + 2.29 1/t

59 cases used 1 cases contain missing values

Predictor          Coef         StDev          T            P
Constant        0.22417       0.04015        5.58        0.000
1/t              2.2892        0.3890        5.88        0.000

S = 0.2457      R-Sq = 37.8%       R-Sq(adj) = 36.7%

Analysis of Variance

Source              DF            SS            MS           F          P
Regression           1        2.0904        2.0904       34.63      0.000
Residual Error      57        3.4405        0.0604
Total               58        5.5308
```

Figure 5.51   Regression output for fitted curved trend model for the moving-range data computed from the fill series.

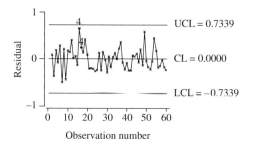

Figure 5.52   Special-cause chart for the moving-range data computed from the fill series.

behavior from moving-range observation 1 to 16. From the perspective of the original process measurements, the implication of this abnormal variation is that there was a period of time where successive fill amounts went through a repeated cycle of being far apart, then close together, then far apart, and so on. Looking closely at the beginning of the fill series as shown in Figure 5.49 reveals this to be the case. Discovery of this localized behavior should initiate an investigation to uncover the underlying cause.

In practice, oscillatory behavior of the process level and/or variation is rarely a *natural* behavior of processes. Data behaving systematically in an oscillatory manner will reflect negative autocorrelation. The typical systematic departure from nonrandomness is positive autocorrelation. For example, trend is a form of positive autocorrelation, as well as the stationary behavior found with the phosphorus series of Section 5.5. With this in mind, any evidence of oscillatory behavior suggests the likelihood that interventions to the process are being made,

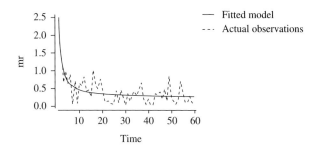

Figure 5.53   Fitted-values chart for the moving ranges from the fill series.

altering the underlying natural behavior. Alwan (1991) shows that certain common types of adjustments found in practice applied to an in-control process with the intent to improve the process result in degraded overall process performance and an induced oscillatory time-series behavior. Thus, the out-of-control signal found in the special-cause chart of Figure 5.52 might be a tip-off that the source of the abnormal variation is a practice of operator tampering in an attempt to stabilize the fill process. If such a practice were found, the out-of-control signal would then be designated as a special cause. In turn, management would benefit from training the operators about the concepts of random variability and why reactions to such variability can be counterproductive.

Finally, the plot of the fitted values along with the original moving ranges is given in Figure 5.53. As can be seen, the underlying evolution of the moving-range process is characterized by a downward-curved trend. These fitted values confirm our first suspicion that variability of the fill measurements is becoming tighter as time passes. Given the nature of a filling process, the systematic reduction in variability may be directly related to the upward trend in the fill levels. As the fill level increases toward the capacity of the container and there is potentially wasted overflow, the fill amounts are physically constrained, and thus the variability is bounded. Efforts should be made to understand both the underlying systematic movements of the process level and variation for improvement purposes.

In conclusion, we strongly recommend that users avoid the construction and interpretation of the standard moving-range chart. Instead, we recommend that the data-analytic methods presented in this and earlier subsections be implemented for proper monitoring of process variability over time.

EXERCISES

5.1. Would you agree with this statement? "In general, process applications associated with individual observations are more common than process applications associated with subgroups of observations."

5.2. From your personal life, give two examples of personal processes on which you could collect variable data in the form of a series of individual measurements.

5.3. What are the two different estimates for process standard deviation that can be used in the implementation of an X chart?

5.4. What is the value of c_4 if the data series has 20 observations? What is its value if the data series has 50 observations?

5.5. Consider the following summary statistics for data series on a variable X:

Variable	N	Mean	Median	TrMean	StDev	SE Mean
x	30	128.38	128.76	128.61	15.63	2.85

Variable	Minimum	Maximum	Q1	Q3
x	89.85	160.89	117.97	139.75

Using one of the two estimates described in this chapter, what is an estimate for the process standard deviation?

5.6. Consider the following summary statistics for a series of moving ranges:

Variable	N	Mean	Median	TrMean	StDev	SE Mean
mr	39	6.436	5.655	6.196	4.399	0.704

Variable	Minimum	Maximum	Q1	Q3
mr	0.283	17.873	2.646	10.100

  a. What is an estimate for the process standard deviation?
  b. If an X chart were constructed, how many observations would be involved?

5.7. Below are measurements of a critical dimension (mm) on 20 consecutive plastic parts produced from an injection mold:

```
0.6474   0.6465   0.6517   0.6504   0.6479   0.6508   0.6487   0.6462
0.6444   0.6480   0.6462   0.6508   0.6468   0.6492   0.6504   0.6516
0.6465   0.6509   0.6480   0.6539
```

  a. Based on the sample standard deviation, what is an estimate for the process standard deviation?
  b. Based on the moving-range method, what is an estimate for the process standard deviation?

5.8. Suppose a series of individual measurements is placed in a column of a Minitab worksheet. Determine a sequence of commands (menu or session) that will "create" and place the values of the moving ranges in another column of the worksheet. [*Hint:* There are at least two ways of accomplishing this task. Consider the options provided in the time-series menu (Stat >> Time Series).]

5.9. Refer to Exercise 5.5 and determine the values for the X chart control limits.

5.10. Refer to Exercise 5.7 and determine the values for the X chart control limits based on the sample standard deviation–based estimate. In addition, construct the X chart by plotting the individual measurements with the computed limits.

5.11. Refer to Exercise 5.7 and determine the values for the X chart control limits based on the moving-range-based estimate. In addition, construct the X chart by plotting the individual measurements with the computed limits.

5.12. Consider the stopwatch data series introduced in Section 2.3. Using the moving-range-based estimate, construct an X chart for the data series.

5.13. Both individual and team sports performance can vary greatly over time. We often say that a team had a "hot streak" for a while, then a "cold streak" or "slump." The possibility that sports performance processes are actually in a state of statistical control runs contrary to popular belief in streaks due to assignable reasons. As a measure of team performance, consider the point spread, which is simply defined as points scored by a team minus points scored by its opponent. Technically, these data are discrete; however, a variable measurement framework works satisfactorily for this application. Below are the consecutive game point spreads for the Chicago Bulls for the first half of the 1997–1998 regular season (point spreads for the entire season are given in "bulls.dat"):

-7	20	4	13	-2	13	-21	-7	13	9
-4	9	15	-1	-11	5	10	22	18	-8
-2	27	7	21	8	5	10	7	6	-4
14	9	11	-27	1	5	10	-10	10	6
31									

    a. Perform randomness and normality checks on the data.

    b. Using the moving-range-based estimate, construct an X chart for the data series. Supplement the control chart with supplementary runs rules to identify any unusual streaks. What is your conclusion about the Bulls' performance over time?

    c. Many sports writers suggested that the Bulls were "revitalized" after the all-star break. Below are the point spreads for the second half of the regular season:

2	-7	18	13	7	-25	39	-8	7	2
2	9	8	37	6	22	-5	15	28	13
15	-7	10	16	6	6	2	23	15	15
17	5	14	15	18	-6	9	-9	-8	7
2									

      Project out the retrospective control computed in part b and plot the new observations. What is your conclusion?

5.14. Refer to the general discussion of Exercise 5.13. Consider now the individual performance of Chicago Bulls player Michael Jordan in terms of points scored per game. Below are the points scored by Jordan in consecutive games for the entire 1997–1998 regular season (data are also given in "jordan.dat"):

30	16	29	29	27	15	19	28	28	27
30	49	33	26	26	29	29	13	29	25
28	11	31	36	24	27	24	47	41	33
44	34	19	26	44	32	40	20	27	45
33	32	32	20	29	14	31	15	40	18
29	37	21	27	16	19	17	33	28	30
42	37	26	30	17	35	24	33	24	17
34	30	26	41	40	30	29	37	27	19
24	44								

   a. Perform randomness and normality checks on the data.
   b. Using the moving-range-based estimate, construct an *X* chart for the data series. Supplement the control chart with supplementary runs rules to identify any unusual streaks. What is your conclusion about Jordan's performance over time?

5.15. Pick a favorite team or player and choose a game statistic (such as, point spread, point scores, rebounds made) of interest to you. Analyze the chosen data series, using the methods of this chapter. The following Web sites might prove useful:

National Basketball Association:	http://www.nba.com
Women's National Basketball Association:	http://www.wnba.com
National Football League:	http://www.nfl.com
National Hockey League:	http://www.nhl.com

5.16. Financial advisers, financial reporters, and individual investors frequently assign special explanations to "perceived" unusual movements in publicly traded stocks. Below are the day-to-day changes in the closing stock prices for General Motors Corporation from September 4, 1998 to November 27, 1998 (data are also given in "gm.dat"):

-0.875	2.437	-1.187	-2.063	0.938	3.000	-0.563	-0.062	-1.375
-0.375	-0.125	0.062	1.750	-1.500	1.250	1.000	-1.500	-3.125
-0.687	0.812	-1.250	-2.500	0.125	-1.125	0.625	1.063	-0.875
1.125	3.875	2.875	0.812	1.875	-0.687	0.562	-1.187	0.437
1.188	0.500	-1.250	2.000	2.000	-0.438	2.250	-0.250	0.313
-1.063	1.250	0.563	-0.813	3.250	2.563	0.312	-0.437	-0.125
-0.688	1.500	-0.500	-0.937	-0.500				

   a. Perform randomness and normality checks on the data.

    b. Using the moving-range-based estimate, construct an *X* chart for the data series. Supplement the control chart with supplementary runs rules to identify any unusual streaks. What is your conclusion about GM stock movements over time?

5.17. The following data represent the weekly accounts receivable balance (in $1000) of a hardware distributor over a 30-week period (data also given in "acct.dat"):

128.703	120.649	126.215	129.426	135.816	125.131	131.122
121.774	125.367	121.076	119.848	127.251	153.923	125.658
122.172	129.268	122.912	125.351	130.572	116.875	122.587
127.677	132.364	130.950	118.662	131.902	132.220	128.522
132.270	122.671					

    a. Using the moving-range-based estimate, construct an *X* chart for the data series. Supplement the control chart with supplementary runs rules.

    b. Were there any out-of-control signals identified in part *a*? If so, assume a special-cause explanation can be given to the associated point(s).

    c. In light of part *b,* what are the appropriate prospective limits for the *X* chart?

5.18. Consider the daily caloric intake data series introduced in Section 3.3. Given the nature of the data series, construct an appropriate individual measurements control chart to determine the presence of special causes in the process.

5.19. Refer to the change order data series of Exercise 3.15. Given the nature of the data series, construct an appropriate individual measurements control chart to determine the presence of special causes in the process.

5.20. Refer to the weekly sales data series of Exercise 3.5.
    a. Construct a trend control chart for the application.
    b. What are the values of the prospective limits on the trend control chart for period 35?
    c. Construct a special-cause chart and fitted-values chart for the application.

5.21. In a very popular basic quality-related training manual (*The Memory Jogger*, 1994, Methuen, MA: GOAL/QPC), there is an illustration of the construction of an individual measurements chart (p. 48). In particular, the data set consists of consecutive measurements of the time (in seconds) to make an intravenous connection for patients being admitted for open heart surgery. The data are shown below (also provided in "iv.dat"):

600	480	540	240	420	450	480	690	240	360
450	300	480	120	240	210	210	180	240	300
300	130	120	300	180	210				

Other than being constructed in Minitab, below is the exact X chart (including highlighted point) shown in the training manual:

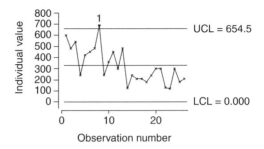

a. Take the data and compute the X chart limits based on the moving-range estimate. Given your computed lower control limit, why is the lower control limit set to zero in the above graph?

b. What is your reaction to the highlighted point in the above graph? Does it legitimately beg isolated special-cause explanation?

c. Fit a simple trend model to the data series. Report the significance of the trend variable and the overall diagnostics of the fitted model.

d. Construct a trend control chart for the application.

e. Construct a special-cause chart and fitted-values chart for the application.

5.22. Refer to the color data series of Exercise 3.30.

a. Using the moving-range-based estimate, construct an X chart for the data series. Supplement the chart with supplementary runs rules, and report the number of special-cause signals based on the standard approach.

b. Fit the data series with an AR(1) model. Report the significance of the lag 1 variable and the overall diagnostics of the fitted model.

c. Construct a special-cause chart. Supplement the chart with supplementary runs rules, and report the number of special-cause signals. Compare your results with the results from part *a*.

d. Construct a fitted-values chart for the application.

5.23. In the book *Time Series Analysis: Forecasting and Control* by Box, Jenkins, and Reinsel (1994), we find a data series associated with a chemical-industry process. In particular, the temperature measurements were made every minute on a chemical reactor process. Below are the data (also given in "boxtemp.dat"):

200	202	208	204	204	207	207	204	202	199
201	198	200	202	203	205	207	211	204	206
204	204	201	198	200	206	207	206	200	204
204	200	200	195	202	204				

a. Using the moving-range based estimate, construct an X chart for the data series. Supplement the chart with supplementary runs rules, and report the number of special-cause signals based on the standard approach.

b. Fit the data series with an AR(1) model. Report the significance of the lag 1 variable and the overall diagnostics of the fitted model.

c. Construct a special-cause chart. Supplement the chart with supplementary runs rules, and report the number of special-cause signals. Compare your results with the results from part *a.*

d. Construct a fitted-values chart for the application.

5.24. Refer to the chemical concentration data series of Exercise 3.32.

   a. Using the moving-range-based estimate, construct an *X* chart for the data series. Supplement the chart with supplementary runs rules, and report the number of special-cause signals based on the standard approach.

   b. Fit the data series with a model based on a lag 1 and lag 2 variable. Report the significance of the two independent variables and the overall diagnostics of the fitted model.

   c. Construct a special-cause chart. Supplement the chart with supplementary runs rules, and report the number of special-cause signals. Compare your results with the results from part *a.*

   d. Construct a fitted-values chart for the application.

5.25. Refer to the global temperature data series of Exercise 3.34.

   a. Using the moving-range-based estimate, construct an *X* chart for the data series. Supplement the chart with supplementary runs rules, and report the number of special-cause signals based on the standard approach.

   b. Fit the data series with a model based on a trend variable and a lag 1. Report the significance of the two independent variables and the overall diagnostics of the fitted model.

   c. Construct a special-cause chart. Supplement the chart with supplementary runs rules, and report the number of special-cause signals. Compare your results with the results from part *a.*

   d. Construct a fitted-values chart for the application.

5.26. For budgeting, warranty, and quality-related issues, a manufacturer of outdoor marine equipment monitors its monthly repair costs (in $1000). Below are 48 consecutive months of data (also given in "repair.dat"):

15.105	26.243	30.271	19.019	27.180	37.921	46.254	35.226
16.980	18.757	24.341	31.498	25.800	13.063	8.041	24.470
41.565	54.392	45.644	38.825	24.413	32.021	19.803	33.203
35.374	38.908	27.030	37.319	20.760	48.538	43.444	39.814
31.745	36.978	36.186	46.496	42.407	63.695	38.056	43.877
34.758	67.865	67.070	53.926	39.491	31.301	45.529	25.233

   a. Using the moving-range-based estimate, construct an *X* chart for the data series. Supplement the chart with supplementary runs rules, and report the number of special-cause signals based on the standard approach.

b. Make a time-series plot of the 48 consecutive observations, labeling the observations by month of the year; refer to the ASQ series illustrated in Section 3.9 to see a time-series plot with monthly labeling. Comment on the type of patterns evidenced in the plot.

c. Create a trend variable and monthly indicator variables. Use stepwise regression on this set of 13 independent variables to sift out the useful predictors. Based on the stepwise regression results, fit the data series with the selected variables.

d. Construct a special-cause chart. Supplement the chart with supplementary runs rules, and report the number of special-cause signals. Compare your results with the results from part *a*.

e. Assume any out-of-control signal(s) identified in part *d* have special-cause explanation. Refit the data series with a special-cause indicator variable along with the earlier chosen independent variables. Report the significance of the independent variables and the overall diagnostics of the fitted model.

f. Present the revised special-cause chart and the fitted-values chart.

5.27. To continue Exercise 5.22, construct and plot the color measurements with time-varying control limits.

5.28. To continue Exercise 5.23, construct and plot the chemical temperature measurements with time-varying control limits.

5.29. To continue Exercise 5.24, construct and plot the concentration measurements with time-varying control limits.

5.30. To continue Exercise 5.25, construct and plot the global temperature measurements with time-varying control limits.

5.31. List the objections to the use of a standard moving-range chart.

5.32. In Section 5.7, it was indicated that the theoretical false-alarm rate for the moving-range chart applied to an iidn process is greater that the standard interpretation of the 3σ limits. To investigate, consider the following Minitab-based simulation experiment:

Step 1   Generate 1001 random observations from the standard normal distribution (Calc $>>$ Random Data $>>$ Normal). (*Note:* by generating 1001 X observations, you will then have 1000 moving-range observations.)

Step 2   Apply Minitab's standard moving-range control chart to the series of individual measurements (Stat $>>$ Control Charts $>>$ Moving Range).

Step 3   Count the number of observations falling outside control limits.

Step 4   Repeat the above three steps ten times over. (*Note:* an alternative is to run the experiment once based on the random generation of 10,000 observations. However, if the Student Edition of Minitab is being used, there is a limitation of 3000 observations in the worksheet.)

   a. Based on your experiment, what is the average number of moving ranges per 1000 observations falling outside control limits?

   b. Refer to Section 5.7 and compare your result with the theoretical false-alarm rate for the standard moving-range chart noted in the section.

5.33.   Consider the plastic part data series of Exercise 5.7.

   a. Construct a half-normal-based control chart for the moving ranges with $\alpha = 0.0027$.

   b. As mentioned in Section 5.7, raise the moving ranges to the 0.40 power and then construct a standard individual observations chart on the transformed moving ranges. Compare the results of this control chart with the control chart from part *a.*

5.34.   Consider the daily changes in GM price data series of Exercise 5.16.

   a. Construct a half-normal-based control chart for the moving ranges with $\alpha = 0.0027$.

   b. As mentioned in Section 5.7, raise the moving ranges to the 0.40 power and then construct a standard individual observations chart on the transformed moving ranges. Compare the results of this control chart with the control chart from part *a.*

5.35.   Consider the accounts receivable data series of Exercise 5.17.

   a. Construct a half-normal-based control chart for the moving ranges with $\alpha = 0.005$.

   b. As mentioned in Section 5.7, raise the moving ranges to the 0.40 power and then construct a standard individual observations chart on the transformed moving ranges. Compare the results of this control chart with the control chart from part *a.*

REFERENCES

Alwan, L. C. (1991): "Time-Series Effects and Degradation of Control Chart Performance Resulting from Overadjustment," *Total Quality Management*, 2, pp. 99–112.

———— and D. Radson (1992): "Time-Series Investigation of Subsample Mean Charts," *IIE Transactions* (Special Issue on Quality in Production), 24, pp. 66–80.

————— and ————— (1993): "Alternative Procedures for the Standard Moving-Range Chart," *Management Research Center Working Paper 93-01*, School of Business Administration, University of Wisconsin-Milwaukee.

Amin, R. W., and R. A. Ethridge (1998): "A Note on Individual and Moving Range Control Charts," *Journal of Quality Technology*, 30, pp. 70–74.

Box, G. E. P., G. M. Jenkins, and G. C. Reinsel (1994): *Time Series Analysis: Forecasting and Control,* 3d ed., Prentice-Hall, Englewood Cliffs, NJ.

————— and G. C. Tiao (1975): "Intervention Analysis with Applications to Economic and Environmental Analysis," *Journal of the American Statistical Association,* 70, pp. 70–79.

Cryer, J. D., and T. P. Ryan (1990): "The Estimation of Sigma for an X Chart: $\overline{MR}/d_2$ or s/c_4?" *Journal of Quality Technology*, 22, pp. 187–192.

Mandel, J. (1969): "The Regression Control Chart," *Journal of Quality Technology*, 1, pp. 1–9.

Montgomery, D. C. (1996): *Introduction to Statistical Quality Control,* 3d ed., Wiley, New York.

Nelson, L. S. (1982): "Control Charts for Individual Measurements," *Journal of Quality Technology*, 14, pp. 172–173.

————— (1990): "Monitoring Reduction in Variation with a Range Chart," *Journal of Quality Technology*, 22, pp. 163–165.

Radson, D., and L. C. Alwan (1995): "Detecting Variance Reductions Using the Moving Range," *Quality Engineering*, 8, pp. 165–178.

Rigdon, S. E., E. N. Cruthis, and C. W. Champ (1994): "Design Strategies for Individuals and Moving Range Control Charts," *Journal of Quality Technology*, 26, pp. 274–287.

Roes, K. C. B., R. J. Does, and Y. Schurink (1993): "Shewhart-Type Control Charts for Individual Observations," *Journal of Quality Technology*, 25, pp. 188–198.

Wadsworth, H. M., K. S. Stephens, and A. B. Godfrey (1986): *Modern Methods for Quality Control and Improvement*, Wiley, New York.

Westgard, J. O., and P. L. Barry (1997), *Cost-Effective Quality Control: Managing the Quality and Productivity of Analytical Processes,* AACC Press, Washington, DC.

Wheeler, D. J., and D. S. Chambers (1992), *Understanding Statistical Process Control,* 2nd ed., SPC Press, Knoxville, TN.

6

Monitoring Subgroup
Variable Measurements

CHAPTER OVERVIEW

In Chapter 5, we considered methods for analyzing and monitoring a series of individual measurements. In such cases, the sample size for each period would be $n = 1$. More generally, control charts for variable measurements involve the collection of samples or *subgroups*. These two terms may be used interchangeably. For each subgroup, pertinent statistics are computed and charted. In particular, the most common subgroup control charts include charts for process variability, using ranges or standard deviations, and charts for process level using means.

In this chapter, we will examine the most popular variety of charts frequently referred to as *Shewhart charts*. Simply stated, Shewhart charts call for control limits—usually, 3σ limits—to be superimposed on a sequence plot of subgroup statistics where the plotted statistic is extracted from a given subgroup in "isolation" of other subgroups. There are other types of control charts which are based on combining current and past subgroup information; we will explore these charts in Chapter 8. Consistent with our approach, we will also discuss and illustrate the practical challenges to the standard methods.

6.1   DESIGNING SUBGROUP CONTROL CHARTS: PRELIMINARY CONSIDERATIONS

A basic requirement for the applicability of a subgroup approach is that the observations within the subgroup naturally represent some appropriately defined unit of time. In a manufacturing setting, an appropriate unit of time is usually defined by the need to have timely, on-line monitoring of the critical processes. If it takes several hours or days to obtain a new process measurement, waiting until a subgroup of five observations is formed would not be reasonable; individual measurement control charts would be preferable. Therefore, subgroup charting techniques are often associated with medium- to high-volume production operations.

Speedy gathering of data is not the only basis for effective application of subgrouping methods. There are many situations in which the appropriate unit of time might be a span of days or weeks. For instance, the tracking of average levels of different external customer satisfaction measures is often done on a monthly basis. As such, all responses in a given month to a customer survey instrument would form an appropriate subgroup.

Thus, the effectiveness and success of control chart procedures for samples or subgroups greatly depend on the manner in which these subgroups are formed. There are three basic "design" issues that need to be addressed: (1) the basis of sampling to obtain the elements used to form the subgroup, (2) the subgroup size, and (3) the timing between subgroups. As will be evident from the following two subsections, the appropriate design should not be arbitrary and requires a careful balance of many competing factors.

RATIONAL SUBGROUPING

It was Shewhart who originally conceptualized a basis for sampling the elements to be used to form a subgroup. His idea was that subgroups should be chosen in such a way that the observations within the sample have been, in all likelihood, measured under the same process conditions. Subgroups selected based on this fundamental principle are commonly referred to as *rational subgroups*. By attempting to have the subgroup elements as homogeneous as possible, special causes are more likely reflected as greater variability of the subgroups themselves. Thus, in the presence of special causes, rational subgrouping maximizes the chance that the plotted statistics, associated with the rational subgroups, will signal the out-of-control situations.

For production processes, the most common scheme for the creation of rational subgroups is to use individual elements taken over a short time. The logic is that observations taken close together have a greater chance of being influenced by similar process conditions. As recommended in the now classic AT&T *Statistical Quality Control Handbook* (1985), a practical implementation of this approach is accomplished by basing the subgroups on consecutive items produced. By allowing too much time between subgroup elements, there are more opportunities for nonhomogenous mixtures of process measurements. For instance, subgroups might consist of observations from stable process and special-cause conditions. As shown later, in the standard construction of control limits for subgroup statistics, within-subgroup variation serves as a basis for the estimation of between-subgroup variation. By virtue of the increased within-subgroup variation due to nonhomogeneity, there will be an inflated view of the between-subgroup variation which results in control limits being placed much farther apart than would be appropriate. Under such situations, the effectiveness of the control chart procedures to detect process change is often severely undermined.

A mixture of special causes and stable process conditions is not the only example of nonhomogeneity. Nonhomogeneous subgroups can be formed by pooling observations sampled across various shifts, operators, or machines that have consistent systematic differences. The problem is that the information on the systematic differences is "buried" within the subgroups, and control charts based on such pooled subgroups will be limited in their ability to detect changes for a given shift, operator, or machine. In Section 6.6, we specifically illustrate the hazards of pooling subgroup observations with no regard to the principles of rational subgrouping. Consistent with the notion of rational subgrouping, one might consider separate control chart procedures for each shift, operator, or machine. Of course, the clear danger is an unmanageable proliferation of control charts. Knowledge of the process should dictate the appropriate balance of control chart activities and should determine which individual factors require more specific monitoring.

The recommendation of sampling items that are close together may indeed ensure the creation of rational subgroups, but it is not without its limitations. Often the underlying process generating the individual measurements is naturally autocorrelated over time, with the correlation increasing as measurements are made more closely. Subgroups based on consecutive items from autocorrelated processes are, in principle, rational in the sense that the subgroup items are influenced by the same natural and predictable variability. The difficulty is that the construction and subsequent interpretation of standard subgroup control chart limits rest on the underlying assumption that the within-subgroup observations behave independently. Some experts have suggested that autocorrelation is simply a special cause which should be eliminated prior to the construction of control chart procedures. Unfortunately, elimination of autocorrelation is often not so simple. As has been pointed out a number of times previously, autocorrelation frequently occurs and represents the natural workings of the process, as opposed to an abrupt process change indicative of a special cause. This, of course, does not oppose the viewpoint that it would be desirable, in most cases, to remove the persistent autocorrelation. However, in practice, often the underlying reasons for the autocorrelated behavior are not fully understood, or it is uneconomical to remove them. In such cases, alternative strategies need to be considered for effective process monitoring.

One possibility is to space the individual measurements farther apart to achieve practical independence, as recommended by Hoerl and Palm (1990) and Coulson (1995). This strategy is viable only for stationary processes, such as the phosphorus process studied in Chapter 5. For nonstationary processes (e.g., trends), there is theoretically no gap between observations that will diminish the between-observations correlation. The "Catch 22" is that a strategy of sampling farther apart to achieve independence among subgroup elements increases the chances of violating the rational-subgroup principle. A second option is to retain the rational-subgroup principle along with the within-subgroup correlation and seek alternative calculations for the placement of subgroup control limits. This option will be explored in greater detail in Section 6.6.

SUBGROUP SIZE

Another important data collection issue concerns the size of the subgroup in terms of the number of individual measurements taken. In practice, subgroup sizes for variable control charts nearly always range from 2 to 25 with a range of 2 to 6 being most typical. Even more specifically, sample sizes of 4 or 5 are predominant in industrial applications. In fact, the American Society for Quality (ASQ) has established control chart forms that are formatted to use up to five measurements. One of the prime reasons for smaller sample sizes is computational ease. Prior to the advent of on-line computing technologies (e.g., computers and programmable calculators), the implementation of control charts depended solely on manual computations. Duncan (1986) notes that the popular choice of $n = 5$ may be influenced by the fact that to find the average of any five numbers, one simply doubles their sum and moves the decimal place one digit to the left. Even today, hand computations are not historical artifacts; many companies (particularly, small companies) still rely on their workforce to manually compute and plot control chart statistics.

But beyond the computational issues, there are several other important considerations in the determination of subgroup size. The most obvious consideration is the general availability of measurement data, which is typically dependent on the process or manufacturing output. For example, if the process generates only one or two measurements per day, it makes little sense to wait one week to form one subgroup. In such a case, it is reasonable to establish a control chart procedure based on a small subgroup size such as $n = 1$ (see Chapter 5). The principle of rational subgrouping can serve as a useful guide with respect to subgroup size and timeliness of data. As long as the time to form a subgroup does not span so wide that it is likely for a special cause to occur within the subgroup, the corresponding subgroup size is justifiable.

Occasionally, attention is paid to the size of process shifts expected to be detected in a timely manner. Intuitively, for a given process shift magnitude, larger sample sizes provide greater sensitivity to the control chart for detecting the shift. If the detection of small changes in the process is critical, then larger sample sizes need to be considered. Alternatively, if small changes are tolerable and the goal is to detect large shifts, smaller sample sizes can be employed. Sensitivity to particular shifts for a given subgroup size can be described by the operating characteristic (OC) curves and the average run length (ARL) curves. In Section 6.3, we will illustrate how an ARL-related measure can help one make decisions with respect to subgroup size and sampling frequency.

In some cases, the determination of subgroup size is influenced by sampling costs. Clearly, for example, destructive testing with large samples would not be an economically feasible combination. However, in many industrial applications, sample measurements are obtained by simple dimensional measuring devices (e.g., gage blocks, micrometers, and dial mechanisms) and via on-line automation which typically have negligible per-unit sampling costs. Note that efforts have been made to embrace the many control chart design issues within an economic framework. For example, Lorenzen and Vance (1986) and Montgomery (1980) detail economic models for control chart design that incorporate cost and time parameters such as the cost of sampling, the cost of investigating out-of-control signals, the cost of correcting special causes, the cost of missing special causes, and the average time between process changes. Unfortunately, given the general complexities of economic models coupled with the difficulties of estimating the various parameters, the implementation of economic models in practice has been limited.

The statistical properties of subgroup statistics can play a key role in the selection of sample size. In particular, a critical assumption made for the construction and interpretation of control charts for subgroup statistics such as the sample mean ($\bar{x}$) and the sample range (R) is that they are well approximated by the normal distribution. With respect to the sample mean, there is reliance on the central limit theorem mechanism to be at work even when the individual measurements are nonnormal. When the underlying individual measurements are not severely nonnormal, the central limit theorem effect on the sample means is usually sufficient for effective control charting, even for sample sizes as small as 4 or 5. Even in the case of severe underlying nonnormality, it is reassuring that in most applications sample sizes larger than 20 are rarely required.

SAMPLING FREQUENCY

A final design issue in the implementation of subgroup charting procedures is the timing between subgroups or, equivalently, the sampling frequency. Obviously, frequent large samples of a process would be advantageous for more timely detection of process changes. However, resource constraints limit such a sampling program. For a given amount of resources, sampling plans are typically conducted in one of two ways, namely, small subgroups on a frequent basis or large subgroups on a less frequent basis. As noted above, one might consider the use of economic design of control chart models to arrive at an appropriate allocation of sampling activities. Minimally, it would be advisable to consider three cost factors: (1) the variable cost of sampling, (2) the fixed cost of sampling, and (3) the cost of missing significant process changes. The variable cost of sampling is the unit cost of testing or measuring an individual item. The fixed cost of sampling includes the required expenses to conduct a test, independent of the subgroup size. Examples of fixed costs include administrative costs (e.g., paperwork or database maintenance), salaries of inspectors and technicians, and the costs associated with holding items for inspection (e.g., downtime and production loss). Probably the most important consideration is the loss due to not recognizing that the process is out of control between samples. Costs associated with this loss may be higher rework and scrap or, ultimately, customer dissatisfaction. Given the overriding concern with the issues associated with the third cost factor, it is not surprising that most organizations adopt the sampling scheme option of small subgroups but on a frequent basis.

Process stability and behavior are also determinants of the frequency of sampling. Processes which exhibit stable *and* acceptable behavior may need less surveillance. As a reminder, process stability does not imply acceptable behavior. The process may be in a state of statistical control void of special causes, but the natural variability is much greater than acceptable standards. In such a situation, frequent monitoring is important to indicate whether needed improvement efforts have been successful. Clearly, processes with erratic behavior would also require more frequent surveillance. The existence of underlying autocorrelated process behavior can have an effect on the appropriate sampling frequency. Alwan and Radson (1992a) show that subgroup means exhibit autocorrelative behavior when the underlying individual measurement series is autocorrelated. Further, they demonstrate that when the underlying process for the individual measurements is stationary, subgroups can be spaced apart to achieve near-random behavior. In general, the required gap between subgroups increases as the underlying autocorrelation increases. Given that the standard control charts rest on the basic assumption of random behavior for the subgroup means, it is essential to assess the underlying process behavior to establish an appropriate sampling frequency.

6.2   SUBGROUP CONTROL CHARTS BASED ON THE RANGE

After the sampling scheme has been designed with consideration given to the basis for rational subgrouping, subgroup size n, and sampling frequency, data should be collected to form the preliminary subgroups. In Chapters 4 and 5, we pointed out that control charts are used retrospectively ("as a judgment") and prospectively ("as an ongoing operation"). Some authors refer to retrospective and prospective control as being a two-phase approach to control chart application. *Phase 1* (or retrospective control) is the analysis of the preliminary samples to determine whether they came from an in-control process. *Phase 2* (or prospective control) deals with the monitoring of future data to determine whether the process remains in control. In our presentation of individual measurement control charts, if the process (original series or residual series) was judged in control from the preliminary data, we simply projected out the retrospective limits to serve as prospective limits. Allowing phase 1 and 2 limits to be the same is justifiable if the number of preliminary samples is sufficiently large, say, 20 to 25. When the number of preliminary samples is not sufficiently large, it is not technically appropriate to use the same set of control limits for each phase; discussion related to this matter can be found in Hillier (1969) and Yang and Hillier (1970). In light of this issue, it is recommended that the number of preliminary subgroups m be 20 or more. In our development of phase 1 limits for various subgroup statistics in this chapter, we will implicitly assume that the value of m is in accordance with this recommendation.

$\bar{x}$ AND R CHARTS

As reflected by the popular subgroup control charts presented in this chapter, the primary emphasis is the detection of change in the level and/or dispersion of the process. As such, it is an established convention to construct two control charts, one to monitor the level (typically, using the subgroup mean) and one to monitor the dispersion (typically, using the subgroup range).

As a starting point, assume that the quality characteristic, denoted by X, is a random variable with mean μ and standard deviation σ. Let now $\bar{X}$ represent the subgroup mean of n observations taken on X. As we showed in Chapter 2, $\bar{X}$ has mean μ and standard deviation $\sigma_{\bar{x}} = \sigma/\sqrt{n}$. Since variations of quality measurements often follow approximately a normal distribution, and since the central limit theorem is at work for subgroup means, even when the measurements are nonnormal, it is assumed that the distribution of subgroup means is well approximated by the normal distribution. Accordingly, the 3σ limits for the subgroup means are given by

$$\text{UCL} = \mu + 3\frac{\sigma}{\sqrt{n}}$$

$$\text{LCL} = \mu - 3\frac{\sigma}{\sqrt{n}}$$

(6.1)

In practice, μ and σ (or $\sigma_{\bar{x}}$) are usually not known, and must be estimated. Denoting the estimates for μ and $\sigma_{\bar{x}}$ as $\hat{\mu}$ and $\hat{\sigma}_{\bar{X}}$, respectively, we see that the control limits of (6.1) become

$$\text{UCL} = \hat{\mu} + 3\hat{\sigma}_{\bar{X}}$$

$$\text{LCL} = \hat{\mu} - 3\hat{\sigma}_{\bar{X}}$$

(6.2)

The most obvious estimate for μ is simply the average of all the observed individual measurements of X. Suppose we let $\bar{x}_1, \bar{x}_2, \ldots, \bar{x}_m$ represent the observed subgroup means for the m preliminary samples. Assuming that the sample sizes used to compute the subgroup means are equal, the average of all individual measurements is equal to the average of the subgroups mean, that is,

$$\bar{\bar{x}} = \frac{\sum\limits_{i=1}^{m} \bar{x}_i}{m}$$

(6.3)

This overall average $\bar{\bar{x}}$ is referred to as the *grand mean* or the *grand average*. In Section 6.4, we will deal with the special case when the subgroup sizes vary.

From (6.2), we now need $\hat{\sigma}_{\bar{X}}$, that is, an estimate of the subgroup standard deviation. There are various possible choices for $\hat{\sigma}_{\bar{X}}$. One possibility is based on the within-sample standard deviation s as an estimate of σ. An alternative procedure, more common in practice, is to base the estimate on the mean of the subgroup ranges R. For now, we will focus on the range method and will discuss the sample standard deviation approach in a subsequent section.

The sample range is defined as the difference between the maximum and minimum observed values in a set of values. If x_{ij} represents the jth individual measurement from the ith preliminary subgroup where $i = 1, 2, \ldots, m$ and $j = 1, 2, \ldots, n$, then the sample range for the ith subgroup is given by

$$R_i = \max_j x_{ij} - \min_j x_{ij}$$

(6.4)

It will prove useful to compute the average range for the m preliminary subgroup ranges:

$$\bar{R} = \frac{\sum\limits_{i=1}^{m} R_i}{m}$$

(6.5)

Based on statistical theory, it can be shown that if we multiply $\bar{R}$ by a factor, labeled A_2, which is a function of the subgroup size n, we then have a reasonable estimate of the term $3\sigma/\sqrt{n}$ found in (6.1); the reader is referred to the end-of-chapter appendix for more technical details. With this in mind, we now have the control limits for the subgroup means:

$$\text{UCL} = \bar{\bar{x}} + A_2 \bar{R}$$

$$\text{CL} = \bar{\bar{x}} \tag{6.6}$$

$$\text{LCL} = \bar{\bar{x}} - A_2 \bar{R}$$

where A_2 is tabulated for various subgroup sizes in Appendix Table V. The above control limits form the basis of what is called the $\bar{x}$ chart (read "x-bar chart"). This $\bar{x}$ chart may be the most well known of all control charts. As a result, many people equate the name *Shewhart control chart* with this chart; we, however, prefer to think of any 3σ-based control chart as a Shewhart control chart.

The focus of the $\bar{x}$ chart is on detection of changes in the process level. To monitor for changes in process variability, a reasonable strategy is to develop a control chart for the plotting of the subgroup ranges. Such a control chart is called an *R chart*. The centerline and the control limits of the *R* chart are given as follows:

$$\text{UCL} = D_4 \bar{R}$$

$$\text{CL} = \bar{R} \tag{6.7}$$

$$\text{LCL} = D_3 \bar{R}$$

where D_3 and D_4 are tabulated for various subgroup sizes in Appendix Table V. Based on their appearance, the control limits of (6.7) do not appear to be Shewhart-type charts in the sense of placing limits plus and minus a certain amount around the centerline. However, embedded in the definitions of D_3 and D_4 is a 3σ structure for the range statistics; the reader is referred to the end-of-chapter appendix for more technical details.

Let us now proceed to an example. However, before we do so, it is important to emphasize an implementation issue with respect to the $\bar{x}$ chart and the *R* chart. At first glance, it would seem that the two control charts might be constructed simultaneously. However, there is a subtlety that makes it advisable to construct the control charts *sequentially*. Namely, control limits for the $\bar{x}$ chart *depend* on a reliable estimate of the process variation given by $\bar{R}$. If the process variability is affected by special causes, then $\bar{R}$ will be "biased," rendering the $\bar{x}$ chart limits less meaningful. As such, we should always study the ranges first, to ensure process variability is in control. If so, then we proceed to the study of the subgroup means.

| | **Example 6.1** |

This subgrouping application is concerned with the computerized tomography (CT) X-ray scanners produced by a major medical equipment manufacturer. Computerized tomography is a sophisticated medical imaging technique that is widely used for emergency trauma care, detailed body studies, and three-dimensional imaging. The CT system consists primarily of an X-ray generation subsystem, an X-ray detector, and an image reconstruction computer. At the

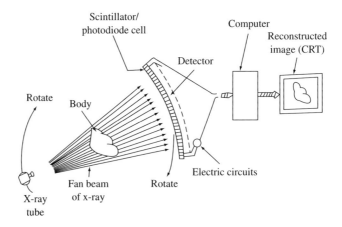

Figure 6.1   Schematic diagram of rotating CT scanner.

heart of the detector are 912 ceramiclike scintillator elements (cells) arranged in an arc to allow for rotational scanning. A schematic diagram of a CT scanner is provided in Figure 6.1. The scintillator elements emit a visible light proportional to the amount of X-rays absorbed in each element. This light is converted to electric signals, nearly 1 million signals per second, and then reconstructed into an image.

One critical property of a CT scintillator element is radiation damage, which can be considered as the loss of light output caused by exposure to a given dose of X-rays. Radiation damage is measured by comparing the response of a scintillator element when exposed to a very large dose of X-rays to a normal level of X-rays. An ideal device would have no change in the response, reflecting robustness to high levels of radiation dosage. Since this is difficult to achieve, levels for the acceptable maximum and cell-to-cell variation of the radiation damage have been determined. By using control charts, ongoing monitoring of this quality characteristic is performed to ensure process stability.

The sampling frequency is based on the production of detectors, which averages about 15 new detectors per week. Associating subgroups with newly produced detectors provides a natural basis for rational subgrouping. Since measuring all 912 scintillator elements on a given detector would take nearly 30 h of testing time, an alternative is to monitor a random sample of scintillators. Given that the cells are uniquely distinguished by location, random sampling can easily be accomplished by using a random number generator. To establish control chart procedures, samples of size 5 for 25 detectors were taken. In terms of the established notation, we have $n = 5$ and $m = 25$. Data of these 25 subgroups along with subgroup statistics are shown in Table 6.1. The individual measurement data can be found in the file "x-ray.dat". The measurements are unitless and represent the ratio of the response of a cell to normal levels of X-rays divided by the response of the same cell to a large dose of X-rays. For ease of operator transcribing and computation, the measurements are "normalized" to 1000 units; note that, in this measurement, detection of unusual observations being either significantly high or low is equally important.

At this stage, our initial step is to study the variability of the detector process. To do so, we consider setting up an R chart. Using the data from Table 6.1, we find the average range to be

$$\bar{R} = \frac{\sum\limits_{i=1}^{25} R_i}{25} = 1.775$$

Table 6.1   X-ray radiation data with subgroup means and ranges

Subgroup Number	Observations					$\bar{x}_i$	R_i
1	998.505	1000.327	999.793	1000.446	1001.268	1000.068	2.763
2	999.986	999.983	999.307	998.721	999.694	999.538	1.265
3	998.720	999.351	999.490	999.636	999.583	999.356	0.916
4	1000.755	999.601	999.229	1000.782	999.513	999.976	1.553
5	1000.381	1000.730	1000.179	1000.363	1000.661	1000.463	0.551
6	999.433	1000.313	999.676	1000.921	1000.410	1000.151	1.488
7	999.550	999.818	1000.215	1001.734	1000.353	1000.334	2.184
8	1001.459	1000.181	1000.310	999.958	999.517	1000.285	1.942
9	999.859	999.653	998.955	999.975	1000.050	999.698	1.095
10	1000.831	999.259	1000.107	999.667	1000.131	999.999	1.572
11	1000.540	1001.213	999.839	998.468	999.215	999.855	2.745
12	999.212	1000.594	1001.613	1000.957	999.893	1000.454	2.401
13	1000.337	1000.103	999.402	1001.348	1000.945	1000.427	1.946
14	999.872	1000.429	1000.409	1000.034	999.309	1000.011	1.120
15	998.773	1000.128	1000.794	1000.363	1000.103	1000.032	2.021
16	1000.043	999.615	999.856	1000.018	1000.509	1000.008	0.894
17	999.538	999.278	1000.810	1001.066	999.439	1000.026	1.788
18	1000.243	998.696	1000.048	1000.070	999.768	999.765	1.547
19	999.929	1000.708	1000.024	999.619	1000.787	1000.213	1.168
20	998.431	1000.260	1000.007	1000.376	1000.448	999.904	2.017
21	999.112	1000.767	1001.017	1000.262	999.541	1000.140	1.905
22	1003.290	1001.916	999.170	999.823	999.080	1000.656	4.210
23	1000.290	1001.004	999.008	999.435	998.937	999.735	2.067
24	1000.370	998.810	1000.610	999.962	1000.102	999.971	1.800
25	1000.673	1000.136	999.571	999.263	1000.445	1000.018	1.410

From Appendix Table V, for subgroup size $n = 5$, the values of D_3 and D_4 are 0 and 2.114, respectively. Using (6.7), we compute the control limits for the R chart as

$$\text{UCL} = D_4\bar{R} = 2.114(1.775) = 3.752$$

$$\text{LCL} = D_3\bar{R} = 0(1.775) = 0$$

The R chart is given in Figure 6.2. On the R chart, subgroup number 22 plots above the upper control limit, and thus, it is deemed an out-of-control point. Had we constructed an $\bar{x}$ chart at this time, the 22nd subgroup would not be viewed as out of the ordinary.

In general, it is not unusual for the R chart to signal while the $\bar{x}$ chart does not, or vice versa. Each chart is looking for different types of violations. If the dispersion of the individual measurements significantly increases around a stable process mean, the R chart, not the $\bar{x}$ chart, will probably pick up this departure. In contrast, it is possible for the dispersion to remain in control and the process mean to shift; this would most likely be seen on the $\bar{x}$ chart

Stat>>Control Charts>>R

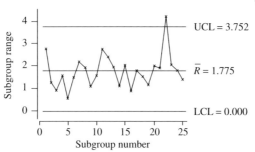

Figure 6.2   Preliminary *R* chart for radiation data.

rather than the *R* chart. There are, of course, situations in which both the process mean and dispersion go out of control together, resulting in signals on both charts.

With respect to the radiation example, an initial inference for the large range value might be that there has been a significant increase in process variability. Under the scenario of increased variability within the subgroup, one expects some observations to deviate farther above the process mean while others are farther below the process mean. A review of the within-subgroup observations (see Table 6.1), however, suggests another explanation; namely, that the large range value appears to be due to an isolated outlier rather than an increased scattering of within-subgroup observations. As a follow-up, a process investigation was then conducted to determine the root cause. Through an analytical study, it was found that a contaminant material was present which can cause a greater risk for radiation damage. The source of the contaminant was such that only a small number of scintillator elements were affected; thus the problem might have gone unnoticed had we ignored the range chart and gone directly to the mean chart. Further investigation brought to light an erratic practice of furnace cleaning which allowed unacceptable levels of the contaminant to enter the production process. Consequently, procedures for regular preventive maintenance were developed, and personnel were provided with the necessary training.

Since a special cause was discovered, the associated range value should be set aside and new retrospective control limits computed. Even though the corresponding subgroup mean is not viewed as an out-of-control signal, common convention is to discard the subgroup outright. Similarly, a subgroup with a special-cause signal on the $\bar{x}$ chart, but not on the *R* chart, is excluded from computations. The inclusion of a subgroup for one chart, but not for the other, tends to be more confusing than helpful. By deleting the 22d subgroup, the revised centerline, based on the remaining 24 subgroups, is

$$\bar{R} = \frac{40.158}{24} = 1.673$$

Accordingly, the revised control limits are

$$\text{UCL} = D_4\bar{R} = 2.114(1.673) = 3.537$$

$$\text{LCL} = D_3\bar{R} = 0(1.673) = 0$$

Now, all the subgroup ranges are found to be in control (see Figure 6.3). If no further subgroups are discarded based on the retrospective analysis of the subgroup means, the above revised control limits serve as the prospective control limits for future on-line process monitoring of future subgroup ranges.

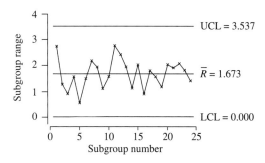

Figure 6.3   Revised R chart for radiation data.

We now turn to the construction of the $\bar{x}$ chart limits. As noted, the control limits are constructed by using all the remaining subgroups except for subgroup number 22. The centerline is computed as

$$\bar{x} = \frac{\sum_{i \neq 22} \bar{x}_i}{24} = \frac{24{,}000.378}{24} = 1000.016$$

The appropriate value of $\bar{R}$ is associated with the revised R chart given in Figure 6.3. To find the $\bar{x}$ chart limits, we use $A_2 = 0.577$ from Appendix Table V for $n = 5$ and Equation (6.6) to find

$$\text{UCL} = \bar{x} + A_2\bar{R} = 1000.016 + 0.577(1.673) = 1000.981$$

$$\text{LCL} = \bar{x} - A_2\bar{R} = 1000.016 - 0.577(1.673) = 999.051$$

Superimposing these computed limits on a sequence plot of the subgroup means gives the $\bar{x}$ chart shown in Figure 6.4. The control chart shows that the subgroup means vary well within the limits with no indications of special causes. Additionally, supplementary rules introduced in Section 4.9 can be implemented. If this were done, no abnormal patterns would be detected. In conclusion, the subgroup means are in a state of statistical control, and the retrospective limits can be projected out for prospective control.

REVISION OF RETROSPECTIVE CONTROL LIMITS AND CONTROL CHART MAINTENANCE

In the previous example, we found an outlying range value which led to the elimination of the subgroup because a special-cause explanation was found. In general, if some subgroup is found to be out of control—range or mean—with special-cause explanation, the convention is to discard the subgroup from both the range and the mean computations. The user then needs to recompute the R chart limits and confirm that the subgroup ranges are still within limits. Thereafter, the $\bar{x}$ chart limits are revised based on the new $\bar{R}$ value, and the remaining subgroup means are assessed for control. Theoretically, one can repeat this iteration several

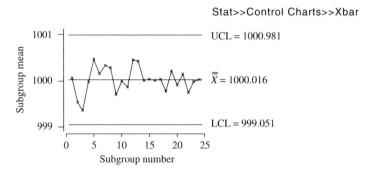

Figure 6.4 x Chart for radiation data.

more times. However, for generally well-behaved data, more than one or two iterations is uncommon. If many iterations are required, it may be that the control limits are not suitable guides because the underlying assumptions are not being appropriately met. In such cases, efforts would be better spent understanding the nature of the process and its incompatibility with standard control charts rather than myopically attempting to seek an ad hoc explanation for each isolated subgroup.

If one is satisfied with both the computed $\bar{x}$ chart limits and R chart limits, then these two sets of limits are used for prospective (i.e., on-line) control of the process for any future changes in the process away from an in-control state. There is always a question of how long prospective limits should be left to use. There is no absolute prescribed answer. There are, however, certain scenarios that "require" the revision of control limits. For example, if there is a change in the subgroup size, this will affect the proper placement of the control limits for both ranges and subgroups. This situation could occur in practice, if it was decided to take smaller samples more frequently, so as to detect a particular type of shift more quickly without increasing the total number of units sampled per day. Or, it may be that smaller samples are taken less frequently because the process has exhibited good stability for an extended period and there is less need to allocate as many resources to the monitoring of the process.

A change in process performance can be another reason for the modification of control limits. For example, if there was a successful effort to reduce process variability, control limits accordingly need to be revised to reflect the new process realities. Beyond these particular scenarios (sample size or process change), it is good practice, as a general rule, to revise centerlines and control limits on a periodic basis. Depending on the application and frequency of sampling, many practitioners establish regular periods for review and revision of control limits, for example, every week, every month, or every 50 or 100 samples. Another common practice is to establish new control limits with each new production run.

Practical Issues Concerning the Range Statistic

As noted earlier, the sample range is widely used in practice in place of the sample standard deviation s to estimate the true process standard deviation. This practice reflects the fact that statistical quality control methods were developed long before the development of the electronic computer or programmable calculators. It is much easier to compute a range by hand than to compute a sample standard deviation. There are a couple of precautions that should be considered in the implementation of the range method. First, it is advisable to restrict the use of range-based methods to small subgroups of size, say, 10 or less. This recommendation is primarily due to the fact that the efficiency of the range-based estimator for σ relative to a standard deviation–based estimator decreases as n increases. Remember, the range statistic is based on only two points—the maximum and the minimum. Thus, all the sample information between these points is ignored. A detailed comparison of the statistical efficiency of the sample standard deviation method relative to the sample range method can be found in Tuprah and Ncube (1987).

OBTAINING AN ESTIMATE FOR σ

We may wish to obtain an estimate for the process standard deviation σ, that is, the standard deviation of individual observations generated by the process. By having an estimate of σ, we can assess the capability of the process in terms of conformance of the individual units produced from the process relative to specification limits. In Chapter 9, we will discuss in detail the issues and computations associated with process capability analysis.

To obtain an estimate for σ based on subgroup range information, we recall that $A_2\bar{R}$ serves as an estimate of $3\sigma/\sqrt{n}$. By simple rearrangement, we obtain the following estimate for σ:

$$\hat{\sigma} = \frac{A_2\bar{R}\sqrt{n}}{3} \tag{6.8}$$

If we define

$$d_2 = \frac{3}{A_2\sqrt{n}}$$

then (6.8) reduces to

$$\hat{\sigma} = \frac{\bar{R}}{d_2} \tag{6.9}$$

Values of d_2 for various subgroup sizes are given in Appendix Table V.

To illustrate the estimation of σ, consider the X-ray data set analyzed in Example 6.1. After elimination of a subgroup due to special-cause influences, we found that $\bar{R} = 1.673$. For a subgroup size of $n = 5$, we find from Appendix Table V that $d_2 = 2.326$. By using these values in (6.9), the estimated process standard deviation is

$$\hat{\sigma} = \frac{1.673}{2.326} = 0.7193$$

Example 6.2

In general, with an estimate for the true process standard deviation of the individual observations, we can estimate the capability of the process in relationship to specifications limits. In Chapter 9, we will explore a variety of measures associated with process capability analysis. For now, we will illustrate, with the next example, the computation of the most common measure of process capability, namely, the proportion of individual units that are expected to meet specifications. Or, we may wish to report the proportion of individual units that are not expected to meet specifications.

Example 6.3

Suppose that the specification limits for the individual scintillators, described in Example 6.1, are 1000 ± 3. From the $\bar{x}$ chart (see Figure 6.4), we can estimate the true mean of the radiation characteristic as $\bar{\bar{x}} = 1000.016$. From the previous example, we found an estimate for σ to be 0.7193. Assuming that the radiation characteristic is normally distributed, with mean 1000.016 and standard deviation 0.7193, we can estimate the proportion of nonconforming scintillators as

$$\hat{p} = P(X < 997) + P(X > 1003)$$

$$= P\left(\frac{X - 1000.016}{0.7193} < \frac{997 - 1000.016}{0.7193}\right) + P\left(\frac{X - 1000.016}{0.7193} > \frac{1003 - 1000.016}{0.7193}\right)$$

$$= P(Z < -4.193) + P(Z > 4.148)$$

$$= 0.0000138 + 0.0000168$$

$$= 0.00003$$

Thus, we estimate about 0.003 percent (30 parts per million) of the scintillators produced will be outside specification limits.

PROBABILITY LIMITS FOR THE $\bar{x}$ CHART

Underlying the development of the $\bar{x}$ chart factor A_2 is the use of a multiple 3 so as to create a 3σ-based control chart. However, as mentioned in Chapter 4, one can define control limits by specifying the false-alarm rate α for the charting procedure. Recall that such limits were called probability limits.

Assuming that subgroup means are approximately normally distributed, we can obtain probability limits for the $\bar{x}$ chart by using $z_{\alpha/2}$ as the multiple in the design of the limits, where $z_{\alpha/2}$ is the upper $\alpha/2$ percentage point of the standard normal distribution. From (6.1) and (6.2), we see that we need an estimate of the standard deviation of $\bar{X}$, or $\sigma/\sqrt{n}$. In the previous subsection, we learned that $\bar{R}/d_2$ serves as an estimate of σ. Thus, an estimate for $\sigma/\sqrt{n}$ is given by

$$\hat{\sigma}_{\bar{X}} = \frac{\bar{R}/d_2}{\sqrt{n}} = \frac{\bar{R}}{d_2\sqrt{n}} \tag{6.10}$$

Using the above estimate with a $z_{\alpha/2}$ multiple, we can see that the estimated probability limits for the $\bar{x}$ chart are given by

$$\text{UCL} = \bar{\bar{x}} + z_{\alpha/2}\left(\frac{\bar{R}}{d_2\sqrt{n}}\right)$$

$$\text{LCL} = \bar{\bar{x}} - z_{\alpha/2}\left(\frac{\bar{R}}{d_2\sqrt{n}}\right)$$

(6.11)

Example 6.4

Consider again the X-ray data set analyzed in Example 6.1. Suppose we wish to construct $\bar{x}$ chart limits based on $\alpha = 0.002$. In this case, we need $z_{0.002/2} = z_{0.001}$. Using either the normal table in the Appendix or Minitab, we find $z_{0.001} = 3.09$. In our previous analysis, we found that the revised $\bar{R} = 1.673$. For a subgroup size of $n = 5$, we find from Appendix Table V that $d_2 = 2.326$. By using these values in (6.11), the estimated probability limits are

$$\text{UCL} = 1000.016 + 3.09\left(\frac{1.673}{2.326\sqrt{5}}\right) = 1001.010$$

$$\text{LCL} = 1000.016 - 3.09\left(\frac{1.673}{2.326\sqrt{5}}\right) = 999.022$$

PROBABILITY LIMITS FOR THE R CHART

As with the $\bar{x}$ chart, the underlying development of the R chart is based on placing control limit at ± 3 standard deviations of the subgroup statistic around its mean level. Even if the underlying population for the individual measurements is normally distributed, it can be shown that the distribution of the ranges is theoretically not normal but rather is asymmetric or positively skewed.

Recall from Chapter 5, we studied the moving-range statistic which is simply a special case of the general range statistic for $n = 2$. It was pointed out that the distribution for the moving-range statistic is highly positively skewed. As a result, the standard moving-range chart's lower control limit, which is based on the symmetric normal distribution, falls unrealistically in a negative region, and thus it is assigned the value of 0. For the same reasons, the lower control limit for the general range statistic is fixed at 0 for all $n \leq 6$. This can be confirmed by noticing that lower control limit factor D_3 is 0 for this range of subgroup sizes (see Appendix Table V). Thus, for these subgroup sizes, the range chart is hampered by its inability to detect variance reductions. Another problem with the application of the symmetric 3σ framework to an inherently skewed population is that the false-alarm rate is not equal to the standard value of 0.0027. For subgroup sizes in the range of 2 to 6, the false-alarm rate for the R chart is typically about 2 or 3 times higher than the standard value of 0.0027. Generally, the severity of skewness in the distribution of the sample range and the false-alarm rate both increase as the sample size decreases.

In one attempt to resolve the issues of a lower control limit set at zero and the inflated false-alarm rate for the moving-range chart, control limits were derived based on the true underlying distribution for the moving-range statistic, namely, the half-normal distribution (see Section 5.7). This approach allows significantly small moving-range values to be revealed below a lower control limit. Furthermore, it enables the user to control the false-alarm rate α to a predetermined level. Since the moving-range statistic is really the range statistic for $n = 2$, the half-normal-based control limits, presented in Section 5.7, directly apply to the range statistic for subgroups of size 2.

By using the results of Harter (1960), it is also possible to construct the probability limits for the R chart for any subgroup size. The estimated probability limits are

$$\text{UCL} = D_{1-\alpha/2}\left(\frac{\bar{R}}{d_2}\right)$$

$$\text{LCL} = D_{\alpha/2}\left(\frac{\bar{R}}{d_2}\right)$$

(6.12)

The pairs of values $(D_{0.025}, D_{0.975})$, $(D_{0.005}, D_{0.995})$, and $(D_{0.001}, D_{0.999})$ for $3 \leq n \leq 10$ are given in Appendix Table VI. Notice that there are no entries in Appendix Table VI relating to $\alpha = 0.0027$, that is, $D_{0.99865}$ and $D_{0.00135}$. The multipliers closest to the 3-σ false-alarm rate are $D_{0.999}$ and $D_{0.001}$ which correspond to $\alpha = 0.002$. The restriction is due to the fact that the table is based on the percentile points originally tabulated by Harter (1960).

For $n = 2$, $D_{1-\alpha/2} = \sqrt{2}z_{\alpha/4}$ and $D_{\alpha/2} = \sqrt{2}z_{1/2-\alpha/4}$, where $z_{\alpha/4}$ and $z_{1/2-\alpha/4}$ are, respectively, the $100(1 - \alpha/4)$ and $100(\frac{1}{2} + \alpha/4)$ percentiles of the standard normal distribution. In this case, given that the $D_{1-\alpha/2}$ and $D_{\alpha/2}$ multipliers can be related to the universally known standard normal distribution, a control chart user has the flexibility to choose practically any value for α.

Example 6.5	In Example 6.4, we constructed probability limits for the $\bar{x}$ chart based on $\alpha = 0.002$. Suppose we wish to construct probability limits for the R chart based on the same specified false-alarm rate. As a comparison with the standard R chart constructed in Example 6.1, let us compute the probability limits from the entire set of 25 subgroups.

By choosing $\alpha = 0.002$, we need to find $D_{0.001}$ and $D_{0.999}$ which from Appendix Table VI are 0.37 and 5.48, respectively. From Example 6.1, the average range based on all 25 subgroups was found to be 1.775. Using (6.12), we see that the estimated probability limits for the original 25 ranges are

$$\text{UCL} = D_{0.999}\left(\frac{\bar{R}}{d_2}\right) = 5.48\left(\frac{1.775}{2.326}\right) = 4.182$$

$$\text{LCL} = D_{0.001}\left(\frac{\bar{R}}{d_2}\right) = 0.37\left(\frac{1.775}{2.326}\right) = 0.282$$

As can be seen, this approach does indeed produce a nonzero lower control limit, allowing significant decreases in variability to be singled out. Furthermore, notice that even with a slightly higher upper control limit threshold, the range value of 4.21 for the 22nd subgroup (see Table 6.1) is still highlighted as an out-of-control point. Again, the subgroup should be eliminated and the above limits recomputed with $\bar{R} = 1.673$.

6.3   PERFORMANCE MEASURES OF $\bar{x}$ AND R CHARTS

In this section, we consider the ability of $\bar{x}$ and R charts to detect shifts in the process away from an in-control state. There are two basic ways to summarize the relative performance of control charts in their detection of process change. Namely, we can construct either operating-characteristic (OC) curves or average run length (ARL). In the subsections to follow, we provide some details on the construction and utilization of these curves.

OPERATING-CHARACTERISTIC CURVE

An OC curve, for instance, is simply a plot of the probability that a single sample will signal out of control against the true value of the process parameter. By this criterion, if two charts have the same probability of signaling when the process is in control, then the chart that has the higher probability of detecting a change in the mean or dispersion parameter is deemed a better chart. In addition to using OC curves for comparing different control charts, the concept of the OC curve was developed to help determine an appropriate subgroup size for a predetermined control chart procedure. By constructing OC curves as a function of process parameters and subgroup size, an analyst can select the minimum required subgroup size to detect, with high probability, process changes—small or large—which are considered important. The standard derivation of OC curves is based on the restrictive assumption that the in-control process is an independently identically distributed normal (iidn) process.

To illustrate the construction of an OC curve, suppose X is a random variable with known in-control parameter values of μ_0 and σ. Suppose further that an $\bar{x}$ chart with subgroup size n is used to monitor the process. Hence, the 3σ control limits are given by

$$\text{UCL} = \mu_0 + 3\left(\frac{\sigma}{\sqrt{n}}\right)$$

$$\text{LCL} = \mu_0 - 3\left(\frac{\sigma}{\sqrt{n}}\right)$$

(6.13)

Now, assume that the process mean shifts from μ_0 to a mean value of μ_1. It is customary to express the increase in the mean value as a multiple k of the process standard deviation σ, that is, $\mu_1 = \mu_0 + k\sigma$. In terms of the subgroup statistic $\bar{X}$, the following are implied:

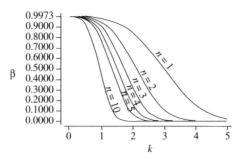

Figure 6.5   OC curves for $\bar{x}$ chart with 3σ limits.

$$\text{In-control state} \qquad \bar{X} \sim N(\mu_0, \sigma^2/n)$$

$$\text{Out-of-control state} \qquad \bar{X} \sim N(\mu_1, \sigma^2/n)$$

As explained in Chapter 4, there is a probability, called the β risk, that a given sample will not signal out of control when the process mean has indeed shifted to μ_1 and is given by

$$\beta = P(\text{LCL} \leq \bar{X} \leq \text{UCL}|\mu = \mu_1)$$

which, in relation to the normal distribution, is

$$\beta = P\left(Z \leq \frac{\text{UCL} - \mu_1}{\sigma / \sqrt{n}}\right) - P\left(Z \leq \frac{\text{LCL} - \mu_1}{\sigma / \sqrt{n}}\right)$$

where Z is the standard normal random variable. After substituting the limits of (6.13) into the above expression and then performing algebraic simplifications, we obtain

$$\beta = P(Z \leq 3 - k\sqrt{n}) - P(Z \leq -3 - k\sqrt{n}) \qquad \textbf{(6.14)}$$

where $k = (\mu_1 - \mu_0) / \sigma$. Since the β risk is the probability that a given subgroup will not signal when a change has occurred, $1 - \beta$ is the probability that a given subgroup will signal that a change has occurred. Recall from Section 4.7, this probability of detection is referred to as the *power*.

For a given subgroup size n, an OC curve is constructed by plotting the β risk or the power versus the shift multiple k. The OC curves for the $\bar{x}$ chart for some typical subgroup sizes are shown in Figure 6.5. (*Note:* The individual measurement control chart is associated with $n = 1$.) The curves reflect the anticipated result that the detection sensitivity increases with the subgroup size. The OC curves can serve as a guide for determining the subgroup size. Suppose, for example, that it is desired to detect a 2σ shift in the mean, with probability 0.90, with a single subgroup. This requirement implies that β is at most 0.10 for $k = 2$. From Figure 6.5, it can be found that a subgroup size of at least $n = 5$ is required.

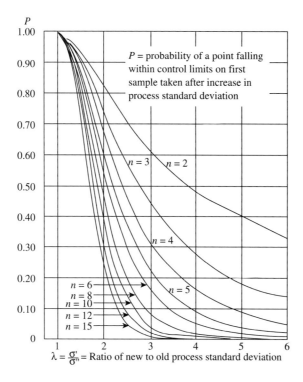

Figure 6.6   OC curves for R chart with 3σ limits. (Adapted from A. J. Duncan, *Quality Control and Industrial Statistics*, 5th ed., Irwin, Homewood, Il, 1986.)

The OC curves can also be constructed for the R chart. The curves would measure the ability of the R chart to detect a shift in the process standard deviation from the in-control value of σ_0 to some other value σ_1. It is an established convention to plot the OC curve against the ratio $\lambda = \sigma_1 / \sigma_0$. To derive OC curves for the R chart, one needs to obtain cumulative probabilities from the distribution of a random variable, known as the *relative range*, defined as $W = R/\sigma$; this random variable is discussed in the end-of-chapter appendix which is devoted to derivation of subgroup control chart factors. For the curious, more details on the required computations can be found in Duncan (1951) and Scheffé (1949). Based on their work, an OC curve for the R chart is presented in Figure 6.6 for various subgroup sizes. Probably the most important message from the graph is that a single subgroup range is very unlikely to fall outside the R chart limits for small to moderate shifts, especially with typical subgroups sizes ($2 \leq n \leq 5$). Thus, greater emphasis should be placed on analyzing the range data for any patterns within the limits that are indicative of change, for example, a drift in the ranges toward a control limit. To this end, supplementary runs rules can be helpful to more quickly detect variance shifts.

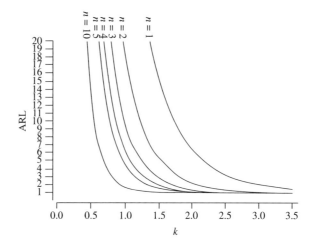

Figure 6.7   ARL curves for x̄ chart wih 3σ limits.

AVERAGE RUN LENGTH FOR THE $\bar{x}$ CHART

First introduced in Section 4.7, another measure of control chart performance is the average run length. The ARL is defined as the expected number of subgroups until an out-of-control signal is given. As noted in Section 4.7, the in-control ARL is given by

$$\mathrm{ARL}_0 = \frac{1}{\alpha} \tag{6.15}$$

while the out-of-control ARL is given by

$$\mathrm{ARL}_1 = \frac{1}{1-\beta} \tag{6.16}$$

In the previous subsection on OC curves, we showed with (6.14) how β can be computed for the $\bar{x}$ chart when the in-control mean has shifted by $k\sigma$. ARL curves for the various subgroup sizes are presented in Figure 6.7. Even though the values in Figure 6.7 are one-to-one functions of the values in Figure 6.5, the ARL curves give a perspective that is not obvious from the OC curves. For instance, if there is a 1σ shift in the mean, the OC curve indicates that for subgroups of size 5, the probability the shift will be detected at the time of the shift is just 0.2225. Although this probability seems low for a single subgroup, the $n = 5$ ARL curve reveals that only four or five subgroups are expected before the shift is detected. If the process is sampled frequently, we can be reassured that the shift will be detected in quick order. However, if subgroups are sampled hours or days apart, the implication of waiting 5 h (or days) is far from timely. Actually, when subgroups are sampled infrequently, the ARL measure may be meaningless since most processes, particularly manufacturing processes, do not remain permanently shifted for 4 or 5 days.

Our discussion of sampling frequency naturally leads us to express performance in terms of the *average time to signal (ATS)*. If subgroups are gathered at fixed intervals of time, say h time units apart, then

$$ATS = h \times ARL \qquad \text{(6.17)}$$

To illustrate the usefulness of the ATS measure, we turn to a simple example.

Example 6.6

Suppose measurements from an in-control manufacturing process have a known mean and standard deviation of 100 and 5, respectively. Given the nature of the application, production personnel have determined a mean shift to 107.5 is not acceptable and must be quickly detected for corrective action. In terms of standard deviations, a mean level of 107.5 is 1.5 standard deviations ($k = 1.5$) away from the in-control mean of 100. Currently, the process is monitored with an $\bar{x}$ chart based on subgroups of size $n = 4$ collected every hour. To begin the analysis of this situation, we compute the β risk for the shift size of concern. Using (6.14), we find

$$\beta = P(Z \le 3 - 1.5\sqrt{4}) - P(Z \le -3 - 1.5\sqrt{4})$$
$$= P(Z \le 0) - P(Z \le -5)$$

Given the mean of the standard normal distribution is zero, we know that $P(Z \le 0)$ is 0.50. The second cumulative probability is essentially equal to 0. Hence, β is equal to 0.50. Therefore, from (6.16), the out-of-control ARL is

$$ARL_1 = \frac{1}{1 - 0.50} = 2$$

Since a subgroup is taken every hour ($h = 1$), the average time to signal is

$$ATS = 1(2) = 2\,h$$

Suppose this average time is considered unacceptably long. The question is, How can we reduce the average time needed to detect the out-of-control condition? There are a few alternatives. One is to consider an alternative charting procedure that has greater power of detecting shifts of size 1.5 standard deviations; for example, consideration might be given to the non-Shewhart schemes outlined in Chapter 8. If, however, personnel wish to continue using an $\bar{x}$ chart, the second alternative is to sample more frequently. For example, if subgroups are gathered every 0.5 h, the average time to signal is ATS $= (\frac{1}{2})2 = 1$ h. A third possibility is to increase the subgroup size which will accordingly increase the power of detection. For example, if we use $n = 10$, then we find $\beta = 0.0406$ which implies

$$ARL_1 = \frac{1}{1 - 0.0406} - 1.04$$

So, if samples are taken every hour, the average time to signal is 1(1.04) = 1.04 h. Thus, for this example, a control chart design based on the larger subgroups ($n = 10$) gathered every hour has about the same performance, in terms of average time to detect the shift, as a control chart design based on smaller subgroups ($n = 4$) taken every 0.5 h.

We close our discussion of the ARL measure by noting that ARL calculations can be extended to evaluate the use of almost any control criteria beyond just the use of control limits on the $\bar{x}$ and R charts. For instance, Champ and Woodall (1987) demonstrate a procedure for determining the ARL for the $\bar{x}$ chart supplemented with any combination of runs rules. As noted in Section 4.9, the notion of a false-alarm rate β is not clearly defined when one is studying control criteria such as runs rules. Since an OC curve is a direct function of β, this implies that an OC curve may not be meaningful in the interpretation of runs rules. In such cases, it is best to evaluate the control criteria using the ARL measure. As will be noted in Chapter 8, the ARL concept also plays a key role in the comparison of the Shewhart method to other control charts, such as CUSUM and EWMA charts.

6.4   SUBGROUP CONTROL CHARTS BASED ON THE SAMPLE STANDARD DEVIATION

As an alternative to the range method, the within-sample standard deviation estimator S can be used to estimate σ. As mentioned earlier, the range statistic is predominantly favored over S because the required calculations are more easily done by hand. Given the advent of inexpensive, small, high-speed computers, the advantage of computational simplicity in the use of the range over the standard deviation is less compelling now. The primary advantage of the sample standard deviation method over the range method is statistical efficiency. With small sample sizes, little information is lost by use of the sample range rather than the sample standard deviation. However, for sample sizes greater than 10, it is generally recommended that the sample standard deviation be utilized because the range method has considerably less statistical efficiency.

$\bar{x}$ AND S CHARTS WITH CONSTANT SUBGROUP SIZE

As was the case with the range method, when one is using the subgroup standard deviations, there is a need to develop an estimate for σ. Suppose that m preliminary subgroups, each of size n, are available, and the sample standard deviations for these subgroups are computed and are denoted by $s_1, s_2, \ldots, s_m$. Consider the simple average of these standard deviations, that is,

$$\bar{s} = \frac{\sum_{i=1}^{m} s_i}{m} \tag{6.18}$$

It can be shown that $\bar{s}/c_4$ is an unbiased estimate of σ where c_4 is a function of subgroup size and is tabulated for various sample sizes in Appendix Table V. Recall that a variant of this estimator was presented in Chapter 5 for estimating the standard deviation of a process of individual measurements. It should be carefully noted that in the case of individual measurements—subgroups of size 1—the value of c_4 is based on the number of total plotted values. From (6.1), to construct an $\bar{x}$ chart, we need an estimate of $\sigma/\sqrt{n}$, that is, the standard deviation of the subgroup mean statistic. Given $\bar{s} / c_4$ is an unbiased estimate of σ, an unbiased estimate of $\sigma/\sqrt{n}$ is given by

$$\hat{\sigma}_{\bar{x}} = \frac{\bar{s}}{c_4\sqrt{n}} \tag{6.19}$$

Accordingly, the control limits for an $\bar{x}$ chart based on the use of subgroup standard deviations are given by

$$\text{UCL} = \bar{\bar{x}} + \frac{3\bar{s}}{c_4\sqrt{n}} = \bar{\bar{x}} + A_3\bar{s}$$

$$\text{LCL} = \bar{\bar{x}} - \frac{3\bar{s}}{c_4\sqrt{n}} = \bar{\bar{x}} - A_3\bar{s} \tag{6.20}$$

where $A_3 = 3 / (c_4\sqrt{n})$ and its values are tabulated for various sample sizes in Appendix Table V.

To monitor changes in process variability, we can plot the subgroup standard deviations with limits

$$\text{UCL} = B_4\bar{s}$$

$$\text{LCL} = B_3\bar{s} \tag{6.21}$$

where B_3 and B_4 are tabulated for various subgroup sizes in Appendix Table V. A control chart for the sample standard deviations is referred to as an *s chart*. More details on the derivation of the *s* chart limits are provided in the end-of-chapter appendix.

Example 6.7

This example reanalyzes the X-ray data from Example 6.1. However, this time we shall construct an *s* chart and an $\bar{x}$ chart based on the sample standard deviation method. As with the range method, we begin with the implementation of the control chart for variability since the $\bar{x}$ chart depends on a reliable estimate for process variability.

In Table 6.2, we show the sample standard deviation values for the 25 preliminary subgroups. The average sample standard deviation is

$$\bar{s} = \frac{\sum_{i=1}^{25} s_i}{25} = 0.7222$$

From Appendix Table V, we find that the values of B_3 and B_4 when $n = 5$ are 0 and 2.089, respectively. Using (6.21), we see that the *s* chart limits based on the average subgroup standard deviation are

$$\text{UCL} = B_4\bar{s} = 2.089(0.7222) = 1.509$$

$$\text{LCL} = B_3\bar{s} = 0(0.7222) = 0$$

A sequence plot of the subgroup standard deviations with the latter set of limits superimposed is shown in Figure 6.8. Once again, subgroup number 22 is highlighted as an out-of-control signal. Since a special-cause explanation can be provided, the subgroup is eliminated and the *s* chart limits are recalculated. The *s* chart, with revised control limits, can be found in Figure

Table 6.2   Subgroup means and standard deviations for X-ray radiation data

Subgroup Number	$\bar{x}_i$	s_i
1	1000.068	1.0209
2	999.538	0.5348
3	999.356	0.3716
4	999.976	0.7365
5	1000.463	0.2280
6	1000.151	0.5974
7	1000.334	0.8449
8	1000.285	0.7222
9	999.698	0.4418
10	999.999	0.5872
11	999.855	1.0778
12	1000.454	0.9318
13	1000.427	0.7550
14	1000.011	0.4598
15	1000.032	0.7566
16	1000.008	0.3278
17	1000.026	0.8424
18	999.765	0.6214
19	1000.213	0.5108
20	999.904	0.8405
21	1000.140	0.8015
22	1000.656	1.8649
23	999.735	0.8907
24	999.971	0.6951
25	1000.018	0.5906

6.9. The intermediate calculations are left as an exercise for the reader. The revised s chart gives no indication of departures from an in-control state. Unless any additional special causes are revealed on the $\bar{x}$ chart, the control limits, along with the centerline shown in Figure 6.9, can be projected for prospective control.

We now turn to the construction of the $\bar{x}$ chart. From Figure 6.9, we can find the average subgroup standard deviation $\bar{s}$ for the 24 usable subgroups to be 0.6746. For $n = 5$, Appendix Table V gives $A_3 = 1.427$. By applying these values to (6.20), the $\bar{x}$ chart limits are computed as

$$\text{UCL} = \bar{\bar{x}} + A_3\bar{s} = 1000.016 + 1.427(0.6746) = 1000.98$$

$$\text{LCL} = \bar{\bar{x}} - A_3\bar{s} = 1000.016 - 1.427(0.6746) = 999.05$$

Notice that the above values are nearly identical to the earlier computed $\bar{x}$ chart limits using the ranges (see Figure 6.4). The lack of disparity between the two sets of limits reflects the fact

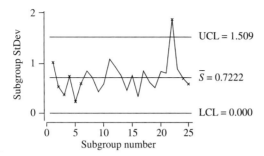

Figure 6.8   Preliminary s chart for radiation damage data.

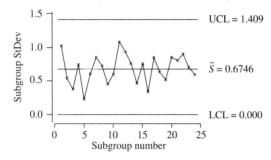

Figure 6.9   Revised s chart for radiation damage data.

that the data are well behaved and in control. For larger subgroup sizes, the differences between the two approaches will be more discernible given the much greater statistical efficiency of the sample standard deviation method over the sample range method.

With the X-ray data set, we have seen that the sample range method and sample standard deviation method reveal the existence of a special cause manifested as an outlier within a particular subgroup. Although both approaches provided similar guidance in this case, note that the sample standard deviation method, especially for large subgroup sizes, tends to be slightly less sensitive than the sample range methods in revealing outliers residing within subgroups. The main reason is that the influence of individual points, including outliers, is "averaged" out in the sample standard deviation calculation. In contrast, the sample range computation, by definition, places its primary weight on any outliers within a subgroup regardless of subgroup size. Surprisingly, the R chart is *not* significantly more responsive than the s chart to within-subgroup outliers as subgroup size increases. This stems from the fact that the expected width of R chart limits increases with subgroup size while the expected width of s chart limits decreases with subgroup size. The net effect is that the two charts are nearly equalized with the R chart having only a slight edge over the s chart in the detection of within-subgroup outliers. Regardless of which charting method is implemented, it is good practice to review directly the individual measurements for any substantial deviations. One way to reveal any unusual observations is simply to construct a histogram of all the individual measurements, preferably on a standardized scale.

$\bar{x}$ AND s CHARTS WITH VARIABLE SUBGROUP SIZE

Implicit in our earlier development of the $\bar{x}$ chart, R chart, and s chart is that the sample size is equal for all subgroups. Our focus on equal sample sizes is primarily due to the fact that unequal subgroup sizes for variable charts occur less frequently in practice. Generally, this is because subgroup sizes for variables tend to be small and thus easily maintained at some predetermined size. Nonetheless, if subgroup sizes do vary, it is less reasonable to use the simple average of the subgroup means and standard deviations since this gives equal weight to each of the subgroups. It is preferable to weight each subgroup by its size and then average them together in a combined or *pooled* estimate. If n_i is the number of observations in the ith subgroup, then the grand mean is computed as follows:

$$\bar{\bar{x}} = \frac{\sum_{i=1}^{m} n_i \bar{x}_i}{\sum_{i=1}^{m} n_i} = \frac{\sum_{i=1}^{m}\sum_{j=1}^{n_i} x_{ij}}{\sum_{i=1}^{m} n_i} \tag{6.22}$$

In other words, the grand mean is the average of all the individual measurements. For the sample standard deviation, we consider the following pooled estimate:

$$s_p = \sqrt{\frac{\sum_{i=1}^{m}(n_i - 1)s_i^2}{\sum_{i=1}^{m}(n_i - 1)}} \tag{6.23}$$

where s_i^2 is the estimated sample variance for subgroup i.

Authors of most SPC textbooks suggest that the pooled grand mean estimate and the pooled standard deviation estimate can simply be substituted into the limits for the $\bar{x}$ chart and s chart given by (6.20) and (6.21). Recall that we mentioned that $\bar{s} / c_4$ is an unbiased estimate of σ. By substituting s_p into (6.20) or (6.21), we are implicitly using s_p / c_4 as an estimate for σ. Even though s_p / c_4 is still an unbiased estimate of σ, it can be demonstrated that an alternative unbiased estimate is associated with better statistical properties. Namely, if we define $c_4(d)$ as the c_4 constant cross-referenced at

$$d = \sum_{i=1}^{m} n_i - m + 1 \tag{6.24}$$

then estimator $s_p / c_4(d)$ can be shown to be a more efficient unbiased estimator of σ than s_p / c_4.

Based on derivations shown in the end-of-chapter appendix, the following s chart control limits are recommended:

$$\text{UCL} = \frac{c_4(n_i)}{c_4(d)} B_4 s_p$$

$$\text{LCL} = \frac{c_4(n_i)}{c_4(d)} B_3 s_p$$

(6.25)

where $c_4(\cdot)$ is the c_4 constant (see Appendix Table V) cross-referenced to the argument found in the parentheses. The dependency of the control limits on subgroup size n_i implies that the width of the control limits will vary as n_i changes. In particular, the limits become tighter as n_i increases.

With respect to the $\bar{x}$ chart, the control limits using the pooled estimates are

$$\text{UCL} = \bar{\bar{x}} + \frac{3s_p}{c_4(d)\sqrt{n_i}}$$

$$LCL = \bar{\bar{x}} - \frac{3s_p}{c_4(d)\sqrt{n_i}}$$

(6.26)

Unfortunately, there is no convenient way to redefine the constants found in (6.26) for a more compact presentation. Let us now turn to an example to illustrate the computation of $\bar{x}$ and s charts with varying subgroup sizes.

Example 6.8

PreciseTech, Inc., produces valves for motorcycle engines using a hot metal-forging process. Prior to machining and grinding operations, manufacturing personnel randomly select a subgroup of valves hourly, and the valve diameters (in centimeters) are recorded. Because past experience has suggested that variability in raw material (such as the supply of metal alloys) can adversely affect the stability of the valve production process, PreciseTech has instituted a process control procedure which calls for tighter surveillance when new raw material inputs the process. In particular, the introduction of new raw material triggers a policy of randomly selecting 10 valves hourly. If five consecutive subgroups show no evidence of process instability, then only five valves are randomly selected every hour. Valve diameter measurement data for $m = 30$ consecutive preliminary subgroups, along with relevant subgroup statistics, are shown in Table 6.3.

We begin by constructing the s chart to determine if variability is in control. The pooled standard deviation is computed from (6.23) as follows:

$$s_p = \sqrt{\frac{\sum_{i=1}^{30}(n_i - 1)s_i^2}{\sum_{i=1}^{30}(n_i - 1)}}$$

$$= \sqrt{\frac{4(0.106395)^2 + \cdots + 9(0.112171)^2 + \cdots + 9(0.118491)^2 + \cdots + 4(0.12780)^2}{4 + \cdots + 9 + \cdots + 9 + \cdots + 4}}$$

$$= \sqrt{\frac{1.5781}{145}}$$

$$= 0.10432$$

Table 6.3 Valve measurement data

Subgroup Number	Observations										$\bar{x}_i$	s_i
1	4.92	4.96	5.00	4.82	5.11						4.962	0.106395
2	4.91	5.02	5.05	4.96	4.90						4.968	0.066106
3	5.17	4.82	4.81	5.15	4.92						4.974	0.175300
4	4.87	5.11	5.06	5.04	4.98						5.012	0.092033
5	4.93	4.98	4.94	5.09	4.99						4.986	0.063482
6	4.91	4.86	5.08	5.20	4.96						5.002	0.137550
7	5.04	4.98	4.96	5.12	5.16						5.052	0.086718
8	4.88	4.99	5.19	5.09	4.88						5.006	0.135019
9	4.98	4.88	5.03	4.98	4.94						4.962	0.055857
10	5.00	4.84	4.79	4.98	5.15						4.952	0.142373
11	5.08	4.99	5.06	5.01	4.92						5.012	0.063008
12	4.95	5.08	4.88	4.88	5.00						4.958	0.084971
13	4.90	4.99	4.90	4.92	5.01						4.944	0.052249
14	5.07	4.93	5.21	4.99	4.98						5.036	0.109453
15	4.98	5.02	5.14	4.93	5.17						5.048	0.103296
16	4.96	4.86	4.87	4.93	4.89						4.902	0.042071
17	4.88	5.11	5.03	5.11	4.98						5.022	0.096799
18	4.99	5.08	4.94	5.12	5.05						5.036	0.071624
19	5.20	4.98	4.99	4.87	5.04	5.00	4.89	5.04	5.05	4.80	4.986	0.112171
20	4.95	4.89	5.08	4.79	4.85	5.09	5.03	5.13	5.04	5.04	4.989	0.113279
21	5.09	4.96	4.97	5.03	4.98	5.12	4.96	5.04	5.11	4.98	5.024	0.063456
22	5.14	4.99	4.90	5.03	5.05	4.78	4.95	4.84	4.94	5.01	4.963	0.105204
23	5.04	4.97	5.10	4.92	4.95	5.01	4.83	5.21	4.83	5.06	4.992	0.118491
24	4.83	5.11	4.98	4.89	4.96						4.954	0.105499
25	5.06	5.00	4.89	5.04	5.27						5.052	0.138456
26	4.93	5.01	4.96	5.18	4.92						5.000	0.106536
27	5.02	4.96	4.99	4.82	4.89						4.936	0.080808
28	4.94	5.20	5.12	5.03	4.89						5.036	0.127004
29	5.22	5.04	5.01	5.08	4.92						5.054	0.109909
30	5.05	4.90	5.06	4.99	4.76						4.952	0.124780

From (6.25), we see that we need the values for $c_4(5)$, $c_4(10)$, and $c_4(d)$ where the value of d is determined from (6.24):

$$d = \sum_{i=1}^{30} n_i - 30 + 1 = 175 - 30 + 1 = 146$$

The values of $c_4(5)$ and $c_4(10)$ can be found from Appendix Table V to be 0.9400 and 0.9727, respectively. To obtain the value for $c_4(146)$, the approximation given in Section 5.1 may be used, namely,

$$c_4(146) = \frac{4(146) - 4}{4(146) - 3} = 0.9983$$

Finally, we need the values of B_3 and B_4 when $n = 5$ and when $n = 10$. From Appendix Table V, we find that $B_3(5) = 0$, $B_4(5) = 2.089$, $B_3(10) = 0.284$, and $B_4(10) = 1.716$. Because the control limits depend on the subgroup size, we have two sets of limits. Using (6.25), we find for $n = 5$

$$\text{UCL} = \frac{0.940}{0.9983}(2.089)(0.10432) = 0.2052$$

$$\text{LCL} = \frac{0.940}{0.9983}(0)(0.10432) = 0$$

For $n = 10$, we have

$$\text{UCL} = \frac{0.9727}{0.9983}(1.716)(0.10432) = 0.1744$$

$$\text{LCL} = \frac{0.9727}{0.9983}(0.284)(0.10432) = 0.0289$$

In Figure 6.10, we give the s chart for the preliminary subgroups, using the above two sets of limits. In terms of variability, the s chart gives no indications of departures from an in-control state. Thus, we can proceed to the computation of the limits for the $\bar{x}$ chart.

First, we use (6.22) to find the pooled grand mean:

$$\bar{\bar{x}} = \frac{\sum_{i=1}^{30} n_i \bar{x}_i}{\sum_{i=1}^{30} n_i} = \frac{5(4.952) + \cdots + 10(4.986) + \cdots + 10(4.992) + \cdots + 5(4.952)}{5 + \cdots + 10 + \cdots + 10 + \cdots + 5}$$

$$= \frac{873.63}{175}$$

$$= 4.992$$

Again, because the control limits depend on the subgroup size, we have two sets of limits. Using (6.26), we find for $n = 5$

$$\text{UCL} = 4.992 + \frac{3(0.10432)}{0.9983\sqrt{5}} = 5.132$$

$$\text{LCL} = 4.992 - \frac{3(0.10432)}{0.9983\sqrt{5}} = 4.852$$

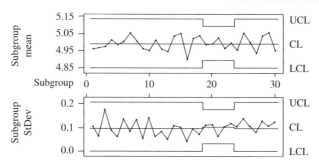

Figure 6.10   *x̄* and *s* Control chart for valve data with variable subgroup size.

For $n = 10$, we have

$$\text{UCL} = 4.992 + \frac{3(0.10432)}{0.9983\sqrt{10}} = 5.091$$

$$\text{LCL} = 4.992 - \frac{3(0.10432)}{0.9983\sqrt{10}} = 4.893$$

The $\bar{x}$ chart can also be found in Figure 6.10. As with the standard deviations, the subgroup means appear to be stable and in control with no indications of special causes being present.

S^2 CHART

A variant of the s chart is the s^2 chart, which monitors the subgroup variances rather than the subgroup standard deviations. Unlike the subgroup range or the subgroup standard deviation statistic, a control chart developed based on ± 3 standard deviations of the subgroup statistics around the centerline is not advisable for the sample variance. The reason is that the asymmetry of the distribution for the sample variance statistic is much more pronounced than that of the distribution for the aforementioned statistics.

Recall from Chapter 2 that if individual observations are random and normally distributed, then the chi-square (χ^2) distribution underlies the behavior of S^2. The idea is then to establish probability limits directly from the underlying chi-square distribution. The estimated control limits for an s^2 control chart are

$$\text{UCL} = s_p^2 \left(\frac{\chi^2_{\alpha/2,\, n-1}}{n - 1} \right)$$

$$\text{LCL} = s_p^2 \left(\frac{\chi^2_{1-\alpha/2,\, n-1}}{n - 1} \right)$$

(6.27)

where $\chi^2_{\alpha/2,\,n-1}$ and $\chi^2_{1-\alpha/2,\,n-1}$ are the $100(1 - \alpha/2)$ and $100(\alpha/2)$ percentile points of a chi-square distribution with $n - 1$ degrees of freedom. In the case of unequal subgroup sizes, the estimate s_p^2 is the square of the pooled standard deviation given by (6.23). When subgroups are equal in size, the square of the pooled standard deviation simplifies to average sample variance, that is,

$$s_p^2 = \frac{\sum_{i=1}^{m} s_i^2}{m} \tag{6.28}$$

6.5   ASSUMPTIONS UNDERLYING DEVELOPMENT OF SUBGROUP CHARTS

There are two fundamental assumptions underlying the development of the R, s, and s^2 charts. First, it is assumed that the individual observations within the subgroups are independent. In the next section, we discuss the impact of nonindependence on the R chart. Second, it is assumed that the individual observations within the subgroups are normally distributed; refer to the end-of-chapter appendix to verify this working assumption of normality for the individual observations in the derivations of all the various control chart constants associated with either subgroup mean or variability charts.

Several investigators have studied the effect of nonnormality on the control chart constants used in the construction of $\bar{x}$ and R charts. Notably, Burr (1967) concluded that the normal-based control chart constants are quite insensitive to small to moderate departures from the normality assumption. In other words, unless the departure from normality is extremely severe, Burr's conclusion is that application of the standard control chart constants is considered satisfactory. In contrast to the range method, even small departures from normality can severely affect the appropriateness of the s and s^2 charts; in a related application, the reader is referred to Hahn (1970) who notes that confidence intervals and hypothesis tests for σ and σ^2 are unduly influenced by small to moderate departures from normality.

When the within-subgroup observations are nonnormal, a data-analytic solution is to transform the data so that the individual observations are approximately normal. Upon transformation, subgroup statistics can be computed from the transformed individual observations, and then R, s, or s^2 charts can be constructed. In the full version of Minitab, there is an option with all the subgroup charts to apply a power transformation (referred to as a Box-Cox transformation by Minitab) to the individual observations prior to construction of the charts.

With respect to monitoring subgroup means, the general assumptions are that the individual measurements within the subgroup are independent and that the *averages* of these individual measurements are approximately normally distributed. The premise is that the central limit theorem mechanism will take hold to give approximate normality to the subgroup means when the individual measurements are not normally distributed. As a point of interest, Shewhart indeed recognized the possibility of nonnormal individual measurements. In Shewhart (1931), we find that he performed a detailed study of the effects of various nonnormal distributions for the individual measurements on the subgroup means based on $n = 4$. He found that the distribution of subgroup means did not differ much from the normal distribution even when the individual observations were generated from a highly nonnormal distribution.

With this said, this does not imply that the standard $\bar{x}$ chart limits do not assume normality of the individual measurements. If we refer to the end-of-chapter appendix, we find that the $\bar{x}$ chart constants of A_2 (range method) and A_3 (standard deviation method) are based on constants d_2 and c_4, respectively, which, in turn, are based on the underlying assumption of normality. However, as we just noted, Burr (1967) demonstrated that small to moderate departures in normality do not adversely impact the usefulness of the range-based constant, and hence, the implementation $\bar{x}$ chart will be satisfactory in the presence of such departures. But this would not be the case for an $\bar{x}$ chart based on the standard deviation–based constant. It would be advisable to deal with any detectable departures from normality before implementing any standard deviation–based charts. Below is a summary of our discussion and general guidelines for the implementation of subgroup charts in the presence of nonnormality:

1. *The $\bar{x}$ chart (range-based) and R chart:* If the individual observations exhibit a small to moderate departure from normality, implementation of the $\bar{x}$ chart and R chart using the original measurements is generally acceptable.

2. *The $\bar{x}$ chart (range-based) and R chart:* If the individual observations exhibit a substantial departure from normality, apply a normalizing transformation to the individual observations and then implement the $\bar{x}$ chart and R chart, using the transformed measurements.

3. *The $\bar{x}$ chart (standard deviation–based) and s chart:* If the individual observations exhibit a small to moderate departure from normality, consider the following:

    a.   When the subgroup size is small ($n \leq 10$) and constant, implement instead an $\bar{x}$ chart (range-based) and R chart using the original measurements. Or, apply a normalizing transformation to the individual observations and then implement the $\bar{x}$ chart (standard deviation–based) and s chart using the transformed measurements.

    b.   When the subgroup size is large ($n > 10$) or not constant, apply a normalizing transformation to the individual observations and then implement the $\bar{x}$ chart (standard deviation–based) and s chart using the transformed measurements.

4. *The $\bar{x}$ chart (standard deviation–based) and s chart:* If the individual observations exhibit a substantial departure from normality, apply a normalizing transformation to the individual observations and then implement the $\bar{x}$ chart (standard deviation–based) and s chart using the transformed measurements.

We have concentrated our discussion on the implications of departures from the normality assumption. However, we also stated that there is another fundamental assumption underlying the development of subgroup charts; namely, it is assumed that the individual measurements within the subgroups are independent. By means of illustrative examples, we study the impact of violations of this assumption in the next section.

6.6   PRACTICAL CHALLENGES TO STANDARD SUBGROUP CHARTS

In this section, we consider two published data sets taken from real settings that nicely illustrate some common problems that can occur with subgroup control charts in practice. We should clearly state that our intent is not to criticize the distinguished authors and pioneers—Ishikawa and Shewhart—but to study the impact of certain scenarios on the effectiveness of standard control chart procedures.

DANGERS OF STUDYING SUBGROUP STATISTICS ONLY

For our first example, we consider a data set which illustrates the dangers of basing analysis only on subgroup statistics without regard to the individual measurements. The data are taken from the widely used quality control manual by Ishikawa (1986, p. 66) and consist of five measurements of moisture content of a textile product taken at successive times: 6:00, 10:00, 14:00, 18:00, and 22:00 for 25 consecutive days. The measurements along with certain subgroup statistics are shown in Table 6.4. In addition, the data can be found in the file "ishikawa.dat".

We will see later that the systematic sampling, with respect to time within days, is the key information to a better understanding of the process. However, for now, we ignore this possibility and follow Ishikawa in his simultaneous construction of an $\bar{x}$ chart and R chart for the 25 subgroups. From the values provided in Table 6.4, we find $\bar{\bar{x}} = 12.94$ and $\bar{R} = 1.352$. For a subgroup size of 5, the values of D_3, D_4, and A_2 are 0, 2.114, and 0.577, respectively. Using (6.6) and (6.7), we calculate the control limits for the $\bar{x}$ chart as UCL = 12.94 + 0.577(1.352) − 13.72 and LCL = 12.94 − 0.577(1.352) = 12.16, and the control limits for the R chart as UCL = 2.114(1.352) = 2.858 and LCL = 0(1.352) = 0. Like Ishikawa, we show the two control charts together in Figure 6.11.

We notice that no points exceed either set of control limits. Thus, on the surface, the control charts for the subgroups suggest a process in a state of statistical control. Ishikawa lists no indications of any unusual process behavior and, implicitly, leaves his readers with the impression that the process is indeed in control. However, upon *close* inspection, there seems to be subtle evidence that the subgroup means are tightly dispersed around the centerline relative to the

Table 6.4   Ishikawa moisture data

Subgroup Number	Time					$\bar{x}_i$	R
	6:00	**10:00**	**14:00**	**18:00**	**22:00**		
1	14.0	12.6	13.2	13.1	12.1	13.00	1.9
2	13.2	13.3	12.7	13.4	12.1	12.94	1.3
3	13.5	12.8	13.0	12.8	12.4	12.90	1.1
4	13.9	12.4	13.3	13.1	13.2	13.18	1.5
5	13.0	13.0	12.1	12.2	13.3	12.72	1.2
6	13.7	12.0	12.5	12.4	12.4	12.60	1.7
7	13.9	12.1	12.7	13.4	13.0	13.02	1.8
8	13.4	13.6	13.0	12.4	13.5	13.18	1.2
9	14.4	12.4	12.2	12.4	12.5	12.78	2.2
10	13.3	12.4	12.6	12.9	12.8	12.80	0.9
11	13.3	12.8	13.0	13.0	13.1	13.04	0.5
12	13.6	12.5	13.3	13.5	12.8	13.14	1.1
13	13.4	13.3	12.0	13.0	13.1	12.96	1.4
14	13.9	13.1	13.5	12.6	12.8	13.18	1.3
15	14.2	12.7	12.9	12.9	12.5	13.04	1.7
16	13.6	12.6	12.4	12.5	12.2	12.66	1.4
17	14.0	13.2	12.4	13.0	13.0	13.12	1.6
18	13.1	12.9	13.5	12.3	12.8	12.92	1.2
19	14.6	13.7	13.4	12.2	12.5	13.28	2.4
20	13.9	13.0	13.0	13.2	12.6	13.14	1.3
21	13.3	12.7	12.6	12.8	12.7	12.82	0.7
22	13.9	12.4	12.7	12.4	12.8	12.84	1.5
23	13.2	12.3	12.6	13.1	12.7	12.78	0.9
24	13.2	12.8	12.8	12.3	12.6	12.74	0.9
25	13.3	12.8	13.0	12.3	12.2	12.72	1.1

control limits. In other words, the control limits seem a bit too wide, giving the appearance that the subgroup means are hugging the centerline. Ironically, the term *hugging,* describing the tendency for points to cluster around the center-line, is suggested by Ishikawa (1986); other authors refer to this phenomenon as *stratification.*

To understand why the control limits can be too wide in general, remember that the standard $\bar{x}$ chart control limits are based only on the within-subgroup variation. If the within-subgroup variation is significantly inconsistent with the between-subgroup variation, control limits will be either too wide or too tight depending on the direction of the inconsistency. The typical cause of this inconsistency is the presence of systematic or nonrandom variability within the

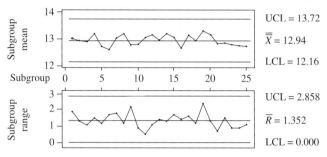

Figure 6.11   x and R Charts for Ishikawa's moisture data.

subgroups. To visualize this disparity, we construct an alternative control chart for the subgroup means with limits based on the variation of the subgroup means themselves, ignoring the within-subgroup information. Thus, we are treating the subgroup mean values as individual measurements.

In Chapter 5, two individual charts for the standard case were presented; one was based on the moving range while the other was based on the sample standard deviation. Either version will illustrate the issue at hand. Let us take, for example, the sample standard deviation approach. The sample standard deviation calculation applied directly to the mean values of the 25 subgroups produces the value of 0.1892. Recall that the correction factor of c_4 is required which, for 25 observations, is found to be 0.9897. Thus, the 3σ limits, based only on the between-subgroup variation, are the grand mean, 12.94, plus and minus 3(0.1892/0.9897), which results in UCL = 13.51 and LCL = 12.37.

In Figure 6.12, we show the sequence plot of the subgroup means with the alternative control limits and with the standard $\bar{x}$ chart control limits. Again, all points are within control limits. However, we notice some difference in the width of control limits; in particular, the limits are wider for the standard $\bar{x}$ chart. If the within-subgroup observations are randomly distributed, the expected width of control limits calculated from within-subgroup information will be the same as limits calculated from the standard deviation of subgroup means.

Rather than stop with the appearance of statistical control for the subgroup means, we pursue the evident inconsistency of the within- and between-subgroup sample variation. It would seem that careful examination of the individual measurements might be rewarding. A time-series plot of the series of 125 individual observations, with points numbered to correspond to the five sampling periods each day, is shown in Figure 6.13. Immediately, we see that a systematic pattern is uncovered! Namely, the first period of each day is high relative to other periods of the day. The first period is maximum for 21 out of 25 days; this is significantly higher than the expected number of $(\frac{1}{5})25 = 5$ days for a random process.

To capture the systematic variation by time of day, regression modeling can be used. Specifically, we set up a seasonal indicator variable period$_1$, which is 1 for period 1 and 0 for periods 2 to 5. Regressing the individual measurements

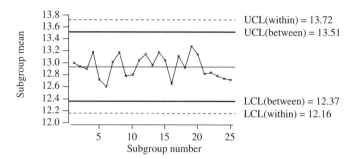

Figure 6.12   Subgroup means for moisture data with different control limits.

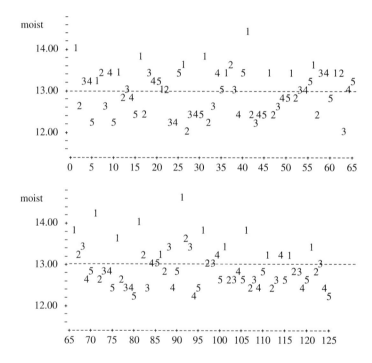

Figure 6.13   Time-series plot of individual moisture measurements.

(moist$_i$) on this independent variable gives the computational results shown in Figure 6.14. The estimated coefficient for period$_1$ is highly significant with a p value < 0.0005. Unlike the original moisture measurements, analysis of the residuals would indicate no substantial evidence against random normal behavior.

The special-cause chart (Figure 6.15) for the residuals shows no outliers. However, the application of supplementary runs rules reveals a rule 5 violation, which implies that two of three successive points (namely, observations 90 to 92)

```
The regression equation is
moist = 12.8 + 0.865 period1

Predictor          Coef         StDev            T            P
Constant        12.7670        0.0408       312.61        0.000
period1         0.86500       0.09132         9.47        0.000

S = 0.4084      R-Sq = 42.2%       R-Sq(adj) = 41.7%

Analysis of Variance

Source             DF           SS           MS            F            P
Regression          1       14.964       14.964        89.72        0.000
Residual Error    123       20.516        0.167
Total             124       35.480
```

Figure 6.14   Regression output for seasonal indicator model for moisture data.

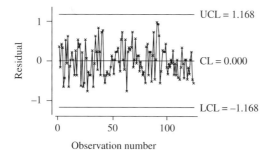

Figure 6.15   Special-cause chart for moisture series.

fall beyond 2σ limits on one side of the centerline (see Section 4.9). It is appropriate to find an explanation for this out-of-control signal which, interestingly, was not brought to light by either of the subgroup charts.

To visualize the implications of the fitted model, we display the fitted-values chart in Figure 6.16. The plot clearly shows that the first observation of each day is systematically higher than that of the other periods. Efforts should be concentrated on discovering the causes of this systematic variation, perhaps due to a "start-up" effect, and, if possible, to remove them. If the systematic variation, modeled by the seasonal model, could be eliminated, we would realize a 42.2 percent (R^2) improvement in the process as measured by the variance.

This example illustrated that vital information, for process understanding and improvement, can be lost by studying only subgroup statistics. In particular, we saw that the subgrouping covered up a systematic nonrandom variation which can only be discovered through analysis of the individual observations. Of course, much of the problem stems from the design of the data collection

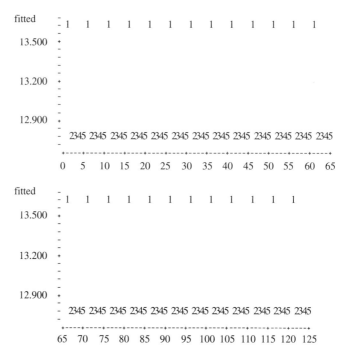

Figure 6.16   Fitted-values chart for moisture series.

scheme. It can be legitimately argued that in this case the principle of rational subgrouping was violated. Is there an alternative approach? As suggested earlier in this chapter, one might form subgroups over time based on closely produced items. Now the question is, How should the subgroups be charted? If the subgroups are plotted on one chart, it is expected that the seasonality, buried within the subgroups, will surface and will be reflected in the behavior of the subgroups themselves. As demonstrated in Chapter 5, control limits applied to a series of nonrandom observations are not necessarily reliable guides for the detection of change over and above those of the natural workings of the process. In the end, it is clear that the broader perspective of a data analysis approach, as opposed to the standard SPC techniques, has provided us with the critical insights for more effective process understanding and monitoring.

CONSEQUENCES OF WITHIN-SUBGROUP CORRELATION

In the previous example, we studied the effects of having a certain type of a systematic component "buried" within the subgroups on the traditional subgroup charts. In particular, we learned that one form of nonrandomness can cause the standard limits to have an inflated view of the true variability of the subgroup means. Other types of nonrandomness can have an opposite effect. To illustrate this possibility, we consider a set of data taken from W. A. Shewhart's (1931) pioneering work, *Economic Control of Quality Manufactured Product* (p. 20). This data set includes 204 consecutive measurements of the electrical resistance of insulation in megohms (MΩ) and can be found in file "shewhart.dat". Our aim is not to criticize Shewhart but to illustrate that even a great statistician, without access to modern time-series methodology or modern computing, can overlook the kind of nonrandom effects we are studying here. Indeed, the fact that Shewhart is the father of quality control adds a unique dimension to our study.

To illustrate the $\bar{x}$ chart, Shewhart proceeded to take successive groups of 4 observations to form 51 subgroups. The data formatted on this basis, along with pertinent subgroup statistics, are shown in Table 6.5. The subgroup size of 4 was arbitrary, as Shewhart indicated. First, we construct the $\bar{x}$ chart using Shewhart's arbitrary subgroup size choice of $n = 4$, resulting in 51 (204/4) subgroups. The average of subgroup means $\bar{\bar{x}}$ is 4498.2, and the average of the subgroup ranges $\bar{R}$ is 658.6. The control limits for the $\bar{x}$ chart require the value for A_2 for subgroups of size 4 to be 0.729. Using (6.6), we see that the $\bar{x}$ chart control limits are UCL = 4498.2 + 0.729(658.6) = 4978.32 and LCL = 4498.2 − 0.729(658.6) = 4018.08.

The $\bar{x}$ chart for the sample data is presented in Figure 6.17. It is immediately apparent from the $\bar{x}$ chart that the control limits are very tight, allowing 10 subgroup means—19.6 percent of the subgroup means—to breach control limits. Matters are made worse when one applies supplementary runs rules; namely, 10 additional "out-of-control" signals are revealed. Collectively, 17 unique subgroups are involved, implying that nearly 35 percent of the total sample is viewed as out of control!

Clearly, there are more fundamental issues at hand, and the standard method is not providing appropriate guidance for detecting potential special causes. Much of this is due to the fact that the observed within-subgroup variation does not accurately reflect the observed between-subgroup variation of the means. As was done for the Ishikawa example, to visualize the inconsistency between the two sources of variation, we can construct an alternative control chart for the subgroup means with limits based on the between-subgroup variation of the subgroup means, ignoring the within-subgroup information. Accordingly, we find the standard deviation of the subgroup means to be 352.3, and the correction factor $c_4 = 0.995$. Thus, the 3σ limits based only on the between-subgroup variation are calculated from 4498.2 ± 3(352.3/0.995) which produces UCL = 5560.41 and LCL = 3435.99, where 4498.2 is the grand mean.

In Figure 6.18, we present the plot of the subgroup means with both types of control limits superimposed on the data series. As can be plainly seen, there is a substantial difference in the width of the control limits. In particular, the control limits based on the between-subgroup variation are much wider than the limits based on the within-subgroup variation. Control limits calculated directly from the subgroup means appear more realistic; however, we must be careful not to place too much trust in these limits either. This is because time-series analysis of the subgroup means would reveal them to be nonrandom in the form of positive autocorrelation.

Table 6.5 Data on electrical resistance of insulation (MΩ)

Subgroup	Observations				$\bar{x}_i$	R
1	5045	4350	3975	4350	4430.00	1070
2	4290	4430	4285	4485	4372.50	200
3	3980	3925	3760	3645	3827.50	335
4	3300	3685	5200	3463	3912.00	1900
5	5100	4635	5450	5100	5071.25	815
6	4635	4720	4565	4810	4682.50	245
7	4410	4065	5190	4565	4557.50	1125
8	4725	4640	4895	4640	4725.00	255
9	4790	4845	4600	4700	4733.75	245
10	4110	4410	4790	4180	4372.50	680
11	4790	4340	5750	4895	4943.75	1410
12	4740	5000	4255	4895	4722.50	745
13	4170	3850	4650	4445	4278.75	800
14	4170	4255	4375	4170	4242.50	205
15	4175	4550	2855	4450	4007.50	1695
16	2920	4375	4355	4375	4006.25	1455
17	4090	5000	5000	4335	4606.25	910
18	4640	4335	4615	5000	4647.50	665
19	4215	4275	5000	4275	4441.25	785
20	4615	4735	4700	4215	4566.25	520
21	4700	4700	4095	4700	4548.75	605
22	4095	3940	3650	3700	3846.25	445
23	4445	4000	5000	4845	4572.50	1000
24	4560	4700	4310	4310	4470.00	390
25	5000	4575	4430	4700	4676.25	570
26	4850	4850	4570	4570	4710.00	280

Subgroup	Observations				$\bar{x}_i$	R
27	4855	4160	4325	4125	4366.25	730
28	4100	4340	4575	3875	4222.50	700
29	4050	4050	4685	4685	4367.50	635
30	4430	4300	4690	4560	4495.00	390
31	3075	2965	4080	4080	3550.00	1115
32	4425	4300	4430	4840	4498.75	540
33	4840	4310	4185	4570	4476.25	655
34	4700	4440	4850	4125	4528.75	725
35	4450	4450	4850	4450	4550.00	400
36	3635	3635	3635	3900	3701.25	265
37	4340	4340	3665	3775	4030.00	675
38	5000	4850	4775	4500	4781.25	500
39	4770	4500	4770	5150	4797.50	650
40	4850	4700	5000	5000	4887.50	300
41	5000	4700	4500	4840	4760.00	500
42	5075	5000	4770	4570	4853.75	505
43	4925	4775	5075	4925	4925.00	300
44	5075	4925	5250	4915	5041.25	335
45	5600	5075	4450	4215	4835.00	1385
46	4325	4665	4615	4615	4555.00	340
47	4500	4765	4500	4500	4566.25	265
48	4850	4930	4700	4890	4842.50	230
49	4625	4425	4135	4190	4343.75	490
50	4080	3690	5050	4625	4361.25	1360
51	5150	5250	5000	5000	5100.00	250

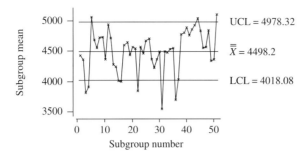

Figure 6.17   x Chart for Shewhart's megohm data.

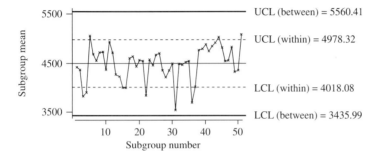

Figure 6.18   Subgroup means for megohm data with different control limits.

The standard $\bar{x}$ chart limits are clearly challenged by the data. It would be instructive to study the sequence of individual measurements, ohms$_t$, which Shewhart used to form his arbitrary subgroups (see Figure 6.19). The series shows some visual evidence of positively autocorrelated behavior, that is, a meandering pattern to the observations. The runs test and ACF, with two standard error limits, are shown in Figure 6.20. If the process were in a state of statistical control, we would expect on average 102 runs above and below the mean, given the observed frequencies of observations above and below the mean. The low count of 49 observed runs confirms the clustering tendency. The associated p value of 0.0000 (that is, < 0.00005) clearly justifies the rejection of the null hypothesis of randomness, for any reasonable level of significance. The ACF reveals several significant autocorrelations which also suggest that the series is nonrandom.

Given the meandering pattern, we should consider the possible use of lagged variables as predictor variables in a regression model. With this in mind, a PACF can serve as a useful guide for determining how many lagged variables are required to describe the nonrandom behavior. The PACF for the ohms series is given in Figure 6.21. PACF clearly cuts off after the first lag. The indication is that the series can be fitted with an AR(1) model.

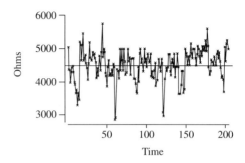

Figure 6.19   Time-series plot of individual
megohm measurements.

```
ohms

K = 4498.1763

The observed no. of runs = 49
The expected no. of runs = 102.0196
112 Observations above K    92 below
       The test is significant at    0.0000
```

(a) **Runs test**

```
ACF of ohms
           -1.0 -0.8 -0.6 -0.4 -0.2 0.0  0.2  0.4  0.6  0.8  1.0
           +----+----+----+----+----+----+----+----+----+----+
   1    0.546                           XXXXXXXXXXXXXXX
   2    0.306                           XXXXXXXXX
   3    0.216                           XXXXXX
   4    0.162                           XXXXX
   5    0.082                           XXX
   6   -0.001                           X
   7   -0.048                          XX
   8   -0.026                          XX
   9    0.016                           X
  10    0.085                           XXX
  11    0.093                           XXX
  12    0.110                           XXXX
  13    0.046                           XX
  14    0.006                           X
  15   -0.004                           X
  16   -0.040                          XX
  17   -0.093                          XXX
  18   -0.057                          XX
  19   -0.052                          XX
  20   -0.053                          XX
```

(b) **ACF**

Figure 6.20   Randomness checks of megohm data series.

```
PACF of ohms
                -1.0 -0.8 -0.6 -0.4 -0.2 0.0  0.2  0.4  0.6  0.8  1.0
                 +----+----+----+----+----+----+----+----+----+----+
    1    0.546                              XXXXXXXXXXXXXXXX
    2    0.012                              X
    3    0.063                              XXX
    4    0.026                              XX
    5   -0.042                            XX
    6   -0.064                           XXX
    7   -0.039                            XX
    8    0.034                              XX
    9    0.051                              XX
   10    0.101                              XXXX
   11    0.015                              X
   12    0.044                              XX
   13   -0.087                           XXX
   14   -0.038                            XX
   15   -0.007                             X
   16   -0.038                            XX
   17   -0.048                            XX
   18    0.063                              XXX
   19   -0.020                             X
   20   -0.023                            XX
```

Figure 6.21   PACF of megohm data series.

By regressing ohms$_t$ on ohms$_{t-1}$, we obtain the estimated AR(1) model shown in Figure 6.22. The autoregression is clearly significant. The t ratio for the independent variable ohms$_{t-1}$ is 9.31 with an associated p value < 0.0005. The runs test and ACF of the residuals provide no evidence against randomness (see Figure 6.23). The histogram and normal probability plot of the standardized residuals are given in Figure 6.24. Generally, the plots look satisfactory; however, there are 3 residuals, or 1.5 percent of the data, that fall beyond 3 standard deviations versus 0.27 percent expected under the normal distribution. Whether these reflect special causes or simply a general departure from normality in the direction of a "fat-tailed" distribution cannot be determined, since there is no way now to conduct an investigation into possible special causes. But regardless of this, the first-order AR model gives a good idea as to what is going on.

The special-cause chart is shown in Figure 6.25. The control limits show the 3σ outliers to be observations 16, 60, and 121. In addition, two unusual runs of points labeled as rule 2 and rule 7 violations are revealed; see Section 4.9 for an explanation of these runs violations. All these noted signals serve as legitimate signals for special-cause investigation. As an interesting comparison, these individual signals correspond to only 4 out of the 17 subgroup means, which were earlier viewed as being out of control by the supplementary runs rules when applied to the sequence of subgroup means.

```
The regression equation is
ohms = 2029 + 0.549 ohms-1

203 cases used 1 cases contain missing values

Predictor          Coef         StDev           T          P
Constant          2028.8         266.3        7.62      0.000
ohms-1           0.54867       0.05892        9.31      0.000

S = 390.4     R-Sq = 30.1%      R-Sq(adj) = 29.8%

Analysis of Variance

Source               DF            SS           MS          F          P
Regression            1      13216531     13216531      86.70      0.000
Residual Error      201      30638896       152432
Total               202      43855427
```

Figure 6.22   Regression output for fitted lag 1 model for megohm data series.

The fitted-values chart, along with the original measurements, is shown in Figure 6.26. This plot gives us a view of the evolution of the estimated process level through time. Ideally, we would like to find and remove the sources of the underlying systematic behavior reflected by the fitted-values chart. Rather than chase down the many signals flagged by the standard approach, efforts would be more reasonably spent trying to understand the causes of the persistent period-to-period carryover effects. If the nonrandom variation, captured by the fitted AR(1) model, could be eliminated, the process variation would be reduced by 30.1 percent (R^2).

This example clearly brings out the point that in the presence of positive autocorrelation, the $\bar{x}$ chart based on the mean of subgroup ranges tends to signal too many outliers, thus misdirecting the search for special causes. The reason is that the positive autocorrelation within the subgroups deflates the estimate of variability, and thus forces control limits too close together around the centerline. This disadvantage, however, does have one saving grace; an alert analyst will recognize that not all the out-of-control points represent special causes and hence must question the appropriateness of the standard assumptions.

GAPPING STRATEGY

The Ishikawa and Shewhart examples illustrated two detrimental effects of within-subgroup nonrandomness on the standard $\bar{x}$ chart implementation. First, depending on the nature of the nonrandomness, the $\bar{x}$ chart limits may be either too wide or too narrow relative to the natural variability of the subgroup means. In both situations, these anomalies misdirect special-cause searching. Second, the subgroup means themselves may exhibit nonrandom behavior, as a carryover of the underlying nonrandomness of the individual measurements. The Ishikawa

```
residual
K = 0.0000
The observed number of runs = 102
The expected number of runs = 102.4778
100 Observations above K      103 below
      The test is significant at 0.9464
      Cannot reject at alpha = 0.05
```

(a) Runs test

```
ACF of residual
              -1.0 -0.8 -0.6 -0.4 -0.2 0.0  0.2  0.4  0.6  0.8  1.0
               +----+----+----+----+----+----+----+----+----+----+
    1   -0.000                          X
    2   -0.026                         XX
    3    0.046                         XX
    4    0.072                         XXX
    5    0.024                         XX
    6   -0.021                         XX
    7   -0.070                        XXX
    8   -0.019                          X
    9   -0.012                          X
   10    0.080                         XXX
   11    0.023                         XX
   12    0.109                         XXXX
   13    0.001                          X
   14   -0.012                          X
   15   -0.001                          X
   16    0.002                          X
   17   -0.095                        XXX
   18   -0.002                          X
   19   -0.020                         XX
   20   -0.044                         XX
```

(b) ACF

Figure 6.23   Randomness checks of residuals from fitted lag 1 model for megohm series.

data set reflected problems with the first issue while the Shewhart data set included problems with both issues.

Let us now consider another example that especially amplifies the issues at hand. The data are the daily levels of the Standard & Poors (S&P) composite index from January 6, 1992, to December 24, 1992, and are found in the file "sp500.dat". As mentioned in Chapter 4, it is well known that almost any market index, such as the S&P and the Dow Jones Industrial Index, exhibits severe non-random behavior closely resembling a random-walk model. To bring in subgrouping, let us define a subgroup as a business week starting on Monday and ending on Friday; for convenience, the date of January 6, 1992, was chosen to start the data series because it is a Monday.

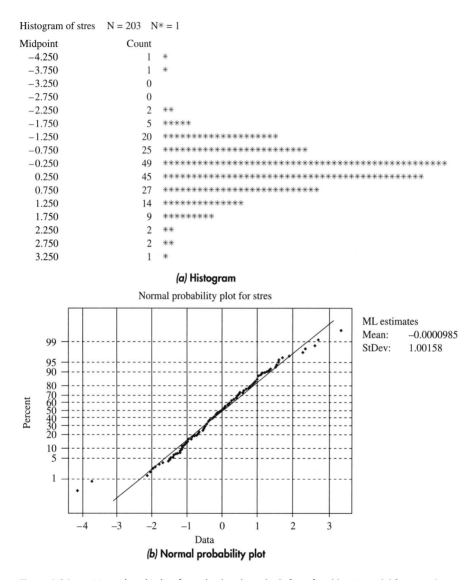

Histogram of stres   N = 203   N* = 1

Midpoint	Count	
−4.250	1	*
−3.750	1	*
−3.250	0	
−2.750	0	
−2.250	2	**
−1.750	5	*****
−1.250	20	********************
−0.750	25	*************************
−0.250	49	***
0.250	45	***
0.750	27	***************************
1.250	14	**************
1.750	9	*********
2.250	2	**
2.750	2	**
3.250	1	*

(a) Histogram

Normal probability plot for stres

ML estimates
Mean: −0.0000985
StDev: 1.00158

(b) Normal probability plot

Figure 6.24   Normality checks of standardized residuals from fitted lag 1 model for megohm data series.

Figure 6.27 shows the $\bar{x}$ and R charts for the 51 formed weekly S&P subgroups. The reader will notice that there are a few places where the control limits vary slightly. This is because some subgroups have missing observations due to holidays, leaving these subgroups with four rather than five observations. For our purposes, the effect is so minor that it can be ignored. A couple of facts can be plainly seen from the subgroup mean chart. First, the time-series behavior of

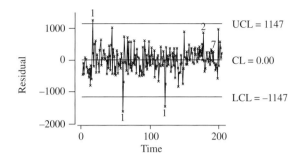

Figure 6.25   Special-cause chart for megohm series.

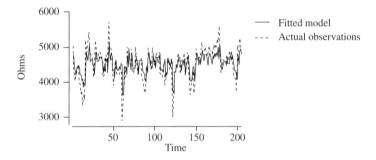

Figure 6.26   Fitted-values chart for megohm series.

the subgroup means is extremely nonrandom. Second, the $\bar{x}$ chart limits are so tight that more than one-half of the subgroups fall outside the limits! Similar to the Shewhart data set, the tightness of the limits stems from the within-subgroup observations being close together, and thus deflating the range estimate and, in turn, the width of the control limits.

Depending on the structure of the nonrandomness, two basic approaches may be considered in the monitoring of subgroup means. One possibility is to model directly the nonrandomness of the subgroup means; the other possibility is to design a sampling scheme that diminishes the correlation between the subgroup means. Alwan and Radson (1992a) show that for certain time-series behaviors when subgroups are spaced far enough apart, the subgroup means are practically independent—in some cases, entirely independent. The critical requirement for such a "gapping" strategy is that the underlying subgroup process be stationary. Theoretically, the correlation between subgroup means for nonstationary processes (e.g., trends and random walks) cannot be washed out even for arbitrarily wide gaps. Even though, theoretically, the correlation cannot be diminished for nonstationary processes, it is sometimes possible to space subgroups apart in a finite sample to give the "appearance" of randomness. This can lull one into placing retrospective control limits on the seemingly well-behaved data. The problem, however, is that the long-run variance of nonstationary

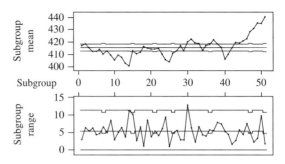

Figure 6.27 $\bar{x}$ and R charts for S&P subgroups.

processes is infinite; thus these processes will *naturally* wander away from the retrospective limits.

To demonstrate, suppose that the S&P subgroups are separated by 3 weeks; that is, form a weekly subgroup, skip 3 weeks, and then form the next subgroup, and so on. This sampling scheme gives us 13 subgroups from the originally formed 51 subgroups. Although not perfectly random, the 13 subgroup means exhibit far less nonrandomness than do the adjacent subgroup means. With this in mind, we proceed to establish the retrospective control limits. We have to be careful to recognize that the gapping of subgroups does not affect the behavior of the within-subgroup observations. Thus, an estimate based on the within-subgroup variability will still substantially underestimate the between-subgroup variability. As an alternative, it would be more reasonable to compute 3σ control limits directly from the observed subgroup mean values. Based on this approach, retrospective limits for the 13 subgroup means were determined and are displayed in Figure 6.28. Additionally, we projected these limits into the future and plotted subgroup means for the postobservation period of December 28, 1992, to December 31, 1993. Consistent with the initial subgroups, the future subgroups were sampled on the skipped basis. As is evident from the plot, the subgroup means rapidly stray from the control limits. The standard interpretation is that the process has shifted, reflecting some potential special cause. However, this is not the case; we are simply witnessing the natural evolution of a nonstationary process. It would be a futile exercise to seek special explanations for the "out-of-control" signals. This logic does not dissuade the stock pundits from suggesting special causes for variations of this index, for example, "profit taking," "concern about the budget deficit," or "lowering of the discount rate." A very simple application of time-series modeling suggests that there are no special causes of the market behavior in the sense of sudden shocks that are incompatible with a state of statistical control. As mentioned earlier, we are observing a random-walk model which relates an observation Y_t at time t to the observation one period prior Y_{t-1} through the simple model:

$$Y_t = \mu + Y_{t-1} + \varepsilon_t$$

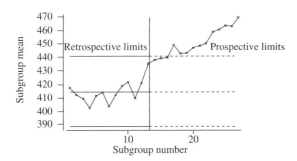

Figure 6.28   Subgroup mean chart for gapped S&P subgroups.

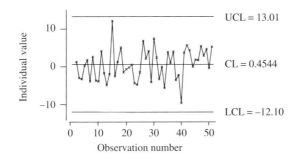

Figure 6.29   X chart applied to the differences in weekly S&P subgroup means.

where ε_t is a random error at time t. The random-walk model is a special case of the AR(1) model with lag 1 autoregressive parameter $\phi = 1$. Although a random-walk process is nonrandom and nonstationary, the implication of this model is that the *differences* or changes from one period to the next period are random.

As an example, an individual measurement control chart for the differences of the original 51 subgroups is presented in Figure 6.29. Visually, the differences appear to be consistent with randomness. Although not shown, the more formal diagnostics of the run test and ACF also confirm this verdict. Furthermore, standard normality checks reveal that the differences are compatible with the normal distribution. Therefore, we can safely interpret the control limits in Figure 6.29 which suggest no abnormal behavior in the market index. Unlike earlier, these control limits can be projected for future on-line monitoring of the process; remember that future successive differences would be plotted, not the values of the S&P index.

The S&P example demonstrated that there are processes—nonstationary—that require direct modeling to unify the picture rather than some inventive sampling scheme. However, for stationary processes, a sampling design based on gapping can provide a suitable alternative to a model-based approach. The

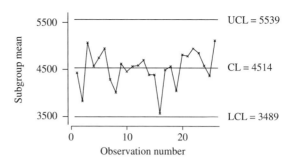

Figure 6.30   Subgroup mean chart for gapped megohm subgroups.

Shewhart data set can be used to illustrate this alternative approach. We found earlier that the individual ohms measurements followed a stationary AR(1) model with ϕ estimated to be 0.5525; in Chapter 3, we pointed out that a stationary process is associated with a ϕ value less than 1 in absolute value. We also noted that an analysis of the 52 subgroup means revealed them to be autocorrelated.

Suppose, instead of using the 52 subgroups, we take every other subgroup as the basis for analysis, which gives us 26 subgroups. It can be shown that these 26 subgroup means closely resemble a random series. The curious reader may want to verify our conclusion. At this stage, we can pursue the construction of control limits for the selected subgroup means. As was stressed before, the gapping of subgroups does not lessen the inadequacy of the observed within-subgroup variation as a predictor for the between-subgroup variation. As such, it would be more appropriate to treat the subgroup means directly as individual observations. Using Minitab's individual measurement control chart option, we obtain the control chart shown in Figure 6.30. In contrast to the standard construction, no out-of-control signals are revealed. Unlike in the S&P example, it would be reasonable to extend these control limits out for future monitoring. This is because the underlying stationarity of the subgroup process implies that there is no tendency for the subgroup means to significantly drift away. Thus, any future departures from the control limits are likely to reflect legitimate signals of process change.

R CHART IN THE PRESENCE OF WITHIN-SUBGROUP CORRELATION

Up to this point, our focus has been on the hazards of autocorrelation on effectively monitoring subgroup means, with little attention given to the R chart. If we look back at the R chart for the S&P data (Figure 6.27), it appears to be much less affected than the counterpart $\bar{x}$ chart by the underlying nonrandomness of the within-subgroup observations. Using simulation methods, Alwan and Radson (1992b) examined various properties of the subgroup range statistic in the presence of autocorrelated data. Their study indicated that subgroup ranges will exhibit near-random behavior even with the severest levels of underlying autocorrelation. However, they also found that the distribution of the sample range

```
Histogram of std(R)    N = 51          Histogram of stdsqrt(R)    N = 51
Midpoint    Count                      Midpoint    Count
  -2.750      0                          -2.750      0
  -2.250      0                          -2.250      2     **
  -1.750      2    **                     -1.750      2     **
  -1.250      5    *****                   -1.250      4     ****
  -0.750      9    *********               -0.750      6     ******
  -0.250     11    ***********             -0.250     11     ***********
   0.250     14    **************           0.250     11     ***********
   0.750      3    ***                      0.750      7     *******
   1.250      2    **                       1.250      4     ****
   1.750      3    ***                      1.750      3     ***
   2.250      1    *                        2.250      1     *
   2.750      1    *                        2.750      0
```

(a) Original ranges   (b) Transformed ranges

Figure 6.31   Histograms of standardized range and transformed range values for S&P data series.

statistic becomes increasingly more skewed as the level of autocorrelation increases. It is instructive to take a closer look at the subgroup ranges from the S&P example.

The R chart (Figure 6.27) for the S&P subgroups highlights one subgroup range (subgroup 30) as an out-of-control signal outlier. The standard procedure would be to search for possible special causes associated with this subgroup. However, before we seek special causes, we do well to examine the validity of the R chart assumptions.

Unlike the original S&P data, a visual inspection of the control chart of subgroup ranges provides no suggestion of nonrandom behavior. Other tests, such as the runs test and ACF, would support this conclusion; the reader is encouraged to confirm our assessment. To check for normality, we present a histogram of the standardized values of the ranges in Figure 6.31a. Visual examination of the histogram clearly shows an apparent nonnormality of the range data in the form of positive skewness. Hence, we proceed to seek a normalizing transformation to resolve the skewness. In light of the positive skewness, plausible choices are square root, cube root, and log transformations. In this application, the square root transformation was found to be the most appropriate. In Figure 6.31b, we also present the histogram of the standardized values of the square root transformed values of range, which are denoted by stdsqrt(R). The transformed values are substantially closer to normality than the original ranges. What is clear is that there are no substantial outliers in either tail, and that the standard R chart limits are not well suited in the presence of the original skewed distribution.

In Figure 6.32, we display a control chart for the transformed ranges using Minitab's individual measurement control chart option. Since the tests of randomness and normality are reasonably satisfactory, this control chart would be more appropriate for special-cause detection than the standard use of the R chart seen earlier. Unlike the standard R chart (Figure 6.27), the control chart of

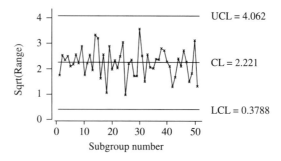

Figure 6.32   *X* chart applied to the transformed ranges computed from S&P series.

Figure 6.32 reveals no outliers, and thus there is no need to search for special causes. As a note of interest, our review of the financial research literature and discussions with finance experts confirm that no notable events or shocks occurred during the year 1992. This is consistent with the control chart of S&P differences (Figure 6.29) and the control chart of transformed ranges (Figure 6.32).

We explored some implications of underlying nonrandom behavior on the standard subgroup charts, and we provided some suggestions on how to cope with this departure. When extra process variability is undesirable (e.g., in target-based manufacturing applications), the best solution is clearly the removal of the underlying nonrandomness. The removal of systematic causes would have two benefits. First, the process would be improved since the overall variation would be reduced, leaving only simple random variation. Second, the process would be more conducive to standard SPC methods for the purposes of monitoring.

Often careful attention and hard work can bring processes to a state of statistical control in the sense of a random process. The Ishikawa example is a good example where the elimination of the systematic time-series component not only will fundamentally improve the process but also will result in a random process which is amenable to standard SPC methods. Preliminary interpretation of the fitted model for this application (see Figure 6.16) is direct and intuitive; namely, there appears to be a start-up effect. In this situation, it is not unrealistic to expect that some team of individuals (e.g., production managers, engineers, technicians, or vendors of process instrumentation and raw material), with background knowledge of the process, can find and eliminate the root cause of the persisting effect. It should be emphasized again that the identification of the start-up effect, and thus the opportunity for process improvement, was revealed not by standard SPC methods but rather by a practical data analysis approach, supported by time-series analysis.

A more difficult interpretation of a fitted-values chart is meandering or "wavelike" autoregressive variation. Our experience suggests that sources of this type of behavior are seldom identified in practice; hence, these processes are not easily brought to a state of statistical control in the sense of an iid process. Thus, the ongoing use of broader data analysis tools is essential for effective monitoring. There are, however, some processes that can be adjusted in real time, giving then the opportunity to compensate for the systematic autocorrelated behavior so that the remaining process variation will be random and in control. In Chapter 10, we will discuss this idea in greater detail.

APPENDIX TO CHAPTER 6

For the technically interested reader, we provide details underlying the derivations of some of the subgroup control charts presented in this chapter. Take particular note that these derivations are all based on a 3σ structure with the underlying assumption that the individual measurements within the subgroups are normally distributed.

DERIVATION OF R CHART LIMITS

The R chart control limits are based on the standard 3σ framework, namely, $\mu_R \pm 3\sigma_R$ where μ_R and σ_R are, respectively, the population mean and population standard deviation of the range statistic R. The population mean of ranges is estimated by the sample average of the ranges, that is, $\bar{R}$. To complete the development of the control limits, an estimate of σ_R must be made. Consider the random variable

$$W = \frac{R}{\sigma} \tag{A6.1}$$

where σ is the standard deviation of the sampled quality characteristic X. The random variable W is known as the *relative range*. Applying expectations to (A6.1), we obtain

$$E(W) = E\left(\frac{R}{\sigma}\right) = \frac{E(R)}{\sigma} \tag{A6.2}$$

The distribution of W is well known when X follows the normal distribution. In particular, it has been determined that $E(W)$ is simply a function of subgroup size n. Symbolically $E(W)$ is represented by the symbol d_2, which is tabulated in Table V. Rearranging (A6.2) and substituting in the term d_2 give the following relationship for σ:

$$\sigma = \frac{E(R)}{d_2} \tag{A6.3}$$

As pointed out earlier, $\bar{R}$ is an estimate for μ_R or, equivalently, $E(R)$. Thus, from (A6.3), we see that an estimate for σ in terms of the ranges is given by

$$\hat{\sigma} = \frac{\bar{R}}{d_2} \tag{A6.4}$$

At this stage, we take the variance of the relative range given in (A6.1) and obtain

$$\text{Var}(W) = \text{Var}\left(\frac{R}{\sigma}\right) = \frac{\text{Var}(R)}{\sigma^2} \tag{A6.5}$$

Rearranging (A6.5) and taking square roots, we have

$$\sigma_R = \sigma \sigma_W \tag{A6.6}$$

As previously noted, the distribution of W is known when X is normally distributed. As was true for the mean of the distribution, the standard deviation σ_W, of the distribution is also a function of subgroup size n. The standard deviation of the relative range is represented by the symbol d_3. Consequently, (A6.6) can be rewritten as

$$\sigma_R = d_3 \sigma \tag{A6.7}$$

Since σ is not known, the estimate given in (A6.4) can be substituted into (A6.7) to get the following estimate for σ_R:

$$\hat{\sigma}_R = d_3 \left(\frac{\bar{R}}{d_2}\right) \tag{A6.8}$$

Now, the theoretical 3σ control limits of $\mu_R \pm 3\sigma_R$ can estimated by

$$\begin{aligned} \text{UCL} &= \bar{R} + 3d_3\left(\frac{\bar{R}}{d_2}\right) \\ \text{LCL} &= \bar{R} - 3d_3\left(\frac{\bar{R}}{d_2}\right) \end{aligned} \tag{A6.9}$$

Since both d_2 and d_3 are functions of n, one can condense (A6.9) by defining the following quantities:

$$D_3 = 1 - \frac{3d_3}{d_2}$$

and

$$D_4 = 1 + \frac{3d_3}{d_2}$$

Replacing these defined quantities in (A6.9), we obtain

$$\text{UCL} = D_4\bar{R}$$

$$\text{LCL} = D_3\bar{R}$$

DERIVATION OF $\bar{x}$ CHART LIMITS BASED ON RANGE METHOD

The theoretical 3σ control limits for the subgroup means are given by $\mu \pm 3\sigma_{\bar{x}}$. The overall average of the subgroups means (that is, $\bar{\bar{x}}$) serves as an appropriate estimate of μ. We now need an estimator for $\sigma_{\bar{x}} = \sigma/\sqrt{n}$. Dividing the estimate for σ given in (A6.4) by the square root of n, we have an estimate for $\sigma_{\bar{x}}$, namely,

$$\hat{\sigma}_{\bar{x}} = \frac{\bar{R}}{d_2\sqrt{n}}$$

Replacing the grand mean and the above estimate in the general control limits of $\mu \pm 3\sigma_{\bar{x}}$, we obtain

$$\text{UCL} = \bar{\bar{x}} + 3\left(\frac{\bar{R}}{d_2\sqrt{n}}\right) = \bar{\bar{x}} + A_2\bar{R}$$

$$\text{LCL} = \bar{\bar{x}} - 3\left(\frac{\bar{R}}{d_2\sqrt{n}}\right) = \bar{\bar{x}} - A_2\bar{R}$$

where $A_2 = 3/(d_2\sqrt{n})$.

DERIVATION OF s CHART LIMITS: EQUAL SUBGROUP SIZES

We begin again by assuming that the random variable X for the individual observations within the subgroups is normally distributed. If S^2 represents the sample standard deviation statistic for X, we can state the following fact originally noted in Section 2.9:

$$\frac{(n-1)S^2}{\sigma^2} \sim \chi^2_{n-1}$$

If we take the square root, then we have the following random variable

$$Y = \frac{S\sqrt{n-1}}{\sigma} \qquad \text{(A6.10)}$$

It turns out that Y follows a distribution known as the *chi distribution* with $n-1$ degrees of freedom. For the chi distribution, it can be shown that

$$E(Y) = \sqrt{2}\,\frac{\Gamma(n/2)}{\Gamma[(n-1)/2]}$$

where $\Gamma(\cdot)$ is the gamma function.[1] Rearranging (A6.10) and applying the above result, we have

$$E(S) = \sqrt{\frac{2}{n-1}}\,\frac{\Gamma(n/2)}{\Gamma[(n-1)/2]}\sigma$$

$$\qquad \text{(A6.11)}$$

$$= c_4\sigma$$

Recognizing that the $E(S^2) = \sigma^2$ and utilizing (A6.11), we consider the following facts:

---

[1] In general, $\Gamma(p) = \int_0^\infty x^{p-1}e^{-x}dx$ where $p > 0$. If p is a positive integer, then $\Gamma(p) = (p-1)(p-2)\cdots(2)(1) = (p-1)!$.

$$\text{Var}(S) = E(S^2) - [E(S)]^2$$
$$= \sigma^2 - (c_4\sigma)^2 \qquad \text{(A6.12)}$$
$$= \sigma^2(1 - c_4^2)$$

As with the other subgroup charts, the s chart control limits are based on the standard 3σ framework, namely, $\mu_S \pm 3\sigma_S$, where μ_S and σ_S are, respectively, the population mean and population standard deviation of the statistic S. Given (A6.11) and (A6.12), the 3σ control limits for the s chart are

$$\text{UCL} = c_4\sigma + 3\sigma\sqrt{1 - c_4^2}$$
$$\text{LCL} = c_4\sigma - 3\sigma\sqrt{1 - c_4^2} \qquad \text{(A6.13)}$$

It is necessary at this stage to develop an estimate for σ. If we rearrange (A6.11), we have

$$\sigma = \frac{E(S)}{c_4} \qquad \text{(A6.14)}$$

Given a set of subgroups with their sample standard deviations, a reasonable estimate for the population mean of S is the average sample standard deviation for the set of subgroups. Taking note of (A6.14) and denoting the average sample standard deviation by $\bar{s}$, we then have the following estimate for σ:

$$\hat{\sigma} = \frac{\bar{s}}{c_4} \qquad \text{(A6.15)}$$

By replacing this estimate in (A6.13), the estimated 3σ limits are

$$\text{UCL} = \bar{s} + 3\frac{\bar{s}}{c_4}\sqrt{1 - c_4^2}$$
$$\text{LCL} = \bar{s} - 3\frac{\bar{s}}{c_4}\sqrt{1 - c_4^2} \qquad \text{(A6.16)}$$

For a condensed presentation, define

$$B_4 = 1 + 3\left(\frac{1}{c_4}\right)\sqrt{1 - c_4^2}$$
$$B_3 = 1 - 3\left(\frac{1}{c_4}\right)\sqrt{1 - c_4^2} \qquad \text{(A6.17)}$$

Accordingly, the control limits given in (A6.16) can be rewritten as

$$\text{UCL} = B_4\bar{s}$$
$$\text{LCL} = B_3\bar{s}$$

(*Note:* The development of $\bar{x}$ chart limits based on the standard deviation method and equal subgroup sizes was described in the chapter.)

DERIVATION OF s CHART LIMITS: UNEQUAL SUBGROUP SIZES

Because of the changing subgroup sizes, we need to replace c_4 with, $c_4(n_i)$, which results in

$$\text{UCL} = c_4(n_i)\sigma + 3\sigma\sqrt{1 - c_4^2(n_i)}$$
$$\text{LCL} = c_4(n_i)\sigma - 3\sigma\sqrt{1 - c_4^2(n_i)}$$

(A6.18)

As discussed in the chapter, when subgroup sizes vary, it is more appropriate to use a pooled standard deviation estimate for σ, namely,

$$\hat{\sigma} = \frac{s_p}{c_4(d)}$$

where s_p and $c_4(d)$ are defined in the chapter. Replacing the above estimate in (A6.18), we get the following:

$$\text{UCL} = c_4(n_i)\frac{s_p}{c_4(d)} + 3\frac{s_p}{c_4(d)}\sqrt{1 - c_4^2(n_i)}$$
$$\text{LCL} = c_4(n_i)\frac{s_p}{c_4(d)} - 3\frac{s_p}{c_4(d)}\sqrt{1 - c_4^2(n_i)}$$

(A6.19)

By using the B_3 and B_4 constants defined in (A6.17) with $c_4(n_i)$ in place of c_4, the control limits can be written as

$$\text{UCL} = \frac{c_4(n_i)}{c_4(d)}B_4 s_p$$

$$\text{LCL} = \frac{c_4(n_i)}{c_4(d)}B_3 s_p$$

EXERCISES

6.1. What is meant by *rational subgrouping*?

6.2. What are the two fundamental assumptions underlying the derivation of standard subgroup control chart statistics?

6.3. Explain the difference in interpretation between a point falling above the upper control limit on an $\bar{x}$ chart and a point falling above the upper control limit on an R chart.

6.4. What is the meaning of phase 1 and phase 2 control limits?

6.5. What is the effect of subgroup size on the location of R and s chart limits?

6.6. Suppose that a standard $\bar{x}$ chart with subgroup size n is being used to detect mean shifts in a manufacturing process. However, it is felt that the current charting scheme is not detecting a particular shift size of concern quickly enough. Given the desire for quicker detection, what are the alternatives to the current charting scheme?

6.7. In this chapter, it was mentioned that the moving-range statistic is a special case of the range statistic for $n = 2$. This fact was pointed out in relation to a discussion about control chart constants. However, what is one aspect that differs between the moving-range statistic and the range statistic for $n = 2$? *Hint:* Supplementary runs rules are not advisable for the moving ranges but may be applied to the subgroup ranges. Explain why.

6.8. Suppose that, in a state of statistical control, a particular quality characteristic is known to have a true mean and standard deviation of 100 and 7, respectively. For an $\bar{x}$ chart with $n = 6$, find the following:
  a. The 3σ limits.
  b. The 0.005 probability limits.

6.9. Tensile strength is a critical characteristic for bridge bolts. Because the testing of tensile strength destroys the bolt and each bolt is expensive, a small sample size of 3 was decided upon. Suppose 20 preliminary subgroups were gathered and measurements (in thousands of pounds per square inch) made on the two bolts per subgroup. Below are two summary statistics:

$$\sum_{i=1}^{20} \bar{x}_i = 1508.67 \qquad \sum_{i=1}^{20} R_i = 86.53$$

Assume that the quality characteristic is approximately normally distributed.
  a. Compute the control limits for the $\bar{x}$ and R charts.
  b. Assuming both charts exhibit stable behavior, estimate the mean and the standard deviation of the process generating the individual measurements.

6.10. Redo part a of Exercise 6.9, using the $\bar{x}$ and R charts with $\alpha = 0.002$.

6.11. List the concerns one might have with the implementation of a standard R chart with $n = 2$.

6.12. A manufacturer of batteries periodically takes random samples of size 7 from a production line that manufactures 1.5 volt (V) AA batteries. The sampled batteries are tested on a voltmeter. Below are two summary statistics based on 35 samples:

$$\sum_{i=1}^{35} \bar{x}_i = 52.185 \qquad \sum_{i=1}^{35} s_i = 0.6097$$

Assume that the quality characteristic is normally distributed.
  a. Compute the control limits for the $\bar{x}$ and s control charts.

b. Assuming both charts exhibit stable behavior, estimate the true mean and the standard deviation of the process generating the individual measurements.

6.13. To continue Exercise 6.12, based on the subgroup-based estimate of the process standard deviation, estimate the proportion of batteries having a voltage capacity below a lower specification limit of 1.425 volts.

6.14. To assess the weight control of a 16-ounce (oz) packaged product, the net weight of the product is monitored by $\bar{x}$ and R charts, using a subgroup size of $n = 5$. Below are data for 25 successive subgroups:

Subgroup	$\bar{x}$	R
1	16.03	0.23
2	15.97	0.28
3	16.00	0.22
4	15.98	0.24
5	15.98	0.15
6	15.92	0.13
7	16.09	0.74
8	16.10	0.16
9	16.01	0.32
10	16.02	0.39
11	15.99	0.29
12	16.01	0.13
13	16.03	0.25
14	15.92	0.33
15	16.04	0.38
16	15.98	0.28
17	15.99	0.34
18	16.02	0.15
19	16.07	0.22
20	15.99	0.28

Assume that the quality characteristic is normally distributed.

a. Set up $\bar{x}$ and R control charts for this application. Assume that any out-of-control signals are associated with special causes. What are the prospective limits for each of these charts?

b. Estimate the true mean and standard deviation of the process generating the individual measurements.

6.15. Redo part *a* of Exercise 6.14, using the $\bar{x}$ and R charts with $\alpha = 0.004$.

6.16. To continue Exercise 6.14, based on the subgroup-based estimate of the process standard deviation, estimate the proportion of packages not conforming to specifications of 16 ± 0.25.

6.17. A dental clinic is studying how long patients wait to see the dentist or dental hygienist. For every patient, waiting times are measured (defined as the number of seconds from registration to first contact with the dentist or dental hygienist). At the end of each day, the times are entered in a row of a spreadsheet. The number of patients varies from day to day. To keep the analysis simple, an option in the spreadsheet software to randomly pick a fixed number of observations can be invoked; in particular, $n = 5$ was chosen. Defining a given day as a subgroup, below are data for 20 consecutive days (data also given in "clinic.dat"):

Day	x_1	x_2	x_3	x_4	x_5
1	839	585	583	613	768
2	706	527	956	644	612
3	322	614	807	282	170
4	1001	904	419	522	430
5	259	811	847	603	572
6	681	523	665	397	873
7	799	874	553	173	296
8	263	585	546	445	455
9	591	637	739	927	825
10	372	997	496	283	641
11	633	371	425	625	752
12	294	522	599	740	468
13	714	600	738	694	610
14	311	829	216	447	506
15	720	446	393	431	707
16	306	550	633	845	394
17	561	574	370	495	580
18	431	500	580	349	626
19	404	791	773	576	1073
20	640	687	754	627	482

  a. Construct $\bar{x}$ and R control charts for these data. Is the process in control? What are the prospective limits for each of these charts?
  b. Estimate the true mean and standard deviation of the process generating the individual measurements.

6.18. To continue Exercise 6.17, experience suggests that people generally become dissatisfied when having to wait more than 15 min. Based on the subgroup-based estimate of the process standard deviation, estimate the probability that a patient will have to wait 15 min or more.

6.19. Redo part *a* of Exercise 6.17, using the $\bar{x}$ and R charts with $\alpha = 0.005$.

6.20. Redo all of Exercise 6.17, using the $\bar{x}$ and s charts.

6.21. An important quality characteristic in a bronze casting is the percentage of copper in the casting. Suppose that copper determinations are made on samples of three castings as shown below:

Subgroup	x_1	x_2	x_3
1	87.15	86.97	85.69
2	87.07	87.20	86.95
3	86.58	87.03	86.54
4	87.62	86.89	87.23
5	86.93	87.34	86.81
6	86.58	86.75	86.55
7	87.43	87.06	86.78
8	87.54	87.29	87.45
9	87.16	86.71	86.87
10	87.25	87.38	87.23
11	87.31	86.52	86.27
12	86.60	86.62	87.49
13	87.22	87.51	87.31
14	86.85	87.99	87.28
15	87.29	86.35	87.13
16	86.85	86.90	87.49
17	87.23	86.81	86.95
18	86.90	86.01	86.52
19	87.18	86.96	86.97
20	87.07	87.06	86.99
21	87.13	87.88	86.56
22	86.77	87.34	86.90
23	86.68	87.74	87.30
24	87.52	87.72	86.94
25	86.96	87.45	87.05

a. Construct $\bar{x}$ and R control charts for these data. Is the process in control? What are the prospective limits for each of these charts?

b. Estimate the true mean and standard deviation of the process generating the individual measurements.

    c. Assuming normality for the individual observations, if the specifications for percentage copper are 87 ± 0.8, what is the estimated number of castings per 1000 that are expected to meet specifications?

6.22. Redo all of Exercise 6.21, using the $\bar{x}$ and s charts.

6.23. The formation of a silicon dioxide layer on silicon wafers is a critical step in the manufacture of integrated circuits. Silicon dioxide is an electrical insulator and is usually created by thermal oxidation. Using special furnaces, oxidation is performed by exposing silicon wafers to oxygen at about 1000°C. During production, there are 250 wafers in every furnace run. The thickness of the oxide layer is of critical importance and is measured in angstroms (Å). Taken from successive furnace runs, below are the data associated with 25 subgroups of size $n = 6$ (data are also given in "wafer1.dat"):

Subgroup	x_1	x_2	x_3	x_4	x_5	x_6
1	299.91	289.15	294.58	292.69	290.59	289.84
2	286.75	294.03	298.97	296.61	294.72	300.84
3	299.36	298.20	294.79	291.72	293.17	291.72
4	291.65	288.21	294.88	293.29	290.44	296.63
5	294.85	300.51	303.41	298.46	289.26	295.94
6	286.17	291.46	287.98	292.66	288.48	292.98
7	295.41	289.84	296.71	301.80	285.38	301.82
8	294.72	301.68	294.19	297.21	293.78	294.91
9	295.35	294.43	295.91	297.93	288.61	289.58
10	295.37	293.65	302.26	294.69	296.31	289.20
11	299.47	287.60	296.23	298.04	293.74	291.52
12	291.65	302.86	293.32	289.79	295.16	297.77
13	295.31	298.85	289.63	289.10	296.58	292.21
14	298.46	291.80	287.00	292.42	296.81	290.13
15	293.82	291.19	296.99	291.68	295.67	295.68
16	293.72	298.63	294.11	287.50	294.82	294.78
17	300.91	296.09	291.52	299.77	292.85	295.85
18	290.04	294.42	299.57	291.73	291.80	296.64
19	294.71	292.57	301.65	294.99	288.94	291.95
20	281.16	296.72	292.48	284.18	288.62	284.01
21	291.35	295.34	301.99	293.68	291.00	296.44
22	290.28	296.21	293.66	292.66	290.93	293.43
23	297.49	302.51	299.86	285.41	291.45	296.64
24	296.61	289.90	296.55	288.73	298.48	290.72
25	288.05	284.72	296.40	300.28	300.57	291.85

a. Assuming any out-of-control signal has an identifiable special-cause explanation, establish the prospective control limits for the $\bar{x}$ and R charts.

b. The target value for the thickness of the oxide layer is 300 Å. Would you say the process is satisfactorily achieving this desired target?

c. Dissatisfied with process performance, manufacturing engineers conducted a study to determine ways to improve the process on two dimensions: (1) To bring the mean level of the process closer to target and (2) to reduce the overall process variation. Following the implementation of new production procedures, 20 new subgroups were collected (data are also given in "wafer2.dat"). Plot the subgroup means and ranges along with the prospective limits determined in part *a*. What are your conclusions?

New Subgroup	x_1	x_2	x_3	x_4	x_5	x_6
1	299.07	298.82	296.10	297.67	298.95	301.75
2	299.69	297.11	300.34	298.53	302.50	299.41
3	302.18	299.96	298.46	297.97	296.99	298.56
4	299.33	298.20	301.32	299.47	299.41	303.66
5	299.67	301.93	298.43	300.38	302.52	298.98
6	297.43	300.80	302.49	298.41	304.85	301.97
7	302.68	300.23	298.66	298.01	304.36	300.66
8	299.67	298.88	304.00	301.33	301.93	297.98
9	300.31	297.96	305.48	297.80	299.76	295.37
10	297.93	301.82	303.07	300.65	298.06	299.61
11	298.09	299.82	301.76	299.07	299.19	302.34
12	294.79	296.79	297.60	299.75	303.22	300.40
13	299.66	302.17	299.84	297.90	303.26	298.99
14	300.87	296.68	301.87	298.09	300.50	301.17
15	305.77	300.67	302.37	296.47	298.28	303.15
16	302.77	303.82	300.75	298.81	301.66	304.40
17	302.22	303.20	303.21	298.53	301.93	301.46
18	300.14	302.64	302.14	298.77	301.57	297.11
19	297.01	298.62	300.43	297.23	299.11	300.12
20	301.94	297.38	302.72	300.13	300.23	301.59

6.24. Veneer bricks ("thin bricks") are manufactured for residential and commercial building applications. An important quality characteristic is the length of the bricks. Twenty subgroups of five bricks each are collected, and the lengths (in inches) are shown below:

Subgroup	x_1	x_2	x_3	x_4	x_5	$\bar{x}$	R
1	7.97	8.03	8.01	7.97	7.99	7.994	0.06
2	7.95	8.01	8.00	8.01	8.00	7.994	0.06
3	8.02	7.99	7.96	8.04	7.99	8.000	0.08
4	8.02	8.00	7.98	7.95	8.00	7.990	0.07
5	7.99	8.00	7.99	7.99	8.02	7.998	0.03
6	7.97	8.04	8.00	8.02	8.02	8.010	0.07
7	8.00	7.99	7.98	8.00	8.00	7.994	0.02
8	8.04	8.02	7.97	8.02	7.96	8.002	0.08
9	7.92	8.05	7.93	8.02	7.99	7.982	0.13
10	7.97	8.01	8.06	8.02	7.99	8.010	0.09
11	8.04	7.98	8.03	8.03	8.01	8.018	0.06
12	8.02	7.98	8.04	7.98	7.94	7.992	0.10
13	8.02	7.99	7.93	7.95	7.99	7.976	0.09
14	7.99	7.98	8.01	7.95	7.99	7.984	0.06
15	8.00	8.02	7.97	8.04	8.04	8.014	0.07
16	7.99	8.01	8.02	7.94	8.02	7.996	0.08
17	8.02	7.96	7.99	8.04	7.97	7.996	0.08
18	8.02	7.96	8.02	8.02	8.02	8.008	0.06
19	7.96	7.93	8.01	7.93	7.99	7.964	0.08
20	7.98	8.03	8.02	8.04	8.02	8.018	0.06

a. Construct $\bar{x}$ and R control charts for these data. Is the process in control? What are the prospective limits for each of these charts?

b. After establishing the control charts in part *a,* 10 new subgroups were collected. Plot the subgroup means and ranges along with the prospective limits determined in part *a.* What are your conclusions?

New Subgroup	x_1	x_2	x_3	x_4	x_5	$\bar{x}$	R
1	7.94	8.06	7.97	8.03	7.99	7.998	0.12
2	8.06	7.96	7.96	7.99	7.93	7.980	0.13
3	7.98	7.97	8.03	7.98	8.00	7.992	0.06
4	7.97	8.06	7.95	8.11	8.03	8.024	0.16
5	7.92	7.92	7.95	8.02	7.99	7.960	0.10
6	7.97	8.03	7.89	7.99	8.12	8.000	0.23
7	7.99	7.96	7.97	8.03	8.02	7.994	0.07
8	8.09	7.99	7.94	8.00	8.04	8.012	0.15
9	7.98	7.95	8.05	8.13	7.94	8.010	0.19
10	7.99	7.96	8.02	8.04	7.86	7.974	0.18

6.25. Redo all of Exercise 6.24, using the $\bar{x}$ and s charts.

6.26. Consider the X-ray data studied in this chapter. Referring to Table 6.2, construct an s^2 chart with $\alpha = 0.002$.

6.27. Consider the waiting time data of Exercise 6.17. Construct an s^2 chart with $\alpha = 0.005$.

6.28. Consider the bronze casting data of Exercise 6.21. Construct an s^2 chart with $\alpha = 0.004$.

6.29. Suppose that, in a state of statistical control, a particular quality characteristic is normally distributed and is known to have a true standard deviation of $\sigma = 4$. If subgroups are of size 5, refer to the end-of-chapter appendix to determine
   a. The values of 3σ limits for the R chart.
   b. The values of 3σ limits for the s chart.

6.30. Below are data on a critical quality characteristic. (*Note:* Some subgroups have fewer observations.)

Subgroup	x_1	x_2	x_3	x_4	x_5	x_6	x_7
1	101.97	100.55	100.55	102.13	95.50	103.61	99.41
2	97.85	101.88	105.88				
3	89.72	99.75	106.91				
4	99.97	98.05	99.44				
5	97.54	103.19	100.23	107.06	101.11	103.52	95.50
6	104.82	104.04	98.11	90.47	99.21	106.15	99.14
7	93.20	87.94	104.53	102.67	92.77	96.05	103.49
8	93.90	110.32	100.46	101.37	98.55	99.75	101.53
9	98.86	104.15	99.09	96.90	100.51	105.69	90.39
10	86.34	97.25	92.26	97.11			
11	105.49	105.62	103.53	98.83			
12	93.45	90.99	100.52	102.92	95.92	104.97	98.49
13	104.43	89.80	102.12	105.01	97.13	92.66	104.39
14	103.64	101.43	102.43	103.97	104.61	110.38	91.38
15	98.81	100.18	100.63	106.13	100.15	96.58	89.24

   a. Construct $\bar{x}$ and s control charts for these data. Is the process in control?
   b. Estimate the true mean and standard deviation of the process generating the individual measurements.

6.31. Assuming normality for the quality characteristic, find the probability that a given subgroup mean will signal out of control if there is a negative shift in the mean equal to 2.5 standard deviations when the subgroup size is

            a. $n = 3$
            b. $n = 4$
            c. $n = 5$
            d. $n = 10$

6.32. Find the ARL for each part of Exercise 6.31.

6.33. Refer to the OC curves of Figure 6.6 to determine
    a. The approximate probability that any given subgroup range, based on a subgroup size of $n = 5$, will signal out of control if the process standard deviation tripled.
    b. The minimum subgroup size needed if it is desired to detect a doubling of process standard deviation with probability 0.30 with a single subgroup range.

6.34. Using a standard $\bar{x}$ chart, suppose that it is desired that a 1σ shift in the mean be detected within 10 samples, on average. Refer to the ARL curves of Figure 6.7 to determine the minimum subgroup size needed to meet this objective.

6.35. Suppose that at a high-volume production facility, a standard $\bar{x}$ chart is used to monitor for permanent mean shifts from an in-control normal process. Furthermore, assume that the current scheme is to collect hourly subgroups of size 3.
    a. For the current scheme, what is the average time to signal (ATS) for the detection of mean shifts of size 1.75 standard deviations?
    b. Suppose that it is desired to detect mean shifts of size 1.75 standard deviations in 0.5 h on average. How could this be accomplished if the subgroup size is left at 3? How could this be accomplished if the sampling frequency remains hourly? For both situations, give the specific details of the control charting scheme.

6.36. Based on an accelerated lifetime testing procedure, consider the following subgroup measurements of the lifetime of an electronic component (data are also given in "lifetime.dat").

Subgroup	x_1	x_2	x_3	x_4	x_5
1	0.473	0.405	0.213	3.187	0.572
2	0.430	2.623	1.415	0.915	2.933
3	0.148	1.938	1.057	2.019	1.256
4	5.209	0.211	1.047	0.492	0.388
5	0.308	0.536	0.570	2.951	1.741
6	0.539	0.229	1.969	0.449	0.461
7	3.763	0.500	0.954	0.407	1.250

8	0.705	0.571	1.038	0.880	0.607
9	13.915	0.300	2.812	1.735	0.644
10	0.666	0.961	0.659	6.590	0.984
11	1.223	1.590	0.828	0.200	0.668
12	0.532	0.343	2.470	8.159	0.996
13	0.721	0.029	1.392	1.073	0.235
14	0.605	1.185	1.706	0.339	3.455
15	1.114	0.323	1.054	0.303	0.716
16	2.535	2.381	2.087	0.463	0.301
17	2.365	0.576	0.680	1.028	0.603
18	0.831	0.994	1.175	0.867	8.310
19	1.937	1.553	0.415	2.581	1.361
20	4.969	0.449	0.424	0.457	0.570
21	0.840	0.323	1.843	2.316	0.360
22	0.852	0.176	0.556	0.988	0.248
23	0.271	0.517	0.592	1.104	0.544
24	1.080	0.344	0.177	0.716	3.526
25	0.805	1.162	1.306	0.733	0.359

a. Without removing any subgroups, construct $\bar{x}$ and R control charts for these data. What do the control charts suggest about the stability of the process?

b. Place all the individual observations in a single column of your worksheet. Perform normality checks on the individual observations. What do you conclude?

c. Transform the individual measurements, using the logarithmic transformation, and then construct $\bar{x}$ and R control charts based on the transformed measurements. What do you now conclude about the stability of the process and possible existence of special-cause variation? (*Note:* Users of the full version of Minitab can simply use the Box-Cox charting option. Users of the student version need to transform the original observations and then place the transformed data in another column before invoking the available charting options.)

6.37. Suppose we are interested in studying the average number of points scored per game by a certain group of basketball players. In particular, below are the game-by-game points scored by Michael Jordan, Scottie Pippen, Toni Kukoc, and Dennis Rodman of the Chicago Bulls for 20 consecutive games during the 1997–1998 regular season.

Game	Jordan	Pippen	Kukoc	Rodman
1	29	23	9	3
2	37	24	6	7
3	21	20	22	0
4	27	25	19	2
5	16	22	9	2
6	19	23	17	4
7	17	13	8	7
8	33	12	11	6
9	28	29	14	9
10	30	24	11	4
11	42	25	8	6
12	37	19	16	2
13	26	18	22	3
14	30	13	21	0
15	17	12	21	4
16	35	15	17	6
17	24	13	16	0
18	33	33	10	9
19	24	27	17	12
20	17	23	10	2

a. Construct an $\bar{x}$ chart for these data based on the range method. Respond to the visual appearance of the subgroup mean series relative to the control limits. Do you have any concerns? Explain.

b. Construct an alternative subgroup mean chart that would not be affected by the phenomenon observed in part *a*.

6.38. An article in *Quality Engineering* (K. C. Gilbert, K. Kirby, and C. R. Hild, vol. 9, 1991, pp. 367–382) presents subgroup data on the charge weights (in grams) of insecticide dispensers; the original source of the data is I. W. Burr, *Statistical Quality Control Methods*, Marcel Dekker, Inc., New York, 1976. Below we show a part of the published data (data are also given in "charge.dat").

Subgroup	x_1	x_2	x_3	x_4	$\bar{x}_i$
1	456	458	439	448	450.25
2	459	462	495	500	479.00
3	443	453	457	458	452.75
4	470	450	478	470	467.00

5	457	456	460	457	457.50
6	434	424	428	438	431.00
7	460	444	450	463	454.25
8	467	476	485	474	475.50
9	471	469	487	476	475.75
10	473	452	449	449	455.75
11	477	511	495	508	497.75
12	458	437	452	447	448.50
13	427	443	457	485	453.00
14	491	463	466	459	469.75
15	471	472	472	481	474.00
16	443	460	462	479	461.00
17	461	476	478	454	467.25

a. Without removing any subgroups, construct an $\bar{x}$ chart for these data. What does the control chart suggest about the stability of the process?

b. Treating the subgroup means as individual observations, construct an X chart for the subgroup means.

c. Superimpose the limits computed for the $\bar{x}$ chart in part a with the limits from part b on a sequence plot of the subgroup means. What suspicions are raised by the plot?

6.39. To continue Exercise 6.38, consider now an investigation of the individual observations. Unlike with the Shewhart megohms series studied in Section 6.6, it cannot be assumed that the time between the last observation of a given subgroup and the first observation of the next subgroup is the same as the time between adjacent observations within a given subgroup. Thus, if we wish to compute the lag 1 autocorrelation in the individual measurements, we must consider x_t as the value for the second, third, or fourth observation within the subgroup versus x_{t-1}, which is the value of the immediate predecessor of x_t within the subgroup. Put x_t and x_{t-1}, in two columns of a Minitab worksheet; that is, in a given row, the first column has the x_t value, and the second column has the x_{t-1} value, as just described.

a. Compute the correlation between successive within-subgroup observations (Stat >> Basic Statistics >> Correlation). Given the reported p value, is the computed correlation significantly different from zero? If so, relate this finding to the results from Exercise 6.38.

b. Using x_t as the dependent variable and x_{t-1} as the independent variable, fit a regression model. Report the significance of the independent variable. What percentage of the variation in the individual observations is attributable to the systematic pattern in the within-subgroup observations?

c. Construct a special-cause chart for the residuals from the fitted model of part *b*. Over and above the fitted model, are there any indications of special causes?

REFERENCES

Alwan, L. C., and D. Radson (1992a): "Time-Series Investigation of Subsample Mean Charts," *IIE Transactions*, 24, pp. 66–80.

——— and D. Radson (1992b): "Time-Series Investigation of Subsample Ranges for Positively Autocorrelated Processes," *Modeling and Simulation: Control, Signal Processing, Robotics, Systems, and Power*, 23, pp. 1791–1798.

AT&T (1985): *Statistical Quality Control Handbook*, 11th printing, AT&T Technologies, Indianapolis, IN.

Burr, I. W. (1967): "The Effect of Non-Normality on Constants for $\overline{X}$ and R Charts," *Industrial Quality Control*, 23, pp. 563–569.

Coulson, S. H. (1995): Discussion of "The Problem of Misplaced Control Limits," *Applied Statistics*, 44, p. 98.

Duncan, A. J. (1951): "Operating Characteristics of R Charts," *Industrial Quality Control*, 7, pp. 40–41.

——— (1986): *Quality Control and Industrial Statistics,* 5th ed., Irwin, Homewood, IL.

Hahn, G. J. (1970): "How Abnormal Is Normality," *Journal of Quality Technology*, 3, pp. 18–22.

Harter, H. L. (1960): "Tables of Range and Studentized Range," *The Annals of Mathematical Statistics*, 31, pp. 1122–1147.

Hillier, F. S. (1969): "$\overline{X}$ and R-Chart Control Limits Based on a Small Number of Subgroups," *Journal of Quality Technology*, 1, pp. 17–26.

Hoerl, R. W., and A. C. Palm (1990): "Discussion: Integrating SPC and APC," *Technometrics*, 34, pp. 268–272.

Ishikawa, K. (1986): *Guide to Quality Control,* 2d ed., Asian Productivity Association, Tokyo, Japan.

Lorenzen, T. J., and L. C. Vance (1986): "The Economic Design of Control Charts: A Unified Approach," *Technometrics*, 28, pp. 3–10.

Montgomery, D. C. (1980): "The Economic Design of Control Charts: A Review and Literature Survey," *Journal of Quality Technology*, 12, pp. 75–87.

Scheffé, H. (1949): "Operating Characteristics of Average and Range Charts," *Industrial Quality Control*, 5, pp. 13–18.

Shewhart, W. (1931): *Economic Control of Quality Manufactured Product*, D. Van Nostrand, New York; reprinted by the American Society for Quality Control in 1980, Milwaukee, WI.

Tuprah, K., and M. Ncube (1987): "A Comparison of Dispersion Quality Control Charts," *Sequential Analysis*, 6, pp. 155–163.

Yang, C. H., and F. S. Hillier (1970): "Mean and Variance Control Chart Limits Based on a Small Number of Subgroups," *Journal of Quality Technology*, 2, pp. 9–16.

7

Monitoring Attribute Data

CHAPTER OVERVIEW

Up to this point, we have primarily studied control charts for variables, that is, for quality-related characteristics that can be quantitatively measured, such as cycle time, customer satisfaction levels, weight, diameter, flow rate, and temperature. This chapter, however, considers control charts for applications in which the data are discrete, or countable. In the typical application a quality characteristic can be classified into one of two categories, depending on whether it meets some specification (or guideline) of quality. Quality measurements of this type are called *attributes*. In some applications, an attribute measure is the only way to define quality, for example, the absence or presence of a required part. In other applications, a precise numerical measure may be possible but it is overly expensive; for example, a go/no-go gage may be used to indicate whether a dimension is outside or within engineering specifications because precise measurement of the dimension would be more time-consuming.

A very common attribute measurement is an evaluation of whether a particular item is defective or nondefective, acceptable or not acceptable. Another common attribute measurement is the number of occurrences of an undesirable quality characteristic per item. The term *defect* refers to an undesirable quality characteristic of an item, such as a flaw, a blemish, or an imperfection. Conceivably, an individual item can have several defects without being classified *defective*. For example, an automobile body with some surface scratches is undesirable, but obviously this does not imply the entire automobile is defective.

Even though the terms *defect* and *defective* seem to make the necessary distinctions, the American Society for Quality (ASQ) recommends replacing these terms with *nonconformity* and *nonconforming,* respectively. The reason is that the term *defective* implies that an item is seriously impaired or unusable while *nonconforming* simply implies that an item does not meet specifications or requirements. Nonconforming items may be repairable or sold as imperfect items at reduced prices. Thus, defective is viewed as a more serious form of nonconforming. For similar reasons and for consistency in terminology, the term *nonconformity* is used in place of the term *defect.* For the most part, we will adopt the recommended terminology in our exposition.

Based on these two different attribute measures, two standard types of attribute control charts are frequently used in practice. One, called a *p chart,* focuses on the proportion or percentage of nonconforming units in individual subgroups. A variation of this chart, an *np chart,* is based on the number of nonconforming units per subgroup. The other type, a *c chart,* focuses on counts of individual nonconformities within some unit which may be defined by area, space, or time. A chart closely related to the *c* chart is the *u chart* which is used in situations where multiple units are gathered and the average number of nonconformities is monitored. There is also a chart, called the *D chart,* that combines nonconformities based on the severity of different types of nonconformities. We will discuss the statistical basis for all these standard attribute charts and explore their effectiveness in practical applications.

7.1 MONITORING BINOMIAL-BASED STATISTICS

Consider the situation in which we observe an item in terms of whether it is conforming (acceptable) or nonconforming (not acceptable). Thus, we are concerned with dichotomous measurements that reflect which of the two basic outcomes, for a given item, has occurred. To quantify these outcomes, we can assign one outcome the value of 1 and the other the value of 0. In most statistical applications, the value 1 corresponds to a "success" while the value 0 corresponds to a "failure." In the quality area, the term *success* is not taken literally since it is an accepted convention to assign to a nonconforming item the value of 1 and to a conforming item the value of 0. In principle, there is no reason why this assignment cannot be reversed; if reversed, we would then be monitoring the percentage conforming.

Since these two outcomes are complements, and if p denotes the probability of a nonconforming item, the probability of a conforming item is $1 - p = q$. Recall from Section 2.7 that a random variable defined on this basis is known as a *Bernoulli random variable*. In practice, we are not simply interested in the result of one Bernoulli trial; rather, we wish to consider a sequence of Bernoulli trials, each of which results in a conforming or nonconforming item.

To determine the capability and stability of a process, it is theoretically possible to study strings of 1s and 0s generated from a series of Bernoulli trials. However, practically speaking, this would be a rather difficult task. Instead, a subgroup of items is collected, and the percent (or number) of nonconforming items in the subgroup is recorded and monitored over time. As presented in Section 2.7, the binomial distribution serves as the theoretical distribution for the number of successes (nonconforming) in repeated equal-size samples of Bernoulli trials. Recall that the binomial distribution is given by

$$P(X = x) = \binom{n}{x} p^x (1 - p)^{n-x} \quad x = 0, 1, \dots, n$$

where X is a random variable representing the number of nonconforming items in a sample of n items with probability p of a nonconforming item. From Section 2.7 we learned that the mean and variance of a binomial random variable are np and $np(1 - p)$, respectively.

The *sample proportion nonconforming* $\hat{p}$ is defined as the ratio of the number of nonconforming items in the sample to the sample size, that is, $\hat{p} = X / n$. As noted in Section 2.7, the mean and standard deviation of $\hat{p}$ are, respectively,

$$\mu_{\hat{p}} = p \tag{7.1}$$

and

$$\sigma_{\hat{p}} = \sqrt{\frac{p(1 - p)}{n}} \tag{7.2}$$

In Section 2.9, we pointed out that $\hat{p}$ is the *average* number of nonconforming items in a subgroup of n items, and thus the central limit theorem takes effect.

Namely, as n increases, the distribution of $\hat{p}$ approaches the normal distribution. Since the binomial distribution is symmetric when $p = 0.5$, the normal approximation is obviously better, for a given sample size n, as p gets closer to 0.5. In a subsection to follow, we discuss in greater detail guidelines for determining subgroup sizes when dealing with $\hat{p}$. Transformations can also be considered for normalizing skewed binomial data. One such transformation is based on the trigonometric arcsin ($\sin^{-1}$) function. Excellent discussions of the implications of this transformation can be found in Box, Hunter, and Hunter (1978) and in Ryan (1989).

CONTROL CHART FOR PROPORTION NONCONFORMING: p CHART

When the binomial distribution is adequately approximated by the normal distribution, control limits for monitoring the proportion nonconforming are easy to derive. Namely, we apply the standard 3σ framework which calls for limits placed at $\mu_{\hat{p}} \pm 3\hat{\sigma}_{\hat{p}}$. If the true proportion nonconforming p is known, then replacing (7.1) and (7.2) in the 3σ control chart model gives

$$\text{UCL} = p + 3\sqrt{\frac{p(1-p)}{n}}$$

$$\text{CL} = p \tag{7.3}$$

$$\text{LCL} = p - 3\sqrt{\frac{p(1-p)}{n}}$$

In most practical applications, we need to obtain an estimate for p. As a first step, we select m preliminary subgroups, each of size n. As in the case of variable subgroups, it is recommended that m be at least 20. This will allow us to extend out the retrospective (or phase 1) control limits as prospective (or phase 2) control limits. For each of these subgroups, we have the observed number of nonconforming items, denoted by $x_1, x_2, \ldots, x_m$. Accordingly, we can compute the sample proportion nonconforming for the subgroups, denoted by $\hat{p}_1, \hat{p}_2, \ldots, \hat{p}_m$. We can now proceed to estimate p, by calculating the proportion nonconforming observed across all subgroups:

$$\bar{p} = \frac{\sum_{i=1}^{m} \hat{p}_i}{m} = \frac{\sum_{i=1}^{m} x_i}{mn} \tag{7.4}$$

where x_i is the observed number of nonconforming items in the ith subgroup. Using this unbiased estimate in place of p in (7.2), we can estimate the standard deviation of the proportion nonconforming by

$$\hat{\sigma}_{\hat{p}} = \sqrt{\frac{\bar{p}(1-\bar{p})}{n}} \tag{7.5}$$

Hence, by substituting the estimates of (7.4) and (7.5) into (7.3), we obtain the following upper and lower control limits:

$$\text{UCL} = \bar{p} + 3\sqrt{\frac{\bar{p}(1 - \bar{p})}{n}}$$

$$\text{LCL} = \bar{p} - 3\sqrt{\frac{\bar{p}(1 - \bar{p})}{n}}$$

(7.6)

with centerline positioned at $\bar{p}$. A sequence plot of the proportion nonconforming in successive subgroups with these limits superimposed is called a *p chart*. Given that we are using a symmetric distribution—normal—to approximate a generally skewed binomial distribution, it is possible for the lower control limit computed from the above formula to have a negative value. In such cases, LCL is set equal to zero which implies, for all practical purposes, that the limit plays no role. When LCL equals zero, it is difficult to know if a subgroup proportion ($\hat{p}$) which is equal to zero reflects a truc special cause or simply occurred because the true value of p is small. In a subsection to follow, we will discuss how increasing the subgroup size can ensure that the lower control limit is positive, and thus allows zero-valued subgroup proportions to stand out.

We need to emphasize *strongly* that an adequate approximation of the subgroup proportions to the normal distribution is not the only requirement for effective use of standard p chart limits. Another assumption is that the subgroup proportions, except for possible local departures reflecting special causes, should behave randomly. As discussed and illustrated in the variable control chart chapters, horizontal limits superimposed on systematic nonrandom behavior do not necessarily provide a reliable means of detecting special causes. There is also an assumption that the estimated standard deviation given in (7.5) accurately reflects the true variability of the subgroup proportions. For this to happen, the within-subgroup observations of 1s and 0s should be independent and the probability of the occurrence of a nonconforming item remains constant. Additionally, the probability of a nonconforming item should remain constant from subgroup to subgroup. If these requirements are violated, the estimated standard deviation given in (7.5) will not be an appropriate estimate for the variability of the subgroup proportions, which, in turn, undermines the applicability of the standard p chart limits. In Section 7.2, we present an application that dramatically illustrates this point.

In addition to the control limits, other control criteria, such as supplementary runs rules, can be applied to detect abnormal variation within the control limits. In Section 4.9, we listed eight control criteria, which are available through Minitab, for application on variable data. Theoretically, all eight of these tests can be applied to the attribute data. However, Minitab has chosen to offer only the first four of the eight listed criteria in Section 4.9 for attribute control charts.

Depending on the direction of an out-of-control signal, the interpretation and implication can be quite different. If an out-of-control point is on the high side, an investigation should be launched with the goal of removing and protecting against any special causes of poor quality. However, if an out-of-control point is on the low side, an investigation should also be conducted to promote any special causes of high quality.

Keep in mind that the basic principles of retrospective and prospective control for attribute charts are the same as those for variable charts. Thus, the initial control limits computed from the preliminary subgroups are regarded as trial limits. Any subgroup found to be unrepresentative due to special causes is set aside, and the trial limits are revised. The process continues until the remaining data are deemed void of special causes; at that time, the data can serve as a basis for prediction and prospective control.

Example 7.1

We now turn to an application that illustrates the construction of a standard p chart. In particular, the data come from the home mortgage Records and Support Center (RSC) of a large bank. The RSC is a centralized department that supports most administrative functions, such as storage and tracking of mortgage loans, filing documents into those loans, and delivering mail to all staff members at several facilities.

One area that was found to be in need of immediate improvement was the *drop-filing* area. Drop-filing is the process of placing all documents into the appropriate mortgage loan file. Misplacing documents contributes to delays in the processing time of mortgages. The implication of processing cycle times is serious since many customers simultaneously apply to more than one bank, waiting for the first approval. In fact, the bank has lost much business due to its long cycle times.

Since there are literally tens of thousands of files and inspection is manual, it is not feasible to perform a 100 percent audit. A practical alternative is to periodically audit a sample of files. A nonconforming item is defined as any file which is missing required documents or any file including misplaced documents intended for some other file. To establish a baseline understanding of the drop-filing error rate, RSC conducted a 2-month pilot study in which 150 active files were audited every other business day (in particular, Monday, Wednesday, and Friday). Since mortgages are coded sequentially by a 10-digit number, random samples are easily obtained with the use of a simple computer program that generates random numbers. In general, for detailed discussions on statistical sampling concepts applied to administrative and support processes, refer to Glasser (1985). Data collected from 24 subgroups (8 weeks $\times$ 3 days), each of size 150, are shown in Table 7.1 and can also be found in the file "drop-file.dat".

For the data in Table 7.1, we can compute the overall average proportion nonconforming as follows:

$$\bar{p} = \frac{\sum_{i=1}^{m} x_i}{mn} = \frac{250}{24(150)} = 0.06944$$

For a subgroup size of 150, we might expect an adequate approximation by the normal distribution for the data set. In Figure 7.1, we display a normal probability plot of the standardized proportions. Since the subgroup proportions will be seen to exhibit no systematic time patterns, we can safely interpret the normality checks. (Recall from Chapter 3 the Deming spring elongation series which illustrated that the normal diagnostic checks should generally follow after the time behavior of the data is assessed.) The normal diagnostics in Figure 7.1 show that the sample distribution of the subgroup proportions is symmetric and consistent with approximate normality.

Table 7.1 Data from drop-filing process

Subgroup	Number of Nonconforming x_i	Number Inspected n_i	Proportion Nonconforming $\hat{p}_i$
1	12	150	0.080
2	11	150	0.073
3	10	150	0.067
4	12	150	0.080
5	6	150	0.040
6	12	150	0.080
7	11	150	0.073
8	12	150	0.080
9	8	150	0.053
10	13	150	0.087
11	11	150	0.073
12	7	150	0.047
13	9	150	0.060
14	15	150	0.100
15	11	150	0.073
16	12	150	0.080
17	6	150	0.040
18	8	150	0.053
19	11	150	0.073
20	9	150	0.060
21	13	150	0.087
22	10	150	0.067
23	8	150	0.053
24	13	150	0.087

Using the value of 0.06944 for $\bar{p}$ in (7.6), we calculate the p chart limits:

$$\text{UCL} = \bar{p} + 3\sqrt{\frac{\bar{p}(1-\bar{p})}{n}} = 0.06944 + 3\sqrt{\frac{0.06944(0.93056)}{150}} = 0.1317$$

and

$$\text{LCL} = \bar{p} - 3\sqrt{\frac{\bar{p}(1-\bar{p})}{n}} = 0.06944 - 3\sqrt{\frac{0.06944(0.93056)}{150}} = 0.007176$$

The p chart with the above computed upper and lower limits is shown in Figure 7.2. The limits appear to be reasonably placed relative to the data. Furthermore, there is no evidence of nonrandom systematic patterns or out-of-control situations; we invoked the runs rules option in Minitab, and no flags resulted. Therefore, the data are reflecting an in-control binomial process.

Even though the process is in a state of statistical control, the proportion of nonconforming files was viewed as unacceptably high, and thus attention was directed to analyzing the system

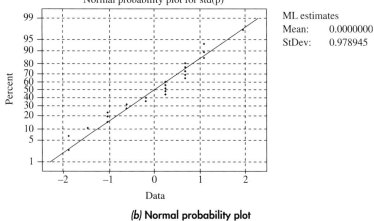

(a) **Histogram**

(b) **Normal probability plot**

Figure 7.1   Normality checks for standardized proportions from drop-filing application.

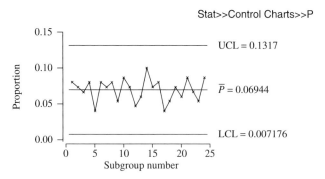

Figure 7.2   *p* Chart for the drop-filing application.

for improvement. Using problem-solving tools, such as flowcharts and cause-and-effect diagrams, a project team identified a simple solution that requires a change in the sorting procedure. Currently, documents are directly dropped in the files based on a unique 10-digit identification number associated with each loan. The improvement suggestion was to first *rough*-sort the documents based only on the first two digits, and then *fine*-sort the documents

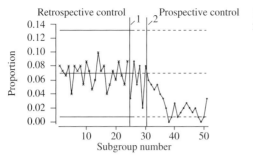

Figure 7.3   Verifying improvement with a *p* chart for the drop-filing application.

into the order that appears on the file shelf. By doing these presorts, the team felt that drop-filing can be done faster and more accurately. As reflected in Figure 7.3, the new drop-filing procedure resulted in a substantially improved process; the average proportion of nonconforming files dropped from about 7 percent to less than 2 percent. Given the fundamental change in the process, new control limits should be established for future monitoring. These new limits would serve as the baseline to determine if any additional efforts result in further improvements.

CONTROL CHART FOR NUMBER OF NONCONFORMING: *np* CHART

An equivalent alternative to plotting the proportion nonconforming is to plot the number of nonconforming in each subgroup. This option, called the *np chart,* is sometimes preferred since personnel may find it easier to understand counts than proportions. Additionally, once control limits are established, on-line computations are eliminated since there is no need to compute proportions, only counting is required.

Using $\bar{p}$ as an estimate for p along with the earlier stated mean and variance of the binomial distribution, we see the control limits for the *np chart* are

$$\text{UCL} = n\bar{p} + 3\sqrt{n\bar{p}(1 - \bar{p})}$$
$$\text{LCL} = n\bar{p} - 3\sqrt{n\bar{p}(1 - \bar{p})}$$

(7.7)

with the centerline positioned at $n\bar{p}$.

Since the plotted statistics are whole numbers, fractional control limit values can be replaced with appropriate integer values for operator ease. In particular, the closest integer in an outward direction away from the centerline should be chosen. For example, if UCL = 20.4 and LCL = 1.6, then the upper and lower control limits may be set at 21 and 1, respectively. If control limits are "rounded" in this manner, it is especially important to remember that points that fall exactly on the control limits should be regarded as out-of-control signals.

Example 7.2	For the data studied in Example 7.1, we could have equivalently implemented an *np* chart. From Example 7.1, we have $n = 150$ and $\bar{p} = 0.06944$. To illustrate, we compute limits based on (7.7):

$$\text{UCL} = 150(0.06944) + 3\sqrt{150(0.06944)(0.93056)} = 19.76$$

$$\text{LCL} = 150(0.06944) - 3\sqrt{150(0.06944)(0.93056)} = 1.076$$

The centerline is set at $150(0.06944) = 10.42$. The *np* chart, with this centerline and the above control limits, is given in Figure 7.4. The same conclusions are drawn with this chart as with the *p* chart shown in Figure 7.2.

SUBGROUP SELECTION

It is useful to recognize that the *p* chart is conceptually similar to the $\bar{x}$ chart because both charts track the average values of the subgroups. In one case the average of a set of 1s and 0s is computed while in the other case the average of a set of continuous (or nearly continuous) measurement values is computed. Thus, many of the subgroup design issues pertinent to the $\bar{x}$ chart (see Section 6.1) apply also to the *p* chart. For instance, the concept of rational subgrouping is equally important for the *p* chart; that is, all items within the subgroups should be influenced by identical conditions. It is not advisable to sample across production runs, operators, processes, work centers, shifts, and so on, since vital information on their differences would be buried. Additionally, the added within-subgroup variation would widen the control limits, rendering the charting procedure less sensitive to detecting out-of-control departures. Lastly, even if an out-of-control signal were revealed, mixing together different processes makes it difficult to pin down the specific source for investigation.

Determining subgroup size is another important design issue. There are various guidelines for determining the value of *n*. In Section 7.1, we mentioned that

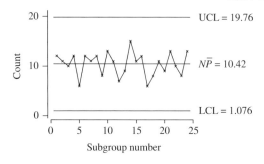

Figure 7.4   *np* Chart for the drop-filing application.

the subgroups should be chosen sufficiently large that the normal approximation to the binomial distribution is adequate. Grant and Leavenworth (1996) suggest a rule of thumb of np greater than 5 for consideration of the normal distribution to approximate the binomial distribution. Note that this rule of thumb implicitly assumes that $p \leq 0.5$. If $p > 0.5$, we then use $1 - p$ in place of p, that is, $n(1 - p) > 5$. To avoid ambiguity, an alternative rule of thumb of $np(1 - p) > 5$ might be considered. Besides the appropriateness of the normal approximation, we discuss below other considerations for subgroup size determination.

One approach is to choose the subgroup size large enough that there is a high probability that some nonconforming items will be observed. A detailed discussion of this approach is also found in Farnum (1994). If the subgroups are too small and the probability p of a nonconforming item is also small, there is a real possibility that most of, if not all, the preliminary subgroup proportions $\hat{p}_i$ will be equal to 0. If all $\hat{p}_i = 0$, the result is a meaningless p chart since the average proportion $\bar{p} = 0$ which, in turn, implies the lower and upper control limits are set to zero. To minimize the possibility of this scenario, one can specify a probability γ of observing a subgroup with at least one nonconforming item to be high (say, 0.90 or 0.95). Since we are dealing with a small value of p, the required probability calculations can be based on the Poisson approximation to the binomial (see Section 2.7). To determine the probability of at least one nonconforming item, we recognize the following fact:

$$\gamma = P(X > 0) = 1 - P(X = 0) \tag{7.8}$$

Using the Poisson approximation with the Poisson parameter $\lambda = np$, the probability of 0 nonconforming items is

$$P(X = 0) \cong \frac{e^{-\lambda}\lambda^0}{0!} = e^{-\lambda} = e^{-np}$$

Substituting this approximate probability in (7.8), we have $\gamma \cong 1 - e^{-np}$. Using the natural logarithm and solving for n, we find

$$n = \frac{-\ln(1 - \gamma)}{p}$$

If, for instance, $p = 0.01$ and it is desired that the probability be at least 0.95 of observing a subgroup with at least one nonconforming item, then the minimum sample size should be

$$n = \frac{-\ln(1 - 0.95)}{0.01} = \frac{-\ln 0.05}{0.01} = \frac{-(-2.996)}{0.01} = 299.6 \text{ or } 300$$

However, we should take note that this minimum sample does not meet the requirement of the normal approximation rule of thumb which calls for $np > 5$. Another consideration for determining the subgroup size is the sensitivity of the chart in detecting a specified increase in the proportion nonconforming. As described in Chapter 6 for variable control charts, the operating-characteristic (OC) and average run length (ARL) curves measure the capability of a control chart to detect shifts in the underlying process parameters. In a similar fashion, the OC and ARL curves can be constructed for the p chart. By constructing OC (or ARL) curves for different subgroup sizes, one can find what subgroup size provides an expected level of performance for the detection of specific changes in p. For illustrations and detailed discussion of the OC and ARL curves for p charts, the interested reader is referred to Duncan (1986).

From another perspective, many quality control textbooks suggest that the subgroup size might be selected such that the lower control limit of the p chart is positive. The idea is to allow the p chart to highlight significant reductions in proportion nonconforming, possibly reflecting improvement efforts. Thus, we are requiring that

$$\text{LCL} = p - 3 \sqrt{\frac{p(1-p)}{n}} > 0$$

which implies

$$n > \frac{9(1-p)}{p} \tag{7.9}$$

Even though this criterion ensures a positive LCL for a given p, it does not guarantee that a positive $\hat{p}_i$ can fall below LCL. In fact, it may be *impossible* for a positive $\hat{p}_i$ to be less than LCL even if the subgroup size is based on this criterion; surprisingly, this fact is not pointed out in most quality control textbooks. Actually, there are several popular textbooks in which the authors wrongly imply that observing an unusually small number of nonconforming items (such as one nonconforming item) is ensured by this criterion. To illustrate, let us assume $p = 0.05$. For $p = 0.05$, we find that (7.9) implies the need for at least 172 items to satisfy the positive LCL criterion. If we choose this minimum value, then the p chart LCL becomes

$$\text{LCL} = p - 3 \sqrt{\frac{p(1-p)}{n}} = 0.05 - 3 \sqrt{\frac{0.05(1-0.05)}{172}} = 0.000145$$

As can be seen, the limit is indeed positive. But notice that if one nonconforming item occurs in some subgroup, the corresponding proportion nonconforming is $\hat{p} = 1/172 = 0.0058$ which is greater than LCL. The only value of $\hat{p}$ that can appear below the lower control limit is zero. Thus, this criterion only guarantees that zero-valued $\hat{p}$ will be viewed as legitimate signals of change. For $p = 0.05$, a subgroup size of 210 is required to ensure that a positive $\hat{p}$ can breach the lower control limit.

Finally, given the interdependence between the recommended subgroup sizes and the underlying proportion nonconforming, subgroup size choice is not a static decision. Presumably, processes targeted for monitoring will be improved upon. As such, new standard p chart limits would be established, and larger subgroup sizes may likely be required so that approximate normality is attained given the smaller value for p.

VARIABLE CONTROL LIMITS

Both the p chart and np chart just presented assume a constant subgroup size n. However, there are situations in which subgroup sizes might vary. For instance, subgroups might be formed based on the complete inspection of hourly or daily production output. For service applications, subgroups may be formed on a day's or week's worth of administrative activities, for example, correct filing or application reporting.

To deal with variable subgroup sizes, one approach is to simply replace n by n_i in the computation of control limits, where n_i is the size of subgroup i. Hence the control limits will vary depending on the subgroup size. Theoretically, we could use this approach for either the p chart or the np chart, given their equivalence. However, it is generally recommended that this adaptation be applied only to the p chart. The problem with the np chart is that its centerline in addition to the control limits will vary with subgroup size. As a result, pattern analysis of any sequence of observations may become difficult. It may, for instance, not be clear if an increase in the number of nonconforming items is due to a true process deterioration or is simply the result of getting more nonconforming items because of an increase in subgroup size. Conversion to proportions allows for a fairer comparison but in certain situations may not be satisfactory, as illustrated soon.

Variable control limits for the p chart are shown below:

$$\text{UCL}(i) = \bar{p} + 3\sqrt{\frac{\bar{p}(1-\bar{p})}{n_i}}$$
$$\text{LCL}(i) = \bar{p} - 3\sqrt{\frac{\bar{p}(1-\bar{p})}{n_i}}$$

(7.10)

Rather than take the simple average of the subgroup proportions, $\bar{p}$ should be computed as the total number of defects divided by the total number of trials, that is,

$$\bar{p} = \frac{\sum_{i=1}^{m} x_i}{\sum_{i=1}^{m} n_i}$$

(7.11)

where x_i is the number of nonconforming items in subgroup i.

An interesting example of the use of the p chart is the monitoring of the percentage of price quotations that are accepted by customers. This percentage is sometimes referred to as a *capture rate* (see Latzko and Dowhin 1991) or as a *kill ratio*—a vernacular term common among sales personnel. Clearly, an understanding and an improvement of the price quotation process are at the heart of being in business for many organizations. Data for the number of weekly price quotations along with the number not accepted from a division of a large manufacturing

Example 7.3

Table 7.2   Data from the quotation process

Week	Number Not Captured x_i	Number of Quotes n_i	Proportion Not Captured $\hat{p}_i$	Week	Number Not Captured x_i	Number of Quotes n_i	Proportion Not Captured $\hat{p}_i$
1	38	59	0.644	21	38	62	0.613
2	44	63	0.698	22	38	63	0.603
3	30	54	0.556	23	34	58	0.586
4	39	60	0.650	24	26	53	0.491
5	38	60	0.633	25	33	59	0.559
6	27	55	0.491	26	41	63	0.651
7	31	56	0.554	27	43	65	0.662
8	35	60	0.583	28	34	60	0.567
9	31	59	0.525	29	31	54	0.574
10	40	63	0.635	30	22	50	0.440
11	27	75	0.360	31	36	63	0.571
12	27	54	0.500	32	37	61	0.607
13	30	57	0.526	33	40	65	0.615
14	28	52	0.538	34	40	63	0.635
15	45	65	0.692	35	34	63	0.540
16	24	56	0.429	36	40	63	0.635
17	35	62	0.565	37	29	55	0.527
18	33	55	0.600	38	31	53	0.585
19	23	51	0.451	39	33	56	0.589
20	41	65	0.631	40	39	63	0.619

company are provided in Table 7.2 and found in the file "capture.dat". To be consistent with the concept of nonconforming items, data are presented in terms of the quotations not captured. The levels of percent not captured, as seen in Table 7.2, are not unusual. In most competitive markets, capture rates of less than 50 percent are quite common. Additionally, it is not surprising that the number of price quotations varies from week to week.

By using (7.11), the average proportion of price quotations not captured is computed as

$$\bar{p} = \frac{\sum\limits_{i=1}^{40} x_i}{\sum\limits_{i=1}^{40} n_i} = \frac{1365}{2373} = 0.5752$$

The upper and lower control limits for each subgroup are given by (7.10):

$$\text{UCL}(i) = 0.5752 + 3\sqrt{\frac{0.5752(0.4248)}{n_i}} = 0.5752 + \frac{1.483}{\sqrt{n_i}}$$

$$\text{LCL}(i) = 0.5752 - 3\sqrt{\frac{0.5752(0.4248)}{n_i}} = 0.5752 - \frac{1.483}{\sqrt{n_i}}$$

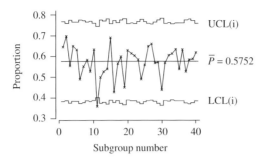

Figure 7.5 p Chart for the price quotation application.

Thus, for week 1

$$\text{UCL}(1) = 0.5752 + \frac{1.483}{\sqrt{59}} = 0.7683$$

$$\text{LCL}(1) = 0.5752 - \frac{1.483}{\sqrt{59}} = 0.3821$$

The p chart, with individual control limits for all 40 weeks, is shown in Figure 7.5. Generally, it appears that the subgroup proportions are reasonably dispersed within the control limits. There is, however, a signal for week 11 which is outside its control limits on the low side; that is, the proportion of noncaptured customers was significantly low or, equivalently, the capture rate was significantly high.

A follow-up study revealed that week 11 was the week of a trade exposition sponsored by certain practitioner associations which in the past the company had limited contact with. Interestingly, not only was the subgroup proportion significantly different for week 11, but also we see from Table 7.2 that the number of price quotations—opportunities—was much larger than that under ordinary conditions. Both these facts provided credible evidence to the company that consideration should be given to greater involvement with similar trade shows and possibly targeting, by way of advertising, members of the associations involved in the week 11 show. At this stage, since a special explanation can be attached to the significant subgroup proportion, this subgroup should be set aside and $\bar{p}$ revised, along with the control limits.

CONTROL CHART FOR STANDARDIZED PROPORTION DATA

A second approach to the issue of variable subgroup sizes is to convert or rescale the proportion data in a way that directly incorporates the subgroup sizes. One recommended rescaling is done as follows:

$$z_i = \frac{\hat{p}_i - \bar{p}}{\sqrt{\dfrac{\bar{p}(1 - \bar{p})}{n_i}}} \tag{7.12}$$

where $\bar{p}$ is calculated from (7.11). Notice that this conversion calls for subtracting from each subgroup proportion the estimated mean, and then dividing by an estimate of the standard deviation of the subgroup proportions. This rescaling is basically a form of standardization, thus the use of the symbol z_i. As previously seen, there are other ways to standardize data. For instance, the most common version is to subtract the sample mean and divide by the sample standard deviation s, where s is estimated directly from the data with no assumptions on the underlying distributions. The standardization done in (7.12) is rooted in a firm belief that the binomial model is applicable to the process in question. We will present an example in Section 7.2 which illustrates the contrary.

Assuming an underlying binomial model, the control limits for the standardized values are placed at ± 3 around a centerline of zero. A control chart for the standardized values is often called a standardized or stabilized p chart. The appeal of this control chart is that the control limits are fixed, not variable.

There are other advantages to standardization or rescaling. From a data analysis perspective, the incorporation of the subgroup sizes properly places the data observations relative to each other which, in turn, allows for appropriate interpretation of diagnostic checks of normality and randomness. Similarly, many authors suggest that the standardized p chart is more suitable to supplementary runs rules and pattern analysis than the np chart or the p chart with varying subgroup sizes. Earlier in this section, we discussed intuitively why pattern analysis seems to be unsuited for np chart data with varying subgroup sizes. However, there can also be misleading situations when one is dealing with subgroup proportions. To illustrate, suppose $\bar{p} = 0.04, \hat{p}_1 = 0.075$, and $\hat{p}_2 = 0.080$ with $n_1 = 200$ and $n_2 = 100$. Apparently, the second subgroup reflects a poorer level of quality than the first subgroup. Before we draw a definitive conclusion, let us examine the standardized values as computed from (7.12):

$$z_1 = \frac{0.075 - 0.040}{\sqrt{\dfrac{0.040(0.960)}{200}}} = 2.53$$

and

$$z_2 = \frac{0.080 - 0.040}{\sqrt{\dfrac{0.040(0.960)}{100}}} = 2.04$$

Thus, the first subgroup is 2.53 standard deviations above the average, while the second subgroup is 2.04 standard deviations above the average. The standardized values paint a different picture than the original proportions; the first subgroup, not the second, is viewed as farther above the average proportion nonconforming. Therefore, changes in the subgroup proportions are more accurately represented when taken relative to their subgroup size. For this reason, many authors strongly recommend against using runs rules and pattern analysis directly on proportion data. Consequently, it is no surprise that Minitab does not allow for the application of runs rules for p or np charts with nonconstant subgroups sizes.

Example 7.4

Consider the price quotation data studied in Example 7.3. In that example, we found that $\bar{p} = 0.5752$. From (7.12), the following standardized values for price quotation data are determined:

Table 7.3   Data for standardized p chart

Week	Proportion Not Captured $\hat{p}_i$	Standardized Value z_i	Week	Proportion Not Captured $\hat{p}_i$	Standardized Value z_i
1	0.644	1.070	21	0.613	0.601
2	0.698	1.978	22	0.603	0.449
3	0.556	−0.292	23	0.586	0.170
4	0.650	1.172	24	0.491	−1.246
5	0.633	0.911	25	0.559	−0.247
6	0.491	−1.265	26	0.651	1.214
7	0.554	−0.327	27	0.662	1.408
8	0.583	0.127	28	0.567	−0.134
9	0.525	−0.773	29	0.574	−0.017
10	0.635	0.959	30	0.440	−1.934
11	0.360	−3.770	31	0.571	−0.061
12	0.500	−1.118	32	0.607	0.495
13	0.526	−0.747	33	0.615	0.655
14	0.538	−0.536	34	0.635	0.959
15	0.692	1.910	35	0.540	−0.570
16	0.429	−2.220	36	0.635	0.959
17	0.565	−0.170	37	0.527	−0.719
18	0.600	0.372	38	0.585	0.143
19	0.451	−1.795	39	0.589	0.213
20	0.631	0.906	40	0.619	0.704

$$z_i = \frac{\hat{p}_i - 0.5752}{\sqrt{\dfrac{0.5752(0.4248)}{n_i}}} = \frac{\sqrt{n_i}(\hat{p}_i - 0.5752)}{0.49431}$$

Table 7.3 gives the original subgroup proportions along with the standardized z values. The standardized p chart is shown in Figure 7.6.[1] Again, week 11 sticks out relative to the control limits. In general, the p chart with exact variable limits and the standardized p chart are equivalent in highlighting individual points which fall outside the control limits. Additionally, as we noted, standardization allows us to be absolutely safe in the application of runs rules. The plot of standardized values in Figure 7.6 gives no evidence of sequences inconsistent with randomness. Thus, only subgroup 11 should be discarded in the revision of limits.

[1]Note: Minitab does not have a standardized p chart option. The user must first use the column of proportions to calculate and store in another column the standardized values; the calculation and storage are accomplished with Minitab's "Calculator" feature (Calc >> Calculator). The user can then implement an individual measurement control chart (Stat >> Control Charts >> Individuals) on the standardized values with mean and standard deviation set to 0 and 1, respectively. With Minitab's individual measurement chart option, the user has the further option of supplementing the chart with runs rules.

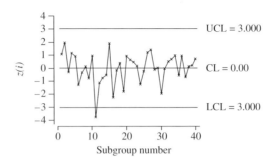

Figure 7.6   Standardized p chart for the price quotation application.

CONTROL LIMITS BASED ON AVERAGE SUBGROUP SIZE

A third approach is to construct approximate but fixed control limits by using the average subgroup size in the computation of p chart limits. This possibility is most suitable when the differences among subgroup sizes are small. As a rule of thumb, this approach is preferred if the subgroup sizes vary by no more than 25 percent around the average subgroup size (see Ford Motor Company, 1984). The average subgroup size is computed as follows:

$$\bar{n} = \frac{\sum_{i=1}^{m} n_i}{m} \tag{7.13}$$

Consequently, the approximate control limits are given by

$$\text{UCL}_{\text{approx}} = \bar{p} + 3 \sqrt{\frac{\bar{p}(1 - \bar{p})}{\bar{n}}}$$

$$\text{LCL}_{\text{approx}} = \bar{p} - 3 \sqrt{\frac{\bar{p}(1 - \bar{p})}{\bar{n}}} \tag{7.14}$$

where $\bar{p}$ is calculated from (7.11). Given the approximate nature of the control limits, points falling near the control limits should be interpreted with caution. It is possible that a point is deemed out of control by the approximate limits but is within exact limits, and vice versa. To help determine when exact computation may be necessary, Besterfield (1986) describes the following four possibilities:

Case 1: $n_i < \bar{n}$ and $\text{LCL}_{\text{approx}} < \hat{p}_i < \text{UCL}_{\text{approx}}$
In this case, the exact limits will be wider than the approximate limits. Since the subgroup proportion falls within the approximate limits, it must necessarily fall within the exact limits. Thus, there is no need to compute exact limits.

Case 2: $n_i > \bar{n}$ and $\text{LCL}_{\text{approx}} < \hat{p}_i < \text{UCL}_{\text{approx}}$
In this case, the exact limits will be narrower than the approximate

limits. Therefore, a subgroup proportion that falls within the approximate limits may not necessarily fall within the exact limits. Thus, if a subgroup proportion is slightly within the approximate limits, it is advisable to compute the exact control limits.

Case 3: $n_i > \bar{n}$ and $\hat{p} < \text{LCL}_{\text{approx}}$ or $\hat{p}_i > \text{UCL}_{\text{approx}}$
In this case, the exact limits will be narrower than the approximate limits. Since the subgroup proportion falls outside the approximate limits, it must necessarily fall outside the exact limits. Thus, there is no need to compute exact limits.

Case 4: $n_i < \bar{n}$ and $\hat{p}_i < \text{LCL}_{\text{approx}}$ or $\hat{p}_i > \text{UCL}_{\text{approx}}$
In this case, the exact limits will be wider than the approximate limits. Therefore, a subgroup proportion that falls outside the approximate limits may not necessarily fall outside the exact limits. Thus, if a subgroup proportion is slightly within the approximate limits, it is advisable to compute the exact control limits.

The appeal of the average subgroup size approach is that it offers a control chart with fixed limits which may be less intimidating to personnel. However, this attractive feature is counterbalanced by the computational "rework" that may be required for exact determination of whether suspected points are legitimate signals.

Example 7.5

Consider again the price quotation data first studied in Example 7.3. In Example 7.3, we constructed a p chart with exact but varying control limits. A drawback of a p chart with varying limits is the jagged look to the chart. So, as an alternative, we constructed a standardized p chart in Example 7.4. However, a drawback with the standardized p chart approach is the loss of reference to the actual proportion nonconforming. To be able to plot the original proportions of nonconforming with the "clean" appearance of fixed control limits, the p chart limits can be approximated by using the average subgroup size. For the price quotation data, the average subgroup size, given by (7.13), is

$$\bar{n} = \frac{\sum_{i=1}^{40} n_i}{40} = \frac{2373}{40} = 59.325$$

The control limits, based on $\bar{n}$, are found by using (7.14) and are

$$\text{UCL}_{\text{approx}} = 0.5752 + 3\sqrt{\frac{0.5752(0.4248)}{59.325}} = 0.7677$$

$$\text{LCL}_{\text{approx}} = 0.5752 - 3\sqrt{\frac{0.5752(0.4248)}{59.325}} = 0.3827$$

Figure 7.7[2] shows the p chart with control limits constructed from the average subgroup size. Notice that subgroup 11 has a proportion nonconforming of 0.360 which falls below the approximate lower control limit. Suppose, however, that we had no knowledge of the exact lower

---

[2]Note: Minitab does not have a p chart option to allow for the use of average subgroup size. An easy way to create the chart is to plot the proportions using the time-series option (Graph >> Time Series Plot). With the control limits hand-calculated, one can have the control limits superimposed by choosing the reference option (Frame >> Reference) and inputting the values of the control limits to the Positions box.

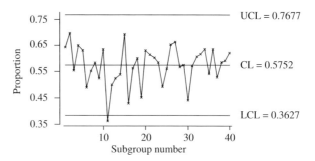

Figure 7.7   *p* Chart based on average subgroup size for the price quotation application.

control limit for subgroup 11. Would the exact limit need to be computed? Not in this case since the actual subgroup size is 75, which is greater than the average subgroup size, implying that the exact limit would be narrower than the approximate limits. Hence, this subgroup will exceed the exact control limits, as we have already seen. Again, since this subgroup has a special explanation, it should be put aside and control limits recomputed based on revised $\bar{p}$ and $\bar{n}$.

BEYOND SPECIAL-CAUSE SEARCHING

Even though the *p* chart helps in the discovery of unusual changes (good or bad) indicative of special causes, this is not necessarily its sole purpose. For instance, a plot of the proportion nonconforming provides management with a sense of the general quality level and of how well the organization is doing relative to some desired objectives. If the process is consistently performing at an unacceptable level (that is, $\bar{p}$ is too high), management should focus on the system of common causes generating this level of poor quality. Sporadic special causes affecting the system become secondary issues. Thus, until the process has been stabilized to more acceptable levels of performance, it is advisable to simply plot the proportion nonconforming *without* distraction from the control limits.

In seeking root causes for high average proportion nonconforming, remember that the observed data are dependent on the *definition* of a nonconforming item. For manufacturing processes, conformance is typically defined in terms of upper and lower specification limits, sometimes referred to as product tolerances. Unacceptable levels of nonconforming product may be due to unrealistic, unnecessary, or improper tolerance limits. A detailed manufacturing review would be a useful activity to sort out the possible culprits. Not only should machines and methods be studied, but also information on customer requirements should be incorporated. An effective review process is necessarily cross-functional and, minimally, should include representatives from marketing, product design, manufacturing engineering, quality, and production. An excellent discussion of the ingredients for a successful manufacturing review can be found in Juran and Gryna (1988).

Inappropriate guidelines for quality are not restricted to the manufacturing sectors. For example, we have personally encountered an effort in a large commercial lending institution to reduce the number of telephone rings prior to the calls being answered. This is a common improvement activity in many service organizations since it is believed that quick answering reflects responsiveness to the customer, which should in turn please customers. At this particular company, changes in methods and procedures were indeed made resulting in significantly decreased answering times such that most calls were caught by the third ring. Embraced by the notion of continual improvement, *good* quality was then defined as answering by the second ring. Daily data on the percentage of calls answered after two rings $\hat{p}_i$ were collected and monitored. The results were disappointing: $\bar{p}$ exceeded 0.30 and there was no apparent solution short of costly additional staffing. The problem was that the definition of quality was overly restrictive and not realistic. As an interesting epilogue, the organization, frustrated by the lack of further improvement, decided to ask customers what their requirements were. Much to the surprise of the company, customers did not notice (or care) if the number of rings was one, two, three, or, for that matter, four. The predominant concern was being transferred or connected correctly to the desired parties. Much too often organizations assume they know what should be improved rather than develop a system to highlight what truly should be improved.

Another purpose of a p chart is to suggest places for more precise measurements. For some applications, one of the primary reasons for an attribute approach is that it is more convenient and less costly than variable measurements. As noted in the introduction of this chapter, in some situations, attribute information is the only type of measurement that makes sense; for example, an electronic device works or does not, or an owner's manual is in the package or is not. If variable measurements are possible but an attribute approach is chosen instead, the disadvantage is that a quality characteristic is classified into only two classes (conforming or nonconforming). There is no information on how close or far the quality characteristic is from acceptable standards; for manufacturing processes, conformance is typically defined in terms of upper and lower specification limits.

With only attribute information, we cannot determine whether the process is off-target or whether the process variability is so great that nonconforming units are inevitable (see Figure 7.8). Variable data can provide the specific information to distinguish between the two scenarios depicted in Figure 7.8, and hence help determine appropriate courses of action for process improvement purposes. Since it is uneconomical to measure precisely all quality characteristics important to an organization, a reasonable strategy, as a starting point, is to monitor many processes by means of attribute data at a fraction of the cost. This will enable the organization to determine the general condition of these processes. Processes shown by the p chart to be generating an intolerable level

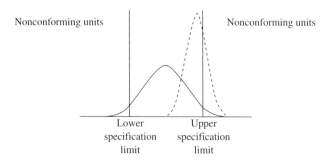

Figure 7.8   Nonconforming units generated from different processes.

of proportion nonconforming or to be hopelessly out of control can be earmarked for detailed troubleshooting that might entail analysis of pertinent variable measurements.

7.2   VIOLATIONS OF p CHART ASSUMPTIONS

The fundamental assumption of all control charts, variable and attribute, is that the individual measurements (and thus also the subgroup statistics obtained from these measurements) are independent and identically distributed. In addition, there is a distribution assumption depending on the particular type of chart. For subgroup variable charts, it is assumed that the subgroup means and ranges approximately follow a normal distribution. The p chart assumes that the subgroup proportions, for successive subgroups of size n_i, are independently and binomially distributed with parameters p and n_i. Furthermore, there is an assumption that the normal distribution adequately approximates the binomial distribution.

If the subgroup proportions do not follow a binomial distribution, then the estimated standard deviation $\sqrt{\bar{p}(1 - \bar{p}) / n_i}$ will not, in general, be applicable. In turn, control limits based on the binomial assumption will be wrong and potentially seriously misleading. Mosteller and Tukey (1977) underscore this phenomenon by stating[3]:

> If among thousands of manufactured piece parts the observed fraction of defective piece parts is p, and 1000 pieces are produced by one operator on one machine in one shift, it is risky business to suppose that the long-run average proportion of defectives will be between p − 3*SQRT(pq/1000) and p + 3*SQRT(pq/1000). The fraction defective is likely to depend on many things: the day of the week (Mondays being notorious), the operator, the machine, the shift, the supervisor, the inspector, and other plant matters we should not detail here. Appreciating this bramble of sources of variability led Walter Shewhart (1931) to devise methods of quality control with limits

---

[3]Recall that q is equal to $1 - p$.

Table 7.4   Data from the glass jar process

Subgroup	Proportion Nonconforming $\hat{p}_i$	Number Inspected n_i	Subgroup	Proportion Nonconforming $\hat{p}_i$	Number Inspected n_i
1	0.0387	116,000	29	0.0454	115,000
2	0.0313	150,000	30	0.0647	238,000
3	0.0233	472,000	31	0.0391	1,814,000
4	0.0251	477,000	32	0.0261	6,468,000
5	0.0354	48,000	33	0.0349	429,000
6	0.0517	433,000	34	0.0257	854,000
7	0.0543	374,000	35	0.0443	158,000
8	0.0340	2,561,000	36	0.0266	638,000
9	0.0373	428,000	37	0.0338	36,000
10	0.0345	318,000	38	0.0532	639,000
11	0.0700	100,000	39	0.0729	431,000
12	0.0381	262,000	40	0.0203	867,000
13	0.0239	92,000	41	0.0450	269,000
14	0.0349	858,000	42	0.0346	3,208,000
15	0.0453	369,000	43	0.0255	869,000
16	0.0577	214,000	44	0.0431	186,000
17	0.0364	314,000	45	0.0373	990,000
18	0.0871	264,000	46	0.0687	422,000
19	0.0303	857,000	47	0.0428	210,000
20	0.0314	261,000	48	0.0378	1,821,000
21	0.0451	864,000	49	0.0410	2,692,000
22	0.0250	132,000	50	0.0329	161,000
23	0.0495	2,120,000	51	0.0318	1,046,000
24	0.0385	1,063,000	52	0.0377	61,000
25	0.0182	650,000	53	0.0227	423,000
26	0.0604	371,000	54	0.0276	115,000
27	0.0555	108,000	55	0.0406	640,000
28	0.0288	316,000	56	0.0342	38,000

p +/− 3*SQRT(pq/n) as an ideal to be nearly achieved only after the most strenuous and sophisticated engineering efforts. What mass production with all its control and measuring ability cannot attain, other fields cannot expect formulas to give. Belief in such formulas may produce fancied security and sad surprises. (p. 123)

Let us now direct our attention to an example that amplifies the inappropriateness of the standard *p* chart limits.

Example 7.6

This example is related to the process of making glass jar caps from a lithographed tin plate. (The source of the data is H. V. Roberts, *Data Analysis for Managers*, Duxbury Press, Belmont, CA, 1991.) Specifically, the data include the proportions of nonconforming jar caps for 56 consecutive production orders over a period of several months. The data are presented in Table 7.4 and can also be found in the file "jarcap.dat".

First, we will construct the standard *p* chart with varying subgroup sizes. For varying subgroup sizes, the average proportion nonconforming is found by dividing the total number of nonconforming items by the total number of items inspected. Since the data in Table 7.4 are in terms of proportions, the average proportion nonconforming can equivalently be found as follows:

$$\bar{p} = \frac{\sum_{i=1}^{56} n_i p_i}{\sum_{i=1}^{56} n_i} = \frac{1,468,590}{40,430,000} = 0.03632$$

Hence, the exact control limits for each subgroup are calculated by

$$\text{UCL}(i) = 0.03632 + 3\sqrt{\frac{0.03632(0.96368)}{n_i}} = 0.03632 + \frac{0.1871}{\sqrt{n_i}}$$

$$\text{LCL}(i) = 0.03632 - 3\sqrt{\frac{0.03632(0.96368)}{n_i}} = 0.03632 - \frac{0.1871}{\sqrt{n_i}}$$

The sequence plot of the proportion nonconforming, with exact limits computed from the above equations, is given in Figure 7.9. No, we did not forget to put the limits; take a second look! The control limits are so tight that essentially every subgroup proportion is outside the control limits. Clearly, the points outside the limits should not, at this stage, be marked for isolated special-cause explanations; suspicion must be directed to the limits themselves.

The control limits are very tight because the binomial variation estimated by $\sqrt{\bar{p}(1 - \bar{p})/n_i}$ only accounts for a small part of the observed variation of $\hat{p}_i$. To understand how this may have occurred, it is useful to take stock of the requirements for the applicability of the binomial model:

1.   The within-subgroup observations (Bernoulli trials) are independent.
2.   The probability of a nonconforming item remains constant for all Bernoulli trials within the subgroup.
3.   The probability of a nonconforming item remains constant from subgroup to subgroup.

A process that behaves under these conditions is said to be a *simple binomial process.*

To different degrees, any violations in these three assumptions can render $\sqrt{\bar{p}(1 - \bar{p})/n_i}$ as a less reliable estimate of the variability of the subgroup proportions. Consider a violation of the Bernoulli independence assumption. This implies that the probability of a nonconforming item depends on the outcome(s) of prior trials. Often in practice, we may find that the long-run proportion of nonconforming items is fairly stable; however, in the short run, conforming and nonconforming items come in "strings." In other words, the probability of a nonconforming item is higher if a nonconforming item occurred recently, and similarly for conforming items. If the subgroups are relatively small, there is a chance that clusters of 1s or 0s may reside within the individual subgroups. This will result in the subgroup proportions being more variable than predicted by the standard estimate.

A second possible violation to the binomial model is a *randomly* changing probability of a nonconforming item for each trial within the subgroups while the average probability of a

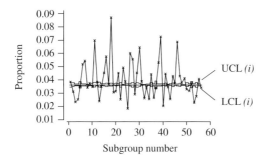

Figure 7.9 p Chart for the glass jar application.

nonconforming from subgroup to subgroup remains constant. To understand the implications of such a scenario, imagine having 100 unfair coins with the probabilities of getting a head (nonconforming item) distributed among the coins as follows:

Number of coins	25	25	25	25
Probability of coin landing as a head	0.10	0.20	0.30	0.40

If we denote the mean of the distribution of probabilities of getting a head as μ_π, it is not difficult to see that $\mu_\pi = 0.25$. The variance of the distribution of probabilities is found

$$\sigma_\pi^2 = \tfrac{1}{4}(0.10 - 0.25)^2 + \tfrac{1}{4}(0.20 - 0.25)^2 + \tfrac{1}{4}(0.30 - 0.25)^2 + \tfrac{1}{4}(0.40 - 0.25)^2 = 0.0125$$

Now, consider a repeated coin experiment where all $n = 100$ coins are thrown at once and the number of heads is recorded, or equivalently the proportion of heads. The idea is to view the 100 coins as a subgroup and the observed heads as nonconforming items. By tossing all the coins together, we are guaranteeing varying probabilities among the subgroup trials. Based on the results of Box, Hunter, and Hunter (1978), the mean and standard deviation of the subgroup proportion, for such a coin experiment, can be shown to be, respectively

$$\mu_{\hat{p}} = \mu_\pi \tag{7.15}$$

and

$$\sigma_{\hat{p}} = \sqrt{\frac{\mu_\pi(1 - \mu_\pi)}{n} - \frac{\sigma_\pi^2}{n}} \tag{7.16}$$

where μ_π and σ_π^2 have been derived above. If we take, for example, $n = 100$ with the distribution of probabilities illustrated above, then we have

$$\mu_{\hat{p}} = 0.25 \qquad \sigma_{\hat{p}} = \sqrt{\frac{0.25(0.75)}{100} - \frac{0.0125}{100}} = 0.0418$$

For comparative purposes, consider the standard case where the probability of a nonconforming item is a constant value p of 0.25; then the mean and standard deviation are

$$\mu_{\hat{p}} = 0.25 \qquad \sigma_{\hat{p}} = \sqrt{\frac{0.25(0.75)}{100}} = 0.0433$$

Interestingly and perhaps surprisingly, the expected variability of the subgroup proportions is slightly *less* than the standard case when the probabilities vary within the subgroups. Thus, probabilities varying only within the subgroups clearly cannot be at the heart of the matter with respect to the glass jar process example we are discussing.

Another possible departure from the standard binomial process is a randomly changing probability of nonconforming items, from subgroup to subgroup. Using the coin example, suppose instead: (1) a coin is taken at random from the 100 biased coins; (2) the particular coin is tossed $n = 100$ times, and the number of heads is recorded; and (3) the coin is returned, and the sequence of steps is repeated. Again, using the results of Box, Hunter, and Hunter (1978), we can show that the mean and variance of the subgroup proportion are

$$\mu_{\hat{p}} = \mu_{\pi} \tag{7.17}$$

and

$$\sigma_{\hat{p}} = \sqrt{\frac{\mu_{\pi}(1 - \mu_{\pi})}{n} + \frac{(n - 1)\sigma_{\pi}^2}{n}} \tag{7.18}$$

Accordingly, for the distribution of probabilities with $\mu_{\pi} = 0.25$ and $\sigma_{\pi}^2 = 0.0125$,

$$\mu_{\hat{p}} = 0.25 \qquad \sigma_{\hat{p}} = \sqrt{\frac{0.25(0.75)}{100} + \frac{99(0.0125)}{100}} = 0.1194$$

Notice that this expected variability, measured in terms of standard deviations, of the subgroup proportions is nearly 3 times larger than the standard case ($\sigma_{\hat{p}} = 0.0433$) with constant probability equal to 0.25.

We have explored, in isolation, the impact of varying probabilities within subgroups versus between subgroups on the expected variability of the subgroup proportions. A coin experiment can be quite easily designed to allow for varying probabilities within and between subgroups. The reader is invited to design such an experiment. Even though the nature of the varying probabilities was restricted to the design of the coin experiment, we are left with some general impressions. Namely, for reasonable conditions, varying probabilities within subgroups seem to have only a minor decreasing effect on the variability of the subgroup proportions while varying probabilities between subgroups result in a relatively larger increase in the variability.

With the above facts, we speculate that the large variability in subgroup proportions for the glass jar cap process is due to changing levels of probability of nonconforming items from production order to production order, rather than changes in the probability of nonconforming for the within-subgroup Bernoulli trials from standard assumptions. Thus, the process is apparently operating under two sources of common-cause variation: (1) a minor component related to within-subgroup or binomial variation and (2) a major component related to the production order to production order variation over and above the binomial variation. If *in control* is strictly defined as adherence to binomial variation, then clearly the process is out of control. However, we might ask, Are the proportions behaving in control on a broader perspective? Additionally, are there any indications of special causes, that is, unusual subgroups relative to the general variability? To answer these questions, we need to study directly the behavior of the subgroup proportion data without distraction from the standard p chart limits.

Given that the subgroup sizes vary, we technically should incorporate the subgroup sizes in our analysis for more accurate assessment of proportion data. As discussed and illustrated in previous sections, this can be accomplished by standardizing the proportion data computed by dividing the deviations from the centerline by the estimated standard deviation, namely, $\sqrt{\bar{p}(1 - \bar{p})/n_i}$. For data generated from a "pure" binomial process, the standardized data are expected to vary between ± 3. Since we know that standardization, computed in this manner,

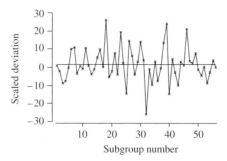

Figure 7.10   Time-series plot of the scaled deviations from the glass jar application.

for the glass jar cap data will not result in data varying between ± 3, there is little need to incorporate the $\bar{p}(1 - \bar{p})$ component in the calculation. We simply need to scale the deviations from the centerline based on the subgroup sizes. Given that $\bar{p} = 0.03632$ for the glass jar data, the scaled deviations are then computed as follows:

$$
\begin{aligned}
d_i &= \frac{\hat{p}_i - \bar{p}}{\sqrt{1/n_i}} \\
&= \sqrt{n_i}(\hat{p}_i - \bar{p}) \\
&= \sqrt{n_i}(\hat{p}_i - 0.03632) \qquad i = 1, 2, \ldots, 56
\end{aligned}
$$

We now proceed to analyze the scaled deviations d_i. In Figure 7.10, a time-series plot of the scaled deviation data is presented. The visual impression of the data supports randomness as do the runs test and ACF presented in Figure 7.11. Figure 7.12 gives the histogram and normal probability plot of the standardized (or centered) deviation values. Even though there is a slight pull in the right tail, an overall assessment of approximate normality is quite reasonable. Thus, the scaled deviations closely behave as a normal random process, that is, a process in a state of statistical control.

As an alternative to the inappropriate standard limits based on the assumption of binomial-only variation, control limits can be constructed by treating the scaled deviations as individual measurements. In Figure 7.13, we show an *X* chart for the scaled deviations. There are no indications of special-cause departures in relation to the control limits. Additionally, we applied supplementary runs rules to the data, and no abnormal variations were revealed. Thus, in a larger sense, the forces behind the variability of the subgroup proportions are *stable*, and there is no evidence of isolated special causes.

If any investigations are to be undertaken, they are best directed toward understanding why a standard binomial process is not at work for this application. The elimination of the additional between-subgroup variation may not be necessarily the desired goal. The underlying forces push the proportion data beyond standard limits not only in an upward direction but also in a downward direction. What needs to be learned is not only the sources for the extra variability but also how the sources can be controlled to produce the lower levels of proportion nonconforming found below the standard *p* chart limits. However, in the meantime, the control chart of Figure 7.13 serves as the most useful guide for general process monitoring.

```
              scaled deviation

                K =   1.4713

                The observed number of runs =   29
                The expected number of runs =   28.4286
                24 Observations above K 32 below
                    The test is significant at 0.8749
                    Cannot reject at alpha = 0.05
```

(a) Runs test

```
ACF of scaled deviation

                  -1.0 -0.8 -0.6 -0.4 -0.2 0.0  0.2  0.4  0.6  0.8  1.0
                  +----+----+----+----+----+----+----+----+----+----+
       1  -0.073                           XXX
       2  -0.115                          XXXX
       3   0.042                            XX
       4  -0.114                          XXXX
       5  -0.035                            XX
       6  -0.176                         XXXXX
       7  -0.024                            XX
       8   0.204                            XXXXXX
       9   0.022                            XX
      10  -0.047                           XX
      11   0.099                           XXX
      12   0.114                           XXXX
```

(b) ACF

Figure 7.11   Randomness checks of the scaled deviations from the glass jar application.

TESTING THE BINOMIAL ASSUMPTION

The glass bottle cap example was a dramatic illustration of the inconsistency of the standard limits with the observed variability of the proportion data. In our experience, such inconsistencies, though not quite as severe as shown in the last example, are quite commonplace. So, it is important to be able to recognize when the assumption of a binomial process may not be appropriate.

One of the simplest checks for departures from the standard assumptions is simply a visual comparison of the p chart (or np chart) limits with the observed data. With the possible exception of an occasional special-cause signal, when the proportion data are random and reasonably dispersed between the standard limits, this is a good indication that the binomial assumptions are being adequately met. As examples, recall the p charts for the drop-filing data (Figure 7.2) and the price quotation data (Figure 7.5). On the other end of the spectrum, a visual check of the p chart for the glass jar cap data was more than enough to recognize

```
Histogram of stddeviation    N = 56
   Midpoint        Count
    -2.750            1   *
    -2.250            0
    -1.750            2   **
    -1.250            4   ****
    -0.750            8   ********
    -0.250           17   *****************
     0.250           10   **********
     0.750            7   *******
     1.250            3   ***
     1.750            1   *
     2.250            2   **
     2.750            1   *
```

(a) **Histogram**

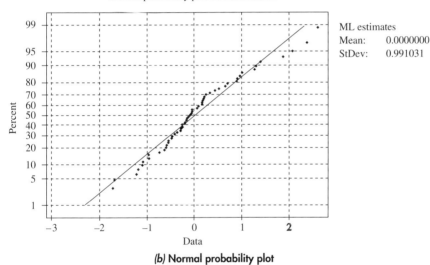

(b) **Normal probability plot**

Figure 7.12   Normality checks of the scaled deviations from the glass jar application.

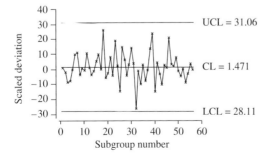

Figure 7.13   *X* chart for the scaled deviations from the glass jar application.

the departure from the binomial assumptions. There are, however, applications, in which the departure is more subtle.

Table 7.5   Data from the fuel injector process

Subgroup	Number of Nonconforming x_i	Number Inspected n_i	Proportion Nonconforming $\hat{p}_i$
1	10	200	0.050
2	1	200	0.005
3	4	200	0.020
4	11	200	0.055
5	8	200	0.040
6	3	200	0.015
7	1	200	0.005
8	8	200	0.040
9	3	200	0.015
10	2	200	0.010
11	12	200	0.060
12	3	200	0.015
13	8	200	0.040
14	2	200	0.010
15	9	200	0.045
16	5	200	0.025
17	6	200	0.030
18	7	200	0.035
19	11	200	0.055
20	2	200	0.010
21	6	200	0.030
22	5	200	0.025
23	8	200	0.040
24	3	200	0.015
25	2	200	0.010

Consider an application in which 200 manufactured fuel injectors are sampled periodically to check for compliance to specifications. The subgroup proportion data are given in Table 7.5 and can also be found in the file "injector.dat". The corresponding p chart for the data is given in Figure 7.14. As an initial reaction, the subgroup proportions appear in a state of statistical control and *seem* to be appropriately monitored by the binomial-based limits. However, upon closer inspection, there is a hint of extra variability in the subgroup proportions, giving a tendency for the points to "crowd" the control limits. A quick check for

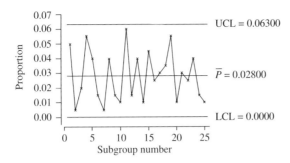

Figure 7.14 p Chart for the fuel injector application.

our suspicion would be to compare the sample standard deviation computed directly from the proportion values with the standard deviation estimate based on the binomial model. In particular, we can compute

$$s_{\hat{p}} = \sqrt{\frac{\sum_{i=1}^{25}(\hat{p}_i - \bar{p})^2}{25 - 1}} = \sqrt{\frac{\sum_{i=1}^{25}(\hat{p}_i - 0.028)^2}{24}} = 0.01708$$

and

$$\hat{\sigma}_{\hat{p}} = \sqrt{\frac{\bar{p}(1 - \bar{p})}{200}} = \sqrt{\frac{0.028(0.972)}{200}} = 0.01167$$

Notice that the sample standard deviation $s_{\hat{p}}$ is nearly 50 percent larger than the standard deviation estimate based on the assumption of binomial variation. Thus, the binomial-based estimate apparently does not fully account for the overall variation of the $\hat{p}$'s. This suggests that the standard p chart control limits may be misleading.

Beyond the visual check or an informal comparison of standard deviation estimates, we can employ a more formal statistical procedure to test for the existence of sources of variability not attributable to the binomial model. It can be shown that if $\hat{p}_1, \hat{p}_2, \ldots \hat{p}_m$ are m subgroups proportions taken from a simple binomial process, then

$$\sum_{i=1}^{m} \frac{(\hat{p}_i - \bar{p})^2}{\bar{p}(1 - \bar{p})/n_i} \overset{\cdot}{\sim} \chi^2_{m-1} \tag{7.19}$$

In other words, assuming the binomial model is true, the sum of the observed proportions minus the estimated mean proportion squared, divided by the estimated variance is approximately chi-square distributed with $m - 1$ degrees of freedom. For equal subgroup sizes, (7.19) becomes

$$\frac{\sum_{i=1}^{m}(\hat{p}_i - \bar{p})^2}{\bar{p}(1 - \bar{p})/n} \overset{\cdot}{\sim} \chi^2_{m-1} \tag{7.20}$$

To gain some insight on the usefulness of the chi-square statistic for our problem, we can rewrite (7.20) as

$$\frac{(m-1)s_{\hat{p}}^2}{\hat{\sigma}_{\hat{p}}^2} \sim \chi_{m-1}^2 \qquad (7.21)$$

Thus, at the heart of chi-square statistic is a ratio of the sample variance, computed directly from the subgroup proportions, to the estimated variance of the subgroup proportions based on the binomial model. If there exists an additional variance component in the subgroup proportions beyond what is expected by the binomial model, the statistic will tend to be large. Therefore, sufficiently *large* chi-square values provide evidence against the hypothesis of a simple binomial process.

For the injector data, since the values of $s_{\hat{p}}$ and $\hat{\sigma}_{\hat{p}}$ were previously found to be 0.01708 and 0.01167, respectively, we use (7.21) to obtain

$$\frac{(25-1)(0.01708)^2}{(0.01167)^2} = 51.41$$

Below is Minitab output showing the probability that a chi-square random variable with 24 degrees of freedom will have a value of 51.41 or less.

```
             Cumulative Distribution Function

             Chisquare with 24 d.f.

                    x        P( X <= x)
             51.4100           0.9991
```

Hence, the probability of observing 51.41 or *larger* on a chi-square distribution with 24 degrees of freedom is close to $1 - 0.9991 = 0.0009$. This value represents the probability of observing a chi-square value of 51.41 or greater due to chance alone if the null hypothesis of binomial model is true; recall that this probability is also referred to as the *statistical significance* or a p value.[4] Given that this probability is much less than the conventional cutoffs of 0.05 or 0.01, we conclude that there is strong evidence to reject the null hypothesis of pure binomial variability underlying the subgroup proportions. Much like the glass jar cap process (Example 7.6), the injector data suggest the existence of external variation not due to a simple binomial process. For purposes of process monitoring, control limits computed directly from the observed proportions would be more appropriate guides for special-cause detection.

As contrast, let us compute the chi-square statistic for the previously studied drop-filing process data. For those data, we found $\bar{p} = 0.06944$. Using subgroup proportions shown in Table 7.1 along with (7.20), we find

$$\frac{\sum_{i=1}^{24}(\hat{p}_i - 0.06944)^2}{0.06944(0.93056)/150} = 13.19$$

---

[4] Be careful not to confuse the p in p value with the p in p chart.

Below is Minitab output showing the probability of observing 13.91 or less on a chi-square distribution with 23 degrees of freedom:

```
Cumulative Distribution Function

Chisquare with 23 d.f.

        x      P( X <= x)
   13.1900        0.0523
```

In this case, the level of significance is $0.9477(= 1 - 0.0523)$ which means that the probability is "large" that we will observe a chi-square value of 13.19 or greater when the underlying process is indeed a simple binomial process. This indicates that the observed variability in the proportions is not unusual, given the assumption of a binomial model implying that the standard p chart analysis is appropriate for the drop-filing process.

There is often a temptation to rely solely on a single-valued statistical test, such as the one presented in this section, but it is risky to do so. The test presented should serve as a supplement to, not a replacement for, the visual study of the proportion data vis-à-vis the standard limits. It is quite possible that the simple binomial process is generally at work for a given application; however, an outlier subgroup proportion influenced by a special cause can potentially throw off the chi-square statistic to be significant, misleading the practitioner to conclude that a nonbinomial force is a consistent influence. A more informed judgment would result from the construction of a p chart and a visual analysis of the data.

7.3 MONITORING POISSON-BASED STATISTICS

In the discussion of the p chart, we were concerned with the occurrence or nonoccurrence of nonconforming items on discrete trials. However, there are many quality-related characteristics for which the notion of a discrete trial is not at all obvious. Examples may include the observance of events that can occur at any point within a continuous interval of time, area, or space. In these cases, the quality observations are then counts of these events within intervals of fixed length. For example, we might be concerned with the number of computer operator errors per day, flaws present in a square foot of woven fabric, emergency service calls per week, or arrivals of airplanes in a given hour.

If such events occur *uniformly at random*, the underlying distribution will be a Poisson distribution. As first introduced in Chapter 2, the Poisson distribution is given by

$$P(X = x) = \frac{\lambda^x e^{-\lambda}}{x!} \quad x = 0, 1, 2, \ldots$$

where X is a random variable representing the count of the number of times some event, such as a nonconformity, has occurred per unit interval of time, area, or space. For example, the Poisson distribution is commonly used in the textile industry (see Kittilitze, 1979) to explain and monitor the number of imperfections in such things as areas of cloth or carpeting and other related products. Recall, from Chapter 2, that the mean and variance of the Poisson distribution are equal to the parameter λ.

In some situations, we may be interested in the sum of two or more Poisson random variables. In such cases, there is an important fact about the sum of Poisson random variables. Recall, from Chapter 2, that the sum $X_1 + X_2 + \ldots + X_k$ of k independent Poisson random variables with parameters $\lambda_1, \lambda_2, \ldots \lambda_k$, respectively, is also a Poisson random variable, with parameter $\lambda = \lambda_1 + \lambda_2 + \ldots + \lambda_k$. As an example, consider the number of cosmetic flaws on an automobile that might arise. For simplicity, assume there are three possible categories of flaws: (1) exterior paint runs, (2) exterior body scratches, and (3) glass bubbles in the windshield. Suppose that the numbers of flaws per automobile from these three sources are independent Poisson random variables with parameters $\lambda_1 = 1.8$, $\lambda_2 = 0.6$, and $\lambda_3 = 3.8$, respectively. We know that the total number of occurrences of flaws from all the sources is a Poisson variable with parameter $\lambda = 1.3 + 0.6 + 3.8 = 5.7$.

As the normal distribution can be used to approximate the binomial distribution, it can also be used to approximate the Poisson distribution. This fact is important since the standard control chart for monitoring Poisson data is based on the 3σ framework, which in turn is rooted in the assumption of approximate normality. As a rule of thumb, the normal approximation to the Poisson is adequate when the mean of the Poisson distribution λ is greater than 5, with the approximation improving as λ increases.

CONTROL CHART FOR NUMBER OF NONCONFORMITIES: c CHART

Assume that the Poisson distribution is adequately approximated by the normal distribution. Given the fact that both the mean and the variance of the Poisson distribution are equal to the parameter λ, a 3σ control chart for the counts of nonconformities is given as follows:

$$\text{UCL} = \lambda + 3\sqrt{\lambda}$$
$$\text{CL} = \lambda \qquad\qquad (7.22)$$
$$\text{LCL} = \lambda - 3\sqrt{\lambda}$$

In most practical applications, we need to obtain an estimate for λ. Suppose m nonoverlapping units are sampled with $c_1, c_2, \ldots, c_m$ observed for the respective units; the use of the symbol c_i instead of x_i to represent observed number of counts is unique to the SPC area. A sample average of these observed count values serves as an unbiased estimate for the mean of the Poisson distribution, that is,

$$\bar{c} = \frac{\sum\limits_{i=1}^{m} c_i}{m} \qquad\qquad (7.23)$$

Since λ represents the mean of the Poisson distribution, $\bar{c}$ serves as its estimate. Replacing $\bar{c}$ for λ in (7.22), we obtain the following control limits:

$$\begin{aligned} \text{UCL} &= \bar{c} + 3\sqrt{\bar{c}} \\ \text{LCL} &= \bar{c} - 3\sqrt{\bar{c}} \end{aligned}$$

(7.24)

with the centerline position at $\bar{c}$. These control limits along with the sequence plot of the number of nonconformities per sampling unit c_i are the basis for a c chart. Similar to the p chart, a computed negative value for the lower control limit for the c chart can arise; since negative counts are impossible, the lower control limit should accordingly be set to zero.

Given the discussed additive property of Poisson random variables, the c chart is used not only for monitoring the occurrences of one type of nonconformity, but also for monitoring collectively the occurrences of several types of nonconformities.

Example 7.7

The general accessibility of books is clearly a critical process for any library. To better understand one aspect of this process, a quality improvement team at a large metropolitan university library was created to analyze the problem of lost books—which includes standard texts and periodicals.[5] Books are viewed as *lost* if they are not easily retrieved after a reasonable search by library staff. Inability to locate a book may be due to several possibilities such as the book's being stolen, misplaced, or not being truly part of the library collection even if the catalog or database indicates so. Lost books not only are an irritation to the library's customers (students and faculty) but also can be quite costly if replacements are required.

To get a basic understanding of the problem, the improvement team decided to record the daily number of pink slip requests for lost-book searches. These requests are filled out by the library staff after several unsuccessful attempts have been made to locate the book with the patron present. Starting at the beginning of the winter/spring semester, data on the number of daily lost-book requests for 46 consecutive days, with the exception of holidays, were collected and shown in Table 7.6. The data are also found in the file "lostbook.dat". The Poisson distribution is a natural candidate to describe this process since we are dealing with the number of occurrences of an event—lost-book request—in the continuous interval of 1 day.

First, we need to find the average number of nonconformities per day, which we calculate to be

$$\bar{c} = \frac{\sum_{i=1}^{46} c_i}{46} = \frac{310}{46} = 6.739$$

By substituting this value in (7.24), the preliminary retrospective control limits for the c chart are

$$\text{UCL} = 6.739 + 3\sqrt{6.739} = 14.53$$
$$\text{LCL} = 6.739 - 3\sqrt{6.739} = -1.05 \rightarrow 0$$

Figure 7.15 shows the sequence plot data provided in Table 7.6 along with the above-computed control limits. Even though there are no individual points outside control limits, a runs

---

[5]The use of process improvement and quality management ideas and techniques has been quite successful in library settings. See, for example, Jurow and Barnard (1993) and Mackey and Mackey (1992).

Table 7.6   Number of formal requests to search for lost books per day

Day i	Nonconformities per Day c_i	Day i	Nonconformities per Day c_i
1	7	24	3
2	8	25	6
3	10	26	5
4	12	27	6
5	9	28	11
6	7	29	4
7	8	30	7
8	7	31	4
9	9	32	3
10	6	33	8
11	9	34	11
12	7	35	2
13	3	36	4
14	1	37	2
15	8	38	8
16	5	39	8
17	4	40	6
18	10	41	9
19	7	42	6
20	8	43	7
21	8	44	5
22	5	45	12
23	7	46	8

Stat>>Control Charts>>C

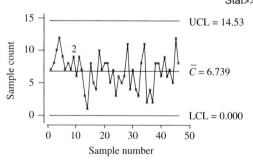

Figure 7.15   c Chart for the lost book request application.

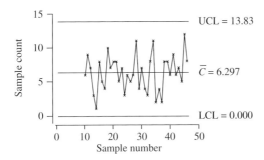

Figure 7.16   Revised *c* chart for the lost-book request application.

violation (rule 2) has been flagged; there are nine consecutive observations above the center-line (see Section 4.9). After careful consideration, the quality improvement team arrived at a plausible explanation for the higher number of lost-book requests. Namely, the consecutive observations occurred at the onset of the academic semester, which is historically the highest demand time during the semester; for example, many students search for suggested readings recommended in course syllabi.

With a special explanation in hand for the runs violation, the nine observations are set aside and the control calculations revised accordingly. In particular, the control limits should be recomputed based on the 37 remaining observations. Once this is done, the revised *c* chart, found in Figure 7.16, is obtained. This chart shows no evidence of isolated or runs violations. From an overall perspective, the data visually appear random with no hint of pervasive patterns, for example, trend or meandering behavior. The other randomness checks, such as the runs test and ACF, would confirm our conclusion; the reader is invited to verify our conclusion. But, before we completely accept the revised *c* chart, it is essential to check the data for approximate normality. The histogram and normality plot of the standardized (using Minitab's center command) data are given in Figure 7.17. As can be seen, approximate conformance to normality is clearly met. The *c* chart limits given in Figure 7.16 can now be extended prospectively for continued monitoring of the data.

Beyond monitoring occurrences of some event in a continuous interval, the *c* chart also comes into control chart applications in a different context. It serves as a handy computational approximation to an *np* chart in applications in which there are many discrete trials and the probability of a nonconforming item is small. This application of the *c* chart stems from the fact that the Poisson distribution is a useful approximation to the binomial distribution when the probability *p* of a defective item on any one trial is small; for any given small *p*, the approximation improves as the subgroup size increases. The Poisson approximation is made by letting the Poisson parameter λ be equal to *np*. Hence, rather than compute binomial limits, the control limits for the number of nonconforming

```
          Histogram of stdcount            N = 37
          Midpoint              Count
           -2.250                 0
           -1.750                 3  ***
           -1.250                 3  ***
           -0.750                 4  ****
           -0.250                 9  *********
            0.250                 5  *****
            0.750                 7  *******
            1.250                 3  ***
            1.750                 2  **
            2.250                 1  *
```

(a) Histogram

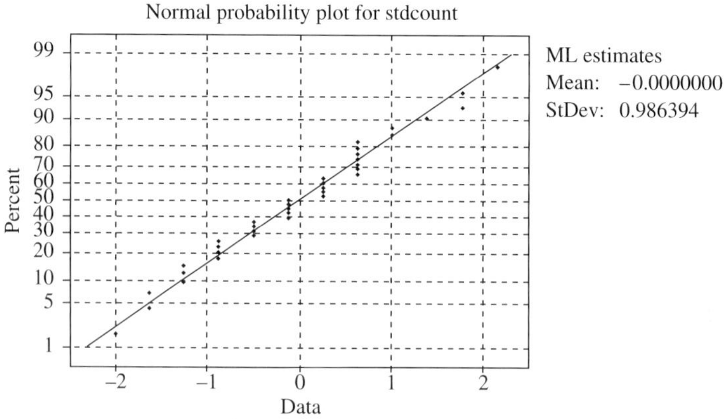

Normal probability plot for stdcount

ML estimates
Mean: -0.0000000
StDev: 0.986394

(b) Normal probability plot

Figure 7.17   Normality checks of the standardized counts from the lost-book request application.

items are more easily calculated by taking the average number of nonconforming items plus and minus the square root of the average number of nonconforming items. A common rule of thumb for the adequacy of the Poisson approximation to the binomial is $n > 20$ and $p < 0.05$.

The fact that the Poisson distribution might adequately approximate a particular binomial application does not imply the c chart is appropriate. Remember that the standard c chart limits additionally assume adequate approximation by the normal distribution. Earlier, we noted that a rule of thumb for adequate approximation of the Poisson by the normal is that $\lambda > 5$. Since $\lambda = np$ in the Poisson approximation of the binomial, the rule of thumb translates to $np > 5$, which by no accident is a suggested rule of thumb for the normal approximation of the binomial noted in Section 7.1. When approximate normality cannot be met, often a simple normalizing transformation (e.g., logarithmic, square root, cube root) will ameliorate the problem. Ryan (1989) provides a good discussion of the use of transformations for Poisson count data.

CONTROL CHART FOR NONCONFORMITIES WHEN UNIT INTERVAL VARIES: u CHART

The c chart assumes that the inspection unit interval length (whether measured in area, time, or volume) from period to period is constant in size. However, in some applications the inspection intervals will vary. For instance, an organization might wish to monitor the total number of nonconformities from production run to production run for all items produced. Since for production scenarios the output inevitably varies, the total period-to-period interval lengths will vary. In some situations, changing interval lengths are unavoidable. Imagine a retail store that monitors the number of customer complaints per day. Because store hours vary from day to day, the time interval of opportunity for complaints or nonconformities changes.

In situations such as described above where the nature of the application is the same as that of the c chart except there are varying interval lengths in which occurrences are counted, a control chart called the u *chart* is used. This chart calls for computing and plotting from period to period the average number of nonconformities per unit, denoted by the symbol u. More specifically, if c_i nonconformities are found in subgroup i, which is made up of n_i inspection units, then the average number of nonconformities per unit for subgroup i is $u_i = c_i / n_i$. Unlike in p chart applications, n_i does *not* have to be a whole number. For instance, suppose the daily number of complaints at a retail store is counted such that 12 h (the typical opening hours) represents 1 unit. Now if Thursdays are open for an extended period of 14 h and Sundays are open only for 6 h, the corresponding values for n_i are 1.167 and 0.5, respectively.

To establish the u chart, m preliminary subgroups are obtained and the centerline of the chart is estimated by

$$\bar{u} = \frac{\sum\limits_{i=1}^{m} c_i}{\sum\limits_{i=1}^{m} n_i} \tag{7.25}$$

There are a few ways to derive the u chart limits. Farnum (1994) provides the most straightforward explanation. Namely, under standard assumptions of counts, we first recognize that c_i $(= n_i u_i)$ follows a Poisson distribution with parameter λ_i. If λ_i is known, 3σ limits for the total number of nonconformities in subgroup i are $\lambda_i \pm 3\sqrt{\lambda_i}$. However, λ_i is not typically known and can be reasonably estimated by $n_i \bar{u}$. This estimate is based on the assumption that the probability of observing a certain number of nonconformities is equal in all intervals of the same length and is proportional to the size of the interval. By using this estimate, the control limits for c_i are estimated by $n_i\bar{u} \pm 3\sqrt{n_i\bar{u}}$. Since $u_i = c_i / n_i$, these limits are divided by n_i to give the u chart limits:

$$\text{UCL}(i) = \bar{u} + 3\sqrt{\frac{\bar{u}}{n_i}}$$

$$\text{LCL}(i) = \bar{u} - 3\sqrt{\frac{\bar{u}}{n_i}}$$

(7.26)

We attach the index i to the control limits to emphasize that the limits depend on n_i and thus are not necessarily constant. As is true for all standard attribute charts, the lower control limit is set at zero if a negative value is computed.

Given that the n_i values are different, the control limits in (7.26) will vary around the constant centerline of $\bar{u}$. As was true in the case of the p chart with variable subgroup sizes, there are also in this case two ways to avoid the jagged control limits. The first possibility is to substitute an average inspection interval size $\bar{n}$ for n_i in (7.26). The result will be a set of approximate control limits given by

$$\text{UCL}_{\text{approx}} = \bar{u} + 3\sqrt{\frac{\bar{u}}{\bar{n}}}$$

$$\text{LCL}_{\text{approx}} = \bar{u} - 3\sqrt{\frac{\bar{u}}{\bar{n}}}$$

(7.27)

where the average inspection interval size is computed from the m preliminary subgroups as follows:

$$\bar{n} = \frac{\sum_{i=1}^{m} n_i}{m}$$

(7.28)

Similar to the p chart based on an average subgroup size, this approach is not recommended if the interval sizes vary by more than 25 percent around the average interval size (see Ford Motor Company, 1984). The fixed control limits are appealing. However, the drawback of the approximate approach is the lack of precise determination for points near (within or outside) the control limits. To determine when exact control limits are needed, simply replace $\hat{p}_i$ with u_i in the general guidelines outlined in Section 7.1.

Analogous to the standardized p chart, a second possible solution to the jagged control limits is to standardize the u_i data and establish a standardized u chart. The standardized value of the number of nonconformities per unit, for the ith sample, may be expressed as

$$z_i = \frac{u_i - \bar{u}}{\sqrt{\dfrac{\bar{u}}{n_i}}}$$

(7.29)

The control limits for the standardized chart would be placed at ± 3 around a centerline of zero. Much of the discussion related to the standardized p chart is equally applicable to the standardized u chart. Most notably, standardization facilitates the application of runs and pattern rules, which may be problematic for the original data since they are based on varying sample sizes.

Before turning to an example to illustrate the computation of u chart limits, we need to underscore that the appropriate interpretation of the u chart and its limits rests on some basic requirements. First, the underlying process for the non-conformities is a simple Poisson process which remains the same from inspection unit to inspection unit and from sample period to sample period. Second, the use of symmetric 3σ control limits requires, at the very least, approximate normality of the u_i data. If these conditions are not met, then direct data analysis of the u_i data potentially provides the opportunity for better insights than the standard application.

Example 7.8

Most retail stores are concerned with the frequency of discrepancies (over or under) of prices, at the point of sale, relative to advertised or shelf prices. Inconsistent pricing—particularly overpricing—is not only a nuisance to customers but also may have a more damaging effect on the customers' perception of the integrity of the store. To diminish such negative implications of overpriced items, many stores have policies such as offering the item for free or giving the customer, in cash, double the difference if this type of discrepancy is spotted. On the other side, underpriced items are also undesirable to a store due to the net accumulated losses.

With these factors in mind, a large Midwestern supermarket chain performs weekly audits on each of its stores to collect data on the number of times price discrepancies occur in the hope that control chart techniques can reveal the frequency of the problem and lead the way to process improvement. Given the large number of products (35,000 to 50,000 different items per store), management has assigned an individual dedicated to the task of auditing the prices. Since nearly all products have a unique Universal Product Code (UPC), the auditor with the aid of a portable photoelectric scanner has the capability of sampling large quantities during the designated audit period.

Table 7.7 shows 30 consecutive audit results which include the number of items audited and the number of price discrepancies caught at one particular store (data also in the file "upc.dat"). Given the fact that we are counting the number of incorrect prices in a finite number of sample product items, this application could be pursued as an application of a p chart with varying subgroup sizes. However, since the subgroup sizes are very large ($> 10,000$) and the probability that a given item has an incorrect price is quite small, we can legitimately pursue this situation as a Poisson application. But the fact that the sample sizes vary requires us to establish a u chart rather than a c chart. To illustrate the construction of a u chart, suppose an inspection unit is defined as 1000 product items. Hence, the number of inspection units n_i is given by the total number sampled, divided by 1000; these values can be found in Table 7.7. Furthermore, the average number of incorrect prices per inspection unit c_i is found by dividing the total number of incorrect prices found in the audit c_i by n_i.

Using (7.25), we find the centerline to be

$$\bar{u} = \frac{1937}{343.745} = 5.635$$

From (7.26), the upper and lower control limits for each week are given by

Table 7.7   Weekly audit of price discrepancies

Week	Total Number of Incorrect Prices c_i	Total Number of Items Audited	Number of Inspection Units n_i	Incorrect Prices per Inspection Unit u_i
1	70	11,688	11.688	5.98905
2	71	11,780	11.780	6.02716
3	76	12,056	12.056	6.30392
4	62	10,695	10.695	5.79710
5	57	11,087	11.087	5.14116
6	66	12,083	12.083	5.46222
7	67	10,518	10.518	6.37003
8	54	10,342	10.342	5.22143
9	79	12,816	12.816	6.16417
10	68	12,958	12.958	5.24772
11	73	12,372	12.372	5.90042
12	69	10,649	10.649	6.47948
13	66	10,813	10.813	6.10376
14	54	11,027	11.027	4.89707
15	45	10,351	10.351	4.34741
16	70	11,941	11.941	5.86216
17	68	10,986	10.986	6.18970
18	59	12,580	12.580	4.68998
19	56	10,151	10.151	5.51670
20	62	11,709	11.709	5.29507
21	75	11,459	11.459	6.54507
22	59	12,366	12.366	4.77115
23	56	11,314	11.314	4.94962
24	60	11,897	11.897	5.04329
25	75	12,963	12.963	5.78570
26	67	10,707	10.707	6.25759
27	67	10,524	10.524	6.36640
28	48	10,461	10.461	4.58847
29	68	10,507	10.507	6.47188
30	70	12,945	12.945	5.40749

$$\text{UCL}(i) = 5.635 + 3\sqrt{\frac{5.635}{n_i}} = 5.635 + \frac{7.121}{\sqrt{n_i}}$$

$$\text{LCL}(i) = 5.635 - 3\sqrt{\frac{5.635}{n_i}} = 5.635 - \frac{7.121}{\sqrt{n_i}}$$

Figure 7.18 shows the u chart for this process. Generally, the points appear well dispersed within the control limits with no indications of special causes. Furthermore, there are

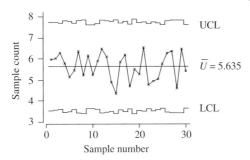

Figure 7.18   *u* Chart for the price discrepancy application.

no apparent patterns such as strings of observations above or below the centerline. If we wish to do a full-blown sequence analysis of the data with no concerns about the effects of the varying sample sizes on pattern analysis, then the u_i should be rescaled by using (7.29). Additionally, other diagnostic checks such as tests for normality are more appropriately interpreted when applied to the rescaled data. The reader can verify that the standard checks for randomness and normality show that the standardized u_i are random and approximately normal. Hence, the standard *u* chart of Figure 7.18 is a suitable guide for monitoring the price discrepancy process. Prospectively, new u_i data are plotted with control limits computed by using (7.26) with $\bar{u} = 5.635$ along with the appropriate values of n_i. Alternatively, new u_i can be standardized by using (7.29) and control limits of ± 3 projected for real-time control.

CONTROL CHART FOR WEIGHTED NONCONFORMITIES: *D* CHART

As noted earlier, a *c* chart can be used to monitor the sum of different types of nonconformities. Similarly, a *u* chart can be used to monitor the average total number of different types of nonconformities per inspection unit. Implicit in the application of *c* charts and *u* charts is that all types of nonconformities are treated equally; that is, they are equally important. In some situations, this may not be a suitable assumption since certain nonconformities may be viewed as minor while others are of more serious consequence. One reasonable solution for such cases is to classify the nonconformities and to develop a rating system that places severity weights on the different nonconformity classifications.

Clearly, a classification scheme depends greatly on an intimate knowledge of the product or service along with customer requirements and expectations. One widely used classification system is defined as follows:

Class 1 defects—very serious. These are nonconformities which lead directly to severe injury or to catastrophic economic loss.

Class 2 defects—serious. These defects lead to significant injury or to significant economic loss.

Class 3 defects—major. These defects are related to major problems that may occur with normal use of a product or service rendered.

Class 4 defects—minor. These nonconformities are related to minor problems that may occur with normal use of a product or a service.

Based on these classifications, we define c_1, c_2, c_3, and c_4 as the number of class 1, class 2, class 3, and class 4 nonconformities, respectively, in a given sample. The next step is to assign weights or demerits for each of the categories, which we can denote by w_1, w_2, w_3 and w_4. For example, we might have demerit weights of $w_1 = 100$, $w_2 = 50$, $w_3 = 10$, and $w_4 = 1$ for the categories listed above. Hence, the total number of demerits D per sample is given by

$$D = 100c_1 + 50c_2 + 10c_3 + c_4$$

In general, when we have k different nonconformity categories, the total number of demerits is

$$D = \sum_{j=1}^{k} w_j c_j \tag{7.30}$$

Assume that each nonconformity category is independent of the others and that the number of nonconformities in any category is represented by a Poisson distribution. Notice that if $w_j = 1$ for all categories, then D is simply the total number of nonconformities in the sample. In this special case, D will follow a Poisson distribution with Poisson parameter equal to the sum of the Poisson parameters associated with the individual categories; hence, one would pursue a standard c chart implementation. However, it can be shown that if any $w_j \neq 1$, then D will not be a Poisson random variable. Nonetheless, the fact that D is a linear combination of independent Poisson random variables allows us to easily determine appropriate control limits for the demerit random variable.

By using basic facts about variance, the variance of D defined in (7.30) is given by

$$\text{Var}(D) = \sum_{j=1}^{k} w_j^2 \text{Var}(c_j) = \sum_{j=1}^{k} w_j^2 \lambda_j \tag{7.31}$$

where λ_j is the associated Poisson parameter for nonconformity category j. To estimate this variance, we need to estimate each of the Poisson parameters λ_j. Since λ_j is the mean (and variance) of a Poisson distribution, a natural estimate is the average number of nonconformities of category j across all samples. In particular, if there are m samples, then an estimate of λ_j is computed from

$$\bar{c}_j = \frac{\sum_{i=1}^{m} c_{ij}}{m} \tag{7.32}$$

where c_{ij} is the number of nonconformities of category j in sample i. Substituting (7.32) for λ_j into (7.31) and taking the square root, we estimate the standard deviation of the demerit random variable by

$$\hat{\sigma}_D = \sqrt{\sum_{j=1}^{k} w_j^2 \bar{c}_j} \qquad (7.33)$$

As is the case for all standard 3σ control charts, the centerline is simply the average of the preliminary values of the relevant plotted statistic, which implies a centerline of

$$\bar{D} = \frac{\sum_{i=1}^{m} D_i}{m} \qquad (7.34)$$

Combining (7.34) and (7.33) in a 3σ framework, we obtain the following control limits:

$$\begin{aligned} \text{UCL} &= \bar{D} + 3\sqrt{\sum_{j=1}^{k} w_j^2 \bar{c}_j} \\ \text{LCL} &= \bar{D} - 3\sqrt{\sum_{j=1}^{k} w_j^2 \bar{c}_j} \end{aligned} \qquad (7.35)$$

These control limits along with a sequence plot of the sample demerits serve as the basis of a *D chart*. Similar to the *c* chart, the *D* chart is formulated on the basis that the sampling interval is constant from sample to sample. For varying sample intervals, one may develop, much like a *u* chart, a variant of the *D* chart (a counterpart to the *u* chart) that monitors the number of demerits per inspection unit; for details, the reader should consult Ryan (1989).

One clear benefit of the *D* chart approach is that the varying weights placed on different types of nonconformities enable more serious problems to influence the variability of the data. But given the aggregation of different categories, it may be difficult to interpret out-of-control points on the *D* chart for isolated corrective action. For example, does an out-of-control point represent a significant increase of a few major nonconformities or of many minor nonconformities or some other combination? As suggested by Grant and Leavenworth (1996), it may be more desirable to maintain separate *c* charts for process improvement efforts while reserving the *D* chart for upper-management reporting.

7.4   VIOLATIONS OF THE *c* CHART ASSUMPTIONS

Like the *p* chart with its assumption of a simple binomial process, the construction of the *c* chart requires some specific conditions to be met:

1. We are counting the number of nonconformities (one or more types) within some defined continuous interval, be it time, area, or space.

2. The number of nonconformities in any given interval is independent of the number in any other nonoverlapping interval.

3. The probability of observing a certain number of nonconformities is equal in all intervals of the same length.

A process that behaves under these conditions is said to be a *simple Poisson process*. Let us turn to examples that challenge the appropriateness of the standard c chart implementation.

Table 7.8   Number of surface nonconformities

Sample i	Nonconformities per Sample c_i	Sample i	Nonconformities per Sample c_i
1	1	16	3
2	2	17	1
3	3	18	6
4	7	19	12
5	8	20	4
6	1	21	5
7	2	22	1
8	13	23	8
9	1	24	7
10	1	25	9
11	6	26	2
12	5	27	3
13	0	28	14
14	6	29	6
15	9	30	8

Example 7.9

In this example, we consider a set of data that pertain to the production of decorative polished brass doorknobs for interior doors. An attribute of particular importance to the customer is the surface quality, in the sense of both appearance and touch. For purposes of process monitoring and understanding, random samples of 50 doorknobs are taken every few hours for inspection. Each knob is visually inspected and hand-felt for surface irregularities, such as cracks and "grainy" bumps. The total number of nonconformities for the collective surface of the 50 knobs is then recorded. The data on 30 consecutive samples are shown in Table 7.8 and given in the file "door.dat".

Let us first proceed by computing the standard c chart limits for the presented data. The estimated mean $\bar{c}$ of the data is 5.133. Thus, the c chart control limits are given by $5.133 \pm 3\sqrt{5.133}$, which produces UCL = 11.93 and LCL = -1.66; thus the LCL is set to zero because of its computed negative value. The c chart, with these limits, is presented in Figure 7.19. As can be seen, three points (observations 8, 19, and 28) lie above the UCL. The standard procedure is to search for possible special causes associated with these signals.

However, before we seek special causes, it is advisable to examine the appropriateness of the c chart assumptions. One quick check for its appropriateness is to compare the estimated variation of the counts based on the belief of an underlying simple Poisson process with the actual observed variability of the counts. Below we can find the pertinent summary statistics of the count data to perform such a comparison:

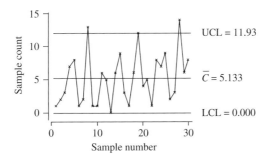

Figure 7.19   c Chart for the brass doorknob application.

Variable	N	Mean	Median	TrMean	StDev	SEMean
c-data	30	5.133	5.000	4.846	3.830	0.699

Variable	Min	Max	Q1	Q3
c-data	0.000	14.000	1.750	8.000

Recall that the Poisson distribution has the property that the variance of the distribution equals the mean of the distribution. Since the mean is typically not known, it would be estimated by $\bar{c}$. For our example, $\bar{c} = 5.133$ which implies that if the underlying process is indeed a simple Poisson process, then the observed variance of the counts should be close to 5.133. From above, we find the estimated standard deviation of the counts is 3.83, and thus the variance is $3.83^2 = 14.67$, which is nearly 3 times larger than the estimated mean of 5.133. Therefore, the Poisson estimate of variation does not fully account for the overall variability of the count data, suggesting that the standard c chart limits may be misleading.

Continuing in our investigation of the appropriateness of the standard c chart, we display the histogram and normal probability plot of the standardized counts in Figure 7.20. A visual examination of the histogram shows clearly a substantial positive skewness. The skewness is also shown in the normal probability plot by the "bowing" of the plotted line. If the process were Poisson, we would expect not substantial skewness but rather closer conformity to symmetry since, for a mean of 5.133, the normal distribution should adequately approximate the Poisson. Again, the data suggest that even though we are dealing with counts, and under certain circumstances counts may follow the Poisson distribution, this is not an application for which the Poisson model is a useful guide in setting control limits.

In light of the positive skewness, a plausible choice is a square root transformation for the data. The logarithmic transformation is unattractive because there is one zero in the data for which logarithmic transformation would be undefined and there is no reason to think that more zeros might not occur with further observation of the process.

In Figure 7.21, we give the histogram and normal probability plot of the standardized values of the square root of the original count data. By comparison to the severe nonnormality of the original data, the transformed data are much closer to approximate normality, even though the histogram does appear a bit rectangular. Other transformations, such as the cube root, did not improve this relatively minor shortcoming. But what is clear is that there are no substantial outliers in either tail, and the original skewed distribution was not a useful guide for directly setting control limits.

For purposes of process monitoring, a more reasonable strategy is to establish a control chart for the square root–transformed data. By treating the transformed data as individual

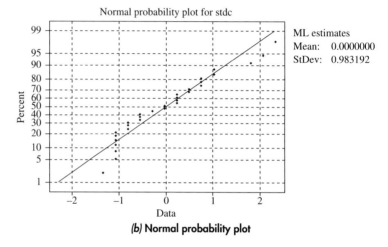

```
          Histogram of stdc      N = 30
       Midpoint              Count
        -1.250                 7    * * * * * * *
        -0.750                 6    * * * * * *
        -0.250                 3    * * *
         0.250                 6    * * * * * *
         0.750                 3    * * *
         1.250                 2    * *
         1.750                 1    *
         2.250                 2    * *
```

(a) Histogram

(b) Normal probability plot

Figure 7.20   Normality checks of the standardized counts from the brass
doorknob application.

observations, we can construct an X chart for the square root data as shown in Figure 7.22. The control limits appear reasonably placed relative to the inherent dispersion of the transformed data. Unlike in the routine application of the c chart, the control chart for the transformed data reveals no outliers. We also applied supplementary runs rules, and no flags were given. Thus, these limits can be projected for on-line process monitoring for detection of any future special-cause departures. In conclusion, the data analysis perspective reveals that the underlying sources do not conform to the standard assumption of a Poisson model but are in a state of statistical control with no need for isolated special-cause searching. Efforts are better spent studying the *system* to understand the sources of the variability not attributable to the Poisson model.

TESTING THE POISSON ASSUMPTION

In Section 7.2, we presented a formal statistical procedure for testing whether observations adhere to a simple binomial process. In a similar fashion, one can

```
Midpoint        Count
  -2.250            1    *
  -1.750            0
  -1.250            6    ******
  -0.750            3    ***
  -0.250            4    ****
   0.250            6    ******
   0.750            5    *****
   1.250            2    **
   1.750            3    ***
```

(a) Histogram

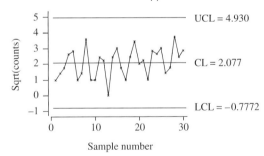

Normal probability plot for stdsqrt(c)

ML estimates
Mean: 0.0000001
StDev: 0.983192

Percent / Data

(b) Normal probability plot

Figure 7.21   Normality checks of the standardized square root of counts from the brass doorknob application.

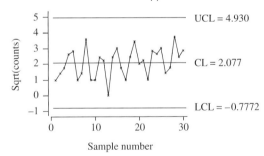

Sqrt(counts) / Sample number

UCL = 4.930
CL = 2.077
LCL = -0.7772

Figure 7.22   *X* chart for the square root of counts from the brass doorknob application.

employ a statistical procedure to compare the consistency of the observed variation of count data with the assumed Poisson distribution.

Other than the occurrence of special causes, two fundamental departures from a simple Poisson process may occur: (1) The counts exhibit nonrandom behavior; and/or (2) the counts exhibit random behavior, but there exists exogenous random variation not explained by the Poisson model. Either of these departures can render the standard c chart limits less effective. In the next section, we will illustrate an example of count data that follow a systematic nonrandom pattern.

The second departure can, sometimes, be detected by visual comparison of the observed variability relative to the c chart limits. If the observed counts are dispersed so much, that an unrealistic percentage of the data falls outside the control limits, doubt is presumably cast on the appropriateness of the limits instead of chasing potentially illusionary special causes. As illustrated earlier, another way to see this discrepancy is to compare the sample variance, computed directly from data, with the sample mean $\bar{c}$; under the assumption of a simple Poisson process these two measures should be close. However, there is also a more formal method of making such a comparison. Namely, if a series of observations $c_1, c_2, \ldots, c_m$ come, hypothetically, from a simple Poisson process, then the statistic

$$\frac{\sum_{i=1}^{m}(c_i - \bar{c})^2}{\bar{c}} \tag{7.36}$$

is approximately chi-square distributed with $m - 1$ degrees of freedom. Large values of this statistic suggest the rejection of the assumption of pure Poisson variability in favor of the belief that there exists exogenous variability beyond the Poisson model.

As an example, let us refer to the lost-book request observations used to establish the revised control chart shown in Figure 7.16, which from Table 7.6 are associated with days 10 through 46; recall that observations 1 through 9 were dropped because they were influenced by special-cause variation. Based on the counts for days 10 through 46, we found $\bar{c} = 6.297$. Using (7.36), we find

$$\frac{\sum_{i=10}^{46}(c_i - 6.297)^2}{6.297} = 40.29$$

To determine where this value falls on a chi-square distribution with 36 degrees of freedom, we find the following cumulative probability:

```
Cumulative Distribution Function
Chisquare with 36 d.f.

        x          P( X <= x)
   40.2900            0.7139
```

Hence, the significance level is 0.2861 ($= 1 - 0.7139$) which indicates that observing a chi-square value of 40.29, or even more extreme, is not that unusual under the assumption of a simple Poisson process. Consequently, the standard c chart can serve as an appropriate means to monitor the application in question.

Let us now take another look at the doorknob surface nonconformities studied in Example 7.9. The average number of nonconformities $\bar{c}$ was found to be 5.133. Using this average value along with the 30 count observations found in Table 7.8, we compute the chi-square statistic to be

$$\frac{\sum_{i=1}^{30}(c_i - 5.133)^2}{5.133} = 82.89$$

To assess the level of significance of this result, we provide the following output:

```
Cumulative Distribution Function
Chisquare with 29 d.f.

        x        P( X <= x)
    82.8900         1.0000
```

Consequently, the probability of observing 82.89, or more extreme, is essentially 0 under the assumption of an underlying Poisson process. Therefore, there is substantial evidence of variation, from sample to sample, not attributable to the Poisson model.

7.5   SYSTEMATIC VARIATION IN ATTRIBUTE PROCESSES

Up to this point, our examples have illustrated p and c chart applications in which either the standard assumptions were basically met or the departure was in the form of added but *random and stable* variability between the subgroup observations. Similar to variable measurements, nonrandom behavior in attribute data is frequently found in practice. In this section, we explore in detail three attribute applications in which, in each case, the associated process exhibits a systematic nonrandom behavior.

Table 7.9   Number of misrouted mail items per day

Day i	Misrouted c_i	Day i	Misrouted c_i
1	9	19	4
2	5	20	5
3	0	21	8
4	1	22	4
5	4	23	0
6	8	24	7
7	3	25	1
8	1	26	5
9	5	27	3
10	3	28	1
11	7	29	2
12	6	30	4
13	0	31	9
14	1	32	7
15	3	33	2
16	5	34	3
17	6	35	5
18	9		

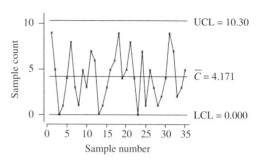

Figure 7.23   *c* Chart for the misrouted mail application.

SEASONAL BEHAVIOR

This application pertains to daily monitoring of misrouted mail for a certain department of a large bank. In general, misrouted mail translates to unnecessary delays. The integrity of the mailing process has become particularly important because the bank, as a convenience, now allows customers to mail in many documents (e.g., loan applications) that in the past needed to be completed in person. To measure the rate of misrouted mail, all employees in the department were asked to place any incorrect mail received as soon as possible and no later than at day's end in a designated bin. Concurrently, employees placed a tally mark on a board found above the bin. In Table 7.9, we present the counts for the daily misrouted mail; the data can also be found in the file "mail.dat".

Since we are dealing with large volumes of mail and the probability that a particular piece of mail will be misrouted is small, it is reasonable to assume that the Poisson model will serve as a suitable guide for understanding the variability of the data. Under this premise, we pursue the construction of a *c* chart. The estimated mean $\bar{c}$ of the data is 4.171. Thus, *c* chart control limits are given by $4.171 \pm 3\sqrt{4.171}$, which produces UCL = 10.30 and LCL = −1.96; thus LCL is set to 0 because its computed value is negative. The *c* chart with these limits is presented in Figure 7.23.

Based on the standard chart, there are no apparent out-of-control conditions. The three points resting on the lower control limit are not flagged by Minitab since the "true" 3σ control limit is at −1.96; hence, in these situations, the lower control set at 0 really serves no purpose. We additionally applied the runs rules available in Minitab's *c* chart option, and no unusual patterns were highlighted. Beyond the simple runs rules, the sequence plot of the data appears random with no obvious patterns. A practitioner would be tempted to judge the process as being in control, with an average daily number of misrouted mail items being slightly more than 4. In fact, this *is* essentially the conclusion reached by the quality analysts at the bank.

However, in light of the doorknob application, Example 7.9, the reader may suspect that the observed variability of the counts, seen in Figure 7.23, is larger than expected under the assumption of a simple Poisson process. Below are the summary statistics for the count data:

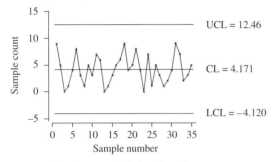

```
   Histogram of stdmail              N = 35
   Midpoint          Count
     -1.750             3    ***
     -1.250             5    *****
     -0.750             2    **
     -0.250             9    *********
      0.250             6    ******
      0.750             2    **
      1.250             5    *****
      1.750             3    ***
```

(a) **Histogram of standardized mail counts**

(b) **X chart applied directly to the counts**

Figure 7.24   Histogram and X chart for the misrouted
mail application.

```
Variable        N      Mean    Median    TrMean    StDev    SEMean
mail           35     4.171     4.000     4.129    2.728     0.461

Variable      Min       Max        Q1        Q3
mail        0.000     9.000     2.000     6.000
```

The estimated variance computed directly from the data is $(2.728)^2 = 7.44$. This value of 7.44 is nearly double the estimated variance of 4.17 expected under Poisson conditions. At this point, we might check the normality of the data, which as shown in Figure 7.24a seems quite satisfactory. In turn, given our experience with the doorknob example, we may be tempted to establish a control chart based directly on the count data, which is given in Figure 7.24b; the computed negative lower control limit is disregarded since the data are, by definition, nonnegative. The message from this control chart is again that there are no apparent special causes. The only difference is the recognition that the counts are influenced by an exogenous variability not accounted for by the Poisson model. Furthermore, the variability appears to be random. For more evidence of randomness, we can conduct the overall runs test and ACF test for the data. A runs test, applied to the counts, finds 19 observed runs versus an expected 18.4 runs for a random process; thus there is no evidence against random behavior. Similarly, the ACF reveals no substantial autocorrelations, further suggesting randomness of the count data.

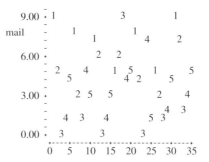

Figure 7.25   Time-series plot of misrouted mail counts labeled by day of the week.

It would seem that there is nothing left to explore. However, we have ignored another possible plotting perspective. Namely, in Figure 7.25, we present a time-sequence plot of the counts with labels identifying the day of the week (1 for Monday, 2 for Tuesday, and so on). Quite dramatically, this plot sheds new light on the application under study. Clearly, the data are nonrandom with discernible day-of-the week effect; Mondays appear high and Wednesdays possibly low. As we see, seasonality can be quite elusive with certain plotting presentations, such as high-resolution graphics. Recall that the runs test and the ACF did not alert us to the nonrandom behavior. This is a good lesson in not relying on one diagnostic tool at the expense of others. Given the realization of nonrandom behavior, neither the horizontal control limits given in Figure 7.23 nor those in Figure 7.24*b* are satisfactory for process monitoring. Moreover, the histogram shown in Figure 7.24*a* has little meaning since it is a mix of nonrandom and random effects.

To capture the systematic pattern in the data, we create indicator variables for the day of the week. Since there does not appear to be a trend in the data and since the ACF reveals no substantial autocorrelations, using only the set of indicator variables as potential predictor variables seems to suffice at this stage. To sift through the five possible predictor variables, we invoke the forward regression[6] stepwise procedure with results shown in Figure 7.26*a*. Consistent with our visual impression, Mondays and Wednesdays exhibit clear nonrandom behavior. The regression fit, based on the corresponding indicator variables, is shown in Figure 7.26*b*. As can be seen, the two independent variables are both significant at the 5 percent level, lending further credence to our premise of seasonal behavior.

With the exception of one point, to be seen in a moment, all the regression diagnostics (randomness and normality) are reasonably satisfactory; the reader may wish to substantiate these facts. With the systematic pattern accounted for, we are now better prepared to detect any potential special causes impinging on the sys-

---

[6]In some counting applications, modeling methods such as logistic and Poisson regression may be needed for very precise results (see Agresti, 1990, and Koch, Atkinson, and Stokes, 1986). Nonetheless, ordinary regression procedures are quite flexible to provide reliable, practical, and useful insights for most applications.

```
Stepwise Regression

F-to-Enter:    4.00  F-to-Remove:    4.00

Response is  mail  on 5 predictors, with N =  35

            Step        1        2
       Constant     3.393    3.905

       mon                   3.89     3.38
       T-Value              4.08     3.63

       wed                           -2.05
       T-Value                       -2.20

       S                    2.26     2.14
       R-Sq                 33.55    42.25
```

(a) Stepwise regression on daily indicator variables

```
The regression equation is
mail = 3.90 + 3.38 mon - 2.05 wed

Predictor        Coef        StDev            T          P
Constant       3.9048       0.4663         8.37      0.000
mon            3.3810       0.9325         3.63      0.001
wed           -2.0476       0.9325        -2.20      0.035

S = 2.137    R-Sq = 42.2%    R-Sq (adj) = 38.6%

Analysis of Variance

Source             DF          SS           MS          F          P
Regression          2     106.876       53.438      11.70      0.000
Residual Error     32     146.095        4.565
Total              34     252.971

Source     DF    Seq SS
mon         1    84.864
wed         1    22.012
```

(b) Regression model with Monday and Wednesday as independent variables

Figure 7.26   Stepwise and regression output for the misrouted mail application.

tem. To do so, we present the control chart for the residuals or special-cause chart in Figure 7.27. Notice that a substantial 3σ signal is revealed that went *unnoticed* by the earlier control charts. In particular, observation 18 (a Wednesday) is significantly higher than expected for this day; this can also be seen in Figure 7.25.

If a special explanation can be given to this particular day, the observation should be set aside and the model, along with control limits, recomputed. However, as discussed in Section 3.9, observations associated with nonrandom processes cannot simply be discarded. If we were to toss out observation 18, then

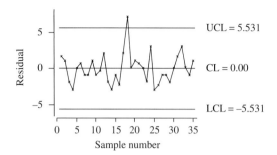

Figure 7.27   Special-cause chart for the misrouted
mail application.

```
The regression equation is
mail = 3.90 + 3.38 mon - 3.24 wed + 8.33 special

Predictor          Coef        StDev           T          P
Constant         3.9048       0.3647       10.71      0.000
mon              3.3810       0.7293        4.64      0.000
wed             -3.2381       0.7736       -4.19      0.000
special          8.333        1.805         4.62      0.000

S = 1.671    R-Sq = 65.8%    R-Sq (adj) = 62.5%

Analysis of Variance

Source                DF          SS          MS          F          P
Regression             3     166.400      55.467      19.86      0.000
Residual Error        31      86.571       2.793
Total                 34     252.971

Source        DF      Seq SS
mon            1      84.864
wed            1      22.012
special        1      59.524
```

Figure 7.28   Regression model with special-cause variable for the misrouted mail
application.

a Tuesday observation and a Thursday observation would inappropriately be side
by side. To extract the unrepresentative effects of observation 18 but preserve the
ordering of the data, we create an indicator variable called special which takes on
the value 1 for observation 18 and the value 0 otherwise. Figure 7.28 shows the
regression model reestimated with the special-cause variable introduced. The
reader can verify that the residual diagnostics for this fitted model are excellent.

The revised special-cause chart is presented in Figure 7.29. The limits of
±4.76, provided in Figure 7.29, can be projected for continued process monitoring. Future points plotted on the special-cause chart will simply be the differences
between what is expected for a given day, as predicted by the regression equation,

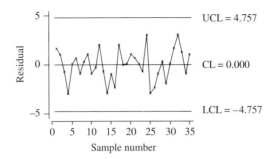

Figure 7.29   Revised special-cause chart for the misrouted mail application.

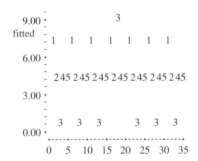

Figure 7.30   Fitted-values chart for the misrouted mail application.

and what is actually observed. Consider, for instance, the monitoring of Monday observations. From Figure 7.28, the regression equation predicts the number of misdirected mail items to be $3.90 + 3.38(1) - 3.24(0) = 7.28$. So, given the control limits of ± 4.76 for the forecast errors, if more than $12.04 (= 7.28 + 4.76)$ or fewer than $2.52 (= 7.28 - 4.76)$ mail items are observed on Monday, a 3σ signal is flagged; or equivalently, a signal is given if the forecast error, the observed number of mail items minus 7.28, is beyond ± 4.76. Additionally, we can use supplementary runs rules (Section 4.9) on the sequence of forecast errors for detection of an unusual run indicative of possible special causes.

To visualize the implications of the estimated model, we display the fitted-values chart in Figure 7.30. The daily pattern is easily seen, and it is quite understandable for personnel of any training background. The plot shows that Mondays are expected to be consistently high, Wednesdays consistently low, and the remaining days (Tuesdays, Thursdays, and Fridays) at an equal level somewhere between the level of Mondays and Wednesdays. The one Wednesday that appears out of place represents the special cause fully accounted for by the special-cause indicator variable. The challenge to management is to discover what specific factors lie behind the systematic pattern.

SEASONAL AND TREND BEHAVIOR

If quality improvement methods can be used to monitor and improve organizational processes, can this approach be applied to our own personal processes? In a thought-provoking book, Roberts and Sergesketter (1993) proposed and demonstrated that indeed many of the ideas of quality management are directly applicable to personal improvement. In their book, Bob Galvin, former CEO of Motorola, underscores the importance of personal aspects of quality by stating:

> We have operated very substantially under the rubric of quality control. Our institutions, our companies have had quality departments. And the old testament was that quality is a company, a department, and an institutional responsibility.
>
> The new truth is radically different. Quality is a very personal obligation. If you can't talk about quality in the first person . . . then you have not moved to the level of involvement of quality that is absolutely essential. [This] . . . is the most useful thing I can say. . . . You must be a believer that quality is a very personal responsibility.

To achieve and improve personal quality, Roberts and Sergesketter (1993) introduce a tool called *personal quality checklist.* A personal quality checklist is a means of recording shortcomings or defects on a collection of key personal activities that an individual wishes to improve. These personal activities might be business-related (e.g., failing to promptly return customer phone calls) or more personal (e.g., eating junk foods). Personal defect data can then be plotted and analyzed for evidence of improvement or special departures.

To demonstrate the use of a personal quality analysis, consider the experience of a student in the Executive MBA Program, School of Business Administration, University of Wisconsin-Milwaukee, who tried the personal quality checklist for a course project[7] in the spring of 1995. The student wrote:

> After consideration and self observation, I put together a personal quality checklist of seven categories. These categories consisted of items which carried over in not only my work, but also in my personal life as well. My personal defects, as well as their definitions, are listed below:
>
> - Failure to return a phone call to a customer request, within 1 hour.
> - Failure to discard incoming "junk" properly (within a day): this includes papers, mail, memos, etc. which are of no use to me.
> - Unnecessary search for something misplaced or lost: even a moment's delay in remembering where a specific document, book, or other business material has been placed.
> - Late for meeting or appointment: even by 1 second.
> - Drinking soda: Pretty self explanatory.
> - Failure to spend at least 30 minutes of quality time with children each day.
> - Failure to do 50 sit-ups each day.

---

[7]As an instructor, I have found the personal quality checklist to be a wonderful basis for projects in statistics and quality management courses for both the regular MBA program as well as the executive MBA program.

Defect Category	Monday	Tuesday	Wednesday	Thursday	Friday	Saturday	Sunday
Late response to phone call	||	|					
Fail to discard incoming junk	|						
Unnecessary searching	⫽||	||||					
Late for meeting or appointment		|					
Drinking soda	||	||					
Not spend half hour with children		|					
Failed to do 50 sit-ups	|						
Total	11	9					

Comments

Figure 7.31   Personal quality checklist.

The method of data collection was quick and simple. I simply carried a small pocket-sized notebook to record daily the occurrences of the noted defects. I made a list of the defects along with the days of the week [see Figure 7.31]. Every time I performed a defect, I recorded a little "tick" in the appropriate category. This became a daily ritual for me for nine weeks.

Notice that by counting the number of defect occurrences per day, we are by definition dealing with attribute data; this is why the personal quality idea is introduced in this chapter. However, personal quality data probably do not lend themselves to analysis by standard attribute control charts, such as p or c charts or any related variants. Here are a few reasons why:

1. Since total defects per day are the result of different categories, each with different probabilities of a defect occurrence, p charts or any standard binomial-based method is not applicable.

2. Even though the data are the number of defect occurrences per day, we may not be dealing with a Poisson distribution. Remember the Poisson is appropriate when there are a large number of trials with a small probability of defect on any given trial, or when it is used to describe random occurrences of some event in a continuous interval. Categories such as "failure to spend time with children" or "failure to do exercise" do not naturally meet these requirements.

3. Even if the specific personal defect categories are Poisson random variables, the total number of defects may not follow a Poisson distribution. Recall that for the sum of nonconformities across different nonconformity types to be a Poisson random variable, the different nonconformity types need to be independent of one another. Often, personal defect categories are not independent of one another. For instance, failing to discard junk mail might increase the likelihood of unnecessary searching for important documents.

4. Total personal defects are frequently not random over time. Trends reflecting improvement are commonplace. Quite often, daily total defects have day-of-week or seasonal effects; that is, certain days are consistently high or low relative to other days.

Table 7.10   Number of daily personal defects

11	9	10	7	7	9	6	9	10	7
9	6	5	7	7	9	7	8	8	6
6	12	6	7	9	9	5	4	9	5
6	7	6	3	4	7	6	6	4	5
2	3	6	2	1	3	2	2	3	4
2	2	2	3	0	2	1	2	4	2
3	2	1							

The general inapplicability of the standard attribute control charts for personal defect data should not be a hindrance for process monitoring and understanding. Without attempting to pigeonhole the application, we can study the process directly by using the simple tools of data analysis. In Table 7.10, we provide the total number of daily personal defects for 63 consecutive days. The data can also be found in the file "personal.dat".

In Figure 7.32, we give a time-sequence plot of total defects with the days of the week labeled. Clearly, the student's personal performance is improving over time, as evidenced by the discernible downward trend. Roberts and Sergesketter (1993) indicate that drops—sometimes substantial—in personal defects often occur just by virtue of using the personal checklist, since it promotes greater awareness of one's personal work. The student concurred and wrote:

> The personal quality project turned out to be a very enjoyable and enlightening experience. The personal quality checklist turned out to be a very effective tool in getting me to become aware of some of my personal defects. Almost immediately I noticed improvements in the categories I selected. I then became constantly conscious of my bad habits and tried to keep them to a minimum.

In addition to the obvious trend effects, there appear to be seasonal effects. Namely, Mondays seem to be high and weekends possibly lower, relative to the rest of the week. Over and above improvement by greater self-awareness, the student attempted the following improvement effort or intervention:

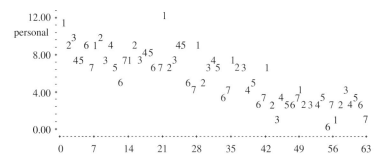

Figure 7.32   Time-series plot of daily personal defects.

On March 3, I decided to try a new organizational system for my desk and for my Day-Timer. The original system consisted of piles of papers on my desk and two corners of my office. These piles were a mixture of different document types. The new system consists of standing files on my desk, each designated by document type. I purchased special plastic "pockets" for my Day-Timer, one for each month, and I place all agendas in the appropriate pocket.

One of the pillars of the scientific method for improving processes is the use of data or facts to provide objective evidence of whether an improvement effort has had any discernible impact. The reader will recall from Chapter 1, the analysis of planned quality improvement efforts is the basis for the check phase of Shewhart's PDCA (plan, do, check, act) cycle. Conventional advice is that first the process should be brought under a state of statistical control—random variation only—and then interventions made. When a process is random before the improvement effort, standard control chart criteria are more suitable to assess whether a change in the process has occurred. As we have illustrated numerous times, many processes are predictable but are not (or cannot be made) random. Additionally, there are processes, such as the personal application, for which random variation is not the most desirable goal. It is clearly preferable to have a nonrandom downward trend in the defects than a random process around some stable mean. However, the difficulty is that the effects of an improvement effort can be hard to discern in the presence of the nonrandom behavior. For the personal application, did the new organization system starting on March 3 (which corresponds to point 40) have an effect on the process? Or, is the lower level of daily defects a continuation of the self-awareness trend? Careful data analysis can help make this distinction.

To evaluate the effects of a deliberate attempt to improve the process, we introduce a regression indicator variable, called organize, that takes on the value 0 for periods prior to the new organizational system (periods 1 to 39) and the value 1 for periods under the new organizational system (periods 40 to 63). In addition to this possible independent variable, we create indicator variables for possible day-of-week effects and a trend variable for the apparent decline in defects. Investigation of the ACF and PACF (both not shown here) does not suggest the need for any lagged variables. In Figure 7.33, we present the results of a stepwise

```
Stepwise Regression

  F-to-Enter:    4.00  F-to-Remove:    4.00

  Response is personal on  9 predictors, with N =  63

Step              1         2         3         4         5
Constant      9.552     9.208     8.485     8.687     8.881

t            -0.131    -0.129    -0.083    -0.084    -0.084
T-Value      -11.67    -12.41     -4.71     -4.97     -5.13

mon                     1.86      1.95      1.75      1.54
T-Value                 3.45      3.86      3.55      3.16

organize                         -2.07     -1.98     -1.92
T-Value                          -3.13     -3.09     -3.09

sat                                        -1.16     -1.38
T-Value                                    -2.36     -2.83

sun                                                  -1.07
T-Value                                              -2.20

S             1.62      1.50      1.40      1.35      1.30
R-Sq         69.06     74.19     77.85     79.79     81.37
```

Figure 7.33   Stepwise regression for daily personal defects application.

regression. Notice that stepwise regression picked the trend variable along with seasonal variables for Monday, Saturday, and Sunday as significant independent variables. Additionally, the indicator variable for the new system was selected as a significant independent variable.

The estimated regression model, based on the stepwise-selected independent variables, is presented in Figure 7.34. As can be seen from the t ratios and the corresponding significance levels, each of the independent variables plays an important role in understanding the process behavior. However, before we read too much into the estimated model, standard checks for model appropriateness need to be performed. In Figure 7.35, we give a snapshot of the diagnostics and leave the full exploration to the reader. The residuals appear random with no evidence of trend or seasonality, as originally seen in the defect data. In addition, the histogram of standardized residuals shows excellent conformity to the normal distribution.

With the regression model deemed satisfactory, we can now make interpretations of the underlying behavior, as suggested by the fitted model. Namely, the model indicates that over the period of observation, personal defects are steadily decreasing at an average pace of 0.0844 per day. Clearly, such a trend cannot last forever. We would soon expect tapering out of the daily defects as the zero level is approached. The goal for the student should be to control the defects at a higher level of personal performance attained from the personal self-improvements. In addition to the trend, we would expect higher levels of defects on Mondays, and lower levels of defects Saturdays and Sundays. A higher level of defects on Mondays is not too surprising since the first day of the week is notorious for catch-up and Blue Monday effects! Lower levels of defects on Saturdays and Sundays

```
The regression equation is
personal = 8.88 - 0.0844 t + 1.54 mon - 1.38 sat - 1.07 sun - 1.92 organize
```

Predictor	Coef	StDev	T	P
Constant	8.8811	0.4236	20.96	0.000
t	-0.08439	0.01644	-5.13	0.000
mon	1.5390	0.4877	3.16	0.003
sat	-1.3814	0.4877	-2.83	0.006
sun	-1.0748	0.4874	-2.20	0.032
organize	-1.9188	0.6211	-3.09	0.003

```
S = 1.304    R-Sq = 81.4%    R-Sq (adj) = 79.7%
```

Analysis of Variance

Source	DF	SS	MS	F	P
Regression	5	423.406	84.681	49.81	0.000
Residual Error	57	96.911	1.700		
Total	62	520.317			

Source	DF	Seq SS
t	1	359.336
mon	1	26.670
sat	1	11.861
sun	1	9.315
organize	1	16.224

Figure 7.34   Regression output for daily personal defects application.

(that is, the weekend) probably reflect the fact that many defect categories are business-oriented which tend to occur during the week.

With respect to the new organizational system, the "organize" variable is highly significant and the corresponding coefficient is negative. This implies that the deliberate improvement effort had a real and favorable impact on the personal process. The fitted-values chart given in Figure 7.36 pictorially summarizes what has been learned from our study of the personal process data. With respect to the PDCA improvement cycle, the confirmation of an improvement by the new approach closes out the check (C) phase. Armed with this knowledge, the student should act (A) on what was observed, by continuing with the new approach, to hold the improvement gains. Finally, in Figure 7.37, we present a special-cause chart for the personal defect process. No residuals are outside the control limits, nor are there any unusual runs of residuals, implying that the residuals are in a state of statistical control. With no special causes at work, primary attention should be given to the information embodied in the fitted model.

AUTOREGRESSIVE BEHAVIOR

For this application, we consider counts of airplane arrivals in successive 10-min intervals in control sector 454LT in the eastern United States from noon to 8

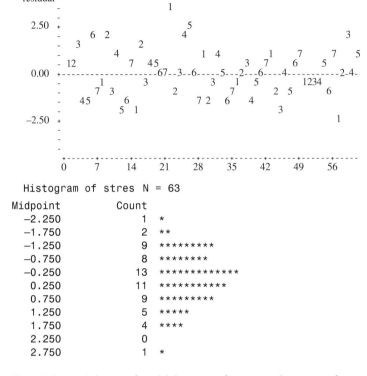

Figure 7.35   Selection of model diagnostics for estimated regression for daily personal defects application.

Table 7-11   Number of airplane arrivals per 10-min interval

1	7	4	5	7	7	4	5	4	6	4	4
6	3	6	4	5	5	2	6	1	5	5	2
3	6	6	9	1	6	7	2	4	5	6	3
4	2	6	4	1	3	6	4	6	2	8	1

p.m., eastern time, on a single day, April 30, 1989.[8] The data are provided in Table 7.11 and can also be found in the file "arrivals.dat".

Given that we are dealing with counts in fixed unit intervals, it is natural to pursue the construction of a c chart. The estimated mean $\bar{c}$ of the data is 4.438. Thus, c chart control limits are given by $4.438 \pm 3\sqrt{4.438}$ which produces UCL = 10.76 and LCL = -1.88; thus LCL is set to 0 because of the computed negative value. The c chart, with these limits, is presented in Figure 7.38.

| [8]Data kindly provided by Der Ann Hsu, University of Wisconsin-Milwaukee.

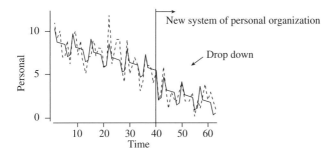

Figure 7.36   Fitted-values chart for daily personal defects application.

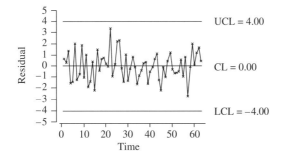

Figure 7.37   Special-cause chart for daily personal defects application.

From the standard chart, we see no apparent out-of-control conditions. We additionally applied the runs rules available in Minitab's c chart option. The supplementary runs rules revealed no unusual patterns in the data series. We might then be tempted to conclude that the counts are in a state of statistical control, in the sense of being an iid process. However, take a closer look at the time-sequence behavior of the counts. It appears that there is a tendency toward *negative* autocorrelation: too much oscillation up and down of the points. The randomness checks shown in Figure 7.39 confirm our suspicion. It appears that it might be rewarding to regress the variable for the count observations, arrivals_t, on its first lag, arrivals_{t-1}. We have done so and found that the autoregression is significant, but the residuals showed evidence of positive skewness. For this reason, we need to consider a simple transformation of the count data. In particular, we will take the square root of arrivals_t; that is, we are now dealing with $\sqrt{\text{arrivals}_t}$.

The reader can confirm that the oscillatory behavior is still evident in the square root of the counts. The regression fit of $\sqrt{\text{arrivals}_t}$ on its first lag, $\sqrt{\text{arrivals}_{t-1}}$, is shown in Figure 7.40. Given that the negative coefficient for the lag variable has a p value of 0.014 (< 0.05), the negative autocorrelation of the arrival counts seems genuine. The reader is left to verify that (1) the runs test and ACF of the residuals provide no significant evidence against randomness and (2)

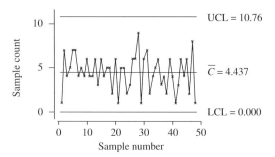

Figure 7.38   *c* Chart for the airplane arrival application.

arrival
  K = 4.4375

  The observed number of runs = 33
  The expected number of runs = 25.0000
  24 Observations above K   24 below
            The test is significant at   0.0196

(a) Runs test

ACF of arrival

```
        −1.0  −0.8  −0.6  −0.4  −0.2   0.0   0.2   0.4   0.6   0.8   1.0
          +----+----+----+----+----+----+----+----+----+----+
 1 −0.330                         X X X X X X X X X
 2  0.053                                  X X
 3  0.026                                  X X
 4  0.037                                  X X
 5  0.048                                  X X
 6 −0.175                            X X X X X
 7 −0.078                              X X X
 8  0.220                                  X X X X X X X
 9 −0.154                            X X X X X
10 −0.028                                  X X
11 −0.044                                  X X
12  0.050                                  X X
```

(b) ACF

Figure 7.39   Randomness checks of the airplane arrival series.

the normality checks of the residuals are satisfactory. Using the individual measurement control chart option in Minitab, we create the special-cause chart
shown in Figure 7.41. Visually, there is a hint of downward trend, or at least of
higher average residuals in the beginning of the series. We should note that the
application of runs rules to the residual series does not suggest any violation.

Finally, we display in Figure 7.42 a fitted-values chart so that we can visualize the estimated systematic tendencies in the count data series. The discovered
negative autocorrelation invites an investigation by management for explanation,
such as a tendency for airplanes to slow down when there is congestion ahead,
and to speed up if the coast seems clear.

```
The regression equation is
sqrt (arr) = 2.80 - 0.355 sqrt (arr-1)

47 cases used 1 cases contain missing values

Predictor          Coef        StDev          T          P
Constant         2.7966       0.2960       9.45      0.000
sqrt (arr-1)    -0.3549       0.1394      -2.55      0.014

S = 0.4783    R-Sq = 12.6%    R-Sq (adj) = 10.7%

Analysis of Variance

Source              DF          SS          MS          F          P
Regression           1      1.4837      1.4837       6.49      0.014
Residual Error      45     10.2936      0.2287
Total               46     11.7773
```

Figure 7.40   Regression output for the fitted lag 1 model for the square root of airplane arrivals.

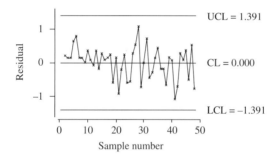

Figure 7.41   Special-cause chart for airplane arrival application.

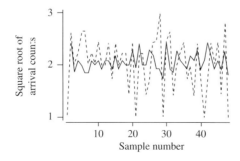

Figure 7.42   Fitted-values chart for airplane arrival application.

EXERCISES

7.1. Explain the difference between a nonconformity and a nonconforming item. Give examples of each.

7.2. How do you describe the sample size for the *c* chart?

7.3. Identify the probability distribution(s) used in the development of *p*, *np*, *c*, and *u* charts.

7.4. Which of the standard attribute control charts are used when the sample size is not constant?

7.5. When the sample size varies, explain why the *np* chart and *c* chart are not appropriate for use.

7.6. For attribute control charts, in general, what is the difference in implications and actions on a process when a point is above an upper control limit and below a lower control limit?

7.7. For most attribute applications, is it really most desirable to have a process in a state of statistical control in the sense of a random process? Give an example of a predictable process, other than a random process, that is more desirable. Be specific.

7.8. List all the conditions required for a process to be regarded as a simple binomial process.

7.9. List all the conditions required for a process to be regarded as a simple Poisson process.

7.10. Under stable conditions, the proportion of rejected product has been found to 0.02. Compute the 3σ control limits for a *p* chart if lots of size 150 are used. What are the 3σ control limits for an *np* chart?

7.11. A *p* chart with 3σ limits has the following limits:

$$\text{UCL} = 0.14424$$

$$\text{CL} = 0.07300$$

$$\text{LCL} = 0$$

What is the subgroup size *n*?

7.12. The Internal Revenue Service (IRS) has offices in numerous regions dedicated to answering questions by telephone from individuals filing income tax forms. Based on a large-scale study of trained IRS representatives, the IRS expects 15 percent of tax questions to be answered incorrectly. To monitor the quality of each of its offices, an IRS quality assurance staff member anonymously calls in and asks a prepared question. For each office, 50 calls are made each week.

    a. Determine the control limits for a *p* chart to determine whether an office is out of control.

b. Determine the control limits for an *np* chart to determine whether an office is out of control.

7.13. To continue Exercise 7.12, brainstorm as many causes as possible of incorrect tax answers by IRS representatives. Construct a cause-and-effect diagram to display the relationships of the enumerated causes.

7.14. A company manufactures string lights for Christmas decorations. A random selection of seven strings, each made up of 100 lights, is done on a periodic basis. Below are the number of defective light sockets out of 700 total sockets for 20 consecutive samples (also given in "socket.dat"):

5	11	7	8	9	12	8	6	10	13
5	9	7	6	10	12	7	8	8	14

Construct a *p* chart for the data. Does the process appear to be in control? What are the prospective limits for the chart?

7.15. Consider the data for Exercise 7.14 used to construct a *p* chart. Find the equivalent *np* chart.

7.16. Consider the data for Exercise 7.14 used to construct a *p* chart. Use the chi-square testing procedure to test whether the proportions are consistent with the assumption of binomial variability. What is the value of the observed chi-square statistic? What is the *p* value for the observed chi-square statistic?

7.17. Throughout his career, professional basketball player Shaquille O'Neal has been criticized for his poor freethrow ability. For 25 consecutive games during the 1997-1998 season, we provide below the number of freethrows missed and the total number of freethrows attempted by O'Neal (also given in "shaq.dat"):

Game	Missed	Attempted	Game	Missed	Attempted
1	3	5	14	6	10
2	2	7	15	5	9
3	4	11	16	7	20
4	11	15	17	7	13
5	4	11	18	4	8
6	2	6	19	6	13
7	5	10	20	7	20
8	5	16	21	6	10
9	5	12	22	6	11
10	7	15	23	8	12
11	6	20	24	4	9
12	3	7	25	7	12
13	3	10			

a. Construct a *p* chart with exact limits for the data. Does the process appear to be in control?

b. Based on the results of part *a*, give the specific details of the design of the control limits for monitoring future proportions of freethrows missed.

7.18. To continue Exercise 7.17, brainstorm as many causes as possible for O'Neal's poor freethrow shooting. Construct a cause-and-effect diagram to display the relationships of the enumerated causes.

7.19. Consider the data for Exercise 7.17 used to construct a *p* chart. Use the chi-square testing procedure to test whether the proportions are consistent with the assumption of binomial variability. What is the value of the observed chi-square statistic? What is the *p* value for the observed chi-square statistic?

7.20. Construct a standardized *p* chart for the data in Exercise 7.17. Also apply all the supplementary runs rules (see Section 4.9) to see if there is any evidence of unusual runs in the data.

7.21. Construct a *p* chart based on average sample size for the data in Exercise 7.17.

7.22. Consider the data on the number of nonconforming glass jars studied in Section 7.2. Use the chi-square testing procedure to test whether the proportions are consistent with the assumption of binomial variability. What is the value of the observed chi-square statistic? What is the *p* value for the observed chi-square statistic?

7.23. The data below give the number of lot-to-lot failures of a plastic-encapsulated resistor to a moisture-resistance test (also given in "resist.dat").

Subgroup Number	Number of Nonconforming	Subgroup Size	Subgroup Number	Number of Nonconforming	Subgroup Size
1	8	250	16	11	125
2	5	125	17	4	125
3	3	125	18	4	125
4	6	210	19	2	150
5	18	250	20	4	210
6	1	150	21	8	150
7	4	125	22	2	150
8	1	125	23	3	150
9	5	150	24	6	125
10	3	250	25	6	210
11	2	150	26	6	125
12	10	250	27	5	150
13	6	125	28	5	125
14	3	125	29	3	150
15	6	150	30	6	150

a. Construct a *p* chart based on the average subgroup size. Which subgroups either are identified as being out of control or are close to be identified as so? Without further calculation, are you able to say with certainty whether these subgroups are out-of-control signals if the *p* chart with exact limits is constructed? Explain.

b. Construct now a *p* chart with exact limits. Which subgroups are identified as being out of control? Compare the results here with part *a*.

c. Assuming that any out-of-control signals from part *b* can be associated with a special-cause explanation, give the specific details of the design of the control limits for monitoring future proportions associated with this application.

7.24. Construct a standardized *p* chart for the data in Exercise 7.23. Also apply all the supplementary runs rules (see Section 4.9) to see if there is any evidence of unusual runs in the data. Assume again that any out-of-control signals can be associated with a special-cause explanation. Explain the specific details of how future proportions will be monitored.

7.25. Consider the following personal process improvement effort undertaken by a former student of mine. The student, an avid golfer who had never been satisfied with his putting, set up an indoor putting green for practice and experimentation. Each day for 15 consecutive days, he attempted 50 putts from a fixed distance. Below are the number of putts missed for each of the 15 days (also given in "putt.dat"):

25	27	28	24	25	25	21	21	29	23
26	20	23	30	24					

a. Construct a *p* chart for the data. Does the process appear to be in control? What are the prospective limits for the chart?

b. The student noticed that nearly 75 percent of the misses were "right" misses. Given this observation, he concluded a major source of the problem was the ball's being positioned too far back in his putting stance. He adjusted his putting process by setting the ball several inches forward, keeping it just inside his left toe. After implementing this modification, he got the following results on 10 new samples:

9	12	11	16	14	13	15	12	14	11

Plot the new subgroup proportions along with the prospective limits determined in part *a*. What are your conclusions? What now should be the values of the prospective limits for future samples?

7.26. The police department of a large metropolitan area is trying to analyze crime rates in certain geographic segments. It has been long recognized that not all offenses are reported for a variety of reasons (e.g., victims feel the offense is too minor to report, or they may not wish to deal with the "hassle" of police reports). For this reason, the police department has contracted out a research firm to randomly contact residences to estimate the incidence of crime offenses. In particular, random samples of 2500 residences are contacted each month. Below are the data for the 18 consecutive months (also given in "crime.dat"):

9	6	12	8	9	3	7	10	6	8
4	6	9	10	7	5	9	14		

a. Construct an *np* chart for the data. Does the process appear to be in control? What are the prospective limits for the chart?

b. Relying on the Poisson distribution as a satisfactory approximation for the binomial distribution, construct a *c* chart for the data. Comment on how well the *c* chart approximates the *np* chart constructed in part *a*.

7.27. To continue Exercise 7.26, brainstorm as many causes as possible for crime rate in a given geographic area. Construct a cause-and-effect diagram to display the relationships of the enumerated causes.

7.28. Consider the data for Exercise 7.26 used to construct an *np* chart. Find the equivalent *p* chart. Use the chi-square testing procedure to test whether the proportions are consistent with the assumption of binomial variability. What is the value of the observed chi-square statistic? What is the *p* value for the observed chi-square statistic?

7.29. Suppose that the true proportion conforming is equal to 0.03. Consider the following subgroup size determination questions:

a. What is the sample size required to ensure a positive *p* chart LCL?

b. What is the sample size required to ensure that a subgroup proportion computed from a subgroup with only one nonconforming will fall below a *p* chart LCL?

7.30. Establish a condition for the value of *p* so that the minimum subgroup size needed to satisfy the normal approximation requirement of $np > 5$ will also result in a positive *p* chart LCL.

7.31. Consider the sample size criterion based on the specification of a probability γ of observing a subgroup with at least one nonconforming item. Establish a general condition for the value of γ so that the minimum subgroup needed to meet this criterion will also satisfy the normal approximation requirement of $np > 5$.

7.32. What must be true about the value of $\bar{c}$ so that the lower control limit of a *c* chart is positive?

7.33. Since the UCL and LCL of a *c* chart typically compute to fractional values and since counts are, by definition, integers, can we simply make the values of control limits integers? How would you suggest conversion of the values of the control limits to integers be done? Explain your decision rule for signaling out of control, given your conversion suggestion.

7.34. Under stable conditions, the mean number of nonconformities has been found to be 12. What are the 3σ control limits for the *c* chart?

7.35. A major automobile maker wishes to monitor and control imperfections present in electropaint coating. Every day, at approximately the same time, 5 car bodies are selected for analysis. The testing procedure calls for a 1-ft^2 frame to be placed in exactly the same position on each car. Using both visual and touch procedures, the inspector then counts the number of surface imperfections (referred to as *inclusions*). Knowledge of the mean number of inclusions produced by the coating process is an important guide for follow-up activities (e.g., the amount of sanding preparation required). Below are the number of inclusions revealed on the set of 5 car bodies for 20 consecutive days (also given in "body.dat"):

12	12	5	8	7	5	2	5	10	5
11	4	7	7	13	9	7	9	4	8

a. After defining the 5 car bodies as a single inspection unit, construct a *c* chart for the data. Does the process appear to be in control? What are the prospective limits for the chart?

b. Use the chi-square testing procedure to test whether the counts are consistent with the assumption of Poisson variability. What is the value of the observed chi-square statistic? What is the *p* value for the observed chi-square statistic?

7.36. Consider the inclusion nonconformity data for Exercise 7.35 used to construct a *c* chart. Suppose future inspection units are redefined as 7 car bodies. What are the centerline and control limits for the *c* chart for monitoring future production based on the newly defined inspection unit?

7.37. An operations manager for an airline at a large airport was interested in studying and monitoring the number of pieces of luggage lost (temporarily or permanently). Below are the numbers of lost-baggage claims recorded per day over a 30-day period (also given in "luggage.dat"):

17	26	37	21	24	8	27	22	26	30
24	27	35	22	30	16	19	21	27	28
34	33	20	21	29	22	24	17	30	23

a. Construct a *c* chart for the data. Does the process appear to be in control?

b. It turns out that on the fourth day there was a major snowstorm that resulted in many canceled flights. In light of this information, establish the prospective control limits for the *c* chart.

7.38. To continue Exercise 7.37, brainstorm as many causes as possible for lost baggage. Construct a cause-and-effect diagram to display the relationships of the enumerated causes.

7.39. Workplace absenteeism is one of the most significant employer issues that human resources practitioners deal with. It creates not only operational problems but also morale problems among coworkers who are called upon to pick up the slack. For this reason, most companies carefully track various measures of absenteeism. Consider, for example, a small manufacturing facility located in Green Bay, Wisconsin. For 35 consecutive weeks (defined as Monday through Friday), we show below the weekly number of frontline employees absent (also given in "absent.dat:)"

13	12	8	14	14	10	12	8	16	12
10	6	11	8	5	12	11	14	14	13
11	9	8	30	11	13	5	15	7	9
13	10	9	15	13					

a. Construct a *c* chart for the data. Does the process appear to be in control?

b. It turns out that during the 24th week, there was a Monday night football game featuring the Green Bay Packers. The desire to attend and watch the Packers is legendary in Green Bay. In light of this information, establish the prospective control limits for the *c* chart.

7.40. To continue Exercise 7.39, brainstorm as many causes as possible for absenteeism at a given company. Construct a cause-and-effect diagram to display the relationships of the enumerated causes.

7.41. A large software development company writes customized computer programs for a variety of industry applications. Each day, completed modules of computer code are run through a diagnostic tool which is able to identify the location of coding errors. Because modules vary in size, a module cannot serve as an inspection unit. Instead, the number of errors must be considered in terms of the number of lines of code. Below are data for 20 consecutive days (also given in "module.dat"):

Day	Total Number of Errors	Total Number of Code Lines
1	43	8,766
2	36	5,605
3	22	5,825
4	39	8,389
5	60	9,715
6	54	8,263
7	30	5,436
8	39	8,431
9	26	5,509
10	61	10,258
11	54	9,802
12	73	10,304
13	33	6,109
14	36	7,735
15	49	10,038
16	51	10,989
17	45	10,221
18	50	8,756
19	40	6,330
`20	52	9,245

a. Define an inspection unit as 1000 lines of code. Construct a u chart with exact limits for the data. Does the process appear to be in control?

b. Based on the results of part a, give the specific details of the design of the control limits for monitoring future data on the total number of coding errors and the total number of code lines.

7.42. Construct a u chart based on average inspection unit size for the data in Exercise 7.41.

7.43. Construct a standardized u chart for the data in Exercise 7.41. Also apply all the supplementary runs rules (see Section 4.9) to see if there is any evidence of unusual runs in the data.

7.44. LaserBond, a manufacturer of paper used in copy machines and laser printers, monitors various aspects of its production by using control charts. Paper is produced in large rolls, 10 ft long and 5 ft in diameter. Each shift, a sample is taken from each completed roll and is checked in the testing laboratory for nonconformities, such as flecks of dirt and discoloration. All these nonconformities are given the same weight in terms of importance. Below are data for 24 consecutive production shifts (also given in "laser.dat"):

Shift	Total Number of Nonconformities	Number of Rolls Produced
1	52	10
2	59	10
3	44	7
4	50	7
5	35	8
6	49	10
7	60	9
8	55	9
9	48	9
10	56	10
11	57	9
12	46	8
13	36	6
14	39	8
15	63	9
16	41	7
17	50	10
18	43	7
19	42	9
20	32	7
21	41	7
22	46	10
23	48	10
24	34	6

      a. Define an inspection unit as one roll of paper. Construct a u chart with exact limits for the data. Does the process appear to be in control?

      b. Based on the results of part a, give the specific details of the design of the control limits for monitoring future data on the total number of nonconformities and the total number of rolls produced.

7.45. Construct a u chart based on average inspection unit size for the data in Exercise 7.44.

7.46. Construct a standardized u chart for the data in Exercise 7.44. Also apply all the supplementary runs rules (see Section 4.9) to see if there is any evidence of unusual runs in the data.

7.47. In Section 7.4, we showed the general formula for how the chi-square statistic, used to test for Poisson variability, is computed from c chart data. Suppose now you have u chart data, that is, u_i and n_i.

      a. In terms of u_i and n_i, write down the general formula for computing the chi-square statistic to test whether the u chart data are consistent with the underlying assumption of Poisson variability.

      b. Consider the data from Exercise 7.44. Based on the formula derived in part a, test whether these u chart data are consistent with the underlying assumption of Poisson variability.

7.48. Nonconformities in the manufacture of motorcycles can be sorted into three categories: minor, major, and serious. Twenty samples of 10 motorcycles are chosen each day, and the total number of nonconformities in each category is recorded. The data are shown below (also given in "cycle.dat"):

Sample	Serious Nonconformity	Major Nonconformity	Minor Nonconformity
1	0	1	3
2	1	3	5
3	1	0	7
`4	1	1	4
5	0	0	4
6	1	0	4
7	1	1	3
8	0	1	4
9	1	1	5
10	1	1	2
11	1	1	6
12	3	0	6
13	0	0	7
14	0	0	4
15	1	0	3
16	0	0	13
17	0	0	6
18	1	0	2
19	1	1	8
20	0	2	5

 a. Assume a weighting scheme of 100, 50, and 10 for serious, major, and minor, respectively. Construct a *D* chart for the data. Does the demerit process appear to be in control? What are the prospective limits for the chart?

 b. Investigate each category of nonconformities separately, possibly using a control chart. What is revealed? In light of your discovery, what do you recommend be done in conjunction with the use of *D* charts?

7.49. An article in *Journal of Quality Technology* (F. C. Kaminsky, J. C. Benneyan, R. D. Davis, and R. J. Burket, 1992, 22, pp. 63–69) presents attribute data on the number of orders for a truck used in the distribution of manufactured product. Below we present the first 35 samples from the data set presented in the article. The data are shown below and are also given in "truck.dat".

22	58	7	39	7	33	8	23	5	26
12	26	10	30	5	24	6	35	6	23
10	17	7	10	6	13	9	21	8	12
4	18	7	17	5					

 a. Construct a *c* chart for the count data. What actions does the chart suggest?

 b. Making no assumption of Poisson variability, compute the sample variance of the count data. Compare the computed sample variance with the sample mean of the count data. What do you conclude about the appropriateness of the application of a standard *c* chart? Explain.

7.50. Continuation of Exercise 7.49: Given the inappropriateness of the standard *c* chart for the data, the authors of the article from which the data were obtained proposed the construction of an alternative control chart based on the geometric distribution. In this problem, we take an alternative "angle" to the analysis of these data.

 a. Perform randomness checks on the count data. What do you conclude?

 b. Fit the data series with a lag 1 variable and a trend variable.

 c. Construct a special-cause chart. What do you find?

 d. Assuming any out-of-control signal revealed in part *c* has an identifiable special-cause explanation, reestimate the fitted model without the effects of the associated data point. Remember, deletion of observations which follow a systematic pattern should be done with an indicator variable. Report the significance of the independent variables for the new fitted model and the overall diagnostics of the fitted model.

 e. Present the revised special-cause chart and the fitted-values chart.

 f. In light of your data analysis, how do feel about the application of an alternative control chart (such as a chart based on the geometric distribution), if the alternative method is grounded in an assumption of iid behavior?

7.51. In the Ford (1984) manual titled *Continuing Process Control and Process Capability Improvement*, count data on the number of "cut flaws" on a blot are presented and used to construct a standard *c* chart. Below are the counts for 25 consecutive samples (also given in "ford.dat"):

9	15	11	8	17	11	5	11	13	7
10	12	4	3	7	2	3	3	6	2
7	9	1	5	8					

   a. Construct a *c* chart for the count data. What actions does the chart suggest?
   b. Making no assumption of Poisson variability, compute the sample variance of the count data. Compare the computed sample variance with the sample mean of the count data. What do you conclude about the appropriateness of the application of a standard *c* chart? Explain.

7.52. Continuation of Exercise 7.51: Instead of using the *c* chart for guidance, consider a data analysis approach.
   a. Perform randomness checks on the count data. What do you conclude?
   b. Fit the data series with a lag 3 variable. Report the significance of the independent variable for the fitted model and the overall diagnostics of the fitted model. What is the model suggesting about the time behavior of the count series?
   c. Construct a special-cause chart. What do you find? Compare this result with the conclusions of the *c* chart from Exercise 7.51*a*.
   d. Present the fitted-values chart.

7.53. In Section 7.5, we studied data from a personal quality checklist. Develop your own personal quality checklist. The idea again is to locate activities that you are not performing well, such as failing to promptly return phone calls or reply to correspondence, or losing papers. The personal quality checklist is then a means of recording your daily performance. Do not be constrained in your thinking, be innovative and have some fun. Keep the number of categories manageable. Six to eight categories are recommended. Each category must be unambiguously defined so that you can recognize a defect when it occurs. After collecting at least 20 consecutive days, perform a complete data analysis on your daily total defect counts.

REFERENCES

Agresti, Alan (1990): *Categorical Data Analysis*, Wiley, New York.

Besterfield, D. H. (1986): *Quality Control,* 2d ed., Prentice-Hall, Englewood Cliffs, NJ.

Box, G. E. P., J. S. Hunter, and W. S. Hunter (1978): *Statistics for Experimenters*, Wiley, New York.

Duncan, A. J. (1986): *Quality Control and Industrial Statistics,* 5th ed., Irwin, Homewood, IL.

Farnum, N. R. (1994): *Statistical Quality Control and Improvement*, Duxbury, Belmont, CA.

Ford Motor Company, Statistical Methods Office (1984): *Continuing Process Control and Process Capability Improvement*, Dearborn, MI.

Glasser, G. J. (1985): "Quality Audits of Paperwork Operations—The First Step toward Quality Control," *Journal of Quality Technology*, 17, pp. 100–107.

Grant, E. L., and R. S. Leavenworth (1996): *Statistical Quality Control,* 7th ed., McGraw-Hill, New York.

Juran, J. M., and F. M. Gryna (1988): *Juran's Quality Control Handbook,* 4th ed., McGraw-Hill, New York.

Jurow, S., and S. Barnard, eds. (1993): Special Issue on Integrating Total Quality Management in a Library Setting, *Journal of Library Administration*, 18, nos. 1–2.

Kittilitze, R. G. (1979): "Poisson Distribution and Textile Mill Problems," *ASQC Annual Quality Congress Transactions*, pp. 126–133.

Koch, G. G., S. S. Atkinson, and M. E. Stokes (1986), "Poisson Regression," pp. 32–41 in *Encyclopedia of Statistical Sciences*, vol. 7, Wiley, New York.

Latzko, W. J., and J. D. Dowhin (1991): "Achieving Service Quality by Charting," *ASQC Annual Quality Congress Transactions*, pp. 2–7.

Mackey, T., and K. Mackey (1992): "Think Quality! The Deming Approach Does Work in Libraries," *Library Journal,* 117, pp. 57–61.

Mosteller, F., and J. W. Tukey (1977): *Data Analysis and Regression*, Addison-Wesley, Reading, MA.

Roberts, H. V., and B. F. Sergesketter (1993): *Quality Is Personal: A Foundation for Total Quality Management*, Free Press, New York.

Ryan, T. P. (1989): *Statistical Methods for Quality Improvement*, Wiley, New York.

Memory Control Charts

CHAPTER OVERVIEW

The control charts discussed thus far are all based conceptually on the control scheme originally devised by Walter A. Shewhart. The key feature of Shewhart charts is that the plotted statistic, at any given period, reflects only the sample information for that period. For this reason, Shewhart control charts are sometimes referred to as *memoryless control charts*. As briefly illustrated in Chapter 4, this memoryless feature renders Shewhart control charts relatively less effective in the detection of smaller process shifts.

To overcome performance deficiencies of Shewhart control charts in detecting smaller shifts, a variety of charts with memory have been developed. In contrast to the Shewhart scheme, the plotted statistic of *memory control charts* incorporates both present and past sample information. The premise of memory control charts is that while a small shift may not lead to an isolated point in the Shewhart scheme to signal out of control, a statistic that considers previous sample outcomes will allow for the shift effect to be amplified through accumulation.

In this chapter, we will present the most common memory charts: the cumulative-sum (CUSUM) control chart, the moving-average chart, and the exponentially weighted moving-average (EWMA) control chart.[1]

8.1   THE CUMULATIVE-SUM CONTROL CHART

Originally introduced by Page (1954), the cumulative-sum control chart is a procedure based on the cumulative sum of the deviations of a sample statistic from some *target value*. The focus on a target value is based on the hypothesis testing framework underlying the procedure. Namely, CUSUM charts have been developed according to the idea of testing between two competing hypotheses, an intended quality level versus a threshold level that is considered important to detect quickly. For example, if the process mean level is being monitored, the hypotheses might be stated as follows:

$$H_0: \mu = \mu_0 \qquad \text{versus} \qquad H_1: \mu = \mu_1 \qquad\qquad \textbf{(8.1)}$$

Depending on the direction of concern, μ_1 is chosen either to be greater than or less than μ_0. We regard μ_0 as the target value or acceptable value for the mean level, while μ_1 is a mean level viewed as unacceptable. For example, the target value might represent the strength of a compound material. For safety considerations, there is a minimum acceptable strength level allowed, and thus this would imply that $\mu_1 < \mu_0$. If the process shifts from the target value, the goal is then to detect the out-of-control mean μ_1 as quickly as possible.

---

[1]It should be pointed out that both the full version and student version of Minitab support options for the construction of moving-average and EWMA charts. However, only the full version of Minitab has CUSUM capabilities.

Equivalently, we can regard $\Delta_\mu = |\mu_1 - \mu_0|$ as the *minimum* shift size that one wishes to detect quickly. In terms of Δ_μ, the hypotheses of (8.1) can be written as

Case of $\mu_1 > \mu_0$:

$$H_0: \mu = \mu_0 \qquad \text{versus} \qquad H_1: \mu = \mu_0 + \Delta_\mu \qquad\qquad \textbf{(8.2)}$$

Case of $\mu_1 < \mu_0$:

$$H_0: \mu = \mu_0 \qquad \text{versus} \qquad H_1: \mu = \mu_0 - \Delta_\mu \qquad\qquad \textbf{(8.3)}$$

The hypotheses stated thus far are concerned with shifts in a single direction, either up or down. In many applications, it is desirable to detect a shift in either direction. If upward and downward shifts of the same magnitude are considered undesirable, we can think of *three* competing hypotheses:

$$H_0: \mu = \mu_0 \qquad H_1: \mu = \mu_0 - \Delta_\mu \qquad H_2: \mu = \mu_0 + \Delta_\mu \qquad \textbf{(8.4)}$$

The alternative hypotheses are symmetric around the null hypothesis H_0 by the amount Δ_μ. It is possible to structure the hypotheses and develop tests for asymmetric alternative hypotheses, that is, to add and subtract different amounts (say, $\Delta_\mu^{(1)}$ and $\Delta_\mu^{(2)}$) to and from μ_0. This would be done for situations where shifts in either direction are important to detect, but smaller shifts below the target (say) might be more critical than comparable-size shifts above the target.

With three competing hypotheses, there must be three decision *zones* associated with the CUSUM procedure. Namely, if the CUSUM statistic falls in a middle zone, the null hypothesis of the process being on target is not rejected. However, if the statistic falls in one of the two other zones, the null hypothesis is rejected in favor of believing that the process has minimally shifted downward to a level given in H_1 or that the process has minimally shifted upward to a level given in H_2.

We have focused our discussion on the situation where the target is equal to the mean of a process generating continuous measurements. However, CUSUM methods are not restricted to detecting departures from target values that are the means of a continuous random variable. For some applications, a possible target (or acceptable) value could be expressed in terms of the percentage of nonconforming items p_0. For more details of this application see Wadsworth, Stephens, and Godfrey (1986). If the intent is to detect a shift in the percentage nonconforming from a level p_0 to another level p_1, the minimum shift level is given by $\Delta_p = |p_1 - p_0|$. Depending on the direction of the shift that one is interested in detecting, the parameters p_0 and Δ_p can then replace μ_0 and Δ_μ, respectively, in the hypothesis-testing framework given by (8.2), (8.3), or (8.4).

Similarly, a specified average number of nonconformities c_0 might constitute a target level with a minimum shift given by $\Delta_c = |c_1 - c_0|$. For more details of this application see Lucas (1985) and White, Keats, and Stanley (1997). Finally, CUSUM methods can also be employed to detect shifts in the process standard deviation, and hence the target of interest could be σ_0. Chang and Gan (1995) and Wadsworth, Stephens, and Godfrey (1986) describe CUSUM techniques for monitoring standard deviations (or ranges) derived from rational subgroups ($n > 1$). In a subsequent subsection, we present an innovative CUSUM technique developed by Hawkins (1993) for detecting changes in process variability in a series of individual observations ($n = 1$).

BASIC IDEA OF A CUSUM-TYPE STATISTIC

Before we develop the decision criteria for testing the hypotheses outlined in this section, it will be useful to develop some insight into the CUSUM-type statistics. Let us assume that a quality characteristic X is normally distributed with a standard deviation known to be $\sigma = 10$. Furthermore, suppose that the target value μ_0 for the process mean level is given by $\mu_0 = 100$. Using Minitab's random number generator, we randomly generate 40 observations, all drawn from a normal distribution with mean 100 and standard deviation 10. These observations are given in Table 8.1 under the column labeled "$x_i^{\text{no shift}}$." Now suppose that we "disturb" the measurements by introducing a sustaining shift starting at observation $i = 20$. In particular, the mean level is increased from $\mu_0 = 100$ to $\mu = 110$. In terms of the standard deviation, the process had shifted by 1σ. The measurements for the shifted process are given under the column labeled "x_i^{shift}."

In Figure 8.1a, both data series are plotted on a Shewhart chart with limits established from the parameters of the in-control process:

$$\text{UCL} = \mu_0 + 3\sigma = 100 + 3(10) = 130$$

$$CL = \mu_0 = 100$$

$$\text{LCL} = \mu_0 - 3\sigma = 100 - 3(10) = 70$$

None of the observations from the slightly shifted process come close to exceeding the upper control limit. If it were not for the simultaneous plotting of both data series, one would be hard pressed to pick up any change in the process by using the Shewhart method alone.

One possibility is to supplement the 3σ rule with additional control criteria, such as the supplementary runs rules introduced in Section 4.9. Figure 8.1b shows the results of applying all the additional control criteria offered by Minitab. As we can see, out-of-control signals are indeed flagged after the occurrence of the shift. In particular, a rule 6 violation (4 of 5 successive observations fall beyond 1 standard deviation on one side of the centerline) occurred at observation 25, and a rule 2 violation (9 successive observations on one side of the centerline) occurred at observations 33, 34, and 35. Even though the additional criteria were successful in alerting us to an out-of-control condition in this example, there are some concerns with the supplementary runs rules approach when the intent is to detect small, but sustaining, process shifts. We list the following three such concerns:

1. As noted previously (Section 4.9), the use of multiple tests can substantially reduce the average run length (ARL) of the control procedure when the

Table 8.1   Unshifted and shifted process data

Sample i	$x_i^{\text{no shift}}$	$x_i^{\text{no shift}} - 100$	$\text{CUSUM}_i^{\text{no shift}}$		x_i^{shift}	$x_i^{\text{shift}} - 100$	$\text{CUSUM}_i^{\text{shift}}$
1	121.12	21.12	21.12		121.12	21.12	21.12
2	98.94	−1.06	20.06		98.94	−1.06	20.06
3	78.38	−21.62	−1.56		78.38	−21.62	−1.56
4	108.62	8.62	7.06		108.62	8.62	7.06
5	88.06	−11.94	−4.88		88.06	−11.94	−4.88
6	94.78	−5.22	−10.10		94.78	−5.22	−10.10
7	106.33	6.33	−3.77		106.33	6.33	−3.77
8	94.58	−5.42	−9.19		94.58	−5.42	−9.19
9	111.53	11.53	2.34		111.53	11.53	2.34
10	97.93	−2.07	0.27		97.93	−2.07	0.27
11	100.40	0.40	0.67		100.40	0.40	0.67
12	86.40	−13.60	−12.93		86.40	−13.60	−12.93
13	105.81	5.81	−7.12		105.81	5.81	−7.12
14	98.15	−1.85	−8.97		98.15	−1.85	−8.97
15	92.64	−7.36	−16.33		92.64	−7.36	−16.33
16	109.64	9.64	−6.69		109.64	9.64	−6.69
17	98.28	−1.72	−8.41		98.28	−1.72	−8.41
18	114.84	14.84	6.43		114.84	14.84	6.43
19	90.23	−9.77	−3.34		90.23	−9.77	−3.34
20	85.59	−14.41	−17.75	**Shift**	95.59	−4.41	−7.75
21	109.67	9.67	−8.08		119.67	19.67	11.92
22	116.59	16.59	8.51		126.59	26.59	38.51
23	108.88	8.88	17.39		118.88	18.88	57.39
24	83.35	−16.65	0.74		93.35	−6.65	50.74
25	104.98	4.98	5.72		114.98	14.98	65.72
26	92.28	−7.72	−2.00		102.28	2.28	68.00
27	101.14	1.14	−0.86		111.14	11.14	79.14
28	91.87	−8.13	−8.99		101.87	1.87	81.01
29	92.98	−7.02	−16.01		102.98	2.98	83.99
30	115.13	15.13	−0.88		125.13	25.13	109.12
31	98.65	−1.35	−2.23		108.65	8.65	117.77
32	108.86	8.86	6.63		118.86	18.86	136.63
33	90.62	−9.38	−2.75		100.62	0.62	137.25
34	96.57	−3.43	−6.18		106.57	6.57	143.82
35	109.20	9.20	3.02		119.20	19.20	163.02
36	80.40	−19.60	−16.58		90.40	−9.60	153.42
37	101.12	1.12	−15.46		111.12	11.12	164.54
38	94.42	−5.58	−21.04		104.42	4.42	168.96
39	103.76	3.76	−17.28		113.76	13.76	182.72
40	117.20	17.20	−0.08		127.20	27.20	209.92

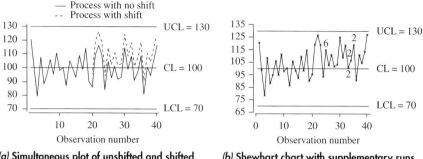

(a) Simultaneous plot of unshifted and shifted processes with Shewhart limits

(b) Shewhart chart with supplementary runs rules applied to shifted process

Figure 8.1   Shewhart chart and supplementary runs rules for the detection of a small shift.

process is in control. A shortened in-control ARL is not desirable because it implies more frequent signaling of out of control when the process is actually in control. It might be more desirable to have a *single* charting procedure which allows the user to control the in-control ARL without sacrificing the ability of the procedure to detect smaller shifts.

2. Intrinsic to their design, runs rules single out short sequences of observations inconsistent with an in-control random process. When a runs rule violation occurs, it may not be clear if the underlying condition is localized or persisting. Clearly, appropriate corrective actions depend greatly on a proper understanding of the nature of the underlying special cause(s). Reacting to the condition as being localized when it is truly persisting can lead to suboptimal corrective action.

3. Investigators have raised a question about how effective runs rules are for picking up small sustaining shifts. Most notably, Champ and Woodall (1987) found that the CUSUM chart is more sensitive to small sustaining shifts than a Shewhart chart supplemented by multiple runs rules.

Given the above concerns with supplementary runs rules, let us focus our attention on another mode of attack on the issue of detecting small sustaining shifts. Consider subtracting first the target value μ_0 from each observed value x_i. If the process is centered on μ_0, we expect the deviations $x_i - \mu_0$ to be a random series of negative and positive numbers with no tendency toward the positive or negative direction; that is, we expect, on average, the deviations to be zero. In contrast, if the underlying mean of the process is located at $\mu_1 \neq \mu_0$, the deviations will have a tendency to be positive if $\mu_1 > \mu_0$ and negative if $\mu_1 < \mu_0$. Now the idea is to amplify the effect of positive or negative tendencies in the deviations. To do so, we can accumulate the deviations over time. If the individual deviations tend to be positive, summing the deviations will result in a statistic that will grow larger and larger. Similarly, if the individual deviations tend to be negative, a cumulative statistic will become larger and larger in magnitude in the negative direction.

Thus, given successive deviations $x_1 - \mu_0, x_2 - \mu_0, x_3 - \mu_0, \ldots, x_k - \mu_0$, we can form the following successive sums:

$$\text{CUSUM}_1 = x_1 - \mu_0$$

$$\text{CUSUM}_2 = x_1 - \mu_0 + x_2 - \mu_0$$

$$\text{CUSUM}_3 = x_1 - \mu_0 + x_2 - \mu_0 + x_3 - \mu_0$$

$$\cdot$$
$$\cdot$$
$$\cdot$$

(8.5)

$$\text{CUSUM}_k = x_1 - \mu_0 + x_2 - \mu_0 + x_3 - \mu_0 + \cdots + x_k - \mu_0$$

It can be recognized that the ith cumulative sum CUSUM_i can be more compactly defined in one of two ways

$$\text{CUSUM}_i = \sum_{j=1}^{i}(x_j - \mu_0) \qquad \text{(8.6a)}$$

$$= x_i - \mu_0 + \text{CUSUM}_{i-1} \qquad \text{(8.6b)}$$

where the initial ("starting") value for the cumulative statistic CUSUM_0 is taken to be zero.

Consider now the application of the cumulative-sum idea to the process data of Table 8.1. Adjacent to the unshifted and shifted data series in columns 2 and 5 of Table 8.1, we provide the deviations from the target value $\mu_0 = 100$. Notice that from sample 20 onward, the deviations are an equal mix of positive and negative values for the unshifted series, while the majority of deviations for the shifted series are positive. Applying either (8.6a) or (8.6b), we obtain the cumulative sums shown for each series in Table 8.1. Starting with sample 20, the cumulative sums for the unshifted series seem to meander negatively and positively with no exceptional tendency to go either way. However, in dramatic contrast, the cumulative sums for the shifted series seem to "explode" in the positive direction.

To visualize the effects on the cumulative sum, Figure 8.2 shows a sequence plot of the CUSUM statistic for each data series. Notice at the time of the shift, the CUSUMs for the shifted series depart rapidly in a trendlike manner away from meandering behavior around the 0 value. This plot brings to bear on the process shift much more dramatically than the plot of the isolated ("memoryless") sample values shown in Figure 8.1a.

Bissell (1994) suggests that there is another interesting advantage to the CUSUM approach over the conventional Shewhart plot. Namely, it can be easier to locate the time of the shift on a CUSUM plot. This is accomplished by looking for *change* or *turning* points in the CUSUM plot. From Figure 8.2, the change point is highlighted and indeed corresponds to the time of the shift. Having a better sense of when the onset of a shift occurred is clearly valuable in searching for special-cause explanations. We caution, however, that the search

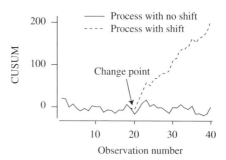

Figure 8.2   Plot of cumulative statistic for unshifted and shifted processes.

for change points should be done only *after* there is a formal signal of a shift because, as we will now see, an in-control CUSUM plot often gives the appearance of change points when in actuality no shift has occurred.

Even though it seemed pretty obvious, from the CUSUM plot, that there exists an out-of-control condition for our example, visual inspection is not a sufficient means for detecting change, in general. In fact, visual inspection alone of a CUSUM plot can be quite deceiving. The reason is that the cumulative statistic, over time, follows nonrandom process which can wander away even if no shift has occurred.

To be more specific, suppose that the process measurements, represented by the random variable X, are random and normally distributed with mean at the target value μ_0 and standard deviation σ. Then the deviations $X_i - \mu_0$ are random and normally distributed with mean 0 and standard deviation σ. Suppose we define a random variable ε_i to be $X_i - \mu_0$. Then from (8.6b) we have

$$\text{CUSUM}_i = \text{CUSUM}_{i-1} + \varepsilon_i \tag{8.7}$$

The reader might recall that this is the model for a random-walk process (see Section 6.6). What is most important to recall is that a random-walk process is a nonstationary meandering process. The property of being nonstationary implies that the cumulative statistic can wander away quite naturally without any provocation. That is, even if the process remains in control and on target at μ_0, CUSUM_i can seemingly drift off.

To demonstrate this fact, we simulated two different series of 200 consecutive random and normal observations with mean $\mu_0 = 100$ and standard deviation $\sigma = 10$ to match the parameters of the example studied previously in this section. Hence, the simulated series are in control around the mean of μ_0 with no underlying shifts. The CUSUM plots for each of these series are depicted in Figure 8.3. In Figure 8.3a, the appearance of a downward trend can lull one into concluding that the process must have shifted in a sustaining manner to a level below $\mu_0 = 100$. While from Figure 8.3b, there is a temptation to conclude that the process shifted to a level above 100. However, such conclusions would be wrong since each data series is in control on target with no underlying sustaining shifts.

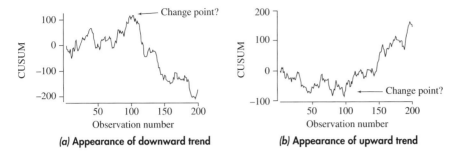

(a) **Appearance of downward trend**   (b) **Appearance of upward trend**

Figure 8.3   Plot of cumulative statistics for randomly generated unshifted processes.

What is needed is a formal statistical test for CUSUM statistics which can help detect when strong and rapid drifts occur, reflecting a *true* sustaining shift. There are two basic approaches, in practice, for monitoring CUSUMs. One approach is to apply a *V mask* (basically, nonparallel control limits) to a CUSUM plot. The other approach, referred to as the *tabular* procedure, represents CUSUMs in a different manner. Historically, both procedures were common among CUSUM users. However, because of recently cited disadvantages with the implementation of the V mask approach, the tabular approach is becoming increasingly the popular choice. We will present both methods, noting their relative advantages and disadvantages and allowing the ultimate choice to rest with the reader.

V Mask CUSUM Procedure for Monitoring the Process Mean

Originally proposed by Barnard (1959), a V mask is a set of control limits applied successively to each cumulative statistic value. For the ith cumulative value $CUSUM_i$, the V mask procedure calls for placing the corner of the V a certain lead distance d ahead of $CUSUM_i$ and extending the legs of the V straight back in such a manner that they are at a distance H units above and below $CUSUM_i$ (see Figure 8.4).

Intuitively, the ever widening V mask arms attempt to account for the natural wanderings of a random walk. If all the previous cumulative sums fall within the arms of the V mask, then there is no evidence of an out-of-control condition. Any cumulative values falling outside the limits are an indication of a shift from target, that is, an out-of-control process. At first glance, it might seem that the arms in Figure 8.4 are incorrectly labeled, namely, the upper arm as the "lower" control limit and the lower arm as the "upper" control limit. The reason for such labeling is that if points fall below the lower arm, this is an indication of an *upward* shift in the underlying process, whereas if points fall above the upper arm, this is an indication of a *downward* shift.

From Figure 8.4, we see that the design of the V mask requires the determination of the lead distance d and H. Or, equivalently, based on principles of

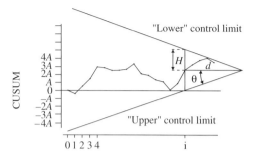

Figure 8.4   Basic V mask design.

geometry, we could determine d and the angle θ as shown in the Figure 8.4. Because we are dealing with issues of geometry, it is important to know the scaling of the vertical axis relative to the horizontal axis. Often the vertical axis scaling is much different than the horizontal axis scaling for purposes of more convenient plotting. As shown in Figure 8.4, the symbol A represents the scale factor between the two axes; that is, A represents the ratio of a vertical scale unit to a horizontal scale unit. If no software is available to automatically construct a V mask, it is much easier to manually draw the mask using the pair (d, H) than the pair (d, θ), because with the pair (d, H) we do not have to pull out a protractor!

Once both the target mean value μ_0 and the smallest shift Δ_μ from the target, for which quick detection is desired, have been determined, the sample statistic underlying the CUSUM statistic needs to be chosen. Individual measurements x_i can be used, as was done earlier. If, however, subgroups ($n > 1$) are collected, we simply replace x_i with $\bar{x}_i$ (the subgroup means) in the computation of the CUSUM statistic. In particular, (8.6a) becomes

$$\text{CUSUM}_i = \sum_{j=1}^{i}(\bar{x}_j - \mu_0) \tag{8.8}$$

Unlike the case with Shewhart charts, Hawkins (1993) points out that there is no real substantial advantage in the performance of CUSUM procedures when the sample size of the subgroups is increased. As such, he argues that sampling resources are better spent spreading the sampling over time rather than at one time. For instance, it would be better to sample a single observation ($n = 1$) every hour than to form a subgroup of size $n = 5$ every 5 h.

Given the chosen sample statistic (individual observation or subgroup mean), the standard deviation of the sample statistic σ_{stat} needs to be identified. If individual measurements x_i are used, then σ_{stat} is simply the standard deviation of the process σ. If sample means are used, then σ_{stat} is given by $\sigma/\sqrt{n}$. For reasons connected with the decision criteria of the CUSUM test procedure, it is necessary to define the following factor:

$$\delta = \Delta_\mu/\sigma_{\text{stat}} \tag{8.9}$$

This factor is interpreted as the smallest shift size one wishes to detect quickly, in terms of the number of standard deviations of the sample statistic used in the CUSUM calculation.

Once these initial steps have been addressed, the most commonly recommended practice in designing the V mask is based on the recommendations of Johnson (1961). Johnson's method is an approximate approach for the formal hypothesis-testing procedure, known as the *sequential likelihood ratio test,* underlying the CUSUM idea originally proposed by Page (1954). Johnson's method requires the user to specify two probabilities: (1) the greatest tolerable probability of getting an out-of-control signal when the process is actually on target (i.e., a false-alarm probability) and (2) the probability of not detecting a shift of size δ (i.e., a type II error risk). Denote these probabilities by α and β, respectively.[2] He recommends that the lead distance d be determined as follows:

$$d = \left(\frac{2}{\delta^2}\right) \ln \left(\frac{1-\beta}{\alpha/2}\right) \qquad \text{(8.10)}$$

Johnson also notes for small β that d is well approximated by

$$d \approx \frac{-2 \ln (\alpha/2)}{\delta^2} \qquad \text{(8.11)}$$

With respect to the height measure, it is given by

$$H = \frac{\Delta_\mu d}{2} \qquad \text{(8.12)}$$

It is important to note that (8.12) always holds true for any CUSUM scheme, whether using Johnson's method or not. In fact, this very same H value, which is the height measure in the V mask design, serves as the cutoff criterion for signaling out of control in the tabular procedure.

Because Johnson greatly simplified the exact, but complex, procedures from sequential likelihood ratio test theory, his method has become a popular choice among most quality control textbooks, references, and software vendors. However, recent works by Adams, Lowry, and Woodall (1992) and Woodall and Adams (1993) bring out serious pitfalls in the interpretation of a V mask based on Johnson's method. They point out that contrary to popular belief, the value of α embedded in Johnson's design is not to be simply interpreted as the probability of a false alarm at any given time. In reality, the probability of a false alarm changes over time with the CUSUM method. It is more appropriate to regard α as a measure of the long-run proportion of CUSUM observations resulting in false alarms. With this said, these same investigators point out another difficulty with Johnson's design. In particular, they demonstrate that the true long-run proportion of CUSUM observations resulting in false alarms is not equal to α with Johnson's design. In particular, they point out that if α for Johnson's method has

---

[2]Given Johnson's original notation, the greatest tolerable probability of a signal is sometimes denoted as 2α. We, however, denote this probability by α to be consistent with the standard notation of type I error risk (probability of rejecting the null hypothesis).

the correct interpretation, then the in-control ARL should be $ARL_0 = 1/\alpha$ (see Section 4.7). In reality, for a given α, for Johnson's method ARL_0 is consistently and substantially larger than $1/\alpha$, on the order of 6 times larger. The consequence is that Johnson's method will be much less sensitive to detecting process shifts than might be expected. These results are important because it is commonly recommended that one choose α as 0.0027 for Johnson's method so that the in-control properties of the CUSUM method are equivalent to the in-control properties of the Shewhart method.

Woodall and Adams (1993) recommend that, rather than rely on some α choice in conjunction with Johnson's method, one establish the CUSUM design values directly on the basis of ARL performance. The suggested use of ARLs for CUSUM design is not new. Goldsmith and Whitfield (1961) evaluated ARLs for a set of V masks so that a choice can be made for d and θ. Other works whose CUSUM design is based on ARL performance include those by Hawkins (1993), Lucas (1976), Woodall (1986), and Yashchin (1987), among many others.

For the most part, the recent efforts to base CUSUM design on ARL performance have focused on determining appropriate design values for the tabular CUSUM procedure. This does not rule out the possibility of using a V mask approach. It is possible to design the V mask so that it will provide *equivalent* performance to a tabular CUSUM design. This would require us[3] to *first* design the decision criteria for the tabular procedure and then establish the correspondence of these criteria with the V mask qualifiers d and H. Is there any compelling reason to design a tabular test and then "hop" to the implementation of the V mask procedure? Yes, because for some, the cumulative statistic for the V mask procedure, as defined in (8.6) or (8.8), is more intuitive and simpler to understand than the cumulative statistic defined for the tabular procedure.

As we will soon learn, the tabular procedure calls for the determination of two design values, k and h. The two approaches are equivalent if

$$d = \frac{h}{k} \tag{8.13}$$

and

$$H = \frac{\Delta_\mu d}{2} \tag{8.14}$$

Notice that (8.14) is (8.12) revisited. For monitoring the process mean, it is usually recommended that k be chosen as one-half the δ factor defined by (8.9); that is, we choose

$$k = \frac{\delta}{2} \tag{8.15}$$

---

[3]This is indeed the approach that Minitab uses if one chooses to implement its V mask procedure. In particular, the user is required to input the tabular design criteria values, and then Minitab creates a V mask that is equivalent in performance to the tabular procedure.

Once k is selected by (8.15), the h factor is selected in such a manner that the pair of values (k, h) is expected to produce a CUSUM procedure with a desired in-control ARL. The most common strategy is to match the in-control ARL of the CUSUM with the in-control ARL of the Shewhart scheme, which is 370.4. By doing so, the interpretation of the control charts (CUSUM or Shewhart) is calibrated to a similar overall false-alarm rate (1 per 370.4 or 0.0027). Adams, Lowry, and Woodall (1992) underscore that, given the nature of the CUSUM procedure, a false-alarm rate for a CUSUM procedure, say 0.0027, must be interpreted *only* as a long-run proportion of false alarms.[4] In contrast, for the Shewhart method, the false-alarm rate could be interpreted as a long-run proportion of false alarms and the probability of a false alarm on any single observation from an in-control process.

In the next section, we discuss more specifically how an h value can be determined given a k value. But first, it is important to illustrate the construction of a V mask procedure by an example.

Example 8.1

We now demonstrate the calculations and construction of a V mask by using the data from Table 8.1. Recall that the target value is $\mu_0 = 100$. Since we are using the individual measurements in the computation of the cumulative statistic, σ_{stat} is equal to the process standard deviation σ, which is known to be 10. Suppose we are interested in detecting a shift in the process mean of 10 units in either direction as quickly as possible. In other words, we are interested in quickly detecting if the mean shifts down to $90 (= 100 - 10)$ or shifts up to $110 (= 100 + 10)$. In terms of earlier-defined notation, this implies $\Delta_\mu = 10$. Using (8.9), we need to convert the magnitude of this shift size into multiples of σ_{stat}:

$$\delta = \frac{10}{10} = 1$$

Thus, the interest is to detect a $1.0\sigma_{stat}$ shift in the process mean, which in this case is equivalent to saying a 1 standard deviation (1σ) shift in the process mean. From (8.15), this implies $k = 0.5$. For a k value of 0.5, h is typically selected in the neighborhood of 4.5 to 5 with $h = 5$ the most common choice. The pair $(0.5, 5)$ gives rise to an in-control ARL roughly the same as that for a Shewhart chart. In the next section, we will learn how to fine-tune the value of h to more closely match a desired in-control ARL.

Applying (8.13) to the selected h and k values, we find the lead distance of the V mask to be

$$d = \frac{h}{k} = \frac{5}{0.5} = 10$$

The height value H is now obtained from (8.14):

$$H = \frac{\Delta_\mu d}{2} = \frac{10(10)}{2} = 50$$

---

[4]So that the notion of a false-alarm rate is properly interpreted, we will be careful, in our discussions of this chapter, to use terms such as *overall false-alarm rate* or *long-run false-alarm rate*.

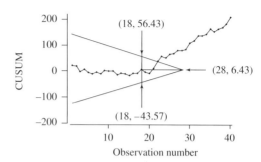

Figure 8.5   "Manually" drawn V mask positioned on the 18th CUSUM.

Fortunately, Minitab[5] will automatically superimpose a V mask with the computed lead distance and height specifications. However, to further solidify our understanding of the V mask design, it is beneficial to demonstrate how one can manually draw a properly proportioned V mask on a CUSUM plot. A simple way of doing so is accomplished by the following steps:

1.   Pick an arbitrary CUSUM observation, $CUSUM_i$. For this observation, recognize that the x axis and y axis coordinate values are given by $(x_i, y_i) = (i, CUSUM_i)$.

2.   Determine the xy point for the tip of the V mask. Since the tip is at a horizontal distance of d units from the plotted cumulative value, the x and y coordinate values are given by $(i + d, CUSUM_i)$.

3.   Determine the xy points which are located H units above and below the plotted cumulative value. Since these points are vertically straight up and down from the plotted cumulative value, the xy coordinate values for the two points are given by $(i, CUSUM_i + H)$ and $(i, CUSUM_i - H)$.

4.   Starting with the tip point determined in step 2, draw two different lines going through the two different points determined from step 3.

Manually drawing does not necessarily imply "hand" drawing. The above steps are particularly useful when you are using a graphics software package (e.g., a spreadsheet package such as Microsoft Excel or Lotus 123) to create the CUSUM chart.

   To demonstrate the mechanics, suppose we pick the 18th cumulative statistic value. From Table 8.1, we find $(i, CUSUM_i) = (18, 6.43)$. The tip of the V mask is then given by $(18 + 10, 6.43) = (28, 6.43)$. The vertically displaced points are given by $(18, 6.43 + 50) = (18, 56.43)$ and $(18, 6.43 - 50) = (18, -43.57)$. In Figure 8.5, we see these points labeled on the graph with rays drawn through the points to form the V mask. Once the V mask has been constructed, it can be positioned on each of the cumulative statistic values. Actually, the proper implementation of a V mask requires that, starting with the first cumulative-sum observation, the V mask be applied successively to *each* subsequent observation. Because the V mask moves with the observations, some authors refer to the mask as a *mobile* V mask.[6] As the V mask is moved to the right, the legs of the V mask should be long enough to reach the period

---

[5]Full version.

[6]Moving a manually created V mask is not difficult in most Windows-based software. In Excel, for example, the four line segments shown in Figure 8.5 can be grouped as one object. Once grouped, the V mask can be moved by grabbing it and dragging with the mouse.

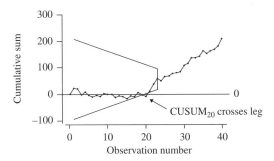

Stat>>Control Charts>>CUSUM

Figure 8.6   Minitab automatically generated V mask positioned on the 23rd CUSUM.

of the first cumulative sum. If the plotting is done by hand, this can be problematic because the CUSUM values might wander off the sheet of graph paper.

From Figure 8.5, we see that when the V mask is positioned on $CUSUM_{18}$, no cumulative values prior to period 18 fall outside the limits. More generally, no points would fall outside the limits if we positioned the V mask on any of the cumulative values for periods 1 to 22. For $CUSUM_{23}$, we use Minitab's generated V mask (see Figure 8.6). Notice that Minitab does not draw the triangle part of the V mask which forms the tip. This is because the triangle part of the V mask is not really used in assessing observations. As opposed to prior periods, the CUSUM graph now crosses one of the legs of the mask, indicating an out-of-control condition. In particular, the 20th cumulative sum crossed the lower leg, and as such, it represents an estimated change point. As we already know, the onset of the shift actually did occur in period 20.

As with any other control chart procedure, an out-of-control signal should be acted on promptly to seek special-cause explanation. Once the special cause is identified and corrective action is taken, the cumulative sum is reset to a starting value of 0 and a new chart is implemented.

It is interesting to note that the sample average of the underlying process over any localized segment can be obtained from the CUSUM data without recourse to the original observations. In particular, the sample average of observations $x_i, x_{i+1}, \ldots, x_j$ can be computed as follows:

$$\hat{\mu} = \bar{x}_{i,j} = \mu_0 + \frac{CUSUM_j - CUSUM_{i-1}}{j - i + 1} \qquad (8.16)$$

To illustrate, suppose for Example 8.1 one desires a quick estimate of the new mean level upon the signal of a shift. One strategy might be to compute the average over the range from the suspected change point period ($i = 20$) to the period associated with the first signal ($j = 23$). From (8.16), the sample average over this range is

$$\bar{x}_{20,23} = 100 + \frac{\text{CUSUM}_{23} - \text{CUSUM}_{19}}{23 - 20 + 1}$$

$$= 100 + \frac{57.39 - (-3.34)}{4}$$

$$= 115.18$$

The reader can verify that the same value would be obtained if the average were computed directly from the individual observations $x_{20}, x_{21}, x_{22},$ and x_{23}. This estimate of the mean somewhat overestimates the true mean, which we know to be 110. However, we have to remember that this estimate is based on only a very small sample of 4 observations. If we compute the sample average over the range of 20 to 40, we find $\bar{x}_{20,40} = 110.16$.

TABULAR CUSUM PROCEDURE FOR MONITORING THE PROCESS MEAN

One obvious disadvantage of the V mask procedure is the "burden" of having to continually reposition a geometric object on the evolving graphical plot. By modifying the representation of the cumulative statistic, a CUSUM procedure with equivalent performance to the V mask procedure can be devised that does not even require a chart. The approach was originally devised to facilitate a hand-only calculation approach to implementing a CUSUM scheme. Since the hand calculations need to be laid out in a table format, this alternative approach is referred to as the tabular CUSUM approach.

Instead of accumulating the deviations $x_i - \mu_0$ to define the cumulative statistic, the tabular approach calls for two adaptations. First, a *reference* value denoted by K is incorporated in the computation. This reference value (sometimes referred to as the *slack* or *allowance* value) is chosen to be one-half of the magnitude of the shift which one is interested in quickly detecting, that is, $K = \Delta_\mu/2$. Given that $\Delta_\mu = \delta\sigma_{\text{stat}}$ [from (8.9)] along with the choice of k from (8.15), we find that

$$K = k\sigma_{\text{stat}} \tag{8.17}$$

The reference value K is added to and subtracted from the target value μ_0 to produce two points of reference, an upper reference point ($\mu_0 + K$) and a lower reference point ($\mu_0 - K$). The second adaptation is to consider two separate cumulative statistics rather than the single cumulative statistic used in the V mask implementation. One statistic, CUSUM^+, only focuses on accumulating deviations relative to the upper reference point, while the other statistic, CUSUM^-, considers only accumulating deviations relative to the lower reference point. As such, the statistics CUSUM^+ and CUSUM^- are referred to as *one-sided* upper and lower CUSUMs, respectively. The exact procedure for their computation is given by the following steps:

1. Initialize the cumulative statistics to 0, that is, $\text{CUSUM}_0^+ = \text{CUSUM}_0^- = 0$.
2. For $i \geq 1$, compute the upper cumulative sum as follows:

$$\text{CUSUM}_i^+ = \max[0, x_i - (\mu_0 + K) + \text{CUSUM}_{i-1}^+] \qquad \textbf{(8.18)}$$

This implies that if the value of $x_i - (\mu_0 + K) + \text{CUSUM}_{i-1}^+$ is negative, CUSUM_i^+ is reset to 0.

3. For $i \geq 1$, compute the lower cumulative sum as follows:

$$\text{CUSUM}_i^- = \min[0, x_i - (\mu_0 - K) + CUSUM_{i-1}^-] \qquad \textbf{(8.19)}$$

This implies that if the value of $x_i - (\mu_0 - K) + \text{CUSUM}_{i-1}^-$ is positive, then CUSUM_i^- is reset to 0.

If subgroups are collected, the individual observations x_i are simply replaced with the subgroup means $\bar{x}_i$. Once computed, the successive values of CUSUM_i^+ and CUSUM_i^- can be entered into a table or displayed as a time-sequence plot.

The decision rule for the modified cumulative statistics is remarkably simple. Namely, as long as the statistics fall within the range $[-H, H]$, one concludes that the process has not departed from the target value μ_0. However, if CUSUM_i^+ exceeds H or if CUSUM_i^- is below $-H$, then the process is considered out of control. Thus, we can implement the control procedure by superimposing two parallel lines a distance H above and below a centerline of 0 on a sequence plot of the modified cumulative statistics. The result is a control chart that is Shewhart-like in appearance with $\text{UCL} = H$ and $\text{LCL} = -H$.

As noted in the previous section, the tabular H value is the same value that defines the height from the cumulative statistic to the V mask legs given by (8.14). By using various facts to this point, it is easy to show that H can be rewritten as

$$H = h\sigma_{\text{stat}} \qquad \textbf{(8.20)}$$

In Example 8.1, we noted that for $k = 0.5$, a common corresponding choice for h is 5 because this pair of values gives roughly an in-control ARL close to a Shewhart chart which is 370.4. What is the actual in-control ARL for this pair? A number of investigators have used different approaches to find the exact ARL of the CUSUM chart. The techniques (such as integral equations or Markov analysis methods) required to obtain very precise ARLs are far from trivial (see Brooks and Evans, 1972; Fellner, 1990; Vance, 1986; and Woodall, 1983). As an alternative to these more "difficult" techniques, Hawkins (1992) provides CUSUM users with a relatively simple but accurate method to generate ARLs for a wide variety of h and k values. To find the desired ARL, Hawkins' method calls for the use of an extensive table of constants. Leaning heavily on simplicity, Woodall and Adams (1993) demonstrate that an approximation method given by Siegmund (1985) is surprisingly accurate considering its simplicity. Siegmund's method approximates the ARL for a one-sided CUSUM implementation, that is, if one is monitoring only CUSUM_i^+ or CUSUM_i^-. The ARL approximation for upper-sided CUSUM (ARL^+) and lower-sided CUSUM (ARL^-) are given by

$$\mathrm{ARL}^+ = \frac{\exp[-2(\delta^\mu - k)(h + 1.166)] + 2(\delta^\mu - k)(h + 1.166) - 1}{2(\delta^\mu - k)} \qquad (8.21)$$

and

$$\mathrm{ARL}^- = \frac{\exp[-2(-\delta^\mu - k)(h + 1.166)] + 2(-\delta^\mu - k)(h + 1.166) - 1}{2(-\delta^\mu - k)^2} \qquad (8.22)$$

where $\delta^\mu = (\mu - \mu_0)/\sigma_{\text{stat}}$; that is, δ^μ is the distance (negative or positive) between the actual mean level of the process μ and the target mean level μ_0 in terms of the number of standard deviations σ_{stat}. Please note that unlike δ of (8.9), δ^μ is not defined in terms of an absolute value. Furthermore, note that ARL^+ is not computable when $\delta^\mu = k$ and ARL^- is not computable when $\delta^\mu = -k$.

As Woodall and Adams (1993) point out, the ARL for a two-sided CUSUM implementation $\mathrm{ARL}^{+,-}$ can be obtained by using the following fact:

$$\frac{1}{\mathrm{ARL}^{+,-}} = \frac{1}{\mathrm{ARL}^+} + \frac{1}{\mathrm{ARL}^-} \qquad (8.23)$$

To illustrate, let us return to our question of determining the in-control ARL for a two-sided CUSUM scheme with $k = 0.5$ and $h = 5$. When the process is in control, it is operating at a mean level $\mu = \mu_0$ which implies $\delta^\mu = 0$. When $\delta^\mu = 0$, $\mathrm{ARL}^+ = \mathrm{ARL}^-$, and hence either (8.21) or (8.22) can be used to find

$$\mathrm{ARL}^+ = \frac{\exp[-2(-0.5)(5 + 1.166)] + 2(-0.5)(5 + 1.166) - 1}{2(-0.5)^2}$$

$$= \frac{e^{6.166} - 6.166 - 1}{0.5} = 938.2$$

From (8.23), the in-control ARL for a two-sided scheme is approximated by

$$\frac{1}{\mathrm{ARL}_0^{+,-}} = \frac{1}{938.2} + \frac{1}{938.2}$$

or

$$\mathrm{ARL}_0^{+,-} = 469.1$$

It turns out that the true in-control ARL value is 465, hence the approximation did quite well.

However, we learn that for $k = 0.5$, the value of $h = 5$ results in a CUSUM procedure with an in-control ARL somewhat bigger than 370.4 for the Shewhart scheme. Thus, there is an opportunity to "fine-tune" the choice of h. We could resort to the very precise methods from the above-noted citations (see also Gan, 1991). It turns out, however, we can again utilize Siegmund's method with fairly accurate results.

Woodall and Adams (1993) show that Siegmund's ARL approximation can be used to develop an iterative procedure for selecting h given a k value and a desired in-control ARL (ARL_0). For a two-sided CUSUM implementation, the iterative procedure is defined by the following recursion:

$$h_n = h_{n-1} - \frac{\exp[2k(h_{n-1} + 1.166)] - 2k(h_{n-1} + 1.166) - 1 - 4k^2\,ARL_0}{2k \exp[2k(h_{n-1} + 1.166)] - 2k} \quad \text{(8.24)}$$

with $n = 1, 2, 3, \ldots$ and h_0 is some starting value which we recommend to be 5. As the number of iterations increases, h_n will converge to a single h value corresponding to a CUSUM procedure with a true in-control ARL close to the desired ARL_0 value.

To illustrate a couple of iterations, suppose $k = 0.5$ and the desired ARL_0 is 370.4. After the first iteration, we have

$$h_1 = 5 - \frac{\exp[2(0.5)(5 + 1.166)] - 2(0.5)(5 + 1.166) - 1 - 4(0.5^2)(370.4)}{2(0.5)\exp[2(0.5)(5 + 1.166)] - 2(0.5)}$$

$$= 5 - \frac{e^{6.166} - 6.166 - 1 - 370.4}{e^{6.166} - 1}$$

$$= 4.792$$

After the second iteration, we have

$$h_2 = 4.792 - \frac{\exp[2(0.5)(4.792 + 1.166)] - 2(0.5)(4.792 + 1.166) - 1 - 4(0.5^2)(370.4)}{2(0.5)\exp[2(0.5)(4.792 + 1.166)] - 2(0.5)}$$

$$= 4.792 - \frac{e^{5.958} - 5.958 - 1 - 370.4}{e^{5.958} - 1}$$

$$= 4.767$$

Woodall and Adams (1993) recommend iterating until successive values are within 0.005 of each other (that is, $|h_n - h_{n-1}| < 0.005$) and then reporting the h value to the second decimal place. When doing so, we converge on an h value of 4.77, which is the same value reported by Hawkins (1993) while using a more sophisticated procedure.

To facilitate the computation process, we have provided a Minitab macro[7] named "hvalue.mac" that takes as input the values for k, ARL_0, and h_0. When invoked, this macro will implement the described Siegmund-based iteration procedure and then will print out for the user the "final" h value. In Figure 8.7, we provide the session commands to obtain the h value. In this example, it is assumed that the macro is on a floppy disk in the a: drive. Alternatively, the macro can be copied to the folder named "Macros" created during the installation of Minitab. If done, the "a:" can be dropped and the session line becomes: "MTB> %hvalue 'k' 'ARL' 'h0' 'hfinal' c1".

We are now in a position to fully illustrate the implementation of a tabular CUSUM scheme by an example.

---

[7]This macro can be run in either the full version or the student version of Minitab. In either version, macros can *only* be run at the session command level, that is, by typing the commands at the "MTB>" prompt.

```
MTB > name k1 'k' k2 'ARL' k3 'h0' k4 'hfinal'
MTB > let 'k' = 0.5
MTB > let 'ARL' = 370.4
MTB > let 'h0' = 5
MTB > %a:hvalue 'k' 'ARL' 'h0' 'hfinal' c1
Executing from file: a:hvalue.MAC

Data Display

hfinal      4.77000
```

c1 is a "dummy" column for computational purposes.

Figure 8.7   Minitab session commands to run macro to determine h value.

Example 8.2

In the process of producing shaft seals used in automobile engines, grease is applied to the area of the seal surface which will be in contact with the shaft. The amount of grease is critical since it provides the initial lubrication in the start-up of a new engine before the oil has a chance to fully circulate. For correct operation, the amount of grease applied to the seal should be 0.10 gram (g) in weight. Too much grease can cause the seal to vulcanize, resulting in malfunction, while too little grease will cause the seal to run dry, resulting in seal damage. Departures in the mean level by more than 0.01 g are considered important enough for immediate corrective action. Hence, the target level μ_0 is 0.10 with $\Delta_\mu = 0.01$.

To monitor the process of grease deposition, current practice is to take two preweighed clean (greaseless) seals and run them through the machine and then have them reweighed. The net change in weight reflects the amount of grease on the seal. Grease deposition is monitored hourly on a routine basis. In light of the earlier-cited argument by Hawkins (1993), consideration might be given to changing the sampling practice away from subgroups to more frequent sampling of individual observations, if a CUSUM scheme is preferred. Notwithstanding this recommendation, this example provides us with a nice opportunity to illustrate the appropriate CUSUM calculations when subgroups of size $n > 1$ are collected.

Table 8.2 shows the results of 35 consecutive hourly samples of size 2, along with the corresponding sample means and range. The data on the individual observations are also provided in file "grease.dat".

The construction of the CUSUM chart requires a value for σ_{stat}. Since the sample means $\bar{x}_i$ underlie the computation of the cumulative statistic, σ_{stat} is the standard deviation of the sample mean statistic $\sigma_{\bar{x}}$ which is equal to $\sigma/\sqrt{n}$, where σ is the standard deviation for the process generating the individual observations. If σ is not known, an estimate is used. In some cases, an estimate may be available from a previous detailed process capability study. If not, an estimate can be derived from the data of the current study. Based on the developments of Chapter 6, one possible estimate for σ can be obtained from the sample range data. From Table 8.2, the mean of the ranges $\bar{R}$ is 0.012286. Hence, for subgroups of size 2, the estimate for σ is

$$\hat{\sigma} = \frac{\bar{R}}{d_2} = \frac{0.012286}{1.128} = 0.01089$$

This implies that an estimate for σ_{stat} is

$$\hat{\sigma}_{stat} = \frac{\hat{\sigma}}{\sqrt{2}} = \frac{0.01089}{1.41421} = 0.0077$$

To compute the upper and lower cumulative statistics, we now need the value of the reference value K. In our situation, Δ_μ is specified to be 0.01 implying K is 0.005 ($= \Delta_\mu/2$).

Table 8.2   Grease data

Sample i	$x_{1,i}$	$x_{2,i}$	$\bar{x}_i$	R_i
1	0.091	0.096	0.0935	0.005
2	0.108	0.105	0.1065	0.003
3	0.117	0.079	0.0980	0.038
4	0.105	0.120	0.1125	0.015
5	0.114	0.083	0.0985	0.031
6	0.110	0.108	0.1090	0.002
7	0.087	0.097	0.0920	0.010
8	0.109	0.099	0.1040	0.010
9	0.105	0.094	0.0995	0.011
10	0.100	0.117	0.1085	0.017
11	0.089	0.091	0.0900	0.002
12	0.076	0.091	0.0835	0.015
13	0.119	0.101	0.1100	0.018
14	0.093	0.111	0.1020	0.018
15	0.115	0.101	0.1080	0.014
16	0.125	0.107	0.1160	0.018
17	0.101	0.107	0.1040	0.006
18	0.093	0.093	0.0930	0.000
19	0.108	0.104	0.1060	0.004
20	0.109	0.112	0.1105	0.003
21	0.092	0.105	0.0985	0.013
22	0.092	0.098	0.0950	0.006
23	0.096	0.075	0.0855	0.021
24	0.082	0.081	0.0815	0.001
25	0.106	0.083	0.0945	0.023
26	0.088	0.104	0.0960	0.016
27	0.085	0.076	0.0805	0.009
28	0.069	0.098	0.0835	0.029
29	0.094	0.099	0.0965	0.005
30	0.073	0.092	0.0825	0.019
31	0.097	0.095	0.0960	0.002
32	0.077	0.087	0.0820	0.010
33	0.084	0.093	0.0885	0.009
34	0.108	0.088	0.0980	0.020
35	0.091	0.098	0.0945	0.007
				$\bar{R} = 0.012286$

There are situations in which the value Δ_μ is not given exactly. Instead, what might be stated is a desire to detect a shift of a certain number of standard deviations as quickly as possible. Then K is computed by using $\hat{\sigma}_{stat}$ in conjunction with (8.17).

With a K of 0.005, $\mu_0 = 0.10$, and the sample mean $\bar{x}_i$ as the underlying statistic applied to (8.18) and (8.19), the equations for CUSUM$_i^+$ and CUSUM$_i^-$ are

$$\text{CUSUM}_i^+ = \max[0, \bar{x}_i - 0.105 + \text{CUSUM}_{i-1}^+]$$

and

$$\text{CUSUM}_i^- = \min[0, \bar{x}_i - 0.095 + \text{CUSUM}_{i-1}^-]$$

Let us compute the cumulative statistics for a couple of periods to firm up our understanding. With the starting values being CUSUM$_0^+$ = CUSUM$_0^-$ = 0 and $\bar{x}_1 = 0.0935$, the first period's cumulative statistic values are

$$\text{CUSUM}_1^+ = \max[0, 0.0935 - 0.105 + 0] = \max[0, -0.0115] = 0$$

$$\text{CUSUM}_1^- = \min[0, 0.0935 - 0.095 + 0] = \min[0, -0.0015] = -0.0015$$

Since $\bar{x}_2 = 0.1065$, the second period's cumulative statistic values are

$$\text{CUSUM}_2^+ = \max[0, 0.1065 - 0.105 + 0] = \max[0, 0.0015] = 0.0015$$

$$\text{CUSUM}_2^- = \min[0, 0.1065 - 0.095 + (-0.0015)] = \min[0, 0.01] = 0$$

The cumulative statistic values for all the periods are provided in Table 8.3. The variables RUN$^+$ and RUN$^-$ are running counters of the number of successive periods for which CUSUM$_i^+$ is positive and CUSUM$_i^-$ is negative. As will be shown, these counters are simple housekeeping measures to help backtrack to the likely beginning of a shift.

At this stage, we need to come up with values for k and h. Recall that from (8.15), k is chosen as one-half of the shift size Δ_μ measured in units of σ_{stat}, or, in this case, $\hat{\sigma}_{stat}$. Since we have $\hat{\sigma}_{stat} = 0.0077$, k is computed to be

$$k = \frac{\delta}{2} = \frac{\Delta_\mu/\sigma_{stat}}{2} = \frac{0.01/0.0077}{2} = 0.649$$

Suppose we wish to design the CUSUM chart such that the in-control ARL is 370.4, as is the case with the Shewhart scheme. Given our requirements, we invoke below the earlier-used macro (see Figure 8.7) to approximate an h value:

```
MTB > name k1 'k' k2 'ARL' k3 'h0' k4 'hfinal'
MTB > let 'k'=0.649
MTB > let 'ARL'=370.4
MTB > let 'h0'=5
MTB > %hvalue 'k' 'ARL' 'h0' 'hfinal' c1
Executing from file: C:\PROGRAM FILES\MTBWIN\MACROS\hvalue.MAC

Data Display

hfinal  3.80000
```

From (8.20), an h value of 3.8 results in an H value of

$$H = 3.8(0.0077) = 0.0293$$

Thus, the tabular CUSUM control limits are $\pm H = \pm 0.0293$. The procedure will signal out-of-control if CUSUM$_i^+$ exceeds 0.0293 or if CUSUM$_i^-$ falls below -0.0293. From Table 8.3,

Table 8.3   Tabular CUSUM for grease data

Sample i	x_i	$x_i - 0.105$	CUSUM$_i^+$	RUN$^+$	$x_i - 0.095$	CUSUM$_i^-$	RUN$^-$
1	0.0935	−0.0115	0.0000	0	−0.0015	−0.0015	1
2	0.1065	0.0015	0.0015	1	0.0115	0.0000	0
3	0.0980	−0.0070	0.0000	0	0.0030	0.0000	0
4	0.1125	0.0075	0.0075	1	0.0175	0.0000	0
5	0.0985	−0.0065	0.0010	2	0.0035	0.0000	0
6	0.1090	0.0040	0.0050	3	0.0140	0.0000	0
7	0.0920	−0.0130	0.0000	0	−0.0030	−0.0030	1
8	0.1040	−0.0010	0.0000	0	0.0090	0.0000	0
9	0.0995	−0.0055	0.0000	0	0.0045	0.0000	0
10	0.1085	0.0035	0.0035	1	0.0135	0.0000	0
11	0.0900	−0.0150	0.0000	0	−0.0050	−0.0050	1
12	0.0835	−0.0215	0.0000	0	−0.0115	−0.0165	2
13	0.1100	0.0050	0.0050	1	0.0150	−0.0015	3
14	0.1020	−0.0030	0.0020	2	0.0070	0.0000	0
15	0.1080	0.0030	0.0050	3	0.0130	0.0000	0
16	0.1160	0.0110	0.0160	4	0.0210	0.0000	0
17	0.1040	−0.0010	0.0150	5	0.0090	0.0000	0
18	0.0930	−0.0120	0.0030	6	−0.0020	−0.0020	1
19	0.1060	0.0010	0.0040	7	0.0110	0.0000	0
20	0.1105	0.0055	0.0095	8	0.0155	0.0000	0
21	0.0985	−0.0065	0.0030	9	0.0035	0.0000	0
22	0.0950	−0.0100	0.0000	0	0.0000	0.0000	0
23	0.0855	−0.0195	0.0000	0	−0.0095	−0.0095	1
24	0.0815	−0.0235	0.0000	0	−0.0135	−0.0230	2
25	0.0945	0.0105	0.0000	0	−0.0005	−0.0235	3
26	0.0960	−0.0090	0.0000	0	0.0010	−0.0225	4
27	0.0805	−0.0245	0.0000	0	−0.0145	−0.0370	5
28	0.0835	−0.0215	0.0000	0	−0.0115	−0.0485	6
29	0.0965	−0.0085	0.0000	0	−0.0015	−0.0470	7
30	0.0825	−0.0225	0.0000	0	−0.0125	−0.0595	8
31	0.0960	−0.0090	0.0000	0	0.0010	−0.0585	9
32	0.0820	−0.0230	0.0000	0	−0.0130	−0.0715	10
33	0.0885	−0.0165	0.0000	0	−0.0065	−0.0780	11
34	0.0980	−0.0070	0.0000	0	0.0030	−0.0750	12
35	0.0945	−0.0105	0.0000	0	−0.0005	−0.0755	13

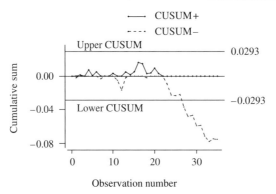

Figure 8.8   Tabular CUSUM chart for grease data.

we find that the lower-side CUSUM for sample 27 is -0.0370 which is less than -0.0293, so an out-of-control signal is flagged.

A simultaneous plot of the upper- and lower-side CUSUM values, along with control limits, can be found in Figure 8.8.[8] The control chart clearly shows that the process has gone out of control on the lower side, indicating a downward process shift relative to the target. Since at the time of the signal ($i = 27$) the counter $RUN^- = 5$, we estimate that the onset of the shift was at period 22 ($= 27 - 5$). A search for special causes is warranted and followed up with appropriate corrective actions.

If corrective actions require an estimate of the new mean level, this can be obtained from the CUSUM information as follows:

$$\hat{\mu} = \mu_0 + K + \frac{CUSUM_i^+}{RUN_i^+} \qquad \text{if } CUSUM_i^+ > H$$

$$= \mu_0 - K + \frac{CUSUM_i^-}{RUN_i^-} \qquad \text{if } CUSUM_i^- < -H \qquad (8.25)$$

For example, the estimate of the mean level at the time of the signal ($i = 27$) is

$$\hat{\mu} = 0.10 - 0.005 + \frac{-0.0370}{5} = 0.0876$$

Adjustments should be attempted to shift the process mean *up* by 0.0124 g to restore the process to the target.

Before we close this section, note that since the tabular procedure is based on the running of two one-sided cumulative sums, the procedure allows for one-sided CUSUM implementation only. If, for example, the critical quality characteristic is strength, then departures below the target are of greatest concern. In

---

[8]In Minitab, the default values for h and k are 4.0 and 0.5, respectively. To change these values, point and click the "Options" button within the CUSUM option box and then type in the desired values for h and k in their dialog boxes.

such a case, a one-sided CUSUM procedure based on plotting only the lower-side cumulative sums ($CUSUM_i^-$) can be implemented. In other situations, the emphasis may be on monitoring the upper-side cumulative sums ($CUSUM_i^+$). Depending on the direction of interest, the decision interval for a one-sided scheme is either $[-H, 0]$ or $[0, H]$. To obtain the appropriate h value for a one-sided CUSUM scheme so that it has a certain in-control ARL, say ARL_0^{one}, the numerical approximation given by (8.24) can be used with ARL_0 replaced by $ARL_0^{one}/2$.

FIR CUSUM

After a shift from target has been detected, the hope is that corrective actions will bring the process back on target. Under the presumption that the process is no longer off target, the cumulative statistics are reset to 0 and new series of cumulative sums are monitored for any possible future shifts. What if, however, the process were not truly brought back to target? This may be due to improper adjustment or the failure to remove all sources of special causes, resulting in the out-of-control condition. In such a situation, we would clearly want to detect the continued lack of control as quickly as possible.

By resetting the cumulative statistics to zero when the process is still out of control, we potentially delay the time to detection. Lucas and Crosier (1982) propose, as a safeguard, that the cumulative statistics be reset not to zero after correction action, but rather to "headstart" values allowing for faster detection of a process still out of control. Their method is referred to as *fast-initial-response CUSUM* or *FIR CUSUM*. They recommend a 50 percent headstart which is accomplished by resetting $CUSUM_0^+$ and $CUSUM_0^-$ to the values $H/2$ and $-H/2$, respectively. Notice that the headstart values are used on both the upper- and lower-side cumulative statistics even though the shift detected before the attempted correction was in one direction (positive or negative). The reason is that we cannot be sure where the process mean ended up after the attempted correction. It is quite possible for the process to be out of control on the other side of the original shift due to overadjustment.

The effect of a headstart can be illustrated with the data in Table 8.4. Hypothetically, imagine that these data represent process measurements after a shift was detected and corrective action attempted. We assume that the target value is 50 and that the standard deviation of the process generating the individual measurements is 2. For convenience, suppose that the objective is to detect a 1 standard deviation shift from the target value with h selected to be 4. This would imply $K = 1$ and $H = 8$. In Table 8.4a, the data are in control with underlying mean equal to the target. Thus, for these data, the corrective action was successful in bringing the process back to target. Ideally, if this were truly known, both cumulative statistics would be reset to zero. Let us consider the relative impact of resetting the cumulative statistics to zero versus to headstart values when the process is in control. Following the recommendation of 50 percent headstart, we

Table 8.4   Cumulative sums with different headstarts for on-target and off-target processes

Sample i	x_i	$x_i - 51$	$x_i - 49$	No Headstart		50% Headstart	
				$CUSUM_i^+$	$CUSUM_i^-$	$CUSUM_i^+$	$CUSUM_i^-$
(a) Process with Mean Equal to 50 (μ_0)							
1	52	1	3	1	0	5	-1
2	51	0	2	1	0	5	0
3	48	-3	-1	0	-1	2	-1
4	48	-3	-1	0	-2	0	-2
5	50	-1	1	0	-1	0	-1
6	49	-2	0	0	-1	0	-1
7	49	-2	0	0	-1	0	-1
8	50	-1	1	0	0	0	0
9	50	-1	1	0	0	0	0
10	48	-3	-1	0	-1	0	-1
(b) Process with Mean Equal to 52 ($\mu_0 + \sigma$)							
1	54	3	5	3	0	7	0
2	53	2	4	5	0	9	0
3	50	-1	1	4	0	8	0
4	50	-1	1	3	0	7	0
5	52	1	3	4	0	8	0
6	51	0	2	4	0	8	0
7	51	0	2	4	0	8	0
8	52	1	3	5	0	9	0
9	52	1	3	6	0	10	0
10	50	-1	1	5	0	9	0

see the headstart values are $CUSUM_0^+ = H/2 = 4$ and $CUSUM_0^- = -H/2 = -4$. For period 1, the cumulative sums without and with headstart are

$$CUSUM_1^+ = \max[0, 52 - 51 + 0] = \max[0, 1] = 1$$
$$CUSUM_1^- = \min[0, 52 - 49 + 0] = \min[0, 3] = 0$$

no headstart

versus

$$CUSUM_1^+ = \max[0, 52 - 51 + 4] = \max[0, 5] = 5$$
$$CUSUM_1^- = \min[0, 52 - 49 - 4] = \min[0, -1] = -1$$

headstart

The results of the remaining calculations are summarized in Table 8.4*a*. As seen, the effect of the headstart values dissipates quickly. As soon as period 2, the effect of the headstart on CUSUM$^-$ is gone, while the effect of the headstart is

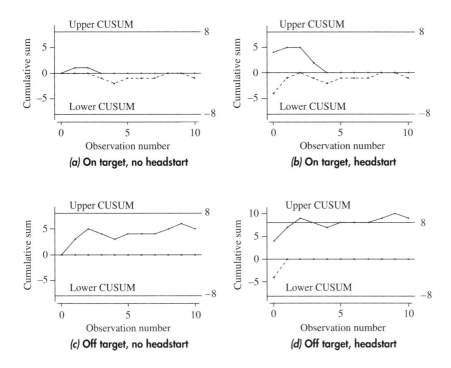

Figure 8.9   Effects of headstart on CUSUM chart performance.

gone by period 4 on CUSUM$^+$. The sequence plots[9] of cumulative sums for the in-control process for no headstart versus headstart are seen in Figure 8.9a and b. The cumulative sums with headstart are clearly converging to the cumulative sums with no headstart.

For an in-control process, the rapidly diminishing effect of a headstart is *good* news. In placing the cumulative statistics closer to the control limits, the obvious concern is that the risk of a false alarm will greatly increase. The expected impact of a 50 percent headstart on in-control and out-of-control processes was formally studied by Lucas and Crosier (1982). In particular, they report the ARLs for a basic CUSUM (no headstart) and for a FIR CUSUM with $k = 0.5$ and $h = 5$ versus various shift sizes δ measured in terms of σ_{stat}. Their results are summarized in columns (a) and (b) of Table 8.5. For an in-control process (no shift), the addition of the FIR feature decreases the ARL slightly from 465 to 430. In terms of a long-run false-alarm rate, it increases a very small amount from 0.0022 ($= 1/465$) for the basic CUSUM to 0.0023 ($= 1/430$) for the FIR CUSUM.

Now, the question turns to the effect of the headstart feature when the process is out of control from the start up. Consider the data presented in Table

---

[9]Minitab offers a FIR CUSUM option. With this option, the user needs to input the headstart in terms of the standard deviation of the statistic plotted. Since $H = h\sigma_{stat}$, a 50 percent headstart of $H/2$ is equivalent to $(h/2)\sigma_{stat}$. Thus, we would enter in Minitab's FIR dialog box the value of $h/2$.

Table 8.5   ARL values for Shewhart and CUSUM schemes with $k = 0.5$ and $h = 5$

δ	(a) CUSUM	(b) FIR CUSUM	(c) Shewhart (Limits at $\pm 3.07\sigma$)	(d) CUSUM-Shewhart (Shewhart Limits at $\pm 3.5\sigma$)	(e) FIR CUSUM-Shewhart (Shewhart Limits at $\pm 3.5\sigma$)
0.00	465	430	465	391	360
0.25	139	122	349.2	130.9	113.9
0.50	38.0	28.7	189.2	37.2	28.1
0.75	17.0	11.2	97.3	16.8	11.2
1.00	10.4	6.35	51.8	10.2	6.32
1.50	5.75	3.37	17.1	5.58	3.37
2.00	4.01	2.36	7.01	3.77	2.36
2.50	3.11	1.86	3.51	2.77	1.86
3.00	2.57	1.54	2.12	2.10	1.54
4.00	2.01	1.16	1.21	1.34	1.16
5.00	1.69	1.02	1.03	1.07	1.02

8.4b, which are the in-control observations shifted up by 2 units or 1 standard deviation. The cumulative sums for these data with and without headstart are shown in the table and plotted in Figure 8.9c and d, respectively. With headstart, notice that the out-of-control condition is caught as quickly as the second sample with $\text{CUSUM}_2^+ = 9$ which is greater than the upper limit of $H = 8$. However, with no headstart, the cumulative sums lag behind, unable to signal even by the 10th observation for this data set.

Comparing columns (a) and (b) of Table 8.5 for $\delta > 0$, we can see the benefit of the headstart feature for detecting out-of-control situations. For example, if the process is shifted out of control by one-half of a standard deviation ($\delta = 0.5$) from the onset, the basic CUSUM is expected to detect the condition in 38 samples on average, while FIR CUSUM is expected to detect the condition in 28.7 samples on average. Thus, in this case, FIR CUSUM is associated with a 25 percent shorter out-of-control ARL. For other values of δ, the out-of-control ARLs for the FIR CUSUM are as much as 40 percent shorter than the corresponding ARLs for the basic CUSUM.

As a technical note, the FIR CUSUM ARLs for $\delta > 0$ are only valid when the δ shift is present at the onset, not when the process is in control initially and the δ shift occurs later. When a process is initially in control, we have learned that the cumulative sums are expected to rapidly drop to zero; hence there is a self-transform of a FIR CUSUM to a basic CUSUM. As such, when the process is initially in control, the FIR CUSUM ARLs for $\delta > 0$ are represented by the basic CUSUM ARLs for $\delta > 0$ found in column (a).

It is important to note that the FIR option was developed for the tabular CUSUM procedure. Since there is only a single cumulative statistic for the V mask procedure, difficulty arises in attempting to give the cumulative series a headstart for both positive and negative shifts. Theoretically, it is possible to implement two V mask charts, one with a headstart for a positive shift and the other with a headstart for a negative shift. However, such an implementation is cumbersome. If the FIR feature is considered an important option, then tabular CUSUM is a clear choice over the V mask CUSUM. Otherwise, the preferable choice is less clear and left to personal preferences.

In summary, the headstart feature greatly enhances the basic CUSUM in the detection of certain out-of-control scenarios without substantially increasing the overall false-alarm rate for the in-control situation. In practical applications, if there is any doubt of whether the process mean level can be adjusted exactly to target after a shift, the user is well advised to consider the FIR CUSUM scheme.

COMBINED SHEWHART-CUSUM SCHEME

Given the better performance of the CUSUM scheme relative to the Shewhart scheme, as illustrated by the first example of this chapter, it is possible to be lulled into believing that the CUSUM scheme must be uniformly dominant over the Shewhart scheme. This is far from the truth. Other than its simplicity, the Shewhart scheme has certain advantages relative to most other charting procedures, including the CUSUM. For one, the Shewhart scheme is much more effective in detecting large process shifts. Some evidence of this fact can be seen by comparing columns (*a*) and (*c*) of Table 8.5. To make a fair comparison, the in-control ARLs are calibrated to the same number, namely, 465. This is accomplished by positioning the Shewhart limits at ± 3.07 standard deviations around the mean rather than the conventional ± 3 standard deviations. Even with the wider Shewhart limits, it can be noticed that somewhere above $\delta = 2.5$ the out-of-control ARLs for the Shewhart scheme are shorter than for the CUSUM scheme with $k = 0.5$ and $h = 5$. The exact crossing point will vary, depending on the CUSUM design values h and k. As a general rule of thumb, Shewhart charts are regarded most effective for detecting 2 standard deviation or larger shifts ($\delta > 2$) and not very effective for detecting 1.5 standard deviation or smaller shifts ($\delta < 1.5$).

Even though it is implicit in ARL studies, an important point is rarely emphasized in the quality-control literature. Namely, comparisons based on ARL measures assume that the nature of the shift is forever sustaining until detected. Thus, when one states that Shewhart is more or less effective than CUSUM for a certain shift size, the shift is assumed to be sustaining. Grant and Leavenworth (1996) poignantly write: "Such [ARL] comparisons, however, may be quite misleading . . . of necessity they assume a shift from an in-control state at some $\overline{X}_0$, to another in-control state at a new level. This state of affairs rarely, if ever, exists in industry" (p. 411). Unfortunately, there are no empirical studies to support or refute these respected authors' conjecture of sustaining shifts being rare in practice. The real importance of their comment is to remind us not to be locked into a mentality that one condition, such as sustaining shifts

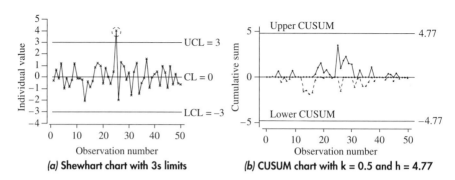

Figure 8.10   Shewhart versus CUSUM chart in the presence of a temporal shock.

with ARL arguments, should serve as the all-powerful litmus test of a charting procedure. From our discussions in Chapter 4, we must remember that departures from an in-control process defined by an independent and identically distributed process come in many shapes and forms, with sustaining shifts being only one part of the picture. Other departures include short-term or sporadic disturbances to the in-control process. And let us not forget the real possibilities of systematic time-series effects shown to be common in industry (see Alwan and Roberts, 1995).

Let us consider the situation of a short-term process shift in discussing the relative advantages of the Shewhart and CUSUM schemes. We simulated 50 random numbers from the normal distribution with mean 0 and standard deviation 1. We then perturbed the 25th observation so that its value is 4.0, that is, 4 standard deviations above the mean. The Shewhart chart of Figure 8.10a clearly brings to our attention the substantial outlier which, in a real application, would call for an immediate search for special-cause explanation.

Now consider implementing a CUSUM chart on the same data series with $\mu_0 = 0$. So as not to cloud the issue, we chose the design parameters to be $k = 0.5$ and $h = 4.77$ which, as noted earlier, give rise to a CUSUM scheme with the same in-control ARL as the 3σ Shewhart chart. The sequence plots of the cumulative statistics (Figure 8.10b) appear well behaved. Furthermore, the CUSUM chart gives no signal of an out-of-control condition; CUSUM_{25}^+ is only bumped up a bit. The failure of the CUSUM chart to detect the isolated shock is not surprising. The CUSUM scheme is designed to exploit a persisting shift by means of accumulation. When the condition is isolated, there is no opportunity to "kick in" the power of the accumulation principle. For CUSUM_{25}^+ to signal, the following would have to happen:

$$x_{25} - 0.5 + \text{CUSUM}_{24}^+ \geq 4.77$$

or

$$x_{25} \geq 5.27 - \text{CUSUM}_{24}^+$$

For the example data series, close inspection of Figure 8.10b reveals that $CUSUM_{24}^+ = 0$. This implies that the 25th observation would have to be an extraordinary value of 5.27 or more standard deviations above the mean for the CUSUM scheme to signal an out-of-control condition. Thus, the advantages of the Shewhart scheme are not just restricted to the detection of sustaining shifts of magnitude 2σ or larger as revealed by a purely ARL mind-set, but are also present for detecting large temporal shocks.

Focused on the advantage of the Shewhart scheme for the detection of larger sustaining shifts, Lucas (1985) suggested the simultaneous implementation of both schemes with a certain modification. He demonstrated that if standard 3σ limits are used with a basic CUSUM design, the in-control ARL drops a little more than might be desired. For example, if a standard Shewhart chart is combined with a basic CUSUM chart with $k = 0.5$ and $h = 5$, the in-control ARL drops from 465 for the basic CUSUM chart alone to 223 for the combined scheme. This translates to more than a doubling of the long-run false-alarm rate. As one way to combat the increased long-run false alarm rate, Lucas recommended positioning the Shewhart limits farther out, at ± 3.5 standard deviations around the mean. From column (d) of Table 8.5, we see that the in-control ARL for a combined scheme with 3.5σ Shewhart limits is 391, which is still a bit lower than the in-control ARL for the CUSUM chart alone. Even after taking into account (albeit informally) this shorter in-control ARL, a comparison of columns (a) and (d) seems to indicate that the combined scheme has an advantage over the CUSUM chain for detecting sustained shifts of 3σ and larger with out of control. Keeping in mind the advantages of the Shewhart scheme for detecting isolated shocks, a combined Shewhart-CUSUM scheme appears to have a lot of merit.

As another point of interest, Lucas (1985) also reported the impact of combining a Shewhart scheme with a FIR CUSUM scheme. The ARLs for this combined scheme are presented in column (e) of Table 8.5. Comparing columns (b) and (e), we see that the addition of the Shewhart scheme makes very little, if any, improvement in the performance of the FIR CUSUM to detect larger shifts. Given this fact, along with the fact that the Shewhart-FIR CUSUM scheme has the shortest in-control ARL of all the schemes shown in Table 8.5, this combined scheme presents itself as the least attractive option.

MONITORING VARIABILITY OF INDIVIDUAL OBSERVATIONS USING CUSUM

A number of investigators have developed procedures for monitoring process variability by using a CUSUM scheme when rational subgroups ($n > 1$) are periodically collected (e.g., see Chang and Gan, 1995). In response to the more frequent application of cumulative sums to individual observations, Hawkins (1981) introduced a simple procedure for monitoring the variability of individual observations (see also Hawkins, 1993).

Suppose that, in an in-control state, a quality characteristic X is normally distributed with mean of μ and standard deviation σ. Consider now the random variable

$$V = \frac{\sqrt{|U|} - 0.822}{0.349}$$

where $U = (X - \mu)/\sigma$. Hawkins (1981) demonstrates that V will be approximately normally distributed with mean equal to 0 and standard deviation equal to 1; that is, V will be approximately a standard normal random variable. If, however, the standard deviation σ shifts to another value, then the *mean* of V will shift away from the in-control mean of 0. Given this fact, Hawkins' idea is that to detect a change in the process standard deviation of the original X variable, we should monitor for a change in mean of the V variable. In terms of observed values of X, Hawkins' procedure calls for the following steps:

1.   Standardize x_i to $u_i = (x_i - \mu)/\sigma$, where μ is the mean or a target value. If μ and σ are unknown, they can be replaced with estimates based on historical in-control process data.

2.   For each u_i, compute the following quantity:

$$v_i = \frac{\sqrt{|u_i|} - 0.822}{0.349} \tag{8.26}$$

3.   Given v_i, compute the tabular cumulative sums

$$\text{CUSUM}_i^+ = \max[0, v_i - k + \text{CUSUM}_{i-1}^+] \tag{8.27}$$

$$\text{CUSUM}_i^- = \min[0, v_i + k + \text{CUSUM}_{i-1}^-] \tag{8.28}$$

where the cumulative sums are initialized to zero unless a FIR CUSUM scheme is implemented. In our original development, the reference value used in the cumulative sum computation is $K = k\sigma_{\text{stat}}$. However, because v_i are standardized values, $\sigma_{\text{stat}} = 1$ and hence $K = k(1) = k$. This is why we see k instead of K in (8.27) and (8.28). Similarly, the decision interval $[-H, H]$ reduces to $[-h, h]$. Hawkins recommends that the h and k values be the same as used in the implementation of a CUSUM for monitoring the process mean; for example, in Hawkins (1993), the proposed technique is illustrated with $k = 0.5$ and $h = 4.77$.

Analogous to the simultaneous implementation of Shewhart mean and variance (or range) charts, Hawkins (1993) recommends the simultaneous implementation of two tabular CUSUM charts, one for the process mean and the other for the process variability.

To illustrate Hawkins' proposal, consider the 40 observations x_i shown in Table 8.6. The first 20 observations were randomly generated from a normal distribution with $\mu = 20$ and $\sigma = 2$, while the remaining 20 observations were randomly generated from a normal distribution with $\mu = 20$ and $\sigma = 3$. A sequence plot of the observations is provided in Figure 8.11a. From this perspective, the increase in variability after observation 20 is not very evident.

Table 8.6   Randomly generated normal data with change in variation

Sample i	x_i	u_i	v_i	Sample i	x_i	u_i	v_i
1	23.5	1.75	1.4352	21	20.7	0.35	−0.6601
2	18.5	−0.75	0.1261	22	23.1	1.55	1.2120
3	15.5	−2.25	1.9427	23	14.2	−2.90	2.5242
4	16.9	−1.55	1.2120	24	21.6	0.80	0.2075
5	18.1	−0.95	0.4375	25	11.4	−4.30	3.5864
6	20.6	0.30	−0.7859	26	18.1	−0.95	0.4375
7	19.8	−0.10	−1.4492	27	18.5	−0.75	0.1261
8	20.6	0.30	−0.7859	28	16.6	−1.70	1.3806
9	17.2	−1.40	1.0350	29	14.4	−2.80	2.4393
10	20.8	0.40	−0.5431	30	20.3	0.15	−1.2456
11	18.0	−1.00	0.5100	31	22.4	1.20	0.7835
12	14.4	−2.80	2.4393	32	21.3	0.65	−0.0452
13	21.9	0.95	0.4375	33	17.9	−1.05	0.5808
14	18.7	−0.65	−0.0452	34	17.2	−1.40	1.0350
15	23.5	1.75	1.4352	35	16.0	−2.00	1.6969
16	22.4	1.20	0.7835	36	24.4	2.20	1.8947
17	17.3	−1.35	0.9739	37	18.2	−0.90	0.3630
18	20.4	0.20	−1.0739	38	21.6	0.80	0.2075
19	21.1	0.55	−0.2303	39	22.4	1.20	0.7835
20	20.5	0.25	−0.9226	40	14.7	−2.65	2.3091

Taking $\sigma = 2$ as the baseline, we standardize the x_i to u_i as follows

$$u_i = \frac{x_i - 20}{2} \qquad i = 1, 2, \ldots, 40$$

All the computed u_i are shown in Table 8.6. By using (8.26), the next step is to convert the u_i to v_i. Below, for instance, is the computation of v_1:

$$v_1 = \frac{\sqrt{|u_1|} - 0.822}{0.349}$$

$$= \frac{\sqrt{|1.75|} - 0.822}{0.349}$$

$$= 1.4352$$

All the computed v_i are shown in Table 8.6. Applying these v_i to a tabular CUSUM scheme, with $k = 0.5$ and $h = 4.77$ and no headstart, we produce the chart shown in Figure 8.11b. The CUSUM chart clearly indicates that a process shift has occurred with all the cumulative sums from $i = 25$ onward exceeding the upper limit. The change point is estimated to be $i = 21$, impressively close to

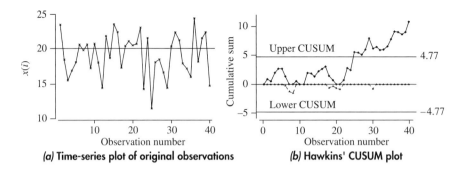

Figure 8.11   Detecting a change in variability of individual observations.

the true time of the shift at $i = 20$. In the next section, we revisit Hawkins' procedure where it is applied in a unique manner.

VIOLATION OF STANDARD CUSUM ASSUMPTIONS

The development and appropriate interpretation of CUSUM charts are based on the underlying assumptions that the process observations are random (or independent) and normally distributed, that is, that the process is iidn. As with Shewhart charts, violations of these assumptions can undermine the effectiveness of a CUSUM chart.

When the observations are random but nonnormal, a viable strategy is to first seek a normalizing transformation, such as a power transformation x^p, and then apply the CUSUM scheme to the normalized data. One important point needs to be considered with transformations and CUSUM. If the target mean level is μ_0 for the original nonnormal data, then the target level for the power-transformed data is not necessarily μ_0^p, but instead, it should be chosen as the long-run average of x^p, that is, $E(x^p)$. When the nonnormal distribution is known, $E(x^p)$ can often be mathematically derived. If the nonnormal distribution is not known, a large sample average of the transformed data when the process is in control can serve as a suitable choice. We will soon illustrate an example of CUSUM implementation in the presence of nonnormality.

With respect to systematic time-series patterns, the interpretation of a CUSUM implementation is often misleading. For instance, Johnson and Bagshaw (1974) demonstrated that the CUSUM procedure is seriously impaired in the presence of autocorrelation. To illustrate, consider the application of a tabular CUSUM to the daily weight series introduced in Chapter 3 and found to follow a stationary meandering AR(1) process with no special causes. To minimize the possibility of an out-of-control signal, we take μ_0 to equal the overall average of the data ($\bar{x} = 181.6$). In Figure 8.12, we present a CUSUM chart applied to the weight series based on Minitab's default options. As can be seen, the cumulative statistics meander around, lingering for a long time above and below the control limits *incorrectly* indicating a lack of control. Part of the problem is

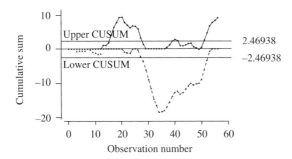

Figure 8.12   CUSUM chart applied to stationary daily weight series.

due to Minitab's default use of the moving-range method to estimate the standard deviation of the process (σ). As demonstrated in Chapter 5, moving-range-based estimators significantly underestimate σ when the process is positively autocorrelated. If we were to use a correct estimate of σ, based on the fitted time-series model, the limits would widen somewhat. However, the CUSUM chart will continue to falsely signal the presence of shifts; demonstration of this fact is left to the reader.

In the presence of meandering behavior, the core problem is that the CUSUM tends to mistakenly view short-term movements as permanent shifts away from the long-run mean, which is constant due to the stationarity of the process. In the same spirit as earlier Shewhart applications, one possibility is to remove the pattern and apply the CUSUM procedure to the residuals and forecast errors, as we will soon illustrate. Other investigators have incorporated a correction to the standard CUSUM procedure in an attempt to reduce the detrimental effects of autocorrelation (e.g., see Yashchin, 1993).

Standard CUSUM applications are not inappropriate in the presence of all types of systematic patterns. The CUSUM procedure works well in signaling the existence of a slowly increasing or decreasing simple trend process. For details on the performance of a CUSUM applied to a trend process and techniques for estimating the trend component from the cumulative sums, refer to Bissell (1984, 1986).

With the next two examples, we illustrate how a CUSUM scheme can be used effectively in the presence of distribution violations or autocorrelation.

In Chapter 7, we presented the most common Shewhart-based control charts for attribute data. When the defect or count rates are very low, say in units per million, most samples for p chart or c chart applications will have zero defects or nonconformity counts. As a result, p charts and c charts will be essentially a plot of a string of zeros with an occasional nonzero observation. As a result, conventional attribute charts become quite ineffective for process monitoring.

As an alternative, Nelson (1994) proposes the monitoring of another dimension associated with the attribute application, namely, the time (or distance) between successive occurrences of

Example 8.3

a defect or nonconformity. For the case of nonconformities (that is, c chart type of data), the distribution assumed to dictate the counts is the Poisson distribution. When a random variable follows a Poisson distribution, the time between occurrences is a random variable which follows the exponential distribution. The exponential distribution was introduced by example in Section 2.6. It is highly skewed, and thus, the application of normal-based Shewhart limits to data on time between occurrences would be inappropriate. To solve this problem, Nelson demonstrates that an exponential random variable when raised to the power of 0.2777 is well approximated by the normal distribution. Hence, we are considering

$$y_i = x_i^{0.2777} \tag{8.29}$$

where x_i represents the ith observation on the time between occurrences. At this point, a standard Shewhart individual measurement control chart can be applied to the y_i data.

Let us consider a hypothetical situation. Suppose the mean time between occurrences of an undesirable event is historically known to be $\mu_0 = 350$. As examples, this may represent the mean time (minutes, hours, etc.) between breakdowns of a critical component, between detection of a containment in a chemical process, or between complaining customers. Or, it may represent the mean distance (inches, feet, etc.) between critical flaws in the production of fiberoptic wire. Furthermore, assume that the individual observations have been found to follow an exponential distribution; in Chapter 9, we show how to test for conformity to an exponential distribution.

Suppose a mean shift down to $\mu_1 = 175$ is viewed as important enough to require quick detection in a CUSUM scheme. In Table 8.7, we provide values of 50 randomly generated observations x_i, with the first 30 coming from an exponential with mean equal to 350 and the remaining 20 coming from an exponential with mean equal to 175. A sequence plot of these observations is provided in Figure 8.13a. Because of the bunching below of observations due to the highly skewed nature of the data, it is rather difficult to see the actuality of the shift.

Using (8.29), we can transform the exponential observations to attain closer conformity to the normal distribution. Below, for example, is the computation of the first transformed value:

$$y_1 = (x_1)^{0.2777}$$
$$= (2325.12)^{0.2777}$$
$$= 8.6072$$

All the transformed values are shown in Table 8.7.

The mean and standard deviation can be estimated from the transformed data to construct the Shewhart control chart, as recommended by Nelson (1994). However, since the mean of the exponential distribution before the shift is known, we can theoretically determine the mean and standard deviation of the distribution for the transformed data. In particular, it can be shown that if an exponentially distributed random variable, with mean μ_0, is raised to the 0.2777 power, then the mean and standard deviation of the transformed random variable are given by

$$\mu_0^{\text{trans}} = 0.901119(\mu_0)^{0.2777} \tag{8.30}$$
$$\sigma^{\text{trans}} = 0.277956(\mu_0)^{0.2777} \tag{8.31}$$

In the in-control state of $\mu_0 = 350$, the mean and standard deviation of the transformed process are

$$\mu_0^{\text{trans}} = 0.901119(350)^{0.2777} = 4.5843$$
$$\sigma^{\text{trans}} = 0.277956(350)^{0.2777} = 1.4140$$

Table 8.7   Exponential and transformed data

Sample i	x_i	y_i	Sample i	x_i	y_i
1	2325.12	8.6072	26	221.84	4.4823
2	66.07	3.2019	27	1524.02	7.6545
3	29.60	2.5620	28	19.52	2.2821
4	505.87	5.6352	29	83.75	3.4199
5	34.10	2.6647	30	793.76	6.3862
6	7.84	1.7714	31	333.90	5.0212
7	1239.74	7.2280	32	101.87	3.6110
8	344.63	5.0655	33	392.49	5.2518
9	80.81	3.3862	34	255.94	4.6638
10	57.50	3.0807	35	1.31	1.0775
11	111.76	3.7052	36	9.75	1.8822
12	343.58	5.0612	37	2.01	1.2143
13	47.15	2.9156	38	466.42	5.5096
14	90.08	3.4898	39	617.01	5.9548
15	272.27	4.7446	40	38.67	2.7594
16	273.66	4.7513	41	47.28	2.9177
17	740.47	6.2641	42	48.37	2.9364
18	338.59	5.0407	43	28.50	2.5353
19	187.60	4.2784	44	222.15	4.4840
20	471.93	5.5276	45	197.19	4.3380
21	326.82	4.9914	46	17.23	2.2044
22	464.92	5.5047	47	336.84	5.0334
23	256.24	4.6653	48	177.97	4.2162
24	78.49	3.3589	49	114.28	3.7282
25	525.67	5.6956	50	23.55	2.4044

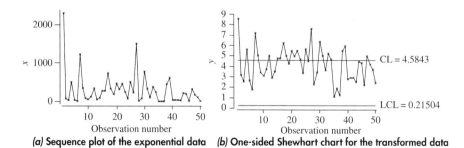

(a) Sequence plot of the exponential data   (b) One-sided Shewhart chart for the transformed data

Figure 8.13   Exponential data with mean shift at observation 30.

Since the interest is only in detecting a shift in one direction, only a one-sided charting scheme need be considered. Suppose the long-run false-alarm rate α is desired to be controlled to a small number such as 0.001. Recall from Chapter 4 that this one-sided rate is associated with placing the Shewhart limit 3.09 standard deviations away from the mean. By using the in-control parameter of $\mu_0 = 350$, the one-sided Shewhart chart limits for the transformed data are

$$CL = \mu_0^{\text{trans}} = 4.5843$$

$$LCL = \mu_0^{\text{trans}} - 3.09\sigma^{\text{trans}} = 4.5843 - 3.09(1.4140) = 0.21504$$

A sequence plot of the transformed data, along with a superimposed lower limit, is shown in Figure 8.13b. The data are clearly more evenly scattered in the transformed scale, but the Shewhart limit fails to detect the downward shift.

The failure of the Shewhart chart to detect the shift is due to the smallness of the shift size. With this in mind, let us explore the application with a CUSUM scheme. Without detailing the computations, a naïve approach would be to apply a normal-based lower-side CUSUM chart to the original exponential data, as is done in Figure 8.14a. Even though the lower-side cumulative statistic is drifting toward the lower limit, the CUSUM chart does not signal the shift during the period of observation. Because of the bunching of lower values in the original exponential data, lower-side cumulative sums have difficulty breaking free to signal the change in a timely fashion. One possible resolution is to develop a CUSUM scheme specifically designed for exponential data (see Gan, 1994). As an alternative, a simple strategy is to apply the normal-based CUSUM to the normalized data.

To pursue the application of a CUSUM scheme for the transformed data, we need to know the mean of the distribution in the transformed scale, associated with the mean in original units, for which we wish to detect quickly. Applying (8.30) to $\mu_1 = 175$, we see the mean in the transformed scale is

$$\mu_1^{\text{trans}} = 0.901119(175)^{0.2777} = 3.7816$$

With this value, the following important values can be determined:

$$\Delta_{\mu^{\text{trans}}} = \left|\mu_0^{\text{trans}} - \mu_1^{\text{trans}}\right| = \left|4.5843 - 3.7816\right| = 0.8027$$

$$\delta = \frac{\Delta_{\mu^{\text{trans}}}}{\sigma^{\text{trans}}} = \frac{0.8027}{1.414} = 0.5677$$

Thus, we have $k = \delta/2 = 0.28385$. Now, it is left to determine the required h value. With a specified overall false-alarm rate of 0.001, the associated in-control ARL is 1000 ($= 1/0.001$).

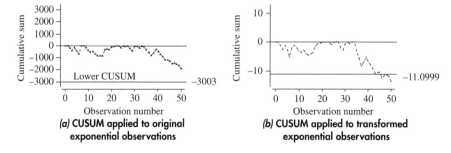

Figure 8.14   CUSUM charts for underlying exponential process.

As noted earlier, when one is dealing with a one-sided scheme, the numerical approximation of (8.24) is used with ARL_0 equal to one-half of the specified one-sided in-control ARL; thus, we use a value of 500 in the approximation. Below, we invoke the supplied macro to determine the necessary h value:

```
MTB > name k1 'k' k2 'ARL' k3 'h0' k4 'hfinal'
MTB > let 'k'=0.28385
MTB > let 'ARL'=500
MTB > let 'h0'=5
MTB > %hvalue k1-k4 c100
Executing from file: C:\MTBWIN\MACROS\hvalue.MAC

Data Display

hfinal   7.85000
```

A plot of the lower-side cumulative sums (with no headstart), along with a lower control limit based on the design parameters $k = 0.28385$ and $h = 7.85$, is given in Figure 8.14b. In this case, the CUSUM chart signals a change with the 45th cumulative sum breaching the control limit. By counting back, the estimated time of the shift is period 33, quite close to the actual period of 30.

Example 8.4

To demonstrate the principles of autoregression, Roberts (1991) studied a data set originally found in an earlier edition of Grant and Leavenworth (1996). The data are measurements of pitch diameter of threads on aircraft fittings. Values are expressed in units of 0.0001 inch (in) in excess of 0.4000 in. The readings are items consecutively produced at times about 1 h apart. Even though the original source treated the application as a subsampling problem, we will treat the data as an individual observation series as in Roberts (1991). We consider only the first 60 observations of a longer series of 100 observations. The data are provided in the column labeled y_i in Table 8.8 and can also be found in the file "pitch.dat".

Let us suppose the first 30 observations form the basis of a retrospective data set. In other words, imagine we have just observed the 30th observation and have yet to observe future observations 31 through 60. Recall that the retrospective data set serves as the information used to estimate the model describing the general workings of the process. As the first step of data analysis, a time-series plot of the pitch diameters needs to be examined (see Figure 8.15a). If we only look at the first 30 observations, it appears as if the time series is exhibiting a meandering type of behavior. Study of the ACF and PACF (not shown) would reveal that the series was indeed not random, with a significant autocorrelation at lag 1. Accordingly, we fit the data with a lag 1 autoregressive—AR(1)—model (see Figure 8.15b). Notice that the lag 1 independent variable is highly significant with a p value of 0.001. The analysis of the residuals (not shown) indicates that the fitted regression model is an appropriate fit.

Thus, the fitted regression model not only serves as a satisfactory summary of past behavior but also can be used to make forecasts of future process outcomes. In Table 8.8 under the column labeled $\hat{y}_i$, the retrospective and prospective fitted values are provided. The retrospective fitted values are estimates for the time-series pattern underlying the 30 observations; there is no fitted value for the first period since the model requires a previous period's observation, which is not known prior to period 1. The prospective fitted values are the one-step-ahead forecasts as one moves into the future observing new observations. The differences

Table 8.8   Pitch diameters of threads on aircraft fittings (continued)

Sample i	y_i	$\hat{y}_i$	e_i	u_i	v_i
1	36	*	*	*	*
2	35	34.5088	0.4912	0.31855	−0.73811
3	34	33.9375	0.0625	0.04053	−1.77844
4	33	33.3662	−0.3662	−0.23748	−0.95896
5	32	32.7949	−0.7949	−0.51550	−0.29804
6	31	32.2236	−1.2236	−0.79351	0.19712
7	31	31.6523	−0.6523	−0.42302	−0.49169
8	34	31.6523	2.3477	1.52250	1.18022
9	32	33.3662	−1.3662	−0.88599	0.34175
10	30	32.2236	−2.2236	−1.44202	1.08551
11	30	31.0810	−1.0810	−0.70104	0.04378
12	30	31.0810	−1.0810	−0.70104	0.04378
13	32	31.0810	0.9190	0.59598	−0.14328
14	30	32.2236	0.7764	0.50350	−0.32212
15	32	31.0810	0.9190	0.59598	−0.14328
16	32	32.2236	−0.2236	−0.14501	−1.26419
17	33	32.2236	0.7764	0.50350	−0.32212
18	33	32.7949	0.2051	0.13301	−1.31030
19	32	32.7949	−0.7949	−0.51550	−0.29804
20	35	32.2236	2.7764	1.80052	1.48950
21	32	33.9375	−1.9375	−1.25649	0.85653
22	34	32.2236	1.7764	1.15201	0.72011
23	37	33.3662	3.6338	2.35655	2.04328
24	37	35.0801	1.9199	1.24507	0.84191
25	35	35.0801	−0.0801	−0.05195	−1.70225
26	32	33.9375	−1.9375	−1.25649	0.85653
27	32	32.2236	−0.2236	−0.14501	−1.26419
28	31	32.2236	−1.2236	−0.79351	0.19712
29	33	31.6523	1.3477	0.87399	0.32343
30	33	32.7949	0.2051	0.13301	−1.31030

between the observations y_i and the fitted values $\hat{y}_i$ are the residuals, given under the column labeled e_i. For the prospective data set, the residuals are better referred to as forecast errors. To illustrate, the forecast for the future period 31 done at period 30 is

$$\hat{y}_{31} = 13.942 + 0.5713 y_{30} = 13.942 + 0.5713(33) = 32.7949$$

When period 31 arrives, the actual process outcome is 33, implying a forecast error e_{31} of 0.2051 (= 33 − 32.7949).

Table 8.8   (concluded)

Sample i	y_i	$\hat{y}_i$	e_i	u_i	v_i
31	33	32.7949	0.2051	0.13301	−1.31030
32	33	32.7949	0.2051	0.13301	−1.31030
33	36	32.7949	3.2051	2.07853	1.77568
34	32	34.5088	−2.5088	−1.62698	1.29951
35	31	32.2236	−1.2236	−0.79351	0.19712
36	23	31.6523	−8.6523	−5.61109	4.43202
37	33	27.0819	5.9181	3.83794	3.25807
38	36	32.7949	3.2051	2.07853	1.77568
39	35	34.5088	0.4912	0.31855	−0.73811
40	36	33.9375	2.0625	1.33755	0.95852
41	43	34.5088	8.4912	5.50661	4.36853
42	36	38.5079	−2.5079	−1.62639	1.29886
43	35	34.5088	0.4912	0.31855	−0.73811
44	24	33.9375	−9.9375	−6.44455	4.91866
45	31	27.6532	3.3468	2.17043	1.86601
46	36	31.6523	4.3477	2.81952	2.45600
47	35	34.5088	0.4912	0.31855	−0.73811
48	36	33.9375	2.0625	1.33755	0.95852
49	41	34.5088	6.4912	4.20960	3.52358
50	41	37.3653	3.6347	2.35713	2.04383
51	34	37.3653	−3.3653	−2.18243	1.87766
52	38	33.3662	4.6338	3.00506	2.61178
53	35	35.6514	−0.6514	−0.42244	−0.49297
54	34	33.9375	0.0625	0.04053	−1.77844
55	38	33.3662	4.6338	3.00506	2.61178
56	36	35.6514	0.3486	0.22607	−0.99293
57	38	34.5088	3.4912	2.26407	1.95611
58	39	35.6514	3.3486	2.17160	1.86714
59	39	36.2227	2.7773	1.80110	1.49012
60	40	36.2227	3.7773	2.44961	2.12929

As long as the conditions remain the same as time evolves, the retrospective residuals and forecast errors will behave as a series of independent and identically distributed series around a mean of 0 and a level of variation remaining constant. Thus, as discussed in Chapter 5, tracking the forecast errors is a reasonable strategy for detection of process change. In Figure 8.16a, we present a time-series plot of the residuals continuing into the forecast errors. By comparing the 30 retrospective residuals and the 30 prospective forecast errors, it is quite clear that there was a significant increase in the level of variation. From a real-time perspective, the question

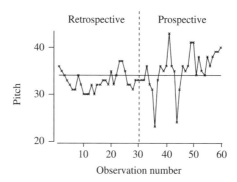

(a) Time-series plot of the pitch diameter measurements

```
The regression equation is
pitch = 13.9 + 0.571 pitch–1

29 cases used 1 case contain missing values

Predictor    Coef     StDev       T      P
Constant   13.942     4.797    2.91  0.007
pitch–1    0.5713    0.1462    3.91  0.001

S = 1.542    R–Sq = 36.1%  R–Sq (adj) = 33.8%

Analysis of variance

Source          DF       SS        MS       F      P
Regression       1   36.324    36.324   15.27 0.001
Residual error  27   64.228     2.379
Total           28  100.552
```

(b) Regression output for fitted AR(1) model

Figure 8.15   Time-series analysis of the pitch diameter series.

is, When would the variation shift be detected starting from period 31? Since the residuals and forecast errors represent a series of individual observations, the earlier-described technique developed by Hawkins can be applied.

Recall that the first step is to standardize the underlying individual observations. The standard deviation of the residuals can be estimated by using the range techniques illustrated in Chapter 5 or can be directly extracted from the regression output. From Figure 8.15b, we find the standard deviation of the residuals is equal to 1.542. Since the mean of the residuals is 0, the standardized values u_i are computed by

$$u_i = \frac{e_i - 0}{1.542} = \frac{e_i}{1.542}$$

The next step is convert the u_i to v_i by using (8.26). The final computed values of v_i, along with the values for the fitted values, residuals, and u_i, are given in Table 8.8. By implementing the CUSUM scheme with Hawkins' recommendation of $h = 4.77$, the CUSUM chart shown in Figure 8.16b results.

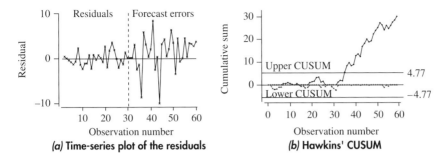

Figure 8.16 Nonconstant variance in the residuals and forecast errors.

The CUSUM chart on the forecast errors signals a variation shift at period 35. If we were to look back at the plot of the original observations (Figure 8.15a), a reasonable question might come to mind. Namely, why not apply Hawkins' technique to the original observations, given that the nonconstant variance is also apparent in that plot? Even though it is true that the nonconstant variance is manifested in the observations, we must not forget that the observations are highly correlated, and thus the u_i and v_i required in the technique will also be correlated. By applying the CUSUM scheme to the correlated data, we place ourselves at risk of obtaining misleading results, as indicated by Johnson and Bagshaw (1974) and illustrated earlier with the weight series (Figure 8.12). Nevertheless, if we ignored these facts and implemented Hawkins' CUSUM to the original observations with no regard to the autocorrelation, an out-of-control signal would be flagged at period 41. Thus, by disregarding the autocorrelation, we have incurred a substantial delay—six periods $(41 - 35)$—in the detection of the out-of-control condition.

8.2 THE MOVING-AVERAGE CONTROL CHART FOR MONITORING THE PROCESS MEAN

We have demonstrated that the CUSUM scheme is a powerful scheme relative to the Shewhart scheme for the detection of small sustaining process shifts. For the same purpose, other memory-based schemes have been developed as alternatives to the Shewhart scheme.

One such alternative is the *moving-average* control chart. Relative to the CUSUM scheme, the moving-average approach is quite basic. Suppose individual observations $(x_1, x_2, \ldots)$ are collected. The *moving average* of width (or span) w at time i is defined as

$$M_i = \frac{x_i + x_{i-1} + \cdots + x_{i-w+1}}{w} = \frac{\sum_{j=i-w+1}^{i} x_j}{w} \qquad i \geq w \qquad \textbf{(8.32)}$$

Thus, the moving-average statistic is simply the average of the w most recent observations. Because the monitored statistic M_i incorporates previous observations, the moving-average method indeed falls under the category of a memory-based control chart. For periods $i < w$, there are not yet w observations to calculate a moving average of width w. For these periods, the average of all observations up to period i defines the moving average. So, for example, if $w = 3$, we have

$$M_1 = \frac{x_1}{1} = x_1$$

$$M_2 = \frac{x_1 + x_2}{2}$$

$$M_i = \frac{x_i + x_{i-1} + x_{i-2}}{3} \qquad i \geq 3$$

Notice that if $w = 1$, the moving averages are just the individual observations, that is, $M_i = x_i$. Thus, the Shewhart scheme is a special case of a moving-average scheme with $w = 1$. Even though for $w > 1$, the plotted statistic is not a Shewhart-type statistic, in the sense of having no memory, the moving-average scheme calls for computing the control limits for the moving averages in much the same manner as for a Shewhart scheme.

To determine moving-average limits, we need to know the variance of the moving-average statistic. Under the assumption that the individual observations are independent, we can use the fact that the variance of a sum of random variables is the sum of the variances of these random variables (see Section 2.6). Furthermore, we need to recall that the variance of a constant times a random variable is the constant squared times the variance of the random variable (see Section 2.6). Hence, the variance of the moving average M_i is

$$\text{Var}(M_i) = \text{Var}\left(\frac{1}{w} \sum_{j=i-w+1}^{i} X_j\right) \qquad i \geq w$$

$$= \frac{1}{w^2} \text{Var}\left(\sum_{j=i-w+1}^{i} X_j\right)$$

$$= \frac{1}{w^2} \sum_{j=i-w+1}^{i} \text{Var}(X_j)$$

$$= \frac{1}{w^2}(w\sigma^2) = \frac{\sigma^2}{w} \qquad \text{(8.33)}$$

For periods $i < w$, the variance of the moving average is

$$\text{Var}(M_i) = \frac{\sigma^2}{i} \qquad \text{(8.34)}$$

If the intent is to monitor the process relative to a mean target value μ_0, then, for periods $i \geq w$, the centerline and 3σ control limits are given by

$$\text{UCL} = \mu_0 + 3\frac{\sigma}{\sqrt{w}} \qquad (8.35)$$

$$\text{CL} = \mu_0$$

$$\text{LCL} = \mu_0 - 3\frac{\sigma}{\sqrt{w}} \qquad (8.36)$$

For periods $i < w$, $\sigma/\sqrt{w}$ is replaced with $\sigma/\sqrt{i}$. We see again that if $w = 1$, the moving-average limits become nothing more than the standard 3σ Shewhart limits.

The implementation of the moving-average chart does not require a target value μ_0 as does the CUSUM scheme. Without a specified target value, standard practice is to replace μ_0 in (8.35) and (8.36) with the overall average of the individual observations $\bar{x}$. When σ is not known, either $\overline{MR}/d_2$ or s/c_4 can serve as an estimate for σ (refer to Section 5.1).

The moving-average scheme can also be applied to subgroups of size $n > 1$. In such cases, one simply replaces x_i with $\bar{x}_i$ in the computation of the moving average, given by (8.32). Since the standard deviation of the sample average is $\sigma/\sqrt{n}$, the moving-average control limits of (8.35) and (8.36) become

$$\text{UCL} = \mu_0 + 3\frac{\sigma}{\sqrt{nw}} \qquad (8.37)$$

$$\text{LCL} = \mu_0 - 3\frac{\sigma}{\sqrt{nw}} \qquad (8.38)$$

For periods $i < w$, w is replaced with i. In place of μ_0, the grand mean $\bar{\bar{x}}$ can be used. Finally, σ might be replaced with subgroup-based estimates such as $\hat{\sigma} = \bar{R}/d_2$ or $\hat{\sigma} = \bar{s}/c_4$ (refer to Sections 6.2 and 6.4).

The performance of the moving-average chart for the detection of sustaining shifts has been investigated by Roberts (1959) and Lai (1974). Generally speaking, the moving-average chart with $w > 1$ has been found to detect small shifts more quickly than the Shewhart scheme, but is less quick in detecting large shifts. Also, the moving-average chart is generally less effective in detecting small shifts than the CUSUM method. Unlike the CUSUM procedure, one serious difficulty with the moving-average chart is the lack of good guidance for addressing design issues. Most notably, how does one determine an appropriate choice of w? The choice is typically made with the following insight. Looking at the moving-average limits, we can see that as w increases, the width of the control limits decreases. Thus, small sustaining shifts are more effectively detected with larger values of w. The cost of larger w is a slower response to large sustaining shifts. Beyond this insight, recommendations for the selection of w are very nonspecific. Part of the problem is the discreteness of w. Unlike in the CUSUM procedure with its design parameter h, the design parameter w cannot be "fine-tuned" so that certain specific performance requirements can be met.

Table 8.9 Moving averages ($w = 5$) for grease data

Sample i	x_i	M_i	Sample i	x_i	M_i
1	0.0935	0.093500	19	0.1060	0.105400
2	0.1065	0.100000	20	0.1105	0.105900
3	0.0980	0.099333	21	0.0985	0.102400
4	0.1125	0.102625	22	0.0950	0.100600
5	0.0985	0.101800	23	0.0855	0.099100
6	0.1090	0.104900	24	0.0815	0.094200
7	0.0920	0.102000	25	0.0945	0.091000
8	0.1040	0.103200	26	0.0960	0.090500
9	0.0995	0.100600	27	0.0805	0.087600
10	0.1085	0.102600	28	0.0835	0.087200
11	0.0900	0.098800	29	0.0965	0.090200
12	0.0835	0.097100	30	0.0825	0.087800
13	0.1100	0.098300	31	0.0960	0.087800
14	0.1020	0.098800	32	0.0820	0.088100
15	0.1080	0.098700	33	0.0885	0.089100
16	0.1160	0.103900	34	0.0980	0.089400
17	0.1040	0.108000	35	0.0945	0.091800
18	0.0930	0.104600			

Example 8.5

To illustrate the construction of a moving-average chart, consider again the shaft seal grease data studied in Example 8.2 and presented in Table 8.2. Recall that subgroups of size $n = 2$ were collected. Thus, sample means would be used in the moving-average calculations. For $w = 5$, the sample means along with the computed moving averages are shown in Table 8.9.

A range-based estimate for σ was found in Example 8.2 to be 0.01089. Given that the target mean μ_0 is 0.10, the moving-average control limits are obtained from (8.37) and (8.38) and are given by

$$\text{UCL} = 0.10 + 3\frac{0.01089}{\sqrt{2(5)}} = 0.11033$$

$$\text{LCL} = 0.10 - 3\frac{0.01089}{\sqrt{2(5)}} = 0.08967$$

These control limits are applicable for periods $i \geq 5$. For periods $1 \leq i < 5$, the number 5 in the above limits is replaced by i.

The moving-average chart is shown in Figure 8.17a. Initially, the limits are wider until they become constant for periods $i \geq 5$. For comparison, a moving-average chart with $w = 1$ (that is, a Shewhart chart) is given in Figure 8.17b.

The Shewhart chart fails to detect the relatively small shift. By contrast, the moving-average chart, with $w = 5$, signals an out-of-control condition. Interestingly, the chart first signals at sample period 27, which is the same sample period in which the CUSUM scheme detected the out-of-control condition (see Figure 8.8). This is not suggestive of equal capability

Stat>>Control Charts>>Moving Average

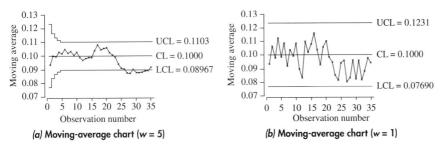

Figure 8.17   Moving-average charts for the grease data.

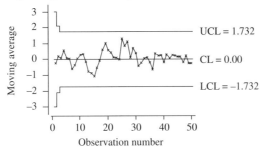

Figure 8.18   Moving-average chart ($w = 3$) concealing temporal shock.

of the two methods, because a CUSUM scheme is generally more effective in the detection of small shifts. Nevertheless, if simplicity is an overriding concern, the moving-average chart might be viewed as an attractive alternative to the CUSUM scheme for the detection of smaller shifts which are not quickly detectable by a Shewhart scheme.

Before we close our discussion of the moving-average chart, two points are worth noting. First, the successive moving averages when $w > 1$ are highly correlated due to the way they are constructed. As a result, the moving averages will exhibit nonrandom patterns even when no out-of-control condition is present, much as the sequence of cumulative sums exhibits nonrandom patterns. Because of this, it is ill advised to supplement the moving-average control limits with other control criteria, such as runs rules. Second, the moving-average chart, like the CUSUM chart and all standard memory charts, is not effective in revealing isolated shifts or shocks to the process. To demonstrate, we present in Figure 8.18 a moving-average chart with $w = 3$ for the data series with a 4 standard deviation observation shown in Figure 8.10. Notice that the averaging of observations completely "buries" the potentially important special-cause signal.

8.3   THE EXPONENTIALLY WEIGHTED MOVING-AVERAGE CONTROL CHART

The moving-average method described in the previous subsection can be regarded as a weighted-average scheme. In this section, we describe an alternative weighted-average scheme, known as the *exponentially weighted moving average (EWMA)*, which has been proved to compete very well with the CUSUM method for the detection of small sustaining shifts.

EWMA CHART FOR MONITORING THE PROCESS MEAN

To begin our development of a new weighted-average scheme, let us look more closely at the weighting system associated with the moving-average method. We can write the expression for M_i given by (8.32) as

$$M_i = \left(\frac{1}{w}\right)x_i + \left(\frac{1}{w}\right)x_{i-1} + \cdots + \left(\frac{1}{w}\right)x_{i-w+1} + 0x_{i-w} + 0x_{i-w-1} + \cdots + 0x_1 \quad \text{(8.39)}$$

Hence, the w most recent observations are equally weighted by $1/w$, while all earlier observations are equally weighted by 0.

Roberts (1959) introduced a control chart based on another weighting scheme. He suggested a statistic be constructed based on varying weights where relatively more weight is assigned to the most recent observation with the remaining weights steadily decreasing as observations become less recent. Because Roberts (1959) chose the weights to decrease geometrically with the age of the observations, he referred to the control chart based on such a weighting system as a *geometric moving-average control chart*. However, in recent times, there is a preference among quality-control professionals to refer to the chart as an *exponentially weighted moving-average (EWMA) chart*. The preferred name is an attempt to reflect that the weighting system underlying the EWMA chart is exactly the same as the weighting system underlying a popular forecasting technique known as exponential smoothing. But, because of this commonality, the name *EWMA* can be a source of confusion. The problem is that exponential smoothing techniques are also used in modeling a process which exhibits systematic time-series patterns, such as meandering autoregressive behavior. In contrast, EWMA charts (like CUSUM charts) have been designed to detect sustaining process shifts from an in-control random process, *not* modeling nonrandom behavior. In Chapter 10, we discuss the role of exponential smoothing techniques as a possible alternative to regression methods for the modeling of nonrandom data. For now, we will assume that the in-control process generates random observations and apply the EWMA chart to detect shifts from this in-control state.

If individual observations $x_1, x_2, \ldots$ are collected, the exponential moving average, at period i, is defined as

$$\text{EWMA}_i = \lambda x_i + (1 - \lambda)\text{EWMA}_{i-1} \quad \text{(8.40)}$$

where λ is a weighting constant ($0 < \lambda \leq 1$). To compute the first value (EWMA_1), a value for EWMA_0 is required. If the intent is to monitor the process for departures away from some target value μ_0, then EWMA_0 is set equal to μ_0. Without a specified target value, standard practice is to initialize the statistic to the average of a preliminary set of data, that is, $\text{EWMA}_0 = \bar{x}$. If rational subgroups of size $n > 1$ are collected, simply replace x_i with $\bar{x}_i$. The starting value is either μ_0 if specified or the grand mean $\bar{\bar{x}}$.

We will soon need to address in some detail the design of an EWMA chart. At this stage, however, some preliminary insights can be gained by simply looking at (8.40). Notice that if $\lambda = 1$, the values of the EWMA statistic are exactly equal to the individual observations x_i. In other words, the EWMA statistic becomes the memoryless statistic for the Shewhart scheme. Indeed, all the subsequent development of the EWMA scheme collapses to the Shewhart scheme when $\lambda = 1$. Since Shewhart schemes work better in the detection of large sustaining shifts, it stands to reason that detection of small sustaining shifts is associated with λ values on the other side of the spectrum, that is, λ closer to 0 than to 1.

By repeatedly using the recursive Equation (8.40), we can see how the EWMA scheme weights all the individual observations:

$$
\begin{aligned}
\text{EWMA}_i &= \lambda x_i + (1 - \lambda)[\lambda x_{i-1} + (1 - \lambda)\text{EWMA}_{i-2}] \\
&= \lambda x_i + \lambda(1 - \lambda)x_{i-1} + (1 - \lambda)^2\text{EWMA}_{i-2} \\
&= \lambda x_i + \lambda(1 - \lambda)x_{i-1} + \lambda(1 - \lambda)^2 x_{i-2} + \lambda(1 - \lambda)^3 x_{i-3} \\
&\quad + \cdots + \lambda(1 - \lambda)^{i-1}x_1 + (1 - \lambda)^i\text{EWMA}_0 \\
&= \sum_{j=0}^{i-1} \lambda(1 - \lambda)^j x_{i-j} + (1 - \lambda)^i\, \text{EWMA}_0 \qquad \textbf{(8.41)}
\end{aligned}
$$

For a λ value between 0 and 1, also $1 - \lambda$ is between 0 and 1. This means that the multiplication of a weight by $1 - \lambda$ will decrease the value of the weight. Hence, the weights $\lambda(1 - \lambda)^j$ decrease—geometrically—with the age of the observations.

For comparative purposes, consider an EWMA scheme with $\lambda = 0.25$ versus a moving-average scheme with $w = 4$. From (8.39) and (8.41), we find the following weighted averages:

$$MA_i = 0.25x_i + 0.25x_{i-1} + 0.25x_{i-2} + 0.25x_{i-3} + 0x_{i-4} + 0x_{i-5} + \ldots$$

versus

$$\text{EWMA}_i = 0.25x_i + 0.1875x_{i-1} + 0.1406x_{i-2} + 0.1055x_{i-3} + 0.0791x_{i-4} + 0.0593x_{i-5} + \cdots$$

Since the EWMA statistic incorporates past information, the corresponding control chart scheme rightfully falls under the category of memory-based control charts.

As with the moving-average statistic, the EWMA statistic is plotted with control limits derived in a Shewhart-like manner. If the individual observations are independent random variables with underlying variance equal to σ^2, then the standard deviation of the exponentially weighted moving-average statistic, at period i, can be shown to be

$$\sigma_{\text{EWMA}(i)} = \sigma \sqrt{\frac{\lambda}{2 - \lambda} [1 - (1 - \lambda)^{2i}]} \tag{8.42}$$

With a specified target mean value, the centerline and control limits for the EWMA control chart, at period i, are as follows:

$$\text{UCL}_i = \mu_0 + L\sigma \sqrt{\frac{\lambda}{2 - \lambda}[1 - (1 - \lambda)^{2i}]} \tag{8.43}$$

$$\text{CL} = \mu_0$$

$$\text{LCL}_i = \mu_0 - L\sigma \sqrt{\frac{\lambda}{2 - \lambda} [1 - (1 - \lambda)^{2i}]} \tag{8.44}$$

where L is the number of standard deviations the control limits are from the centerline. If rational subgroups of size $n > 1$ are used, σ is replaced with $\sigma/\sqrt{n}$ in (8.42), (8.43), and (8.44). When μ_0 is not specified, the convention is to replace it in the control limit construction with $\bar{x}$ for the individual observation case and $\bar{\bar{x}}$ for the subgroup case.

From (8.43) and (8.44), we see that the width of control limits changes with the value of the sample period i through the factor $(1 - \lambda)^{2i}$. But since $(1 - \lambda)^{2i}$ converges to 0 as i increases, the exact EWMA limits converge to

$$\text{UCL} = \mu_0 + L\sigma \sqrt{\frac{\lambda}{2 - \lambda}} \tag{8.45}$$

$$\text{LCL} = \mu_0 - L\sigma \sqrt{\frac{\lambda}{2 - \lambda}} \tag{8.46}$$

Theoretically, these control limit values, known as *asymptotic* limits, are attained when i approaches infinity. However, for typical choices of λ, $(1 - \lambda)^{2i}$ converges to 0 rapidly after a practical number of periods, allowing then for the use of asymptotic limits in place of exact limits. For small values of i, it is best to stick with the exact limit computation, as already incorporated in most software packages, including Minitab.

In most applications, the process standard deviation σ that appears in the exact and asymptotic limits must be estimated from the data. When one is dealing with subgroups, σ is typically estimated as $\bar{R}/d_2$ or $\bar{s}/c_4$ (refer to Sections 6.2 and 6.4). For the case of individual measurements, either $\overline{MR}/d_2$ or s/c_4 can be used as an estimate for σ (refer to Section 5.1).

With respect to the choice of L, one of the commoner choices for L is obviously 3, to mimic the standard Shewhart chart. To compete with a CUSUM chart for the detection of small shifts, a λ value is commonly chosen in the range of 0.05 to 0.25. Minitab's default choices are $L = 3$ and $\lambda = 0.20$. For a more refined approach, the EWMA chart can be designed to have certain performance characteristics. As was the case with the CUSUM method, an EWMA chart can

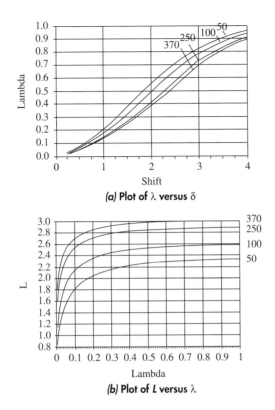

Lambda

Shift

(a) **Plot of** λ **versus** δ

L

Lambda

(b) **Plot of L versus** λ

Figure 8.19   Design of EWMA scheme. Adapted from Crowder (1989) with permission from the American Society for Quality.

be designed to a specified in-control ARL. Since there are an infinite number of (L, λ) combinations which give rise to an EWMA chart with a particular in-control ARL, some additional user specifications are required. Crowder (1989) and Lucas and Saccucci (1990) have developed an EWMA design strategy akin to the design of a CUSUM chart. Their design strategy requires the user to specify two requirements: (1) a desired in-control ARL and (2) the magnitude of a shift in the process mean that is desired to be detected as quickly as possible.

One nice feature of Crowder's work is that he has developed simple plots that enable the user to easily obtain the appropriate (L, λ) combination, given the specified requirements. To design an EWMA chart relative to a target mean μ_0, the following steps are suggested:

Step 1   Specify the in-control ARL (ARL_0). In Crowder (1989), plots are given for ARL_0 choices of 50, 100, 250, 370, 500, 750, 1000, 1500, and 2000. We will present only a subset of these choices for illustration of the procedure.

Step 2   Specify the critical shift size $\Delta_\mu = |\mu_1 - \mu_0|$ away from the target level that is desired to be detected as quickly as possible. As done with the CUSUM design, the shift magnitude needs to be converted to standardized units

$$\delta = \frac{\Delta_\mu}{\sigma_{stat}}$$

where σ_{stat} is σ if individual observations (x_i) underlie the computation of the EWMA statistic and is $\sigma/\sqrt{n}$ if subgroup means $(\bar{x}_i)$ are used in the computation.

Step 3   Refer to the plot given in Figure 8.19a, and project a vertical line from the value of δ determined in step 2 up to the curve associated with the specified ARL_0. From that point on the curve, project a horizontal line to the left to obtain the value for λ.

Step 4   Refer to the plot given in Figure 8.19b, and project a vertical line from the value of λ determined in step 3 up to the curve associated with the specified ARL_0. From that point on the curve, project a horizontal line to the left to obtain the value for L.

With a finely scaled ruler, accurate values of L and λ can be easily obtained to the second decimal place.

Example 8.6

To illustrate the construction of an EWMA chart, consider the grease data first studied in Example 8.2. Recall from Example 8.2 that the desire is to detect a shift of 0.01 g from a target value of $\mu_0 = 0.10$ as quickly as possible. Since subgroups of size 2 were collected, σ_{stat} is given by $\sigma/\sqrt{2}$. In Example 8.2, σ_{stat} was estimated to be 0.0077. Thus, the critical shift size in standardized units (δ) is estimated to be 1.3 $(= 0.01/0.0077)$. Suppose it is desired to have an EWMA scheme with similar in-control properties to a 3σ Shewhart scheme in terms of ARL. This would imply an in-control ARL of about 370. Using Figure 8.19a and b, with a fine-scale ruler, in conjunction with the above outlined steps, we find that the combination of $(\lambda, L) = (0.20, 2.86)$ should be a suitable choice for detecting shifts in the neighborhood of 1.3 standardized units. We should point out that Crowder's procedure is based on a known value of δ. In our example, this would mean the standard deviation σ needs to be exactly known. Of course, in practice, parameters are rarely known. With reliable estimates for the necessary parameters, theoretical studies, such as Crowder's, play the important role of providing us with a good "sense" of likely performance of the control chart scheme, be it EWMA, CUSUM, or Shewhart. Remember also that theoretical studies, such as ARL studies, optimize theoretical performance for very specific conditions, such as a sustaining shift. In reality, the process will not be affected in as precise a manner as assumed in a theoretical development. With this perspective, we proceed with the selected λ and L values as reasonable values for meeting the stated intentions.

Let us now illustrate the EWMA calculations for the first couple of periods. With the starting value of $EWMA_0 = 0.10$ and $\bar{x}_1 = 0.0935$, we can apply (8.40) to find the first period's EWMA value, which is equal to

$$EWMA_1 = \lambda \bar{x}_1 + (1 - \lambda)EWMA_0$$

$$= 0.20(0.0935) + 0.80(0.10)$$

$$= 0.0987$$

Since $\bar{x}_2 = 0.1065$, the second period's EWMA value is

$$EWMA_2 = \lambda \bar{x}_2 + (1 - \lambda)EWMA_1$$

$$= 0.20(0.1065) + 0.80(0.09870)$$

$$= 0.10026$$

The summary of the remaining computations is given in Table 8.10.

Using (8.43) and (8.44) with the estimate for the subgroup mean standard in place of σ, the control limits for the first period are

$$UCL_1 = 0.10 + 2.86\left(\frac{\hat{\sigma}}{\sqrt{2}}\right) \sqrt{\frac{0.20}{2 - 0.20}[1 - (1 - 0.20)^{2(1)}]}$$

$$= 0.10 + 2.86(0.0077) \sqrt{\frac{0.20}{2 - 0.20}[1 - (1 - 0.20)^{2(1)}]}$$

$$= 0.10440$$

and

Table 8.10   Exponentially weighted moving averages ($\lambda = 0.20$) for grease data

Sample i	x_i	$EWMA_i$	Sample i	x_i	$EWMA_i$
1	0.0935	0.098700	19	0.1060	0.102858
2	0.1065	0.100260	20	0.1105	0.104386
3	0.0980	0.099808	21	0.0985	0.103209
4	0.1125	0.102346	22	0.0950	0.101567
5	0.0985	0.101577	23	0.0855	0.098354
6	0.1090	0.103062	24	0.0815	0.094983
7	0.0920	0.100849	25	0.0945	0.094886
8	0.1040	0.101479	26	0.0960	0.095109
9	0.0995	0.101084	27	0.0805	0.092187
10	0.1085	0.102567	28	0.0835	0.090450
11	0.0900	0.100053	29	0.0965	0.091660
12	0.0835	0.096743	30	0.0825	0.089828
13	0.1100	0.099394	31	0.0960	0.091062
14	0.1020	0.099915	32	0.0820	0.089250
15	0.1080	0.101532	33	0.0885	0.089100
16	0.1160	0.104426	34	0.0980	0.090880
17	0.1040	0.104341	35	0.0945	0.091604
18	0.0930	0.102073			

$$\text{LCL}_1 = 0.10 - 2.86\left(\frac{\hat{\sigma}}{\sqrt{2}}\right)\sqrt{\frac{0.20}{2 - 0.20}[1 - (1 - 0.20)^{2(1)}]}$$

$$= 0.10 - 2.86(0.0077)\sqrt{\frac{0.20}{2 - 0.20}[1 - (1 - 0.20)^{2(1)}]}$$

$$= 0.09560$$

For the second period, the limits are:

$$\text{UCL}_2 = 0.10 + 2.86(0.0077)\sqrt{\frac{0.20}{2 - 0.20}[1 - (1 - 0.20)^{2(2)}]}$$

$$= 0.10564$$

and

$$\text{LCL}_2 = 0.10 - 2.86(0.0077)\sqrt{\frac{0.20}{2 - 0.20}[1 - (1 - 0.20)^{2(2)}]}$$

$$= 0.09436$$

Notice that the control limits are widening as i increases. But the difference in control limit width, from one period to the next, diminishes as i increases. After a sufficient number of periods, the exact limit values are indistinguishable from the asymptotic limits, which from (8.45) and (8.46) are computed to be

$$\text{UCL} = 0.10 + 2.86(0.0077)\sqrt{\frac{0.20}{2 - 0.20}}$$

$$= 0.10734$$

and

$$\text{LCL} = 0.10 - 2.86(0.0077)\sqrt{\frac{0.20}{2 - 0.20}}$$

$$= 0.09266$$

In Figure 8.20, we show the EWMA control chart as produced by Minitab.[10] From the display, the exact limits became indistinguishable from the asymptotic limits by the seventh or eighth sample period. The control chart signals an out-of-control condition at observation 27, matching the earlier CUSUM results.

Lucas and Saccucci (1990) demonstrated that the EWMA scheme can be made nearly as powerful as the CUSUM scheme for detecting small sustaining shifts. Given its simplicity for implementation, the EWMA chart positions itself as an attractive alternative to the CUSUM chart. When designed to quickly detect small shifts, the EWMA will, logically, perform less well in detecting large shifts compared to a Shewhart scheme. As suggested with the CUSUM, a possible solution could be to combine the EWMA chart with a Shewhart chart where the Shewhart chart has limits placed a bit wider than 3σ so as to reduce the risk of

---

[10]In Minitab, the default value for L is 3. To change this factor to 2.86, point and click the "S Limits" option button within the EWMA option box and then type in the numbers −2.86 and 2.86 in the dialog box labeled "Sigma limit positions."

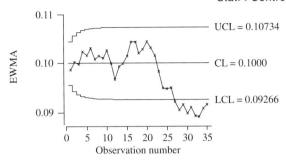

Figure 8.20   EWMA chart for the grease data with $\lambda = 0.20$ and $L = 2.86$.

false alarm. Additionally, the EWMA can be adapted with a headstart to more quickly detect a still out-of-control process after an attempted correction. For details on FIR EWMA schemes, consult Lucas and Saccucci (1990) and Rhoads, Montgomery, and Mastrangelo (1996).

Finally, we wish to address a common misconception about the relationship between the moving-average chart and the exponentially weighted moving-average chart. Some quality control authors suggest that these two schemes are equivalent, if

$$\lambda = \frac{2}{w + 1} \tag{8.47}$$

This conjecture of equivalence is not entirely true. What is true is that if λ and w are related by (8.47), the stabilized EWMA control limit values equal the stabilized moving-average control limit values. However, the values of the plotted EWMA versus moving-average statistics can be very different, and thus different results are bound to occur. To demonstrate, consider the moving-average chart shown earlier in Figure 8.17a based on $w = 5$ applied to the grease data. Substituting $w = 5$ into (8.47) results in $\lambda = 0.333\overline{3}$. Selecting the L factor to be 3, as in the standard implementation of the moving-average chart, and $\lambda = 0.333\overline{3}$, we obtain the EWMA chart shown in Figure 8.21. Comparing Figures 8.17a and 8.21, we see that the values of the stabilized upper and lower control limits for the two charts are indeed identical. However, close inspection of the plots reveals different movements of the plotted statistic values. In fact, the charts differ in their signaling of out of control. In particular, the 31st moving-average value is well below the lower control limit, whereas the 31st EWMA value is well above the lower control limit. Thus, the two procedures are not equivalent even when the condition of (8.47) is met.

MONITORING PROCESS VARIABILITY WITH EWMA

A variety of EWMA methods have also been developed to monitor process variability. For example, Ng and Case (1989) proposed a number of EWMA-based

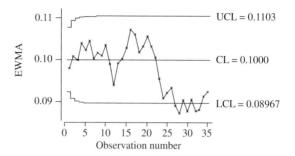

Figure 8.21   EWMA chart for grease data with $\lambda = 0.333\overline{3}$ and $L = 3$.

charts for monitoring variability when dealing with rational subgroups ($n > 1$). Also for subgroups, Crowder and Hamilton (1992) introduced a logarithm-based EWMA procedure that can be designed to detect variance shifts of specified magnitude as quickly as possible.

As in the case of our CUSUM presentation, we will focus on techniques for monitoring variability of individual observations. Even though originally proposed for the CUSUM scheme, one possibility is to apply an EWMA chart to the v_i statistic considered by Hawkins (1993). Recall that if the process is in control, v_i will approximately follow the standard normal distribution. Any changes in the underlying standard deviation will result in a mean shift in the v_i data. Using an EWMA chart to detect such a shift would seem to be a reasonable strategy. We will soon illustrate this approach by example. Now we consider control charts specifically designed for monitoring variability based on an EWMA-related approach.

MacGregor and Harris (1993) proposed two possible EWMA-based charts for monitoring variability of individual observations. Each of their proposed charts assumes that the individual observations x_i arise from an independent and normally distributed process. The first of their proposed charts is based on a statistic called the *exponentially weighted mean square error (EWMS)*, which is computed each period as follows:

$$\text{EWMS}_i = \lambda(x_i - \mu)^2 + (1 - \lambda)\text{EWMS}_{i-1} \qquad \textbf{(8.48)}$$

where μ is the mean of the in-control process. The statistic is initialized to $\text{EWMS}_0 = \sigma_0^2$, where σ_0^2 is taken to be a specified target value or is obtained from historical data when the process is in control. Rather than plot EWMS_i, MacGregor and Harris (1993) recommend that the square root of EWMS_i ($\sqrt{\text{EWMS}_i}$) be plotted. Accordingly, they call the corresponding control chart an *exponentially weighted root mean square (EWRMS) chart*. The limits for the EWRMS chart are approximated by

$$\text{UCL} = \sigma_0 \sqrt{\frac{\chi_{\phi/2,f}^2}{f}} \qquad \textbf{(8.49)}$$

and

$$\text{LCL} = \sigma_0 \sqrt{\frac{\chi^2_{1-(\phi/2),f}}{f}} \qquad \textbf{(8.50)}$$

where $\chi^2_{\phi/2,f}$ and $\chi^2_{1-(\phi/2),f}$ denote the $100(1 - \phi/2)$ and $100(\phi/2)$ percentile points of the chi-square distribution with f degrees of freedom, respectively. The number of degrees of freedom is related to the exponential weighting constant λ as follows:

$$f = \frac{2 - \lambda}{\lambda} \qquad \textbf{(8.51)}$$

It should be pointed out that ϕ does not represent an overall false-alarm rate. Using simulation methods, MacGregor and Harris (1993) determined the in-control ARL (that is, the inverse of the overall false-alarm rate) for a handful of (ϕ, λ) combinations. The closest reported combination to the popular choice of 370.4 for an in-control ARL is $(\phi, \lambda) = (0.01, 0.05)$. This combination is expected to give an in-control ARL of 436, or an overall false-alarm rate of 0.0023, which is not too far from 0.0027.

The EWRMS statistic will react not only to shifts in the process variance but also shifts in the process mean. The Hawkins v_i statistic similarly tends to react to both mean and variance changes. Thus, if an out-of-control signal occurs, it may not be clear with only the EWRMS statistic or the Hawkins v_i statistic which process dimension has undergone a possible change. In response to this issue, MacGregor and Harris (1993) devised a statistic closely related to the EWRMS, called the *exponentially weighted moving variance (EWMV)*, which allows for a changing mean. Thus, an out-of-control signal from EWMV is most likely an indication of a variance change. Even though control limits are developed for the EWMV statistic, limited details are provided in their article for control chart design.

Another reasonable solution is to simultaneously monitor the EWRMS statistic or Hawkins' statistic and a statistic dedicated to detecting changes in the process mean. Thus, if only the EWRMS statistic signals, this suggests a change in variance. However, if both charts signal, we suspect that a change in the mean has *minimally* occurred with the possibility of both mean and variance having changed.

To illustrate the EWMA approaches to monitoring process variability, consider the data from Table 8.6 which were used to demonstrate Hawkins' statistic in a CUSUM framework. Recall that observations 1 to 20 were generated from a normally distributed process with $\mu = 20$ and $\sigma = 2$, and observations 21 to 40 were generated from a normally distributed process with $\mu = 20$ and $\sigma = 3$. The data observations, along with the values of Hawkins' v_i statistic, are summarized in Table 8.11.

As suggested, a standard EWMA chart can be applied to the v_i data. Since we will construct an EWRMS chart with design parameters $(\phi, \lambda) = (0.01, 0.05)$, which give rise to a chart

Example 8.7

Table 8.11   Summary of computations for EWMA-related variability monitoring schemes

Sample i	x_i	v_i	$EWMA_i(v)$	$(x_i - 20)^2$	$EWMS_i$	$EWRMS_i$
1	23.5	1.4352	0.28704	12.25	4.4125	2.1006
2	18.5	0.1261	0.25485	2.25	4.3044	2.0747
3	15.5	1.9427	0.59242	20.25	5.1017	2.2587
4	16.9	1.2120	0.71634	9.61	5.3271	2.3081
5	18.1	0.4375	0.66057	3.61	5.2412	2.2894
6	20.6	−0.7859	0.37128	0.36	4.9972	2.2354
7	19.8	−1.4492	0.00718	0.04	4.7493	2.1793
8	20.6	−0.7859	−0.15144	0.36	4.5298	2.1283
9	17.2	1.0350	0.08585	7.84	4.6953	2.1669
10	20.8	−0.5431	−0.03994	0.64	4.4926	2.1196
11	18.0	0.5100	0.07005	4.00	4.4679	2.1137
12	14.4	2.4393	0.54390	31.36	5.8126	2.4109
13	21.9	0.4375	0.52262	3.61	5.7024	2.3880
14	18.7	−0.0452	0.40906	1.69	5.5018	2.3456
15	23.5	1.4352	0.61428	12.25	5.8392	2.4164
16	22.4	0.7835	0.64813	5.76	5.8353	2.4156
17	17.3	0.9739	0.71328	7.29	5.9080	2.4306
18	20.4	−1.0739	0.35585	0.16	5.6206	2.3708
19	21.1	−0.2303	0.23862	1.21	5.4001	2.3238
20	20.5	−0.9226	0.00637	0.25	5.1426	2.2677
21	20.7	−0.6601	−0.12692	0.49	4.9099	2.2158
22	23.1	1.2120	0.14086	9.61	5.1449	2.2682
23	14.2	2.5242	0.61753	33.64	6.5697	2.5631
24	21.6	0.2075	0.53552	2.56	6.3692	2.5237
25	11.4	3.5864	1.14570	73.96	9.7487	3.1223
26	18.1	0.4375	1.00406	3.61	9.4418	3.0728
27	18.5	0.1261	0.82847	2.25	9.0822	3.0137
28	16.6	1.3806	0.93889	11.56	9.2061	3.0342
29	14.4	2.4393	1.23898	31.36	10.3138	3.2115
30	20.3	−1.2456	0.74206	0.09	9.8026	3.1309
31	22.4	0.7835	0.75035	5.76	9.6005	3.0985
32	21.3	−0.0452	0.59124	1.69	9.2050	3.0340
33	17.9	0.5808	0.58915	4.41	8.9652	2.9942
34	17.2	1.0350	0.67832	7.84	8.9089	2.9848
35	16.0	1.6969	0.88204	16.00	9.2635	3.0436
36	24.4	1.8947	1.08457	19.36	9.7683	3.1254
37	18.2	0.3630	0.94026	3.24	9.4419	3.0728
38	21.6	0.2075	0.79370	2.56	9.0978	3.0163
39	22.4	0.7835	0.79166	5.76	8.9309	2.9885
40	14.7	2.3091	1.09515	28.09	9.8889	3.1447

with an in-control ARL of 436, we will also like to select EWMA parameters (L, λ) giving rise to a similar ARL, for comparative purposes. By interpolation of the plots shown in Crowder (1989), we found one possible combination for these parameters to be $(L, \lambda) = (2.9, 0.2)$.

As noted earlier, when the process generating the x_i observations is on target for some mean and variance, the v_i observations will approximately follow a normal distribution with mean 0 and standard deviation 1. Any change in the variance of the process for the observations will result in a change in the *mean* of the distribution for the v_i observations. Thus, we implement a standard EWMA chart on the v_i observations with $\mu_0 = 0$ and $\sigma_v = 1$. Given $v_1 = 1.4352$ and an initial value of $\text{EWMA}_0^v = \mu_0 = 0$, we use (8.40) to obtain the value of the first period's EWMA:

$$\text{EWMA}_1^v = 0.2(1.4352) + 0.8(0) = 0.28704$$

The summary of the remaining computations is given in Table 8.11 under the column titled $\text{EWMA}_i(v)$. Using (8.45) and (8.46), we see the stabilized EWMA limits for v_i are computed to be

$$\text{UCL} = \mu_0 + 2.90\sigma_v \sqrt{\frac{0.20}{2 - 0.20}}$$

$$= 0 + 2.90(1) \sqrt{\frac{0.20}{2 - 0.20}}$$

$$= 0.9667$$

and

$$\text{LCL} = \mu_0 - 2.90\sigma_v \sqrt{\frac{0.20}{2 - 0.20}}$$

$$= 0 - 2.90(1) \sqrt{\frac{0.20}{2 - 0.20}}$$

$$= -0.9667$$

The EWMA chart for the v_i observations is shown in Figure 8.22a. The 25th observation (EWMA_{25}), along with many subsequent plotted values, exceeds the upper control limit indicating a potential out-of-control condition which we know began at period 20. So, the chart was successful in identifying an underlying change in the process variance for the x_i observations.

To implement the EWRMS chart, first we need to establish a baseline or target value σ_0^2 for the variance. In this case, we take the variance of the process prior to the shift as the target variance. Since the standard deviation prior to the shift is 2, $\sigma_0^2 = 4$. As for the values of the design parameters (ϕ, λ), let us select the reported values of $(\phi, \lambda) = (0.01, 0.05)$ in MacGregor and Harris (1993) to roughly match the in-control ARL of the EWMA chart just illustrated. We use (8.48) with $\mu = 20$ and $\text{EWMS}_0 = \sigma_0^2 = 4$ to obtain the first period's EWMS:

$$\text{EWMS}_1 = 0.05(23.5 - 20)^2 + (1 - 0.05)(4) = 4.4125$$

Taking the square root of the above-computed value gives the first period's plotted value for the EWRMS chart, namely, $\text{EWRMS}_1 = \sqrt{4.4125} = 2.1006$. The summary of the remaining computations for the EWRMS chart can be found in Table 8.11.

To construct the control limits, the chi-square degrees of freedom f must be determined. From (8.51), we find

$$f = \frac{2 - 0.05}{0.05} = 39$$

With $\phi = 0.01$, we see from (8.49) and (8.50) that the values of $\chi^2_{39, \phi/2} = \chi^2_{39, 0.005}$ and $\chi^2_{39, 1 - \phi/2} = \chi^2_{39, 0.995}$ are needed. The tables in Appendix Table III can be used, or we can use Minitab to

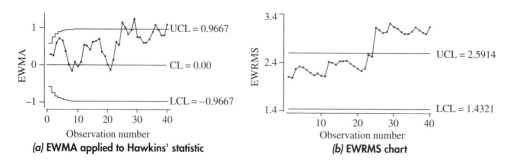

(a) EWMA applied to Hawkins' statistic

(b) EWRMS chart

Figure 8.22   EWMA-related scheme for monitoring variability.

accomplish this task. Using Minitab, we find that $\chi^2_{39,0.005} = 65.4756$ and $\chi^2_{39, 0.995} = 19.9959$. We are now prepared to compute the control limits, using (8.49) and (8.50), and the limits are

$$\text{UCL} = \sigma_0 \sqrt{\frac{\chi^2_{39,0.005}}{39}} = 2 \sqrt{\frac{65.4756}{39}} = 2.5914$$

$$\text{LCL} = \sigma_0 \sqrt{\frac{\chi^2_{39,0.005}}{39}} = 2 \sqrt{\frac{19.9959}{39}} = 1.4321$$

The EWRMS chart is shown in Figure 8.22*b*. This chart quite definitively indicates the out-of-control condition with all the plotted values from period 25 onward falling beyond the upper control limit.

VIOLATIONS OF STANDARD EWMA ASSUMPTIONS

No different from the CUSUM or Shewhart charts, the development of the standard EWMA charts is based on the assumptions of randomness and normality for the in-control process. The question is, How sensitive is a given method, in terms of reduced effectiveness, to departures from these assumptions? We have demonstrated in previous chapters the pitfalls of the Shewhart scheme to randomness and/or normality departures. In a subsection of Section 8.1, we provided evidence that the CUSUM scheme is also susceptible to misleading effects when either one of these standard assumptions is violated.

For an EWMA scheme, when λ approaches 1, the chart approaches a Shewhart chart, and thus inherits the problems associated with Shewhart charts in the presence of assumption violations. On the other end of the spectrum, small values for λ (that is, closer to 0) produce EWMA charts nearly identical, in performance, to CUSUM charts designed for detecting small sustaining shifts (see Lucas and Saccucci, 1990). Actually, this is not an unexpected result. When λ is very small, it turns out that EWMA values divided by a constant are equal to the

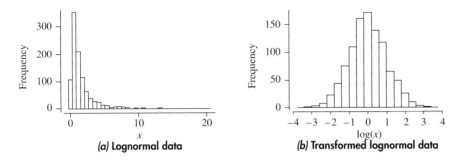

Figure 8.23   Histograms of lognormal and transformed lognormal data.

CUSUM statistic values [as originally defined by (8.6)]. Thus, an EWMA scheme, based on a small λ, will inherit the type of problems associated with violations in assumptions that are unique to the CUSUM scheme. As Mittag and Rinne (1993, p. 501) point out: "the EWMA chart lies between the Shewhart and the CUSUM chart and one can actually choose its position between those two charts."

With respect to the normality assumption, there is a perception that the EWMA is robust to nonnormality. The rationale for this belief is based on an appeal to the central limit theorem mechanism. This argument is reflected by Montgomery (1996, p. 333) who writes, "Since the EWMA can be viewed as a weighted average of all past and current observations, it is very insensitive to the normality assumption." The notion of "very insensitive" implies robustness to even the most extreme cases.

In actuality, EWMA charts can be quite sensitive to departures from normality for reasonable choices of λ. Let us illustrate this point with a set of simulated nonnormal data. Figure 8.23a shows a histogram of 1000 randomly generated observations from a skewed distribution known as the lognormal distribution. (The data are provided in the file "lognormal.dat".) A lognormal random variable is a random variable which, when transformed by the logarithm, is normally distributed. This can be seen from the histogram of the logged transformed data shown in Figure 8.23b.

Since the skewed data are all generated randomly from a single distribution, an appropriate control chart scheme should indicate that the process is in control with the exception of some false alarms expected by design. We purposely simulated a large number of observations to get a better sense of the false-alarm issue in the presence of nonnormality. Suppose we consider applying EWMA charts to the skewed data, for different values of λ, where the corresponding L factor was selected so that a long-run false-alarm rate of 0.0027 is *expected*. On one end of the extreme, if an EWMA chart with $\lambda = 1$ (that is, a Shewhart chart with 3σ limits) were applied, we would find 130 out-of-control signals, all above the upper control limit! This is around 40 times more false alarms than the expected 3 in about 1000. But it can be argued that with $\lambda = 1$, no averaging of the observations takes place to allow for any normalizing effect. To make sure that

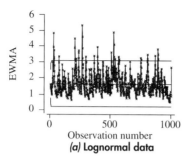

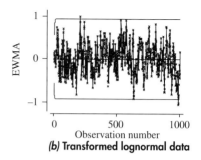

Figure 8.24   EWMA charts applied to lognormal and transformed lognormal data.

more averaging occurs, let us "hop" down to a λ value of 0.2. In Figure 8.24a, the corresponding EWMA chart is presented. The effects of nonnormality are clearly still substantial with a bunching up of the observations below the center-line and many stragglers above the upper control limit. In this case, there are 39 out-of-control signals, again all above one limit. This translates to 13 times more false alarms than expected. Granted, the averaging associated with a smaller λ did ameliorate the detrimental effects somewhat, but obviously not enough to make standard interpretations with the EWMA chart. It is worth pointing out that even smaller values of λ result in little additional improvement.

Obviously, not only does nonnormality have the potential of greatly mis-leading the user in terms of false alarms, but also it can make the chart inef-fective in detecting real out-of-control conditions. Given the bunching of observations on the low side of the present EWMA chart (Figure 8.24a), it is not hard to imagine the difficulty the chart will have in detecting downward shifts.

As emphasized throughout the text, the correct course of action should be based on the basic principles of data analysis. Had we been confronted with the illustrated nonnormal data, we would have sought a normalizing transformation. The likelihood is that the logarithmic, or a closely related, transformation would have been identified to best normalize the data. Once normalized, the EWMA (or any standard chart) can be appropriately applied. In Figure 8.24b, we present the EWMA chart with $\lambda = 0.2$, that is, the same λ value as used in Figure 8.24a. Un-like in Figure 8.24a, the data are now evenly dispersed. There are three points beyond the control limits, which exactly matches expectations. This perspective allows us to correctly conclude that this process is in control.

EXERCISES

8.1. Explain why supplementary runs rules would not be appropriate to apply to the plotted statistics of any of the memory charts discussed in this chapter.

8.2. How should the concept of a false-alarm rate be interpreted in the con-text of memory-based charts?

8.3. Discuss the relative advantages and disadvantages of a Shewhart-based chart versus a memory-based chart.

8.4. Explain how it is possible for a CUSUM chart to signal out of control even if the process is perfectly stable, in the sense of being random and normally distributed over time.

8.5. Discuss the relative advantages and disadvantages of implementing a V mask CUSUM scheme versus a tabular CUSUM scheme.

8.6. It has been suggested in the quality control literature that the 3σ Shewhart chart is a special case of the CUSUM chart when $k = 3$ and $h = 0$. Make up a set of a few numbers to demonstrate, by counterexample, that this is not the case; in particular, give an example in which the CUSUM chart would signal but the Shewhart chart would not. For convenience, you can assume that $\mu_0 = 0$ and $\sigma_{stat} = 1$.

8.7. For a given k value, what is the impact of increasing the h value on the properties of a CUSUM chart?

8.8. For a given k value, what is the impact of increasing the specified in-control ARL on the associated h value?

8.9. Suppose that $\hat{\sigma}_{stat} = 3$ and $h = 5$. If one wanted to implement a FIR CUSUM with a 75 percent headstart, what would be the values of $CUSUM_0^+$ and $CUSUM_0^-$?

8.10. Why is it not reasonable to make headstart values for $CUSUM_0^+$ and $CUSUM_0^-$ greater than H in absolute value?

8.11. Suppose you wish to monitor individual measurements with a two-sided CUSUM chart. Given $k = 0.5$, use the numerical procedure based on Siegmund's method to determine an h value for the following in-control ARLs:

ARL	250	300	370	500	1000
h					

8.12. Consider the shifted data series found in Table 8.1. In Example 8.1, we illustrated the construction of a V mask CUSUM chart with $h = 5$ and $k = 0.5$ for this data series. We found that the chart first signaled out of control at the 23rd sampling period (see Figure 8.6). Using the same design parameters, implement a tabular CUSUM chart and demonstrate that it also first signals out of control at the 23rd sampling period.

8.13. Suppose that a V mask scheme was implemented for a target value of 30. If $CUSUM_{10} = 14.755$ and $CUSUM_{14} = 40.758$, what is the average of the individual observations for periods 11 through 14?

8.14. The chemical diacetyl is an important factor in the flavor and aroma of beer. As such, most brewers routinely sample beer from production to measure for the diacetyl content. For a popular beer produced by a large

Midwestern brewery, the target value is 0.05 mg/L. Based on large-scale consumer testing, it has determined that departures in the mean level by more than 0.03 mg/L are important enough to warrant immediate corrective action. The following data show diacetyl contents for 25 consecutive samples taken at a rate of one sample per shift (data also given in "diacetyl1.dat"):

0.057	0.043	0.021	0.027	0.034	0.049	0.037	0.047	0.075
0.064	0.031	0.030	0.047	0.030	0.049	0.100	0.059	0.027
0.062	0.005	0.053	0.049	0.035	0.058	0.038		

a. Using the moving-range method, estimate the process standard deviation. What is the estimated value of δ?

b. Implement a tabular CUSUM scheme for the process mean such that the in-control ARL is expected to be around 370. Interpret the chart results.

8.15. Continuation of Exercise 8.14: Shortly following the data collected in Exercise 8.14, a new supplier of yeast for the product line was established. Below are 20 new observations (data also given in "diacetyl2.dat"):

0.104	0.063	0.081	0.060	0.085	0.076	0.044	0.045	0.074
0.044	0.028	0.092	0.057	0.057	0.071	0.116	0.047	0.065
0.037	0.086							

Using the estimate for the process standard deviation determined in Exercise 8.14, implement a tabular CUSUM scheme, with the same design parameters of Exercise 8.14, to determine if the process is in control around the target value with the new supplier. Interpret the results of the chart.

8.16. Rework Exercise 8.15, using a FIR CUSUM with 50 percent headstart. Compare the results of the FIR CUSUM with the results of Exercise 8.15.

8.17. Consider the diacetyl data of Exercise 8.15 for the construction of a V mask CUSUM. Design an equivalent V mask scheme to the tabular chart of Exercise 8.15. What are the values of the lead distance design parameter d and the height design parameter H? Plot the cumulative sums and place the V mask on the sampling period that first signaled out of control with the tabular chart of Exercise 8.15. Demonstrate that the V mask also signals out of control at this same sampling period.

8.18. Reconsider the diacetyl data in Exercises 8.14 and 8.15 for establishing Shewhart-based control charts.

a. Using the moving-range method to estimate σ and taking $\mu_0 = 0.05$, establish prospective 3σ limits for the data given in Exercise 8.14.

b. Plot the data given in Exercise 8.15 against the Shewhart limits determined in part a. For these data, compare the performance results of the Shewhart chart with those of the CUSUM chart.

8.19. Deep vein thrombosis, or DVT, describes a condition in which blood clots form in the deep blood vessels of the legs and groin. These blood clots can block the flow of blood from the legs back to the heart. Sometimes, a piece of blood clot is carried by the bloodstream through the heart to a blood vessel, where it lodges and reduces, or blocks, the flow of blood. This life-threatening condition is called an *embolism*. To minimize the possibility of these conditions, physicians prescribe for high-risk patients a daily intake of an anticoagulant medication, the most common being Coumadin®. One measure of the effectiveness of an anticoagulant therapy is called the *prothrombin time* (or *protime*). For high-risk patients, a protime target value of 18 s is commonly recommended. Beyond this target value, protimes that are too high are associated with an unacceptable risk of hemorrhaging, while too low protimes are associated with an unacceptable risk of blood clotting. Suppose that it has been determined that departures in the mean level by more than 1 s are important enough to merit immediate corrective action, such as dosage changes.

Recognizing that there is a general tendency for health care providers to overreact to variation in patient measurements, a medical clinic has decided to embrace an innovative idea of implementing SPC procedures to help detect legitimate changes in patient measurements. You have been hired to implement such procedures. Given the need to detect small shifts in protimes, you have decided to implement a CUSUM scheme. Consider the case of patient X. Because of a chronic condition, this patient is required to have his protime measured once a week. Below are the protime measurements for this patient for 45 consecutive weeks (data also given in "protime.dat"):

17.7	18.4	18.7	18.2	17.6	19.0	18.3	17.9	17.8	18.2
17.0	18.1	17.5	18.3	17.9	17.2	17.4	17.9	18.5	17.9
18.0	18.0	17.2	18.6	17.2	18.5	18.3	17.2	18.2	18.8
17.7	17.9	16.9	17.5	17.6	17.4	16.8	16.5	18.0	17.2
17.1	18.2	16.4	17.7	17.2					

   a. Based on a baseline study of this patient's measurements over an extended period, σ has been estimated to equal 0.61. What is the estimated value of δ?

   b. Implement a tabular CUSUM scheme for the process mean such that the in-control ARL is expected to be around 370. Interpret the chart results. If the CUSUM chart signals out of control, is there a suggestion to decrease or increase the dosage of the anticoagulant medication?

   c. If the CUSUM chart signals out of control, estimate the time of the onset of the shift.

8.20. Reconsider the protime data in Exercise 8.19 for establishing a Shewhart-based control chart:

   a. Using the historically based estimate of σ and taking $\mu_0 = 18.0$, establish 3σ limits for the data given in Exercise 8.19.

   b. Plot the data given in Exercise 8.19 against the Shewhart limits determined in part *a*. For these data, compare the performance results of the Shewhart chart with those of the CUSUM chart.

8.21. Given that Hawkins' statistic v_i is approximately normally distributed with mean 0 and standard deviation 1, what does that imply about the distribution of $\sqrt{|u_i|}$?

8.22. The target value of a part dimension is 5 mm. Based on much past experience, it can be assumed that $\sigma = 0.40$. Below are 30 successive observations from the process (data also given in "part1.dat"):

4.83	5.39	5.90	4.93	5.04	5.12	4.97	5.41	4.52	5.08
5.22	5.17	5.44	4.83	4.52	5.27	5.19	5.50	4.66	4.73
4.85	4.93	5.49	4.88	5.15	5.07	5.55	4.59	5.72	4.85

  a. Implement a tabular CUSUM scheme for the process mean, using $k = 0.50$ and $h = 4.77$. Interpret the results of the chart.
  b. Use Hawkins' CUSUM scheme with $k = 0.50$ and $h = 4.77$ to assess the stability of process variability over time. Interpret the results of the chart.

8.23. Continuation of Exercise 8.22: Not satisfied with the level of variability in the part dimension process, manufacturing personnel conducted a study to reduce overall process variation around the target level. Following the implementation of new production procedures, the following 20 new observations were collected (data also given in "part2.dat"):

5.09	4.95	4.95	5.27	4.80	4.92	4.91	5.01	5.23	5.04
5.01	5.10	4.61	5.00	5.16	4.86	4.92	4.65	4.91	5.21

  a. Implement a tabular CUSUM scheme for the process mean, using $k = 0.50$ and $h = 4.77$ to determine if the process is in control around the target value with the new production system. Note that since there was an attempt to change the process variability, it can no longer be assumed that $\sigma = 0.40$. Use a moving-range-based estimate of σ in your CUSUM construction.
  b. Use Hawkins' CUSUM scheme with $k = 0.50$ and $h = 4.77$ to determine if the intervention to reduce overall process variation was successful.

8.24. One of the responsibilities of a manager of a certain manufacturing facility is to control costs of manufactured goods. Of particular interest to the manager is the weekly unit cost, which is defined as the total cost in dollars of goods manufactured for the week divided by the number of units produced during the week. Below are the weekly unit costs for the 30 most recent weeks (data also given in "unitcost.dat"):

7.03	7.26	6.85	7.25	7.14	6.73	6.91	7.42	7.30	7.25
7.39	7.31	6.70	7.31	6.86	6.47	7.04	6.77	7.43	7.18
6.98	6.83	6.82	6.94	6.54	7.32	6.99	7.36	7.11	6.91

  a. Using the moving-range method, estimate the process standard deviation.

b. Suppose that the manager wishes to hold the weekly unit cost to a target of $7.00. Furthermore, because of slim per-unit profit margins, the manager is concerned with small shifts *upward* in the average unit cost. In particular, the manager wishes to detect quickly a mean shift upward of $0.50 from target. Implement a one-sided CUSUM scheme for the process mean such that the in-control ARL is expected to be around 370. Interpret the chart results.

8.25. For its high-volume copying machine, a 24-h copy center wants to set up a control chart for monitoring the occurrence of failures which require technician response. Based on long experience, it is known that the mean time between failures is 250 h. Below is the number of hours between failures for the last 40 failures (data also given in "copier.dat"):

418.3	384.5	302.2	19.6	160.4	63.0	795.1	84.9	317.1
61.8	114.4	94.7	189.5	112.4	144.9	393.9	110.2	141.1
349.0	69.0	15.3	35.2	239.3	256.0	2.4	31.1	98.6
108.5	27.6	79.0	83.6	73.7	189.9	66.4	18.5	93.8
237.7	49.8	184.6	160.1					

a. Assume that the underlying distribution for failure times is the exponential distribution. Furthermore, suppose that a mean shift down to 150 h is considered unacceptable, and thus requiring quick detection. Using Example 8.3 as a guide, implement a one-sided tabular CUSUM scheme, with an in-control ARL of 500, for this application. Interpret the results of the chart.

b. Based on the same in-control ARL specified in part *a*, implement a one-sided Shewhart chart on the appropriately transformed failure times. Compare the performance results of the Shewhart chart with those of the CUSUM chart constructed in part *a*.

8.26. The following 40 observations represent viscosity measurements (in Krebs) on a chemical process taken every 15 min (data also given in "krebs.dat"):

62.67	64.49	60.44	56.49	52.06	55.57	50.40	57.77	57.43
62.86	62.10	66.06	62.96	64.88	67.80	59.69	60.88	60.24
64.29	63.40	59.80	58.35	61.39	67.24	70.90	67.08	62.91
65.83	71.65	76.61	66.11	66.02	68.38	63.77	58.23	62.60
63.53	63.43	64.67	69.38					

a. In a stable state, viscosity measurements taken every 15 min have been estimated to follow an AR(1) model given by $\hat{x}_t = 24 + 0.6x_{t-1}$. Given this fitted model, determine the fitted values and residual values for observations 2 through 40.

b. If the standard deviation of the residuals for the fitted model is 4.0, then implement a tabular CUSUM scheme for the process mean of the residuals, using $k = 0.50$ and $h = 4.77$ Interpret the chart results.

c. If the CUSUM chart signals out of control, estimate the time of onset of the shift.

8.27. Minitab's default design parameters for the EWMA chart are $\lambda = 0.2$ and $L = 3$. Does such a design offer the same in-control properties as a 3σ Shewhart chart? Explain.

8.28. For a given L value, what are the effects of increasing the λ value on the EWMA scheme?

8.29. Suppose you wish to design an EWMA chart with an in-control ARL of about 370. Refer to an appropriate figure in this chapter to determine the approximate L values for the following λ values:

λ	0.05	0.10	0.20	0.30	0.40
L					

8.30. Suppose you wish to monitor individual measurements with an EWMA chart. Furthermore, suppose you desire an EWMA chart with an in-control ARL of about 370. Determine the values of design parameters λ and L for the following process mean shift sizes (in terms of process standard deviation):

δ	λ	L
1.00		
1.25		
1.50		
1.75		
2.00		

8.31. In Ishikawa (*Guide to Quality Control,* 2nd ed., Asian Productivity Association, Tokyo, Japan, 1986), we find the following subgroup data on the hole diameters for a certain mechanical part (data also given in "ishpart.dat"):

Subgrp	x_1	x_2	x_3	x_4	x_5	$\bar{x}$	R
1	7	24	24	20	25	20.0	18
2	17	37	28	16	26	24.8	21
3	12	22	40	36	34	28.8	28
4	52	35	29	36	24	35.2	28
5	28	28	34	29	48	33.4	20
6	39	27	48	32	25	34.2	23
7	36	21	31	22	28	27.6	15

8	5	33	15	26	42	24.2	37
9	50	34	37	27	34	36.4	23
10	21	17	20	25	16	19.8	9
11	34	18	29	43	24	29.6	25
12	18	35	26	23	17	23.8	18
13	10	28	19	26	21	20.8	18
14	21	23	35	28	38	29.0	17
15	27	41	15	22	23	25.6	26
16	37	19	39	21	38	30.8	20
17	37	46	22	26	25	31.2	24
18	13	32	35	56	45	36.2	43
19	9	51	25	37	39	32.2	42
20	14	27	34	37	52	32.8	38
21	30	51	34	36	28	35.8	23
22	54	31	35	29	25	34.8	29
23	45	21	38	38	31	34.6	24
24	19	31	27	25	38	28.0	19
25	25	45	41	36	43	38.0	20
26	30	24	44	48	38	36.8	24
27	64	32	32	42	42	42.4	32
28	8	58	65	33	39	40.6	57
29	38	37	50	37	33	39.0	17
30	64	38	47	49	41	47.8	26

    a. With these subgroups, Ishikawa presents an $\bar{x}$ chart which first signals out of control at subgroup number 30, that is, the last subgroup. He attributed the out-of-control condition to the introduction of new raw material starting with subgroup number 15. Using a range-based estimate for the process standard deviation, construct an $\bar{x}$ chart and verify Ishikawa's result.

    b. Using a range-based estimate for the process standard deviation, implement an EWMA chart with $\lambda = 0.3$ and $L = 2.9$. Compare the results with those of part *a*.

8.32. Reconsider the diacetyl data in Exercises 8.14 and 8.15 for implementing an EWMA chart.

    a. Given the estimated value of δ determined in Exercise 8.14a, determine the values of design parameters λ and L so that the EWMA chart is expected to have an in-control ARL of about 370.

    b. Combine the data from Exercises 8.14 and 8.15 into a single data series. Given the values of λ and L determined in part *a*, implement an EWMA chart for the data series. Compare the results obtained with the CUSUM chart in Exercise 8.15.

8.33. Reconsider the protime data in Exercise 8.19 for implementing an EWMA chart.

  a. Given the estimated value of δ determined in Exercise 8.19a, determine the values of design parameters λ and L so that the EWMA chart is expected to have an in-control ARL of about 370.

  b. Given the values of λ and L determined in part a, implement an EWMA chart for the data series. Compare the results obtained with the CUSUM chart in Exercise 8.19.

8.34. Analysis of the monthly returns, defined as the monthly percentage changes, of the Dow Jones Industrial Index over the 10-year period of October 1977 to September 1987 would reveal the returns to be random over time. For this 10-year span, the average monthly return was 0.009333 (that is, slightly less than 1 percent). Skipping over the market crash month of October 1987, below are the monthly returns over the subsequent 10-year period of November 1987 to October 1997 (data also given in "dow8797.dat"):

−0.08365	0.05583	0.00995	0.05630	−0.04117	0.02202	−0.00060
0.05302	−0.00608	−0.04668	0.03922	0.01677	−0.01602	0.02524
0.07707	−0.03649	0.01548	0.05314	0.02505	−0.01630	0.08655
0.02839	−0.01637	−0.01789	0.02287	0.01719	−0.06090	0.01407
0.02998	−0.01881	0.07952	0.00140	0.00847	−0.10548	−0.06392
−0.00415	0.04692	0.02850	0.03827	0.05191	0.01093	−0.00896
0.04722	−0.04070	0.03982	0.00619	−0.00885	0.01720	−0.05851
0.09049	0.01707	0.01364	−0.00990	0.03750	0.01118	−0.02334
0.02243	−0.04103	0.00438	−0.01397	0.02416	−0.00123	0.00270
0.01820	0.01890	−0.00220	0.02872	−0.00322	0.00663	0.03109
−0.02668	0.03468	0.00091	0.01886	0.05802	−0.03748	−0.05252
0.01250	0.02061	−0.03614	0.03777	0.03880	−0.01811	0.01675
−0.04418	0.02514	0.00245	0.04258	0.03591	0.03859	0.03275
0.02017	0.03290	−0.02101	0.03799	−0.00704	0.06493	0.00837
0.05294	0.01660	0.01834	−0.00324	0.01322	0.00203	−0.02248
0.01567	0.04627	0.02472	0.07849	−0.01132	0.05503	0.00944
−0.04373	0.06263	0.04492	0.04556	0.06921	−0.07579	0.04148
−0.06543						

An interesting question to ask is, Did the average monthly return of Dow Jones change during the 10-year period of November 1987 to October 1997 relative to the prior 10-year period of October 1977 to September 1987? To investigate this question, use the mean level of the first 10-year period as a mean target value in the implementation of an EWMA chart. In implementation of the EWMA chart, use $\lambda = 0.3$ and $L = 2.9$. Interpret the results of the chart.

8.35. Consider the shifted data series introduced in Section 8.1 and found in Table 8.1. Implement a moving-average control chart with $w = 5$. Compare the results with the CUSUM chart presented in Example 8.1.

8.36. Consider the Ishikawa data in Exercise 8.31, and construct a moving-average control chart with $w = 5$. Compare the results with the EWMA chart constructed in Exercise 8.31.

8.37. Consider the protime data in Exercise 8.19, and construct a moving-average control chart with $w = 3$. Compare the results with the CUSUM chart constructed in Exercise 8.19.

8.38. Reconsider the part dimension data in Exercise 8.23. Implement an EWRMS chart with $\sigma = 0.40$ and $(\phi, \lambda) = (0.01, 0.05)$. Compare the results with the Hawkins CUSUM procedure implemented in Exercise 8.23.

REFERENCES

Adams, B. M., C. Lowry, and W. H. Woodall (1992): "The Use (and Misuse) of False Alarm Probabilities in Control Chart Design," in *Frontiers in Statistical Quality Control 4*, edited by H. J. Lenz, G. B. Wetherill, and P. Th. Wilrich, Physica-Verlag, Heidelberg, Germany.

Alwan, L. C., and H. V. Roberts (1995): "The Pervasive Problem of Misplaced Control Limits," *Applied Statistics*, 44, pp. 269–278.

Barnard, G. A. (1959): "Control Charts and Stochastic Processes," *Journal of Royal Statistical Society,* series B, 16, pp. 151–174.

Bissell, D. (1984): "The Performance of Control Charts and CUSUMs under Linear Trend," *Applied Statistics*, 33, pp. 145–151.

———— (1986): "Corrigendum (to Bissell, 1984)," *Applied Statistics*, 35, p. 214.

———— (1994): *Statistical Methods for SPC and TQM*, Chapman & Hall, London.

Brooks, D., and D. A. Evans (1972): "An Approach to the Probability Distribution of CUSUM Run Length," *Biometrika*, 59, pp. 539–550.

Champ, C. W., and W. H. Woodall (1987): "Exact Results for Shewhart Control Charts with Supplementary Runs Rules," *Technometrics*, 29, pp. 393–399.

Chang, T. C., and F. F. Gan (1995): "A Cumulative Sum Control Chart for Monitoring Process Variance," *Journal of Quality Technology*, 27, pp. 109–119.

Crowder, S. V. (1989): "Design of Exponentially Weighted Moving Average Schemes," *Journal of Quality Technology*, 21, pp. 155–162.

————, and M. D. Hamilton (1992): "An EWMA for Monitoring a Process Standard Deviation," *Journal of Quality Technology,* 24, pp. 12–21.

Fellner, W. H. (1990): "Average Runs Lengths for Cumulative Sum Schemes," *Applied Statistics*, 39, pp. 402–412.

Gan, F. F. (1991): "An Optimal Design for CUSUM Quality Control Charts," *Journal of Quality Technology*, 23, pp. 279–286.

———— (1994): "Design of Optimal Exponential CUSUM Control Charts," *Journal of Quality Technology*, 26, pp. 109–124.

Goldsmith, P. L., and H. Whitfield (1961): "Average Runs Lengths in Cumulative Chart Quality Control Schemes," *Technometrics*, 3, pp. 11–20.

Grant, E. L., and R. S. Leavenworth (1996): *Statistical Quality Control,* 7th ed., McGraw-Hill, New York.

Hawkins, D. M. (1981): "A CUSUM for a Scale Parameter," *Journal of Quality Technology*, 13, pp. 228–231.

———— (1992): "A Fast Accurate Approximation of Average Runs Lengths of CUSUM Control Charts," *Journal of Quality Technology*, 24, pp. 37–42.

———— (1993): "Cumulative Sum Control Charting: An Underutilized SPC Tool," *Quality Engineering*, 5, pp. 463–477.

Johnson, N. L. (1961): "A Simple Theoretical Approach to Cumulative Sum Charts," *Journal of American Statistical Association*, 56, pp. 835–840.

Johnson, R. A., and M. Bagshaw (1974): "The Effects of Serial Correlation on the Performance of CUSUM Tests," *Technometrics*, 16, pp. 103–112.

Lai, T. L. (1974): "Control Charts Based on Weighted Sums," *Annals of Statistics*, 2, pp. 134–147.

Lucas, J. M. (1976): "The Design and Use of Cumulative Sum Quality Control Schemes," *Journal of Quality Technology*, 8, pp. 1–12.

———— (1985): "Counted Data CUSUM's," *Technometrics*, 27, pp. 129–144.

————, and R. B. Crosier (1982): "Fast Initial Response for CUSUM Quality-Control Schemes," *Technometrics*, 24, pp. 109–206.

————, and M. S. Saccucci (1990): "Exponentially Weighted Moving Average Schemes: Properties and Enhancements," *Technometrics*, 32, pp. 1–12.

MacGregor, J. F., and T. J. Harris (1993): "The Exponentially Weighted Moving Variance," *Journal of Quality Technology*, 25, pp. 106–118.

Mittag, H. J., and H. Rinne (1993): *Statistical Methods of Quality Assurance*, Chapman & Hall, London.

Montgomery, D. C. (1996): *Introduction to Statistical Quality Control* (3rd ed.), Wiley, New York.

Nelson, L. S. (1994): "A Control Chart for Parts-per-Million Nonconforming," *Journal of Quality Technology*, 26, pp. 239–240.

Ng, C. H., and K. E. Case (1989): "Development and Evaluation of Control Charts Using Exponentially Weighted Moving Averages," *Journal of Quality Technology,* 21, pp. 242–250.

Page, E. S. (1954): "Continuous Inspection Schemes," *Biometrika*, 41, pp. 100–115.

Rhoads, T. R., D. C. Montgomery, and C. M. Mastrangelo (1996): "A Fast Initial Response Scheme for the Exponentially Weighted Moving Average Control Chart," *Quality Engineering*, 9, pp. 317–327.

Roberts, H. V. (1991): *Data Analysis for Managers with Minitab*, (2nd ed.), Duxbury Press: Belmont, CA.

Roberts, S. W. (1959): "Control Chart Tests Based on Geometric Moving Averages," *Technometrics*, 1, pp. 239–250.

Siegmund, D. (1985): *Sequential Analysis: Tests and Confidence Intervals*, Springer-Verlag, New York.

Vance, L. C. (1986): "Average Runs Lengths of Cumulative Sum Control Charts for Controlling Normal Means," *Journal of Quality Technology*, 18, pp. 180–193.

Wadsworth, H. M., K. S. Stephens, and A. B. Godfrey (1986): *Modern Methods for Quality Control and Improvement*, Wiley, New York.

White, C. H., J. B.Keats, and J. Stanley (1997): "Poisson CUSUM versus *c* Chart for Defect Data," *Quality Engineering*, 9, pp. 673–679.

Woodall, W. H. (1983): "The Distribution of the Run Length of One-Sided CUSUM Procedures for Continuous Random Variables," *Technometrics*, 25, pp. 295–301.

——— (1986): "The Design of CUSUM Quality Control Charts," *Journal of Quality Technology*, 18, pp. 99–102.

———, and B. M. Adams (1993): "The Statistical Design of CUSUM Charts," *Quality Engineering*, 5, pp. 559–570.

Yashchin, E. (1987): "Some Aspects of the Theory of Statistical Control Schemes," *IBM Journal of Research and Development*, 31, pp. 199–205.

——— (1993): "Performance of CUSUM Control Schemes for Serially Correlated Observations," *Technometrics*, 35, pp. 37–52.

Process Capability

CHAPTER OVERVIEW

The concept of process capability was first introduced briefly in Chapter 4. Recall that process capability involves a determination of the extent to which a process can achieve certain operational criteria, taking into account the natural variation within the process. For example, in manufacturing applications, there is often an interest in knowing the likelihood that a process can meet certain quality requirements, defined by a target level along with specification or tolerance limits. The notion of process capability is not restricted to the engineering requirements for some product characteristics. For other processes, such as service processes, measures of response time might reflect their operational capability.

In this chapter, we will present a variety of measures that attempt to indicate the ability of the process to meet specifications. We will also investigate the impact of violations of assumptions underlying process capability measures. Since process capability studies greatly depend on accurate and precise measurements, this chapter will also deal with basic methods for evaluating measurement system performance.

9.1   PROCESS CAPABILITY STUDIES

A *process capability study* is a carefully planned study designed to provide specific information about the performance of a process under certain operating conditions. Depending on the application, a number of potential questions might be asked from a process capability study, such as

- Is the process centered to some prespecified target level?
- Is the level of variability acceptable?
- What proportion of items is acceptable relative to the specifications or to some go/no go or pass/fail assessment of quality?
- What is the rate of occurrence of faults, accidents, or other undesirable events over a unit interval?
- What factors contribute to the variability, defective process outcomes, or other undesirable events?

From this set of common questions, it can be concluded that some process capability studies are based on attribute data, while others are based on variable data.

Attribute process capability studies determine process performance in terms of measures such as the proportion of nonconforming or defective items produced by the process or the rate of occurrence of some undesirable event in some sampling interval. Attribute process capability studies are naturally linked to the information provided from attribute control charts—*p*, *np*, *c*, or *u* charts.

In the case of a p chart, if the data are in control, then process performance can be defined as the average proportion of nonconforming items $\bar{p}$. If we define capability as the achievement of satisfactory performance, then $1 - \bar{p}$, or $100(1 - \bar{p})$ percent, would be reported. With dealing with the capability of a process measured in terms of event data, we turn to c or u charts and their associated statistics of $\bar{c}$ (the average number of occurrences per unit) and $\bar{u}$ (the average number of occurrences per unit based on varying sampling sizes), respectively. Unlike in the p chart case, the average measures of $\bar{c}$ and $\bar{u}$ are not to be interpreted in percentage terms; hence $1 - \bar{c}$ and $1 - \bar{u}$ have little meaning.

One of the limitations of attribute-based capability studies is the lack of specific information on the degree to which an item failed to meet some underlying operational criteria. For this reason, *variable process capability studies* are particularly important to product designers and manufacturing engineers in the study of their own organization's processes or perhaps those of suppliers. Variable-based capability information allows personnel to compare natural variability to specifications and to predict quantitatively how well a process will likcly meet specifications. This helps management in many actions, such as deciding whether to invest in new technologies, planning production schedules, establishing sampling and inspection strategies, and selecting capable suppliers.

At the core of variable process capability studies are three essential components: (1) design specifications, (2) the centering of the process, and (3) the range of the natural variation. Design specifications are typically given by an ideal target level along with specification limits. As stated earlier (see Section 4.8), specification limits define the conformance boundaries for an individual unit from a manufacturing or service operation. These limits may be of the two-sided type with both an upper specification limit (USL) and a lower specification limit (LSL), or the one-sided type with either an upper limit or a lower limit. A critical question is, How does the distribution of the quality characteristic compare with the target level and specification limits? To be more specific, does the mean of the distribution coincide with the target level? Also, how is the distribution spread relative to the distance between specification limits (two-sided case) or the distance from a single specification limit to the target (one-sided case)?

If we can assume normality for the individual measurements, it is customary to summarize the centering of the process, along with the range of natural variation, in terms of *natural tolerance limits*. Natural tolerance limits are symmetric around the mean and chosen to be wide enough that "most" of the distribution is encompassed. The convention is to set the natural tolerance limits (NTLs) at ± 3 standard deviations around the mean. That is,

$$\text{UNTL} = \mu + 3\sigma$$

$$\text{LNTL} = \mu - 3\sigma$$

(9.1)

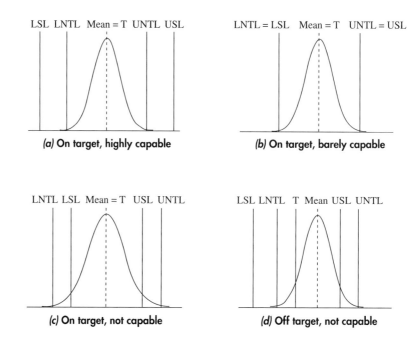

Figure 9.1   Natural tolerance limits versus specification limits.

In practice, parameters μ and σ are typically not known and must be replaced by estimates $\hat{\mu}$ and $\hat{\sigma}$. The choice of estimates for the mean and standard deviation often depends on the nature of the data collection scheme. For instance, if a single random sample is collected, the sample average $\bar{x}$ and sample standard deviation s are reasonable choices for $\hat{\mu}$ and $\hat{\sigma}$, respectively. If relevant data have been already collected for control chart purposes, estimates can be extracted from the control chart information. For example, if subgroups are collected for constructing subgroup control charts, we might use $\hat{\sigma} = \bar{R} / d_2$ or $\hat{\sigma} = \bar{s} / c_4$ as the estimate for σ.

Even though natural tolerance limits are computed in the same manner as Shewhart control limits for *individual* measurements, these sets of limits should be viewed as having separate purposes. Control limits help identify process changes while natural tolerance limits provide an interval associated with most of the distribution. The choice of 3 standard deviations for the different types of limits is, to a certain extent, coincidental. We emphasize the word *individual* because natural tolerance limits and specification limits should be compared only to measurements of the *individual* units of output. To compare an average of a sample of individual units against such limits can be seriously misleading. As an example, consider a case of two individual units both of which are defective, such that one unit is above the upper specification limit and the other unit is below the lower specification limit, each an equal distance away from the target. If we were to look at the average value, it would be exactly on target!

In a sense, natural tolerance limits define the range of *actual* process performance, while specification limits define the range of *allowable* process performance. Figure 9.1 illustrates four possible scenarios that can arise between natural tolerance limits and two-sided specifications. For each scenario, it is assumed that the target level T is equidistant from the upper and lower specification limits. Before we discuss each scenario, it is important to realize that the overall observed variability in the measurements of a quality characteristic will be due to the variability in the product or service itself and due to measurement error variability. Measurement errors cloud the true values of the quality characteristic. Variability in measuring comes from a number of sources, most notably human error and measurement equipment error. In Section 9.7, we will discuss, in greater detail, some approaches for estimating the variability due to measurement error. Until then, we will assume that the measurement errors are negligible, allowing us to focus on the process capability in terms only of the true variability of the quality characteristic in question.

In Figure 9.1*a,* the distribution of the process outcomes is centered on the target level, and the specification limits are wider than the natural tolerance limits. This represents a highly desirable scenario since the items are centered to the target and the probability of a nonconforming item is small, that is, the total area outside of specification limits is small. It is conceivable that management could tighten the specification limits and report its product as being more uniform or consistent than that of its competitors. The organization can rightfully claim that there is little rework and that its customers should experience less difficulty. All this should translate to higher profits and competive edge for the company.

In Figure 9.1*b,* the process is on target with its spread equal to the specification spread; thus the process is capable of meeting specifications, but barely so. Given the normal model, the process is expected to produce approximately 0.27 percent nonconforming items. A primary concern with this situation is that if the process mean moves to the right or left even a little bit, a significant amount of output will exceed one of the specification limits. Similarly, increases in the process variability can easily lead to unacceptable levels of rejected items. Hence, it is critical to monitor the process closely for any shifts in the mean or variance. Under proper conditions, control charts are obviously excellent tools to keep an eye on the process for any shifts.

In Figure 9.1*c,* the range of natural variability is greater than the specification range. Such a process is not capable of producing an acceptable amount of nonconforming product, regardless of whether the process is centered on target. Immediate actions must be taken. The first thing to do is to confirm that the specification limits are valid. Excessively tight specification limits often result from inadequate communication and coordination between product designers and operations personnel. When there is a lack of process capability information, designers have a tendency to be overly cautious and thus set specifications too tightly. Had a system been in place to allow for process data gathering and feedback to the designers from operations, it might be revealed that relaxing specifications will not adversely affect the assembly or use of the product.

If careful examination and communication reveal that the specifications are realistic in the presence of much wider natural tolerance limits, other actions must be considered. The primary goal is to dramatically reduce variability. Opportunities for variation reduction might be revealed from a *design-of-experiments* study. Design of experiments (DOE) is a systematic statistical approach for discovering what variables are most influential in terms of affecting the mean and variability of the observed output. By isolating and controlling the major sources of variability, the overall process variability can be dramatically reduced. Box, Hunter, and Hunter (1978) and Montgomery (1997) provide excellent coverage of introductory DOE principles. Even though DOE can prove to be a powerful approach for improvement efforts, successful improvement stories are most often supported with the use of "basic" quality improvement tools, such as ones described in Chapter 1. Gitlow et al. (1995) illustrate numerous examples in which the application of flowcharting, Pareto charts, and cause-and-effect diagrams within a PDCA framework resulted in bringing an incapable process into one operating well within specification limits.

If improvement efforts (basic or DOE-supported) are not capable of transforming an incapable process to a capable process, management may need to overhaul the system with possible consideration of investing in new technology, new material, or more experienced operators to meet the strict, but required, specifications. Needless to say, a system overhaul can be a financial burden to the organization. Management must carefully assess how critical to the overall customer expectations and satisfaction is the specific quality characteristic and must weigh this criticality against the investment costs. Until substantial improvements are made, the organization might have to contend with the *temporary* possibility of setting up a sorting system that attempts to sort out items that fail to meet specifications.

Finally, in Figure 9.1*d,* the range of natural variability is equal to the specification range, but the process average is off-center relative to the target specification. A common explanation is a faulty machine setting. When measurement errors are not negligible, an off-center distribution might also be due to poorly calibrated measuring equipment. If no corrective action is taken, an unacceptable portion of items will fall outside specification limits.

In Figure 9.1, the renderings of the data distribution curves were idealized as population distribution curves. In practice, data are summarized in the form of a histogram to assess process variation and concentration relative to specification limits. Examples of histograms with specification limits were first introduced in Chapter 2. Since Minitab automatically provides histograms with superimposed specification limits for its process capability routines, we will soon see more examples of the use of histograms in the general assessment of process capability.

9.2   THE NEED FOR PROCESS STABILITY

As a prerequisite for any process capability study, it is a standard recommendation that the process be brought to a state of statistical control, in the classical sense of being random and void of special causes. Clearly, statistics for attribute capability studies, such as $\bar{p}$, $\bar{c}$, and $\bar{u}$, mean little when the process is unstable in the sense of being erratic in mean or variability. Likewise for variable capability studies, estimates for the mean level $\hat{\mu}$ and standard deviation $\hat{\sigma}$ have little interpretative value when the process is unstable.

The next question is, What about a predictable process that is not being influenced by erratic special causes? Recall the Deming elongation example from Chapter 3 in which we witnessed a process trending downward, clearly on a predictable path toward unacceptable product. Deming (1986, pp. 312–313) notes that this application is seriously misleading because the "distribution is fairly symmetric, and both tails fall well within specifications." In other words, the estimated natural tolerance limits fall inside the specification limits, leaving one to believe, quite incorrectly, that the process is capable. Because of the trend, the data descriptors, such as the sample mean and sample standard deviation, do not reflect past or future performance of the process. If, on the other hand, the process is a random process, these single measures of location or spread are appropriate for summarizing process behavior and distribution.

Given the Deming example, it seems that both the conditions of randomness and no special causes present are required for process capability analysis. We, however, would suggest that pure randomness is not a necessary condition. Many predictable nonrandom processes can be viewed as stable and capable. Consider, for instance, the daily weight process of the masters in business administration student, which was found to follow a nonrandom stationary meandering pattern, in particular, an autoregressive lag 1 process or AR(1) process (see Section 3.8). As discussed earlier, this nonrandom component of the process is natural and nonremovable. Now suppose that the individual has a target weight within the neighborhood of 180 to 185 lb and wishes not to be heavier than 190 lb (USL) or lighter than 175 lb (LSL). In Figure 9.2a, the time-sequence plot of the weight series with upper and lower specifications is shown. Because of the stationary nature of the weight series, we can easily see that it would be highly unlikely for the process to drift toward either one of the specification limits. Instead, the process is expected to mcandcr *around* the mean level which is estimated to be 181.61 lb. Hence, the process is highly capable.

It is also interesting to look at this application from the perspective of a distribution summary. In earlier chapters, adamant warnings were given not to construct histograms as a first step of analysis. The primary reason is that a histogram tells nothing about the time-behavior of the process. There are situations, such as the Deming elongation example, in which the histogram may be absolutely misleading because of the long-run implications of the camouflaged nonrandom behavior. However, there are some nonrandom data series which

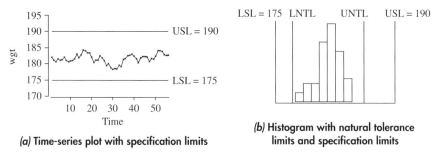

(a) Time-series plot with specification limits

(b) Histogram with natural tolerance limits and specification limits

Figure 9.2   Process capability and stationary nonrandom weight series.

allow for the interpretation of a histogram. In particular, if a nonrandom process is *stationary*, then data can be collapsed into a histogram to answer only *certain* questions.

For example, it can be shown that the overall distribution of observations x_t from a stationary process, such as an AR(1), is the normal distribution if the underlying random errors ε_t are normally distributed. In the next section, we will discuss this result in more specific detail. In Figure 9.2*b,* a histogram of the weight observations is given along with estimated natural tolerance limits and specification limits. This histogram provides a perspective on the *long-run* behavior of the process in terms of location and spread. It appears that the overall distribution, as reflected by the histogram, is relatively symmetric, and the normal distribution can serve as a satisfactory benchmark for overall capability assessments. Given that the natural tolerance limits are well within the specification limits, we again see that the process is highly capable. Indeed, the normal distribution with estimates for overall mean and standard deviation can be used to estimate the rate of nonconforming observations over the long run.

What the histogram cannot reveal is the short-term capability of this process. If the previous day's weight observation is above the mean (in this case, the target), then the next observation is expected to be off-target also. In particular, the probability of a nonconformance is smaller if the process is currently near the overall mean level than if it is at one of the peaks or troughs of the meandering pattern. Hence, the short-term probability of an observation's being out of specification is dependent on the short-term location of the process.

It is worth emphasizing that the fact that the histogram cannot provide clues for short-term forecasts of a nonrandom process is the very reason why we should never start process data analysis with a histogram, as we previously recommended. Effective real-time process monitoring requires an understanding of the time-dependent—predictable—movements of the process. In summary, a process can be viewed as capable from an overall or long-term perspective when the basic condition of stationarity is met. For the special case of a random process, which is a stationary process, the short-term forecast and the long-term forecast for the process level and variability are the same. Hence, the comparison of specification limits with the histogram or natural tolerance limits reflects the expected process capability at any time.

9.3   REPORTING NONCONFORMANCE RATES IN PARTS PER MILLION UNITS

Imagine having the natural tolerance equal to the design tolerance. Under the assumptions of an in-control normally distributed process, this means that only 0.27 percent of the output is expected to fall outside specifications or, equivalently, that 99.73 percent of the output is acceptable. In many organizations, a product error-free rate of 99.73 percent is regarded as quite good. Consider what this level of quality would really mean in societal measures:

- At least 20,000 wrong drug prescriptions each year.
- More than 15,000 newborn babies accidentally dropped by doctors or nurses each year.
- No telephone service for nearly 10 min each week.
- Nearly 500 incorrect surgical operations per week.

Would you be happy with this level of quality? Clearly no, and neither was Motorola. In the 1980s, Motorola was rapidly losing market share in the semiconductor industry to Japanese companies capable of producing similar products at higher quality levels with lower costs. In response, Motorola committed to an ambitious plan in the mid-1980s, called Six Sigma, which is a systematic means of achieving a minimal error-free rate of 99.999966 percent, or 3.4 defects per million. Motorola's ambitious goal has been met in several of its operations, helping Motorola reclaim its stature as one of the exceptional world-class companies in terms of quality and customer satisfaction. The program continues to be the guiding principle at Motorola, and its demonstration that such high levels of quality are possible has inspired hundreds of companies to emulate Motorola's achievements.

There is a practical consequence to reporting rates of nonconformance for highly capable processes. Instead of reporting the nonconformance rates in very small fractions, it is more convenient to report the rates in units of parts per million (ppm). Hence, a nonconformance rate of 0.0027 or 0.27 percent is reported as 2700 ppm. As we will soon see from Minitab's process capability output, Minitab reports nonconformance rates in ppm units.

Before we end our discussion of changing the base of reporting from percentages to ppm units, it is interesting to see how Motorola's 6σ level quality is associated with a minimal nonconformance rate of 3.4 ppm. Motorola characterizes an "ideal" process as one that is on target and has variability so small that the upper or lower specification would be 6 standard deviations away from the mean. The total area under a normal curve outside of 6 standard deviations from the mean is 0.000000002, implying 2 defects per billion parts. This probability is so small that even Minitab doesn't have the built-in precision to compute it.

We still have not answered the question, Where does the value of 3.4 ppm come from? The answer requires us to know that Motorola is willing to tolerate small shifts in the process mean away from the target level. In particular, Motorola accepts the possibility that the process mean might shift as much as 1.5 standard deviations away from the target. Motorola's reasoning is based on its choice to employ only standard SPC charts, such as Shewhart charts, which have better detection capabilities for shifts of 2 standard deviations or more. Thus, it is not unusual for smaller shifts, such as 1.5 standard deviations, to go unnoticed.

Let us now calculate the probability of a nonconformance under the "worst"-case scenario of a process mean shift of 1.5 standard deviations from the target. Assume, for instance, that the process shifted in the direction of the upper specification limit. It is convenient to recognize two facts. First, since the lower and upper specifications are both 6σ from the target level, $USL - LSL = 12\sigma$, implying $(USL - LSL)/2 = 6\sigma$.

Second, when the process is on target, the process mean μ is located halfway between specifications; thus $\mu = (USL + LSL)/2$, which implies that a 1.5σ shift in the direction of USL results in $\mu = (USL + LSL)/2 + 1.5\sigma$. With these facts in mind, the probability of nonconformance (NCR) is

$$NCR = P(X < LSL) + P(X > USL)$$

$$= P\left\{Z < \frac{LSL - \left[\frac{(LSL + USL)}{2} + 1.5\sigma\right]}{\sigma}\right\} + P\left\{Z > \frac{USL - \left[\frac{(LSL + USL)}{2} + 1.5\sigma\right]}{\sigma}\right\}$$

$$= P\left\{Z < \frac{-\left[\frac{USL - LSL}{2}\right] - 1.5\sigma}{\sigma}\right\} + P\left|Z > \frac{\frac{USL - LSL}{2} - 1.5\sigma}{\sigma}\right|$$

$$= P\left(Z < \frac{-6\sigma - 1.5\sigma}{\sigma}\right) + P\left(Z > \frac{6\sigma - 1.5\sigma}{\sigma}\right)$$

$$= P(Z < -7.5) + P(Z > 4.5)$$

We can use Minitab[1] (or Appendix Table I) to infer that $P(Z < -7.5)$ is 0 to at least seven decimal places. We also find that $P(Z > 4.5)$ is 0.0000034. Thus, to the seventh decimal place, the probability of nonconformance is 0.0000034 (=0.0000000 + 0.0000034), or 3.4 ppm.

All our preceding computations of ppm relied on the normal model to be perfectly precise in describing the distribution of the process outcomes at an

---

[1]As we have done often in this book, Minitab can be used to compute the necessary cumulative probabilities. We have done so by choosing the "Cumulative probability" option within the normal distribution option area (Calc >> Probability Distributions >> Normal). Within this option area, there are two ways to input the z values (in this case, -4.5 or -7.5): (1) The z values can be typed into the "Input constant" dialog box. Or (2) first the z values can be placed into a worksheet column, and then the name of this column is placed in the "Input column" dialog box. The second way of inputting ("Input column") should be chosen if the user wishes to have the computations reported beyond the fourth decimal place, as was necessary for us in our 6σ computations.

extreme number of standard deviations from the mean. Bissell (1994, p. 239) comments that "it is ludicrous to stretch the normal model to this extent." Consistent with the Motorola philosophy, a more reasonable interpretation of very highly capable processes is that a small to moderate shift in the process mean will not cause much of a problem. It is more of an interpretation of high safety than a specific number of defects per million or billion.

9.4   PROCESS CAPABILITY INDICES

At its core, process capability analysis is basically an assessment of the relationship between the natural variability of a process and the design specifications. Earlier, we summarized the range of natural variability by constructing natural tolerance limits. The location and width of natural tolerance limits were then visually compared with the design specifications (see Figure 9.1). A common desire of many practitioners is to be able to summarize the relationship of actual process performance relative to allowable process performance in terms of a single summary statistic. Such statistics are referred to as process capability indices or process capability ratios.

Numerous process capability indices have been developed and reported in the quality control literature. Kotz and Johnson (1993) devote an entire highly technical textbook to the wide variety of proposed indices found in the literature. For the less technically inclined, their book can still serve as an excellent source for an extensive bibliography of process capability–related references. In this chapter, we consider five process capability indices most commonly used and supported by Minitab: C_p, C_{pl}, C_{pu}, C_{pk}, and C_{pm}.

Any discussion of process capability indices must be preceded by some warnings. First, computing and reporting a process capability index of any type are meaningless if the process is not stable. This suggests that the process is free of any special-cause influences which result in unpredictable behavior. Additionally, with respect to natural process workings, we have demonstrated that stationarity can serve as a baseline criterion for stability. Second, all standard process capability indices are constructed with an assumption of normality for the process measurements. Substantial departures from the normal distribution undermine the meaningful interpretation of the indices. We discuss further the issue of nonnormality and process capability indices in Section 9.5.

Finally, Berezowitz and Chang (1995) remind us, nicely but convincingly, that even if the statistical assumptions are met, process capability indices are of little value to an organization if the specifications, incorporated in the calculations, do not reflect the ultimate customer needs and satisfaction.

POTENTIAL CAPABILITY INDEX C_p

The most basic version of an index for process capability considers the width of specification limits (allowable process spread) relative to the range of natural variability (actual process spread). Denoted C_p, the index is defined by

$$C_p = \frac{\text{USL} - \text{LSL}}{\text{UNTL} - \text{LNTL}} = \frac{\text{USL} - \text{LSL}}{6\sigma} \qquad (9.2)$$

If the process is not normal, then the estimated natural spread, given by 6σ, is not likely to reflect a range expected to encompass 99.7 percent of possible process observations. In general, proper interpretation of standard process capability indices rests on an assumption of normality.

In Figure 9.3, we show the correspondence between C_p and three scenarios of process performance. Because each of the three scenarios reflects a centered process, they reflect the best possible scenarios, given their respective levels of natural variability. In the top case, the centered process is quite capable since the specification spread is considerably greater than the range of natural variability, implying that the relative ratio of these ranges (C_p) would be greater than 1. In the second case, when $C_p = 1$, the range of natural variability coincides with the specification spread; that is, allowable tolerance equals natural tolerance. Under the assumption of normality, such a process is expected to produce 0.27 percent nonconforming items, or 2700 ppm. Finally, $C_p < 1$ implies that the range of natural variability is greater than the width of specifications. This is the least desirable situation since the process is expected to produce a "high" fraction of nonconforming items.

Clearly, higher values of C_p are more desirable. Montgomery (1996) recommends minimum values for C_p to be 1.33 for existing processes and 1.50 for new processes. Indeed, a C_p value of 1.33 is commonly used by many companies, most notably in the automotive industry, as a minimum contractual requirement of process capability expected from suppliers. For processes associated with issues such as safety, Montgomery recommends that the thresholds be raised to 1.50 for existing processes and 1.67 for new processes.

For centered processes with an underlying normal distribution, one can determine the implied rate of nonconformance for a given value of C_p. First, notice from (9.2) that the distance between specification limits is given by $\text{USL} - \text{LSL} = 6\sigma C_p$. Hence, for a centered process, $\mu - \text{LSL} = 3\sigma C_p$. Using this fact along with a symmetry argument, we find the rate of nonconformance in terms of C_p as

$$\text{NCR} = P(X < \text{LSL}) + P(X > \text{USL})$$

$$= 2P(X < \text{LSL})$$

$$= 2P\left(Z < \frac{\text{LSL} - \mu}{\sigma}\right)$$

$$= 2P\left(Z < \frac{-3\sigma C_p}{\sigma}\right)$$

$$= 2P(Z < -3C_p) \qquad (9.3)$$

For example, the common requirement of 1.33 for C_p implies NCR = $2P(Z < -3.99)$. From Appendix Table I, we find $P(Z < -3.99) = 0.00003304$.

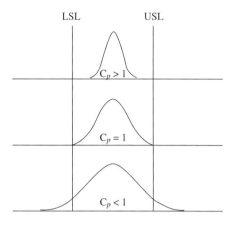

Figure 9.3   Illustrations of C_p and different capability levels.

Thus, for a centered process with a C_p of 1.33, the nonconforming rate is 0.0000661 ($= 2 \cdot 0.00003304$), or 66.1 ppm.

ESTIMATING PROCESS STANDARD DEVIATION

Referring to (9.2), we see that the computation of C_p requires a value for σ. When process capability analysis is done in conjunction with subgroup control charts, estimates such as $\bar{R}/d_2$ or $\bar{s}/c_4$ can be used. These estimates of standard deviation are based solely on the within-subgroup variation.

An alternative method of estimating the process standard deviation is to ignore the subgrouping of the data and to simply estimate the standard deviation from the single set of all individual measurements. For example, say we have 25 subgroups, each of size 5. Rather than compute $\bar{R}/d_2$ we derive an estimate from the set of 125 individual measurements. When one is dealing with a single set of individual measurements, the most common choice of an estimate for σ is the sample standard deviation s. Another possible choice is the unbiased estimate s/c_4, where the correction factor is based on the sample size n. As noted in Chapter 5, for sample sizes greater than 20, c_4 is well approximated by $(4n - 4)/(4n - 3)$.

If the process is in control and the within-subgroup observations are consistent with the normal distribution, then the standard deviation estimates ($\bar{R}/d_2$ or $\bar{s}/c_4$) based on subgrouping will give nearly equivalent results to the standard deviation estimates (s or s/c_4) based on the single set of individual measurements. On the other hand, if the process is out of control, the single-set estimates will typically be larger than the subgroup-based estimates because the single-set estimates are capturing the extra variation associated with the mean shifts occurring *between* the subgroups. Given this fact, there are some practitioners who prefer the reporting of process capability indices based on a single-set estimate. Their feeling is that such computed indices will more accurately reflect the actual process performance than indices based only on the within-subgroup information.

As we will see, Minitab computes and reports process capability indices based on within-subgroup and single-set estimates of process standard deviation. Minitab refers to indices based on the within-subgroup estimates as *short-term* indices, whereas indices based on the single-set estimates are referred to as *long-term* indices.

It is this author's opinion that the notion of computing a process capability index so as to summarize an out-of-control process with a single number is misguided. If a process is not stable, then the standard deviation estimates (subgroup-based or single-set-based) have no predictive value. If a process is out of control, then as Wheeler and Chambers (1992, p. 134) state, "The computation of the following values [long-term indices] is essentially an exercise in frustration. . . ." The summarization of process capability should occur only after the process is brought to a stable state. When done so, the issue of whether to base a process capability index on a within-subgroup estimate or a single-set estimate becomes a moot point since either type will produce nearly equal results. With this in mind, let us now turn to a couple of examples to illustrate the computation of the process capability index C_p.

Example 9.1

Optimal Optics Inc. is a manufacturer of numerous optical products, such as beam splitters, lenses, mirrors, and prisms. Mirrors are critical components for many optical systems. They can be used to bend, fold, and focus optical light beams. Within its mirror product line, some of the most popular mirrors are the spherical concave mirrors. Coated with aluminum, spherical concave mirrors are typically used for laser beams and concentration imaging. A number of characteristics are important to the overall imaging capability of the spherical mirror, for example, surface quality, coating quality, and various dimensional characteristics. One dimension, with a specified target and limits, is the center thickness (CT), as depicted in Figure 9.4.

One of the spherical concave mirrors produced by Optimal Optics is the SM100A. The SM100A is a 100-mm radius-of-curvature mirror with target center thickness of 6.8 mm and specification limits ±0.2 mm around the target. To assess process capability, center thickness data were collected on 100 randomly chosen spherical mirrors. The data are shown in Table 9.1 and also found in file "mirror.dat". For these data, we find that

$$\bar{x} = 6.7983 \qquad s = 0.079426$$

From these summary statistics, the natural tolerance limits are estimated as:

$$\text{UNTL} = \hat{\mu} + 3\hat{\sigma} = 6.7983 + 3(0.079426) = 7.03658$$

$$\text{LNTL} = \hat{\mu} - 3\hat{\sigma} = 6.7983 - 3(0.079426) = 6.56002$$

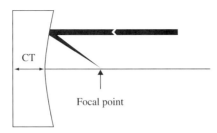

Figure 9.4   Sketch of spherical mirror.

Table 9.1   Center thickness measurements for 100 spherical mirrors (mm)

6.904	6.732	6.594	6.808	6.795	6.802	6.826	6.787	6.665	6.765
6.844	6.706	6.799	6.714	6.767	6.827	6.962	6.702	6.829	6.769
6.767	6.833	6.978	6.713	6.827	6.644	6.853	6.635	6.917	6.761
6.830	6.764	6.832	6.778	6.767	6.711	6.682	6.726	6.891	6.769
6.784	6.922	6.671	6.762	6.977	6.751	6.617	6.822	6.843	6.758
6.785	6.814	6.764	6.773	6.797	6.908	6.715	6.831	6.942	6.781
6.740	6.884	6.820	6.855	6.848	6.797	6.835	6.798	6.832	6.730
6.914	6.829	6.782	6.740	6.999	6.859	6.797	6.802	6.811	6.869
6.910	6.792	6.807	6.870	6.730	6.843	6.773	6.817	6.903	6.767
6.768	6.754	6.668	6.871	6.804	6.776	6.716	6.898	6.858	6.672

These values appear to be "fairly" close to the specification limits of 6.8 ± 0.2, USL $= 7.0$ and LSL $= 6.6$. To gain a more complete perspective of how the data distribution compares with the specifications, we can invoke Minitab's "Capability Analysis" option. In addition to reporting various statistics, this option creates a histogram of the data, along with a number of process capability measures, as shown in Figure 9.5. As can be seen from the histogram, it is reasonable to assume normality for the CT measurements.

Given the estimate for standard deviation ($s = 0.079426$), the estimated value for C_p is

$$\hat{C}_p = \frac{7.0 - 6.6}{6(0.079426)} = 0.8394$$

Minitab's reported $\hat{C}_p$ of 0.84, under the table labeled "Potential (ST) Capability," is indeed the same value we computed, when rounded to the second decimal place. As discussed earlier, Minitab computes short-term (ST) and long-term (LT) indices. The LT indices are based on s/c_4 as an estimate for the process standard deviation. When the sample size is large, as in this example, c_4 will be nearly equal to 1, and hence s and s/c_4 will be close in value.[2] In turn, an

---

[2]Note that Minitab does not have a built-in option for computing the process capability indices based on the sample standard deviation s as the estimate for σ. However, there is an easy solution to this "problem." The user simply needs to first compute s and then type in its value in the dialog box labeled "Historical sigma" found within the general capability analysis option box.

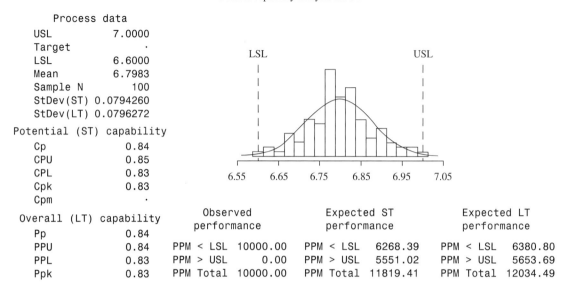

Figure 9.5 Minitab's process capability output for the spherical mirror data.

ST index value will be essentially equal to the corresponding LT index value. To distinguish an LT index from an ST index, Minitab replaces the C symbol with a P symbol. For instance, from Figure 9.5, we find that the short-term index $\hat{C}_p$ and the long-term index $\hat{P}_p$ are both equal to 0.84.

Given the benchmark of $C_p = 1$, an estimated C_p less than 1 would imply an estimated rate of nonconformance greater than 0.27 percent. One way to estimate the nonconformance rate is to substitute $\hat{C}_p$ into (9.3). We suggest, however, a direct approach of estimating the nonconformance rate for each specification. Such an approach does not depend on the process being exactly centered, as required by (9.3).

Given the assumption of normality, the estimate of the nonconformance rate expected from the process is found as follows:

$$\widehat{NCR} = P(X < 6.6) + P(X > 7.0)$$

$$= P\left(Z < \frac{6.6 - 6.7983}{0.079426}\right) + P\left(Z > \frac{7.0 - 6.7983}{0.079426}\right)$$

$$= P(Z < -2.496664) + P(Z > 2.539471)$$

Since the z entries in Appendix Table I are only to two decimal places, we use Minitab to find that $P(Z < -2.496664) = 0.006268$ and $P(Z > 2.539471) = 0.005551$. Referring to Figure 9.5, we notice that Minitab reports the probabilities of nonconformance in terms of parts per million (ppm). For example, we expect 5551.02 items per million to exceed the USL; or equivalently, the probability of this type of nonconformance is $5551.2/1,000,000 = 0.00555102$, which is the number we just found, to more decimal places. Similarly, we expect 6268.39 items per million to exceed the USL; or equivalently, the probability of this type of nonconformance is 6268.39/1,000,000 = 0.00626839, which again is consistent with our computation.

Given the two estimated probabilities, we can add them to obtain an overall estimated rate of nonconformance, which is 0.00555102 + 0.00626839 = 0.01181941, or 11,819.41 ppm. In addition to the estimated rate of nonconforming items, Minitab reports the observed number of nonconforming items in the data set, in terms of ppm. The reported ppm of 0.00 implies no observations fell above USL, while the reported ppm of 10,000 implies that 1 observation out of 100 fell below the LSL.

The previous example demonstrated the computation of the process capability index from a single random sample of n observations. Many times, data used to compute process capability measures originate from control chart applications; that is, the data are collected in the form of subgroups or individual observations over some period.

| | **Example 9.2** |

To illustrate the use of control chart data in process capability index calculations, consider an application taken from a manufacturer of a wide variety of seals, gaskets, and O-rings used in a number of industries, such as automotive, chemical processing, oil refining, medical, and aerospace. Within the O-ring product family, the company produces an aerospace industry class of O-rings known as AS568A. The AS568A O-rings are classified by two important characteristics: cross-sectional width and inside diameter. Visually an O-ring looks like a doughnut. Its cross-sectional width is the thickness of the doughnut ring, and its inside diameter is the diameter distance for the inner circle, that is, the hole. For example, the AS568A no. 427 is specified to have a cross-sectional width of 0.275 in with an inside diameter of 4.725 in. The lower and upper specifications for the inside diameter are 4.692 and 4.758 in, or equivalently 4.725 ± 0.033 in.

The company has established strict quality control procedures for all its O-ring processes. In particular, subgroups of O-rings are gathered and monitored by using standard $\bar{x}$ and R charts. Table 9.2 presents the inside-diameter measurements for 30 subgroups of size 5. These data are also provided in the file "O-ring.dat". Before we assess the capability of the process relative to the specification limits, it is important to establish whether the process is stable. The procedure of constructing $\bar{x}$ and R charts was discussed in Chapter 6. By using Minitab, these subgroup charts can be created from its control chart options. However, when there is an interest in performing process capability in conjunction with SPC analysis, Minitab offers an alternative option called "Probability Capability Sixpack." The output from this option, as applied to the O-ring data, is shown in Figure 9.6. Our attention should first turn to the control charts. For both charts, the data series is stable around the centerline with no signals of out-of-control behavior. Therefore, we can proceed to the question of process capability.

Since process capability analysis is concerned with the individual measurements, as opposed to averages, relative to the specifications, Minitab provides a plot that reveals the individual measurements within each subgroup. There does not appear to be any single point below LSL = 4.692 or above USL = 4.758. Normality checks—a histogram and a normal probability plot—of the 150 (= 5 · 30) individual measurements are also provided because the interpretation of standard process capability indices rests on the assumption of normality for the individual measurements. The data indeed appear consistent with the normal distribution assumption.

Given the control chart results and normality checks, it is now appropriate to consider the process capability summaries found under the title "Potential (ST)" in Figure 9.6. Using the control chart results, we can estimate the parameters of the process as

Table 9.2   Inside diameter of O-rings (in)

Row	d_1	d_2	d_3	d_4	d_5
1	4.7227	4.7274	4.7189	4.7201	4.7201
2	4.7204	4.7133	4.7111	4.7335	4.7279
3	4.7129	4.7219	4.7222	4.7384	4.7235
4	4.7275	4.7302	4.7430	4.7183	4.7221
5	4.7346	4.7314	4.7355	4.7237	4.7293
6	4.7228	4.7197	4.7383	4.7188	4.7114
7	4.7235	4.7289	4.7265	4.7290	4.7214
8	4.7253	4.7197	4.7313	4.7312	4.7197
9	4.7293	4.7449	4.7285	4.7369	4.7324
10	4.7268	4.7306	4.7381	4.7279	4.7259
11	4.7213	4.7340	4.7077	4.7353	4.7381
12	4.7224	4.7356	4.7304	4.7166	4.7213
13	4.7357	4.7241	4.7267	4.7288	4.7363
14	4.7229	4.7160	4.7274	4.7295	4.7221
15	4.7145	4.7231	4.7349	4.7288	4.7253
16	4.7311	4.7262	4.7278	4.7322	4.7304
17	4.7096	4.7229	4.7329	4.7201	4.7299
18	4.7231	4.7262	4.7296	4.7289	4.7167
19	4.7386	4.7366	4.7288	4.7321	4.7183
20	4.7251	4.7400	4.7191	4.7162	4.7288
21	4.7297	4.7125	4.7137	4.7262	4.7206
22	4.7223	4.7120	4.7294	4.7372	4.7200
23	4.7242	4.7282	4.7362	4.7401	4.7347
24	4.7233	4.7259	4.7243	4.7275	4.7360
25	4.7236	4.7243	4.7296	4.7225	4.7368
26	4.7319	4.7255	4.7251	4.7283	4.7251
27	4.7100	4.7234	4.7156	4.7233	4.7370
28	4.7275	4.7161	4.7281	4.7256	4.7263
29	4.7212	4.7198	4.7334	4.7225	4.7241
30	4.7263	4.7379	4.7124	4.7308	4.7200

$$\hat{\mu} = \bar{\bar{x}} = 4.726$$

$$\hat{\sigma} = \frac{\bar{R}}{d_2} = \frac{0.01732}{2.326} = 0.0074463$$

Under the title "Potential (ST)," we see the same value for $\hat{\sigma}$. Thus, the C_p index is estimated to be

$$\hat{C}_p = \frac{4.758 - 4.692}{6(0.00745)} = 1.476$$

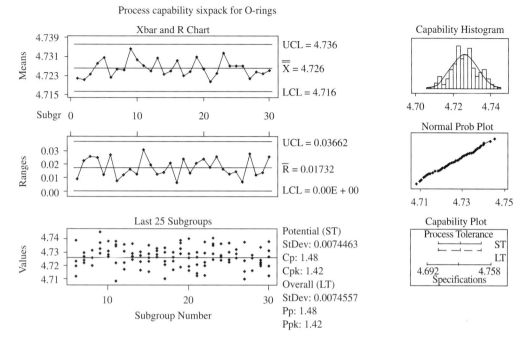

Figure 9.6   Control chart and process capability sumary of the O-ring data.

which, when rounded, is equal to 1.48, as reported by Minitab. In terms of an estimated non-conformance rate, we find

$$\widehat{\text{NCR}} = P\left(Z < \frac{4.692 - 4.726}{0.0074463}\right) + P\left(Z > \frac{4.758 - 4.726}{0.0074463}\right)$$

$$= P(Z < -4.56603) + P(Z > 4.29744)$$

Using Minitab to compute the above probabilities, we find that $P(Z < -4.56603) = 0.0000025$ and $P(Z > 4.29744) = 0.0000086$. Thus, the overall nonconformance rate is estimated to be 0.0000111, or 11.1 ppm.

ONE-SIDED CAPABILITY INDICES C_{pl} AND C_{pu}

The C_p index is constructed on the basis that two-sided specifications are associated with the process in question. Suppose only a single specification limit is given. Then we can define one-sided indices as follows:

$$C_{pu} = \frac{\text{USL} - \mu}{3\sigma} \tag{9.4}$$

$$C_{pl} = \frac{\mu - \text{LSL}}{3\sigma} \tag{9.5}$$

The change in denominator from 6 to 3 standard deviations is to reflect the focus on one specification rather than two. As we will soon see, the denominator change is also convenient because it allows other process indices to be developed in such a manner that comparisons can easily be made among them.

Example 9.3

In Example 9.1, the data were acquired by a random sampling scheme, hence we safely proceeded into the distribution computations without any overriding concerns about possible complications from nonrandom or correlated observations. In Example 9.2, the data were collected over time, but we demonstrated the data to be random and stable by time-series plot and control charts.

Let us now consider an example for which the observations are not random but probability assessments of capability can still be made. The example is taken from a plastics manufacturer which is part of the larger continuous process industry. In continuous process industries, properties of materials—temperatures, pressures, and concentrations—are typically monitored for control purposes. Because of the tendency of these measures to "flow," observations sampled closely together in time tend to also be close together in values; hence correlated observations are commonplace in many continuous process industry applications (see Harris and Ross, 1991).

The plastics company in question manufactures a wide range of extruded plastic sheet products based on a polypropylene polymer. Their main product line is a twin-wall plastic sheet which can be used as an alternative to materials such as cardboard, plywood, and rigid metal. Among the general specifications of the twin-wall product is its tensile strength. In the plastics industry, the testing of polymer properties, such as tensile strength, is typically guided by the American Society for Testing and Materials (ASTM) procedures. Tension is analogous to pulling from opposite directions. The strength of a material is simply defined as the force needed to deform the material. For many materials, a simple operational definition for strength is to determine how much force is required to pull apart the material. This is known as *catastrophic failure*. As covered under ASTM D628, the procedure to test the tensile strength of polymers is generally done by cutting out samples shaped like "dogbones" so that the central third is narrower than the ends. Upon pulling the ends, this ensures that nearly all the deformation will take place in the narrower section.

Table 9.3   Tensile strength measurements (lb/in^2)

3914.38	3854.79	3912.18	3897.53	3877.20	3833.65	3729.96	3766.62
3766.73	3737.39	3699.82	3746.74	3760.01	3707.80	3708.43	3680.99
3745.77	3763.10	3778.25	3726.20	3821.15	3758.56	3784.79	3782.43
3753.51	3762.91	3824.81	3851.97	3796.46	3869.74	3892.00	3833.36
3872.16	3846.46	3821.53	3794.25	3784.44	3756.38	3762.56	3728.69
3759.07	3802.58	3835.13	3835.00	3858.04	3920.79	4027.29	4046.88
3936.88	3956.61	3927.77	4013.05	3845.62	4006.66	3951.08	4010.44
4000.81	3958.64	3977.04	3928.18	3935.04	3870.91	3881.00	3933.67
3941.20	3823.59	3807.01	3783.23	3709.49	3821.28	3768.46	3838.35
3804.91	3786.37	3830.39	3832.24	3800.78	3897.73	3813.81	3748.56
3764.20	3816.62	3831.21	3821.32	3846.30	3866.01	3848.33	3896.03
3901.06	3891.94						

To study the nature of the process and its capability with respect to tensile strength, the company designed a study which called for pulling off sample plastic sheets every hour from a continuous production process. Strength measurements (in pounds per square inch) for 90 consecutive samples are provided in Table 9.3 and are also given in the file "tensile.dat". The product specification calls for a minimal tensile strength of 3700 lb/in^2. Since the data are consecutive over time, the first step is to examine a time-series plot which is given in Figure 9.7a. The figure clearly shows that the data series is not random and seems to exhibit a meandering pattern around the centerline, positioned at the average. The reader is encouraged to further confirm the nonrandomness by conducting a runs test and an ACF test.

The meandering pattern around the mean level suggests that the process is likely to be stationary with no tendency to stray too far from the mean level. Determining whether the process is stationary is of critical importance. If one cannot conclude stationarity, then the implication is that the nonstationary process will most likely drift away, with no tendency around any single mean level. In such a case, long-run process capability measures based on estimates of overall mean and variation levels are not computable. Thus, once we estimate the appropriate model for a data series, it will be important to check if stationarity is implied from the fit.

As demonstrated in previous examples, meandering is reflective of a carryover effect from past observation values. Recall that the PACF indicates what lags are most useful in the prediction of a data series. From Figure 9.7b, significant correlations at lags 1 and 2 are revealed for the data series, tensile$_t$, while all other lags are insignificant. This implies that tensile$_{t-1}$ and tensile$_{t-2}$ together serve as important predictors of tensile$_t$. A model developed on this basis is called a *second-order autoregressive* model, or AR(2) model, and is simply an extension of the AR(1) model found in several earlier applications. In general, an AR(2) model is given by the following:

$$X_t = \tau + \phi_1 X_{t-1} + \phi_2 X_{t-2} + \varepsilon_t \qquad \text{(9.6)}$$

where ε_t is a zero-mean random error process and ϕ_1 and ϕ_2 are constants representing the carryover effect from the previous two periods, respectively. For an AR(1) model, stationarity is implied if the value of the coefficient for the lag 1 term is in absolute value less than 1 (see Section 3.8). The stationarity for an AR(2) model depends on the following conditions for ϕ_1 and ϕ_2:

$$\phi_1 + \phi_2 < 1 \qquad \phi_2 - \phi_1 < 1 \qquad -1 < \phi_2 < 1 \qquad \text{(9.7)}$$

To determine the stationarity conditions for the larger class of time-series models, AR(p) models, for which AR(1) and AR(2) models are special cases, refer to Box, Jenkins, and Reinsel (1994) and Abraham and Ledolter (1983). By using a regression approach, the parameters of the AR(2) model can be estimated by regressing tensile$_t$ on the two independent variables tensile$_{t-1}$ and tensile$_{t-2}$. The estimated regression equation is shown in Figure 9.8a. The t ratios of 5.40 and 2.92 are highly significant with associated p values less than 0.0001 and 0.005, respectively. Residual analysis would show that the residuals are in control, in the sense of random and normal behavior with no outliers. Hence, the model is deemed satisfactory. Notice that the estimates $\hat{\phi}_1 = 0.5563$ and $\hat{\phi}_2 = 0.3002$ satisfy the stationarity conditions of (9.7):

$$0.5563 + 0.3002 = 0.8565 < 1 \qquad 0.3002 - 0.5563 = -0.2256 < 1 \qquad -1 < 0.3002 < 1$$

Thus, given our visual impression of the data meandering along with the AR(2) model stationarity conditions being met, it is safe to assume the tensile strength process is stationary.

Given that a process is stationary, it is possible to summarize the long-run properties of the process by means of a single distribution. Earlier in this chapter, we demonstrated that collapsing the nonrandom, but stationary, weight series into a histogram is acceptable (see

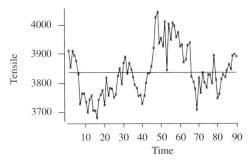

(a) Time-series plot of the tensile strengths

```
PACF of tensile
            -1.0 -0.8 -0.6 -0.4 -0.2   0.0   0.2   0.4   0.6   0.8   1.0
             +----+----+----+----+----+----+----+----+----+----+
  1   0.784                             XXXXXXXXXXXXXXXXXXXXX
  2   0.299                             XXXXXXXX
  3  -0.040                          XX
  4   0.004                           X
  5  -0.084                         XXX
  6  -0.092                         XXX
  7   0.029                          XX
  8  -0.012                           X
  9   0.032                          XX
 10  -0.095                         XXX
 11   0.023                          XX
 12  -0.040                         XX
```

(b) Partial autocorrelation function

Figure 9.7   Time-series checks of the tensile strength series.

Figure 9.2). Theoretically, if the underlying errors (ε_t) are normally distributed, then the overall or long-run distribution of X_t is also normal with mean and variance being some function of the time-series parameters. For an AR(2) process, the mean and standard deviation are given by, respectively,

$$\mu = \frac{\tau}{1 - \phi_1 - \phi_2} \tag{9.8}$$

and

$$\sigma = \sigma_\varepsilon \sqrt{\frac{1}{1 + \phi_2} \frac{1 - \phi_2}{(1 - \phi_2)^2 - \phi_1^2}} \tag{9.9}$$

where σ_ε is the standard deviation of the underlying random error process. Since the residuals of the fitted AR(2) model serve as estimates for the true random errors and are indeed compatible with normality, it is appropriate to assume that the long-run distribution of the tensile process is normal. We have already reported the coefficient estimates for lagged variables, in particular, $\hat{\phi}_1 = 0.5563$ and $\hat{\phi}_2 = 0.3002$. We now need the estimates for τ and σ_ε. From the regression output (Figure 9.8a), the estimate for τ is given by the estimated constant term, which is 550.9. Finally, the standard deviation of residuals which is reported to be 49.56 serves as an appropriate estimate for σ_e. Replacing all the estimates in (9.8) and (9.9), we obtain the following estimates:

$$\hat{\mu}_{\text{tensile}} = \frac{550.9}{1 - 0.5563 - 0.3002} = 3839.02$$

$$\hat{\sigma}_{\text{tensile}} = 49.56 \sqrt{\frac{1}{1 + 0.3002} \frac{1 - 0.3002}{(1 - 0.3002)^2 - 0.5563^2}} = 85.64$$

What if the sample mean and standard deviation were computed directly from the tensile data using standard methods? Below are the standard summary statistics:

Variable	N	Mean	Median	Tr Mean	StDev	SE Mean
tensile	90	3837.9	3830.8	3835.0	83.6	8.8

Variable	Min	Max	Q1	Q3
tensile	3681.0	4046.9	3766.7	3896.4

The sample mean of 3837.9 is nearly identical to the AR(2) model-based estimate of 3839.02. For stationary processes, it turns out that either mean estimate is appropriate and both are expected to be nearly identical. With respect to the estimate for the overall process standard deviation, the sample standard deviation estimate of 83.6 is quite close to the model-based estimate of 85.64. The fact that the sample standard deviation is less than the model-based estimate is not solely due to chance. Alwan, Champ, and Maragah (1994) have demonstrated that, for positively autocorrelated stationary time series, the standard sample standard deviation will slightly underestimate the true standard deviation of the process. Hence, we recommend that the model-based estimate of the process standard deviation be used instead of the standard sample standard deviation estimate. Estimates of the process standard deviation, based on ranges or moving ranges, should also be avoided in the presence of nonrandom process behavior. Recall that from the control chart applications of Chapters 5 and 6, range-based estimates of variability can severely underestimate true variability. For example, the moving-range estimator $\overline{MR}/d_2$ would produce an estimate of 37.20 for the process standard deviation, more than 50 percent smaller than the appropriate estimate of 85.64.

Using (9.5), along with $\hat{\mu}_{\text{tensile}} = 3839.02$ and $\hat{\sigma}_{\text{tensile}} = 85.64$, we estimate the one-sided lower process capability index to be

$$\hat{C}_{pl} = \frac{3839.02 - 3700}{3(85.64)} = 0.54102$$

This value, rounded to the second decimal place, is found in Figure 9.8b next to the label "CPL." In general, the rate of nonconformance from a one-sided index in terms of the relevant one-sided capability index value is found as follows:

$$\text{NCR} = P(Z < -3C_{pl}) \quad \text{or} \quad P(Z < -3C_{pu}) \tag{9.10}$$

In this case, we estimate NCR to be $P[Z < -3(0.54102)] = P(Z < -1.62306) = 0.05228803$, or 52,288.03 ppm, which is essentially equal to the ppm number of 52,261.91 seen in Figure 9.8b.[3] Assuming continuation of the basic conditions now being observed, the implication is that the *long-run* rate of nonconforming is estimated to be around 5.23 percent. In the short

---

[3]We have to remember that we are comparing 0.05228803 with 0.05226191. Thus, we are dealing with differences in the probability values at the fifth decimal. Such minor differences are due to our hand rounding of various values used to determine the estimated probability.

```
The regression equation is
tensile = 551 + 0.556 tensile-1 + 0.300 tensile-2

88 cases used 2 cases contain missing values

Predictor     Coef    StDev       T       P
Constant     550.9    256.7     2.15   0.035
tensile-1   0.5563   0.1030     5.40   0.000
tensile-2   0.3002   0.1028     2.92   0.004

S = 49.56    R - Sq = 66.1%  R - Sq (adj) = 65.3%

Analysis of variance

Source           DF      SS       MS       F      P
Regression        2   406755   203378   82.79  0.000
Residual error   85   208810     2457
Total            87   615565

Source           DF   Seq SS
tensile-1         1   385820
tensile-2         1    20936
```

(a) Regression output for fitted AR(2) model

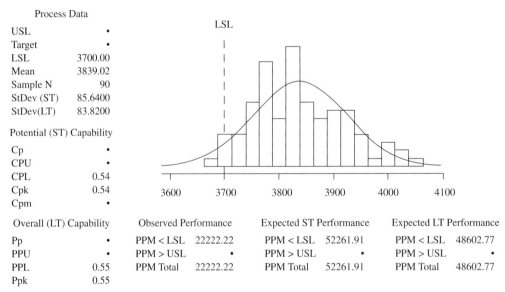

(b) Process capability output

Figure 9.8   Process capability of tensile process based on fitted AR(2) model.

run, the rate of nonconformance changes with the meandering pattern of the process. Hence, the short-run rates may be higher or lower than the long-run rate with the short-run rates averaging out to the long-run rate.

Computation of nonconformance rates from a process capability index value brings to bear a common misconception. Some practitioners implicitly assume that equivalent values for different process capability indices imply equivalent levels of capability. For instance, many practitioners and authors suggest that it is desirable to have the two-sided index or the one-sided index exceed 1. As a result of this recommendation, one may be led to believe that a $C_p = 1$ is associated with similar process performance as a C_{pl} or $C_{pu} = 1$. In terms of nonconformance rates, this inference is incorrect. A centered process with two-sided specifications and with $C_p = 1$ is associated with NCR = 0.0027, or 2700 ppm, while a process with a one-sided specification (say, upper) and with $C_{pu} = 1$ is associated with NCR = 0.00135, or 1350 ppm, that is, one-half the rate of nonconformance.

In addition to being used for applications with only one specification, one-sided capability indices serve in the development of an alternative two-sided capability index C_{pk}, which attempts to address a certain deficiency with the C_p index. In particular, C_p does not take into account the location of the process mean relative to the specifications. This implies that two processes with the same level of natural variability will have the same C_p value regardless of the process mean location.

For illustration, assume for convenience that $\sigma = 1$ and that the specification limits are positioned at ± 6 around a midpoint and target value of 0. Figure 9.9 illustrates four processes with equal C_p values of 2.0, with only one process centered between the specifications while the others are centered off-target and are closer to the upper specification limit. Even though all processes have the same C_p index, the true capability of the other processes is substantially lower. However, the off-center processes have the *potential* of matching the high capability of the first process if the mean can be adjusted to the center of the specification interval. For this reason, C_p is said to be a measure of the potential capability in a process. Only when the process is centered does C_p appropriately represent the actual capability of the process.

PERFORMANCE CAPABILITY INDEX C_{pk}

When the process mean is off-center, the focus of attention is on the one specification that is closest to the process mean, since nonconformities are most likely to occur relative to that specification. With this in mind, a modified capability index C_{pk} was developed to reflect the undesirable condition of a process mean

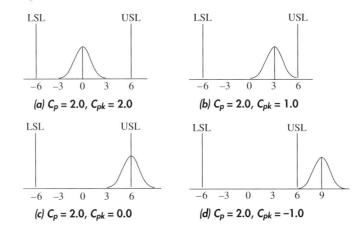

Figure 9.9   Processes with identical C_p values but different performance levels.

positioned away from the center of the specification interval and toward one of the specification limits. To determine which specification limit comes most into play, the computation of C_{pk} requires two calculations, one reflecting the process performance relative to the lower specification limit (C_{pl}) and the other relative to the upper specification limit (C_{pu}). The two-sided index C_{pk} is then defined as the smaller of the two one-sided indices, that is, the worse-case capability:

$$C_{pk} = \min\left(\frac{\mu - \text{LSL}}{3\sigma}, \frac{\text{USL} - \mu}{3\sigma}\right) = \min(C_{pl}, C_{pu}) \tag{9.11}$$

To understand of the relationship between C_p and C_{pk}, consider again the four process scenarios depicted in Figure 9.9. Even though C_p remains constant, we see that the more off-center the process is, the smaller C_{pk} becomes. Hence, whereas C_p measures process potential, C_{pk} attempts to measure the *actual* process capability or performance. To illustrate the required computations, C_{pk} for the process in Figure 9.9*b* is found as follows:

$$C_{pk} = \min\left[\frac{3 - (-6)}{3(1)}, \frac{6 - 3}{3(1)}\right] = \min(3, 1) = 1$$

For any given process, a few general points can be inferred from the panels of Figure 9.9:

1. $C_{pk} \le C_p$ always.
2. $C_p = C_{pk}$ only when the process is centered at the midpoint of the specification interval.
3. $C_{pk} = 0$ whenever the process is centered on either the upper or the lower specification limit.

4. $C_{pk} < 0$ implies that the process is centered below the lower specification limit or above the upper specification limit.

Except when $C_{pk} \leq 0$, it cannot be determined if a process is off-center from the C_{pk} value alone. For $C_{pk} > 0$, the process being off-center is only recognized when its value is found to be different from the C_p value. In fact, it can be shown that C_{pk} can be defined in terms of C_p and a measure of displacement of the process mean from the midpoint of the specification interval. The measure of displacement is captured in the k factor found in the subscript of C_{pk}, where k is given by

$$k = \frac{|m - \mu|}{(USL - LSL)/2} \tag{9.12}$$

where m is the midpoint of the specification interval, that is, $m = (USL + LSL)/2$. With this definition of k, the relationship between C_{pk} and C_p is given by

$$C_{pk} = (1 - k)C_p \tag{9.13}$$

Since $k \geq 0$, it can now be formally seen that C_{pk} can never exceed C_p. Furthermore, we see that $C_{pk} = C_p$ only when $k = 0$, which from (9.12) implies that the process mean μ is positioned at the midpoint of the specification interval m. As an illustration of (9.12) and (9.13), below we show the computation of k and C_{pk} for the four processes with $C_p = 2$ shown in Figure 9.9:

Figure 9.9a: $k = \dfrac{|0 - 0|}{[6 - (-6)] \div 2} = 0$ $C_{pk} = (1 - 0)2 = 2$

Figure 9.9b: $k = \dfrac{|0 - 3|}{[6 - (-6)] \div 2} = 0.5$ $C_{pk} = (1 - 0.5)2 = 1$

Figure 9.9c: $k = \dfrac{|0 - 6|}{[6 - (-6)] \div 2} = 1$ $C_{pk} = (1 - 1)2 = 0$

Figure 9.9d: $k = \dfrac{|0 - 9|}{[6 - (-6)] \div 2} = 1.5$ $C_{pk} = (1 - 1.5)2 = -1$

It is useful to notice that if $LSL \leq \mu \leq USL$, then $0 \leq k \leq 1$.

Because C_{pk}, as opposed to C_p, incorporates a measure of the process location and process variation, many companies prefer to use C_{pk} as a measure of process performance. Many practitioners and authors suggest that a minimal value of 1 for C_{pk} is desirable for internal or supplier processes. Such a recommendation is not without logic. A recommendation for $C_{pk} \geq 1$ ensures that the process has a C_p minimally equal to 1 and ensures that the process mean is within specifications. Given these two facts, the rate of nonconformance is expected to be low. Actually, for a given positive C_{pk}, a bound on the rate of nonconformance can be determined. Using (9.10), we can express the rate of nonconformance for two-sided specifications in terms of one-sided indices as follows:

$$\text{NCR} = P(Z < -3C_{pl}) + P(Z < -3C_{pu}) \qquad \text{(9.14)}$$

Since $C_{pk} = \min(C_{pl}, C_{pu})$, it follows that $C_{pk} \leq C_{pl}$ and $C_{pk} \leq C_{pu}$, which implies that $-3C_{pk} \geq -3C_{pl}$ and $-3C_{pk} \geq -3C_{pu}$. Hence, a bound on (9.14) is given by

$$\text{NCR} \leq P(Z < -3C_{pk}) + P(Z < -3C_{pk}) = 2P(Z < -3C_{pk}) \qquad \text{(9.15)}$$

Therefore, a minimum requirement of $C_{pk} \geq 1$ implies a rate of nonconforming no greater than $2P(Z < -3 \times 1) = 0.0027$.

TARGET-FOCUSED CAPABILITY INDEX C_{pm}

We have found that C_p failed to distinguish between processes of equal variation but different means. In an attempt to address this case, the C_{pk} index was developed. However, as is true with any single numerical measure of a process, some situations can call into question the appropriateness of the C_{pk} measure. For example, consider the two process distributions depicted in Figure 9.10. Each process has the identical C_{pk} value of 1; however, it is clear that we are dealing with two very different processes. Process 1 is centered at target with greater inherent variation than the off-target process 2.

By comparing C_{pk} with C_p, the differences between the two processes would have been revealed. Therein lies the problem: The location of the process cannot be revealed by C_{pk} alone, but only in conjunction with C_p. The pitfalls in C_{pk} as a single measure of process centering stem from the way that location is incorporated into the index. Namely, C_{pk} considers the location of the process in a worst-case scenario, relative to the specifications and not relative to the target value. In an attempt to directly account for the location of the target value, the following index was developed, independently, by Hsiang and Taguchi (1985) and Chan, Cheng, and Spiring (1988):

$$C_{pm} = \frac{\text{USL} - \text{LSL}}{6\sqrt{E[(X - T)^2]}} \qquad \text{(9.16)}$$

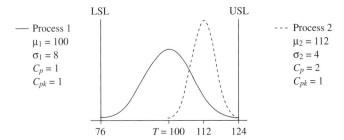

Figure 9.10   Two processes with identical C_{pk} values.

where T is a constant representing the target value. The denominator of the index is proportional to the square root of the expected squared deviation from the target value T. It is useful to rewrite the expected-value term of (9.16). Using the rules of expectation and the definition of variance, we can show that

$$E[(X - T)^2] = E[(X - \mu)^2] + (\mu - T)^2$$
$$= \sigma^2 + (\mu - T)^2$$

Then (9.16) can be rewritten as

$$C_{pm} = \frac{USL - LSL}{6\sqrt{\sigma^2 + (\mu - T)^2}} \tag{9.17}$$

It is clear from (9.17) that the greater the discrepancy between the process mean and the target, the smaller the overall index will be. One nice feature of this index is that there is no implicit assumption that the target is equidistant from the specifications. For some applications, off-center target values are required. The most common situation involves off-center targeting as related to "mating" part dimensions, as noted by Kane (1986). There may be some applications for which we do not wish to have the penalty factor that is symmetric and quadratic around the target. In such cases, Kotz and Johnson (1993) offer a modified C_{pm} which allows for different penalty schemes.

With respect to estimating C_{pm}, Chan, Cheng, and Spiring (1988) and Kotz and Johnson (1993) discuss various alternative estimators and their properties. Based on the recommendation of Spiring (1989), Minitab calculates the following estimate:

$$\hat{C}_{pm} = \frac{USL - LSL}{6\sqrt{\dfrac{\left[\displaystyle\sum_{i=1}^{n}(x_i - T)^2\right]}{(n - 1)}}} \tag{9.18}$$

Let us now see how C_{pm} "views" the two processes with identical C_{pk}, shown in Figure 9.10. We find for process 1 that

$$C_{pm} = \frac{124 - 76}{6\sqrt{8^2 + (100 - 100)^2}} = 1.0$$

Notice that $C_{pm} = C_{pk} = C_p$ for this process. In general, if $\mu = T$, it is seen that (9.17) collapses to the definition of C_p given by (9.2). If, in addition, T is the midpoint of the specification interval, as in this situation, then C_{pm} and C_{pk} will be the same. For process 2, we find that

$$C_{pm} = \frac{124 - 76}{6\sqrt{4^2 + (112 - 100)^2}} = 0.632$$

Thus, in contrast to C_{pk}, C_{pm} reacted to the off-center process as reflected by a substantially smaller index value.

If deviations from target within the specification range are not particularly relevant and, instead, the simple measure of meeting specification is the only concern, then C_{pm} is inappropriate. To demonstrate, below is the computation of the nonconformance rates for the two illustrated processes:

$$\text{NCR}_1 = P\left(Z < \frac{76 - 100}{8}\right) + P\left(Z > \frac{124 - 100}{8}\right) = P(Z < -3) + P(Z > 3) = 0.0027$$

$$\text{NCR}_2 = P\left(Z < \frac{76 - 112}{4}\right) + P\left(Z > \frac{124 - 112}{4}\right) = P(Z < -9) + P(Z > 3) = 0.00135$$

From a perspective of purely meeting the specifications, the off-target process outperforms the centered process. Clearly, the conclusion is that the choice of an appropriate measure depends on the application and objective. Schneider, Pruett, and Lagrange (1995) provide a thorough discussion of uses and misuses of process capability indices in different settings.

9.5   NONNORMALITY AND PROCESS CAPABILITY INDICES

We have noted that the interpretation of process capability indices is based on the assumption of normality for the quality characteristic in question. Numerous studies have demonstrated that the standard interpretation of C_{pk} is inappropriate in the presence of a variety of nonnormal distributions; see Gunter (1989) and Somerville and Montgomery (1996). English and Taylor (1993) expressed similar concerns with the C_p index. An extensive discussion of the effects of nonnormality on process capability indices can be found in Kotz and Johnson (1993).

There are two basic approaches to dealing with nonnormality. One approach is to seek a normalizing transformation so that the standard process capability computations can be carried out on the transformed data relative to the transformed specifications.

The other approach is to directly deal with the nonnormal distribution. If the particular probability distribution is known, given the context of the application, then relevant process capability measures can be calculated directly from the given nonnormal distribution. The other possibility is to fit a mathematical function to the original data such that the function has the properties of a probability density function. Based on the data, the parameters of the function are estimated. This approach is typically based on the use of flexible families of density functions that can be "molded" to the observed data distribution. Some popular families of functions include the gamma, Weibull, Pearson, and Johnson systems. The procedures for selecting and fitting density functions are beyond the scope of this book, but interested readers are referred to Hahn and Shapiro (1967) and Lawless (1982). Rodriguez (1992) and Farnum (1996) provide extensive discussions of density function fitting and the construction of process capability indices.

When one is dealing directly with the nonnormal distribution (known or estimated), the recommendation by numerous practitioners and authors is to replace the normal-based 99.7 percent natural tolerance range ($\mu \pm 3\sigma$) with a generalized natural tolerance range constructed from the percentiles $X_{0.00135}$ and $X_{0.99865}$, which are defined as the points on the nonnormal distribution such that 0.00135 and 0.99865 of the distribution are to the *right* of the points, respectively. Thus, a generalized C_p^* can be defined as

$$C_p^* = \frac{\text{USL} - \text{LSL}}{X_{0.00135} - X_{0.99865}} \qquad (9.19)$$

Because the median M is viewed as a preferable measure of central tendency in the presence of nonnormality, it is commonly recommended (e.g., see Bissell, 1994; and Farnum, 1996) that the one-sided generalized indices be constructed as follows:

$$C_{pl}^* = \frac{M - \text{LSL}}{M - X_{0.99865}} \qquad (9.20)$$

$$C_{pu}^* = \frac{\text{USL} - M}{X_{0.00135} - M} \qquad (9.21)$$

Finally, a generalized C_{pu}^* has been defined as the minimum of the generalized one-sided indices.

UNKNOWN UNDERLYING NONNORMAL DISTRIBUTION | Example 9.4

Consider the following interesting application of a process capability analysis. Plumbing World Inc. runs a highly automated plant in the Midwest. The plant manufactures a wide range of plumbing components, including, but not limited to, sump/ejector pumps, toilet seats, replacement tanks, fill valves, and faucet stems. During the early 1990s, Plumbing World converted its production philosophy to one based on a *just-in-time* (*JIT*) production system, an innovation first introduced by Toyota Motor Co. JIT is a production system in which the movement of goods during production and deliveries from suppliers are carefully timed so that the right parts, in the right quantities, are provided at the right time. The success of a JIT approach depends on the capability of the company's suppliers to meet several criteria on quality, quantity, and delivery performance. Because of this dependency, JIT companies, including Plumbing World, typically have extensive supplier evaluation programs.

Among the performance measures collected on its suppliers, Plumbing World gathers data on the timeliness of deliveries for certain critical parts. Ideally, a supplier should have shipments arrive precisely on a scheduled date. Any shipment made later than the scheduled time is viewed as a late delivery and potentially disruptive to Plumbing World's JIT production schedules. Hence, the scheduled time serves as both a target *and* an upper specification

Table 9.4   Delivery times from supplier

−2.20833	−2.16667	−1.91667	−3.83333	−2.79167	−1.33333	1.58333
−3.45833	−1.70833	−4.95833	−3.83333	−3.70833	−2.62500	−0.62500
−2.45833	−4.95833	−1.87500	−4.29167	−2.25000	−2.04167	−2.16667
−4.54167	−3.33333	−4.00000	−2.54167	0.87500	−4.83333	−3.58333
−2.00000	−3.29167	−1.66667	−3.29167	1.66667	−1.62500	−3.41667
−3.87500	−3.08333	−3.70833	0.66667	6.66667	0.37500	−1.00000
−4.08333	−1.75000	1.79167	−1.54167	−4.37500	−2.20833	−3.41667
−3.50000	−2.79167	2.66667	−3.54167	−3.45833	−2.75000	−3.83333
−3.58333	−3.33333	−3.87500	−4.50000	−4.20833	0.33333	−1.58333
−2.75000	−1.08333	−3.91667	−2.54167	0.95833	−0.45833	−3.20833
−4.45833	0.25000	−2.41667	−2.25000	−4.79167	−3.70833	−2.12500
−2.66667	0.50000	0.91667	−2.83333	−4.20833	−2.41667	−2.50000
−1.08333	−3.87500	−3.12500	−3.95833	−1.50000	−1.16667	−3.75000
−2.79167	−0.25000	−3.83333	−2.50000	3.75000	−4.66667	−3.66667
−0.12500	5.91667					

limit. On the other side, deliveries made early are discouraged and viewed as "bad" timeliness. The reason is the fact that early deliveries translate to extra inventory holdings for Plumbing World and unnecessary costs. In general, Plumbing World has a supplier policy calling for shipments not to be made 5 or more days *prior* to the scheduled date. This early-shipment restriction can be viewed as a lower specification limit.

In Table 9.4, delivery time data on a random sample of 100 different orders from a particular supplier are provided and can be found in the file "jit.dat". The measurements are in units of days, early or late, relative to the scheduled time; negative values represent early shipments, and positive values represent late shipments. Because all shipments are bar-coded, exact arrival times are recorded. The arrivals are compared to 7 a.m. of the scheduled date and can be accepted until closing of the shipping dock at 7 p.m., while production is an around-the-clock operation. The data provided are measured to the nearest 0.5 h. Consider, for example, the first entry in Table 9.4, which is equal to −2.20833. Since −2.20833 is equal to −(2 + 2.5/12), this value represents a shipment that was 2 days and 5 half-hours early, or 2 days and 2.5 h early. So, if the scheduled date is Friday at 7 a.m., the shipment arrived around 4:30 p.m. on Wednesday. Given our explanation above, LSL equals −5 and USL equals 0.

If normality is assumed and standard process capability computations are made, then we have $\hat{C}_{pk} = 0.34$ (see Figure 9.11) and a nonconformance rate estimate of 0.256 (= 0.1574 + .0986). However, the histogram is not consistent with the superimposed normal curve; instead the data are consistent with a nonnormal positively skewed distribution.

Since the particular form of the distribution is not known, a normalizing transformation for the data is a logical alternative. Consider the power transformations introduced in Chapter 3 which are of the form x^p for $p \neq 0$ and $\ln x$ for $p = 0$. Because power transformations can be undefined for zero or for negative values of x, this general class of power transformations is

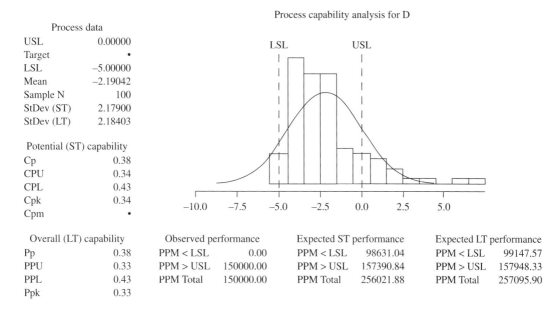

Process data

USL	0.00000
Target	•
LSL	−5.00000
Mean	−2.19042
Sample N	100
StDev (ST)	2.17900
StDev (LT)	2.18403

Potential (ST) capability

Cp	0.38
CPU	0.34
CPL	0.43
Cpk	0.34
Cpm	•

Overall (LT) capability		Observed performance		Expected ST performance		Expected LT performance	
Pp	0.38	PPM < LSL	0.00	PPM < LSL	98631.04	PPM < LSL	99147.57
PPU	0.33	PPM > USL	150000.00	PPM > USL	157390.84	PPM > USL	157948.33
PPL	0.43	PPM Total	150000.00	PPM Total	256021.88	PPM Total	257095.90
Ppk	0.33						

Figure 9.11   Normal-based process capability assessments of the delivery time process.

meaningful only for positive-valued data. However, this stipulation is not a serious restriction, because a single constant K can be added to the data if there are some negative values. In our case, the minimum observation is −4.958; hence a value of 5 or more needs to be added to all the observations before an appropriate power transformation is sought.

The specific choice of the constant value K is not necessarily arbitrary. It turns out that different values of K added to the data result in different choices for the appropriate power transformation. For instance, if 5 is added to the data, a power of 0.337 is found most appropriate. We found by trial and error that $K = 6$ provides the best ultimate results, for the transformed data to be as compatible with the normal distribution as possible. Invoking Minitab's Box-Cox procedure[4] for the shifted data, we find the power of $p = 0$ is optimal, that is, the logarithmic function (see Figure 9.12a). If we define D as the random variable representing delivery time, we are suggesting that the random variable ln $(D + 6)$ approximately follows the normal distribution with estimated mean and standard deviation found below:

```
Variable       N     Mean    Median    Tr Mean      StDev    SE Mean
ln(d+6)      100   1.1985    1.2102     1.1959     0.5249     0.0525

Variable     Min      Max        Q1         Q3
ln(d+6)   0.0408   2.5390    0.8293     1.5018
```

In the transformed scale, the lower specification limit is ln $(−5 + 6) = 0$, and the upper specification limit is ln $(0 + 6) = 1.7918$. We can now estimate C_{pk} as follows

---

[4]There is an option with the "Capability Analysis" option that allows the user to apply a power transformation to the data prior to the computation of the process capability indices.

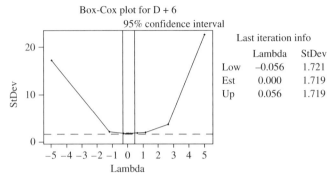

(a) Box-Cox output for choosing power transformation

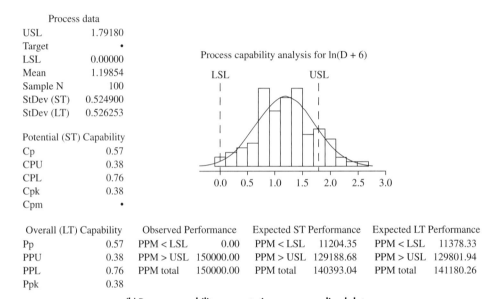

Process data

USL	1.79180
Target	•
LSL	0.00000
Mean	1.19854
Sample N	100
StDev (ST)	0.524900
StDev (LT)	0.526253

Potential (ST) Capability

Cp	0.57
CPU	0.38
CPL	0.76
Cpk	0.38
Cpm	•

Overall (LT) Capability		Observed Performance		Expected ST Performance		Expected LT Performance	
Pp	0.57	PPM < LSL	0.00	PPM < LSL	11204.35	PPM < LSL	11378.33
PPU	0.38	PPM > USL	150000.00	PPM > USL	129188.68	PPM > USL	129801.94
PPL	0.76	PPM total	150000.00	PPM total	140393.04	PPM total	141180.26
Ppk	0.38						

(b) Process capability computations on normalized data

Figure 9.12   Analysis of normalized delivery time data.

$$\hat{C}_{pk} = \min\left[\frac{1.1985 - 0}{3(0.5249)}, \frac{1.7918 - 1.1985}{3(0.5249)}\right] = \min(0.761, 0.377) = 0.377$$

This estimated value, along with the two estimated probabilities of nonconformance relative to the specification limits in the transformed scale, can be found in Figure 9.12b. Namely, the estimated probability of exceeding the upper limit is 0.1292, and the estimated probability of falling below the lower limit is 0.0112, implying that the NCR is estimated to be 0.1404 (= 0.1292 + 0.0112). This estimated rate differs substantially from the value of 0.256 attained from the inappropriate assumption of normality of the original data.

We now turn to another example in which the underlying distribution is known, leading one to possibly consider the generalized nonnormal indices recommended in the literature. Given the gaining popularity of nonnormal process capability indices, it is interesting to put these recommendations to the "test."

KNOWN UNDERLYING NONNORMAL DISTRIBUTION

Example 9.5

For many manufactured products, an important measure of product quality is the probability that the product will perform at an acceptable level for a minimum specified time. This measure of product quality is referred to as the *reliability* of the product. Depending on the nature of the product and conditions of use, a number of probability distributions might be used to model the lifetime of a product, or equivalently the failure rate. One of the simplest distributions for reliability modeling is the exponential distribution first encountered in Chapter 2. In reliability modeling, the exponential distribution is best when it represents the lifetime of systems or products, such as electronic components, that do not seem to age or fatigue over time but rather are subject to a constant chance of failure. In its general form, the exponential density function for lifetime t is given by

$$f(t) = \frac{1}{\mu} e^{-t/\mu} \qquad t \geq 0$$

where the parameter μ is the mean of the distribution and is called the *mean time to failure* and $1/\mu$ is called the *failure rate* of the product or system.

To assess the reliability of a certain electronic component, a life-testing procedure was undertaken, which basically calls for the continuous operation of a random sample of components and the recording of the failure time for each component. For this component a minimum duration requirement of 200 h is expected; that is, the lower specification is 200. In Table 9.5, the lifetimes in hours for a sample of 250 components are presented. The data are also found in the file "lifetime.dat".

Even though the application lends itself to the application of the exponential model, it is always a good idea to confirm the applicability of this model, much as we have done when normality was assumed for the analysis. There are a number of methods for checking whether a set of data is compatible with a particular theoretical distribution. For normality checking, we relied on two graphical procedures: the histogram and normal probability plot. Recall that if the data are from the normal distribution, the normal probability plot should approximate a straight line. The normal probability plot for the lifetime data is provided in Figure 9.13a. It clearly indicates the data to be highly nonnormal, but this does not necessarily imply the use of the exponential distribution.

The concept of the normal probability plot can be extended to any theoretical distribution. For instance, an exponential probability plot is also a plot of the cumulative distribution found in the data set, but now the vertical axis is adjusted to produce a straight line only if the data followed an exact exponential distribution. The exponential probability plot for the lifetime data is presented in Figure 9.13b. In this case, the data are compatible with the theoretical straight line established from the exponential distribution; hence our analysis can safely proceed with the assumption that the exponential distribution models the lifetime data. For more discussions of the testing of data for compatibility with the exponential distribution and other common reliability-related distributions, refer to Hooper and Amster (1990) and Shapiro (1990).

Table 9.5   Lifetimes of an electronic component (h)

1274.9	807.0	578.1	3563.8	57.3	2897.1	7566.5
1252.8	1643.6	301.9	1418.6	10360.5	8793.3	862.8
841.7	203.5	1953.3	660.7	102.4	1092.9	1236.6
47.3	2655.9	2263.6	648.4	1347.0	1283.6	312.3
245.1	3712.9	2440.4	4302.3	5527.4	2510.3	3831.8
6144.4	860.6	745.2	717.7	4117.2	1957.0	1455.0
3096.5	3244.7	5713.1	1939.1	1073.9	2403.5	338.3
774.6	214.1	1417.7	3949.6	37.9	737.0	4928.0

Given the acceptability of an exponential distribution assumption, the question is, What is the estimated proportion of components failing to meet the minimum specification of 200 h? First, we need an estimate of the population mean:

```
Variable     N     Mean    Median   Tr Mean    StDev    SE Mean
lifetime    56     2223     1382      1952      2252      301

Variable    Min     Max      Q1        Q3
lifetime     38   10360     722      3208
```

Hence, the estimated mean lifetime is $\hat{\mu} = 2223$ h. It might be noticed that the sample standard deviation of 2252 is curiously close to the sample mean. This is not accidental because, as we have stated in Chapter 3, the theoretical standard deviation and mean of the exponential distribution are equal.

Let us now consider computing a one-sided lower capability index according to the recommended construction given by (9.20). To do so, we need the estimated sample median $\tilde{x}$ which, from the summary statistics, is 1382 (see above). Below we use Minitab's inverse CDF command to find the required percentile:

```
MTB >   InvCDF 0.00135;
SUBC>     Exponential 2223.

Inverse Cumulative Distribution Function

Exponential with mean = 2223.00

P( X <= x)       x
     0.0014    3.0031
```

Remember, we want the point such that the area to the right is 0.99865, which is equivalent to finding, as Minitab does, the point such that the area to the left is 0.00135. Using (9.20), we find the estimated generalized index value for the lower-specification limit to be

$$\hat{C}_{pl}^* = \frac{1382 - 200}{1382 - 3.0031} = 0.857$$

Given that the nonnormal indices are based on a natural tolerance range defined by the same percentiles as the normal-based counterparts, the *hope* is that the value of a nonnormal-based index can be interpreted comparably had the value been associated with a normal-based index

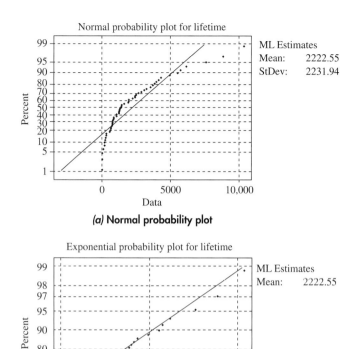

Figure 9.13   Probability plots for lifetime data.

derived from normal data. From (9.10), if the data were normally distributed, a lower-specification index value of 0.857 implies a capability level measured in terms of an estimated nonconformance rate of $P(Z < -3 \times 0.857) = 0.005$, or 0.5 percent.

An exponential distribution with a mean of 2223 along with the region of interest specified above is shown in Figure 9.14a. We can use Minitab's cumulative distribution command with exponential option to directly estimate the nonconformance rate:

```
MTB >   CDF 200;
SUBC>     Exponential 2223.

Cumulative Distribution Function

Exponential with mean = 2223.00

          x       P( X <= x)
   200.0000          0.0860
```

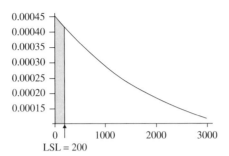

(a) Percent nonconforming and exponential distribution

Process capability analysis for transformed lifetimes

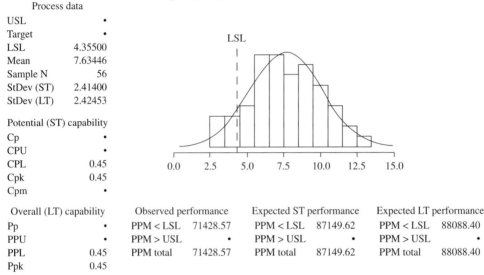

Process data	
USL	•
Target	•
LSL	4.35500
Mean	7.63446
Sample N	56
StDev (ST)	2.41400
StDev (LT)	2.42453

Potential (ST) capability	
Cp	•
CPU	•
CPL	0.45
Cpk	0.45
Cpm	•

Overall (LT) capability		Observed performance		Expected ST performance		Expected LT performance	
Pp	•	PPM < LSL	71428.57	PPM < LSL	87149.62	PPM < LSL	88088.40
PPU	•	PPM > USL	•	PPM > USL	•	PPM > USL	•
PPL	0.45	PPM total	71428.57	PPM total	87149.62	PPM total	88088.40
Ppk	0.45						

(b) Process capability computations on normalized data

Figure 9.14   Estimating the rate of nonconformance.

Hence, the estimated rate of nonconformance is 0.086, or 8.6 percent. Under the assumption of normality, an estimated nonconformance rate of 0.086 is associated with a C_{pl} of 0.455, which differs considerably from 0.857. Thus, in this example, the nonnormal-based index value gives the impression of a much more capable process than the one that actually exists. Our experience suggests that this example is not an anomaly, particularly for one-sided indices. Quite often, interpreting the values of nonnormal-based one-sided indices in a comparable fashion to values from normal-based one-sided indices can be highly misleading. Hence, requiring a nonnormal-based capability index to be greater than 1, as commonly recommended for normal-based indices, may be questionable. Since the nonnormal version of C_{pk} is based on the nonnormal one-sided indices, similar concerns apply to this two-sided index. For more discussions of alternative C_{pk} indices, in the presence of nonnormality, refer to Kotz and Johnson (1993).

To complete this example, let us consider not computing the process capability index directly from the nonnormal exponential data, but instead transforming the data first so that normal-based methods can be used, as was done in Example 9.4. With no knowledge of the application, a power transformation could be sought, based on either the median/quantile approach (see Chapter 3) or the Box-Cox method. Using the Box-Cox method, we would find the power $p = 0.337$ to be recommended. However, given background knowledge and data evidence of an exponential distribution at work, Nelson (1994) demonstrated that $p = 0.2777$ is more appropriate. Given the closeness of the two possibilities, either power choice is bound to give similar results. Using $p = 0.2777$, we present in Figure 9.14b Minitab's process capability output. Note that the lower specification of 200 in the transformed scale is $(200)^{0.2777} = 4.355$. From the output, we see that C_{pl} is reported to be 0.45 (more precisely, 0.452) with an estimated nonconformance rate of 8.71 percent, which is nearly equal to the estimated rate of 8.60 percent obtained directly from the exponential distribution.

We believe this example raises serious questions about the appropriateness of the nonnormal indices. Fortunately, most cases of nonnormality in practice are well handled by an appropriate normalizing transformation; in that case, standard process capability index computations can be employed on the normalized data, usually with good results. However, for "pathological" cases, in which transformations cannot rectify the nonnormality, one must resort to direct computations based on the nonnormal distribution (either known or fitted). In such cases, our advice is that nonnormal-based indices be used only for comparative purposes, between similar nonnormal situations, and *not* for the estimation of nonconformance rates.

9.6   CAPABILITY INDICES AND SAMPLING VARIATION

When one is computing a capability index value from a given sample, it is important to recognize that this computed value is an estimate of the process capability. Even if process conditions do not change, we can expect that an estimate derived from another sample will differ due to sampling variation. Hence, $\hat{C}$ estimators should be viewed as random variables with some underlying probability or sampling distribution. Kushler and Hurley (1992) note that many practitioners tend to rely solely on the single point estimate of process capability, overlooking the importance "to properly account for the uncertainty due to sampling variability in order to make inferences about the true value of the capability index" (p. 189). Rather than simply report a point estimate, it may be more reasonable to consider reporting an interval estimate (i.e., a confidence interval) for the process capability index in question. If an interval estimate for an index is constructed, an interval estimate for a nonconformance rate can also be determined by using the relationships between C indices and NCR.

Because the computation of $\hat{C}_p$ depends on only one estimate (namely, $\hat{\sigma}$), the construction of a confidence interval for C_p is relatively straightforward. If we use the sample standard deviation S as an estimator for σ, then we can utilize the fact (see Section 2.9) that the random variable $(n - 1)S^2/\sigma^2$ follows a chi-square distribution with $n - 1$ degrees of freedom. In Section 2.10, we used this fact to make the following probability statement:

$$P\left(\chi^2_{1-\alpha/2,\,n-1} \le \frac{(n - 1)S^2}{\sigma^2} \le \chi^2_{\alpha/2,\,n-1}\right) = 1 - \alpha$$

where $\chi^2_{1-\alpha/2,\,n-1}$ and $\chi^2_{\alpha/2,\,n-1}$ denote the $100(\alpha/2)$ and $100(1 - \alpha/2)$ percentile points of the chi-square distribution with $n - 1$ degrees of freedom, respectively. Subjecting the above probability statement to some algebraic manipulations, we find

$$1 - \alpha = P\left(\frac{\chi^2_{1-\alpha/2,\,n-1}}{n - 1} \le \frac{S^2}{\sigma^2} \le \frac{\chi^2_{\alpha/2,\,n-1}}{n - 1}\right)$$

$$= P\left(\sqrt{\frac{\chi^2_{1-\alpha/2,\,n-1}}{n - 1}} \le \frac{S}{\sigma} \le \sqrt{\frac{\chi^2_{\alpha/2,\,n-1}}{n - 1}}\right)$$

$$= P\left(\frac{\text{USL} - \text{LSL}}{6}\sqrt{\frac{\chi^2_{1-\alpha/2,\,n-1}}{n - 1}} \le \frac{\text{USL} - \text{LSL}}{6\sigma}S \le \frac{\text{USL} - \text{LSL}}{6}\sqrt{\frac{\chi^2_{\alpha/2,\,n-1}}{n - 1}}\right)$$

$$= P\left(\frac{\text{USL} - \text{LSL}}{6}\sqrt{\frac{\chi^2_{1-\alpha/2,\,n-1}}{n - 1}} \le C_p S \le \frac{\text{USL} - \text{LSL}}{6}\sqrt{\frac{\chi^2_{\alpha/2,\,n-1}}{n - 1}}\right)$$

$$= P\left(\frac{\text{USL} - \text{LSL}}{6S}\sqrt{\frac{\chi^2_{1-\alpha/2,\,n-1}}{n - 1}} \le C_p \le \frac{\text{USL} - \text{LSL}}{6S}\sqrt{\frac{\chi^2_{\alpha/2,\,n-1}}{n - 1}}\right)$$

Thus, it can be seen that for a given data set, the $100(1 - \alpha)$ percent confidence interval is given by

$$\frac{\text{USL} - \text{LSL}}{6s}\sqrt{\frac{\chi^2_{1-\alpha/2,\,n-1}}{n - 1}} \le C_p \le \frac{\text{USL} - \text{LSL}}{6s}\sqrt{\frac{\chi^2_{\alpha/2,\,n-1}}{n - 1}}$$

where s is the observed sample standard deviation. Since $\hat{C}_p = (\text{USL} - \text{LSL})/(6s)$, the above confidence interval is, equivalently, given by

$$\hat{C}_p \sqrt{\frac{\chi^2_{1-\alpha/2,\,n-1}}{n - 1}} \le C_p \le \hat{C}_p \sqrt{\frac{\chi^2_{\alpha/2,\,n-1}}{n - 1}} \qquad (9.22)$$

Two critical assumptions underlie (9.22). First, the sample standard deviation s is used as an estimate for the process standard deviation σ. If σ is estimated from control chart subgroup data (e.g., using the range method), then an alternative and approximate method that incorporates correction factors, depending on subgroup size, needs to be considered (see Bissell, 1994). Second, the validity of (9.22) rests on the assumption that the quality characteristic being monitored is normally distributed. It is well known that the sampling distribution behavior of S can change dramatically away from the chi-square distribution for even moderate departures from normality.

Suppose hypothetically a process is centered approximately at the midpoint of a specification interval bounded below by LSL = 100 and bounded above by USL = 120, thus allowing for proper interpretation of a computed C_p index value. Suppose also that, based on a random sample of 30 observations, the sample standard deviation s was computed to be 2.35. Hence, we find that

Example 9.6

$$\hat{C}_p = \frac{120 - 100}{6(2.35)} = 1.42$$

If the company owning this process required a capability of 1.33 for C_p, the estimated value of 1.42 would seem quite reassuring. Let us now consider going beyond a point estimate to the construction of a 95 percent confidence interval for C_p. From (9.22), we need the values for $\chi^2_{0.975, 29}$ and $\chi^2_{0.025, 29}$, which are the points on the chi-square distribution such that the areas to the right are 0.975 and 0.025, respectively. Below we use Minitab[5] to find the required percentile points:

```
MTB > name c1 'probability' c2 'chi-value'
MTB >    InvCDF c1 c2;
SUBC>      Chisquare 29.
MTB > print c1 c2

Data Display

Row    probability    chi-value

 1          0.025       16.0471
 2          0.975       45.7223
```

Therefore, the 95 percent confidence interval for C_p is given by

$$1.42\sqrt{\frac{16.0471}{29}} \leq C_p \leq 1.42\sqrt{\frac{45.7223}{29}}$$

$$1.06 \leq C_p \leq 1.78$$

The confidence interval provides another "feel" of the situation versus simply reporting a point estimate of 1.42. Once taking sampling variation into account, we now learn that the possibility that the true C_p is as low as 1.06 or as high as 1.78 cannot be rejected based on a 5 percent level of significance. In particular, there is no evidence against the possibility that the true C_p equals certain values less than 1.33, for example, 1.1 or 1.2. On the other hand, it is quite plausible that the true C_p equals some value greater than 1.33, but not greater than 1.78.

For any of the other indices (C_{pl}, C_{pu}, C_{pk}, or C_{pm}), there is no definitive method to obtain a confidence interval. Many different authors have suggested a variety of approximate confidence intervals, each with pros and cons. See Kotz and Johnson (1993) for a summary of proposed confidence intervals along with

| [5]Remember Minitab's inverse CDF option finds points associated with areas to the *left* of the points.

relevant references. One version applicable to C_{pl}, C_{pu}, and C_{pk} that has gained some general acceptance was proposed by Bissell (1990). Under the conditions of normality for the process measurements and the use of the sample standard deviation estimate s, an approximate $100(1 - \alpha)$ percent confidence interval is given by

$$\hat{C}\left(1 - z_{\alpha/2}\sqrt{\frac{1}{9n\hat{C}^2} + \frac{1}{2n-2}}\right) \leq C \leq \hat{C}\left(1 + z_{\alpha/2}\sqrt{\frac{1}{9n\hat{C}^2} + \frac{1}{2n-2}}\right) \quad \text{(9.23)}$$

where $z_{\alpha/2}$ is the upper $\alpha/2$ percentage point on the standard normal distribution and C represents C_{pl}, C_{pu}, or C_{pk}.

Example 9.7

If a process is estimated to have a $\hat{C}_{pk} = 2.0$ from a sample of size $n = 50$, then an approximate 95 percent confidence interval for C_{pk} is given by

$$2\left[1 - 1.96\sqrt{\frac{1}{9(50)2^2} + \frac{1}{2(50)-2}}\right] \leq C_{pk} \leq 2\left[1 + 1.96\sqrt{\frac{1}{9(50)2^2} + \frac{1}{2(50)-2}}\right]$$

$$1.59 \leq C_{pk} \leq 2.41$$

With this interval, it is safe to say that the process capability is at a level higher than the common standard of $C_{pk} = 1.33$.

As is true with any confidence interval type, the width of the interval is dependent, among other things, on the sample size n. The sensitivity of the interval to n varies among different types of confidence intervals. Unfortunately, confidence intervals for capability indices when n is "small" are too wide for meaningful insight about the true process capability. For example, if the sample size had been 20 in Example 9.7 which gave rise to the estimate of $\hat{C}_{pk} = 2.0$, the 95 percent confidence interval would be $1.06 \leq C_{pk} \leq 2.94$. Now, the best we can say is that it is quite possible for C_{pk} to be near 1, implying that the process is not so great (relative to, say, 1.33), or the process is highly capable with C_{pk} near 3. In other words, we have a very limited understanding about the actual process capability. As a general rule of thumb, Kotz and Johnson (1993) recommend a minimal sample size of 50 be used to obtain useful confidence interval results.

9.7   MEASUREMENT ERROR

To emphasize the importance of data for quality management or process improvement efforts, there is a popular motto which applies: "You don't know what you can't measure!" We offer a modification to this motto: "How much you know depends on how well you measure." In this chapter, we have been concerned with characterizing the capability of a process, that is, summarizing the

inherent variability in the quality characteristic in question relative to the specifications. Clearly, any error in measuring the true value of the quality characteristic can affect the ability to judge conformance of a particular item and more
generally the capability of the overall process.

In general, any observed value x_{obs} can be assumed to be equal to the sum of
two parts: the "true" value of the product characteristic x_{prod} and the measurement
error ε. Thus, in equation form, we have

$$x_{\text{obs}} = x_{\text{prod}} + \varepsilon \qquad (9.24)$$

Recall from the stopwatch example, introduced in Chapter 2, that a measurement system is assessed on the basis of two broad categories, generally
known as accuracy and precision. The accuracy of a measurement system was
defined as the extent to which the average of numerous repeated measurements
on the same item differed from the true value. Irrespective of the average value
of the repeated measurements, precision is a measure of the extent of variation
among the repeated measurements, that is, the ability of a measurement system
to replicate identical measurements for the same item. Refer to Figure 2.3 to visualize the concepts of accuracy and precision. In the subsections which follow,
we consider in greater detail the estimation of the accuracy and precision of a
given measurement system.

ESTIMATING ACCURACY

Suppose one takes n measurements $x_{\text{obs},1}, x_{\text{obs},2}, \ldots, x_{\text{obs},n}$ of a benchmark unit of
known value x_{known}. Benchmark units are reference-unit "yardsticks" used for
calibration purposes. Summing over (9.24) and dividing by n, we obtain

$$\frac{\sum_{i=1}^{n} x_{\text{obs},i}}{n} = \frac{\sum_{i-1}^{n} x_{\text{known}}}{n} + \frac{\sum_{i=1}^{n} \varepsilon_i}{n} \qquad (9.25)$$

where ε_i represents the measurement error associated with the ith measurement.
Rearranging and combining terms in (9.25), we have the following expression
for the average measurement error:

$$\bar{\varepsilon} = \frac{\sum_{i=1}^{n}(x_{\text{obs},i} - x_{\text{known}})}{n} = \bar{x}_{\text{obs}} - x_{\text{known}} \qquad (9.26)$$

If there is no tendency for the measurements to be above or below the true
value, we expect $\bar{\varepsilon}$ to be close to 0. Such a measurement system is viewed as
accurate. An inaccurate or biased measurement system would produce measurements systematically below or above the true value and, thus an average
measurement error significantly different from zero. Inaccuracy is often due to

measuring devices that are out of calibration. Inadequate training in the proper use of measuring devices can be another cause.

Even if the measurement system is accurate, we can expect the sample average $\bar{\varepsilon}$ to differ from zero due to sampling variation. If we represent the underlying population mean of the measurement errors by μ_ε, the underlying question becomes, Can the sample average error be viewed as significantly different from $\mu_\varepsilon = 0$?

For process capability studies, a practical consequence of an inaccurate measurement system is a mean shift of the observed measurements away from the true mean of the quality characteristic. Hence, a process generating a quality characteristic which is intrinsically centered on the target will be viewed as being off-target with displacement equal to the mean of the measurement errors.

Example 9.8

Using a caliper device, 30 repeated width measurements of a benchmark unit of 5-mm width were obtained and are shown below:

5.05	5.06	5.08	5.06	4.91	4.91	4.91	4.97	4.83	4.89
4.94	4.86	4.89	4.88	4.97	4.93	5.00	4.99	4.88	4.95
4.86	4.95	5.03	5.10	4.93	5.06	4.96	4.96	4.95	4.93

The individual measurement errors can be found simply by subtracting 5 from each measurement, which gives the following data:

0.05	0.06	0.08	0.06	-0.09	-0.09	-0.09	-0.03	-0.17	-0.11
-0.06	-0.14	-0.11	-0.12	-0.03	-0.07	0.00	-0.01	-0.12	-0.05
-0.14	-0.05	0.03	0.10	-0.07	0.06	-0.04	-0.04	-0.05	-0.07

A negative error implies that the caliper reading was below the true value, while a positive error implies that the caliper reading was above the true value. Below are the summary statistics for the measurement errors:

Descriptive Statistics

Variable	N	Mean	Median	Tr Mean	StDev	SE Mean
error	30	-0.0437	-0.0500	-0.0454	0.0718	0.0131

Variable	Min	Max	Q1	Q3
error	-0.1700	0.1000	-0.0950	0.0075

To make allowance for sampling variability, we can construct a confidence interval for the underlying population mean of the measurement errors. As shown in Section 2.10, the general $100(1 - \alpha)$ percent confidence interval for the population mean was given by (2.31). In our case, the 95 percent confidence interval for μ_ε is

$$-0.0437 - t_{0.025,\,29}\frac{0.0718}{\sqrt{30}} \leq \mu_\varepsilon \leq -0.0437 + t_{0.025,\,29}\frac{0.0718}{\sqrt{30}}$$

or

$$-0.0437 - 2.045\frac{0.0718}{\sqrt{30}} \leq \mu_\varepsilon \leq -0.0437 + 2.045\frac{0.0718}{\sqrt{30}}$$

which reduces to

$$-0.0705 \leq \mu_\varepsilon \leq -0.0169$$

Since the confidence interval does not include $\mu_\varepsilon = 0$, it appears that the caliper is biased in the negative direction, implying an inaccurate caliper. Formally, we say that the null hypothesis of $\mu_\varepsilon = 0$ is rejected at a 5 percent level of significance. Practically speaking, there is clear evidence that the measuring device is out of calibration.

In the previous example, the accuracy of the measuring device was assessed relative to one particular reference unit. Namely, for a reference unit of 5-mm width, the estimated bias of the caliper is -0.0437 mm. It is quite possible that the magnitude of the bias will change for different reference unit sizes. For instance, the caliper has a negative bias for measuring small units but a positive bias for measuring large units. In measurement system terminology, the caliper is said to exhibit *nonlinearity*. Linearity occurs in a situation in which the bias remains the same over the operating range of the measuring device. The extent of linearity (or lack of) clearly has a direct impact on calibration procedures.

Finally, measurement system accuracy can change over time even when the reference unit remains fixed. A measurement system is said to have *stability* if there is no discernible change in overall accuracy as time progresses. Drifts in calibration, or cyclic effects, due to ambient temperature or humidity are common causes of instability. Detailed discussions of the issues of linearity, stability, and other features of a measurement system can be found in *Measurement Systems Analysis: Reference Manual* (1995), jointly published by Chrysler, Ford, and General Motors.

ESTIMATING PRECISION COMPONENTS: REPEATABILITY AND REPRODUCIBILITY

Questions of accuracy deal with the mean level of the measurement errors, while precision deals with the variability of the measurement errors. The variability observed in measured values is due, in part, to the variability of the product and, in part, to the variability inherent in the measurement system.

Assuming that the random variables for the product quality and the measurement error given in (9.24) are independent (which is likely to be the case), we can write

$$\sigma^2_{\text{overall}} = \sigma^2_{\text{prod}} + \sigma^2_\varepsilon \tag{9.27}$$

where $\sigma^2_{\text{overall}}$ is the overall total observed variance, σ^2_{prod} is the variance due to the product (process), and σ^2_ε is the variance due to the measurement system.

The primary goal is to design a study to separate and estimate the two variance components of the right-hand side of (9.27). This will enable us to assess the general capability of the measurement system, in terms of what proportion of the overall variation is due to the measurement errors relative to the inherent variation in the process. We then have an opportunity to assess the true capability of the process, in terms of the product itself, unclouded by the measurement errors.

Obtaining a realistic estimate of σ_ε^2 requires an understanding of two potential sources for measurement errors:

1. *Repeatability:* Even if a particular individual, using the same measuring device, measures the same item, variation in the measurement readings is expected. This variation is referred to as *repeatability,* or *within-operator variation.* Typically, poor repeatability is due to the measuring device itself; hence repeatability is sometimes called *equipment variation.*

2. *Reproducibility*: When different operators use the same measuring device on the same item, variation again can be expected. This type of variation, called *reproducibility,* or *between-operator variation,* reflects the inability of operators to reproduce or match the results of other operators. Training problems and unclear measurement procedures are common explanations for reproducibility problems.

Given these two sources of variation, measurement error variance can be expressed as the combined effect of repeatability and reproducibility errors, namely,

$$\sigma_\varepsilon^2 = \sigma_{\text{repeat}}^2 + \sigma_{\text{repro}}^2 \tag{9.28}$$

The planning for a statistical study of measurement errors requires that we address three basic questions:

- How many operators (k) will be involved in making measurements?
- How many similar-type parts (m) will be measured?
- How many repeat measurements (n) will be made by each operator?

The number of operators, parts, and measurements can vary. For example,

- *Experiment A:* One operator, one or more parts, several measurements per part ($k = 1, m \geq 1, n > 1$). Such an experiment will provide a measure of consistency of readings on particular parts taken by one person. Since different operators are not considered, this experiment can be referred to as the *repeatability-only* experiment. If operator effects do indeed exist, a repeatability-only experiment will underestimate the measurement error variance.

- *Experiment B:* Several operators, one or more parts, one measurement per part ($k \geq 1, m \geq 1, n = 1$). Here the focus is on the consistency of readings among operators. With $n = 1$, the experiment, called a *reproducibility-only* experiment, does not permit the estimation of repeatability. If significant repeatability exists, a *reproducibility-only* experiment will underestimate the measurement error variance.

- *Experiment C:* Several operators, several parts, several measurements per part ($k > 1, m > 1, n > 1$). This experimental design is most popular since it permits for the estimation of both repeatability and reproducibility. Statistical studies attempting to estimate the separate effects of repeatability and reproducibility are called *gage R&R* studies. Even though a gage (or gauge) is a specific type of measuring device, a gage R&R study generically applies to the study of any type of measurement system.

Once the experimental design has been chosen, there is then the question of deciding what statistical methodology should be used in the analysis of the experimental data. There are basically two approaches for the analysis of measurement data:

1. *Analysis-of-variance (ANOVA) method:* ANOVA is a general framework of statistical techniques for estimating and testing the potential different sources of variability underlying some random variable under study. In a gage R&R study, it is typically assumed that the operators and parts involved in the study are samples taken from a larger population of operators and parts. With this assumption, the appropriate ANOVA model to be considered is called a *two-factor random effects* model, with operators and parts being the two factors. Montgomery and Runger (1993a, b) provide a detailed discussion of the use of the two-factor random effects model for gage R&R studies. Even though an ANOVA method might be considered the definitive approach, these authors also illustrate potential pitfalls with the ANOVA approach when certain estimation procedures are used. An excellent reference, concerning the issues related to the estimation of the components of variance in a two-factor random effects model, is by Searle, Casella, and McCulloch (1992).

2. *Range method:* For much the same historical reasons as for the implementation of many SPC methods, gage R&R studies are typically based on the easy-to-hand-compute range statistic. The most obvious shortcoming of a range-based approach is the less efficient use of the data relative to an ANOVA approach. Ranges only consider the largest and smallest of a given set of data points, while ANOVA estimators (akin to the sample standard deviation) utilize all the observations in a data set. However, as will be demonstrated by the next example, the range approach is attractive because it lends itself naturally to the construction of a control chart which has an interesting interpretation.

Even with the less efficient use of the data by the range approach, both methods will give very similar results for most applications. There is, however, one situation that only the ANOVA approach can handle appropriately, and this is the study of a subcomponent of variability known as an *interaction effect*. In gage R&R studies, this would mean an operator-part interaction. Such interaction reflects some sort of nonindependence between operators and parts. For example, an operator might tend to measure small parts on the high end, whereas large parts tend to be measured on the low end. If significant interaction effects are present, the range method will overlook these effects, resulting in a significant underestimation of the true measurement-error variance. Any suspicion of interaction effects should be grounds for pursing the use of ANOVA methodology. We now turn to an example to flesh out the methods for estimating variance components associated with measurement errors.

Table 9.6   Gage R&R data

| | Operator 1 | | | | Operator 2 | | | | Operator 3 | | | |
| | Measurements | | | | Measurements | | | | Measurements | | | |
Part	1	2	$\bar{x}$	R	1	2	$\bar{x}$	R	1	2	$\bar{x}$	R
1	31.83	31.67	31.75	0.16	32.55	32.07	32.31	0.48	31.99	32.23	32.11	0.24
2	30.31	30.47	30.39	0.16	30.63	30.39	30.51	0.24	30.15	30.15	30.15	0.00
3	29.51	29.83	29.67	0.32	29.83	29.67	29.75	0.16	29.51	29.83	29.67	0.32
4	29.43	29.27	29.35	0.16	29.83	29.35	29.59	0.48	29.27	29.51	29.39	0.24
5	29.91	30.23	30.07	0.32	30.07	29.67	29.87	0.40	29.83	29.27	29.55	0.56
6	30.47	30.31	30.39	0.16	30.87	30.87	30.87	0.00	30.31	30.31	30.31	0.00
7	28.79	28.79	28.79	0.00	29.03	28.71	28.87	0.32	28.23	28.47	28.35	0.24
8	29.99	29.67	29.83	0.32	30.07	29.91	29.99	0.16	29.67	29.83	29.75	0.16
9	28.55	28.63	28.59	0.08	28.95	28.79	28.87	0.16	28.55	28.23	28.39	0.32
10	29.35	29.11	29.23	0.24	29.83	29.59	29.71	0.24	29.59	29.51	29.55	0.08
Overall averages:			29.806	0.192			30.034	0.264			29.722	0.216
			$(\bar{O}_1)$	$(\bar{R}_1)$			$(\bar{O}_2)$	$(\bar{R}_2)$			$(\bar{O}_3)$	$(\bar{R}_3)$

Example 9.9

Consider the experiment in which three operators ($k = 3$) were asked to make two measurements ($n = 2$) on each of 10 different parts ($m = 10$). The data from this experiment are provided in Table 9.6.

The first step is to investigate the question of repeatability. Ideally, an operator has no repeatability problems, and hence each measurement on a given part will be identical. That is, there will be no variation among the repeated measurements for a given part. By using the range statistic as a measure of variability, the range values shown in Table 9.6 represent the difference (in absolute value) between the two measurements made on the same part by a given operator. The average ranges for the three operators are given by $\bar{R}_1$, $\bar{R}_2$, and $\bar{R}_3$. The overall range $\bar{R}$ of the repeated measurements can be found by averaging the 30 individual ranges or by taking the average of the three operator average ranges:

$$\bar{R} = \frac{\bar{R}_1 + \bar{R}_2 + \bar{R}_3}{3}$$
$$= \frac{0.192 + 0.264 + 0.216}{3}$$
$$= 0.224$$

Using this overall average range as a centerline, the individual ranges can be plotted on a standard R chart, with limits calculated in the typical manner:

$$\text{UCL}_R = D_4 \bar{\bar{R}} = 3.267(0.224) = 0.7318$$
$$\text{LCL}_R = D_3 \bar{\bar{R}} = 0(0.224) = 0$$

Note that the control chart factors (D_3 and D_4) are found in Appendix Table V with the sample size of 2 since the individual ranges are calculated from 2 repeated measurements.

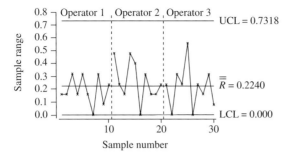

Figure 9.15 Repeatability range chart.

The repeatability range chart is shown in Figure 9.15. All the ranges are in control, which implies that repeatability levels for all the operators appear to be the same, within sampling variation. An out-of-control point would be an indication that the associated operator was having difficulty in making consistent measurements. This may be due to some difficulty by the operator in using the measurement device. In such cases, it is advisable to rectify the problem and obtain new measurement data for a gage R&R assessment.

Recall that our goal is to estimate each of the variance components σ^2_{prod}, σ^2_{repeat}, and σ^2_{repro} contributing to the overall observed variation $\sigma^2_{overall}$. In Chapter 6, we showed that the sample average of the ranges divided by factor d_2 serves as an estimator for the standard deviation of the observations. The scale factor d_2 depends on the size of the sample from which the ranges are calculated. Although not explicitly stated earlier, d_2 is an appropriate factor when the number of samples—as opposed to the sample size—is large. When the number of samples is small, the appropriate scale factor is d_2^*, originally derived by Duncan (1958). As listed in Table 9.7, d_2^* is cross-tabulated by sample size and number of samples. When the number of samples is larger than 15, essentially d_2^* is equal to the factor d_2. Since it was recommended that control chart limits be established with 30 or more samples, there was little need to make the distinction between d_2^* and d_2 during our control chart developments. As we will soon see, the distinction will be important in gage R&R computations.

Following our comments, an estimate for the repeatability component is given by

$$\hat{\sigma}^2_{repeat} = \left(\frac{\bar{\bar{R}}}{d_2^*}\right)^2 \qquad (9.29)$$

For this example, there are 30 sample ranges, each based on a sample size of 2. Since 30 is sufficiently large, we can use here the large-sample d_2 factor which, from Appendix Table V, is 1.128. Therefore,

$$\hat{\sigma}^2_{repeat} = \left(\frac{0.224}{1.128}\right)^2 = 0.03943$$

As defined, gage reproducibility represents the variability among operators. If this variability does exist, we can expect the individual operator overall averages $\bar{O}$ to differ. Based on results from the general ANOVA approach to gage R&R, it can be shown that the estimate for the reproducibility component is given by

$$\hat{\sigma}^2_{repro} = \hat{\sigma}^2_O - \frac{\hat{\sigma}^2_{repeat}}{mn} \qquad (9.30)$$

Table 9.7 d_2^* values

Number of Samples	Sample Size													
	2	3	4	5	6	7	8	9	10	11	12	13	14	15
1	1.41	1.91	2.24	2.48	2.67	2.83	2.96	3.08	3.18	3.27	3.35	3.42	3.49	3.55
2	1.28	1.81	2.15	2.40	2.60	2.77	2.90	3.02	3.13	3.22	3.30	3.38	3.45	3.51
3	1.23	1.77	2.12	2.38	2.58	2.75	2.89	3.01	3.11	3.21	3.29	3.37	3.43	3.50
4	1.20	1.75	2.10	2.37	2.57	2.74	2.88	3.00	3.10	3.20	3.28	3.36	3.43	3.49
5	1.19	1.74	2.10	2.36	2.56	2.73	2.87	2.99	3.10	3.19	3.28	3.35	3.42	3.49
6	1.18	1.73	2.09	2.35	2.56	2.73	2.87	2.99	3.10	3.19	3.27	3.35	3.42	3.49
7	1.17	1.73	2.09	2.35	2.55	2.72	2.86	2.99	3.09	3.19	3.27	3.35	3.42	3.48
8	1.17	1.72	2.08	2.35	2.55	2.72	2.86	2.98	3.09	3.19	3.27	3.35	3.42	3.48
9	1.16	1.72	2.08	2.34	2.55	2.72	2.86	2.98	3.09	3.18	3.27	3.35	3.42	3.48
10	1.16	1.72	2.08	2.34	2.55	2.72	2.86	2.98	3.09	3.18	3.27	3.34	3.42	3.48
11	1.16	1.71	2.08	2.34	2.55	2.72	2.86	2.98	3.09	3.18	3.27	3.34	3.41	3.48
12	1.15	1.71	2.07	2.34	2.55	2.71	2.86	2.98	3.09	3.18	3.27	3.34	3.41	3.48
13	1.15	1.71	2.07	2.34	2.54	2.71	2.86	2.98	3.09	3.18	3.27	3.34	3.41	3.48
14	1.15	1.71	2.07	2.34	2.54	2.71	2.86	2.98	3.09	3.18	3.26	3.34	3.41	3.48
15	1.15	1.71	2.07	2.34	2.54	2.71	2.85	2.98	3.08	3.18	3.26	3.34	3.41	3.48

where $\hat{\sigma}_{\bar{O}}^2$ is an estimate of the variance of the operator averages. Since $\hat{\sigma}_{\bar{O}}^2$ includes the influences of equipment variation, it must be adjusted by subtracting a fraction of the repeatability component. Using the range method, $\hat{\sigma}_{\bar{O}}^2$ is given by

$$\hat{\sigma}_{\bar{O}}^2 = \left(\frac{R_{\bar{O}}}{d_2^*}\right)^2 \qquad (9.31)$$

where $R_{\bar{O}}$ is the range of the overall operator averages. Combining (9.30) and (9.31), we have

$$\hat{\sigma}_{\text{repro}}^2 = \left(\frac{R_{\bar{O}}}{d_2^*}\right)^2 - \frac{\hat{\sigma}_{\text{repeat}}^2}{mn} \qquad (9.32)$$

In the unusual situation of a negative value computed for $\hat{\sigma}_{\text{repro}}^2$, the recommendation by *Measurement Systems Analysis: Reference Manual* (1995) is to set the reproducibility component to zero.[6]

From Table 9.6, we find that the overall operator averages are $\bar{O}_1 = 29.806$, $\bar{O}_2 = 30.034$, and $\bar{O}_3 = 29.722$. This implies that $R_{\bar{O}}$ is 0.0312 $(= 30.034 - 29.722)$. Since there is only *one* range calculation from a sample of size 3 (i.e., number of samples = 1 and sample size = 3), d_2^* is taken from Table 9.7 to be 1.91. Note that this factor differs considerably from the large-sample d_2 factor of 1.693. Thus, we find from (9.32) that

$$\hat{\sigma}_{\text{repro}}^2 = \left(\frac{0.312}{1.91}\right)^2 - \frac{0.03943}{10(2)} = 0.02471$$

It is worth pointing out that if the large-sample d_2 factor is used, the reproducibility estimate is 0.032, that is, an estimate which is nearly 30 percent larger than the one above. We point this out in light of the fact that some quality control practitioners and authors base gage studies on the d_2 factor, rather than on the appropriate d_2^* factor. In general, the practice of using the d_2 factor leads to inflated estimates of the gage components (see Asokan, 1997).

Having the estimates of both the repeatability and reproducibility variance components, we can now arrive at an overall estimate of the measurement error variance

$$\hat{\sigma}_{\varepsilon}^2 = \hat{\sigma}_{\text{repeat}}^2 + \hat{\sigma}_{\text{repro}}^2$$
$$= 0.03943 + 0.02471$$
$$= 0.06414$$

We now need an estimate for the product variance σ_{prod}^2. It might be tempting to estimate the overall total variance $\sigma_{\text{overall}}^2$ by directly applying a standard variance estimator (such as S^2) to the individual measurement data and then subtracting $\hat{\sigma}_{\varepsilon}^2$ from this overall estimate to obtain the estimate for product variance. However, this approach, which is followed by some practitioners, is not appropriate when the gage study is based on repeated measurements. The problem is that the repeated measurements are correlated, hence violating the assumption of independence underlying the standard variance estimators. The appropriate approach is to estimate the product variance based on estimates taking into account the nature of the experimental design. By using the range method, the appropriate estimate is given by

$$\hat{\sigma}_{\text{prod}}^2 = \left(\frac{R_{\bar{P}}}{d_2^*}\right)^2 - \frac{\hat{\sigma}_{\text{repeat}}^2}{kn} \qquad (9.33)$$

where $R_{\bar{P}}$ is the range of the overall part averages. The need to subtract an appropriate "correction" factor is based on the same argument as the need to incorporate a correction

---

[6]The underlying reason for a possible negative estimate for a nonnegative variance component stems from the standard ANOVA estimation procedures. Montgomery and Runger (1993b) provide a discussion of the use of alternative parameter estimation procedures, as applied to ANOVA gage R&R analysis.

factor in the reproducibility estimate given by (9.30). The correction factors for (9.30) and (9.33) are directly adapted from the formal ANOVA setup. We emphasize these points because the influential book *Measurement Systems Analysis: Reference Manual* (1995) fails to incorporate the correction factor in its suggested estimate for σ_{prod}^2; that is, the book recommends to the reader (9.33) *without* the subtracted term. By failing to incorporate the correction factor, the estimate for σ_{prod}^2 is biased upward. This will suggest to the user that the product variation constitutes a higher percentage of the total observed variation. On the other hand, the percentage variation due to the measurement error relative to the total variation will be deflated downward, hence implying a better performance of the measurement system than may truly be the case. Because Minitab's gage R&R range estimation procedure is based on the uncorrected estimate for σ_{prod}^2, we do not recommend the use of Minitab for range-based gage computations. Users are advised to only invoke Minitab's ANOVA option for gage R&R analysis.

Continuing with our example, we see there are 10 part averages for the 10 parts. For instance, to obtain the average for part 1, we can compute either the average of the 6 individual part 1 measurements or the average of the 3 operator averages for part 1. From Table 9.6,

$$\bar{P}_1 = \frac{31.75 + 32.31 + 32.11}{3} = 32.057$$

By computing all 10 averages, it can be found that $\bar{P}_1$ is the largest with a value of 32.057 and $\bar{P}_9$ is the smallest with a value of 28.617. This implies that $R_{\bar{P}}$ is 3.44 (= 32.057 − 28.617). Since there is only one range calculation from a sample of size 10, d_2^* is taken from Table 9.7 to be 3.18. Then from (9.33), we have

$$\sigma_{\text{prod}}^2 = \left(\frac{3.44}{3.18}\right)^2 - \frac{0.03943}{3(2)} = 1.16364$$

Below is a summary of the estimated variance components:

Source	Variance $(\hat{\sigma}^2)$	Standard Deviation $(\hat{\sigma})$
Total Gage R&R	0.06414	0.25326
Repeatability	0.03943	0.19857
Reproducibility	0.02471	0.15719
Product	1.16364	1.07872
Total	1.22778	1.10805

The total variance estimate of 1.22778 is simply the sum of the total measurement error variance estimate (0.06414) and the estimate of the product variance (1.16364). Adjacent to the variance column is a column of standard deviation estimates obtained by taking the square root of the variance estimates.

By using the estimated results, there are a number of ways to make a final judgment on the capability of the measurement system. One commonly used measure relates the variation due to the measurement errors, in standard deviation units, to the specification width (USL − LSL). The most popular version of this measure is known as the *precision-to-tolerance, or P/T, ratio,* defined as follows:

$$\frac{P}{T} = \frac{6\hat{\sigma}_\varepsilon}{\text{USL} - \text{LSL}} \tag{9.34}$$

Assuming that the measurement system has no bias ($\mu_\varepsilon = 0$) and that the measurement errors are normally distributed, $6\hat{\sigma}_\varepsilon$ captures most of the measurement error distribution. Hence, the *P/T* ratio may be interpreted as the percent of the specification width used by the measurement error distribution.

If the product in the previous example has LSL = 25 and USL = 35, the precision-to-tolerance ratio is estimated to be

$$\frac{P}{T} = \frac{6\sqrt{0.06414}}{35 - 25} = 0.152$$

A common rule of thumb claims that if *P/T* is less than 0.10, the measurement system is deemed acceptable. Montgomery and Runger (1993a, p. 119) explain that this rule "is often justified on the widely used engineering rule that suggests that a measurement device be calibrated in units one-tenth as large as the accuracy required in the final measurement."

One disadvantage of the *P/T* ratio is that for a given product, when faced with different customer specifications, the ratio will differ. To overcome this dependency on the specifications, a frequently recommended alternative (see *Measurement Systems Analysis: Reference Manual*, 1995) is to consider the ratio of $\hat{\sigma}_\varepsilon$ to $\hat{\sigma}_{\text{total}}$. When this ratio is multiplied by 100, we are then expressing the measurement error component as a percentage of the total variability. This percentage is typically referred to as percent R&R. For Example 9.9, we have

$$100 \times \frac{\hat{\sigma}_\varepsilon}{\hat{\sigma}_{\text{total}}} = 100 \times \frac{\sqrt{0.06414}}{\sqrt{1.22778}} = 22.9\%$$

Many industries follow the guideline that a percent R&R of less than 10 percent is acceptable, 10 to 30 percent may be acceptable depending on the application and relevant costs, and beyond 30 percent is unacceptable, requiring immediate improvement of the measurement system.

9.8   MEASUREMENT ERROR VARIABILITY AND PROCESS CAPABILITY

In our development of process capability indices, it was assumed that the measurement errors were negligible. An interesting question to ask is, What is the impact on the process capability computations if the measurement errors are not negligible but not accounted for? To gain some insight, consider the C_p index, which we can express as follows:

$$C_p = \frac{\text{USL} - \text{LSL}}{6\sigma_{\text{obs}}} = \frac{\text{USL} - \text{LSL}}{6\sqrt{\sigma_{\text{prod}}^2 + \sigma_\varepsilon^2}} \tag{9.35}$$

Supose that product variation approaches zero. If there is no measurement error variation, C_p will get arbitrary large (approach infinity), reflecting a process approaching perfection or an ideal state. However, from (9.35), when measurement error variation is present, C_p cannot approach an arbitrarily large value and is bound from above, namely,

$$C_p \leq \frac{\text{USL} - \text{LSL}}{6\sigma_\varepsilon} \tag{9.36}$$

If, for example, the specification width is 10 and $\sigma_\varepsilon = 1$, then C_p cannot be reported to be higher than 1.67 (discounting sampling variation) regardless of how capable the true underlying process may be! Hence, any goal to achieve a C_p level higher than 1.67 cannot be realized when the computation is based solely on observed measurements from the current measurement system. Ultimately, the wisest course of action is to fundamentally improve the measurement system (equipment, training, etc.). Until then, process capability calculations (indices and nonconformance rates) can be based on $\hat{\sigma}_{\text{prod}}$. This estimate can be directly extracted from the gage R&R study as shown in Example 9.9. Or, if an estimate for total observed variance $\hat{\sigma}_{\text{obs}}^2$ was found from a data study outside the gage study, $\hat{\sigma}_{\text{prod}}$ can then be taken to be $\sqrt{\hat{\sigma}_{\text{obs}}^2 - \sigma_\varepsilon^2}$.

EXERCISES

9.1. What is process capability analysis? What are some of its objectives and benefits?

9.2. Explain the difference between natural tolerance limits and specification limits.

9.3. Why must a process be stable before capability is assessed? Is stability in the classical sense of a random in-control process required for meaningful capability analysis? Explain.

9.4. Assume that the target value is located halfway between specifications with the upper and lower specification limits 6 standard deviations away from the target. If the following shift sizes away from the target level are tolerated, what are "worst"-case nonconformance rates in terms of ppm?

Shift size	1.00σ	1.25σ	1.50σ	1.75σ	2.00σ
ppm					

9.5. Under what conditions can 5σ quality be associated with 3.4 defects per million?

9.6. What distribution assumption underlies the interpretation of standard capability indices?

9.7. Why is C_p referred to as a *potential capability* index whereas C_{pk} is referred to as an *actual capability* index?

9.8. Under what conditions would C_p equal C_{pk}? What does it mean if C_{pk} is negative?

9.9. Is it true that C_p equals C_{pk} if the process is centered on the target value? Explain.

9.10. Under what conditions would C_p equal C_{pm}? Under what conditions would C_p, C_{pk}, and C_{pm} all be equal?

9.11. Suppose the process specifications are 100 ± 10 and $\hat{C}_p = 1.5$. What is the estimated standard deviation of the process?

9.12. Given Motorola's definition of an ideal process, what is its associated C_{pk} value? Again, as defined by Motorola, what is the value of C_{pk} in the worst-case scenario?

9.13. Suppose a process is centered on target with the target being halfway between the specification limits. If $\hat{C}_p = 0.80$, what is the estimated nonconformance rate in terms of ppm?

9.14. Suppose a process is centered on target with the target being halfway between the specification limits. If the estimated nonconformance rate is 1700 ppm, what is the value of $\hat{C}_p$?

9.15. Estimate process capability, using the summary statistics for the battery data in Exercise 6.12. If the lower-specification limit is 1.425 volts (V), estimate the appropriate process capability index.

9.16. Estimate process capability, based on the $\bar{x}$ and R chart calculations, for the net weight data in Exercise 6.14. If the specifications are at 16 ± 0.025, estimate C_p, C_{pk}, and C_{pm}.

9.17. Estimate process capability, based on the $\bar{x}$ and R chart calculations, for the patient waiting data in Exercise 6.17. If the upper specification is 15 min, estimate the appropriate process capability index.

9.18. Estimate process capability, based on the $\bar{x}$ and R chart calculations, for the copper percentage data in Exercise 6.21. If the specifications are at 87 ± 0.8, estimate C_p, C_{pk}, and C_{pm}.

9.19. Piston rings used in automobiles are coated with hard chrome plating. For a particular stage of production, the requirement of the plating thickness is 195 ± 20 micrometers (μm). A random sample of 50 piston rings was obtained. The data are shown below (also given in "ring.dat"):

189	181	183	191	180	182	187	188	189	189
177	191	192	180	184	190	179	198	197	179
171	188	165	191	176	172	199	185	183	183
187	189	186	191	185	172	182	184	185	201
189	187	186	183	178	173	172	193	184	183

    a. Perform normality checks on the data.
    b. Using the sample standard deviation s as an estimate for σ, estimate C_p, C_{pk}, and C_{pm}.
    c. Estimate the rate of nonconformance.

9.20. To continue Exercise 9.19, construct 95 percent confidence intervals for C_p and C_{pk}.

9.21. The voltage of an electron gun is a critical characteristic associated with the quality of a TV picture. If the voltage is too high, the image will be too bright, whereas if voltage is too low, the image will be too dark. The specifications on voltage are 55 ± 10 V. Based on past empirical study, it was determined that voltage measurements are compatible with the normal distribution. Random samples of 7 electron guns are chosen on a periodic basis. For each subgroup, the sample mean $\bar{x}$ and the sample standard deviation s are found. After 30 subgroups, the following summary information is obtained:

$$\sum_{i=1}^{30} \bar{x}_i = 1497.6 \qquad \sum_{i=1}^{30} s_i = 68.751$$

a. Estimate C_p, C_{pk}, and C_{pm}. Comment on their values.
b. Estimate the rate of nonconformance.
c. How much will the rate of nonconformance be reduced if the process mean is adjusted to be on target?

9.22. An article in *Quality Engineering* (by K. S. Krishnamoorthi, 1991, 3, pp. 41–47) presents individual measurement data on sand compactability. Below are the individual measurements given in sequential order (also given in "sand.dat"):

46	43	41	42	40	44	40	41	40	42
41	41	43	43	40	38	45	42	41	43
42	43	39	44	44	45	43	42	41	46
41	39	40	40	42	44	42	40	43	

a. Using the moving-range method, construct an X chart for the data series. What do you conclude about the stability of the process?
b. As reported by the author of the article, the lower and upper specifications for sand compactability are 38 and 46, respectively. Based on computations from part *a*, estimate C_p and C_{pk}.
c. Estimate the rate of nonconformance.

9.23. Refer to Exercise 9.22. Redo parts *b* and *c*, using the sample standard deviation s as an estimate for σ.

9.24. To continue Exercise 9.23, construct 95 percent confidence intervals for C_p and C_{pk}.

9.25. Show that $E[(X - T)^2]$ indeed equals $\sigma^2 + (\mu - T)^2$. (*Hint: $X - a = X - a + b - b$.*)

9.26. Express $\hat{C}_p$ in terms of the sample variance s^2. (*Hint: $x_i - a = x_i - a + b - b$.*) Given this general expression, determine the value of $\hat{C}_p$ if LSL $= 95$, $T = 100$, USL $= 105$, $n = 50$, $\bar{x} = 100.5$, and $s^2 = 0.965$.

9.27. What is the value of C_{pm} if $C_p = 1.7$, $\sigma^2 = 4$, $\mu = 50$, and $T = 47$?

9.28. Refer to the change order data series of Exercise 3.15. Suppose that a change order time of more than 60 days is considered unacceptable.
   a. Ignoring the severe nonnormality of the data, estimate the one-sided capability index C_{pu}. If we assume normality, what is the estimated rate of nonconformance?
   b. Transform the data with an appropriate normalizing transformation and then redo part *a*.

9.29. Suppose that X_t follows an AR(1) process, that is,

$$X_t = \tau + \phi_1 X_{t-1} + \varepsilon_t$$

For an AR(1) process, it can be shown that the mean and standard deviation are given by, respectively,

$$\mu = \frac{\tau}{1 - \phi_1}$$

$$\sigma = \frac{\sigma_\varepsilon}{\sqrt{1 - \phi_1^2}}$$

Refer to Figure 3.39 to find the estimated AR(1) model for the daily weight series. If the student has personal specification limits of 180 ± 5 lb, estimate C_p, C_{pk}, and C_{pm}.

9.30. Consider the stopwatch data series introduced in Section 2.3. Recall that this series consists of my personal attempts to measure 10 s with an electronic stopwatch. Based on $\alpha = 0.05$, test whether my personal measurement system is accurate.

9.31. A manufacturer of postal weight scales randomly selects scales on a periodic basis for tests of accuracy and precision. Below are 25 repeated weight measurements of a benchmark unit of known weight equal to 96 ounces (oz):

96.0	95.9	96.0	96.0	96.0	96.0	96.0	95.9	96.0	95.9
95.9	95.8	96.0	96.0	95.8	96.0	96.1	96.1	96.0	95.9
95.8	95.9	95.9	96.0	96.0					

Based on $\alpha = 0.05$, what is your conclusion about the accuracy of the tested scale?

9.32. An article in *Quality Engineering* (by C. Lin, C. Hong, and J. Lai, 1997, 9, pp. 561–573) presents data for an R&R analysis based on a measuring device known as an optical comparator. Using an optical comparator, two operators took two measurements (in millimeters) of a particular dimension on 10 gear parts. The results are shown below:

	Operator 1		Operator 2	
Part	Test 1	Test 2	Test 1	Test 2
1	3.045	3.041	3.048	3.046
2	3.037	3.038	3.031	3.038
3	3.021	3.024	3.023	3.022
4	3.017	3.012	3.011	3.016
5	3.048	3.049	3.042	3.048
6	3.046	3.045	3.042	3.043
7	3.039	3.034	3.033	3.035
8	3.033	3.035	3.038	3.037
9	3.048	3.041	3.034	3.032
10	3.041	3.049	3.046	3.045

    a. Construct a repeatability range chart for the data. Are there any out-of-control points indicating operator difficulty?

    b. Estimate gage repeatability and reproducibility.

    c. Obtain an estimate for gage percent R&R. Given the standard guidelines, what is your assessment of the measurement system?

9.33. Continuation of Exercise 9.32: If the specifications for the measured dimension are 3.035 ± 0.05, what can you say about gage capability in terms of a *P/T* ratio?

9.34. The American Society for Testing and Materials (ASTM) has established a standard of performance for fiberglass shingles, ASTM D3462, "Standard Specification for Fiber Glass Asphalt Shingles." This test measures the force, in grams, required to tear a shingle in a specialized measuring device known as the *Elmendorf tear tester.* To conduct a gage R&R study with the Elmendorf tester, four shingle specimens were cut out of a large piece of shingle. For each large piece of shingle, its cutout specimens were then randomly assigned so that each operator had two specimens. Below are the data results associated with seven large sample shingle pieces:

	Operator 1		Operator 2	
Sample	Test 1	Test 2	Test 1	Test 2
1	2139	2127	2109	2118
2	2426	2427	2363	2394
3	2105	2144	2079	2105
4	2157	2155	2113	2112
5	2119	2112	2091	2092
6	2378	2412	2371	2371
7	2170	2186	2153	2162

> *a.* Construct a repeatability range chart for the data. Are there any out-of-control points indicating operator difficulty?
>
> *b.* Estimate gage repeatability and reproducibility.
>
> *c.* Obtain an estimate for gage percent R&R. Given the standard guidelines, what is your assessment of the measurement system?

9.35. Explain the impact of an increased measurement error variation on the determination of process capability.

9.36. Demonstrate that if $\sigma_{prod} = \sigma_{\varepsilon}$, then the "reported" C_p would be about 71 percent of the "true" C_p.

9.37. Establish a condition for the value of σ_{ε}, in relation to two-sided specification limits, such that C_p would always be computed to be less than 1 no matter the *true* capability of the process.

REFERENCES

Abraham, B., and J. Ledolter (1983): *Statistical Methods for Forecasting,* Wiley, New York.

Alwan, L. C., C. W. Champ, and H. D. Maragah (1994): "Study of Average Run Lengths for Supplementary Runs Rules in the Presence of Autocorrelation," *Communications in Statistics: Simulation and Methods,* 23, pp. 373–391.

Asokan, M. V. (1997): "On the Estimation of Gauge Capability," *Quality Engineering,* 10, pp. 263–266.

Berezowitz, W. A., and T. H. Chang (1995): "Capability Indices— Somewhere the Point Got Lost," Technical Paper MS95-148, Society of Manufacturing Engineers.

Bissell, D. (1990): "How Reliable Is Your Capability Index?" *Applied Statistics,* 39, pp. 331–340.

——— (1994): *Statistical Methods for SPC and TQM,* Chapman & Hall, London.

Box, G. E. P., J. S. Hunter, and W. J. Hunter (1978): *Statistics for Experimenters,* Wiley, New York.

———, G. M. Jenkins, and G. C. Reinsel (1994): *Time Series Analysis: Forecasting and Control,* 3d ed., Prentice-Hall, Englewood Cliffs, NJ.

Chan, L. K., S. W. Cheng, and F. A. Spiring (1988): "A New Measure of Process Capability, C_{pm}," *Journal of Quality Technology,* 20, pp. 160–175.

Chrysler Corporation, Ford Motor Company, and General Motors Corporation (1995): *Measurement Systems Analysis: Reference Manual,* Automotive Industry Action Group, Troy, MI.

Deming, W. E. (1986): *Out of the Crisis,* MIT Center for Advanced Engineering Study, Cambridge, MA.

Duncan, A. J. (1958): "Design and Operation of a Double-Limit Variables Sampling Plan," *Journal of American Statistical Association*, 53, pp. 543–550.

English, J. R., and G. D. Taylor (1993): "Process Capability Analysis: A Robustness Study," *International Journal of Production Research*, 31, pp. 1621–1635.

Farnum, N. R. (1996): "Using Johnson Curves to Describe Non-Normal Process Data," *Quality Engineering*, 9, pp. 329–336.

Gitlow, H., A. Oppenheim, and R. Oppenheim (1995): *Quality Management: Tools and Methods for Improvement,* 2d ed., Irwin, Burr Ridge, IL.

Gunter, B. H. (1989): "The Use and Abuse of C_{pk}," *Quality Progress*, 22(3), pp. 108–109; (5), pp. 79–80.

Hahn, G. J., and S. S. Shapiro (1967): *Statistical Models in Engineering*, Wiley, New York.

Harris, T. J., and W. H. Ross (1991): "Statistical Process Control Procedures for Correlated Observations," *Canadian Journal of Chemical Engineering*, 69, pp. 48–57.

Hooper, J. H., and S. J. Amster (1990): "Analysis and Presentation of Reliability Data," in *Handbook of Statistical Methods for Engineers and Scientists*, edited by H. M. Wadsworth, McGraw-Hill, New York.

Hsiang, T. C., and G. Taguchi (1985): "A Tutorial on Quality Control and Assurance—Taguchi Methods," Presentation at the American Statistical Association Meeting, Las Vegas, NV.

Kane, V. E. (1986): "Process Capability Indices," *Journal of Quality Technology*, 18, pp. 41–52.

Kotz, S., and N. L. Johnson (1993): *Process Capability Indices*, Chapman & Hall, London.

Kushler, R., and P. Hurley (1992): "Confidence Bounds for Capability Indices," *Journal of Quality Technology*, 24, pp. 188–195.

Lawless, J. F. (1982): *Statistical Models and Methods for Lifetime Data,* Wiley, New York.

Montgomery, D. C. (1996): *Introduction to Statistical Quality Control,* 3d ed., John Wiley, New York.

——— (1997): *Design and Analysis of Experiments,* 4th ed., John Wiley, New York.

——— and G. C. Runger (1993a): "Gauge Capability Analysis and Designed Experiments, I, Basic Methods," *Quality Engineering*, 6, pp. 115–135.

——— and ——— (1993b): "Gauge Capability Analysis and Designed Experiments, II, Experimental Design Models and Variance Component Estimation," *Quality Engineering*, 6, pp. 289–305.

Nelson, L. S. (1994): "A Control Chart for Parts-per-Million Nonconforming," *Journal of Quality Technology*, 26, pp. 239–240.

Rodriguez, R. N. (1992): "Recent Developments in Process Capability Analysis," *Journal of Quality Technology*, 24, pp. 176–186.

Schneider, H., J. Pruett, and C. Lagrange (1995): "Uses of Process Capability Indices in the Supplier Certification Process," *Quality Engineering*, 8, pp. 225–235.

Searle, S. R., G. Casella, and C. E. McCulloch (1992): *Variance Components*, Wiley, New York.

Shapiro, S. S. (1990): "Selection, Fitting, and Testing Statistical Models," in *Handbook of Statistical Methods for Engineers and Scientists*, edited by H. M. Wadsworth, McGraw-Hill, New York.

Somerville, S. E., and D. C. Montgomery (1996): "Process Capability Indices and Non-Normal Distributions," *Quality Engineering*, 9, pp. 305–316.

Spiring, F. A. (1989): "An Application of C_{pm} to the Toolwear Problem," *Transactions ASQC Congress*, Toronto, pp. 123–128.

Wheeler, D. J. and D. S. Chambers (1992): *Understanding Statistical Process Control,* SPC Press, Knoxville, TN.

10

Related Special Topics

CHAPTER OVERVIEW

Over the decades since Shewhart's pioneering work, hundreds of statistical techniques have been developed for process monitoring. The previous chapters detailed the most dominant of these techniques found in practice. This chapter is reserved for a certain set of special topics. The notion of a special topics chapter should not be viewed as a miscellaneous chapter of less important topics.

All this chapter's topics could have been incorporated in earlier chapters. However, there are a few reasons for our treatment of a topic as special. For one, we consider in this chapter certain methods particularly specialized for certain sectors of industrial practice. For instance, we begin with the presentation of SPC methods for short production runs often associated with job-shop environments. We also reserved this chapter for methods which are more technical relative to earlier presentations. Multivariate quality control and adaptive control are examples of relatively more sophisticated topics. When discussing these more advanced methods, we will give a basic description of the method and then provide appropriate references for more complete coverage. Additionally, this chapter gives us an opportunity to discuss some natural extensions or variants of approaches discussed earlier. For instance, as an alternative to regression modeling, we will discuss the use of the exponentially weighted moving-average (EWMA) technique as a basis for modeling time-series data. We will also discuss and illustrate how modeling a single time series can be extended to incorporate information from other time series. We close the chapter and the book by listing in "bulletlike" fashion a number of other specialized topics with references for interested readers to explore independently.

Given that this chapter is a collection of special topics, the flow of the chapter is much different from that of previous chapters. In particular, there is no necessity to approach the sections of this chapter in any particular order. Thus, the reader can pick and choose sections based on his or her interests.

10.1   SPC METHODS FOR SHORT-RUN PRODUCTION

Historically, traditional SPC methods were developed for application to high-volume manufacturing processes, that is, mass production environments. Indeed, many companies with high-volume production have enjoyed the benefits of these methods. We have also emphasized and demonstrated, throughout the text, that SPC methods or variants, based on broader data analysis techniques, can be successfully applied to a wide variety of nonmanufacturing process applications, ranging from sales data to personal process data.[1] In all these applications (manufacturing or not), there is an assumption that the process generates a "sufficient"

---

[1]Given the nature of the special topics in this chapter, the emphasis leans more toward manufacturing applications. However, the reader can take note that our presentation of modeling a time series using information from other time series revolves on a nonmanufacturing type of process.

amount of data measurements to establish the control chart limits or the fitted statistical model. For example, the implementation of traditional Shewhart charts is typically based on at least 20 to 25 samples. Anything less is viewed as insufficient for adequate process control. This viewpoint is underscored by Shewhart (1939), who wrote, "Control of this kind cannot be reached in a day. It can not be reached in the production of a product in which only a few pieces are manufactured. It can, however, be approached scientifically in a continuing mass production" (p. 46).

Since Shewhart's era, many techniques, technologies, and philosophies have been introduced, creating a wide array of manufacturing environments other than just a mass production one. Among them are just-in-time (JIT) systems, lean and agile manufacturing, flexible manufacturing, computer-integrated manufacturing (CIM), and robotics, to name a few. Interestingly, for some environments, one consequence of technological advances is the ability to obtain great amounts of data. In particular, automatic test equipment integrated within a CIM environment can be utilized to gather and download an abundance of process data for immediate data analysis.

On the other hand, manufacturing advances have resulted in environments in which relevant data are more likely to be scarce than abundant. In particular, manufacturing environments tailored for the flexibility of producing a wide variety of products tend to process only small batches of a particular product. As such, not enough items in a single production run are available to generate data for the establishment of a traditional control chart. A job shop is one such environment characterized by short production runs or small lot sizes. Because short-run processes represent a challenge for traditional SPC methods, many manufacturers feel that they cannot utilize the control chart methodology. In actuality, simple adaptations of the implementation of traditional control charts can be made to allow for their use in low-volume manufacturing environments.

In a low-volume environment, some general approaches (see Farnum, 1992; and Woodall, Crowder, and Wade, 1995) for process control include

1. Using control charts with greater ability to pick up process changes in fewer samples. For example, CUSUM and EWMA control charts, discussed in Chapter 8, might be considered.

2. Adjusting the placement of standard Shewhart limits. Hillier (1969) has demonstrated that the true false-alarm rate of a Shewhart control chart constructed from a small number of data points is much different from the expected rate of 0.0027 associated with 3σ limits. Given this fact, a number of investigators have proposed techniques to adjust the limits of $\bar{x}$ and individual charts so that a desired false-alarm rate can be achieved. An extensive reference list of works related to this approach is found in Woodall, Crowder, and Wade (1995).

3. Monitoring and controlling secondary operating variables rather than the product characteristics. For example, control of ambient temperature, pressure, humidity, or machine revolutions (rpm) may be sufficient for control of

the ultimate product output. Since data on secondary operating variables are easily obtained and plentiful by nature, standard charting procedures can be implemented.

4. Designing control charts based on the pooling of data on similar product characteristics from different batches. Even though any given batch is small in size, pooling many batches together can overcome the data limitations associated with small lot sizes inherent in short-run production environments. With enough data, standard Shewhart charts can be implemented.

In the following subsections, we expand on the idea of pooling data (approach 4) for purposes of control charting.

SHORT-RUN VARIABLE CHARTS: CONSTANT VARIATION ACROSS BATCHES

The most common approach to the data limitations in a short-run production environment is to pool data across different batches. To do so, the idea is to consider the *deviation* of a product's measurement from its nominal (or target) value, that is, actual observation minus target value. For example, if the target of the specification is 0.475, actual measurements of 0.477 and 0.471 are recorded as $+0.002$ and -0.004, respectively.

The benefit of using deviations, rather than actual measurements, is clear in small batches of products where the nominal values vary from batch to batch. Namely, deviation data can allow for the charting of several different part numbers on the same chart. A single chart of deviations from nominal values is commonly referred to as a *DNOM chart*.

A key initial step in setting up a DNOM chart is to identify the family of products that can be combined. As pointed out by Winchell and Millis (1990), the grouping process of *group technology* may be used for establishing families for the purpose of statistical process control applications, when batch sizes are small. Group technology involves the identifying of products with common traits in either design characteristics or manufacturing characteristics (for more details, see Stevenson, 1999).

As with any other control chart, the construction of a DNOM chart requires an estimate of the process variability. If it can be assumed that the process variation is the same for all items regardless of nominal values, then all the deviations can be used to arrive at a single estimate of process variation. With the constant-variation assumption, the idea of the DNOM chart is pretty basic. Namely, one treats the deviation data as one would treat original measurements in the construction of standard control charts. In Table 10.1, we give the general layout for the case of k batches of n observations, where the ith batch has an associated nominal value of T_i. If $n = 1$, the sequence of individual deviations can be used to construct standard individual charts, such as Shewhart individual charts (Chapter 5) or memory-based charts (Chapter 8). In the case of the CUSUM chart, recall that there is a need for the selection of a target value. Since the measurements

Table 10.1   General conversion to DNOM control chart data

Batch	Original Measurements	Nominal	Deviations
1	$x_{11}, x_{12}, \ldots, x_{1n}$	T_1	$x_{11} - T_1, x_{12} - T_1, \ldots, x_{1n} - T_1$
2	$x_{21}, x_{22}, \ldots, x_{2n}$	T_2	$x_{21} - T_2, x_{22} - T_2, \ldots, x_{2n} - T_2$
.	.	.	.
.	.	.	.
.	.	.	.
k	$x_{k1}, x_{k2}, \ldots, x_{kn}$	T_k	$x_{k1} - T_k, x_{k2} - T_k, \ldots, x_{kn} - T_k$

are deviations, the logical choice for the target value is 0. When $n > 1$, subgroup statistics such as averages, ranges, and standard deviations can be computed and monitored by using the standard charting procedures of Chapters 6 and 8. To illustrate the construction of a DNOM chart, we turn now to an example.

Example 10.1

In this example, we consider data taken from a manufacturer of a wide array of coatings and resin products, including house paints, appliance coatings, and automotive coatings. One of the main product lines is a base-coat paint used for automotive exterior plastic parts such as bumpers, trim around door handles, and mirror housings. Based on customer specifications, base-coat paints can be matched to a wide variety of solid or metal-based colors. One of the key characteristics of a paint is the percentage of solids content. Basically, paints are made up of three components: binder (or resin), pigment, and solvent. Binder and pigment constitute the solids. In the base-coat product family, the percentage of solids content typically falls in the range of 40 to 55 percent. Certain solid colors, such as reds and blues, are associated with lower percentage while metallics and white have high solids content. For the interested reader, the reason why the color white has a higher percentage of solids is because more pigments must be added to hide the primer undercoat, which is typically gray. The percentage of solids is such a key characteristic that paint manufacturers are often required by customers to contractually agree to a solids content percentage.

Production of paint is done in a batch processing environment with batch sizes running from 1 gallon (gal) to 40,000 gal. The processing time of a batch of base-coat paint is around 4 h. On any given day, there are 20 to 30 batches of base-coat paint being processed. In the facility at large, batches of other paint lines are also being processed. Because essentially all batches have different solids content target values, a standard control chart applied to measurements on solids content would be inappropriate. The natural alternative is to consider deviations from the target values.

There is a question of what batches can be combined for a single deviation-based chart. As stated earlier, the DNOM approach assumes that the underlying variances for the deviations are approximately equal regardless of nominal values. Based on historical data and background experience, production managers have observed relatively constant variation around targets within product families but not necessarily across product families. For example, product families with much lower solids percentage targets (say, range of 10 to 15 percent) are known to have much greater deviations around the targets than product families with higher solids percentage targets. Lower percentage of solids content means a higher percentage of solvent content. The process of adding solvent is inherently more variable when it is added in large amounts, and thus the explanation for greater deviations in solids content. With these facts in mind, we consider the implementation of a DNOM approach to only one product family, namely, base-coat paints.

To measure solids content, three paint samples from a batch are obtained. Each sample of wet paint is placed on a foil pan and weighed (in grams) to the fourth decimal place. The samples are baked at 150° F until only solids remain. The remaining solids are weighed. Given the prebake weights, the percentage of solids can then be computed for each sample.

Table 10.2 shows the subgroup information, along with other relevant statistics for 25 batches processed on a particular day. The columns labeled with the d symbol are the deviations; specifically, $d_1 = x_1 -$ nominal, $d_2 = x_2 -$ nominal, and $d_3 = x_3 -$ nominal. Treating the deviation data as original measurements, we can construct, for instance, Shewhart subgroup average and range charts. The subgroup averages and ranges for the deviation data are found in the rightmost two columns of Table 10.2. Remember, the first step is to analyze the range data prior to the construction of the subgroup average chart. From Appendix Table V, for subgroup size $n = 3$, the range chart factors D_3 and D_4 are 0 and 2.574, respectively. The preliminary control limits for the R chart are then given by

$$\text{UCL}_R = D_4\bar{R} = 2.574(1.2376) = 3.186$$

$$\text{LCL}_R = D_3\bar{R} = 0(1.2376) = 0$$

The R chart with these limits is given in Figure 10.1a. The chart shows no indication of an out-of-control condition; hence, the range estimate can be used to construct the subgroup average chart. From Appendix Table V, we find the A_2 factor to be 1.023. As a result, the control chart limits for the deviation averages are

$$\text{UCL}_{\bar{d}} = \bar{\bar{d}} + A_2\bar{R} = 0.02 + 1.023(1.2376) = 1.286$$

$$\text{LCL}_{\bar{d}} = \bar{\bar{d}} - A_2\bar{R} = 0.02 - 1.023(1.2376) = -1.246$$

The DNOM average chart is shown in Figure 10.1b. The chart shows that the deviation averages are well within limits, with no indications of special causes or abnormal patterns.

TESTING THE EQUALITY OF VARIANCES

The assumption of equal variances across different nominal values is fundamental to the construction of the DNOM charts just described. There are several formal tests for checking whether this assumption is reasonable. We shall consider two tests, the *variance ratio test* and the *Barlett test*. The variance ratio test is used to test the equality of variances between two populations, while the Barlett test can be used for studying whether r populations have equal variances. Both tests make the assumption that the underlying distribution for observations is the normal distribution. If normality cannot be assumed, then tests which do not depend on a distribution assumption (known as nonparametric tests) might be considered; references for nonparametric tests can be found in Glaser (1982).

Variance Ratio Test   Consider first the case of two normal populations with means and variances μ_1 and σ_1^2 and μ_2 and σ_2^2. Each population represents the underlying

Table 10.2   Measurements on solid percentages and deviations from nominal values

Batch	x_1	x_2	x_3	Nominal	d_1	d_2	d_3	$\bar{d}$	R
1	45.10	43.22	43.27	43.78	1.32	−0.56	−0.51	0.0833	1.88
2	51.69	52.26	52.04	52.17	−0.48	0.09	−0.13	−0.1733	0.57
3	43.57	44.06	42.20	43.14	0.43	0.92	−0.94	0.1367	1.86
4	42.31	42.16	43.09	42.98	−0.67	−0.82	0.11	−0.4600	0.93
5	51.95	52.43	51.12	51.44	0.51	0.99	−0.32	0.3933	1.31
6	48.23	50.97	48.63	48.56	−0.33	2.41	0.07	0.7167	2.74
7	41.65	40.04	39.94	40.32	1.33	−0.28	−0.38	0.2233	1.71
8	43.00	43.04	42.25	42.97	0.03	0.07	−0.72	−0.2067	0.79
9	48.12	47.91	48.33	48.57	−0.45	−0.66	−0.24	−0.4500	0.42
10	44.66	45.53	43.79	44.03	0.63	1.50	−0.24	0.6300	1.74
11	46.45	46.24	45.86	45.96	0.49	0.28	−0.10	0.2233	0.59
12	50.31	51.16	50.24	50.39	−0.08	0.77	−0.15	0.1800	0.92
13	47.80	47.99	47.89	47.66	0.14	0.33	0.23	0.2333	0.19
14	53.64	53.54	53.90	54.23	−0.59	−0.69	−0.33	−0.5367	0.36
15	48.77	48.47	48.03	49.12	−0.35	−0.65	−1.09	−0.6967	0.74
16	46.08	43.61	46.52	45.07	1.01	−1.46	1.45	0.3333	2.91
17	42.44	43.00	43.15	43.18	−0.74	−0.18	−0.03	−0.3167	0.71
18	49.56	48.54	49.47	48.93	0.63	−0.39	0.54	0.2600	1.02
19	51.52	50.95	52.95	51.45	0.07	−0.50	1.50	0.3567	2.00
20	53.94	52.74	53.23	53.24	0.70	−0.50	−0.01	0.0633	1.20
21	55.11	54.20	55.49	54.52	0.59	−0.32	0.97	0.4133	1.29
22	40.32	41.31	41.96	41.64	−1.32	−0.33	0.32	−0.4433	1.64
23	42.27	43.26	42.09	42.92	−0.65	0.34	−0.83	−0.3800	1.17
24	50.35	50.99	50.98	50.48	−0.13	0.51	0.50	0.2933	0.64
25	51.97	50.86	50.36	51.44	0.53	−0.58	−1.08	−0.3767	1.61
								$\bar{\bar{d}} = 0.02$	$\bar{R} = 1.2376$

distribution of process measurements at different nominal values. Suppose that two independent random samples of sizes n_1 and n_2 are drawn from each of these populations.

The statistical procedure for comparing population variances σ_1^2 and σ_2^2 is to make an inference about the ratio σ_1^2/σ_2^2. If the null hypothesis of equal variances is true, the ratio σ_1^2/σ_2^2 equals 1 whereas the ratio differs from 1 when the variances are unequal. Hence, the competing hypotheses may be stated as

$$H_0: \frac{\sigma_1^2}{\sigma_2^2} = 1 \qquad (\sigma_1^2 = \sigma_2^2)$$

$$H_1: \frac{\sigma_1^2}{\sigma_2^2} \neq 1 \qquad (\sigma_1^2 \neq \sigma_2^2)$$

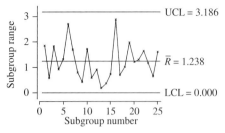

(a) Range chart for deviation data

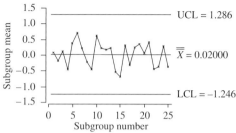

(b) Subgroup mean chart for deviation data

Figure 10.1   DNOM Shewhart charts.

Logically, the test statistic is the ratio of sample variances obtained from the sample data:

$$F = \frac{s_1^2}{s_2^2} \tag{10.1}$$

The random variable F can only take on nonnegative values because sample variances can never be negative.[2] If the value of F is "around" 1, we have no reason to doubt the null hypothesis that the underlying true variances are equal. However, small or large values for the ratio would provide evidence that the true variances are different.

How small or large must the ratio be before we reject the null hypothesis? To answer this question, we need to know the sampling distribution of s_1^2/s_2^2 when the null hypothesis is true. The distribution was first tabulated by Sir Ronald Fisher; the distribution that accounts for the variability of s_1^2/s_2^2, when the null hypothesis is true, is called the *F distribution*. In general, the F distribution is characterized by two parameters ν_1 and ν_2, called the numerator and denominator degrees of freedom, respectively. In our case, the degrees of freedom associated with s_1^2 and s_2^2 are $n_1 - 1$ and $n_2 - 1$, respectively.

To use the F distribution to test the hypotheses, we need the critical values of F such that the combined area in the right and left tails of the distribution is some specified level of significance α. As is done with other two-sided statistical tests (including the standard control chart), the overall α value is divided by 2, with $\alpha/2$ associated with a lower critical point and $\alpha/2$ associated with an upper critical point. Denote $F_{\alpha/2,n_1-1,n_2-1}$ and $F_{1-\alpha/2,n_1-1,n_2-1}$ as the $100(1 - \alpha/2)$ and $100(\alpha/2)$ percentile points of the F distribution with $n_1 - 1$ numerator and $n_2 - 1$ denominator degrees of freedom, respectively. The decision rule is then given by

Do not reject H$_0$ if $F_{1-\alpha/2,n_1-1,n_2-1} \leq F \leq F_{\alpha/2,n_1-1,n_2-1}$; otherwise reject H$_0$ **(10.2)**

Percentage points for the F distribution for different (ν_1, ν_2) combinations and common levels of significance α are provided in Appendix Table IV. Note that

---

[2] In honor of Sir Ronald Fisher, the test statistic was named F by George W. Snedecor.

the tables show only percentage points for the upper tail. But there is a useful property about the F distribution which allows us to determine lower percentage points from upper percentage points. Namely, to find the value of F such that the area in the lower tail is p, we can use the following fact:

$$F_{1-p,\,\nu_2,\nu_1} = \frac{1}{F_{p,\,\nu_1,\,\nu_2}} \qquad\qquad (10.3)$$

Note that the degrees of freedom in the subscripts are interchanged. A more flexible alternative to tables is to employ Minitab to determine percentage points for any combination of (ν_1, ν_2) and value of α.

Example 10.2

A production manager at a machining job shop wishes to determine what range of parts can be combined for establishing a single process control chart. The need for a combined chart exists because parts being produced vary according to customer order; hence there is no single series of observations on a particular part. One measurement of interest is the hole diameter, in millimeters, of a certain part type. For this part type, hole diameters typically range from 14 to 30 mm.

There is some suspicion that hole precision worsens for larger hole sizes; that is, the deviations from nominal values are smaller for small hole sizes than for large hole sizes. If this is true, variability for small holes is not representative of variability for large holes, and vice versa. To investigate the issue, a planned experiment was conducted on the two most extreme hole sizes, namely, 14-mm and 30-mm holes. Suppose 20 parts of each hole size are manufactured and the sample variances are found to be $s_{14\,\mathrm{mm}}^2 = 0.235$ and $s_{30\,\mathrm{mm}}^2 = 0.315$. Before we proceed with the test, it is important that the data be checked for conformance to normality. Assuming that the normality assumption is reasonable, from (10.1), we compute the test statistic to be

$$F = \frac{s_{14\,\mathrm{mm}}^2}{s_{30\,\mathrm{mm}}^2} = \frac{0.235}{0.315} = 0.746$$

The numerator degrees of freedom are $\nu_1 = n_1 - 1 = 20 - 1 = 19$, and the denominator degrees of freedom are $\nu_2 = n_2 - 1 = 20 - 1 = 19$. Choosing $\alpha = 0.05$, we need to find $F_{0.025,19,19}$ and $F_{0.975,19,19}$. These points can be found from Minitab as follows:

```
MTB > InvCDF 0.975;     Calc>>Probability Distributions>>F
SUB C> F 19 19.

Inverse Cumulative Distribution Function

F distribution with 19 DF in numerator and 19 DF in denominator

P( X <= x)          x
   0.9750        2.5265    ◄──────── F₀.₀₂₅,₁₉,₁₉

MTB > InvCDF 0.025;
SUB C> F 19 19.

Inverse Cumulative Distribution Function

F distribution with 19 DF in numerator and 19 DF in denominator

P( X <= x)          x
   0.0250        0.3958    ◄──────── F₀.₉₇₅,₁₉,₁₉
```

Since $s^2_{14\ mm}/s^2_{30\ mm}$, with its value of 0.746, falls within the range of 0.3958 to 2.5265, we cannot reject the null hypothesis of equal variances. As such, there is no evidence against the pooling of these different nominal sizes for the construction of a DNOM chart.

Barlett Test   We saw that the null hypothesis that variances of two normal populations are equal can be tested with the F test. To test the more general case that the variances of r populations are equal, alternative tests need to be considered. If all r populations are approximately normal, then the *Barlett test for homogeneity of variances* can be used to test the null hypothesis

$$H_0: \sigma^2_1 = \sigma^2_2 = \cdots = \sigma^2_r$$

against the alternative hypothesis that at least two variances differ.

The first step of the Barlett test procedure is to obtain independent random samples from the r populations. From these samples, the sample variances s^2_1, $s^2_2, \ldots, s^2_r$ are then computed. The Barlett test statistic is

$$B = \frac{\left[\sum_{i=1}^{r}(n_i - 1)\right]\ln \bar{s}^2 - \sum_{i=1}^{r}(n_i - 1) \ln s^2_i}{1 + \dfrac{1}{3(r - 1)}\left[\sum_{i=1}^{r}\dfrac{1}{(n_i - 1)} - \dfrac{1}{\sum_{i=1}^{r}(n_i - 1)}\right]} \tag{10.4}$$

where $n_i = $ size of ith sample

$$\bar{s}^2 = \text{weighted average of } r \text{ sample variances} = \frac{\sum_{i=1}^{r}(n_i - 1)s^2_i}{\sum_{i=1}^{r}(n_i - 1)}$$

$\ln x = $ natural logarithm (i.e., log to base e) of quantity x

Barlett showed that under the null hypothesis of equal variances, the test statistic B has a sampling distribution which is closely approximated by a chi-square (χ^2) distribution with $r - 1$ degrees of freedom. He further demonstrated that departures from the null hypothesis tend to result in larger values for B. Thus, for a level of significance α, the appropriate decision rule is:

$$\text{If } B \leq \chi^2_{\alpha,r-1}, \text{ do not reject } H_0$$
$$\text{If } B > \chi^2_{\alpha,r-1}, \text{ reject } H_0 \tag{10.5}$$

where $\chi^2_{\alpha,r-1}$ is the upper α percentage point of the χ^2 distribution with $r - 1$ degrees of freedom.

| Example 10.3 | In Example 10.1, we combined data from a product line of paint coatings having target percentages of solids content falling within the range of 40 to 55 percent because of the need for |

Table 10.3   Calculations for Barlett test for equality of three population variances

Population i	s_i^2	n_i	$n_i - 1$	$(n_i - 1)s_i^2$	$\ln s_i^2$	$(n_i - 1)\ln s_i^2$
(15%) 1	0.984	25	24	23.616	-0.016129	-0.3871
(45%) 2	0.506	35	34	17.204	-0.681219	-23.1614
(75%) 3	0.392	30	29	11.368	-0.936493	-27.1583
Sum total			87	52.188		-50.7068

$$\bar{s}^2 = \frac{52.188}{87} = 0.59986$$

$$\ln \bar{s}^2 = -0.51106$$

constancy of variance around nominal values. Table 10.3 contains summary data from a study on the percentage of solids content for three widely separated target specifications, namely, 15, 45, and 75 percent. With these data, we can formally investigate whether constancy of variance can reasonably be assumed for these specification values. In other words, we wish to test the null hypothesis

$$H_0: \sigma_{15\%}^2 = \sigma_{45\%}^2 = \sigma_{75\%}^2$$

Substituting into (10.4) the needed calculations shown in Table 10.3, we obtain the following value for B:

$$B = \frac{87(-0.51106) - (-50.7068)}{1 + \frac{1}{3(2)}\left[\left(\frac{1}{24} + \frac{1}{34} + \frac{1}{29}\right) - \frac{1}{87}\right]} = 6.148$$

Choosing $\alpha = 0.05$, we need to find $\chi_{0.05,3-1}^2$. Below, we employ Minitab to find this point:

```
MTB > InvCDF 0.95;
SUBC> Chisquare 2.

Inverse Cumulative Distribution Function

Chi-Square with 2 DF

P( X <= x)      x
0.9500       5.9915    ◄――――――   χ²₀.₀₅,₂
```

Since $B = 6.148 > 5.9915$, we reject the hypothesis that all three population variances are equal. Hence, it is not appropriate to combine these nominal values to construct a DNOM chart based on an assumption of equal variances.

STANDARDIZED SHORT-RUN CHARTS

If variances are different for different nominal values, the "deviation from the nominal value" approach described earlier is not advisable. An alternative approach is to standardize the subgroup statistics, relative to the variability associated with their nominal values.

With different variances for different nominal values, a single reliable estimate for the average range cannot be obtained by pooling different part types or batches. Instead, a standardized approach requires an estimate of variability at each nominal value. For instance, let $\bar{R}_i$ be the average range value for a product type with a nominal value of T_i. Since, due to the short-run conditions, there will not typically be enough data for any given nominal value at the time of control chart construction, the standardized approach requires estimates of each $\bar{R}_i$ to be obtained from prior history. Hence, the standardized approach requires greater background data collection efforts than those required by the constant-variance case.

Given an average range value $\bar{R}_i$, the traditional range chart control limits for the ith product type are given by $\text{UCL}_R = D_4 \bar{R}_i$ and $\text{LCL}_R = D_3 \bar{R}_i$. Notice that plotting a subgroup range value R relative to these limits is equivalent to plotting $R/\bar{R}_i$ relative to an upper control limit equal to D_4 and a lower control limit equal to D_3. For example, a plotted point is not a signal of out of control if

$$D_3 \bar{R}_i < R < D_4 \bar{R}_i \qquad \text{or} \qquad D_3 < \frac{R}{\bar{R}_i} < D_4$$

The statistic $R/\bar{R}_i$ is referred to as a *standardized range*. Correspondingly, the control chart for the standardized ranges is referred to as a *standardized R chart*. Because the limits of a standardized R chart are fixed at $\text{UCL} = D_4$ and $\text{LCL} = D_3$ regardless of the nominal value, standardized ranges associated with different nominal values can be plotted together under a single control chart.

Similarly, a *standardized $\bar{x}$ chart* can be developed to allow for the construction of a single control chart in the presence of different nominal values. For a given average range value $\bar{R}_i$ and a nominal target value T_i, the subgroup mean chart control limits for the ith product type are given by $\text{UCL}_{\bar{x}} = T_i + A_2 \bar{R}_i$ and $\text{LCL}_{\bar{x}} = T_i - A_2 \bar{R}_i$. An out-of-control signal is not flagged if

$$T_i - A_2 \bar{R}_i < \bar{x} < T_i + A_2 \bar{R}_i \qquad \text{or} \qquad -A_2 < \frac{\bar{x} - T_i}{\bar{R}_i} < A_2$$

Thus, in general, a standardized $\bar{x}$ chart calls for the plotting of *standardized means* $(\bar{x} - T_i)/\bar{R}_i$ with control limits fixed at $\text{UCL} = -A_2$ and $\text{LCL} = A_2$ regardless of nominal value. The centerline is set at the value 0.

As with the DNOM approach, combining of part types or batches, to be used in a single standardized chart, should be done in a judicious and logical manner. A common recommended application for standardized short-run charts is for a process in which the disturbances of central tendency or variability are potentially related to some common denominator regardless of the nominal value, for example, monitoring of a particular part characteristic produced from a particular machining center such as drilling or blanking.

The standardized mean and range charts make no assumption about the relationship between variation and nominal value. The only assumption is that the inherent level of variation is not equal across different nominal values. In some applications, it has been observed that variation tends to change in a systematic fashion as nominal value changes; for example, variation increases proportionally as the nominal value increases. In the proportional case, Farnum (1992) demonstrates that a control chart can be implemented to monitor $\bar{x}_i/T_i$ rather than $(\bar{x} - T_i)/\bar{R}_i$; the reader is referred to his article for the details of the proposed control chart construction.

Table 10.4   Standardized attribute statistics for short-run applications

Attribute	Average Value (Part Type i)	Plotted Statistic
Proportion nonconforming	$\bar{p}_i$	$z_i = \dfrac{\hat{p}_i - \bar{p}_i}{\sqrt{\bar{p}_i(1 - \bar{p}_i)/n}}$
Number of nonconforming	$n\bar{p}_i$	$z_i = \dfrac{n\hat{p}_i - n\bar{p}_i}{\sqrt{n\bar{p}_i(1 - \bar{p}_i)}}$
Number of nonconformities	$\bar{c}_i$	$z_i = \dfrac{c_i - \bar{c}_i}{\sqrt{\bar{c}_i}}$
Number of nonconformities per unit	$\bar{u}_i$	$z_i = \dfrac{u_i - \bar{u}_i}{\sqrt{\bar{u}_i}}$

The notion of standardizing variable data for a short-run application can be carried over to deal with attribute data in a short-run environment. Standardizing of attribute data was discussed in Chapter 7 as a means of compensating for variable sample sizes so that constant-value control limits can be used. Similarly, standardizing allows for different part types with different rates or proportions of nonconformities to be plotted on the same chart. As with standardized charts for variable data, short-run attribute charts require historical data to obtain good estimates of average rate or proportion levels by part type; for example, $\bar{p}_i$ represents the estimate of the average proportion defective for the ith part type. A summary of some common standardized attribute statistics is provided in Table 10.4. For all cases, a standardized attribute control chart is constructed with a center line at 0 and control limits fixed at UCL $= 3$ and LCL $= -3$.

10.2   MULTIVARIATE QUALITY CONTROL

In all our analyses thus far, the emphasis was on the use of data analysis techniques (from standard control charts to time-series modeling) to study the behavior of data series associated with a single variable. There are many practical situations for which process understanding and control require the simultaneous study of two or more quality characteristics. For example, in the manufacture of critical bolts used in the assembly of airplanes, the properties of hardness and tensile strength together determine the integrity of a bolt. A critical bolt that might be used to hold a wing to the main body needs to have high tensile strength to handle the high load requirement. It is known that the harder the material, the higher the tensile strength *tends* to be; that is, a positive statistical relationship between the two quality characteristics exists. But the existence of a relationship does not imply that only one variable needs to be studied and used as a proxy for the other variable. In this case, hardness must be measured and controlled along with tensile strength. Softer material is associated with low tensile strengths, but very hard material, though associated with higher tensile strength, is problematic because a very hard material suffers from brittleness and susceptibility to cracking. Hence, both quality characteristics of hardness and tensile strength must be monitored simultaneously.

Table 10.5　　Bivariate correlated data

Sample	x_1	x_2	Sample	x_1	x_2
1	11.09	20.28	16	9.67	20.16
2	9.33	19.01	17	9.94	21.40
3	10.95	20.96	18	8.77	19.33
4	10.53	21.33	19	9.40	17.45
5	11.95	22.51	20	11.41	23.57
6	9.60	18.94	21	8.75	18.02
7	10.05	19.72	22	12.52	24.86
8	11.36	22.05	23	9.85	20.88
9	8.43	15.69	24	11.07	21.29
10	9.26	16.93	25	11.60	17.37
11	9.79	20.01	26	10.42	20.60
12	9.13	17.76	27	8.53	17.27
13	10.05	20.19	28	9.36	18.19
14	10.73	21.92	29	10.20	21.29
15	9.44	20.23	30	11.02	20.93

One possible approach for the monitoring of two or more related quality characteristics is to independently monitor the characteristics with separate control charts. For example, suppose that a product is defined by two measurable characteristics x_1 and x_2, and measurements of both characteristics are taken at regular intervals. Suppose that the mean values are known to be $\mu_1 = 10$ and $\mu_2 = 20$, and the standard deviations are known to be $\sigma_1 = 1$ and $\sigma_2 = 2$. Furthermore, the characteristics are positively correlated with the underlying correlation ρ equal to 0.80. Recall that correlation is a measure of linear association ranging from -1 to 1. In Table 10.5, we present observations for x_1 and x_2 for 30 successive sampling periods.

As illustrated in Figure 10.2, two separate 3σ Shewhart charts can then be constructed. Examining these charts separately, we see that each data series appears random and stable with no indication of unusual observations. Hence, from this perspective, the process would be deemed in control.

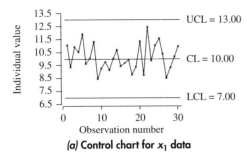
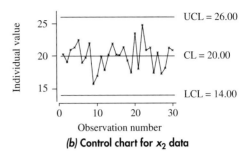

(a) **Control chart for x_1 data**   (b) **Control chart for x_2 data**

Figure 10.2   Implementing separate control charts.

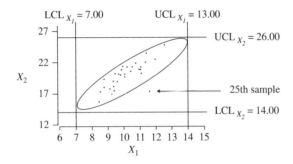

Figure 10.3   Elliptical control chart versus independent
control chart limits.

Because of the positive correlation between the quality characteristics, it can be noticed that there is a similarity in the movements of the x_1 and x_2 data. Namely, at any given time point, both plotted points tend to be on one side of their respective centerlines, and the relative distances to the centerline are about the same on each chart. There appears to be one exception to this general tendency. At sampling period 25, the positions of the two plotted points, relative to their respective centerlines, differ greatly; the x_1 observation is well above its centerline while the x_2 observation is well below its centerline. This sample period is inconsistent with the general workings of the process. In technical terms, we say that the *bivariate* process is out of control. However, by viewing the process univariately (i.e., one variable at a time) the inconsistency escapes attention.

Figure 10.3 provides a perspective on the issues at hand. By using separate univariate control charts (as shown in Figure 10.2), the process will be considered in control if both points fall within their respective control limits. This is equivalent to concluding that the process is in control if the pair of sample observations falls within the rectangle shown in Figure 10.3. On the other hand, the process will be considered out of control if the pair of sample observations falls outside the rectangle. One difficulty with using the rectangular region for control decisions is that it fails to take into account the joint tendencies between the individual variables. From Figure 10.3, we see that the paired observations are concentrated within an elliptical region. This concentration reflects the general workings of the process. As such, any substantial departure from this concentration reflects a potential out-of-control condition.

Given that the superimposed ellipse serves as a more appropriate boundary for natural variability, we see that the point associated with the observations for the 25th sampling period falls well beyond the elliptical control limits. However, this legitimate out-of-control signal was not viewed as unusual from the perspective of the rectangular control region. This is indeed the same sampling period which we noted earlier as escaping attention of the separate control charts shown in Figure 10.2.

Susceptibility to not detecting bivariate out-of-control conditions represents one of the weaknesses in the use of a rectangular control region or separate control charts. The difficulty to maintain a desired overall type I error rate (α) constitutes another problem with the simultaneous implementation of separate control charts. As an example, consider the situation of implementing two separate control charts to monitor two quality characteristics which are independent. If 3σ control limits are used on each characteristic, the probability that both charts will simultaneously signal the process is in control is $0.9973(0.9973) = 0.994607$. Thus, the probability that one or both charts will signal out of control, when the process is truly in control, is 0.005393 ($= 1 - 0.994607$). Thus, the overall type I error rate is nearly twice what is expected with a single 3σ based control chart.

When the quality characteristics are not independent, the overall type I error rate associated with the simultaneous implementation of separate control charts is less straightforward to calculate. However, we can gain some insight about the overall type I error rate for the dependent case from Figure 10.3. As noted earlier, the separate control charts are 3σ based. The ellipse in the figure was constructed in such a manner that the probability of a joint observation's falling within the ellipse is 0.9973; or, equivalently, the probability of a joint observation's falling outside the ellipse is 0.0027. As we have explained, the use of two separate control charts is equivalent to using the inner rectangular region, seen in Figure 10.3, for control purposes. Thus, when we are using two separate control charts, an out-of-control signal is associated with a bivariate point's falling *outside* the rectangular region. Since the region outside the rectangle is a subset of the larger region outside the ellipse, we realize that the overall type I error rate with the implementation of simultaneous separate control charts must necessarily be *less* than 0.0027. Notice that this result is in contrast to the independent case where we showed that the overall false-alarm rate is greater than 0.0027 when separate charts are maintained. In either case, we learn that the overall false-alarm rate is not controlled to 0.0027 with the separate chart implementation.

Given the scatter of points in Figure 10.3, our use of an elliptical region as a control region seems visually appealing. There are, however, also sound theoretical reasons for the appropriateness of an ellipse. In problems concerning the study of two random variables X and Y, observations are regarded in pairs, where

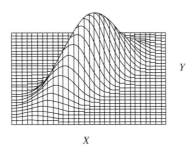

Y

X

Figure 10.4   Bivariate normal distribution.

each pair consists of one observation on *X* and one on *Y*. A population consisting of all possible paired values on two random variables is called a *bivariate population* or *distribution*. One extremely important bivariate distribution is the bivariate normal distribution. The bivariate normal distribution is an extension of the normal distribution for a single random variable. Because two random variables are involved, the mathematical equation for the bivariate density function is of the form $Z = f(X, Y)$; readers interested in more details of the bivariate normal function are referred to Johnson and Wichern (1998). Graphing such a function produces a three-dimensional figure such as that shown in Figure 10.4. The precise shape of the figure depends on the variances and means of *X* and *Y* and on the correlation between *X* and *Y*. Probabilities are represented by volumes instead of by areas, as is the case with the univariate normal distribution. Thus, the total volume under the bivariate density curve is 1.

An important fact about the bivariate normal distribution is that if the three-dimensional solid of Figure 10.4 is cut by any plane parallel to the *XY* plane, the curve formed is an ellipse centered at the point (μ_X, μ_Y) where μ_X and μ_Y are the mean values of *X* and *Y*, respectively. As such, ellipses serve as a basis for many testing and estimation procedures associated with the bivariate normal distribution. It is also important to know that the probabilities of an (*X*, *Y*) pair's falling within or outside an ellipse can easily be computed from well-known univariate distributions, such as the chi-square or *F* distributions. As we will soon see, the use of a chi-square or an *F* distribution depends on whether parameters are known or need to be estimated.

In summary, one possible approach for monitoring two correlated quality characteristics which follow a bivariate normal distribution is to plot paired observations on a scatter diagram with a superimposed ellipse. This ellipse would be constructed in such a manner that the probability of a false signal (type I error) is controlled to some desired α level. If a plotted point falls within the ellipse, the process is deemed in control while points outside the ellipse are signals of potential out-of-control conditions. There are, however, two primary disadvantages to an elliptical control chart approach. First, scatter-type diagrams provide no indication of time or sequence order. Hence, it is not possible from the elliptical plot alone to determine the time at which a given observation occurred. Second, the plotting of data, along with a control ellipse, becomes difficult or impossible for more general cases of jointly monitoring three or more quality characteristics. If three characteristics are considered, then the scatter plot is a three-dimensional plot with the elliptical control region being a three-dimensional ellipsoid (that is, an egglike object). Fortunately, both disadvantages can be overcome with procedures described in the following subsections.

CHI-SQUARE MULTIVARIATE CONTROL CHART: PARAMETERS KNOWN

When one is concerned with the analysis of measurements made on several variables or characteristics, it is important to have a structured framework which allows for a clearer presentation of the data. Matrix notation provides the basis for such a framework. Matrices also provide a convenient medium for manipulation of the data. Through the use of certain algebraic operations that can be performed on matrices, the matrix approach permits large arrays of data to be denoted compactly and operated upon efficiently. In an end-of-chapter appendix, we provide a brief and elementary presentation of matrices and matrix algebra. A fuller treatment of matrix algebra, with application to statistical methods, can be found in specialized books such as those by Johnson and Wichern (1998) and by Searle (1982).

Suppose now that we are interested in simultaneously monitoring p (≥ 2) quality characteristics. As a generalization of the bivariate normal distribution, we assume that the p characteristics $X_1, X_2, \ldots, X_p$ are jointly distributed according to a *multivariate* normal distribution. Let $\mu_1, \mu_2, \ldots, \mu_p$ be the mean values and $\sigma_1^2, \sigma_2^2, \ldots, \sigma_p^2$ be the variances of the respective quality characteristics when the process is in control. Furthermore, we can assume that $n \geq 1$ measurements of each quality characteristic are taken in each sampling period. Hence statistics, such as the sample mean, can be computed for each of the p variables.

Since the characteristics may be interdependent, a measure of dependence between each of the characteristics is required. The measure used in the development of the multivariate procedure is known as *covariance*. Covariance measures the extent to which two random variables "vary together." The theoretical or population covariance of X_i and X_j is given by

$$\sigma_{ij} = E[(X_i - \mu_i)(X_j - \mu_j)] \tag{10.6}$$

If X_i tends to vary in the same direction from its mean (above or below) as does X_j from its mean, then the products of those deviations, on average, will be positive. Conversely, if X_i tends to vary in the opposite direction from its mean relative to X_j and its mean, then the products of those deviations, on average, will be negative. If there are little or no tendencies for the variables to vary together, the covariance will be close to zero. It is useful to recognize that variance is a special case of covariance. Namely, variance represents how a random variable varies with itself. Thus, the covariance of X_i and X_i is simply the variance of X_i as shown below:

$$\sigma_{ii} = E[(X_i - \mu_i)(X_i - \mu_i)] = E[(X_i - \mu_i)^2] = \sigma_i^2$$

The notion of covariance may seem quite similar to a correlation measure. Indeed, covariance and correlation are directly related. If we divide the covariance by the standard deviations of the two random variables, we get a measure constrained in value between -1 and $+1$. The formula by which the covariance can be converted to the correlation measure is

$$\rho_{ij} = \frac{\sigma_{ij}}{\sigma_i \sigma_j} \qquad (10.7)$$

It is now convenient to organize the details of our discussion into a matrix framework. If sample means are collected at each period on each quality characteristic, we are then concerned with the joint behavior of p random variables: $\bar{X}_1, \bar{X}_2, \ldots, \bar{X}_p$. These random variables can be organized as follows:

$$\bar{\mathbf{X}}_k = \begin{bmatrix} \bar{X}_{1k} \\ \bar{X}_{2k} \\ \vdots \\ \bar{X}_{pk} \end{bmatrix} \qquad (10.8)$$

where k indicates the sampling period, $k = 1, 2, \ldots$. The use of boldface type indicates that the defined symbol refers to a matrix. In this case, the matrix has p rows and 1 column. Such a matrix is often called a *vector*. Since $\bar{\mathbf{X}}_k$ contains elements that are random variables, $\bar{\mathbf{X}}_k$ is referred to as a *random vector*. Since the expected value of $\bar{X}_i$ equals the expected value of X_i (see Section 2.9), the expected value of $\bar{\mathbf{X}}_k$ is given by

$$E(\bar{\mathbf{X}}_k) = \begin{bmatrix} E(\bar{X}_{1k}) \\ E(\bar{X}_{2k}) \\ \vdots \\ E(\bar{X}_{pk}) \end{bmatrix} = \begin{bmatrix} \mu_1 \\ \mu_2 \\ \vdots \\ \mu_p \end{bmatrix} = \boldsymbol{\mu} \qquad (10.9)$$

The vector $\boldsymbol{\mu}$ is then a population mean vector.

By considering every pairwise combination of X_i and X_j, we can organize all the covariances into the following matrix:

$$\boldsymbol{\Sigma} = \begin{bmatrix} \sigma_{11} & \sigma_{12} & \cdots & \sigma_{1p} \\ \sigma_{21} & \sigma_{22} & \cdots & \sigma_{2p} \\ \cdots & \cdots & \cdots & \cdots \\ \sigma_{p1} & \sigma_{p2} & \cdots & \sigma_{pp} \end{bmatrix} \qquad (10.10)$$

The matrix $\boldsymbol{\Sigma}$ has p rows and p columns. This matrix of covariances is often written in a different way by using the following two facts. First, as earlier demonstrated, the covariance of a random variable on itself is the variance of the random variable; thus $\sigma_{ii} = \sigma_i^2$. Second, the covariance of X_i and X_j is clearly equal to the covariance of X_j and X_i, thus $\sigma_{ij} = \sigma_{ji}$. With these facts in mind, (10.10) can be conveniently presented in terms of its diagonal and its upper diagonal part, that is,

$$\Sigma = \begin{bmatrix} \sigma_1^2 & \sigma_{12} & \cdots & \sigma_{1p} \\ & \sigma_2^2 & \cdots & \sigma_{2p} \\ & & \ddots & \vdots \\ & & & \sigma_p^2 \end{bmatrix} \qquad (10.11)$$

Implicitly, the lower diagonal part is a symmetric reflection of the upper diagonal part. Since Σ is a matrix of variances and covariances, it is commonly referred to as a *variance-covariance matrix*.

Given the dimensions of the defined matrices, certain matrix operators, and the rules of matrix algebra (see the end-of-chapter appendix), the matrices can be combined to define the following random variable:

$$\chi_k^2 = n(\overline{\mathbf{X}}_k - \boldsymbol{\mu})'\Sigma^{-1}(\overline{\mathbf{X}}_k - \boldsymbol{\mu}) \qquad (10.12)$$

where $(\overline{\mathbf{X}}_k - \boldsymbol{\mu})'$ is the transpose of $\overline{\mathbf{X}}_k - \boldsymbol{\mu}$ and Σ^{-1} is the inverse of Σ. This random variable is one-dimensional; that is, the right-hand side (10.12) reduces to a scalar or single number. If the means, variances, and covariances remain at their in-control values (i.e., no change in process parameters), the random variable of (10.12), as reflected by its notation, follows a chi-square distribution with p degrees of freedom. If one or more of the mean values change, then there will be a tendency for the values of the random variable to be *larger*, on average, than when the process is in control. Hence, to construct a control chart for values of χ_k^2, we need only an upper control limit. In particular, for a specified α, the upper control limit is

$$\text{UCL} = \chi_{\alpha,p}^2 \qquad (10.13)$$

For the important case of two quality characteristics, the chi-square statistic of (10.12) can be conveniently expressed without the use of matrices. Namely, if $\bar{x}_{1,k}$ and $\bar{x}_{2,k}$ represent the sample averages of the two quality characteristics for sampling period k, the chi-square statistic is computed as follows:

$$\chi_k^2 = \frac{n}{\sigma_1^2\sigma_2^2 - \sigma_{12}^2}[\sigma_2^2(\bar{x}_{1,k} - \mu_1)^2 + \sigma_1^2(\bar{x}_{2,k} - \mu_2)^2 - 2\sigma_{12}(\bar{x}_{1,k} - \mu_1)(\bar{x}_{2,k} - \mu_2)] \qquad (10.14)$$

where σ_{12}^2 is the square of the covariance σ_{12}. The value of this statistic would be compared with an upper control limit value of $\chi_{\alpha,2}^2$. In the previous subsection, we demonstrated that bivariate normal data can be monitored by superimposing an ellipse, constructed in such a manner that the probability of a sample mean pair's being plotted outside the elliptical region is α, when the process is in control. Plotting of the statistic defined by (10.14), with an upper control limit of $\chi_{\alpha,2}^2$, is an equivalent procedure in that if χ_k^2 is less than the upper control limit, then $(\bar{x}_{1,k}, \bar{x}_{2,k})$ will fall within the elliptical region. Similarly, if χ_k^2 is greater than the upper control limit, then $(\bar{x}_{1,k}, \bar{x}_{2,k})$ will fall outside the elliptical region. The advantages of the chi-square approach over the ellipse approach are the ability to plot the data sequentially in time order and the ease of handling more than two dimensions (characteristics).

To illustrate the construction of a chi-square chart for multivariate data, we consider the data studied in the previous subsection and presented in Table 10.5. Recall that the means for the characteristics x_1 and x_2 are known to be $\mu_1 = 10$ and $\mu_2 = 20$, respectively, and the standard deviations are known to be $\sigma_1 = 1$ and $\sigma_2 = 2$. Additionally, the underlying correlation ρ between the characteristics is equal to 0.80. From (10.7), we can determine the covariance to be $\sigma_{12} = \rho_{12}\sigma_1\sigma_2 = 0.80(1)(2) = 1.6$. Hence, the mean vector and variance-covariance matrix are, respectively,

$$\boldsymbol{\mu} = \begin{bmatrix} \mu_1 \\ \mu_2 \end{bmatrix} = \begin{bmatrix} 10 \\ 20 \end{bmatrix}$$

Example 10.4

and

$$\boldsymbol{\Sigma} = \begin{bmatrix} 1^2 & 1.6 \\ 1.6 & 2^2 \end{bmatrix} = \begin{bmatrix} 1 & 1.6 \\ 1.6 & 4 \end{bmatrix}$$

Since only one measurement on each characteristic is taken each sampling period, the subgroup size n is equal to 1. For the first sampling period, we find from Table 10.5 that the observed sample vector is

$$\bar{\mathbf{x}}_1 = \begin{bmatrix} x_{11} \\ x_{21} \end{bmatrix} = \begin{bmatrix} 11.09 \\ 20.28 \end{bmatrix}$$

From (10.12), we see the need to subtract the mean vector from the sample vector, which results in

$$\bar{\mathbf{x}}_1 - \boldsymbol{\mu} = \begin{bmatrix} 11.09 \\ 20.28 \end{bmatrix} - \begin{bmatrix} 10 \\ 20 \end{bmatrix} = \begin{bmatrix} 1.09 \\ 0.28 \end{bmatrix}$$

Consequently, applying (10.12), we have

$$\chi_1^2 = [1.09 \ 0.28] \begin{bmatrix} 1 & 1.6 \\ 1.6 & 4 \end{bmatrix}^{-1} \begin{bmatrix} 1.09 \\ 0.28 \end{bmatrix}$$

To determine the inverse of $\boldsymbol{\Sigma}$, we can use Minitab.[3] Below are the Minitab session commands and entries leading to the determination of the inverse:

```
MTB > name m1 'Sigma'
MTB > read 2 2 m1                    Calc>>Matrices>>Read
DATA> 1.0 1.6
DATA> 1.6 4.0
        2 rows read.
MTB > inverse m1 m2                  Calc>>Matrices>>Invert
MTB > name m2 'Sigma Inverse'
MTB > print m2

Data Display

 Matrix Sigma Inverse

    2.77778   -1.11111
   -1.11111    0.69444
```

---

[3]Matrix capabilities are found only in the full versions of Minitab, not in the student versions. As an alternative, Microsoft Excel can be used to accomplish all the necessary matrix operations.

Thus, we now have

$$\chi_1^2 = [1.09\ \ 0.28]\begin{bmatrix} 2.77778 & -1.11111 \\ -1.11111 & 0.69444 \end{bmatrix}\begin{bmatrix} 1.09 \\ 0.28 \end{bmatrix}$$

Since the dimensions of the above vectors and matrix are small, hand computation of their product is relatively straightforward by using the rules of matrix multiplication found in the end-of-chapter appendix. As a continuation of the just-presented sequence of Minitab session commands, below we show how Minitab can be employed to complete the remaining calculations:

```
MTB > read 1 2 m3
DATA> 1.09 0.28
      1 rows read.
MTB > name m3 'xtranspose'
MTB > read 2 1 m4
DATA> 1.09
DATA> 0.28
      2 rows read.
MTB > name m4 'x'
MTB > multiply 'xtranspose' 'Sigma Inverse' m5
MTB > multiply m5 'x' m6          Calc>>Matrices>>Arithmetic
Answer = 2.6765
```

Since this application deals with two quality characteristics, the value for χ_1^2 can alternatively be determined from (10.14) as follows:

$$\chi_1^2 = \frac{1}{1(4) - 1.6^2}[4(11.09 - 10)^2 + 1(20.28 - 20)^2 - 2(1.6)(11.09 - 10)(20.28 - 20)] = 2.6765$$

For period 2, we find

$$\chi_2^2 = \frac{1}{1(4) - 1.6^2}[4(9.33 - 10)^2 + 1(19.01 - 20)^2 - 2(1.6)(9.33 - 10)(19.01 - 20)] = 0.4536$$

If we choose $\alpha = 0.0027$, the upper control limit for the computed statistics is $\chi_{0.0027,2}^2$, which, as seen below, equals 11.829:

```
MTB > InvCDF 0.9973;
SUBC> ChiSquare 2.

Inverse Cumulative Distribution Function

Chi-Square with 2 DF

P( X <= x)               x
0.9973              11.8290
```

A sequence plot of the bivariate control chart data with this control limit is shown in Figure 10.5.[4] Consistent with the elliptical chart (Figure 10.3), the chi-square chart views the 25th sample as an out-of-control point. The challenge is to discover the causes, if any exist, for the bivariate departure in the process.

---

[4]Minitab does not have options for multivariate control charts. This multivariate control chart, along with others presented in this chapter, was constructed by using Minitab's time-series plotting option.

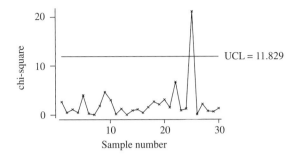

Figure 10.5   Chi-square control chart for $p = 2$ quality characteristics.

HOTELLING T^2 CONTROL CHART: PARAMETERS UNKNOWN

Since process parameters are typically unknown, there is a need to estimate thcm from thc data. As in the case of univariate charts, preliminary samples are gathered for purposes of parameter estimation and control chart construction. For the multivariate case, we need to estimate the elements of the mean vector μ and the elements of the variance-covariance matrix Σ.

Technically, the chi-square multivariate control chart presented in the previous section is applicable only when the variance and covariance parameters are known. This is theoretically equivalent to saying that an infinite number of samples have been observed so that there is no uncertainty about the true values of the parameters. In practical application, it is often sufficient to have a "large" number of samples to justify the application of infinite samples–based methods. For example, when parameters are unknown, the standard Shewhart chart is based on the normal distribution when technically the t distribution is the appropriate baseline distribution. The assumption is made that the number of samples used in the construction of the Shewhart chart is large enough, say, 25 or 30, to take advantage of the fact that the normal distribution is close enough to the t distribution for such sample sizes.

For multivariate control chart applications, the F distribution is the appropriate distribution to use in control chart construction when μ and Σ need to be estimated. Unlike in the univariate case, there is no simple rule of thumb for the minimum number of samples required to justify using the chi-square distribution in place of the F distribution. The difficulty is that the number of samples necessary for chi-square distribution–based control limits to adequately approximate the F distribution–based limits depends on the number of quality characteristics (p) and the sample size (n). Lowry and Montgomery (1995) provide tables indicating the recommended minimum number of samples for $p = 2, 3, 4, 5, 10$, and 20 and for $n = 3, 5$, and 10. It is worth noting that for all investigated combinations of p and n, the recommended minimum values are always greater than 20 samples, and for many combinations greater than 50 samples.

When the number of preliminary samples used in the estimation of parameters is not sufficiently large, there are a couple of considerations that should be incorporated in the implementation of a multivariate control chart scheme. First, it is advisable to use the *F* distribution, not the large-sample–based chi-square distribution. Second, it is necessary to consider a two-phase approach to control chart implementation. As first discussed in Chapter 6, *phase 1* is the analysis of the preliminary samples to determine whether they came from an in-control process. Phase 1 analysis is what we earlier referred to as retrospective control. *Phase 2* (or prospective control) deals with the monitoring of future data to determine whether the process remains in control. In our presentation of univariate control charts, if the process was judged to be in control from the preliminary data, we simply projected out the retrospective limits to serve as prospective limits. Allowing phase 1 and phase 2 limits to be the same is justifiable if the number of preliminary samples is sufficiently large, say, 20 or 25, as is often the case. When the number of preliminary samples is not sufficiently large, it is not technically appropriate to use the same set of control limits for each phase. In this section, we will present the multivariate extension of the phase 1 and 2 approach as presented by Alt and Smith (1990). As is shown below, it is more convenient to separate our presentation into two cases: (1) sample size *n* is greater than 1, and (2) sample size *n* is equal to 1.

Sample Size $n > 1$ Suppose that *m* preliminary samples of size n (> 1) are drawn for purposes of estimating process parameters and determining whether the process is in control. As before, we will assume that *p* quality characteristics are simultaneously being measured and jointly monitored. For each of the *m* samples, a vector of sample means can be computed:

$$
\begin{bmatrix} \bar{x}_{11} \\ \bar{x}_{21} \\ \vdots \\ \bar{x}_{p1} \end{bmatrix},
\begin{bmatrix} \bar{x}_{12} \\ \bar{x}_{22} \\ \vdots \\ \bar{x}_{p2} \end{bmatrix},
\dots,
\begin{bmatrix} \bar{x}_{1m} \\ \bar{x}_{2m} \\ \vdots \\ \bar{x}_{pm} \end{bmatrix}
\tag{10.15}
$$

where $\bar{x}_{jk}$ represents the sample mean of the *j*th quality characteristic for the *k*th sample, that is,

$$
\bar{x}_{jk} = \frac{\sum\limits_{i=1}^{n} x_{ijk}}{n} \qquad \begin{array}{l} j = 1, 2, \dots, p \\ k = 1, 2, \dots, m \end{array}
\tag{10.16}
$$

where x_{ijk} is the *i*th observation on the *j*th quality characteristic in the *k*th sample. An overall estimate of the mean level for the *j*th quality characteristic is simply obtained by averaging the sample means over all *m* samples:

$$
\bar{\bar{x}}_{j} = \frac{\sum\limits_{k=1}^{m} \bar{x}_{jk}}{m} \qquad j = 1, 2, \dots, p
\tag{10.17}
$$

Placing the individual mean estimates in a vector, we obtain an unbiased estimate for the mean vector $\boldsymbol{\mu}$:

$$\hat{\boldsymbol{\mu}} = \begin{bmatrix} \bar{\bar{x}}_1 \\ \bar{\bar{x}}_2 \\ \vdots \\ \bar{\bar{x}}_p \end{bmatrix} \tag{10.18}$$

We now need an estimate of the variance-covariance matrix $\boldsymbol{\Sigma}$. For the kth sample, an estimate of the variance for quality characteristic j is obtained from the usual sample variance calculation:

$$s_{jk}^2 = \frac{\sum_{i=1}^{n}(x_{ijk} - \bar{x}_{jk})^2}{n-1} \qquad \begin{array}{l} j = 1, 2, \ldots, p \\ k = 1, 2, \ldots, m \end{array} \tag{10.19}$$

An estimate of the covariance between quality characteristic j and quality characteristic h from the kth sample is

$$s_{jhk} = \frac{\sum_{i=1}^{n}(x_{ijk} - \bar{x}_{jk})(x_{ihk} - \bar{x}_{hk})}{n-1} \qquad \begin{array}{l} j = 1, 2, \ldots, p \\ j \neq h \end{array} \tag{10.20}$$

As was done with the mean estimates, the variance and covariance estimates can be averaged over all m samples to obtain overall estimates:

$$\bar{s}_j^2 = \frac{\sum_{k=1}^{m} s_{jk}^2}{m} \qquad j = 1, 2, \ldots, p \tag{10.21}$$

and

$$\bar{s}_{jh} = \frac{\sum_{k=1}^{m} s_{jhk}}{m} \qquad j \neq h \tag{10.22}$$

Placing all the estimates from (10.21) and (10.22) in a matrix structure, we obtain an unbiased estimate of the true variance-covariance matrix $\boldsymbol{\Sigma}$. Denoting the matrix estimate by $\mathbf{S}$, we have

$$\mathbf{S} = \begin{bmatrix} \bar{s}_1^2 & \bar{s}_{12} & \cdots & \bar{s}_{1p} \\ & \bar{s}_2^2 & \cdots & \bar{s}_{2p} \\ & & \ddots & \vdots \\ & & & \bar{s}_p^2 \end{bmatrix} \tag{10.23}$$

By replacing both the mean vector $\boldsymbol{\mu}$ with $\hat{\boldsymbol{\mu}}$ and the variance-covariance matrix $\boldsymbol{\Sigma}$ with $\mathbf{S}$ in (10.12), we have the following statistic value, for sampling period k:

$$T_k^2 = n(\bar{\mathbf{x}}_k - \hat{\boldsymbol{\mu}})' \mathbf{S}^{-1}(\bar{\mathbf{x}}_k - \hat{\boldsymbol{\mu}}) \tag{10.24}$$

Due to the pioneering work of Hotelling (1947), the above statistic is often referred to as a *Hotelling T^2 statistic*. Analogous to (10.14), for the bivariate case ($p = 2$), the statistic can be computed in such a manner:

$$T_k^2 = \frac{n}{\bar{s}_1^2 \bar{s}_2^2 - \bar{s}_{12}^2}[\bar{s}_2^2(\bar{x}_{1k} - \bar{\bar{x}}_1)^2 - \bar{s}_1^2(\bar{x}_{2k} - \bar{\bar{x}}_2)^2 - 2\bar{s}_{12}(\bar{x}_{1,k} - \bar{\bar{x}}_1)(\bar{x}_{2,k} - \bar{\bar{x}}_2)] \quad \textbf{(10.25)}$$

Just as with the chi-square statistic for the parameters-known case, there is only an upper control limit involved in monitoring the T^2 statistic. As shown by Alt (1982), the phase 1 upper control limit applied to the preliminary samples is given by

$$UCL = \frac{p(m-1)(n-1)}{m(n-1) - p + 1}F_{\alpha, p, m(n-1)-p+1} \quad \textbf{(10.26)}$$

If any of the initial T_k^2 signal out of control and a special-cause explanation can be assigned, the sample is discarded and the sample mean vector $\hat{\boldsymbol{\mu}}$ and sample variance-covariance matrix $\mathbf{S}$ are recomputed. Note that the control limit given by (10.26) must be recomputed with m replaced by m^*, which represents the number of remaining samples. The iterative procedure continues until the remaining data are deemed not to be influenced by special causes.

As recommended by Alt and Smith (1990), for phase 2, future sample vectors are mapped to T^2 values and monitored with an upper control limit set at

$$UCL = \frac{p(m^* + 1)(n - 1)}{m^*(n - 1) - p + 1}F_{\alpha, p, m^*(n-1)-p+1} \quad \textbf{(10.27)}$$

If no samples are discarded in phase 1, then $m^* = m$. Even if $m^* = m$, we notice that the phase 1 limit and phase 2 limit will still differ because of the replacement of the term $m - 1$ in (10.26) with the term $m + 1$.

Example 10.5

A manufacturer of pencil lead for mechanical pencils views three quality characteristics as important enough for simultaneous control: (1) diameter of the lead, (2) force, in pounds, required to break the lead extended 1 mm out of the pencil, angled at 40°, and (3) a proprietary measure of glide, that is, how smoothly the lead moves on the paper (a point value around 5 is desirable, too smooth or too much drag is less desirable). Table 10.6 shows data for these three characteristics for the 0.5-mm-diameter product line. Data are also found in the file "pencil.dat". For computational convenience, the diameter measurements are in hundredths of millimeters relative to 0.5 mm; thus a measurement of 2.3 implies a diameter of 0.523 mm. In terms of our notation, we have $p = 3$, $m = 20$, and $n = 3$.

As a first step in the construction of the Hotelling T^2 control chart, we need to compute the sample means, sample variances, and sample covariances. The sample means of each quality characteristic for each sample are obtained by using (10.16). For example, the mean of the diameter readings for sample 1 is

$$\bar{x}_{11} = \frac{2.3 + (-0.1) + (-1.4)}{3} = 0.267$$

Similarly, the mean of the breaking force data for sample 1 ($\bar{x}_{21}$) is 2.827, and the mean of the glide data for sample 1 ($\bar{x}_{31}$) is 5.280.

Table 10.6   Mechanical lead pencil data

Sample	Diameter (mm) j = 1			Breaking Force (lb) j = 2			Glide (Points) j = 3		
1	2.3	−0.1	−1.4	3.48	2.72	2.28	5.85	5.53	4.46
2	−1.7	0.0	0.1	2.19	2.90	2.74	4.25	5.51	4.24
3	0.4	−0.1	0.4	2.89	2.81	3.24	6.83	5.01	4.65
4	−0.6	0.5	−0.7	2.66	2.98	2.51	6.00	6.40	4.47
5	0.2	0.4	−0.9	2.76	2.52	2.69	4.60	4.70	5.55
6	1.7	−0.4	0.5	3.29	2.76	3.20	5.52	4.51	6.14
7	−0.4	1.3	−1.0	2.94	3.40	2.40	5.32	6.19	5.07
8	−1.1	−0.9	−1.9	2.13	1.85	2.81	5.43	4.24	4.91
9	−0.8	−0.7	0.6	2.59	2.65	2.85	4.70	4.98	5.57
10	0.3	1.5	−0.5	2.87	3.21	2.51	5.63	6.16	4.48
11	−0.3	0.7	−0.7	3.00	2.81	2.32	6.64	4.75	5.21
12	−1.0	−0.8	−0.9	2.03	1.95	2.13	3.94	4.41	5.14
13	0.2	1.4	−0.8	2.42	2.78	2.57	4.99	5.94	4.90
14	0.8	1.4	−0.3	2.57	3.42	2.53	5.22	4.99	5.07
15	−1.1	0.5	−1.0	2.56	2.93	2.53	5.38	5.55	4.47
16	−0.2	0.6	−0.3	2.51	2.79	2.90	4.94	4.92	5.64
17	−0.6	0.4	1.4	2.34	2.72	3.45	5.41	6.19	5.89
18	0.2	1.1	2.1	3.10	2.70	3.46	5.32	4.85	6.13
19	0.0	−1.7	1.6	2.89	2.31	2.66	5.06	4.12	4.50
20	0.2	−0.2	−0.1	3.13	2.38	3.41	5.17	5.04	6.24

The sample variances of each characteristic for each sample can be found by using (10.19). For example, the sample variance of the diameter measurement for sample 1 is

$$s_{11}^2 = \frac{(2.3 - 0.267)^2 + (-0.1 - 0.267)^2 + (-1.4 - 0.267)^2}{3 - 1} = 3.523$$

For each sample, three sample covariances need to be calculated by using (10.20): the covariance between the diameter and the break force, the covariance between the diameter and the glide, and the covariance between the break force and the glide. For sample 1, we find

$$s_{121} = \frac{(2.3 - 0.267)(3.48 - 2.827) + (-0.1 - 0.267)(2.72 - 2.827) + (-1.4 - 0.267)(2.28 - 2.827)}{3 - 1}$$

$$= 1.139$$

and

$$s_{131} = \frac{(2.3 - 0.267)(5.85 - 5.28) + (-1 - 0.267)(5.53 - 5.28) + (-1.4 - 0.267)(4.46 - 5.28)}{3 - 1}$$

$$= 1.217$$

and

Table 10.7 Mechanical lead pencil statistics for construction of Hotelling's T^2 chart

Sample k	$\bar{x}_{1k}$	$\bar{x}_{2k}$	$\bar{x}_{3k}$	s^2_{1k}	s^2_{2k}	s^2_{3k}	s_{12k}	s_{13k}	s_{23k}	T^2_k
1	0.267	2.827	5.280	3.523	0.369	0.530	1.139	1.217	0.397	0.2796
2	−0.533	2.610	4.667	1.023	0.139	0.533	0.364	0.333	0.182	2.3602
3	0.233	2.980	5.497	0.083	0.052	1.366	0.043	0.122	−0.129	1.5352
4	−0.267	2.717	5.623	0.443	0.058	1.038	0.155	0.485	0.211	2.0964
5	−0.100	2.657	4.950	0.490	0.015	0.272	−0.032	−0.355	0.009	0.4671
6	0.600	3.083	5.390	1.110	0.080	0.677	0.270	0.474	0.199	3.0084
7	−0.033	2.913	5.527	1.423	0.251	0.346	0.568	0.701	0.276	1.5417
8	−1.300	2.263	4.860	0.280	0.244	0.356	−0.260	−0.082	0.104	6.6017
9	−0.300	2.697	5.083	0.610	0.019	0.197	0.105	0.335	0.060	0.3874
10	0.433	2.863	5.423	1.013	0.123	0.738	0.349	0.819	0.295	0.6823
11	−0.100	2.710	5.533	0.520	0.123	0.971	0.128	−0.327	0.184	1.1304
12	−0.900	2.037	4.497	0.010	0.008	0.366	−0.004	0.024	0.036	11.4242
13	0.267	2.590	5.277	1.213	0.033	0.332	0.124	0.586	0.091	2.3363
14	0.633	2.840	5.093	0.743	0.253	0.014	0.344	−0.018	−0.043	2.0981
15	−0.533	2.673	5.133	0.803	0.050	0.337	0.198	0.300	0.087	1.1077
16	0.033	2.733	5.167	0.243	0.040	0.168	0.014	−0.122	0.058	0.0330
17	0.400	2.837	5.830	1.000	0.318	0.155	0.555	0.240	0.102	2.7758
18	1.133	3.087	5.433	0.903	0.145	0.419	0.181	0.399	0.242	4.7416
19	−0.033	2.620	4.560	2.723	0.085	0.224	0.296	0.326	0.134	3.2077
20	−0.033	2.973	5.483	0.043	0.284	0.434	0.053	−0.025	0.272	2.3667
Means	−0.0067	2.7355	5.2153	0.9102	0.1343	0.4736	0.2295	0.2716	0.1384	

$$s_{231} = \frac{(3.48 - 2.827)(5.85 - 5.28) + (2.72 - 2.827)(5.53 - 5.28) + (2.28 - 2.827)(4.46 - 5.28)}{3 - 1}$$

$$= 0.397$$

The sample means, variances, and covariances for all the preliminary samples are shown in Table 10.7.

The overall averages of the sample means, sample variances, and sample covariances are also shown in Table 10.7. Hence, the estimated mean vector and variance-covariance matrix are

$$\hat{\boldsymbol{\mu}} = \begin{bmatrix} \bar{x}_1 \\ \bar{x}_2 \\ \bar{x}_3 \end{bmatrix} = \begin{bmatrix} -0.0067 \\ 2.7355 \\ 5.2153 \end{bmatrix} \qquad \mathbf{S} = \begin{bmatrix} \bar{s}_1^2 & \bar{s}_{12} & \bar{s}_{13} \\ \bar{s}_{21} & \bar{s}_2^2 & \bar{s}_{23} \\ \bar{s}_{31} & \bar{s}_{32} & \bar{s}_3^2 \end{bmatrix} = \begin{bmatrix} 0.9102 & 0.2295 & 0.2716 \\ 0.2295 & 0.1343 & 0.1384 \\ 0.2716 & 0.1384 & 0.4736 \end{bmatrix}$$

For the first sample ($k = 1$), the observed sample vector is

$$\bar{\mathbf{x}}_1 = \begin{bmatrix} \bar{x}_{11} \\ \bar{x}_{21} \\ \bar{x}_{31} \end{bmatrix} = \begin{bmatrix} 0.267 \\ 2.827 \\ 5.280 \end{bmatrix}$$

From (10.24), we see the need to subtract the estimated mean vector, from the sample vector which results in

$$(\bar{\mathbf{x}}_1 - \hat{\boldsymbol{\mu}}) = \begin{bmatrix} 0.267 \\ 2.827 \\ 5.280 \end{bmatrix} - \begin{bmatrix} -0.0067 \\ 2.7355 \\ 5.2153 \end{bmatrix} = \begin{bmatrix} 0.2737 \\ 0.0915 \\ 0.0647 \end{bmatrix}$$

Consequently, by applying (10.24), the T^2 value for the first sample is found from the following matrix operation:

$$T_1^2 = 3[0.2737 \quad 0.0915 \quad 0.0647] \begin{bmatrix} 0.9102 & 0.2295 & 0.2716 \\ 0.2295 & 0.1343 & 0.1384 \\ 0.2716 & 0.1384 & 0.4736 \end{bmatrix}^{-1} \begin{bmatrix} 0.2737 \\ 0.0915 \\ 0.0647 \end{bmatrix}$$

Below are the Minitab session commands and entries leading to the determination of $\mathbf{S}^{-1}$:

```
MTB > name m1 'S'
MTB > read 3 3 m1
DATA> 0.9102 0.2295 0.2716
DATA> 0.2295 0.1343 0.1384
DATA> 0.2716 0.1384 0.4736
        3 rows read.
MTB > inverse m1 m2
MTB > name m2 'S inverse'
MTB > print m2

Data Display

 Matrix S inverse

    1.9444    -3.1103   -0.2062
   -3.1103    15.6298   -2.7838
   -0.2062    -2.7838    3.0432
```

Thus, the following matrix computation needs to be performed:

$$T_1^2 = 3[0.2737 \quad 0.0915 \quad 0.0647] \begin{bmatrix} 1.944 & -3.1109 & -0.2062 \\ -3.1103 & 15.629 & -2.7838 \\ -0.2062 & -2.7838 & 3.0432 \end{bmatrix} \begin{bmatrix} 0.2737 \\ 0.0915 \\ 0.0647 \end{bmatrix}$$

Continuing with the above Minitab session, we can find the value of T_1^2 as follows:

```
MTB > read 1 3 m3
DATA> 0.2737 0.0915 0.0647
      1 rows read.
MTB > name m3 'xtranspose'
MTB > read 3 1 m4
DATA> 0.2737
DATA> 0.0915
DATA> 0.0647
      3 rows read.
MTB > name m4 'x'
MTB > multiply 3 'xtranspose' m5
MTB > multiply m5 'S inverse' m6
MTB > multiply m6 'x' m7
Answer =           0.2796    ←————— T_1^2
```

The values of T^2, for all the preliminary samples, are given in Table 10.7.

The phase 1 upper control limit for the T^2 chart, with $\alpha = 0.0027$, is found by using (10.26):

$$UCL = \frac{3(20-1)(3-1)}{20(3-1)-3+1} F_{0.0027,3,20(3-1)-3+1}$$

$$= 3F_{0.0027,3,38}$$

The value of $F_{0.0027,3,38}$ is found below:

```
MTB > InvCDF 0.9973;
SUBC>   F 3 38.

Inverse Cumulative Distribution Function

F distribution with 3 DF in numerator and 38 DF in denominator

 P( X <= x)            x
 0.9973            5.6327
```

Thus, the phase 1 upper control limit is 3(5.6327), or 16.898. A sequence plot of the T^2 values, with this control limit, is shown in Figure 10.6. Notice that no points exceed the upper control; thus we would conclude the process is in control from a multivariate context. For prospective control, new T^2 values would be plotted against a phase 2 upper control limit. Since no preliminary samples were removed, m^* of (10.27) is equal to 20. As such, the phase 2 upper control limit is

$$UCL = \frac{3(20+1)(3-1)}{20(3-1)-3+1} F_{0.0027,3,20(3-1)-3+1}$$

$$= 3.3158 F_{0.0027,3,38}$$

$$= 3.3158(5.6327)$$

$$= 18.677$$

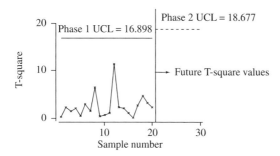

Figure 10.6   Hotelling T^2 control chart for mechanical lead pencil data.

As seen from Figure 10.6, this control limit is projected outward for on-line monitoring of future T^2. In our example, the distinction between the phase 1 and 2 limits is visually apparent. However, as the number of preliminary samples increases, the phase 1 and 2 limits become nearly equivalent in value.

Sample Size $n = 1$ The Hotelling T^2 chart just presented is applicable when subgrouping is used. As discussed in Chapter 5, some situations do not lend themselves to subgrouping, and thus only an individual measurement ($n = 1$) of a given quality characteristic is collected at a given sampling period. Since no subgrouping is involved, the observed vector is a vector of individual measurements and not a vector of subgroup means, namely,

$$\mathbf{x}_k = \begin{bmatrix} x_{1k} \\ x_{2k} \\ \vdots \\ x_{pk} \end{bmatrix} \qquad k = 1, 2, \ldots, m \tag{10.28}$$

where x_{jk} represents the individual observation on the jth quality characteristic for the kth sample. The mean vector is then estimated by averaging the individual observations for a given quality characteristic over m samples, that is,

$$\hat{\boldsymbol{\mu}} = \begin{bmatrix} \bar{x}_1 \\ \bar{x}_2 \\ \vdots \\ \bar{x}_p \end{bmatrix} \tag{10.29}$$

where

$$\bar{x}_j = \frac{\sum\limits_{k=1}^{m} x_{jk}}{m} \qquad j = 1, 2, \ldots, p \tag{10.30}$$

In the case of $n > 1$, m variance and covariance estimates are computed, and then the estimates are averaged across the m samples to arrive at the final estimates of the variances and covariances. However, when $n = 1$, the first step of obtaining variance and covariance estimates from each sample is not possible. Thus, the variance and covariance estimates must be obtained by going across the preliminary samples. In particular, the variances are determined as follows:

$$s_j^2 = \frac{\sum_{k=1}^{m} (x_{jk} - \bar{x}_j)^2}{m - 1} \qquad j = 1, 2, \ldots, p \tag{10.31}$$

An estimate of the covariance between quality characteristic j and quality characteristic h is given by

$$s_{jh} = \frac{\sum_{k=1}^{m} (x_{jk} - \bar{x}_j)(x_{hk} - \bar{x}_h)}{m - 1} \qquad \begin{array}{l} j = 1, 2, \ldots, p \\ j \neq h \end{array} \tag{10.32}$$

Mapping the estimates given by (10.31) and (10.32) into a matrix, we obtain an estimate of the variance-covariance matrix, namely,

$$\mathbf{S} = \begin{bmatrix} s_1^2 & s_{12} & \cdots & s_{1p} \\ & s_2^2 & \cdots & s_{2p} \\ & & \ddots & \vdots \\ & & & s_p^2 \end{bmatrix} \tag{10.33}$$

Note: As with the true variance-covariance matrix given by (10.11), we do not need to present the lower triangle of the sample variance-covariance matrix because of its symmetry. Now with the sample vector $\mathbf{x}_k$, estimated mean vector $\hat{\boldsymbol{\mu}}$, and sample variance-covariance matrix $\mathbf{S}$ defined by (10.28), (10.29), and (10.33), respectively, the Hotelling T^2 of (10.24) becomes

$$T_k^2 = (\mathbf{x}_k - \hat{\boldsymbol{\mu}})' \mathbf{S}^{-1} (\mathbf{x}_k - \hat{\boldsymbol{\mu}}) \tag{10.34}$$

As with the subgrouping case, the distinction between phase 1 and phase 2 limits must also be considered for individual observations. Referring to the phase 1 and 2 limits with subgrouping given by (10.26) and (10.27), we notice that the only difference is the slight modification in the factor in front of the similar F factor. Namely, if no preliminary samples are deleted, then $m - 1$ in the phase 1 limit is replaced with $m + 1$ for the phase 2 limit. For the case of $n = 1$, the distinction is not so slight. For individual observations, Tracy, Young, and Mason (1992) demonstrated that a distribution, known as a beta distribution, should serve as the basis for the determination of the appropriate phase 1 limits. Even though the beta distribution is available on Minitab for the determination of percentile points, it turns out that the beta distribution can be expressed as a function of the F distribution. Based on the results of Tracy, Young, and Mason (1992), the phase 1 upper control limit for the individual-based T^2 statistic is

$$UCL = \frac{(m-1)^2[p/(m-p-1)]F_{\alpha,p,m-p-1}}{m + [mp/(m-p-1)]F_{\alpha,p,m-p-1}} \qquad (10.35)$$

This limit would be used to retrospectively analyze the preliminary samples. As usual, if any out-of-control points are given special-cause explanation(s), then the associated preliminary samples are removed, and new mean vector and variance-covariance matrix estimates are obtained with the remaining samples.

For the monitoring of future process values, the appropriate phase 2 upper control limit for the individual-based T^2 statistic is

$$UCL = \frac{p(m^*+1)(m^*-1)}{(m^*)^2 - m^*p}F_{\alpha,p,m^*-p} \qquad (10.36)$$

where m^* is the number of remaining preliminary samples, that is, the number of preliminary samples not discarded during phase 1.

Table 10.8   Chemical process data

Sample	% Impurities	Temperature	Concentration	$T_k^2(m = 14)$	$T_k^2(m = 13)$
1	14.92	85.77	42.26	10.93	*
2	16.90	83.77	43.44	2.04	1.84
3	17.38	84.46	42.74	5.58	5.33
4	16.90	86.27	43.60	3.86	3.58
5	16.92	85.23	43.18	0.04	0.23
6	16.71	83.81	43.72	2.25	2.17
7	17.07	86.08	43.33	1.44	1.46
8	16.93	85.85	43.41	1.21	1.05
9	16.71	85.73	43.28	0.68	1.91
10	16.88	86.27	42.59	2.17	5.16
11	16.73	83.46	44.00	4.17	3.84
12	17.07	85.81	42.78	1.40	1.65
13	17.60	85.92	43.11	2.33	7.00
14	16.90	84.23	43.48	0.90	0.77
Means	16.830	85.190	43.209		

Example 10.6

To illustrate the implementation of a T^2 chart with $n = 1$, we consider the data application presented in Tracy, Young, and Mason (1992). As explained by the authors, the data are taken from a chemical industrial process for which individual measurements on three quality characteristics are gathered each sampling period: percentage of impurities X_1, temperature X_2, and concentration X_3. The authors provide data on $m = 14$ preliminary samples. These data are shown in Table 10.8 and also found in the file "chemical.dat".

By averaging the individual observations across preliminary samples, we obtain an estimated mean vector of

$$\hat{\boldsymbol{\mu}} = \begin{bmatrix} \bar{x}_1 \\ \bar{x}_2 \\ \bar{x}_3 \end{bmatrix} = \begin{bmatrix} 16.830 \\ 85.190 \\ 43.209 \end{bmatrix}$$

From (10.31), the sample variances are then

$$s_1^2 = \frac{(14.92 - 16.83)^2 + (16.90 - 16.83)^2 + \cdots + (16.90 - 16.83)^2}{14 - 1} = 0.3641$$

$$s_2^2 = \frac{(85.77 - 85.19)^2 + (83.77 - 85.19)^2 + \cdots + (84.23 - 85.19)^2}{14 - 1} = 1.0366$$

$$s_3^2 = \frac{(42.26 - 43.209)^2 + (43.44 - 43.209)^2 + \cdots + (43.48 - 43.209)^2}{14 - 1} = 0.2250$$

From (10.32), the sample covariances are

$$s_{12} = \frac{(14.92 - 16.83)(85.77 - 85.19) + \cdots + (16.90 - 16.83)(84.23 - 85.19)}{14 - 1} = -0.0214$$

$$s_{13} = \frac{(14.92 - 16.83)(42.26 - 43.209) + \cdots + (16.90 - 16.83)(43.48 - 43.209)}{14 - 1} = 0.1004$$

$$s_{23} = \frac{(85.77 - 85.19)(42.26 - 43.209) + \cdots + (84.23 - 85.19)(43.48 - 43.209)}{14 - 1} = -0.2444$$

Hence, the estimated variance-covariance matrix is

$$\mathbf{S} = \begin{bmatrix} s_1^2 & s_{12} & s_{13} \\ & s_2^2 & s_{23} \\ & & s_3^2 \end{bmatrix} = \begin{bmatrix} 0.3641 & -0.0214 & 0.10040 \\ & 1.03660 & -0.2444 \\ & & 0.22500 \end{bmatrix}$$

The reader can verify that the inverse of this matrix is

$$\mathbf{S}^{-1} = \begin{bmatrix} 3.23274 & -0.36748 & -1.84168 \\ & 1.33858 & 1.61797 \\ & & 7.02372 \end{bmatrix}$$

To compute now the T^2 value for the first sample, we first subtract the estimated mean vector from the observed sample vector,

$$\mathbf{x}_1 - \hat{\boldsymbol{\mu}} = \begin{bmatrix} 14.92 \\ 85.77 \\ 42.26 \end{bmatrix} - \begin{bmatrix} 16.830 \\ 85.190 \\ 43.209 \end{bmatrix} = \begin{bmatrix} -1.910 \\ 0.580 \\ -0.949 \end{bmatrix}$$

Applying (10.34), we find the T^2 value for the first sample as follows:

$$T_1^2 = \begin{bmatrix} -1.910 & 0.580 & -0.949 \end{bmatrix} \begin{bmatrix} 3.23274 & -0.36748 & -1.84168 \\ & 1.33858 & 1.61797 \\ & & 7.02372 \end{bmatrix} \begin{bmatrix} -1.910 \\ 0.580 \\ -0.949 \end{bmatrix}$$

The T^2 values, based on 14 preliminary samples, are found in the next-to-last column of Table 10.8.

From (10.35), the phase 1 upper control limit with $\alpha = 0.0027$ is

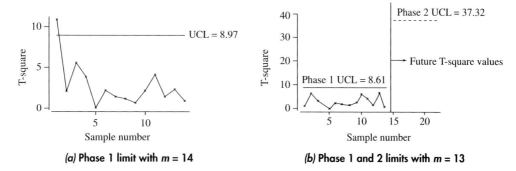

(a) **Phase 1 limit with** $m = 14$ (b) **Phase 1 and 2 limits with** $m = 13$

Figure 10.7   Multivariate control charts when $n = 1$.

$$
\begin{aligned}
\text{UCL} &= \frac{(14-1)^2[3/(14-3-1)]F_{0.0027,3,14-3-1}}{14 + [14(3)/(14-3-1)]\,F_{0.0027,3,14-3-1}} \\[6pt]
&= \frac{50.7F_{0.0027,3,10}}{14 + 4.2F_{0.0027,3,10}} \\[6pt]
&= \frac{50.7(9.6267)}{14 + 4.2\,(9.6267)} \\[6pt]
&= 8.97
\end{aligned}
$$

A sequence plot of the T^2 values, along with this upper control, is shown in Figure 10.7a. Notice that observation 1 exceeds the upper control limit. Tracy, Young, and Mason (1992) report that this unusual observation was due to a "sampling error" with the impurities measurement. Given that a special-cause explanation was found, it is appropriate to remove the first preliminary sample.

Once the sample is removed, the mean vector and variance-covariance matrix need to be reestimated with the remaining $m = 13$ samples. We leave the recalculations to the reader as a chapter exercise. Applying the revised $\hat{\boldsymbol{\mu}}$ and $\mathbf{S}$ to the remaining 13 samples, we obtain the T^2 values shown in the rightmost column of Table 10.8. The corresponding phase 1 upper control limit (now with $m = 13$) is

$$
\begin{aligned}
\text{UCL} &= \frac{(13-1)^2[3/(13-3-1)]F_{0.0027,3,13-3-1}}{13 + [13(3)/(13-3-1)]F_{0.0027,3,13-3-1}} \\[6pt]
&= \frac{48F_{0.0027,3,9}}{13 + 4.3333F_{0.0027,3,9}} \\[6pt]
&= \frac{48(10.4879)}{13 + 4.3333(10.4879)} \\[6pt]
&= 8.61
\end{aligned}
$$

Replotting the T^2 values, we see from Figure 10.7b that none of the preliminary sample observations signals out of control. We can now proceed to prospective analysis using the revised $\hat{\boldsymbol{\mu}}$ and $\mathbf{S}$ in the computation of future T^2 values.

For future T^2 values, we use the upper control limit given by (10.36), which is equal to

$$\text{UCL} = \frac{3(13 + 1)(13 - 1)}{13^2 - 13(3)} F_{0.0027,3,13-3}$$

$$= 3.877 F_{0.0027,3,10}$$

$$= 3.877(9.6267) = 37.32$$

This limit is projected outward for future values on Figure 10.7*b*. Notice that the disparity between the phase 1 and phase 2 limits is substantial. In general, to reduce the disparity, the number of preliminary samples *m* needs to be increased. As a rough rule of thumb, if the number of quality characteristics *p* is in the range of 2 to 5, then at least 100 preliminary samples are required so that the retrospective and prospective limits will be approximately the same.

SPECIAL-CAUSE SEARCHING AND MULTIVARIATE SIGNALS

In our presentation of the chi-square multivariate chart, we illustrated the procedure in Example 10.4 with a data set for which we already knew the reason for the out-of-control signal shown in Figure 10.5. Namely, we learned from earlier plots and discussion that the signal was not due to an unusual x_1 value alone or an unusual x_2 value alone, but rather due to the pair (x_1, x_2) of values being inconsistent with the general workings of the process. In our presentation of the Hotelling T^2 chart for individual measurements, we illustrated the procedure in Example 10.6 with a data set for which we were told by Tracy, Young, and Mason (1992) the reason for the out-of-control signal shown in Figure 10.7. Namely, the multivariate signal was due to an unusual x_1 value alone, not due to some unusual behavior in terms of possible two-way or three-way combinations of the x_1, x_2, x_3 values, that is, (x_1, x_2), (x_1, x_3), (x_2, x_3), or (x_1, x_2, x_3).

In practical applications, when confronted with a multivariate signal based on simultaneous monitoring of *p* quality characteristics, where should the process analyst turn for special-cause explanation? The answer is not so clear. For example, it may be that the signal is due to one of the *p* variables going out of control. Or, the signal may be due to a combination of two or more variables, which is inconsistent with the general process behavior. In the presence of a significant T^2 observation, some practitioners respond by plotting univariate charts for each quality characteristic. Examining characteristics individually can be helpful in tracking down the source of the problem, if the multivariate signal was due to a highly unusual observation, in the sense of being too large or too small, on one of the characteristics. However, as we have illustrated, the study of univariate charts alone does not capture all possible explanations for multivariate signals.

Jackson (1980, 1991) recommended the use of the multivariate statistical technique called *principal-component analysis* to help determine the sources of a significant T^2 observation. With principal-component analysis, a set of interrelated variables is transformed to a new set of uncorrelated variables. The new variables are called *principal components*. Each principal component is a linear combination of the original variables. Jackson's idea is to decompose the T^2

statistic into a sum of principal components and then to analyze the components with control charts to identify which components lead to the out-of-control signal. One difficulty with this approach is that the principal components can be hard to interpret in how they relate to the original variables. There are, however, some instances when the principal components represent meaningful combinations of the original variables, allowing the analyst to properly identify the sources of the multivariate out-of-control signal; for example, Jeffery and Young (1993) report a successful interpretation of the principal components in the electrochemical industry.

One of the more promising approaches in interpreting multivariate T^2 signals is based on the works of Mason, Tracy, and Young (1995, 1997). Their idea is to decompose the T^2 statistic into independent components in such a manner that it can be determined if a signal is the result of an individual variable's being unusual or a combination among certain variables being unusual. For the bivariate case ($p = 2$), their approach calls for decomposing the overall T^2 statistic in one of two ways:

$$T^2 = T_1^2 + T_{1|2}^2 \qquad \text{and} \qquad T^2 = T_2^2 + T_{2|1}^2$$

The terms T_1^2 and T_2^2 represent *unconditional* contributions of the x_1 and x_2 values, respectively, to the overall T^2 value. Basically, an unconditional term focuses on the value of the individual variable relative to its mean (or sample mean) with no regard to the other variable. The terms $T_{1|2}^2$ and $T_{2|1}^2$ represent conditional contributions. For example, $T_{1|2}^2$ is a measure of how far the x_1 value is from the predicted value of x_1, given the observed value of x_2. Significant conditional terms suggest that a significant overall T^2 value is due in part to an inconsistent relationship between the two variables. Mason, Tracy, and Young (1995, 1997) provide the details of computing and testing unconditional and conditional terms for the bivariate case and for the more general multivariate cases ($p > 2$).

MULTIVARIATE MONITORING OF PROCESS VARIABILITY

The multivariate control charts, presented in the previous subsections, focused on the detection of unusual changes in individual or mean values of one or more quality characteristics. In a sense, the chi-square and Hotelling T^2 charts represent multivariate extensions of the univariate Shewhart mean chart. As we have learned with the univariate case, it is also important to monitor the process for any instabilities in variability. Recall that monitoring process variability was important not only in itself, but also for the implementation of the Shewhart mean chart since it depends on a reliable estimate of variability which is not distorted by special causes. The need to have a reliable measure of variability and to monitor for changes in process variability is just as important in the multivariate case.

In the multivariate case, process variability is given by the variance-covariance matrix Σ when the parameters are known or by the estimated variance-covariance matrix $\mathbf{S}$ when the parameters are unknown. Several multivariate charts have been developed for monitoring process variability; for example, see Alt (1985) and Alt and Smith (1988). All these charts are based on a statistic known as the *generalized sample variance*, denoted by $|\mathbf{S}|$. The generalized sample variance is a single numerical value for the variation expressed by the variance-covariance matrix $\mathbf{S}$. In particular, the generalized sample variance is defined to be the *determinant* of $\mathbf{S}$. One of the important concepts in matrix algebra is that of the determinant of a matrix. The determinant is used in a variety of ways in matrix algebra. For our purposes, it is sufficient to know that the determinant of a matrix is a single number that is associated with a square matrix, that is, a matrix with an equal number of rows and columns. It is useful to note that the determinant of a matrix composed of one row and one column is defined as the value of the single entry in the matrix. Now notice that if we are dealing with only one quality characteristic ($p = 1$), the sample variance-covariance matrix is simply the matrix $\mathbf{S} = [s_1^2]$. Hence, $|\mathbf{S}|$ (that is, the generalized sample variance) appropriately reduces to the usual sample variance of a single characteristic.

As an illustration of a multivariate control chart for variability, we now present a procedure discussed in Montgomery (1996). Assume that at each sampling period more than one observation is taken on each quality characteristic, that is, $n > 1$. With this assumption of subgrouping, sample variances and sample covariances can be computed at each sampling period. Thus, each sampling period k is associated with an observed sample variance-covariance matrix which we can denote by $\mathbf{S}_k$. The idea then is to compute the determinant of $\mathbf{S}_k$ and plot this value on a control chart with appropriate limits. One reasonable approach for the development of control limits for $|\mathbf{S}_k|$ is to first determine the mean and variance of $|\mathbf{S}_k|$, that is, $E(|\mathbf{S}_k|)$ and $\text{Var}(|\mathbf{S}_k|)$. The next step is to bank on the property that most of the probability distribution of $|\mathbf{S}_k|$ is within 3 standard deviations of $E(|\mathbf{S}_k|)$; that is, set limits at $E(|\mathbf{S}_k|) \pm 3\sqrt{\text{Var}(|\mathbf{S}_k|)}$. It can be shown that:

$$E(|\mathbf{S}_k|) = b_1|\mathbf{\Sigma}| \tag{10.37}$$

and

$$\text{Var}(|\mathbf{S}_k|) = b_2|\mathbf{\Sigma}|^2 \tag{10.38}$$

where

$$b_1 = \frac{1}{(n-1)^p} \prod_{i=1}^{p}(n-i)$$

and

$$b_2 = \frac{1}{(n-1)^{2p}} \prod_{i=1}^{p}(n-i)\left[\prod_{i=1}^{p}(n-i+2) - \prod_{i=1}^{p}(n-i)\right]$$

Analogous to the summation notation, the symbol Π is a shorthand notation for successive multiplication based on an index, i in this case. Given (10.37) and (10.38), the control chart limits for $|\mathbf{S}_k|$ could be set at

$$\text{UCL} = |\mathbf{\Sigma}|(b_1 + 3\sqrt{b_2})$$
$$\text{LCL} = |\mathbf{\Sigma}|(b_1 - 3\sqrt{b_2}) \tag{10.39}$$

If the lower control limit is computed to be a negative value, it is set to zero.

Given how the control limits are defined in terms of b_1 and b_2, there is a subtle restriction imposed on the sample size n; namely, n must be greater than p. If n equaled p, then b_1 and b_2 would equal 0; this occurs because among the terms multiplied together is the last term, $n - p$. If $b_1 = b_2 = 0$, then the upper control limit and the lower control limit are both equal to 0, implying no interval for control. Such a restriction is applicable not only to the multivariate variability chart presented here, but also to other versions discussed in Alt (1985) and Alt and Smith (1988). This means that we could not implement a multivariate chart based on the generalized sample variance for the data set studied in Example 10.6, because $p = n = 3$ in that application. However, this limitation does not preclude us from implementing univariate charts for variability on each quality characteristic. In fact, even if a multivariate chart for variability could be implemented, we still suggest that univariate charts be considered in light of a limitation of the generalized variance measure that we will soon point out.

The control limits of (10.39) assume that the true variance-covariance matrix Σ is known. More realistically, Σ is unknown and needs to be estimated with the sample variance-covariance matrix **S**, based on pooling the preliminary samples as explained earlier. Since the expected value result of (10.37) also applies to **S**, we see that after rearrangement $|\mathbf{S}|/b_1$ is an unbiased estimate of $|\Sigma|$. Substituting $|\mathbf{S}|/b_1$ for $|\Sigma|$ in (10.39), we obtain the following control chart limits:

$$\text{UCL} = \frac{|\mathbf{S}|}{b_1}(b_1 + 3\sqrt{b_2})$$

$$\text{LCL} = \frac{|\mathbf{S}|}{b_1}(b_1 - 3\sqrt{b_2})$$

(10.40)

Suppose two quality characteristics (X_1, X_2) are simultaneously monitored. With a sample size of $n = 10$, suppose $m = 10$ preliminary samples are gathered on these characteristics. The variance and covariance summary statistics are provided in Table 10.9. Using the overall averages of the variances and covariances, we have the following sample variance-covariance matrix:

Example 10.7

$$\mathbf{S} = \begin{bmatrix} 4.435 & 1.991 \\ 1.991 & 2.029 \end{bmatrix}$$

We now need the determinant of **S**. For a 2×2 matrix, the determinant is easily found from the following operation on the elements of **S**:

$$|\mathbf{S}| = 4.435(2.029) - 1.991(1.991) = 5.035$$

For larger matrices, manual computation of the determinant is tedious. In the appendix of this chapter, we show how Minitab can be used to facilitate this computation.

For $n = 10$ and $p = 2$, the constants b_1 and b_2 are

$$b_1 = \frac{1}{9^2}(9)(8) = 0.8889$$

$$b_2 = \frac{1}{9^4}(9)(8)[11(10) - 9(8)] = 0.4170$$

Table 10.9 Data for Example 10.7

Sample k	s_{1k}^2	s_{2k}^2	s_{12}	$|S_k|$
1	4.86	2.23	2.46	4.79
2	1.79	0.82	0.31	1.37
3	4.49	2.42	2.67	3.74
4	6.91	4.00	4.06	11.16
5	4.76	2.39	2.68	4.19
6	4.10	1.35	1.51	3.25
7	3.09	2.21	1.06	5.71
8	6.48	1.48	2.43	3.69
9	4.72	1.59	2.20	2.66
10	3.15	1.80	0.53	5.39
Means	4.435	2.029	1.991	

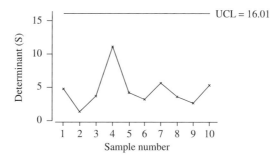

Figure 10.8 Control chart for generalized sample variance.

Substituting the values for b_1, b_2, and $|S|$ in (10.40), we find the control limits to be

$$\text{UCL} = \frac{5.035}{0.8889}(0.8889 + 3\sqrt{0.4170}) = 16.01$$

$$\text{LCL} = \frac{5.035}{0.8889}(0.8889 - 3\sqrt{0.4170}) = -5.94 \quad \rightarrow \quad 0$$

To determine the plotted value for the first sample, we need the determinant of the first sample variance-covariance matrix, which is

$$|S_1| = \begin{vmatrix} 4.86 & 2.46 \\ 2.46 & 2.23 \end{vmatrix} = 4.86(2.23) - 2.46(2.46) = 4.79$$

The values of $|S_k|$ for all the samples are given in the rightmost column of Table 10.9. The control chart for these generalized sample variances is provided in Figure 10.8.

Even though the generalized sample variance is a common summary measure for multivariate variability, the fact that it is a single-value representation of a multivariate scheme necessarily implies some inherent weaknesses. To illustrate a deficiency, suppose a special cause results in an increase in the variance of one quality characteristic and a decrease in the variance of the other characteristic in such a way that the correlation coefficient for the two characteristics remains unchanged. Consider, for example, the two sample variance-covariance matrices shown below:

$$|\mathbf{S}_1| = \begin{bmatrix} 1 & 0.7071 \\ 0.7071 & 2 \end{bmatrix} \qquad |\mathbf{S}_2| = \begin{bmatrix} 2 & 0.7071 \\ 0.7071 & 1 \end{bmatrix}$$

Each of these variance-covariance matrices has the same generalized variance, namely, $|\mathbf{S}_1| = |\mathbf{S}_2| = 1.5$, and yet they reflect distinctly different variance structures. It is because of such situations that we highly recommend the use of univariate control charts for variability as a supplement to any multivariate control based on the generalized variance.

ASSUMPTIONS FOR STANDARD MULTIVARIATE CHARTS

Before closing our presentation of standard multivariate control charts, we must not lose sight of the underlying assumptions made in their development. Similar to the standard univariate charts, the multivariate procedures are based on the assumptions of normality and independence of the observations over time.

With respect to the normality assumption, this translates, more specifically, to the assumption that each vector observation $\mathbf{X}_k$ comes from a multivariate normal distribution. Everitt (1979) studied the sensitivity of the Hotelling T^2 statistic to departures from the assumption of multivariate normality, and found the statistic to be quite sensitive to skewness, with the problem worsening as the vector dimension p increases. It is imperative, then, that diagnostics be available for detecting cases where data exhibit moderate to extreme departures from multivariate normality. Johnson and Wichern (1998) provide some rough-and-ready strategies for assessing multivariate normality. More sophisticated methods are given by Koziol (1982) and Royston (1983). If multivariate nonnormality is detected, power transformations—possibly different transformations for each variable—can be utilized, as we have shown in earlier chapters for univariate applications. See Johnson and Wichern (1998) for discussion and illustration of the multivariate case.

The second assumption is one of independence or randomness, over time, for the sample vectors $\mathbf{X}_k$. In the univariate case, we checked for randomness by using three diagnostics: time-series plot, runs test, and autocorrelation function (ACF). Recall that the ACF is used to determine whether observations of a given variable are significantly correlated with previous observations of the *same* variable. In a multivariate framework, the question is whether vectors are in any way correlated with previous vectors. It may be possible that previous observations of a given variable are significantly correlated with previous observations of the same variable or with previous observations of another variable. Consider the following bivariate vectors separated by one sampling period (i.e., lag 1):

$$\begin{bmatrix} X_{1,t-1} \\ X_{2,t-1} \end{bmatrix} \rightleftarrows \begin{bmatrix} X_{1,t} \\ X_{2,t} \end{bmatrix}$$

The arrows represent the possible lag 1 correlations between the random vectors. If the vectors are generated from a process which is random over time, then these correlations theoretically equal 0. It is important to note that, in general, the correlation between $X_{1,t-1}$ and $X_{2,t}$ is not equal to the correlation between $X_{2,t-1}$ and $X_{1,t}$ (see Box, Jenkins, and Reinsel, 1994). Beyond lag 1, the correlations between all the vector elements, at lags 2, 3, etc., also have to equal 0 for a random process.

To determine whether an observed series of bivariate vectors is significantly correlated across quality characteristics over time, we consider a generalization of the sample autocorrelation function known as the sample *cross-correlation function (CCF)*. For a given set of time-series data x_t and y_t, $1 \le t \le n$, the cross-correlation coefficient, at lag k, is given by

$$r_{xy}(k) = \begin{cases} \dfrac{(1/n)\displaystyle\sum_{t=1}^{n-k}(x_t - \bar{x})(y_{t+k} - \bar{y})}{s_x s_y} & k \ge 0 \\[4mm] \dfrac{(1/n)\displaystyle\sum_{t=1}^{n+k}(y_t - \bar{y})(x_{t-k} - \bar{x})}{s_x s_y} & k < 0 \end{cases}$$

(10.41)

When $k > 0$, $r_{xy}(k)$ represents the sample cross-correlation of the x variable lagging behind the y variable by k periods. Similarly, when $k < 0$, $r_{xy}(k)$ represents the sample cross-correlation of the y variable lagging behind the x variable by k periods. The case of $k = 0$ represents the coincident correlation between the x series and y series, that is, the correlation between the two variables at the same time point.

To illustrate a sample CCF, consider the bivariate series of data presented earlier in Table 10.5. Let us ignore the fact that the data were generated with full knowledge of the underlying parameters. Using Minitab, we obtain the sample cross-correlation function shown in Figure 10.9. First, notice the correlation spike of 0.768 at $k = 0$. This is to be expected since it represents the coincident correlation between x_1 and x_2 at the same time. In other words, this is the inter-correlation between the two characteristics *within* the bivariate vector. Recall that we indicated that the data were generated from a bivariate process for which the two characteristics are positively correlated with an underlying correlation of 0.80. The sample correlation of 0.768 is indeed consistent with the true correlation of 0.80. For lag 1, the cross-correlation between $x_{2,t-1}$ and $x_{1,t}$ is -0.178 while the cross-correlation between $x_{1,t}$ and $x_{2,t+1}$ (equivalently, $x_{1,t-1}$ and $x_{2,t}$) is -0.139. To test whether the sample cross-correlations are significantly different from zero, we need the standard deviation of the sample cross-correlation func-

Stat>>Time Series>>Cross Correlation

```
CCF - correlates x1 (t) and x2 (t+k)

          -1.0 -0.8 -0.6 -0.4 -0.2  0.0  0.2  0.4  0.6  0.8  1.0
           +----+----+----+----+----+----+----+----+----+----+
 -4 -0.140                          XXXX
 -3  0.096                           XXX
 -2  0.299                        XXXXXXXX
 -1 -0.178                       XXXXX
  0  0.768                           XXXXXXXXXXXXXXXXXXXXX
  1 -0.139                       XXXX
  2 -0.079                        XXX
  3 -0.082                        XXX
  4 -0.106                       XXXX
```

Figure 10.9   Sample cross-correlation function for Table 10.5 data.

tion (referred to as the standard error) under the hypothesis of a random bivariate process. It can be shown (see Box, Jenkins, and Reinsel, 1994) that the standard error is approximately given by

$$\sigma_{r_{xy}(k)} = \frac{1}{\sqrt{n - |k|}} \qquad (10.42)$$

Hence, the standard deviation increases as the magnitude of the lag increases. It can further be shown that the sample cross-correlations, computed from a random process over time, will behave approximately as normal random variables with mean equal to zero. Thus, if the bivariate process is random, the probability is about 95 percent that an individual sample cross-correlation will fall between $\pm 2/\sqrt{n - |k|}$ around 0. For lag 1, the approximate standard error is $1/\sqrt{30 - 1} = 0.186$. Therefore, the null hypothesis of randomness cannot be rejected at a 5 percent significance level, if the lag 1 ($k = -1$ or $+1$) sample cross-correlation falls within $\pm 2(0.186) = \pm 0.372$, which is indeed the case here. In Figure 10.9, we superimposed 95 percent limits for all the presented lags. Given the sample correlations at the varying lags, there is no evidence of cross-correlations being significantly different from 0. To make a final conclusion of bivariate randomness, one needs to examine the x_1 data series and x_2 data series separately. For instance, an ACF for each of the two data series should be constructed (not shown here) and examined; this will indicate whether past values of a given variable are correlated.

For dimension cases higher than the bivariate case, one strategy for checking for multivariate randomness over time is to examine the ACF for each variable and the CCF for each pairwise combination. However, one disadvantage of this strategy is that there is an increased risk of falsely rejecting the null hypothesis of randomness (i.e., committing a type I error) because of the use of multiple tests. An alternative approach is to use a single test for multivariate randomness so that the type I error rate can be controlled to a specified level. For more information on single tests for multivariate randomness, see Hosking (1980).

10.3   ALL-PURPOSE MODEL STRATEGY

From the onset of this book, our aim has been to present a strategy for the effective use of data analysis techniques in process applications. A data analysis approach does not mean that the process analysis must be taken out of the hands of the relatively untrained users. Our emphasis has been on a relatively elementary approach to data analysis, relying on statistical techniques accessible to individuals with even a basic analytical background.

Having said that, we note that a data analysis approach does require more expertise than does the implementation of traditional control charts, which can be prepared from worksheets with only simple arithmetic calculations. However, given that time-series effects cannot be ignored for effective process monitoring and control, is there an alternative approach to data modeling which relies on less statistical training? One possibility is to use a simple flexible statistical approach that would work well in many applications. For example, Alwan (1989) and Montgomery and Mastrangelo (1991) suggest the adoption of a simple exponentially weighted moving-average (EWMA) model as an all-purpose model.

When one is talking about the EWMA as a model for process data, an important point needs to be underscored so as to avoid confusion. Recall that the EWMA scheme was earlier introduced in Chapter 8 for purposes of developing a control chart designed to detect small to moderate shifts from a *random* process. In contrast, we are considering here the EWMA method as a means for *modeling nonrandom* processes.

EWMA methods are widely employed for forecasting purposes in many production and operations management applications, largely because of their simplicity and frequent credibility. Its simplest form is the single or simple exponential smoothing model which can be expressed as:

$$F_{t+1} = \lambda x_t + (1 - \lambda)F_t \tag{10.43}$$

where x_t is the actual observation at time t and F_t is the forecast of time t made at time $t - 1$. The constant λ is called the *smoothing constant*. For the EWMA *control chart* developed in Chapter 8, we assumed that the smoothing constant was restricted to the range $0 < \lambda \leq 1$. When restricted to this range, the EWMA *model* has the distinguishable feature that recent observations are weighted most heavily while the weights placed on older observations decay exponentially.

As opposed to the EWMA control chart of Chapter 8, we do not need to restrict λ to the range of 0 to 1. As we will soon illustrate, Minitab has an option to find the optimal λ, which is defined as the λ associated with an EWMA model with minimum sum of residuals squared—which is equivalent to the idea of least-squares estimation in regression. In using the optimization option, there can be cases in which Minitab reports a λ value greater than 1. The allowance for λ values to be greater than 1 stems from the fact that the EWMA model is derived from a more general class of models known as *autoregressive integrated moving average (ARIMA)* models. In particular, it can be shown that a first-order integrated moving average, denoted by ARIMA(0, 1, 1), underlies the simple EWMA model. We will not delve into the workings of ARIMA models; it suffices to mention that, given the parametric relationship between the EWMA and ARIMA(0, 1, 1) models along with a necessary condition for fitting an ARIMA(0, 1, 1) model, known as the *invertibility* condition, the allowable range for λ is from 0 to 2; for a more detailed explanation consult Box, Jenkins, and Reinsel (1994).

To illustrate an all-purpose modeling approach to time-series implementation, consider a previously analyzed positively autocorrelated series, namely, the phosphorus data series studied in Section 5.5. In our original analysis, we identified the fitted model to be an AR(1) model, namely,

Example 10.8

$$\text{Fitted phos}_t = 0.0205 + 0.521 \text{ phos}_{t-1}$$

The diagnostics for this fitted model were satisfactory in the sense that associated residuals were consistent with a normal random process.

To fit a simple EWMA with Minitab, the user has many options, including the option of selecting the optimal smoothing constant. Below Minitab reports optimal EWMA fit for the phosphorus data series:

```
MTB > %ses 'phos'              Stat>>Time Series>>Single Exp Smoothing

Macro is running ... please wait

Single Exponential Smoothing

Data        phos
length      90.0000
NMissing    0

Smoothing Constant
Alpha: 0.635972   ◄──────── λ_optimal

Accuracy Measures
MAPE: 6.15079
MAD:  0.00261
MSD:  0.00001
```

Note from above that Minitab refers to the smoothing constant by the name *Alpha* (instead of lambda, as we are using in our presentation). Applying the results of this output to (10.43), we find that the EWMA fitted values, for the phosphorus series, are generated by

$$F_{t+1} = 0.635972x_t + 0.364028F_t$$

where the fitted value for period 1 (F_1) is often taken to equal the first actual value (phos_1). Minitab allows the user to select other initial values (e.g., the sample mean of part or the whole

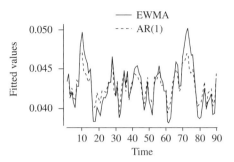

Figure 10.10   Comparison of EWMA and AR(1) fitted models for phosphorus series.

data series); for a general discussion of the initialization of an EWMA model, refer to Gardner (1985).

To see how closely the fitted EWMA model tracks the series relative to the originally fitted AR(1) model, we simultaneously plot, in Figure 10.10, the fitted values from each model. As can be seen, the EWMA closely follows the AR(1) model fitted values. Thus, in this example, the all-purpose EWMA does reasonably well in capturing the nonrandom component underlying the process. At this stage, a control chart can be constructed for the residuals to help detect unusual observations over and above the estimated nonrandom behavior.

Generally, an EWMA modeling approach might work reasonably well if the data series resembles a positive autocorrelated series, dominated only by a meandering component. In many industries, this indeed may be the case. For instance, the chemical and process industries commonly encounter meandering autocorrelated processes (see Harris and Ross, 1991). In a clinical laboratory setting, Alwan and Bissell (1988) found that more than 60 percent of analyzed quality control series followed an ARIMA model which is well suited for approximation by the EWMA model. In these cases, it may be possible to use the EWMA all-purpose model as a substitute for the need for broader time-series training.

However, as we have seen throughout this book, many processes in practice do not fall exclusively in the category of meandering positive autocorrelation. The simple EWMA model performs poorly in the presence of other nonrandom behaviors, such as trends and seasonality. Since many time series cannot be adequately described by the simple EWMA model, improvements can be made to account for different types of time-series behaviors. For instance, Holt (1957) extended the simple EWMA model to double and triple exponential smoothing models to account for different types of trend behavior. Because many time series are influenced by seasonal variation, Winters (1960) describes an EWMA model approach that attempts to account for additive and multiplicative seasonal components as well as trend. In addition to these methods, other researchers have proposed similar models based on the concept of exponential smoothing. Some examples are Trigg and Leach's adaptive exponential smoothing, Chow's adaptive smoothing, Brown's one-parameter adaptive smoothing, and Harrison's harmonic smoothing. For a detailed description and comparison of each of these smoothing methods see Madkridakis, Wheelwright, and McGee (1983).

The need to consider a number of alternative smoothing models defeats the original motivation of the all-purpose model strategy, in particular, an approach that avoids general modeling skills. From a training perspective, the modeling approach developed in this book, which relies on regression, is much more intuitive and understandable to a general audience than alternative smoothing models are. For example, it is this author's experience (in teaching and training) that students and trainees quickly comprehend how to fit and interpret a regression-based trend model but often have difficulty grasping the concept and fitting of a smoothing model designed for trends, such as a double exponential smoothing model.

We must also remember that the ability to model systematic behavior is not the only skill potentially needed in the challenge to understand and monitor processes. Additionally, there is a need to have the ability to diagnostically check for normality and draw upon data analysis techniques to rectify any substantial departures from this standard SPC assumption. In the final analysis, we believe that a strategy based on the elementary data analysis tools presented in this book not only provides a powerful scheme for handling real life applications but also is accessible to a broad audience of process users.

10.4   RELATIONSHIP BETWEEN PROCESSES: A BROADER MODELING PERSPECTIVE

Any process can be viewed as a series of "responses" to one or more series of "inputs." We have provided in this book the techniques necessary to better understand the statistical nature of the response or output series. In particular, we have demonstrated how standard control charts can be used to summarize common-cause variation and identify special-cause variation for many circumstances, while more general data analysis techniques are used for process characterization in other circumstances. Once the process is properly summarized, efforts can be made to backtrack through the process in an attempt to identify the underlying sources of variation reflected in the common-cause variation (random or systematic) or in the special-cause variation.

If we view the primary process for monitoring as an output process and the input processes as the sources of variation, it is useful in the study of common- and special-cause variation to determine the relationship between these two types of processes. Given past data on each of the processes, the data modeling approach illustrated in this book can easily be extended to include the study of two or more processes.

In particular, let us denote the output process by y_t and, for simplicity, assume a single input series x_t. There may, of course, be more than one input series x_t related to y_t. The x_t variable might represent a measurement of an environmental condition, such as temperature, humidity, or pressure, which might have an effect on y_t; or it might be an indicator variable for raw materials, operators, or equipment. Up to this point, we have concentrated on modeling of a single series y_t based on its past, and possibly on some deterministic independent variables, such as trends and seasonal variables. The idea is now to allow for the incorporation of the x_t series to help us better predict and understand y_t. In other words, we are considering a *time-series regression* of y_t on x_t (or several x_t values).

By relating the primary series to an input series, time-series regression provides a potentially valuable tool in helping to explain the common causes and special causes seen in y_t. For instance, a common cause in the form of a trend in the y_t series may be reflecting a trend in an input series x_t. In another example, an outlier in the y_t series may be a reflection of a change in an input series, such as a raw material or operator series. In both cases, the time-series regression suggests a possible avenue for the improvement of the primary series y_t, in particular, a close study of the input series x_t.

Finally, it is important to underscore that by extending the modeling domain to include variables on other processes, we place ourselves in potentially a better position to explain movements in the primary process. As such, we may reveal unusual movements, begging special-cause explanations, in the primary process that would have otherwise escaped our attention if we had studied the primary process alone; we will illustrate this very phenomenon in an example to follow. Hence, time-series regression of y_t on x_t naturally fits as an important tool for SPC efforts.

We should be careful to distinguish a time-series regression of y_t on x_t from a direct regression of y_t on the independent variable x_t. Roberts (1991) explains the reason:

> If we try the regression of Y on X for two time series that are themselves nonrandom, we are likely to obtain a confusing picture that is not easily interpreted. Autocorrelation or other nonrandom effects within the individual series can create this confusion. They can even create an impression of a relationship between the two series when in fact there is none, a situation that has been called "nonsense correlation between time series." (p. 203)

To avoid the danger of nonsense correlation, Roberts suggests that one should consider as independent variables not only the instantaneous or coincident input variable x_t, but also lags of y_t and x_t. As with the univariate data modeling, we might also include deterministic variables in the model, such as trends and seasonal indicator variables. We now turn to an example to illustrate how the incorporation of data from another process can help us better understand the movements of the primary process in question.

Table 10.10 Weekly sales and quotation data

Week	Sales ($000)	Quotes	Week	Sales ($000)	Quotes
1	765.14	682	19	1216.46	1156
2	955.41	846	20	1143.63	898
3	905.13	967	21	1182.72	887
4	468.34	723	22	1306.78	1302
5	778.10	937	23	1300.18	1267
6	1259.52	347	24	690.40	329
7	644.42	991	25	965.84	1089
8	1148.95	1907	26	1425.66	443
9	1438.96	582	27	1296.98	1268
10	948.57	1056	28	1756.99	416
11	1019.70	109	29	1196.34	1957
12	685.79	1454	30	1732.45	373
13	1165.73	1301	31	1192.54	1060
14	1261.17	317	32	1656.40	1070
15	1093.98	898	33	1359.73	554
16	1163.07	1448	34	1060.31	831
17	1424.02	1508	35	1425.68	1244
18	953.39	742	36	1732.28	981

Example 10.9

For this example, the primary process is company sales.[5] In Table 10.10, we show data on weekly sales (thousands of dollars) for a period of 36 weeks. Along with sales data, we provide data on the number of sales quotations given each week by the company's sales personnel to potential customers. The data are also found in the file "quotes.dat".

Studying the sales series alone, we would find that the data were trending upward and well fitted with a simple linear trend model. In Figure 10.11a, we show the fitted trend line superimposed on the sales series. Over and above the systematic trend movement, there appears to be no unusual behavior in the data as reflected by the special-cause chart shown in Figure 10.11b. The basic idea is to now investigate whether the quotation series (X series) can *build upon* or *improve* the analysis of the sales series (Y series). There is no guarantee, in general, that another process will explain the movements in the Y series of interest better. It is possible that the Y series itself contains all the statistical information that is needed for process understanding and monitoring.

If we consider the price quotations data series provided in Table 10.10 as an independent variable, then we are considering only the possibility that the number of price quotations for a given week is a predictor of sales for the same week. This may be the case; however, it is quite plausible that the number of price quotations is a leading indicator of future sales; that is, not only coincident price quotations but lagged price quotations might help in prediction of sales movements. To investigate what possible lags of price quotations might be useful, we can utilize the cross-correlation function (CCF) introduced in Section 10.2. The CCF between price quotations and sales with significance limits is presented in Figure 10.12a.

[5]For confidentiality reasons, the name of the company and nature of the product are omitted. The data were kindly provided by a past executive MBA student of mine.

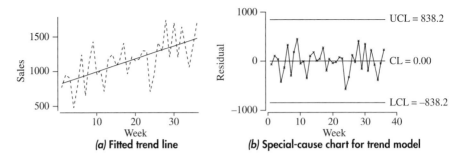

(a) Fitted trend line   **(b)** Special-cause chart for trend model

Figure 10.11   Analysis of sales series alone.

```
CCF - correlates quotes (t) and sales (t+k)

         -1.0 -0.8 -0.6 -0.4 -0.2  0.0  0.2  0.4  0.6  0.8  1.0
          +----+----+----+----+----+----+----+----+----+----+
 -5 -0.225                          XXXXXXX
 -4 -0.033                            XX
 -3  0.021                            XX
 -2  0.020                           XXXX
 -1  0.020                            X
  0 -0.034                            XX
  1  0.470                            XXXXXXXXXXXXXX
  2 -0.319                    XXXXXXXXX
  3  0.073                          XXX
  4 -0.069                          XXX
  5  0.027                           XX
```

(a) CCF of quotation and sales series

```
The regression equation is
sales = 539 + 0.314 quotes-1 + 17.6 time

35 cases used 1 cases contain missing values

Predictor         Coef       StDev        T        P
Constant         539.1       104.9      5.14    0.000
quotes-1       0.31434     0.08171      3.85    0.001
time            17.635       3.514      5.02    0.000

S = 209.6    R-Sq = 57.0%    R-Sq(adj) = 54.3%

Analysis of Variance

Source            DF          SS          MS        F        P
Regression         2     1865543      932772    21.24    0.000
Residual Error    32     1405518       43922
Total             34     3271061
```

(b) Estimated regression equation

Figure 10.12   Incorporating quotations in the analysis of sales.

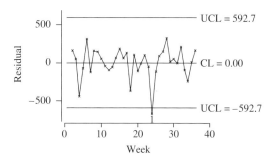

Figure 10.13   Special-cause chart after incorporation of quotations.

For our purposes, we are concerned only with the cross-correlations computed for lags $k \geq 0$ since these correlations are associated with coincident or lagged values of price quotations as related to sales. Because our interest is not in predicting price quotations from past sales, we are not concerned with $k < 0$. From the CCF, we see that the cross-correlation at lag 1 ($k = 1$) is positive and significant. This suggests that the number of price quotations for any given week may be a useful predictor for the following week's sales value. After some thought, this relationship is not too surprising. The number of price quotations is in a sense a measure of interaction between the sales personnel and customers. Hence, it stands to reason that increased (decreased) interaction will be associated with increased (decreased) sales levels, and thus the positive correlation. The lag effect also is quite plausible since, typically, a customer will respond with a purchase sometime *after* obtaining a price quotation.

By regressing the sales data on two independent variables, lag 1 of quotations and time index, we obtain the regression output shown in Figure 10.12b. As reflected by the reported p values, both independent variables contribute significantly in explaining the sales variable. Diagnostics are satisfactory, as the reader can verify. It is interesting to recognize that the continued need for a trend term in the latest model implies that the trend effect is not accounted for by the quotation series.

After allowance for the trend effect and the relationship between number of quotations and sales, we find from Figure 10.13 that there is a 3σ special-cause signal underlying the sales series that was *not* identified by the study of the sales series alone (see Figure 10.11b). In particular, the sales amount for week 24 is substantially lower than expected, given the general trend and the number of quotations in week 23. Management should be concerned and look into this situation to see whether the unusually poor sales are due to internal management problems or to forces external to the company.

10.5   REDUCING PROCESS VARIABILITY BY ADJUSTMENT

In the monitoring and control of manufacturing-type processes, the basic objective is to produce a process having the smallest possible variation around a target value. As pointed out in Section 4.1, two approaches to process control have been developed to help meet such an objective: statistical process control and *adaptive process control* (*APC*).

SPC, or, as we more broadly have called it, statistical process monitoring, is oriented toward the identification and separation of special causes and common causes. The aim is then to ultimately *remove* the effects of the special causes—process shifts and systematic nonrandom patterns—from the process output, thus reducing process variability. In contrast, the aim of APC (also referred to as engineering process control, dynamic control, on-line control, or stochastic control) is not to remove the underlying special causes but rather to tolerate their existence and attempt to *compensate* for the effects of the special causes on the process output, again reducing process variability.

For some practitioners, there appears to be a conflict between statistical and adaptive process control, but this appearance is an illusion. Both approaches lead to process improvement by the reduction of variation about a target value. It is ultimately a question of economics as to whether it is less costly to seek out certain special causes and remove them or to tolerate the causes and compensate for them. Box and Luceño (1997) nicely illustrate the economics of choice between the two approaches:

> The two approaches may be illustrated by considering the dilemma of someone who lives in Madison, Wisconsin, where the ambient temperature (the disturbance) can vary from $-30°F$ to $+95°F$. The problem of keeping the body at a comfortable temperature can be solved by eliminating the special cause, for example, moving to California. But, if this is not possible, convenient, or economical, it can be dealt with by installing central heating and air conditioning in the home with the temperature adjusted by a process of feedback control. (p. 130)

Having said this, we emphasize that the use of APC techniques does not preclude the use of SPC techniques. As a general strategy, SPC techniques should be used as a first line of attack to identify any special causes that can easily and economically be removed. Such efforts can help bring the process closer to a state of statistical control by removing easily resolved sources of variations, (e.g., inconsistent operators due to improper training or differences in raw materials). However, even after such efforts to remove sources of special-cause variation, there may still exist movements in the process due to the influences of naturally occurring phenomena, such as variation in ambient temperature, humidity, and raw material quality. As described in the quotation above, such phenomena may not be economical to eliminate, and thus some system of process adjustment may be called for.

Once adaptive control is underway, SPC techniques can still play a vital and complementary role. In particular, the adjusted process can be monitored for any special-cause impingement that may occur while the process is being regulated. For more technical discussions of the advantages of the complementary use of SPC and APC techniques, see Box and Kramer (1992), Box and Luceño (1997), MacGregor (1987), and Vander Weil et al. (1992). In our data example for this section, we will illustrate when SPC methods can be applied within an APC scheme.

Only a very simple adaptive control scheme will be discussed here. Fortunately, however, this simple scheme is important in practice and serves to bring out many important ideas, including the idea that readjustments, for certain process conditions, can be beneficial, as opposed to the counterproductive effects of readjusting a random process (refer to Section 4.5). The scheme we will introduce is discussed in greater technical detail in Box, Jenkins, and Reinsel (1994) and in Box and Luceño (1997). To develop the scheme, we need to make some assumptions and establish some initial notation:

1.   All APC schemes (simple or sophisticated) presuppose that there exists some mechanism that enables one to alter the observed level of the process output. The standard assumption is that there is another variable (X, say)—referred to as a *compensatory* or *manipulable* variable—that can be adjusted to compensate for movements in the output variable. As an example, Box and Luceño (1997) cite an application in which the output variable (i.e., the quality characteristic of interest) is the thickness of a thin metallic film and the compensatory variable X is the metal deposition rate. The idea is to make regular adjustments to the compensatory variable so that the process output is kept close to some desired target level.

2.   We will assume that y_t represents the value of the output variable, at time t, if *no adjustment is made*. Similarly, we can define d_t as the deviation of the output variable from its target value T if *no adjustment is made*; that is, $d_t = y_t - T$. The deviation d_t is often referred to as the *disturbance* at time t. Since adjustments will be made to the process, the disturbances are not directly observed. If no adjustments are made, then the disturbances are directly observed. Thus, under our APC scenario, the actual output readings will reflect both the disturbances (not directly observed) and the adjustments (known).

3.   We assume that the disturbance process d_t is a nonrandom but predictable process. Recall from Section 4.5, if a process is random, then any adjustments made will have the counterproductive effect of increasing process variation. One of the most common assumptions is that the output (or disturbance) process follows a meandering autocorrelated pattern.

4.   We assume that the underlying disturbance process is *not* affected, regardless of whether adjustments are made. It is helpful to think of a rifle with adjustable sights where the goal is to hit the center of a target. For simplicity, consider only horizontal deviations from the bull's-eye. Suppose the disturbances reflect the effects of a crosswind. Assuming that wind velocity follows an autocorrelated pattern (e.g., the wind velocities close in time tend to be close in value), the disturbances will then vary in an autocorrelated fashion through time. The compensatory variable X is the sight adjustment on the rifle. But recognize that adjusting the rifle sights does not intrinsically change the disturbance process, that is, the wind velocity. Rather, we recognize that we are dealing with the following relationship:

$$\begin{matrix} \text{Observed} \\ \text{deviation from} \\ \text{bull's-eye} \end{matrix} = \text{Disturbance} + f(\text{Rifle adjustment}) \qquad \textbf{(10.44)}$$

The term $f(\ )$ is just some function that relates the rifle adjustment scale to the bull's-eye scale. The objective is then to make sight adjustments, after each shot, so as to compensate for the disturbances in such a manner that the observed deviations from the bull's-eye are minimized.

5.   We assume that the full effect of an adjustment to the compensatory variable X will be reflected in the next period's output reading. Furthermore, we assume that a unit change in X will produce g units of change in the output reading. The constant g is usually call the *process gain*. We can assume, without loss of generality, that a set point of $X = 0$ implies no adjustment. Thus, if at time t, X is set equal to X_t, the output reading will be altered by gX_t units relative to what the output reading would have been if no adjustment had been made.

6.   We assume that there is no cost in making an adjustment. Box and Luceño (1997) note that this is a reasonable assumption in most continuous process industries (e.g., chemical processes) where adjustments are basically costless "knob" turning. However, in some industries, particularly parts industries, there may be fixed costs for making adjustments. For example, there may be costs associated with machine stoppage for setups and tool changing. The reader is referred to Box and Jenkins (1963) and Box and Luceño (1997) for discussion of APC schemes when adjustment costs are nontrivial.

With the above assumptions and discussions, we are now in a position to discuss the remaining details of the APC scheme. At time t, we are faced with a decision as to the set point value for the compensatory variable X_t. Using the rifle relationship of (10.44) as a guide and remembering that there is a process gain of g units and that the effect of adjustment is not felt until the next period, we recognize that at time $t + 1$, the *observed* deviation from the target d_{t+1}^{adjust} after adjustment would be

$$d_{t+1}^{\text{adjust}} = d_{t+1} + gX_t \qquad (10.45)$$

Ideally, if we knew the value of the disturbance at time $t + 1$, we could then set X_t to perfectly counteract the disturbance. In particular, we would set X_t equal to $-d_{t+1}/g$. As a result, the observed deviation from the target would be exactly zero, and the process output would be perfectly on target.

Unfortunately, we do not know precisely the value of a future disturbance. However, we can do the next best thing, namely, we can make a forecast of d_{t+1}. If we denote the forecast by $\hat{d}_{t+1}$, it is useful to write the following expression:

$$d_{t+1} = \hat{d}_{t+1} + e_{t+1} \qquad (10.46)$$

In other words, the actual disturbance equals the forecast of the disturbance plus the forecast error. This breakdown is analogous to the regression decomposition of an observation into a fitted value and a residual. Substituting (10.46) into (10.45), we obtain

$$d_{t+1}^{\text{adjust}} = e_{t+1} + \hat{d}_{t+1} + gX_t \qquad (10.47)$$

Intuitively, the idea is to make adjustments according to the principle of offsetting the *predicted disturbances*. This implies setting X_t so that $gX_t = -\hat{d}_{t+1}$, or equivalently,

$$X_t = \frac{-\hat{d}_{t+1}}{g} \tag{10.48}$$

Since the last two terms of (10.47) cancel, this adjustment policy results in the observed deviation from target being equal to the forecast error, that is,

$$d_{t+1}^{\text{adjust}} = e_{t+1} \tag{10.49}$$

The above fact is true for any time period, not just period $t + 1$.

Equation (10.48) indicates the set point for the compensatory variable, at time t. The adjustment we need to make in the compensatory variable, from its previous value at time $t - 1$, is

$$X_t - X_{t-1} = \frac{-\hat{d}_{t+1}}{g} - \frac{-\hat{d}_t}{g} = \frac{-(\hat{d}_{t+1} - \hat{d}_t)}{g} \tag{10.50}$$

Notice that in our developments we have not made any assumptions about the time-series behavior of the disturbance d_t. For many APC applications, the underlying disturbance process is an autocorrelated meandering process. Given our discussion in Section 10.3, a simple EWMA model can serve as a reasonable predictive model for such processes. Applying the EWMA forecasting scheme [see (10.43)] to the disturbance process, we have

$$\hat{d}_{t+1} = \lambda d_t + (1 - \lambda)\hat{d}_t \tag{10.51}$$

where d_t is the actual disturbance at time t, $\hat{d}_t$ is the forecast of time t made at time $t - 1$, and λ is the smoothing constant. Given (10.51), it will prove useful to notice the following:

$$
\begin{aligned}
\hat{d}_{t+1} - \hat{d}_t &= \lambda d_t + (1 - \lambda)\hat{d}_t - \hat{d}_t \\
&= \lambda d_t + \hat{d}_t - \lambda\hat{d}_t - \hat{d}_t \\
&= \lambda d_t - \lambda\hat{d}_t \\
&= \lambda(d_t - \hat{d}_t)
\end{aligned}
\tag{10.52}
$$

The term $d_t - \hat{d}_t$ is simply the forecast error for time t; thus (10.52) can be rewritten as

$$\hat{d}_{t+1} - \hat{d}_t = \lambda e_t \tag{10.53}$$

But from (10.49), the forecast errors are equal to the deviations from the target for the adjusted process. Hence,

$$\hat{d}_{t+1} - \hat{d}_t = \lambda d_t^{\text{adjust}} \tag{10.54}$$

If we substitute (10.54) into (10.50), we obtain the period-to-period adjustment equation for the compensatory variable

$$X_t - X_{t-1} = \frac{-\lambda d_t^{\text{adjust}}}{g} \qquad (10.55)$$

Thus, determining how much to change X from its previous setting only requires the values for λ and g and the deviation of the current output reading from the target value T.

Example 10.10

To illustrate the APC approach, we consider a data series (known as series A) found in Box, Jenkins, and Reinsel (1994). The data series is composed of 197 concentration readings of a chemical process, single readings taken every 2 h. The data are shown in Table 10.11 and provided in the file named "boxja.dat". Figure 10.14 shows the sequence plot of the series itself (y_t). Suppose it is desired to maintain the concentration characteristic as close as possible to a target value of $T = 17$; the target level is indicated by the superimposed horizontal line in Figure 10.14. On the right-hand side of Figure 10.14 is the scale for the disturbance $d_t (= y_t - 17)$. It is quite clear that the unadjusted process is autocorrelated, as evidenced by its meandering behavior.

Suppose that the sources for the meandering behavior are unknown and thus not immediately removable. However, it is known that the movements in the concentration variable can be compensated for by making adjustments to the feed rate of a catalyst involved in forming

Table 10.11   Chemical concentration data (series A)

17.0	16.6	16.3	16.1	17.1	16.9	16.8	17.4	17.1	17.0
16.7	17.4	17.2	17.4	17.4	17.0	17.3	17.2	17.4	16.8
17.1	17.4	17.4	17.5	17.4	17.6	17.4	17.3	17.0	17.8
17.5	18.1	17.5	17.4	17.4	17.1	17.6	17.7	17.4	17.8
17.6	17.5	16.5	17.8	17.3	17.3	17.1	17.4	16.9	17.3
17.6	16.9	16.7	16.8	16.8	17.2	16.8	17.6	17.2	16.6
17.1	16.9	16.6	18.0	17.2	17.3	17.0	16.9	17.3	16.8
17.3	17.4	17.7	16.8	16.9	17.0	16.9	17.0	16.6	16.7
16.8	16.7	16.4	16.5	16.4	16.6	16.5	16.7	16.4	16.4
16.2	16.4	16.3	16.4	17.0	16.9	17.1	17.1	16.7	16.9
16.5	17.2	16.4	17.0	17.0	16.7	16.2	16.6	16.9	16.5
16.6	16.6	17.0	17.1	17.1	16.7	16.8	16.3	16.6	16.8
16.9	17.1	16.8	17.0	17.2	17.3	17.2	17.3	17.2	17.2
17.5	16.9	16.9	16.9	17.0	16.5	16.7	16.8	16.7	16.7
16.6	16.5	17.0	16.7	16.7	16.9	17.4	17.1	17.0	16.8
17.2	17.2	17.4	17.2	16.9	16.8	17.0	17.4	17.2	17.2
17.1	17.1	17.1	17.4	17.2	16.9	16.9	17.0	16.7	16.9
17.3	17.8	17.8	17.6	17.5	17.0	16.9	17.1	17.2	17.4
17.5	17.9	17.0	17.0	17.2	17.0	17.2	17.4	17.4	17.0
18.0	18.2	17.6	17.8	17.7	17.2	17.4			

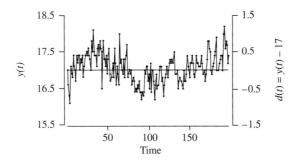

Figure 10.14   Unadjusted chemical concentration series, target value = 17.

the output chemical product. Suppose that the gain is 1.5; that is, an increase of 1 unit in the feed rate increases the concentration by 1.5 units.

Now imagine that we go back in time and implement an APC scheme starting just prior to the first time period. Hence, we are standing from a perspective that the 197 disturbances shown in Figure 10.14 are future values and unknown to us at the present time. Let us assume that an earlier study of a large sample from the process has suggested the appropriateness of an EWMA forecasting model with λ estimated to be 0.30. Before the first reading, we assume that the set point for the compensatory variable is 0, that is, $X_0 = 0$. From (10.48), if $X_0 = 0$, we are implicitly assuming that the forecast value of the first disturbance reading is $\hat{d}_1 = 0$. Since the unadjusted output value for the first period would be 17, the first actual disturbance d_1 is 0.0 (= 17 − 17). Thus, the observed deviation from the target value for the first period, which is also the forecast error [see (10.49)], is

$$d_1^{\text{adjust}} = d_1 - \hat{d}_1 = 0.0 - 0.0 = 0.0$$

This implies that the first-period concentration reading for the adjusted process is

$$y_1^{\text{adjust}} = 17 + d_1^{\text{adjust}} = 17 + 0.0 = 17.0$$

From (10.55), the set point of the feed rate is adjusted by the following amount:

$$X_1 - X_0 = \frac{-0.30 d_1^{\text{adjust}}}{1.5} = -\frac{1}{5}(d_1^{\text{adjust}}) = -\frac{1}{5}(0.0) = 0.0$$

This implies that the set point for the feed rate is

$$X_1 = X_0 + 0.0 = 0.0 + 0.0 = 0.0$$

The picture to date is as follows:

Time	Unobserved Disturbance	Predicted Disturbance	Observed Deviation	Adjusted Output	X Set Point	X Adjustment
1	0.0	0.00	0.00	17.00	0.000	0.000

The action so far seems trivial because we are just starting things off. The next couple of steps, however, bring out the effects of the APC scheme.

Applying (10.51), we can get a forecast for the second-period disturbance:

$$\hat{d}_2 = 0.30(d_1) + (1 - 0.30)\hat{d}_1$$
$$= 0.30(0.0) + 0.70(0.0)$$
$$= 0.0$$

The actual disturbance for the second period d_2 is -0.4 ($= 16.6 - 17$). Thus, the observed deviation from target for the second period is

$$d_2^{\text{adjust}} = d_2 - \hat{d}_2 = -0.4 - 0.0 = -0.4$$

This implies that the second-period concentration reading for the adjusted process is

$$y_2^{\text{adjust}} = 17 + d_2^{\text{adjust}} = 17 + (-0.4) = 16.6$$

The set point of the feed rate is adjusted by the following amount:

$$X_2 - X_1 = -\frac{1}{5}d_2^{\text{adjust}} = -\frac{1}{5}(-0.4) = 0.08$$

This implies that the set point for the feed rate is

$$X_2 = X_1 + 0.08 = 0.0 + 0.08 = 0.08$$

The picture to date is as follows:

Time	Unobserved Disturbance	Predicted Disturbance	Observed Deviation	Adjusted Output	X Set Point	X Adjustment
1	0.0	0.00	0.00	17.00	0.000	0.000
2	−0.4	0.00	−0.40	16.60	0.080	+0.080

Our first action has taken place; namely, we need to adjust the compensatory variable set point *up* by 0.08 unit. But we still have not observed any impact on the output process since the values of the adjusted output for the first two periods are the same as the unadjusted process values shown in Table 10.11. Let us now proceed to the next period.

The forecast for the third-period disturbance is

$$\hat{d}_3 = 0.30(d_2) + 0.70\hat{d}_2$$
$$= 0.30(-0.4) + 0.70(0.0)$$
$$= -0.12$$

The actual disturbance for the third period d_3 is -0.7 ($= 16.3 - 17$). Thus, the observed deviation from the target, for the third period, is

$$d_3^{\text{adjust}} = d_3 - \hat{d}_3 = -0.7 - (-0.12) = -0.58$$

This implies that the third-period concentration reading for the adjusted process is

$$y_3^{\text{adjust}} = 17 + d_3^{\text{adjust}} = 17 + (-0.58) = 16.42$$

The set point of the feed rate is adjusted by the following amount:

$$X_3 - X_2 = -\frac{1}{5}d_3^{\text{adjust}} = -\frac{1}{5}(-0.58) = 0.116$$

This implies that the set point for the feed rate is

$$X_3 = X_2 + 0.116 = 0.08 + 0.116 = 0.196$$

Now the picture to date is as follows:

Time	Unobserved Disturbance	Predicted Disturbance	Observed Deviation	Adjusted Output	X Set Point	X Adjustment
1	0.0	0.00	0.00	17.00	0.000	0.000
2	−0.4	0.00	−0.40	16.60	0.080	+0.080
3	−0.7	−0.12	−0.58	16.42	0.196	+0.116

At this time, that we can see what the APC procedure is starting to accomplish. By making the X adjustment in the second period (i.e., *prior* to the third period's output), we have "made" the concentration reading for the third period equal to 16.42. The concentration reading of 16.42 is *closer* to the target value than the reading of 16.30, which would have resulted had we made no adjustment. In Table 10.12, we show the APC results for the first 15 periods.

Figure 10.15 plots all 197 values of the chemical concentration resulting from the APC procedure. In contrast to the unadjusted output (see Figure 10.14), the output process now appears to be in a state of statistical control in the sense that the chemical concentrations appear random with no tendency for nonrandom meandering. By neutralizing the underlying nonrandom component, the amount of variability in the adjusted process must necessarily be less than

Table 10.12   Results from APC scheme (first 15 periods)

Time	Unobserved Disturbance	Predicted Disturbance	Observed Deviation	Adjusted Output	X Set Point	X Adjustment
1	0.0	0.00	0.00	17.00	0.000	0.000
2	−0.4	0.00	−0.40	16.60	0.080	+0.080
3	−0.7	−0.12	−0.58	16.42	0.196	+0.116
4	−0.9	−0.29	−0.61	16.39	0.317	+0.121
5	0.1	−0.48	0.58	17.58	0.202	−0.115
6	−0.1	−0.30	0.20	17.20	0.161	−0.041
7	−0.2	−0.24	0.04	17.04	0.153	−0.008
8	0.4	−0.23	0.63	17.63	0.027	−0.126
9	0.1	−0.04	0.14	17.14	−0.001	−0.028
10	0.0	0.00	0.00	17.00	−0.001	0.000
11	−0.3	0.00	−0.30	16.70	0.059	+0.060
12	0.4	−0.09	0.49	17.49	−0.038	−0.098
13	0.2	0.06	0.14	17.14	−0.067	−0.028
14	0.4	0.10	0.30	17.30	−0.127	−0.060
15	0.4	0.19	0.21	17.21	−0.169	−0.042
.	.	.	.	.	.	.
.	.	.	.	.	.	.
.	.	.	.	.	.	.

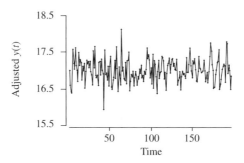

Figure 10.15   Adjusted chemical concentration series, target value = 17.

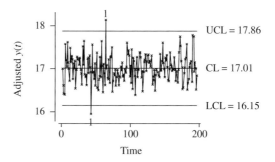

Figure 10.16   Control chart for adjusted output.

that in the unadjusted process. Since we are dealing with a target value, an appropriate criterion for comparing processes in their overall performance must consider the variability relative to the target value. Thus, we do not want to report the sample variance of the output readings (unadjusted or adjusted). A more appropriate measure is the average of the squared deviations from the target value. Suppose the deviations from the target for the unadjusted process (i.e., the disturbances) and the deviations from target for the adjusted process are placed in columns c1 and c2 of Minitab, respectively. Below we use Minitab to compute the average squared deviations for the two processes:

```
MTB > let c3=c1*c1
MTB > let c4=c2*c2
MTB > name c3 'squared disturbances' c4 'squared deviations'
MTB > mean c3

Column Mean

   Mean of squared disturbances = 0.16249
MTB > mean c4

Column Mean

   Mean of squared deviations = 0.10093
```

Thus, in terms of average squared deviations, the use of an APC scheme reduced the process variability around the target value by nearly 40 percent.

Since the adjusted process observations follow an approximately random process and are also compatible with the normal distribution (left for reader to verify), standard SPC methods can be applied to the data series. For example, Figure 10.16 presents a standard Shewhart chart for individual measurements. As can be seen, two outliers are highlighted (observations 43 and 64). If it is found that these outliers are indeed the result of special causes, then efforts should be made to eliminate the underlying sources to prevent future unwanted effects.

In the previous example, we used an EWMA model with λ equal to 0.30 to track the underlying nonrandom component in the output process. As stated, this choice was based on an empirical study of previous observations generated by the process. The fact that the adjusted output series was random implies that the EWMA model, with $\lambda = 0.30$, did a good job of tracking the nonrandom component, thus allowing us to offset the nonrandom effects. If, however, the adjusted series exhibited discernible nonrandom effects, this would be suggestive of a poor predictive model and the need to pursue an alternative model.

There can be situations in which no history is available to estimate an "optimal" forecasting model. If we wished to use an EWMA model as the basis for forecasting in such situations, the lack of history would require us to make an arbitrary choice for the smoothing constant λ. An interesting question is then, What is the effect on the APC results when a less-than-optimal smoothing constant is chosen? Recall from Section 10.3, the optimal λ is defined as the smoothing constant that minimizes the sum of squared residuals (i.e., forecast errors).

Box and Luceño (1997) provide us with some important results, thus enabling us to gain some insight into the effects of using a less-than-optimal smoothing constant. Their results are based on the assumption that the underlying nonrandom component for the disturbance is adequately modeled by an EWMA model. Suppose the true optimum value for the smoothing constant is λ_0. In practice, we obviously do not know λ_0, so we either estimate it based on available data or make an arbitrary choice. Let λ_{choice} represent the value of smoothing constant chosen for implementation of the APC scheme.

Define $\sigma_{\lambda_0}^2$ as the variance about the target for the adjusted processes if λ_0 is used in the APC procedure. Similarly, define $\sigma_{\lambda_{\text{choice}}}^2$ as the variance about the target for the adjusted processes if λ_{choice} is used in the APC procedure. Theoretically (i.e., over the long run), an APC based on the optimum smoothing constant will produce a process with least variability around the target, hence, $\sigma_{\lambda_0}^2 \leq \sigma_{\lambda_{\text{choice}}}^2$. Or, equivalently, the ratio $\sigma_{\lambda_{\text{choice}}}^2 / \sigma_{\lambda_0}^2$ must be greater than or equal to 1. This ratio can be interpreted as an inflation factor in terms of how much the variance of the process output around the target value will be inflated if λ_{choice} is chosen instead of λ_0. Box and Luceño (1997) show that

$$\frac{\sigma_{\lambda_{\text{choice}}}^2}{\sigma_{\lambda_0}^2} = 1 + \frac{(\lambda_{\text{choice}} - \lambda_0)^2}{\lambda_{\text{choice}}(2 - \lambda_{\text{choice}})} \tag{10.56}$$

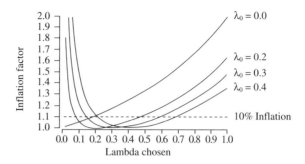

Figure 10.17   Inflation in the variance of the adjusted process arising from a choice of λ different from λ_0.

For a selection of different λ_0 values, Figure 10.17 shows graphs of what the inflation factor would be for λ_{choice} in the range from 0 to 1. First, let us consider the $\lambda_0 = 0$ curve. If the true process situation is associated with $\lambda_0 = 0$, then the process must be in a state of statistical control as defined by a random process. Based on the lessons of Deming, we learned in Section 4.5 that any adjustments made to a random process will have the counterproductive effect of increasing the process variance. Indeed, from Figure 10.17, we see that if any value for λ_{choice} other than 0 is chosen, the process variance is greater than if λ_{choice} were chosen to be 0. In the most extreme case with $\lambda_{\text{choice}} = 1$, an adjustment (apart from the gain factor g) is exactly equal to the current deviation from target. The reader might recall this to be a rule 2 adjustment policy from Nelson's funnel experiment (Section 4.5), which we demonstrated will double the variance of an in-control process. We see from the figure a verification of this fact, namely, the inflation factor is 2 when $\lambda_0 = 0$ and λ_{choice} is taken to be 1.

Notice that for the curves associated with $\lambda_0 > 0$, the curves are quite flat around the optimal value. For moderate departures from the theoretically optimal value, this implies that there is little penalty in terms of increase in variability of the adjusted process. For instance, suppose the optimal value is 0.3, as appropriate for the chemical series just studied. For this λ_0 value, if we choose λ_{choice} to be anywhere within the approximate range of 0.15 to 0.60, the variance of the adjusted process will be at most 10 percent larger than the minimum variance achievable with the optimum value.

Thus, we have learned that the APC procedure is robust to different possible λ_{choice} values for a *given* λ_0 value. This insight is reassuring if we have a reasonable estimate for λ_0 because we know that even if the estimate is a bit off, the impact is minimal. However, what can we learn from the curves when no estimate for λ_0 is available and we have to make a purely arbitrary choice? Box and Luceño (1997) suggest that a safe initial choice for λ_{choice} is somewhere in the neighborhood of 0.2 to 0.4. If λ_{choice} is set to 0.2, for example, then we see from Figure 10.17 that the variance of the adjusted process will increase by at worst roughly 10 percent for any value λ_0 in the range from 0 to 0.4, which is a range wide enough to capture many processes encountered in practice. The "at most" increase of 10 percent represents a worst-case scenario if one sticks to the initial choice for λ_{choice} over the long run. But there is no need to be bound to λ_{choice} since it is possible to revise the smoothing constant with another estimate once a number of adjustments are made based on the initial choice. By recording all the adjustments made and recording all the deviations from the target value for the adjusted process, one can easily "reconstruct" the values of the disturbance process (i.e., the values of the unadjusted process). The reconstructed data series can then be used to estimate the smoothing constant.

10.6   FURTHER OPPORTUNITIES

We have provided in this book a comprehensive presentation of standard SPC methods. We have demonstrated with numerous examples the richness and effectiveness of the standard methods in a variety of applications. We have also armed the reader with basic data analysis techniques for situations when the standard methods are less effective or even misleading for process monitoring and provided necessary insight about the general workings of the process. As emphasized continuously from the beginning of this book, knowledge of this pool of techniques, coupled with an understanding of how to implement these techniques with a user-friendly software package (such as Minitab), gives any user a powerful approach for tackling almost any process challenge encountered in practice.

Nevertheless, as with any book, there are many techniques that may have potential application which are not formally covered. We close this chapter and the book with a list of potentially useful process-control related techniques with references for the interested reader:

Multivariate Extensions of Non-Shewhart Control Chart Schemes   In Section 10.2, we presented chi-square and Hotelling T^2 charts for detecting changes in a multivariate mean vector. These charts can be regarded as Shewhart-type charts in the sense that they use information from only the current sample vector. As a result, they inherit the similar property of univariate Shewhart charts of being less sensitive in detecting small to moderate mean shifts. From Chapter 8, recall that CUSUM and EWMA charts were developed to detect more quickly small to moderate mean shifts in the univariate case. Similarly, investigators have proposed multivariate versions of the CUSUM and EWMA schemes. Crosier (1988) and Pignatiello and Runger (1990) have developed multivariate CUSUM procedures (MCUSUM) while Lowry et al. (1992) and Prabhu and Runger (1997) have developed a multivariate version of the EWMA chart (MEWMA). In addition to these references, Wierda (1994) and Lowry and Montgomery (1995) provide excellent overview reviews of both Shewhart and non-Shewhart type of multivariate control charts.

Multiple Time-Series Modeling   Both the Shewhart-type and non-Shewhart-type multivariate control charts assume that the vectors are independent of one another over time. In Section 10.2, we introduced the cross-correlation function as a possible diagnostic tool for testing such an assumption. When vectors are not independent over time, a strategy of modeling the time behavior of the vectors might be considered. The modeling of vectors over time is sometimes referred to as vector autoregression or multiple time-series modeling. For details on multiple time-series modeling refer to the texts by Brockwell (1996), Reinsel (1993), and Wei (1990). Multiple time-series modeling can be computationally intensive. With the notable exceptions of AUTOBOX,[6] ITSM,[7] SAS,[8] and SCA,[9] few other statistical software include in the necessary algorithms for appropriate model estimation.

Nonparametric Methods   As emphasized throughout this book, the standard SPC methods rest on the assumption of approximate normality for the plotted statistic. When nonnormality is encountered, we have demonstrated that in most cases a simple normalizing transformation can adequately resolve the violation. Another approach proposed by some investigators is to utilize a class of statistical methods, known as nonparametric methods, which allows for inferences with no dependency on a specific distribution assumption such as normality. There are a number of papers in the quality-control literature dedicated to the development of control charts based on nonparametric techniques. They include those by Alloway and Ragnavachari (1991), Bakir and Reynolds (1979), Bhattacharyya and Freierson (1981), Farnum and Stanton (1986), and Hackl and Ledolter (1991).

Sophisticated Outlier Detection Methods   In our data analysis excursions, we have used "simple" techniques for detecting outliers or shifts embedded within a time-series process, for example, the application of standard SPC methods such as Shewhart, CUSUM, or EWMA charts to the residuals. More formally, the detection of outliers in time-series data was first studied by Fox (1972) who introduced two different types of outliers: the additive outlier (AO) and innovational outlier (IO). Simply stated, an AO affects only the level of some particular period's observation whereas an IO affects all observations from some particular period and thereafter. Several approaches to outlier analysis and detection have been proposed in the time-series literature, for example, Bruce and Martin (1989); Chang, Tiao, and Chen (1988); Tsay (1988), and the references therein. As with multiple time-series modeling, few statistical packages integrate the necessary algorithms to perform sophisticated outlier detection. The notable exceptions are the above-noted software vendors AUTOBOX and SCA.

---

[6]Automatic Forecasting Systems, P.O. Box 563, Hataboro, PA 19040 (http://www.autobox.com).
[7]Brockwell and Davis, *ITSM for Windows*, Springer-Verlag, New York, 1994.
[8]SAS Institute Inc., SAS Campus Drive, Cary, NC 27513-2414 (http://www.sas.com).
[9]Scientific Computing Associates Corporation, 913 West Van Buren Street, Suite 3H, Chicago, IL 60607-3528 (http://www.scausa.com).

Advanced APC Schemes   In Section 10.5, we introduced a simple APC scheme based on a number of assumptions. There are, however, practical situations in which these assumptions can be challenged, and there is, therefore, a need to develop alternative APC schemes. We noted, in our earlier discussions, one possible challenge. Namely, when an adjustment is associated with an appreciable cost, it is often not economical to make adjustments on a frequent basis. In our simple APC scheme, we assumed that the full effects of the adjustment will be felt in the next period. However, some adjustment processes can be subject to *dead time;* that is, there is a period of delay before the effects on the output are experienced. For detailed discussion and development of a variety of APC schemes based on such issues as costs, dead time, and other practical departures from the assumptions of a simple APC scheme, the reader should consult Box, Jenkins, and Reinsel (1994) and Box and Luceño (1997).

Economic Design of Control Charts   The practical implementation of a control chart requires the determination of three key *design* parameters: (1) the sample size at a given period, (2) the sampling frequency or time interval between samples, and (3) the specific placement of the control limits. Typically, the selection of these design parameters is based on a combination of statistical criteria and the user's background knowledge of the process environment. For instance, the determination of sample size and control limit placement might be based on some statistical performance measure, such as average run length, of the control chart in detecting a process shift of a particular size. On the other hand, frequency of sampling is typically determined without analytical formality; instead, practitioners tend to arrive at a sampling frequency based on a general assimilation of information on a variety of factors, such as production rates and expected frequencies of process shifts.

In a pioneering article, Duncan (1956) proposed that economic considerations be given to the design of a control chart. Duncan's model focuses on finding the optimum design for an $\bar{x}$ chart given four cost factors: (1) cost of sampling, (2) cost of searching for a special cause when none exists, (3) cost of detecting and removing a special cause, and (4) cost of producing nonconforming units. Subsequent to Duncan's work, many investigators tailored the model to accommodate a variety of specific situations. Montgomery (1980) provides an excellent summary of the earlier developments in the area of economic design of control charts. Subsequent to Montgomery's review paper, Lorenzen and Vance (1986) provided a unified approach to economic design of control charts that generalized Duncan's model, allowing for the optimization of almost any type of control chart, not just an $\bar{x}$ chart. Furthermore, their model is more expansive than Duncan's model, in terms of considering many more cost and operating factors. Lorenzen and Vance's work, along with other more recent contributions to the area, is summarized in the literature papers by Ho and Case (1994) and Svoboda (1991).

Even though the notion of integrating cost facts into the design of a process control scheme seems quite reasonable, the use of economically designed control charts has been limited, at best, in industry. Woodall (1986, 1987) offers two compelling reasons for their lack of implementation in practice. First, Woodall believes that economically designed control charts fall short of properly integrating many important cost factors for business competitiveness. For example, Woodall argues that economic models that attempt to include the costs of customer dissatisfaction or lost sales due to poor process performance cannot possibly capture the full picture because such costs are too difficult to measure correctly. In a related issue, Woodall (1997) argues that economic approaches only consider the costs when the process goes bad and fail to consider the cost benefits of process improvement. The second major concern Woodall has with economically designed control charts is their lack of consideration of statistical criteria. As a result, economically designed control charts can often have poor statistical properties. Most notably, many economic designs result in control charts with an unacceptably high false-alarm rate. Woodall's criticisms mark an important point in the evolution of economically designed control charts. In fact, it is fair to say that his concern about the statistical properties gave rise to a new class of control charts known as *economic statistically designed control charts*. Basically, these designs are economically designed control charts subject to constraints on desired statistical properties. As a first contribution to this new class of control charts, Saniga (1989) introduced economically designed control charts subject to constraints on type I error rate, type II error rate, and the average time to signal a process shift.

Whether the continuing developments of economic designed charts will be embraced by industry is still uncertain. The reader is referred to Keats et al. (1997) for a recent review of the status of economic modeling for statistical process control and suggestions for reducing the barriers to the implementation of economic models in practice.

APPENDIX TO CHAPTER 10

In this appendix, we provide a brief review of matrices along with illustrations on how to use Minitab for basic matrix operations. As noted earlier in this chapter, readers interested in a complete treatment of matrices and matrix algebra can refer to Searle (1982).

DEFINITION

A matrix is a rectangular array of real numbers arranged in rows and columns. Just as there is an algebra of numbers and symbols representing numbers, there is an algebra of matrices. A number of operations can be performed on matrices as mathematical entities.

A matrix can be thought of as a convenient device for arranging real numbers. Usually, the array of numbers is enclosed between brackets (or parentheses), as illustrated by

$$\begin{bmatrix} a_{11} & a_{12} & \cdots & a_{1n} \\ a_{21} & a_{22} & \cdots & a_{2n} \\ \cdots & \cdots & \cdots & \cdots \\ a_{m1} & a_{m2} & \cdots & a_{mn} \end{bmatrix}$$

The numbers which compose a matrix are called *elements* of the matrix. Matrices are typically denoted by boldface letters, and the elements (entries) of the matrix are denoted by lowercase letters. A general element of a matrix is referred to by a_{ij}, where, by convention, the first subscript, i, is used for the row location and the second, j, is used for the column. To identify the size of a matrix, the boldface letter identifying the matrix can be accompanied by the numbers of rows and columns. For example, **A** is an $m \times n$ (read "*m* by *n*") matrix.

Example

$$\mathbf{A} = \begin{bmatrix} 5 & -1 & 6 & 3 & 6 \\ 2 & -3 & 2 & 1 & 5 \\ 0 & 7 & 5 & 1 & 8 \end{bmatrix}$$

is a 3×5 matrix. Creating and storing the above matrix in Minitab (full version only) is done as follows:

```
MTB > read 3 5 m1              Calc>>Matrices>>Read
DATA> 5   -1   6   3   6
DATA> 2   -3   2   1   5
DATA> 0    7   5   1   8
       3 rows read.
MTB > name m1 'A'
MTB > print 'A'

Data Display

 Matrix A

   5  -1  6  3  6
   2  -3  2  1  5
   0   7  5  1  8
```

SPECIAL MATRICES

Several special matrices are used repeatedly in applications. Some of these are the following:

Square Matrix   A matrix with an equal number of rows and columns ($m = n$) is called a square matrix.

Identity Matrix (Unit Matrix)   A square matrix whose principal (NW-SE) diagonal elements are 1 and whose other elements are all 0s is an identity matrix and is denoted by **I**.

Example

$$\mathbf{I} = \begin{bmatrix} 1 & 0 & 0 \\ 0 & 1 & 0 \\ 0 & 0 & 1 \end{bmatrix}$$

is a 3×3 identity matrix. The identity matrix has the same use in the multiplication of matrices as the number 1 has in the multiplication of real numbers. If $\mathbf{A}$ is an $m \times m$ matrix, then

$$\mathbf{IA} = \mathbf{AI} = \mathbf{A} \qquad \mathbf{I} \text{ is an } m \times m \text{ identity matrix}$$

Null Matrix (Zero Matrix)   A matrix whose elements are all zero is called the *null matrix* and is denoted by $\mathbf{0}$. In contrast to an identity matrix, a null matrix does not have to be a square matrix.

Example

$$\mathbf{0} = \begin{bmatrix} 0 & 0 \\ 0 & 0 \end{bmatrix}$$

is a 2×2 zero matrix. The zero matrix plays the same role in matrix addition as the number 0 plays in the system of real numbers. If $\mathbf{A}$ is an $m \times n$ matrix, then

$$\mathbf{0} + \mathbf{A} = \mathbf{A} + \mathbf{0} = \mathbf{A} \qquad \mathbf{0} \text{ is the } m \times n \text{ zero matrix}$$

Symmetric Matrix   A square matrix is said to be *symmetric* if its elements are such that $a_{ij} = a_{ji}$ for all i and j. The variance-covariance matrix, described in this chapter, is a general example of a symmetric matrix.

Example

$$\mathbf{A} = \begin{bmatrix} 4 & 1 & 6 \\ 1 & 7 & 5 \\ 6 & 5 & 2 \end{bmatrix}$$

is a symmetric matrix while

$$\mathbf{B} = \begin{bmatrix} 3 & 2 \\ 1 & 5 \end{bmatrix}$$

is not symmetric.

Row Vector   A row vector of n components is a $1 \times n$ matrix. That is, the row vector $\mathbf{A} = (a_{11}, a_{12}, \ldots, a_{1n})$ is a matrix with 1 row and n columns.

Column Vector   A column vector of m components is an $m \times 1$ matrix. For example,

$$\mathbf{A} = \begin{bmatrix} a_{11} \\ a_{21} \\ \vdots \\ a_{m1} \end{bmatrix}$$

has m rows and 1 column.

MATRIX OPERATIONS

Multiplication by a Scalar   A matrix $\mathbf{A}$ is multiplied by a scalar (constant) s when each element of $\mathbf{A}$ is multiplied by s.

$$s\mathbf{A} = \begin{bmatrix} sa_{11} & sa_{12} & \cdots & sa_{1n} \\ sa_{21} & sa_{22} & \cdots & sa_{2n} \\ \cdots & \cdots & \cdots & \cdots \\ sa_{m1} & sa_{m2} & \cdots & sa_{mn} \end{bmatrix}$$

Example   Let

$$\mathbf{A} = \begin{bmatrix} 3 & 2 \\ 5 & 7 \end{bmatrix} \quad \text{and} \quad s = 2$$

then

$$s\mathbf{A} = \begin{bmatrix} 6 & 4 \\ 10 & 14 \end{bmatrix}$$

Below we show how Minitab can be used to accomplish the above operation:

```
MTB > print 'A'

Data Display

 Matrix A

   3    2
   5    7
MTB > multiply 2 'A' m2        Calc>>Matrices>>Arithmetic
MTB > name m2 '2A'
MTB > print '2A'

Data Display

 Matrix 2A

   6    4
  10   14
```

Addition of Matrices   The sum of two or more matrices of the same order (having equal numbers of rows and columns) is found by adding their corresponding elements.

Example   Let

$$\mathbf{A} = \begin{bmatrix} -1 & 2 & 1 \\ 0 & -4 & 3 \end{bmatrix} \quad \mathbf{B} = \begin{bmatrix} 5 & 1 & 7 \\ 4 & 6 & 2 \end{bmatrix}$$

then

$$\mathbf{A} + \mathbf{B} = \begin{bmatrix} -1+5 & 2+1 & 1+7 \\ 0+4 & -4+6 & 3+2 \end{bmatrix} = \begin{bmatrix} 4 & 3 & 8 \\ 4 & 2 & 5 \end{bmatrix}$$

Below we show how Minitab can be used to accomplish the above operation:

```
MTB > print 'A' 'B'

Data Display

  Matrix A

   -1    2    1
    0   -4    3

  Matrix B

    5    1    7
    4    6    2

MTB > add 'A' 'B' m3        Calc>>Matrices>>Arithmetic
MTB > name m3 'A+B'
MTB > print 'A+B'

Data Display

  Matrix A+B

    4    3    8
    4    2    5
```

Matrix Multiplication   Let A be a matrix of order $m \times p$ and B be a matrix of order $p \times n$. Then the product $AB = C$ is a third matrix of order $m \times n$ whose elements are defined as

$$c_{ij} = a_{i1}b_{1j} + a_{i2}b_{2j} + \cdots + a_{ip}b_{pj} = \sum_{k=1}^{p} a_{ik}b_{kj}$$

where $i = 1, 2, \ldots, m$ and $j = 1, 2, \ldots, n$. In other words, the element in the ith row and the jth column of C is equal to the sum of the products of the elements of the ith row of A multiplied by the elements of the jth column of B as illustrated below:

$$
\begin{bmatrix}
c_{11} & \cdots & c_{1n} \\
\vdots & \vdots & \vdots \\
c_{i1} & \boxed{c_{ij}} & c_{in} \\
\vdots & \vdots & \vdots \\
c_{m1} & \cdots & c_{mn}
\end{bmatrix}
=
\begin{bmatrix}
a_{11} & \cdots & a_{1p} \\
\vdots & \vdots & \vdots \\
\hline
a_{i1} & \cdots & a_{ip} \\
\vdots & \vdots & \vdots \\
a_{m1} & \cdots & a_{mp}
\end{bmatrix}
\begin{bmatrix}
b_{11} & \cdots & b_{1j} & \cdots & b_{1n} \\
\vdots & \vdots & \vdots & \vdots & \vdots \\
\cdot & & \cdot & & \cdot \\
\vdots & \vdots & \vdots & \vdots & \vdots \\
b_{p1} & \cdots & b_{pj} & \cdots & b_{pn}
\end{bmatrix}
$$

It is important to note that matrix multiplication is defined only if the number of columns of the first matrix is equal to the number of rows of the second matrix. Furthermore, matrix multiplication is not commutative; that is, AB is not necessarily equal to BA, even when both operations are possible.

Example   Given

$$A = \begin{bmatrix} 1 & 2 \\ 0 & 1 \end{bmatrix} \qquad B = \begin{bmatrix} 3 & 1 & 2 \\ 4 & 2 & 5 \end{bmatrix}$$

then

$$\mathbf{AB} = \begin{bmatrix} 1 & 2 \\ 0 & 1 \end{bmatrix} \begin{bmatrix} 3 & 1 & 2 \\ 4 & 2 & 5 \end{bmatrix}$$

$$= \begin{bmatrix} 1\cdot3+2\cdot4 & 1\cdot1+2\cdot2 & 1\cdot2+2\cdot5 \\ 0\cdot3+1\cdot4 & 0\cdot1+1\cdot2 & 0\cdot2+1\cdot5 \end{bmatrix} = \begin{bmatrix} 11 & 5 & 12 \\ 4 & 2 & 5 \end{bmatrix}$$

Example   Given

$$\mathbf{A} = \begin{bmatrix} 3 & 2 \\ 1 & 5 \end{bmatrix} \qquad \mathbf{B} = \begin{bmatrix} 1 \\ 2 \end{bmatrix}$$

then

$$\mathbf{AB} = \begin{bmatrix} 3\cdot1+2\cdot2 \\ 1\cdot1+5\cdot2 \end{bmatrix} = \begin{bmatrix} 7 \\ 11 \end{bmatrix}$$

Notice that in each of the above examples **BA** does not exist.

Example   Given

$$\mathbf{A} = \begin{bmatrix} 1 & 0 \\ 1 & 1 \end{bmatrix} \qquad \mathbf{B} = \begin{bmatrix} 2 & 1 \\ 4 & 6 \end{bmatrix}$$

then

$$\mathbf{AB} = \begin{bmatrix} 1\cdot2+0\cdot4 & 1\cdot1+0\cdot6 \\ 1\cdot2+1\cdot4 & 1\cdot1+1\cdot6 \end{bmatrix} = \begin{bmatrix} 2 & 1 \\ 6 & 7 \end{bmatrix}$$

$$\mathbf{BA} = \begin{bmatrix} 2\cdot1+1\cdot1 & 2\cdot0+1\cdot1 \\ 4\cdot1+6\cdot1 & 4\cdot0+6\cdot1 \end{bmatrix} = \begin{bmatrix} 3 & 1 \\ 10 & 6 \end{bmatrix}$$

Therefore **AB** ≠ **BA**. For this last example, we show how Minitab can be used to accomplish the multiplication operation of **A** times **B**:

```
MTB > print 'A' 'B'
Data Display
 Matrix A
  1  0
  1  1
 Matrix B
  2  1
  4  6
MTB > multiply 'A' 'B' m3        Calc>>Matrices>>Arithmetic
MTB > name m3 'AB'
MTB > print 'AB'
Data Display
 Matrix AB
  2  1
  6  7
```

Transpose of a Matrix The transpose of a matrix **A** is obtained from **A** by interchanging the rows and columns, and is usually denoted by $\mathbf{A}'$ or $\mathbf{A}^T$.

Example Consider

$$\mathbf{A} = \begin{bmatrix} 4 & 0 & 1 \\ 7 & 10 & 5 \end{bmatrix} \quad \text{then} \quad \mathbf{A}' = \begin{bmatrix} 4 & 7 \\ 0 & 10 \\ 1 & 5 \end{bmatrix}$$

Below we use Minitab:

```
MTB > print 'A'

Data Display

 Matrix A

   4   0   1
   7  10   5

MTB > transpose 'A' m2              Calc>>Matrices>>Transpose
MTB > name m2 'A-transpose'
MTB > print 'A-transpose'

Data Display

 Matrix A-transpose

   4   7
   0  10
   1   5
```

The following are some rules regarding the operations of transposition:

1. $(\mathbf{A}')' = \mathbf{A}$

    The transpose of the transpose of a matrix is the matrix itself.

2. $(\mathbf{A} + \mathbf{B})' = \mathbf{A}' + \mathbf{B}'$

    The transpose of a sum is the sum of the transposes.

3. $(s\mathbf{A})' = s\mathbf{A}'$

    The transpose of a product of a scalar and a matrix is equal to the scalar times the transpose of the matrix.

4. $(\mathbf{AB})' = \mathbf{B}'\mathbf{A}'$

    The transpose of a product is the product of the transposes in reverse order.

5. The transpose of a row vector ($1 \times n$ matrix) is a column vector ($n \times 1$ matrix). Similarly, the transpose of a column vector ($m \times 1$ matrix) is a row vector ($1 \times m$ matrix).

6. A symmetric matrix is equal to its transpose; that is, **A** is symmetric if and only if $\mathbf{A} = \mathbf{A}'$.

DETERMINANT OF A MATRIX

It has been mentioned that a matrix has no numerical value. However, for each square matrix $\mathbf{A}$ of order $n \times n$, there exists a number called the *determinant* of the matrix, which is denoted by $|\mathbf{A}|$. The determinant of an $n \times n$ matrix is referred to as a determinant of order n.

How do we evaluate a determinant? The process of finding the value of determinants of order 2 and 3 can be easily illustrated schematically. For a 2×2 matrix, the second-order determinant is computed as follows:

$$|\mathbf{A}| = \begin{vmatrix} a_{11} & a_{12} \\ a_{21} & a_{22} \end{vmatrix} = a_{11}a_{22} - a_{21}a_{12}$$

The determinant is the product of the elements along the arrow going downward minus the product of the elements along the arrow going upward.

To evaluate a third-order determinant,

$$|\mathbf{A}| = \begin{vmatrix} a_{11} & a_{12} & a_{13} \\ a_{21} & a_{22} & a_{23} \\ a_{31} & a_{32} & a_{33} \end{vmatrix}$$

the first two columns of the matrix are appended to the right of the matrix, and then the procedure of the 2×2 case is followed; that is, find the products diagonally, as illustrated in the scheme

$$|\mathbf{A}| = \begin{vmatrix} a_{11} & a_{12} & a_{13} & a_{11} & a_{12} \\ a_{21} & a_{22} & a_{23} & a_{21} & a_{22} \\ a_{31} & a_{32} & a_{33} & a_{31} & a_{32} \end{vmatrix}$$

$$|\mathbf{A}| = a_{11}a_{22}a_{33} + a_{12}a_{23}a_{31} + a_{13}a_{21}a_{32} - a_{31}a_{22}a_{13} - a_{32}a_{23}a_{11} - a_{33}a_{21}a_{12}$$

Notice that each term in the sum is a product of three elements whose sign is plus if these elements are along the arrows going down and whose sign is minus if the elements are along the upward-sloping arrows.

Example Let

$$|\mathbf{A}| = \begin{vmatrix} 3 & 4 & 1 \\ 2 & 0 & 3 \\ -2 & 5 & 7 \end{vmatrix}$$

Rewrite and append the first two columns

$$\begin{vmatrix} 3 & 4 & 1 & 3 & 4 \\ 2 & 0 & 3 & 2 & 0 \\ -2 & 5 & 7 & -2 & 5 \end{vmatrix}$$

$$|\mathbf{A}| = 3 \cdot 0 \cdot 7 + 4 \cdot 3 \cdot (-2) + 1 \cdot 2 \cdot 5 - (-2) \cdot 0 \cdot 1 - 5 \cdot 3 \cdot 3 - 7 \cdot 2 \cdot 4 = -115$$

The above schematic method is simple and practical for manually finding the value of a determinant of order 2 or 3. However, this approach is not applicable if the order of the determinant is greater than 3. Thus, a more general method of computing the value of the determinant of any order is needed.

It can be shown that the determinant of a square $n \times n$ matrix is

$$|\mathbf{A}| = \sum_{j=1}^{n} a_{1j} |\mathbf{A}_{1j}| (-1)^{1+j}$$

where $\mathbf{A}_{1j}$ is the $(n-1) \times (n-1)$ matrix obtained by deleting the first row and the jth column of $\mathbf{A}$. This method of computation is called the *method of cofactors.*

Example   Consider again the matrix from the previous example. Then we have:

$$\begin{vmatrix} 3 & 4 & 1 \\ 2 & 0 & 3 \\ -2 & 5 & 7 \end{vmatrix} = 3 \begin{vmatrix} 0 & 3 \\ 5 & 7 \end{vmatrix} (-1)^{1+1} + 4 \begin{vmatrix} 2 & 3 \\ -2 & 7 \end{vmatrix} (-1)^{1+2} + 1 \begin{vmatrix} 2 & 0 \\ -2 & 5 \end{vmatrix} (-1)^{1+3}$$

$$= 3(-15)(1) + 4(20)(-1) + 1(10)(1)$$

$$= -115$$

This determinant value is the same as the one found using the schematic approach; and it should always be so, because the determinant of a matrix is unique, when it exists, and does not depend on the method of computation used.

Even though the method of cofactors allows one to compute the determinant of any order, it can be quite tedious to manually implement for large orders. Unfortunately, Minitab does not have a direct procedure for determining the determinant of any square matrix. Nonetheless, we can use another computation capability of Minitab to quickly determine the determinant of a special class of matrices which happen to be particularly relevant to our interests. In particular, Minitab can be employed to facilitate the computation of the determinant of any *symmetric* matrix. In our discussion of multivariate quality control procedures, we are interested in finding the determinant of a variance-covariance matrix which, by definition, is symmetric.

To find the determinant of a symmetric matrix in Minitab, we can use Minitab's capability of finding eigenvalues of a matrix. Very briefly, an eigenvalue, denoted by λ, for a square matrix $\mathbf{A}$ is a real number such that $|\mathbf{A} - \lambda \mathbf{I}| = 0$, that is, a number such that when multiplied by the identity matrix and then subtracted from $\mathbf{A}$, it will result in a matrix with determinant equal to 0. For an $n \times n$ matrix, there may be as many as n distinct eigenvalues. Even though we will not study the general role of eigenvalues, note that they play an extremely important role in multivariate statistical analysis (see Johnson and Wichern, 1998).

For our purposes, we need to know only the following fact. The determinant of a symmetric matrix is always equal to the product of its eigenvalues.[10] In other words, if $\mathbf{A}$ is a symmetric $n \times n$ matrix, its determinant is

$$|\mathbf{A}| = \prod_{i=1}^{n} \lambda_i$$

---

[10]This result actually applies to any square matrix. However, Minitab only computes eigenvalues for symmetric matrices.

Example Consider the following 4×4 symmetric matrix

$$\mathbf{A} = \begin{bmatrix} 3 & 5 & 8 & 1 \\ 5 & 2 & 4 & 2 \\ 8 & 4 & 5 & 7 \\ 1 & 2 & 7 & 6 \end{bmatrix}$$

If we were to use the method of cofactors, the first step would be to write out the following:

$$|\mathbf{A}| = 3 \begin{vmatrix} 2 & 4 & 2 \\ 4 & 5 & 7 \\ 2 & 7 & 6 \end{vmatrix} (-1)^2 + 5 \begin{vmatrix} 5 & 4 & 2 \\ 8 & 5 & 7 \\ 1 & 7 & 6 \end{vmatrix} (-1)^3 + 8 \begin{vmatrix} 5 & 2 & 2 \\ 8 & 4 & 7 \\ 1 & 2 & 6 \end{vmatrix} (-1)^4 + 1 \begin{vmatrix} 5 & 2 & 4 \\ 8 & 4 & 5 \\ 1 & 2 & 7 \end{vmatrix} (-1)^5$$

Then the determinants for each of the above 3×3 matrices would have to be computed. Alternatively, let us utilize Minitab. Below we used Minitab to determine the eigenvalues of **A:**

```
MTB > print 'A'

Data Display

  Matrix A

    3   5   8   1
    5   2   4   2
    8   4   5   7
    1   2   7   6

MTB > eigen 'A' c1                    Calc>>Matrices>>Eigen Analysis
MTB > name c1 'eigenvalues'
MTB > print 'eigenvalues'

Data Display

eigenvalues
  18.324565    -5.798359    4.614025    -1.140232
```

Now applying the fact that the product of the eigenvalues is the determinant, we find

$$|\mathbf{A}| = 18.324565(-5.798359)(4.614025)(-1.140232) = 559.00$$

The reader can verify that the same value would have been obtained by using the more tedious method of cofactors.

INVERSE OF A MATRIX

Let **A** be a square matrix of order $n \times n$. If there exists a square matrix $\mathbf{A}^{-1}$ of order $n \times n$ such that $\mathbf{AA}^{-1} = \mathbf{A}^{-1}\mathbf{A} = \mathbf{I},$ where **I** is the identity matrix whose order is the same as that of **A**, then the matrix $\mathbf{A}^{-1}$ is called the *inverse* of **A**. It is possible that $\mathbf{A}^{-1}$ does not exist, just as there is no multiplicative inverse for the number 0. When that is the case, **A** is said to be *singular.* Matrices whose inverses exist are called *nonsingular.* Nonsingular matrices are characterized by having a nonzero determinant.

Computation of an Inverse   We will describe one method[11] for finding the inverse of an $n \times n$ square matrix $\mathbf{A}$. First, form what is called a *matrix of cofactors:*

$$\mathbf{C} = \begin{bmatrix} c_{11} & \cdots & c_{1n} \\ c_{21} & \cdots & c_{2n} \\ \cdots & \cdots & \cdots \\ c_{n1} & \cdots & c_{nn} \end{bmatrix}$$

where c_{ij} is the cofactor of a_{ij} in the original matrix $\mathbf{A}$. A cofactor is computed as follows:

$$c_{ij} = |\mathbf{A}_{ij}|(-1)^{i+j}$$

where $\mathbf{A}_{ij}$ is the $(n-1) \times (n-1)$ matrix obtained by deleting the ith row and the jth column of $\mathbf{A}$. If we now take the transpose of C and divide each element by $|\mathbf{A}|$, we have the desired inverse

$$\mathbf{A}^{-1} = \frac{1}{|\mathbf{A}|} \mathbf{C}'$$

Example   Suppose

$$\mathbf{A} = \begin{bmatrix} 2 & 4 & 0 \\ 0 & 2 & 1 \\ 3 & 0 & 2 \end{bmatrix}$$

The matrix of cofactors is

$$\mathbf{C} = \begin{bmatrix} 4 & 3 & -6 \\ -8 & 4 & 12 \\ 4 & -2 & 4 \end{bmatrix}$$

The transpose of the matrix of the cofactors is

$$\mathbf{C}' = \begin{bmatrix} 4 & -8 & 4 \\ 3 & 4 & -2 \\ -6 & 12 & 4 \end{bmatrix}$$

The determinant of $\mathbf{A}$ can be found by any of the previously discussed methods to be

$$|\mathbf{A}| = 20$$

[11]Most computer software use a more efficient method, based on a technique known as Gaussian elimination for finding the inverse, than that described here; for details on more computationally efficient methods for dealing with matrices, refer to G. W. Stewart, *Introduction to Matrix Computations*, Academic Press, New York, 1973. Our purpose here is to show a basic method that can be employed quite simply when the inverse of small-size matrices (2 × 2 or 3 × 3 matrices) is done by hand.

Therefore,

$$\mathbf{A}^{-1} = \frac{\begin{bmatrix} 4 & -8 & 4 \\ 3 & 4 & -2 \\ -6 & 12 & 4 \end{bmatrix}}{20}$$

$$= \begin{bmatrix} 0.20 & -0.40 & 0.20 \\ 0.15 & 0.20 & -0.10 \\ -0.30 & 0.60 & 0.20 \end{bmatrix}$$

The reader can verify that indeed $\mathbf{AA}^{-1} = \mathbf{A}^{-1}\mathbf{A} = \mathbf{I}$.

Conveniently, Minitab has a direct command for computation of a matrix inverse. Below we use Minitab to arrive at the same result:

```
MTB > print 'A'

Data Display

  Matrix A

   2  4  0
   0  2  1
   3  0  2

MTB > inverse 'A' m2
MTB > name m2 'A-inverse'          Calc>>Matrices>>Invert
MTB > print 'A-inverse'

Data Display

  Matrix A-inverse

   0.20  -0.40   0.20
   0.15   0.20  -0.10
  -0.30   0.60   0.20
```

EXERCISES

10.1. List and briefly explain the different approaches for process control in a short-run production environment.

10.2. What is the basic requirement underlying the proper implementation of a DNOM chart?

10.3. Below are data for two parts (P1 and P2) that were made from the same material on the same machine by the same operator (data are also given in "dnom1.dat"). The nominal dimensions for the two parts are $T_{P1} = 5$ and $T_{P2} = 7$.

Sample	Part	x_1	x_2	x_3	x_4
1	P1	5.02	4.99	4.99	4.95
2	P1	5.01	4.98	4.97	5.04
3	P1	5.02	5.00	5.10	4.94
4	P1	5.01	4.95	4.95	4.89
5	P1	5.15	5.00	4.99	5.06
6	P1	5.02	4.94	4.96	4.98
7	P1	4.99	4.99	4.96	4.96
8	P1	5.02	5.04	5.03	5.01
9	P2	6.99	7.10	6.97	6.90
10	P2	6.99	7.05	7.02	6.97
11	P2	6.95	7.00	6.99	7.07
12	P2	6.95	6.97	7.03	7.00
13	P2	6.98	7.03	7.02	7.07
14	P2	6.93	6.98	7.00	7.02
15	P2	7.06	6.99	6.96	7.03
16	P2	7.01	6.99	7.03	7.02
17	P2	7.01	6.96	6.97	6.90
18	P2	7.06	7.00	6.93	7.03
19	P2	7.03	6.96	7.03	7.03
20	P2	7.04	7.01	7.03	6.97

Construct $\bar{x}$ and R charts based on the DNOM approach. Based on the range information, what is the estimated standard deviation for the process regardless of the nominal target value?

10.4. Continuation of Exercise 10.3: Verify that the underlying assumption of constant variance across the two parts is satisfied. Combine all the part 1 observations into one data set, and combine all the part 2 observations into another data set. With $\alpha = 0.05$, use the variance ratio statistic to test the homogeneity of variances.

10.5. Below are data for three machined parts (P1, P2, and P3) that were combined based on being in a same "family" of parts (data are also given in "dnom2.dat"). The nominal dimensions for the three parts are: $T_{P1} = 150$, $T_{P2} = 200$, and $T_{P3} = 350$.

Sample	Part	x_1	x_2	x_3
1	P1	153	150	144
2	P1	146	149	146
3	P1	152	146	148

4	P1	151	151	158
5	P1	142	157	151
6	P2	210	201	194
7	P2	197	196	197
8	P2	209	194	201
9	P2	197	197	205
10	P2	193	204	193
11	P2	192	207	202
12	P2	202	201	199
13	P2	200	193	205
14	P2	205	194	204
15	P3	343	353	342
16	P3	353	347	347
17	P3	353	349	349
18	P3	348	349	340
19	P3	350	356	353
20	P3	346	346	351

Construct $\bar{x}$ and R charts based on the DNOM approach. Based on the range information, what is the estimated standard deviation for the process regardless of the nominal target value?

10.6. Continuation of Exercise 10.5: Verify that the underlying assumption of constant variance across the three parts is satisfied. To do so, combine part observations into three separate data sets. With $\alpha = 0.05$, use the Barlett statistic to test the homogeneity of variances.

10.7. Based on sample sizes of $n = 12$, below are the sample means of two quality characteristics sampled simultaneously:

Sample	$\bar{x}_1$	$\bar{x}_2$
1	99.8	27.9
2	97.9	30.4
3	98.1	29.5
4	106.6	31.8
5	98.0	28.0
6	105.3	29.9
7	100.4	29.4
8	103.1	31.4
9	104.6	30.0
10	99.8	31.5

11	99.7	30.2
12	97.9	28.0
13	96.3	29.4
14	93.1	27.3
15	101.9	30.9

a. Compute the estimated mean vector $\hat{\boldsymbol{\mu}}$.

b. Assume that the estimated variance-covariance matrix is

$$\mathbf{S} = \begin{bmatrix} 160 & 44 \\ 44 & 25 \end{bmatrix}$$

Construct a T^2 chart for the preliminary samples. What is the prospective limit for the T^2 chart?

10.8. A manufacturer of thick acrylic sheets views two quality characteristics as important enough for simultaneous control: the thickness of the sheet and the tensile strength. Twenty samples, each of size 2, were obtained from the process. The data are shown below (data are also given in "acrylic.dat"):

Sample	Thickness (in)		Tensile Strength (000/lb/in²)	
1	2.002	1.994	14.915	14.948
2	1.985	2.037	14.836	15.013
3	2.013	2.023	14.881	14.939
4	1.974	1.982	14.858	14.829
5	1.956	1.997	14.878	14.915
6	1.937	2.003	14.848	14.904
7	1.985	1.983	14.789	14.744
8	1.986	1.965	14.902	14.915
9	1.992	1.998	14.877	14.930
10	2.006	1.995	14.921	14.846
11	1.991	2.025	14.881	14.910
12	1.963	1.962	14.845	14.842
13	2.056	1.997	14.929	14.901
14	2.026	2.000	14.971	14.935
15	2.020	1.990	14.935	14.904
16	2.041	2.001	14.890	14.855
17	2.012	2.026	14.905	14.908
18	1.983	1.991	14.797	14.960
19	1.986	1.996	14.921	14.970
20	1.992	2.002	14.898	14.860

a. Based on $\alpha = 0.005$, construct a T^2 chart for these preliminary samples.

b. What is the prospective limit for the T^2 chart?

10.9. An article in *Quality Engineering* (by D. S. Holmes and A. E. Mergen, 1995, 8, pp. 137–143) presents multivariate data on particle sizes of abrasive grit. In this article, the full data set consists of individual observations on 5 different particle sizes for 35 samples. Below, we consider only a portion of the full data set (data are also given in "grit.dat"):

Sample	A	B	C
1	6.37	17.28	25.87
2	4.51	15.58	19.09
3	7.87	17.03	22.98
4	8.76	16.67	20.69
5	8.04	17.66	20.68
6	7.63	16.15	19.39
7	6.63	16.25	20.80
8	7.10	15.96	21.71
9	8.19	16.34	22.69
10	6.44	17.72	23.29
11	7.06	17.77	23.41
12	7.87	17.06	21.03
13	7.44	17.76	20.46
14	5.10	15.14	18.50
15	4.89	14.78	17.83

a. Based on $\alpha = 0.001$, construct a T^2 chart ($n = 1$) for these preliminary samples.

b. What is the prospective limit for the T^2 chart?

c. Below are some future samples:

Sample	A	B	C
1	6.61	11.74	21.01
2	6.78	11.38	21.50
3	6.12	9.99	21.44

What do you conclude?

CHAPTER 10 • Related Special Topics

10.10. Continuation of Exercise 10.9: For the first 15 preliminary samples, obtain the ACF for each of the variables separately, and superimpose 95 percent limits on each ACF. In addition, obtain the CCF for all pairwise combinations, applying 95 percent limits to each CCF. Based on the review of the ACFs and CCFs, what do you conclude?

10.11. In general, what is the meaning of the computed correlation on the CCF at lag 0 for two variables x and y?

10.12. How could you use a CCF to obtain the same correlations found with an ACF for a given variable?

10.13. Suppose two quality characteristics are simultaneously monitored. Using a sample size of 7, suppose 15 preliminary samples were gathered on these two characteristics. The variance and covariance summary statistics for the samples are given below:

Sample (k)	s_{1k}^2	s_{2k}^2	s_{12}
1	13.13	1.57	2.26
2	9.49	0.18	0.42
3	19.98	1.78	4.99
4	21.95	2.10	6.42
5	16.13	8.30	7.23
6	22.72	0.89	3.61
7	26.27	1.99	4.23
8	26.25	2.83	7.89
9	35.80	1.66	5.21
10	23.33	1.52	5.15
11	30.88	0.91	3.08
12	63.06	5.47	16.89
13	9.61	0.98	0.79
14	10.92	2.02	2.84
15	35.47	1.81	6.17

Implement a multivariate variability chart based on the generalized sample variance. Interpret the results of the chart.

10.14. Continuation of Exercise 10.13: Refer to Section 6.4 and implement separate univariate s^2 charts with $\alpha = 0.0027$. Interpret the results of the two univariate charts. Given the results of these charts and the results of the multivariate chart in Exercise 10.13, what lesson is learned?

10.15. What would the result be if a data series were fitted with an EWMA model with $\lambda = 0$?

10.16. Consider the daily weight series introduced in Section 3.8.

    a. A common rule of thumb that many practitioners use is to choose a value for λ somewhere in the range of 0.1 to 0.5. Minitab's default value for λ is 0.2. Fit the daily weight series with an EWMA model using $\lambda = 0.2$. Obtain the residuals from this model, and test the appropriateness of this fitted model.

    b. Now, use Minitab's option to find the optimal value of λ, and fit the data series with an EWMA model based on this optimal value. Obtain the residuals from this model, and test the appropriateness of this fitted model.

    c. Unlike with fitted models based on least squares, the average of the residuals from an EWMA cannot be assumed to equal zero. Find the average of the residuals, and construct a special-cause chart from this model. Interpret the results of the chart.

10.17. Consider the vehicle dimension measurements series ($kleft_t$) studied in Section 5.6.

    a. In Section 5.6, we found that this data series is well described by a multiple regression model based on a lag 1 variable and a simple trend variable. Fit this model again so as to obtain the residuals from the model. Construct a special-cause chart and supplement the chart with all the supplementary runs rules to identify any unusual variation in the residuals. Report your results.

    b. Now, fit the data series with an EWMA model based on an optimal value of λ. Obtain the residuals from this model, and demonstrate that the model is not completely satisfactory based on model diagnostics. Based on the fitted EWMA model, construct a special-cause chart and supplement the chart with all the supplementary runs rules. Compare the results with part *a*. For this application, explain why viewing any out-of-control signal in the residuals from the EWMA model as a legitimate signal is suspect.

10.18. Other than the data series studied in Example 10.17, pick any data set in the book (whether in chapter sections or chapter exercises) that clearly illustrates the failure of a simple EWMA model to capture satisfactorily the underlying systematic pattern. Support your demonstration with relevant diagnostics.

10.19. Consider the sales data series studied in Section 10.4. Based on the fitted regression model given in Figure 10.12*b*, what is your prediction of sales in week 37, that is, 1 week into the future?

10.20. To study her weight process, a former MBA student collected data on her daily weight (measured each day at 10 p.m. prior to sleeping) and her daily caloric intake for 5 full weeks with the first observation starting on a Monday. Below are the observations (data are also given in "wgtcal.dat"):

Day	Weight	Calories	Day	Weight	Calories
1	152.2	1862	19	153.2	2312
2	152.8	2159	20	153.1	1875
3	152.9	1972	21	153.6	2184
4	152.6	1762	22	153.3	1967
5	151.8	1809	23	153.1	1688
6	151.8	1969	24	153.6	2305
7	152.4	2226	25	153.3	2088
8	151.6	1936	26	153.5	2046
9	151.6	2024	27	153.7	2065
10	151.7	2071	28	154.1	2196
11	152.0	2134	29	153.6	2105
12	151.3	1955	30	153.7	2034
13	151.7	1736	31	153.2	1596
14	152.6	1748	32	153.2	1877
15	152.6	2046	33	153.3	1891
16	152.8	1960	34	153.1	1730
17	153.5	2384	35	154.2	1783
18	152.8	2015			

a. Find an appropriate fitted model for the weight data series. In developing your model, consider not only caloric intake as an independent variable but also lags (possibly on both weight and caloric intake), indicator variables for the day of the week, and a time trend. Report the significance of the independent variables and the overall diagnostics of the fitted model.

b. Construct a special-cause chart. Interpret the results of the chart.

c. Construct a fitted-values chart. Interpret the fitted model in terms of what it is suggesting about the movements of the individual's daily weight series.

10.21. The "loan by phone" department of a large banking institution (serving Wisconsin, Illinois, and Indiana) is concerned with the abandonment of customers who do not wish to wait any longer on the phone for a customer representative. For the business days of Monday through Saturday, daily logs are kept of the number of calls per day and the number of abandoned calls. Below are data for 30 successive business days with the first observation being on a Monday (data are also given in "abandon.dat"):

Day	Abandoned	Total	Day	Abandoned	Total
1	100	1127	16	62	936
2	113	1094	17	69	963
3	63	973	18	4	105
4	96	1027	19	27	677
5	29	608	20	73	1073
6	50	612	21	99	1147
7	37	627	22	92	985
8	102	1071	23	98	1235
9	103	1196	24	46	677
10	131	1285	25	127	1218
11	242	1984	26	116	1207
12	28	418	27	74	969
13	69	873	28	77	1007
14	106	1068	29	59	899
15	46	883	30	12	316

    a. Construct a p chart with exact limits for these data (refer to Section 7.1). What does the chart suggest about the process? Are there any out-of-control signals?

    b. Consider now a general data analysis perspective. Find an appropriate fitted model for the abandoned call series. In developing your model, consider the total number of calls as a possible independent variable; also consider the possibility of the need for indicator variables for the day of the week. Report the significance of the independent variables and the overall diagnostics of the fitted model.

    c. Construct a special-cause chart. Interpret the results of the chart, and compare these results with part *a*.

    d. Construct a fitted-values chart. Interpret the fitted model in terms of what it is suggesting about the movements of the daily abandoned call series.

10.22. Measurement of customers' attitudes and satisfaction is becoming an important element in the quality movement of U.S. organizations. In this exercise, we consider a medium to large service company that monitors the satisfaction ratings of its service by means of monthly satisfaction surveys. Marketing research consultants have suggested to company management that customer satisfaction is influenced by the amount of advertising, among other things. Starting with the month of January, below are data for 36 successive months on monthly percentages of overall satisfied customers (based on a random survey of 1000 customers) and on monthly dollars (in units of $100,000) of advertising (data are also given in "advertise.dat"):

Month	Percentage Satisfied	Advertising	Month	Percentage Satisfied	Advertising
1	0.903	12.5186	19	0.854	11.8586
2	0.900	10.4877	20	0.859	10.7435
3	0.857	12.2691	21	0.826	12.9474
4	0.860	12.2545	22	0.920	12.7544
5	0.894	12.8754	23	0.899	14.5234
6	0.901	11.5503	24	0.822	14.3044
7	0.893	12.8012	25	0.938	12.0964
8	0.855	11.5150	26	0.867	13.2716
9	0.891	11.8585	27	0.935	11.5866
10	0.883	11.4798	28	0.860	13.4995
11	0.861	13.0187	29	0.950	12.4087
12	0.941	12.9914	30	0.913	11.7259
13	0.893	12.0480	31	0.862	12.2054
14	0.830	11.8939	32	0.881	13.5666
15	0.879	11.9322	33	0.909	13.4749
16	0.898	12.3110	34	0.927	13.4345
17	0.838	12.3696	35	0.904	13.6262
18	0.909	11.8493	36	0.941	12.5188

a. First consider the customer satisfaction series alone. Construct an individual measurement control chart (X chart) for these data. What does the chart suggest about the process? Are there any out-of-control signals?

b. Take into consideration that market researchers have postulated that the effects of advertising for a particular month may not be reflected in the same month, but rather that the effects may carry over into future months. Investigate this possibility by obtaining the CCF, with 95 percent limits superimposed, of advertising versus satisfaction.

c. Given the results of part *b,* find an appropriate fitted model for the customer satisfaction series. Report the significance of the independent variable(s) and the overall diagnostics of the fitted model.

d. Construct a special-cause chart. Interpret the results of the chart, and compare these results with part *a.*

    e. Construct a fitted-values chart. Interpret the fitted model in terms of what it is suggesting about the movements of the customer satisfaction series.

10.23. Explain why the mean of the squared deviations from the target is a more appropriate measure for the effectiveness of an APC scheme than the sample variance s^2.

10.24. Refer to the color data series in Exercise 3.30. Suppose that there is a compensatory variable X (initially set to 0) which has a gain of 0.5 on the output series. Assume that you are "sitting" in time just prior to the first time period; that is, you have not yet observed the color observations given in Exercise 3.30. Using an EWMA model with $\lambda = 0.6$, implement the APC scheme described in Section 10.5 to reduce the variability of the process around a target value of 0.74. Fill in the following table.

Time	Unobserved Disturbance	Predicted Disturbance	Observed Deviation	Adjusted Output	X Set Point	X Adjustment
1						
2						
.						
.						
.						
35						

    a. Report the mean of the squared deviations from the target for the unadjusted and adjusted processes. What was the amount of reduction in percentage terms due to the implementation of the APC scheme?

    b. Implement a standard Shewhart chart on the adjusted series. Report your results.

10.25. Refer to the temperature data series in Exercise 5.23. Suppose that there is a compensatory variable X (initially set to 0) which has a gain of 1 on the output series. Assume that you are "sitting" in time just prior to the first time period; that is, you have not yet observed the temperature observations given in Exercise 5.23. Using an EWMA model with $\lambda = 0.7$, implement the APC scheme described in Section 10.5 to reduce the variability of the process around a target value of 204. Fill in the following table.

Time	Unobserved Disturbance	Predicted Disturbance	Observed Deviation	Adjusted Output	X Set Point	X Adjustment
1						
2						
.						
.						
.						
36						

a. Report the mean of the squared deviations from the target for the unadjusted and adjusted processes. What was the amount of reduction in percentage terms due to the implementation of the APC scheme?

b. Implement a standard Shewhart chart on the adjusted series. Report your results.

10.26. An article in *Quality Engineering* (by B. Dodson, 1995, 7, pp. 757–768) presents a data series of the daily individual measurements of the viscosity of a coolant used in a sheet metal rolling production process. Below are 50 consecutive daily readings (data also given in "coolant.dat"):

```
3.04   3.14   3.07   3.15   2.97   3.04   3.14   3.21   3.07   3.21
2.88   3.02   3.08   3.41   3.00   2.87   2.80   2.81   2.85   2.81
2.83   2.87   2.81   2.85   2.91   2.81   2.83   2.84   2.91   2.78
2.80   2.63   2.54   2.61   2.70   2.54   2.60   2.59   2.60   2.63
2.68   2.70   2.53   2.53   2.47   2.47   2.56   2.56   2.60   2.37
```

Suppose that there is a compensatory variable X (initially set to 0) which has a gain of 1.5 on the output series. Assume that you are "sitting" in time just prior to the first time period; that is, you have not yet observed the viscosity observations given above. Using an EWMA model with $\lambda = 0.45$, implement the APC scheme described in Section 10.5 to reduce the variability of the process around a target value of 3.00. Fill in the following table:

Time	Unobserved Disturbance	Predicted Disturbance	Observed Deviation	Adjusted Output	X Set Point	X Adjustment
1						
2						
.						
.						
.						
50						

  a. Report the mean of the squared deviations from the target for the unadjusted and adjusted processes. What was the amount of reduction in percentage terms due to the implementation of the APC scheme?

  b. Implement a standard Shewhart chart on the adjusted series. Report your results.

10.27. In the development of the APC scheme presented in this chapter, we assumed that making an adjustment is costless. Suppose that there is a fixed cost associated with any given adjustment. Without resorting to any derivations, argue intuitively what the general nature of the adjustment decision rule would be.

REFERENCES

Alloway Jr., J. A., and M. Ragnavachari (1991): "Control Chart Based on the Hodges-Lehmann Estimator," *Journal of Quality Technology*, 4, pp. 336–347.

Alt, F. B. (1982): "Multivariate Quality Control: State of the Art," in *ASQC Quality Congress Transactions*, pp. 886–893.

——— (1985): "Multivariate Quality Control," in *Encyclopedia of Statistical Sciences*, vol. 6, edited by N. L. Johnson and S. Kotz, Wiley, New York.

——— and N. D. Smith (1988): "Multivariate Process Control," in *Handbook of Statistics: Quality Control and Reliability*, vol. 7, edited by P. R. Krishnaiah and C. R. Rao, North-Holland, Amsterdam.

——— and ——— (1990): "Multivariate Quality Control," in *Handbook of Statistical Methods for Engineers and Scientists*, edited by H. M. Wadsworth, Jr., McGraw-Hill, New York.

Alwan, L. C. (1989): "Time-Series Modeling for Statistical Process Control," unpublished Ph.D. dissertation, Graduate School of Business, University of Chicago.

——— and M. G. Bissell (1988), "Time-Series Modeling for Quality Control in Clinical Chemistry," *Clinical Chemistry*, 34, pp. 1396–1406.

Bakir, S. T., and M. R. Reynolds, Jr. (1979): "A Nonparametric Procedure for Process Control Based on Within-Group Ranking," *Technometrics*, 21, pp. 175–183.

Bhattacharyya, P. K., and D. Freierson, Jr. (1981): "A Nonparametric Control Chart for Detecting Small Disorders," *Annals of Statistics*, 9, pp. 544–554.

Box, G. E. P., and G. M. Jenkins (1963): "Further Contributions to Adaptive Quality Control: Simultaneous Estimation of Dynamics: Non-Zero Costs," *ISI Bulletin*, 34th Session, Ottawa, Canada, pp. 943–974.

——— and T. Kramer (1992): "Statistical Process Monitoring and Feedback Adjustment: A Discussion," *Technometrics*, 34, pp. 251–285.

——— and A. Luceño (1997): *Statistical Control by Monitoring and Feedback Adjustment*, Wiley, New York.

———, G. M. Jenkins, and G. C. Reinsel (1994): *Time Series Analysis*, 3d ed., Prentice-Hall, Englewood Cliffs, NJ.

Brockwell, P. J. (1996): *An Introduction to Time Series and Forecasting*, Springer-Verlag, New York.

Bruce, A. G., and R. D. Martin (1989): "Leave *k*-out Diagnostics for Time Series," *Journal of the Royal Statistical Society*, Series B, 51, pp. 363–424.

Chang, I., G. C. Tiao, and C. Chen (1988): "Estimation of Time Series Parameters in the Presence of Outliers," *Technometrics*, 30, pp. 193–204.

Crosier, R. B. (1988): "Multivariate Generalizations of Cumulative Sum Quality Control Schemes," *Technometrics*, 30, pp. 291–303.

Duncan, A. J. (1956): "The Economic Design of $\overline{X}$ Control Charts to Maintain Current Control of a Process," *Journal of the American Statistical Association*, 51, pp. 228–242.

Everitt, B. S. (1979): "A Monte Carlo Investigation of the Robustness of Hotelling's One- and Two-Sample T^2 Tests," *Journal of the American Statistical Association*, 74, pp. 48–51.

Farnum, N. R. (1992): "Control Charts for Short Runs: Nonconstant Process and Measurement Error," *Journal of Quality Technology*, 24, pp. 138–144.

——— and L. W. Stanton (1986): "Using Counts to Monitor a Process Mean," *Journal of Quality Technology*, 18, pp. 22–28.

Fox, A. J. (1972): "Outliers in Time Series," *Journal of the Royal Statistical Society*, Series B, 43, pp. 350–363.

Gardner, E. S. (1985): "Exponential Smoothing: The State of the Art," *Journal of Forecasting*, 4, pp. 1–28.

Glaser, R. E. (1982): "Barlett's Test of Homogeneity of Variances," in *Encyclopedia of Statistical Sciences*, vol. 1, edited by N. L. Johnson and S. Kotz, Wiley, New York.

Hackl, P., and J. Ledolter (1991): "A Control Chart Based on Ranks," *Journal of Quality Technology*, 23, pp. 117–124.

Harris, T. J., and W. H. Ross (1991): "Statistical Process Control Procedures for Correlated Observations," *Canadian Journal of Chemical Engineering*, 69, pp. 48–57.

Hillier, F. S. (1969): "$\overline{X}$ and *R*-Chart Control Limits Based on a Small Number of Subgroups," *Journal of Quality Technology*, 1, pp. 17–26.

Ho, C., and K. E. Case (1994): "Economic Design of Control Charts: A Literature Review for 1981–1991," *Journal of Quality Technology*, 26, pp. 39–53.

Holt, C. C. (1957): "Forecasting Seasonal and Trends by Exponential Weighted Moving Averages," Office of Naval Research, Memorandum 52.

Hosking, J. R. M. (1980): "The Multivariate Portmanteau Statistic," *Journal of the American Statistical Association*, 75, pp. 602–608.

Hotelling, H. (1947): "Multivariate Quality Control," in C. Eisenhart, M. W. Hastay, and W. A. Wallis, eds., *Techniques of Statistical Analysis*, McGraw-Hill, New York.

Jackson, J. E. (1980): "Principal Components and Factor Analysis: Part I—Principal Components," *Journal of Quality Technology*, 12, pp. 201–213.

——— (1991): *A User's Guide to Principal Components*, Wiley, New York.

Jeffery, T. C., and J. C. Young (1993): "Monitoring a Chlorine Production Unit by Multivariate Statistical Procedures," *Proceedings of the Symposia on Chlor-alkali and Chlorate Production and New Mathematical Methods and Computational Methods in Electrochemical Engineering*, The Electrochemical Engineering Society, Inc., Pennington, NJ, pp. 136–147.

Johnson, R. A., and D. W. Wichern (1998): *Applied Multivariate Statistical Analysis,* 4th ed., Prentice-Hall, Englewood Cliffs, NJ.

Keats, J. B., E. D. Castillo, V. C. Elart, and E. M. Saniga (1997): "Economic Modeling for Statistical Process Control," *Journal of Quality Technology*, 29, pp. 144–147.

Koziol, J. A. (1982): "A Class of Invariant Procedures for Assessing Multivariate Normality," *Biometrika*, 69, pp. 423–427.

Lorenzen, T. J., and L. C. Vance (1986): "The Economic Design of Control Charts: A Unified Approach," *Technometrics*, 28, pp. 3–10.

Lowry, C. A., and D. C. Montgomery (1995): "A Review of Multivariate Control Charts," *IIE Transactions*, 27, pp. 800–810.

———, W. H. Woodall, C. W. Champ, and S. E. Rigdon (1992): "Multivariate Exponentially Weighted Moving Average Control Chart," *Technometrics*, 34, pp. 46–53.

MacGregor, J. F. (1987): "Interfaces between Process Control and On-Line Statistical Process Control," *Computing and Systems Technology Division Communications*, 10, pp. 9–20.

Madkridakis, S., S. C. Wheelwright, and V. E. McGee (1983): *Forecasting: Methods and Applications,* 2nd ed., Wiley, New York.

Mason, R. L., N. D. Tracy, and J. C. Young (1995): "Decomposition of T^2 for Multivariate Control Chart Interpretation," *Journal of Quality Technology*, 27, pp. 99–108.

———, ———, and ——— (1997): "A Practical Approach for Interpreting Multivariate T^2 Control Chart Signals," *Journal of Quality Technology*, 29, pp. 396–406.

Montgomery, D. C. (1980): "The Economic Design of Control Charts: A Review and Literature Survey," *Journal of Quality Technology*, 12, pp. 75–87.

——— (1996): *Introduction to Statistical Quality Control,* 3d ed., Wiley, New York.

——— and C. M. Mastrangelo (1991): "Some Statistical Process Control Methods for Autocorrelated Data," *Journal of Quality Technology*, 23, pp. 179–193.

Pignatiello, Jr., J. J., and G. C. Runger (1990): "Comparisons of Multivariate CUSUM Charts," *Journal of Quality Technology*, 22, pp. 173–186.

Prabhu, S. S., and G. C. Runger (1997): "Designing a Multivariate EWMA Control Chart," *Journal of Quality Technology*, 29, pp. 8–15.

Reinsel, G. C. (1993): *Elements of Multivariate Time Series Analysis*, Springer-Verlag, New York.

Roberts, H. V. (1991): *Data Analysis for Managers with Minitab,* 2d ed., Duxbury Press, Belmont, CA.

Royston, J. P. (1983): "Some Techniques for Assessing Multivariate Normality Based on the Shapiro-Wilk *W,*" *Applied Statistics*, 32, pp. 121–133.

Saniga, E. M. (1989): "Economic Statistical Control Chart Designs with an Application to $\overline{X}$ and *R* Control Charts," *Technometrics*, 31, pp. 313–320.

Searle, S. R. (1982): *Matrix Algebra Useful for Statistics*, Wiley, New York.

Shewhart, W. (1939): *Statistical Method from the Viewpoint of Quality Control*, Graduate School of the Department of Agriculture, Washington, DC, reprinted by Dover Publications in 1986, Mineola, NY.

Stevenson, W. J. (1999): *Production/Operations Management,* 6th ed., McGraw-Hill/Irwin, Burr Ridge, IL.

Svoboda, L. (1991): "Economic Design of Control Charts: A Review and Literature Survey (1979–1989)," in *Statistical Process Control in Manufacturing*, edited by J. B. Keats and D. C. Montgomery, Marcel Dekker, New York.

Tracy, N. D., J. C. Young, and R. L. Mason (1992): "Multivariate Control Charts for Individual Observations," *Journal of Quality Technology*, 24, pp. 88–95.

Tsay, R. S. (1988): "Outliers, Level Shifts, and Variance Changes in Time Series," *Journal of Forecasting*, 7, pp. 1–20.

Vander Weil, S. A., W. T. Tucker, F. W. Faltin, and N. Doganaksoy (1992): "Algorithmic Statistical Process Control: Concepts and Applications," *Technometrics*, 34, pp. 286–297.

Wei, W. W. (1990): *Time Series Analysis: Univariate and Multivariate Methods*, Addison-Wesley, Redwood City, CA.

Wierda, S. J. (1994): "Multivariate Statistical Process Control—Recent Results and Directions for Future Research," *Statistica Neerlandica*, 48, pp. 147–168.

Winchell, W., and L. A. Millis (1990): "Factors Facilitating Statistical Process Control for Small Batch Sizes," *Quality Engineering*, 2, pp. 331–352.

Winters, P. R. (1960): "Forecasting Sales by Exponentially Weighted Moving Averages," *Management Science*, 6, pp. 324–342.

Woodall, W. H. (1986), "Weakness of the Economic Design of Control Charts" (Letter to the Editor), *Technometrics*, 28, pp. 408–409.

——— (1987): "Conflicts between Deming's Philosophy and the Economic Design of Control Charts," in *Frontiers in Statistical Quality Control,* vol. 3, edited by H.-J. Lenz, G. B. Wetherill, and P.-Th. Wilrich, Physica-Verlag, Heidelberg, Germany.

——— (1997): "A Discussion on Statistically-Based Process Monitoring and Control (Individual Contributions)," *Journal of Quality Technology*, 29, pp. 155–156.

———, S. V. Crowder, and M. R. Wade (1995): "Discussion on Properties of Poisson Q Charts for Attributes," *Journal of Quality Technology*, 18, pp. 328–332.

Table I: Standard Normal Distribution (Right-Tail Probabilities)

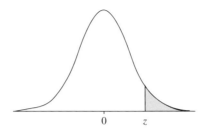

z	P(Z ≥ z)	z	P(Z ≥ z)	z	P(Z ≥ z)	z	P(Z ≥ z)	z	P(Z ≥ z)
0.00	0.500000	0.30	0.382089	0.60	0.274253	0.90	0.184060	1.20	0.115070
0.01	0.496011	0.31	0.378280	0.61	0.270931	0.91	0.181411	1.21	0.113139
0.02	0.492022	0.32	0.374484	0.62	0.267629	0.92	0.178786	1.22	0.111232
0.03	0.488034	0.33	0.370700	0.63	0.264347	0.93	0.176186	1.23	0.109349
0.04	0.484047	0.34	0.366928	0.64	0.261086	0.94	0.173609	1.24	0.107488
0.05	0.480061	0.35	0.363169	0.65	0.257846	0.95	0.171056	1.25	0.105650
0.06	0.476078	0.36	0.359424	0.66	0.254627	0.96	0.168528	1.26	0.103835
0.07	0.472097	0.37	0.355691	0.67	0.251429	0.97	0.166023	1.27	0.102042
0.08	0.468119	0.38	0.351973	0.68	0.248252	0.98	0.163543	1.28	0.100273
0.09	0.464144	0.39	0.348268	0.69	0.245097	0.99	0.161087	1.29	0.098525
0.10	0.460172	0.40	0.344578	0.70	0.241964	1.00	0.158655	1.30	0.096800
0.11	0.456205	0.41	0.340903	0.71	0.238852	1.01	0.156248	1.31	0.095098
0.12	0.452242	0.42	0.337243	0.72	0.235763	1.02	0.153864	1.32	0.093418
0.13	0.448283	0.43	0.333598	0.73	0.232695	1.03	0.151505	1.33	0.091759
0.14	0.444330	0.44	0.329969	0.74	0.229650	1.04	0.149170	1.34	0.090123
0.15	0.440382	0.45	0.326355	0.75	0.226627	1.05	0.146859	1.35	0.088508
0.16	0.436441	0.46	0.322758	0.76	0.223627	1.06	0.144572	1.36	0.086915
0.17	0.432505	0.47	0.319178	0.77	0.220650	1.07	0.142310	1.37	0.085343
0.18	0.428576	0.48	0.315614	0.78	0.217695	1.08	0.140071	1.38	0.083793
0.19	0.424655	0.49	0.312067	0.79	0.214764	1.09	0.137857	1.39	0.082264
0.20	0.420740	0.50	0.308538	0.80	0.211855	1.10	0.135666	1.40	0.080757
0.21	0.416834	0.51	0.305026	0.81	0.208970	1.11	0.133500	1.41	0.079270
0.22	0.412936	0.52	0.301532	0.82	0.206108	1.12	0.131357	1.42	0.077804
0.23	0.409046	0.53	0.298056	0.83	0.203269	1.13	0.129238	1.43	0.076359
0.24	0.405165	0.54	0.294599	0.84	0.200454	1.14	0.127143	1.44	0.074934
0.25	0.401294	0.55	0.291160	0.85	0.197663	1.15	0.125072	1.45	0.073529
0.26	0.397432	0.56	0.287740	0.86	0.194895	1.16	0.123024	1.46	0.072145
0.27	0.393580	0.57	0.284339	0.87	0.192150	1.17	0.121000	1.47	0.070781
0.28	0.389739	0.58	0.280957	0.88	0.189430	1.18	0.119000	1.48	0.069437
0.29	0.385908	0.59	0.277595	0.89	0.186733	1.19	0.117023	1.49	0.068112

Table I: Standard Normal Distribution (Right-Tail Probabilities) *(continued)*

z	$P(Z \geq z)$	z	$P(Z \geq z)$	z	$P(Z \geq z)$	z	$P(Z \geq z)$	z	$P(Z \geq z)$
1.50	0.0668072	1.88	0.0300540	2.26	0.0119106	2.64	0.00414530	3.02	0.00126387
1.51	0.0655217	1.89	0.0293790	2.27	0.0116038	2.65	0.00402459	3.03	0.00122277
1.52	0.0642555	1.90	0.0287166	2.28	0.0113038	2.66	0.00390703	3.04	0.00118289
1.53	0.0630084	1.91	0.0280666	2.29	0.0110107	2.67	0.00379256	3.05	0.00114421
1.54	0.0617802	1.92	0.0274289	2.30	0.0107241	2.68	0.00368111	3.06	0.00110668
1.55	0.0605708	1.93	0.0268034	2.31	0.0104441	2.69	0.00357260	3.07	0.00107029
1.56	0.0593799	1.94	0.0261898	2.32	0.0101704	2.70	0.00346697	3.08	0.00103500
1.57	0.0582076	1.95	0.0255881	2.33	0.0099031	2.71	0.00336416	3.09	0.00100078
1.58	0.0570534	1.96	0.0249979	2.34	0.0096410	2.72	0.00326410	3.10	0.00096760
1.59	0.0559174	1.97	0.0244192	2.35	0.0093867	2.73	0.00316672	3.11	0.00093544
1.60	0.0547993	1.98	0.0238518	2.36	0.0091375	2.74	0.00307196	3.12	0.00090426
1.61	0.0536989	1.99	0.0232955	2.37	0.0088940	2.75	0.00297976	3.13	0.00087403
1.62	0.0526161	2.00	0.0227501	2.38	0.0086563	2.76	0.00289007	3.14	0.00084474
1.63	0.0515507	2.01	0.0222156	2.39	0.0084242	2.77	0.00280281	3.15	0.00081635
1.64	0.0505026	2.02	0.0216917	2.40	0.0081975	2.78	0.00271794	3.16	0.00078885
1.65	0.0494715	2.03	0.0211783	2.41	0.0079763	2.79	0.00263540	3.17	0.00076219
1.66	0.0484572	2.04	0.0206752	2.42	0.0077603	2.80	0.00255513	3.18	0.00073638
1.67	0.0474597	2.05	0.0201822	2.43	0.0075494	2.81	0.00247707	3.19	0.00071136
1.68	0.0464787	2.06	0.0196993	2.44	0.0073436	2.82	0.00240118	3.20	0.00068714
1.69	0.0455140	2.07	0.0192262	2.45	0.0071428	2.83	0.00232740	3.21	0.00066367
1.70	0.0445655	2.08	0.0187628	2.46	0.0069469	2.84	0.00225568	3.22	0.00064095
1.71	0.0436329	2.09	0.0183089	2.47	0.0067557	2.85	0.00218596	3.23	0.00061895
1.72	0.0427162	2.10	0.0178644	2.48	0.0065691	2.86	0.00211821	3.24	0.00059765
1.73	0.0418151	2.11	0.0174292	2.49	0.0063872	2.87	0.00205236	3.25	0.00057703
1.74	0.0409295	2.12	0.0170030	2.50	0.00620967	2.88	0.00198838	3.26	0.00055706
1.75	0.0400592	2.13	0.0165858	2.51	0.00603656	2.89	0.00192621	3.27	0.00053774
1.76	0.0392039	2.14	0.0161774	2.52	0.00586774	2.90	0.00186581	3.28	0.00051904
1.77	0.0383636	2.15	0.0157776	2.53	0.00570313	2.91	0.00180714	3.29	0.00050094
1.78	0.0375380	2.16	0.0153863	2.54	0.00554262	2.92	0.00175016	3.30	0.00048342
1.79	0.0367270	2.17	0.0150034	2.55	0.00538615	2.93	0.00169481	3.31	0.00046648
1.80	0.0359303	2.18	0.0146287	2.56	0.00523361	2.94	0.00164106	3.32	0.00045009
1.81	0.0351479	2.19	0.0142621	2.57	0.00508493	2.95	0.00158887	3.33	0.00043423
1.82	0.0343795	2.20	0.0139034	2.58	0.00494002	2.96	0.00153820	3.34	0.00041889
1.83	0.0336250	2.21	0.0135526	2.59	0.00479880	2.97	0.00148900	3.35	0.00040406
1.84	0.0328841	2.22	0.0132094	2.60	0.00466119	2.98	0.00144124	3.36	0.00038971
1.85	0.0321568	2.23	0.0128737	2.61	0.00452711	2.99	0.00139489	3.37	0.00037584
1.86	0.0314428	2.24	0.0125455	2.62	0.00439649	3.00	0.00134990	3.38	0.00036243
1.87	0.0307419	2.25	0.0122245	2.63	0.00426924	3.01	0.00130624	3.39	0.00034946

Table I: Standard Normal Distribution (Right-Tail Probabilities) *(continued)*

z	P(Z ≥ z)	z	P(Z ≥ z)	z	P(Z ≥ z)	z	P(Z ≥ z)	z	P(Z ≥ z)
3.40	0.00033693	3.77	0.00008162	4.14	0.00001737	4.51	0.00000324	4.88	0.00000053
3.41	0.00032481	3.78	0.00007841	4.15	0.00001662	4.52	0.00000309	4.89	0.00000050
3.42	0.00031311	3.79	0.00007532	4.16	0.00001591	4.53	0.00000295	4.90	0.00000048
3.43	0.00030179	3.80	0.00007235	4.17	0.00001523	4.54	0.00000281	4.91	0.00000046
3.44	0.00029086	3.81	0.00006948	4.18	0.00001458	4.55	0.00000268	4.92	0.00000043
3.45	0.00028029	3.82	0.00006673	4.19	0.00001395	4.56	0.00000256	4.93	0.00000041
3.46	0.00027009	3.83	0.00006407	4.20	0.00001335	4.57	0.00000244	4.94	0.00000039
3.47	0.00026023	3.84	0.00006152	4.21	0.00001277	4.58	0.00000232	4.95	0.00000037
3.48	0.00025071	3.85	0.00005906	4.22	0.00001222	4.59	0.00000222	4.96	0.00000035
3.49	0.00024151	3.86	0.00005669	4.23	0.00001168	4.60	0.00000211	4.97	0.00000033
3.50	0.00023263	3.87	0.00005442	4.24	0.00001118	4.61	0.00000201	4.98	0.00000032
3.51	0.00022405	3.88	0.00005223	4.25	0.00001069	4.62	0.00000192	4.99	0.00000030
3.52	0.00021577	3.89	0.00005012	4.26	0.00001022	4.63	0.00000183	5.00	0.00000029
3.53	0.00020778	3.90	0.00004810	4.27	0.00000977	4.64	0.00000174		
3.54	0.00020006	3.91	0.00004615	4.28	0.00000934	4.65	0.00000166		
3.55	0.00019262	3.92	0.00004427	4.29	0.00000893	4.66	0.00000158		
3.56	0.00018543	3.93	0.00004247	4.30	0.00000854	4.67	0.00000151		
3.57	0.00017849	3.94	0.00004074	4.31	0.00000816	4.68	0.00000143		
3.58	0.00017180	3.95	0.00003908	4.32	0.00000780	4.69	0.00000137		
3.59	0.00016534	3.96	0.00003747	4.33	0.00000746	4.70	0.00000130		
3.60	0.00015911	3.97	0.00003594	4.34	0.00000712	4.71	0.00000124		
3.61	0.00015310	3.98	0.00003446	4.35	0.00000681	4.72	0.00000118		
3.62	0.00014730	3.99	0.00003304	4.36	0.00000650	4.73	0.00000112		
3.63	0.00014171	4.00	0.00003167	4.37	0.00000621	4.74	0.00000107		
3.64	0.00013632	4.01	0.00003036	4.38	0.00000593	4.75	0.00000102		
3.65	0.00013112	4.02	0.00002910	4.39	0.00000567	4.76	0.00000097		
3.66	0.00012611	4.03	0.00002789	4.40	0.00000541	4.77	0.00000092		
3.67	0.00012128	4.04	0.00002673	4.41	0.00000517	4.78	0.00000088		
3.68	0.00011662	4.05	0.00002561	4.42	0.00000494	4.79	0.00000083		
3.69	0.00011213	4.06	0.00002454	4.43	0.00000471	4.80	0.00000079		
3.70	0.00010780	4.07	0.00002351	4.44	0.00000450	4.81	0.00000075		
3.71	0.00010363	4.08	0.00002252	4.45	0.00000429	4.82	0.00000072		
3.72	0.00009961	4.09	0.00002157	4.46	0.00000410	4.83	0.00000068		
3.73	0.00009574	4.10	0.00002066	4.47	0.00000391	4.84	0.00000065		
3.74	0.00009201	4.11	0.00001978	4.48	0.00000373	4.85	0.00000062		
3.75	0.00008842	4.12	0.00001894	4.49	0.00000356	4.86	0.00000059		
3.76	0.00008496	4.13	0.00001814	4.50	0.00000340	4.87	0.00000056		

Source: Computed using Minitab.

Table II: Percentage Points of *t* Distribution

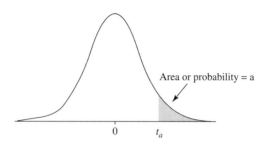

Entries in the table give t_a values, where a is the area or probability in the upper tail of the *t* distribution.

Area or probability = a

df	$t_{0.100}$	$t_{0.050}$	$t_{0.025}$	$t_{0.010}$	$t_{0.005}$	$t_{0.001}$
1	3.078	6.314	12.706	31.821	63.657	318.309
2	1.886	2.920	4.303	6.965	9.925	22.327
3	1.638	2.353	3.182	4.541	5.841	10.215
4	1.533	2.132	2.776	3.747	4.604	7.173
5	1.476	2.015	2.571	3.365	4.032	5.893
6	1.440	1.943	2.447	3.143	3.707	5.208
7	1.415	1.895	2.365	2.998	3.499	4.785
8	1.397	1.860	2.306	2.896	3.355	4.501
9	1.383	1.833	2.262	2.821	3.250	4.297
10	1.372	1.812	2.228	2.764	3.169	4.144
11	1.363	1.796	2.201	2.718	3.106	4.025
12	1.356	1.782	2.179	2.681	3.055	3.930
13	1.350	1.771	2.160	2.650	3.012	3.852
14	1.345	1.761	2.145	2.624	2.977	3.787
15	1.341	1.753	2.131	2.602	2.947	3.733
16	1.337	1.746	2.120	2.583	2.921	3.686
17	1.333	1.740	2.110	2.567	2.898	3.646
18	1.330	1.734	2.101	2.552	2.878	3.610
19	1.328	1.729	2.093	2.539	2.861	3.579
20	1.325	1.725	2.086	2.528	2.845	3.552

Table II: Percentage Points of *t* Distribution *(concluded)*

df	$t_{0.100}$	$t_{0.050}$	$t_{0.025}$	$t_{0.010}$	$t_{0.005}$	$t_{0.001}$
21	1.323	1.721	2.080	2.518	2.831	3.527
22	1.321	1.717	2.074	2.508	2.819	3.505
23	1.319	1.714	2.069	2.500	2.807	3.485
24	1.318	1.711	2.064	2.492	2.797	3.467
25	1.316	1.708	2.060	2.485	2.787	3.450
26	1.315	1.706	2.056	2.479	2.779	3.435
27	1.314	1.703	2.052	2.473	2.771	3.421
28	1.313	1.701	2.048	2.467	2.763	3.408
29	1.311	1.699	2.045	2.462	2.756	3.396
30	1.310	1.697	2.042	2.457	2.750	3.385
35	1.306	1.690	2.030	2.438	2.724	3.340
40	1.303	1.684	2.021	2.423	2.704	3.307
45	1.301	1.679	2.014	2.412	2.690	3.281
50	1.299	1.676	2.009	2.403	2.678	3.261
55	1.297	1.673	2.004	2.396	2.668	3.245
60	1.296	1.671	2.000	2.390	2.660	3.232
65	1.295	1.669	1.997	2.385	2.654	3.220
70	1.294	1.667	1.994	2.381	2.648	3.211
75	1.293	1.665	1.992	2.377	2.643	3.202
80	1.292	1.664	1.990	2.374	2.639	3.195
85	1.292	1.663	1.988	2.371	2.635	3.189
90	1.291	1.662	1.987	2.368	2.632	3.183
95	1.291	1.661	1.985	2.366	2.629	3.178
100	1.290	1.660	1.984	2.364	2.626	3.174
105	1.290	1.659	1.983	2.362	2.623	3.170
110	1.289	1.659	1.982	2.361	2.621	3.166
115	1.289	1.658	1.981	2.359	2.619	3.163
120	1.289	1.658	1.980	2.358	2.617	3.160
∞	1.282	1.645	1.960	2.326	2.576	3.090

Source: Computed using Minitab.

Table III: Percentage Points of χ^2 Distribution

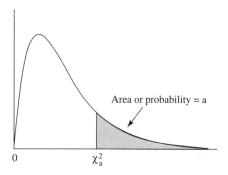

Area or probability = a

0 χ^2_a

						a				
df	0.9995	0.999	0.99865	0.975	0.95	0.05	0.025	0.00135	0.001	0.0005
1	0.0000	0.0000	0.0000	0.0010	0.0039	3.841	5.024	10.273	10.828	12.116
2	0.0010	0.0020	0.0027	0.0506	0.1026	5.991	7.378	13.215	13.816	15.202
3	0.0153	0.0243	0.0297	0.2158	0.3518	7.815	9.348	15.630	16.266	17.730
4	0.0639	0.0908	0.1058	0.4844	0.7107	9.488	11.143	17.800	18.467	19.997
5	0.1581	0.2102	0.2380	0.8312	1.1455	11.070	12.833	19.821	20.515	22.105
6	0.2994	0.3811	0.4234	1.2373	1.6354	12.592	14.449	21.739	22.458	24.103
7	0.4849	0.5985	0.6562	1.6899	2.1673	14.067	16.013	23.580	24.322	26.018
8	0.7104	0.8571	0.9306	2.1797	2.7326	15.507	17.535	25.361	26.124	27.868
9	0.9717	1.1519	1.2413	2.7004	3.3251	16.919	19.023	27.093	27.877	29.666
10	1.2650	1.4787	1.5837	3.2470	3.9403	18.307	20.483	28.785	29.588	31.420
11	1.5868	1.8339	1.9544	3.8157	4.5748	19.675	21.920	30.442	31.264	33.137
12	1.9344	2.2142	2.3499	4.4038	5.2260	21.026	23.337	32.070	32.909	34.821
13	2.3051	2.6172	2.7679	5.0088	5.8919	22.362	24.736	33.671	34.528	36.478
14	2.6967	3.0407	3.2060	5.6287	6.5706	23.685	26.119	35.250	36.123	38.109
15	3.1075	3.4827	3.6624	6.2621	7.2609	24.996	27.488	36.808	37.697	39.719
16	3.5358	3.9416	4.1354	6.9077	7.9616	26.296	28.845	38.347	39.252	41.308
17	3.9802	4.4161	4.6237	7.5642	8.6718	27.587	30.191	39.870	40.790	42.879
18	4.4394	4.9048	5.1260	8.2307	9.3905	28.869	31.526	41.377	42.312	44.434
19	4.9123	5.4068	5.6413	8.9065	10.1170	30.144	32.852	42.871	43.820	45.973

Table III: Percentage Points of χ^2 Distribution *(concluded)*

df	0.9995	0.999	0.99865	0.975	0.95	0.05	0.025	0.00135	0.001	0.0005
					a					
20	5.3981	5.9210	6.1685	9.5908	10.8508	31.410	34.170	44.352	45.315	47.498
21	5.8957	6.4467	6.7069	10.2829	11.5913	32.671	35.479	45.820	46.797	49.011
22	6.4045	6.9830	7.2558	10.9823	12.3380	33.924	36.781	47.278	48.268	50.511
23	6.9237	7.5292	7.8144	11.6886	13.0905	35.172	38.076	48.725	49.728	52.000
24	7.4527	8.0849	8.3822	12.4012	13.8484	36.415	39.364	50.163	51.179	53.479
25	7.9910	8.6493	8.9586	13.1197	14.6114	37.652	40.646	51.591	52.620	54.947
26	8.5379	9.2221	9.5431	13.8439	15.3792	38.885	41.923	53.011	54.052	56.407
27	9.0932	9.8028	10.1353	14.5734	16.1514	40.113	43.195	54.423	55.476	57.858
28	9.6563	10.3909	10.7348	15.3079	16.9279	41.337	44.461	55.828	56.892	59.300
29	10.2268	10.9861	11.3412	16.0471	17.7084	42.557	45.722	57.225	58.301	60.735
30	10.8044	11.5880	11.9541	16.7908	18.4927	43.773	46.979	58.615	59.703	62.162
35	13.7875	14.6878	15.1070	20.5694	22.4650	49.802	53.203	65.477	66.619	69.199
40	16.9062	17.9164	18.3854	24.4330	26.5093	55.758	59.342	72.209	73.402	76.095
45	20.1366	21.2507	21.7668	28.3662	30.6123	61.656	65.410	78.835	80.077	82.876
50	23.4610	24.6739	25.2347	32.3574	34.7643	67.505	71.420	85.374	86.661	89.561
55	26.8658	28.1731	28.7766	36.3981	38.9580	73.311	77.380	91.837	93.168	96.163
60	30.3405	31.7383	32.3828	40.4817	43.1880	79.082	83.298	98.236	99.607	102.695
65	33.8767	35.3616	36.0454	44.6030	47.4496	84.821	89.177	104.577	105.988	109.164
70	37.4674	39.0364	39.7582	48.7576	51.7393	90.531	95.023	110.867	112.317	115.578
75	41.1072	42.7573	43.5158	52.9419	56.0541	96.217	100.839	117.112	118.599	121.942
80	44.7910	46.5199	47.3139	57.1532	60.3915	101.879	106.629	123.317	124.839	128.261
85	48.5151	50.3203	51.1488	61.3888	64.7494	107.522	112.393	129.484	131.041	134.540
90	52.2758	54.1552	55.0173	65.6466	69.1260	113.145	118.136	135.617	137.208	140.782
95	56.0702	58.0220	58.9167	69.9249	73.5198	118.752	123.858	141.720	143.344	146.990
100	59.8957	61.9179	62.8445	74.2219	77.9295	124.342	129.561	147.793	149.449	153.167

Source: Computed using Minitab.

Table IV: Percentage Points of F Distribution $(F_{0.10, \nu_1, \nu_2})$

	Numerator Degrees of Freedom								
ν_1 / ν_2	**1**	**2**	**3**	**4**	**5**	**6**	**7**	**8**	**9**
1	39.86	49.50	53.59	55.83	57.24	58.20	58.91	59.44	59.86
2	8.53	9.00	9.16	9.24	9.29	9.33	9.35	9.37	9.38
3	5.54	5.46	5.39	5.34	5.31	5.28	5.27	5.25	5.24
4	4.54	4.32	4.19	4.11	4.05	4.01	3.98	3.95	3.94
5	4.06	3.78	3.62	3.52	3.45	3.40	3.37	3.34	3.32
6	3.78	3.46	3.29	3.18	3.11	3.05	3.01	2.98	2.96
7	3.59	3.26	3.07	2.96	2.88	2.83	2.78	2.75	2.72
8	3.46	3.11	2.92	2.81	2.73	2.67	2.62	2.59	2.56
9	3.36	3.01	2.81	2.69	2.61	2.55	2.51	2.47	2.44
10	3.29	2.92	2.73	2.61	2.52	2.46	2.41	2.38	2.35
11	3.23	2.86	2.66	2.54	2.45	2.39	2.34	2.30	2.27
12	3.18	2.81	2.61	2.48	2.39	2.33	2.28	2.24	2.21
13	3.14	2.76	2.56	2.43	2.35	2.28	2.23	2.20	2.16
14	3.10	2.73	2.52	2.39	2.31	2.24	2.19	2.15	2.12
15	3.07	2.70	2.49	2.36	2.27	2.21	2.16	2.12	2.09
16	3.05	2.67	2.46	2.33	2.24	2.18	2.13	2.09	2.06
17	3.03	2.64	2.44	2.31	2.22	2.15	2.10	2.06	2.03
18	3.01	2.62	2.42	2.29	2.20	2.13	2.08	2.04	2.00
19	2.99	2.61	2.40	2.27	2.18	2.11	2.06	2.02	1.98
20	2.97	2.59	2.38	2.25	2.16	2.09	2.04	2.00	1.96
21	2.96	2.57	2.36	2.23	2.14	2.08	2.02	1.98	1.95
22	2.95	2.56	2.35	2.22	2.13	2.06	2.01	1.97	1.93
23	2.94	2.55	2.34	2.21	2.11	2.05	1.99	1.95	1.92
24	2.93	2.54	2.33	2.19	2.10	2.04	1.98	1.94	1.91
25	2.92	2.53	2.32	2.18	2.09	2.02	1.97	1.93	1.89
26	2.91	2.52	2.31	2.17	2.08	2.01	1.96	1.92	1.88
27	2.90	2.51	2.30	2.17	2.07	2.00	1.95	1.91	1.87
28	2.89	2.50	2.29	2.16	2.06	2.00	1.94	1.90	1.87
29	2.89	2.50	2.28	2.15	2.06	1.99	1.93	1.89	1.86
30	2.88	2.49	2.28	2.14	2.05	1.98	1.93	1.88	1.85
40	2.84	2.44	2.23	2.09	2.00	1.93	1.87	1.83	1.79
60	2.79	2.39	2.18	2.04	1.95	1.87	1.82	1.77	1.74
120	2.75	2.35	2.13	1.99	1.90	1.82	1.77	1.72	1.68
∞	2.71	2.30	2.08	1.94	1.85	1.77	1.72	1.67	1.63

Denominator Degrees of Freedom (along the left vertical axis)

				Numerator Degrees of Freedom					
10	12	15	20	24	30	40	60	120	∞
60.19	60.71	61.22	61.74	62.00	62.26	62.53	62.79	63.06	63.33
9.39	9.41	9.42	9.44	9.45	9.46	9.47	9.47	9.48	9.49
5.23	5.22	5.20	5.18	5.18	5.17	5.16	5.15	5.14	5.13
3.92	3.90	3.87	3.84	3.83	3.82	3.80	3.79	3.78	3.76
3.30	3.27	3.24	3.21	3.19	3.17	3.16	3.14	3.12	3.10
2.94	2.90	2.87	2.84	2.82	2.80	2.78	2.76	2.74	2.72
2.70	2.67	2.63	2.59	2.58	2.56	2.54	2.51	2.49	2.47
2.54	2.50	2.46	2.42	2.40	2.38	2.36	2.34	2.32	2.29
2.42	2.38	2.34	2.30	2.28	2.25	2.23	2.21	2.18	2.16
2.32	2.28	2.24	2.20	2.18	2.16	2.13	2.11	2.08	2.06
2.25	2.21	2.17	2.12	2.10	2.08	2.05	2.03	2.00	1.97
2.19	2.15	2.10	2.06	2.04	2.01	1.99	1.96	1.93	1.90
2.14	2.10	2.05	2.01	1.98	1.96	1.93	1.90	1.88	1.85
2.10	2.05	2.01	1.96	1.94	1.91	1.89	1.86	1.83	1.80
2.06	2.02	1.97	1.92	1.90	1.87	1.85	1.82	1.79	1.76
2.03	1.99	1.94	1.89	1.87	1.84	1.81	1.78	1.75	1.72
2.00	1.96	1.91	1.86	1.84	1.81	1.78	1.75	1.72	1.69
1.98	1.93	1.89	1.84	1.81	1.78	1.75	1.72	1.69	1.66
1.96	1.91	1.86	1.81	1.79	1.76	1.73	1.70	1.67	1.63
1.94	1.89	1.84	1.79	1.77	1.74	1.71	1.68	1.64	1.61
1.92	1.87	1.83	1.78	1.75	1.72	1.69	1.66	1.62	1.59
1.90	1.86	1.81	1.76	1.73	1.70	1.67	1.64	1.60	1.57
1.89	1.84	1.80	1.74	1.72	1.69	1.66	1.62	1.59	1.55
1.88	1.83	1.78	1.73	1.70	1.67	1.64	1.61	1.57	1.53
1.87	1.82	1.77	1.72	1.69	1.66	1.63	1.59	1.56	1.52
1.86	1.81	1.76	1.71	1.68	1.65	1.61	1.58	1.54	1.50
1.85	1.80	1.75	1.70	1.67	1.64	1.60	1.57	1.53	1.49
1.84	1.79	1.74	1.69	1.66	1.63	1.59	1.56	1.52	1.48
1.83	1.78	1.73	1.68	1.65	1.62	1.58	1.55	1.51	1.47
1.82	1.77	1.72	1.67	1.64	1.61	1.57	1.54	1.50	1.46
1.76	1.71	1.66	1.61	1.57	1.54	1.51	1.47	1.42	1.38
1.71	1.66	1.60	1.54	1.51	1.48	1.44	1.40	1.35	1.29
1.65	1.60	1.55	1.48	1.45	1.41	1.37	1.32	1.26	1.19
1.60	1.55	1.49	1.42	1.38	1.34	1.30	1.24	1.17	1.00

Table IV: Percentage Points of F Distribution $(F_{0.05, \nu_1, \nu_2})$ *(continued)*

Numerator Degrees of Freedom

ν_2 \ ν_1	1	2	3	4	5	6	7	8	9
1	161.45	199.50	215.71	224.58	230.16	233.99	236.77	238.88	240.54
2	18.51	19.00	19.16	19.25	19.30	19.33	19.35	19.37	19.38
3	10.13	9.55	9.28	9.12	9.01	8.94	8.89	8.85	8.81
4	7.71	6.94	6.59	6.39	6.26	6.16	6.09	6.04	6.00
5	6.61	5.79	5.41	5.19	5.05	4.95	4.88	4.82	4.77
6	5.99	5.14	4.76	4.53	4.39	4.28	4.21	4.15	4.10
7	5.59	4.74	4.35	4.12	3.97	3.87	3.79	3.73	3.68
8	5.32	4.46	4.07	3.84	3.69	3.58	3.50	3.44	3.39
9	5.12	4.26	3.86	3.63	3.48	3.37	3.29	3.23	3.18
10	4.96	4.10	3.71	3.48	3.33	3.22	3.14	3.07	3.02
11	4.84	3.98	3.59	3.36	3.20	3.09	3.01	2.95	2.90
12	4.75	3.89	3.49	3.26	3.11	3.00	2.91	2.85	2.80
13	4.67	3.81	3.41	3.18	3.03	2.92	2.83	2.77	2.71
14	4.60	3.74	3.34	3.11	2.96	2.85	2.76	2.70	2.65
15	4.54	3.68	3.29	3.06	2.90	2.79	2.71	2.64	2.59
16	4.49	3.63	3.24	3.01	2.85	2.74	2.66	2.59	2.54
17	4.45	3.59	3.20	2.96	2.81	2.70	2.61	2.55	2.49
18	4.41	3.55	3.16	2.93	2.77	2.66	2.58	2.51	2.46
19	4.38	3.52	3.13	2.90	2.74	2.63	2.54	2.48	2.42
20	4.35	3.49	3.10	2.87	2.71	2.60	2.51	2.45	2.39
21	4.32	3.47	3.07	2.84	2.68	2.57	2.49	2.42	2.37
22	4.30	3.44	3.05	2.82	2.66	2.55	2.46	2.40	2.34
23	4.28	3.42	3.03	2.80	2.64	2.53	2.44	2.37	2.32
24	4.26	3.40	3.01	2.78	2.62	2.51	2.42	2.36	2.30
25	4.24	3.39	2.99	2.76	2.60	2.49	2.40	2.34	2.28
26	4.23	3.37	2.98	2.74	2.59	2.47	2.39	2.32	2.27
27	4.21	3.35	2.96	2.73	2.57	2.46	2.37	2.31	2.25
28	4.20	3.34	2.95	2.71	2.56	2.45	2.36	2.29	2.24
29	4.18	3.33	2.93	2.70	2.55	2.43	2.35	2.28	2.22
30	4.17	3.32	2.92	2.69	2.53	2.42	2.33	2.27	2.21
40	4.08	3.23	2.84	2.61	2.45	2.34	2.25	2.18	2.12
60	4.00	3.15	2.76	2.53	2.37	2.25	2.17	2.10	2.04
120	3.92	3.07	2.68	2.45	2.29	2.18	2.09	2.02	1.96
∞	3.84	3.00	2.60	2.37	2.21	2.10	2.01	1.94	1.88

Denominator Degrees of Freedom

Numerator Degrees of Freedom									
10	**12**	**15**	**20**	**24**	**30**	**40**	**60**	**120**	**∞**
241.88	243.91	245.95	248.01	249.05	250.10	251.14	252.20	253.25	254.31
19.40	19.41	19.43	19.45	19.45	19.46	19.47	19.48	19.49	19.50
8.79	8.74	8.70	8.66	8.64	8.62	8.59	8.57	8.55	8.53
5.96	5.91	5.86	5.80	5.77	5.75	5.72	5.69	5.66	5.63
4.74	4.68	4.62	4.56	4.53	4.50	4.46	4.43	4.40	4.36
4.06	4.00	3.94	3.87	3.84	3.81	3.77	3.74	3.70	3.67
3.64	3.57	3.51	3.44	3.41	3.38	3.34	3.30	3.27	3.23
3.35	3.28	3.22	3.15	3.12	3.08	3.04	3.01	2.97	2.93
3.14	3.07	3.01	2.94	2.90	2.86	2.83	2.79	2.75	2.71
2.98	2.91	2.85	2.77	2.74	2.70	2.66	2.62	2.58	2.54
2.85	2.79	2.72	2.65	2.61	2.57	2.53	2.49	2.45	2.40
2.75	2.69	2.62	2.54	2.51	2.47	2.43	2.38	2.34	2.30
2.67	2.60	2.53	2.46	2.42	2.38	2.34	2.30	2.25	2.21
2.60	2.53	2.46	2.39	2.35	2.31	2.27	2.22	2.18	2.13
2.54	2.48	2.40	2.33	2.29	2.25	2.20	2.16	2.11	2.07
2.49	2.42	2.35	2.28	2.24	2.19	2.15	2.11	2.06	2.01
2.45	2.38	2.31	2.23	2.19	2.15	2.10	2.06	2.01	1.96
2.41	2.34	2.27	2.19	2.15	2.11	2.06	2.02	1.97	1.92
2.38	2.31	2.23	2.16	2.11	2.07	2.03	1.98	1.93	1.88
2.35	2.28	2.20	2.12	2.08	2.04	1.99	1.95	1.90	1.84
2.32	2.25	2.18	2.10	2.05	2.01	1.96	1.92	1.87	1.81
2.30	2.23	2.15	2.07	2.03	1.98	1.94	1.89	1.84	1.78
2.27	2.20	2.13	2.05	2.01	1.96	1.91	1.86	1.81	1.76
2.25	2.18	2.11	2.03	1.98	1.94	1.89	1.84	1.79	1.73
2.24	2.16	2.09	2.01	1.96	1.92	1.87	1.82	1.77	1.71
2.22	2.15	2.07	1.99	1.95	1.90	1.85	1.80	1.75	1.69
2.20	2.13	2.06	1.97	1.93	1.88	1.84	1.79	1.73	1.67
2.19	2.12	2.04	1.96	1.91	1.87	1.82	1.77	1.71	1.65
2.18	2.10	2.03	1.94	1.90	1.85	1.81	1.75	1.70	1.64
2.16	2.09	2.01	1.93	1.89	1.84	1.79	1.74	1.68	1.62
2.08	2.00	1.92	1.84	1.79	1.74	1.69	1.64	1.58	1.51
1.99	1.92	1.84	1.75	1.70	1.65	1.59	1.53	1.47	1.39
1.91	1.83	1.75	1.66	1.61	1.55	1.50	1.43	1.35	1.25
1.83	1.75	1.67	1.57	1.52	1.46	1.39	1.32	1.22	1.00

Table IV: Percentage Points of F Distribution ($F_{0.01, \nu_1, \nu_2}$) (continued)

	Numerator Degrees of Freedom								
ν_1	**1**	**2**	**3**	**4**	**5**	**6**	**7**	**8**	**9**
ν_2									
1	4,052.18	4,999.50	5,403.35	5,624.58	5,763.65	5,858.99	5,928.36	5,981.07	6,022.47
2	98.50	99.00	99.17	99.25	99.30	99.33	99.36	99.37	99.39
3	34.12	30.82	29.46	28.71	28.24	27.91	27.67	27.49	27.35
4	21.20	18.00	16.69	15.98	15.52	15.21	14.98	14.80	14.66
5	16.26	13.27	12.06	11.39	10.97	10.67	10.46	10.29	10.16
6	13.75	10.92	9.78	9.15	8.75	8.47	8.26	8.10	7.98
7	12.25	9.55	8.45	7.85	7.46	7.19	6.99	6.84	6.72
8	11.26	8.65	7.59	7.01	6.63	6.37	6.18	6.03	5.91
9	10.56	8.02	6.99	6.42	6.06	5.80	5.61	5.47	5.35
10	10.04	7.56	6.55	5.99	5.64	5.39	5.20	5.06	4.94
11	9.65	7.21	6.22	5.67	5.32	5.07	4.89	4.74	4.63
12	9.33	6.93	5.95	5.41	5.06	4.82	4.64	4.50	4.39
13	9.07	6.70	5.74	5.21	4.86	4.62	4.44	4.30	4.19
14	8.86	6.51	5.56	5.04	4.69	4.46	4.28	4.14	4.03
15	8.68	6.36	5.42	4.89	4.56	4.32	4.14	4.00	3.89
16	8.53	6.23	5.29	4.77	4.44	4.20	4.03	3.89	3.78
17	8.40	6.11	5.18	4.67	4.34	4.10	3.93	3.79	3.68
18	8.29	6.01	5.09	4.58	4.25	4.01	3.84	3.71	3.60
19	8.18	5.93	5.01	4.50	4.17	3.94	3.77	3.63	3.52
20	8.10	5.85	4.94	4.43	4.10	3.87	3.70	3.56	3.46
21	8.02	5.78	4.87	4.37	4.04	3.81	3.64	3.51	3.40
22	7.95	5.72	4.82	4.31	3.99	3.76	3.59	3.45	3.35
23	7.88	5.66	4.76	4.26	3.94	3.71	3.54	3.41	3.30
24	7.82	5.61	4.72	4.22	3.90	3.67	3.50	3.36	3.26
25	7.77	5.57	4.68	4.18	3.85	3.63	3.46	3.32	3.22
26	7.72	5.53	4.64	4.14	3.82	3.59	3.42	3.29	3.18
27	7.68	5.49	4.60	4.11	3.78	3.56	3.39	3.26	3.15
28	7.64	5.45	4.57	4.07	3.75	3.53	3.36	3.23	3.12
29	7.60	5.42	4.54	4.04	3.73	3.50	3.33	3.20	3.09
30	7.56	5.39	4.51	4.02	3.70	3.47	3.30	3.17	3.07
40	7.31	5.18	4.31	3.83	3.51	3.29	3.12	2.99	2.89
60	7.08	4.98	4.13	3.65	3.34	3.12	2.95	2.82	2.72
120	6.85	4.79	3.95	3.48	3.17	2.96	2.79	2.66	2.56
∞	6.63	4.61	3.78	3.32	3.02	2.80	2.64	2.51	2.41

Denominator Degrees of Freedom

| **Numerator Degrees of Freedom** | | | | | | | | | |
10	12	15	20	24	30	40	60	120	∞
6,055.85	6,106.32	6,157.28	6,208.73	6,234.63	6,260.65	6,286.78	6,313.03	6,339.39	6,365.86
99.40	99.42	99.43	99.45	99.46	99.47	99.47	99.48	99.49	99.50
27.23	27.05	26.87	26.69	26.60	26.50	26.41	26.32	26.22	26.13
14.55	14.37	14.20	14.02	13.93	13.84	13.75	13.65	13.56	13.46
10.05	9.89	9.72	9.55	9.47	9.38	9.29	9.20	9.11	9.02
7.87	7.72	7.56	7.40	7.31	7.23	7.14	7.06	6.97	6.88
6.62	6.47	6.31	6.16	6.07	5.99	5.91	5.82	5.74	5.65
5.81	5.67	5.52	5.36	5.28	5.20	5.12	5.03	4.95	4.86
5.26	5.11	4.96	4.81	4.73	4.65	4.57	4.48	4.40	4.31
4.85	4.71	4.56	4.41	4.33	4.25	4.17	4.08	4.00	3.91
4.54	4.40	4.25	4.10	4.02	3.94	3.86	3.78	3.69	3.60
4.30	4.16	4.01	3.86	3.78	3.70	3.62	3.54	3.45	3.36
4.10	3.96	3.82	3.66	3.59	3.51	3.43	3.34	3.25	3.17
3.94	3.80	3.66	3.51	3.43	3.35	3.27	3.18	3.09	3.00
3.80	3.67	3.52	3.37	3.29	3.21	3.13	3.05	2.96	2.87
3.69	3.55	3.41	3.26	3.18	3.10	3.02	2.93	2.84	2.75
3.59	3.46	3.31	3.16	3.08	3.00	2.92	2.83	2.75	2.65
3.51	3.37	3.23	3.08	3.00	2.92	2.84	2.75	2.66	2.57
3.43	3.30	3.15	3.00	2.92	2.84	2.76	2.67	2.58	2.49
3.37	3.23	3.09	2.94	2.86	2.78	2.69	2.61	2.52	2.42
3.31	3.17	3.03	2.88	2.80	2.72	2.64	2.55	2.46	2.36
3.26	3.12	2.98	2.83	2.75	2.67	2.58	2.50	2.40	2.31
3.21	3.07	2.93	2.78	2.70	2.62	2.54	2.45	2.35	2.26
3.17	3.03	2.89	2.74	2.66	2.58	2.49	2.40	2.31	2.21
3.13	2.99	2.85	2.70	2.62	2.54	2.45	2.36	2.27	2.17
3.09	2.96	2.81	2.66	2.58	2.50	2.42	2.33	2.23	2.13
3.06	2.93	2.78	2.63	2.55	2.47	2.38	2.29	2.20	2.10
3.03	2.90	2.75	2.60	2.52	2.44	2.35	2.26	2.17	2.06
3.00	2.87	2.73	2.57	2.49	2.41	2.33	2.23	2.14	2.03
2.98	2.84	2.70	2.55	2.47	2.39	2.30	2.21	2.11	2.01
2.80	2.66	2.52	2.37	2.29	2.20	2.11	2.02	1.92	1.80
2.63	2.50	2.35	2.20	2.12	2.03	1.94	1.84	1.73	1.60
2.47	2.34	2.19	2.03	1.95	1.86	1.76	1.66	1.53	1.38
2.32	2.18	2.04	1.88	1.79	1.70	1.59	1.47	1.32	1.00

Table IV: Percentage Points of F Distribution $(F_{0.005, \nu_1, \nu_2})$ (continued)

	ν_1	1	2	3	4	5	6	7	8	9
ν_2										
	1	16,210.72	19,999.50	21,614.74	22,499.58	23,055.80	23,437.11	23,714.57	23,925.41	24,091.00
	2	198.50	199.00	199.17	199.25	199.30	199.33	199.36	199.37	199.39
	3	55.55	49.80	47.47	46.19	45.39	44.84	44.43	44.13	43.88
	4	31.33	26.28	24.26	23.15	22.46	21.97	21.62	21.35	21.14
	5	22.78	18.31	16.53	15.56	14.94	14.51	14.20	13.96	13.77
	6	18.63	14.54	12.92	12.03	11.46	11.07	10.79	10.57	10.39
	7	16.24	12.40	10.88	10.05	9.52	9.16	8.89	8.68	8.51
	8	14.69	11.04	9.60	8.81	8.30	7.95	7.69	7.50	7.34
	9	13.61	10.11	8.72	7.96	7.47	7.13	6.88	6.69	6.54
	10	12.83	9.43	8.08	7.34	6.87	6.54	6.30	6.12	5.97
	11	12.23	8.91	7.60	6.88	6.42	6.10	5.86	5.68	5.54
	12	11.75	8.51	7.23	6.52	6.07	5.76	5.52	5.35	5.20
Denominator Degrees of Freedom	13	11.37	8.19	6.93	6.23	5.79	5.48	5.25	5.08	4.94
	14	11.06	7.92	6.68	6.00	5.56	5.26	5.03	4.86	4.72
	15	10.80	7.70	6.48	5.80	5.37	5.07	4.85	4.67	4.54
	16	10.58	7.51	6.30	5.64	5.21	4.91	4.69	4.52	4.38
	17	10.38	7.35	6.16	5.50	5.07	4.78	4.56	4.39	4.25
	18	10.22	7.21	6.03	5.37	4.96	4.66	4.44	4.28	4.14
	19	10.07	7.09	5.92	5.27	4.85	4.56	4.34	4.18	4.04
	20	9.94	6.99	5.82	5.17	4.76	4.47	4.26	4.09	3.96
	21	9.83	6.89	5.73	5.09	4.68	4.39	4.18	4.01	3.88
	22	9.73	6.81	5.65	5.02	4.61	4.32	4.11	3.94	3.81
	23	9.63	6.73	5.58	4.95	4.54	4.26	4.05	3.88	3.75
	24	9.55	6.66	5.52	4.89	4.49	4.20	3.99	3.83	3.69
	25	9.48	6.60	5.46	4.84	4.43	4.15	3.94	3.78	3.64
	26	9.41	6.54	5.41	4.79	4.38	4.10	3.89	3.73	3.60
	27	9.34	6.49	5.36	4.74	4.34	4.06	3.85	3.69	3.56
	28	9.28	6.44	5.32	4.70	4.30	4.02	3.81	3.65	3.52
	29	9.23	6.40	5.28	4.66	4.26	3.98	3.77	3.61	3.48
	30	9.18	6.35	5.24	4.62	4.23	3.95	3.74	3.58	3.45
	40	8.83	6.07	4.98	4.37	3.99	3.71	3.51	3.35	3.22
	60	8.49	5.79	4.73	4.14	3.76	3.49	3.29	3.13	3.01
	120	8.18	5.54	4.50	3.92	3.55	3.28	3.09	2.93	2.81
	∞	7.88	5.30	4.28	3.72	3.35	3.09	2.90	2.74	2.62

Numerator Degrees of Freedom

Numerator Degrees of Freedom									
10	**12**	**15**	**20**	**24**	**30**	**40**	**60**	**120**	**∞**
24,224.49	24,426.37	24,630.21	24,835.97	24,939.57	25,043.63	25,148.15	25,253.14	25,358.57	25,464.46
199.40	199.42	199.43	199.45	199.46	199.47	199.47	199.48	199.49	199.50
43.69	43.39	43.08	42.78	42.62	42.47	42.31	42.15	41.99	41.83
20.97	20.70	20.44	20.17	20.03	19.89	19.75	19.61	19.47	19.32
13.62	13.38	13.15	12.90	12.78	12.66	12.53	12.40	12.27	12.14
10.25	10.03	9.81	9.59	9.47	9.36	9.24	9.12	9.00	8.88
8.38	8.18	7.97	7.75	7.64	7.53	7.42	7.31	7.19	7.08
7.21	7.01	6.81	6.61	6.50	6.40	6.29	6.18	6.06	5.95
6.42	6.23	6.03	5.83	5.73	5.62	5.52	5.41	5.30	5.19
5.85	5.66	5.47	5.27	5.17	5.07	4.97	4.86	4.75	4.64
5.42	5.24	5.05	4.86	4.76	4.65	4.55	4.45	4.34	4.23
5.09	4.91	4.72	4.53	4.43	4.33	4.23	4.12	4.01	3.90
4.82	4.64	4.46	4.27	4.17	4.07	3.97	3.87	3.76	3.65
4.60	4.43	4.25	4.06	3.96	3.86	3.76	3.66	3.55	3.44
4.42	4.25	4.07	3.88	3.79	3.69	3.58	3.48	3.37	3.26
4.27	4.10	3.92	3.73	3.64	3.54	3.44	3.33	3.22	3.11
4.14	3.97	3.79	3.61	3.51	3.41	3.31	3.21	3.10	2.98
4.03	3.86	3.68	3.50	3.40	3.30	3.20	3.10	2.99	2.87
3.93	3.76	3.59	3.40	3.31	3.21	3.11	3.00	2.89	2.78
3.85	3.68	3.50	3.32	3.22	3.12	3.02	2.92	2.81	2.69
3.77	3.60	3.43	3.24	3.15	3.05	2.95	2.84	2.73	2.61
3.70	3.54	3.36	3.18	3.08	2.98	2.88	2.77	2.66	2.55
3.64	3.47	3.30	3.12	3.02	2.92	2.82	2.71	2.60	2.48
3.59	3.42	3.25	3.06	2.97	2.87	2.77	2.66	2.55	2.43
3.54	3.37	3.20	3.01	2.92	2.82	2.72	2.61	2.50	2.38
3.49	3.33	3.15	2.97	2.87	2.77	2.67	2.56	2.45	2.33
3.45	3.28	3.11	2.93	2.83	2.73	2.63	2.52	2.41	2.29
3.41	3.25	3.07	2.89	2.79	2.69	2.59	2.48	2.37	2.25
3.38	3.21	3.04	2.86	2.76	2.66	2.56	2.45	2.33	2.21
3.34	3.18	3.01	2.82	2.73	2.63	2.52	2.42	2.30	2.18
3.12	2.95	2.78	2.60	2.50	2.40	2.30	2.18	2.06	1.93
2.90	2.74	2.57	2.39	2.29	2.19	2.08	1.96	1.83	1.69
2.71	2.54	2.37	2.19	2.09	1.98	1.87	1.75	1.61	1.43
2.52	2.36	2.19	2.00	1.90	1.79	1.67	1.53	1.36	1.00

Table IV: Percentage Points of F Distribution $(F_{0.0027, \nu_1, \nu_2})$ *(continued)*

					Numerator Degrees of Freedom				
ν_1 ν_2	1	2	3	4	5	6	7	8	9
1	55,593.95	68,586.61	74,125.71	77,160.08	79,067.50	80,375.13	81,326.60	82,049.63	82,617.51
2	368.87	369.37	369.54	369.62	369.67	369.70	369.73	369.75	369.76
3	84.98	75.86	72.19	70.20	68.94	68.08	67.44	66.96	66.58
4	43.83	36.49	33.58	31.99	31.00	30.31	29.81	29.42	29.12
5	30.33	24.13	21.69	20.36	19.52	18.94	18.52	18.20	17.94
6	24.05	18.54	16.38	15.21	14.46	13.95	13.57	13.29	13.06
7	20.52	15.47	13.49	12.41	11.73	11.26	10.91	10.64	10.43
8	18.29	13.55	11.69	10.69	10.05	9.60	9.28	9.03	8.83
9	16.76	12.25	10.49	9.53	8.92	8.50	8.19	7.95	7.76
10	15.66	11.32	9.63	8.71	8.12	7.71	7.42	7.18	7.00
11	14.82	10.62	8.98	8.09	7.52	7.13	6.84	6.62	6.44
12	14.17	10.08	8.49	7.62	7.07	6.68	6.40	6.18	6.01
13	13.65	9.65	8.09	7.24	6.70	6.32	6.05	5.83	5.66
14	13.22	9.29	7.77	6.93	6.40	6.04	5.76	5.55	5.38
15	12.86	9.00	7.50	6.68	6.16	5.80	5.53	5.32	5.15
16	12.56	8.76	7.28	6.47	5.95	5.60	5.33	5.12	4.96
17	12.30	8.55	7.08	6.29	5.78	5.42	5.16	4.96	4.80
18	12.08	8.36	6.92	6.13	5.63	5.28	5.02	4.82	4.66
19	11.88	8.21	6.78	6.00	5.50	5.15	4.89	4.69	4.53
20	11.71	8.07	6.65	5.88	5.38	5.04	4.78	4.58	4.43
21	11.56	7.94	6.54	5.77	5.28	4.94	4.68	4.49	4.33
22	11.42	7.83	6.44	5.68	5.19	4.85	4.60	4.40	4.25
23	11.30	7.73	6.35	5.59	5.11	4.77	4.52	4.33	4.17
24	11.19	7.64	6.27	5.52	5.04	4.70	4.45	4.26	4.10
25	11.09	7.56	6.19	5.45	4.97	4.64	4.39	4.19	4.04
26	10.99	7.49	6.13	5.38	4.91	4.58	4.33	4.14	3.98
27	10.91	7.42	6.07	5.33	4.85	4.52	4.28	4.09	3.93
28	10.83	7.36	6.01	5.27	4.80	4.47	4.23	4.04	3.89
29	10.76	7.30	5.96	5.23	4.76	4.43	4.18	3.99	3.84
30	10.69	7.25	5.91	5.18	4.71	4.39	4.14	3.95	3.80
40	10.23	6.88	5.58	4.87	4.42	4.10	3.86	3.67	3.52
60	9.80	6.54	5.27	4.58	4.14	3.83	3.59	3.41	3.27
120	9.39	6.22	4.99	4.31	3.88	3.58	3.35	3.17	3.03
∞	9.00	5.91	4.72	4.06	3.64	3.34	3.12	2.95	2.81

Denominator Degrees of Freedom

Source: Computed using Minitab.

				Numerator Degrees of Freedom					
10	**12**	**15**	**20**	**24**	**30**	**40**	**60**	**120**	**∞**
83,075.26	83,767.57	84,466.59	85,172.22	85,527.47	85,884.33	86,242.78	86,602.80	86,964.37	87,327.48
369.77	369.79	369.80	369.82	369.83	369.84	369.85	369.85	369.86	369.87
66.28	65.81	65.34	64.86	64.62	64.38	64.13	63.88	63.64	63.38
28.87	28.50	28.12	27.74	27.54	27.35	27.15	26.95	26.75	26.54
17.73	17.42	17.10	16.77	16.60	16.43	16.27	16.09	15.92	15.75
12.87	12.59	12.30	12.01	11.86	11.71	11.56	11.40	11.24	11.09
10.26	10.00	9.74	9.46	9.32	9.18	9.04	8.90	8.75	8.60
8.67	8.42	8.17	7.91	7.78	7.65	7.51	7.37	7.23	7.09
7.61	7.37	7.13	6.88	6.75	6.62	6.49	6.36	6.22	6.09
6.85	6.63	6.39	6.15	6.03	5.90	5.77	5.64	5.51	5.37
6.29	6.07	5.84	5.61	5.49	5.36	5.24	5.11	4.98	4.85
5.86	5.65	5.42	5.19	5.07	4.95	4.83	4.70	4.57	4.44
5.52	5.31	5.09	4.86	4.74	4.62	4.50	4.38	4.25	4.12
5.25	5.04	4.82	4.59	4.48	4.36	4.24	4.11	3.99	3.85
5.02	4.81	4.60	4.37	4.26	4.14	4.02	3.90	3.77	3.64
4.83	4.62	4.41	4.19	4.07	3.96	3.84	3.71	3.59	3.45
4.67	4.46	4.25	4.03	3.92	3.80	3.68	3.56	3.43	3.30
4.53	4.32	4.11	3.90	3.78	3.67	3.55	3.42	3.30	3.16
4.40	4.20	3.99	3.78	3.67	3.55	3.43	3.31	3.18	3.05
4.30	4.10	3.89	3.67	3.56	3.45	3.33	3.20	3.08	2.94
4.20	4.00	3.80	3.58	3.47	3.36	3.24	3.11	2.99	2.85
4.12	3.92	3.72	3.50	3.39	3.27	3.16	3.03	2.90	2.77
4.04	3.85	3.64	3.43	3.32	3.20	3.08	2.96	2.83	2.69
3.98	3.78	3.58	3.36	3.25	3.14	3.02	2.89	2.76	2.63
3.91	3.72	3.52	3.30	3.19	3.08	2.96	2.83	2.70	2.57
3.86	3.66	3.46	3.25	3.14	3.02	2.90	2.78	2.65	2.51
3.81	3.61	3.41	3.20	3.09	2.97	2.85	2.73	2.60	2.46
3.76	3.57	3.37	3.15	3.04	2.93	2.81	2.69	2.55	2.41
3.72	3.52	3.32	3.11	3.00	2.89	2.77	2.64	2.51	2.37
3.68	3.49	3.28	3.07	2.96	2.85	2.73	2.60	2.47	2.33
3.40	3.21	3.01	2.81	2.69	2.58	2.46	2.33	2.19	2.05
3.15	2.96	2.76	2.56	2.44	2.33	2.21	2.07	1.93	1.77
2.91	2.73	2.53	2.32	2.21	2.09	1.96	1.83	1.67	1.47
2.69	2.51	2.31	2.10	1.99	1.87	1.73	1.58	1.40	1.00

Table V: Control Chart Constants

Sample Size n	d_3	D_3	D_4	B_3	B_4	A_2	A_3	c_4	d_2
2	0.853	0.000	3.267	0.000	3.267	1.881	2.659	0.7979	1.128
3	0.888	0.000	2.574	0.000	2.568	1.023	1.954	0.8862	1.693
4	0.880	0.000	2.282	0.000	2.266	0.729	1.628	0.9213	2.059
5	0.864	0.000	2.114	0.000	2.089	0.577	1.427	0.9400	2.326
6	0.848	0.000	2.004	0.030	1.970	0.483	1.287	0.9515	2.534
7	0.833	0.076	1.924	0.118	1.882	0.419	1.182	0.9594	2.704
8	0.820	0.136	1.864	0.185	1.815	0.373	1.099	0.9650	2.847
9	0.808	0.184	1.816	0.239	1.761	0.337	1.032	0.9693	2.970
10	0.797	0.223	1.777	0.284	1.716	0.308	0.975	0.9727	3.078
11	0.787	0.256	1.744	0.321	1.679	0.285	0.927	0.9754	3.173
12	0.778	0.283	1.717	0.354	1.646	0.266	0.886	0.9776	3.258
13	0.770	0.307	1.693	0.382	1.618	0.249	0.850	0.9794	3.336
14	0.763	0.328	1.672	0.406	1.594	0.235	0.817	0.9810	3.407
15	0.756	0.347	1.653	0.428	1.572	0.223	0.789	0.9823	3.472
16	0.750	0.363	1.637	0.448	1.552	0.212	0.763	0.9835	3.532
17	0.744	0.378	1.622	0.466	1.534	0.203	0.739	0.9845	3.588
18	0.739	0.391	1.609	0.482	1.518	0.194	0.718	0.9854	3.640
19	0.734	0.404	1.597	0.497	1.503	0.187	0.698	0.9862	3.689
20	0.729	0.415	1.585	0.510	1.490	0.180	0.680	0.9869	3.735
21	0.724	0.425	1.575	0.523	1.477	0.173	0.663	0.9876	3.778
22	0.720	0.435	1.566	0.534	1.466	0.168	0.647	0.9882	3.819
23	0.716	0.443	1.557	0.545	1.455	0.162	0.633	0.9887	3.858
24	0.712	0.452	1.548	0.555	1.445	0.157	0.619	0.9892	3.895
25	0.708	0.459	1.541	0.565	1.435	0.153	0.606	0.9896	3.931

For $n > 25$,

$$c_4 \approx \frac{4n - 4}{4n - 3}$$

Source: Values generated from Minitab.

Table VI: Factors for *R* Chart Probability Limits

Sample Size *n*	$D_{0.001}$	$D_{0.005}$	$D_{0.025}$	$D_{0.975}$	$D_{0.995}$	$D_{0.999}$	
3	0.06	0.13	0.30	3.68	4.42	5.06	
4	0.20	0.34	0.59	3.98	4.69	5.31	
5	0.37	0.55	0.85	4.20	4.89	5.48	
6	0.53	0.75	1.07	4.36	5.03	5.62	
7	0.69	0.92	1.25	4.49	5.15	5.73	
8	0.83	1.08	1.41	4.60	5.25	5.82	
9	0.97	1.21	1.55	4.70	5.34	5.90	
10	1.08	1.33	1.67	4.78	5.42	5.97	

For $n = 2$,

$$D_{1-\alpha/2} = \sqrt{2}z_{\alpha/4}$$
$$D_{\alpha/2} = \sqrt{2}z_{1/2-\alpha/4}$$

Source: Reproduced by permission from E. L. Grant and R. S. Leavenworth, *Statistical Quality Control*, 7th ed., McGraw-Hill, New York, 1996.

B

Basic Overview of Minitab

INTRODUCTION

Minitab is a powerful and very user-friendly statistical software package that enables the user to do quick but insightful data analysis. This appendix will present a basic overview of this package. Our discussion is brief because Minitab is quite easy to learn quickly on one's own.

The learning process is made even easier given its context-sensitive online Help file. The Help file provides comprehensive, easy-to-understand documentation that gives step by step instructions for every procedure. In addition, Minitab provides for nearly all procedures an example of how to implement the given procedure.

OVERVIEW

There are four permanent windows in Minitab: Worksheet, Session, History, and Info.

1. The *Worksheet* window contains the data associated with your Session. Minitab's worksheet is similar in look to a spreadsheet. The worksheet has columns and rows. The columns are labeled C1, C2, etc. Each column contains the values for a particular variable in your data set.

2. The *Session* window is where the user can type and execute commands at a prompt symbolized by "MTB>". Typing directly commands in the Session window is one of two ways to execute Minitab's functional capabilities. The other way is through the use of pull down menus and dialogue boxes. In the Session window, Minitab keeps a running record of all the commands typed and lists the results of your Minitab commands, except for high-resolution graphics. If you decide to run Minitab with menus, Minitab "dumps" the equivalent session commands in the Session window.

3. The *History* window is a running list of all the commands from the Session window. In essence, the History window is the Session window without the results of the commands. One use of this window is to copy commands from the window and paste them into the Session window to execute commands previously executed during your session.

4. The *Info* window contains the basic summary of the worksheet. In particular, it indicates which columns have data entries, the variable names associated with these columns, and the number of data entries in these columns. Alternatively, the user can obtain this equivalent information by typing "info" at the session prompt, that is, "MTB> info".

In addition to the four permanent windows described above, additional windows are created for every high-resolution graph you generate.

There are nine main menus from which you can choose—File, Edit, Manip, Calc, Stat, Graph, Editor, Window, and Help:

1. The *File* menu allows you to open previously saved worksheets and projects, to save current work in the form of a worksheet or project, to print files and windows, and to exit Minitab.

2. The *Edit* menu allows you to edit cells in the worksheet. It also gives you options to edit the most recent dialog box so that the procedure can be quickly re-invoked. Finally, you can find the "Preferences" option within Edit menu. This option lets you customize various aspects of Minitab based on personal preferences. For the most part, you should find the default preferences as satisfactory. The only exception is that you may wish to change the default preference for the prompt in the Session window. The default is to have no "MTB>" prompt appear. Thus, you can not type session commands in the Session window. To change this default, choose "Preferences" (Edit >> Preferences) and then highlight and select "Session Window". At which point, you click the "Enable" option found under "Command Language".

3. The *Manip* menu lets you manipulate entire columns in the worksheet. For example, you have options to stack columns, delete columns, and change the data type of the entries found in a given column.

4. The *Calc* menu gives you the "Calculator" option (Calc >> Calculator) to apply various mathematical functions to the columns in the worksheet. You also find in the Calc menu options to compute simple statistics, either by column or by row. There is also an option enabling you standardize the data. Other important options within this menu include the option to generate random data from a variety of distributions and the option to compute areas or percentile points for a variety of distributions.

5. The *Stat* menu is the core of Minitab in terms of providing you with a large variety of statistical procedures. Here we find the options for obtaining basic summary statistics, estimating regression equations, creating control charts, and performing time-series analysis.

6. The *Graph* menu gives you options to create a wide variety of graphs (high-resolution or character-mode). It should be noted that not all graphs are created from options within the Graph menu. For instance, control charts are created from options within the Stat menu noted above.

7. The *Editor* menu provides some additional editing capabilities, primarily related to the look of the Session Window.

8. The *Window* menu, among other things, allows you to choose which window (Worksheet, Session, History, Info, or a created high-resolution graph) you want displayed and active.

9. The *Help* menu provides the user with a comprehensive on-line help session. Within the Help menu, some first-time users might wish to choose "Getting Started" (Help >> Getting Started). This component is a detailed introduction to Minitab and even provides the user with an example of a typical

Minitab session. To find a detailed description on how to invoke any procedure by means of pull-down menus and dialog boxes, simply choose "Search for Help on" (Help >> Search for Help on). For a complete listing and detailed descriptions of Session commands, you should choose "Session Command Help" (Help >> Session Command Help).

COMMON INPUT AND OUTPUT COMMANDS

As already noted, Minitab can be run either by directly typing session commands or by using pull down menus and dialogue boxes. Below we briefly describe some of the more common input and output Minitab procedures you will likely invoke.

Data can be entered in the worksheet either by directly typing the data entries into the worksheet or by reading in a data file. There are basically three types of data files: (1) Minitab saved file, (2) ASCII or text file, or (3) Other Windows-based software (e.g., Excel spreadsheet).

Suppose as an example we wish get the five columns of measurement data shown in Table 6.1 into a Minitab worksheet. As noted in the Preface, we have provided all data sets in Minitab and ASCII format, namely, "x-ray.mtw" and "x-ray.dat", respectively. (Note: the filename extensions are viewed by Minitab or most any software without regard to case, thus, ".mtw" is equivalent to ".MTW" and ".dat" is equivalent to ".DAT".)

Let us consider the retrieving of a file saved as a Minitab worksheet. First, we should note that worksheets can be saved in one of two ways. The first way is using the "Save Current Worksheet As" option under "File". This saves *only* the current worksheet with the contained data and column names. The resulting file will have a file extension of ".mtw". The files in Minitab format given on the accompanying disk have been saved in such a manner, thus, they all have the extension of ".mtw".

The other saving option is "Save Project As" under "File". This saves all aspects of your current session with Minitab, namely, worksheet, session window of previously typed commands, and any current graphs. The resulting file will have a file extension of ".mpj".

Below is how you retrieve a Minitab worksheet (as opposed to a project):

a. Choose File >> Open Worksheet.
b. Select the directory in which the worksheet file resides.
c. Highlight the worksheet file (for example, "x-ray.mtw") and then click Open.

If you wish to read an ASCII file into Minitab, you proceed as follows:

a. Choose File >> Other Files >> Import Special Text.
b. In the dialog box labeled "Store data in column(s):", enter the columns into which Minitab should place the data (for example, "x-ray.dat" has five columns of data, so you might choose C1-C5). Then click OK.

   c.   Select the directory in which the ASCII file resides.

   d.   Highlight the ASCII file and then click Open.

Once the data are in Minitab, you can proceed with your data analysis. As noted, there are two saving options before you exit Minitab. You can save only the worksheet which is done as follows:

   a.   Choose File >> Save Current Worksheet As.

   b.   Select the directory in which you wish the worksheet file to reside.

   c.   In the dialog box labeled "File name", type in the name you wish to call the worksheet file and then click Save.

The other option is to save all the contents of all the windows (Worksheet, Session, History, and Info). To do so, you do the following:

   a.   Choose File >> Save Project As.

   b.   Select the directory in which you wish the project file to reside.

   c.   In the dialog box labeled "File name", type in the name you wish to call the project file and then click Save.

In a subsequent session, you can retrieve a project file as follows:

   a.   Choose File >> Open Project.

   b.   Select the directory in which the project file (filename extension ".mpj") resides.

   c.   Highlight the project file and then click Open.

INDEX

NAME

INDEX

SUBJECT